THE MAGNETOTELLURIC METHOD
Theory and practice

The magnetotelluric method is a technique for imaging the electrical conductivity and structure of the Earth, from the near-surface down to the 410 km transition zone and beyond. It is increasingly used in geological applications and the petroleum industry. This book forms the first comprehensive overview of magnetotellurics, from the salient physics and its mathematical representation, to practical implementation in the field, data processing, modeling, and geological interpretation.

Electromagnetic induction in 1D, 2D, and 3D media is explored, building from first principles, and with thorough coverage of the practical techniques of time-series processing, distortion, numerical modeling and inversion. The fundamental principles are illustrated with a series of case histories describing geological applications. Technical issues, instrumentation and field practices are described for both land and marine surveys.

This book provides a rigorous introduction to the magnetotelluric method for academic researchers and advanced students, and will be of interest to industrial practitioners and geoscientists wanting to incorporate rock conductivity into their interpretations.

ALAN D. CHAVE is a Senior Scientist at Woods Hole Oceanographic Institution. He has also been a Chartered Statistician (UK) since 2003, and has taught a graduate-level course in statistics in the MIT/WHOI Joint Program for 20 years. For over 30 years, he has conducted research utilizing the magnetotelluric method, primarily in the oceans, and has pioneered research into producing modern magnetotelluric processing methods. Dr Chave has also designed instrumentation for optical and chemical measurements in the ocean, and has played a leadership role in developing long-term ocean observatories worldwide. He has been an editor of *Journal of Geophysical Research* and editor-in-chief of *Reviews of Geophysics*.

ALAN G. JONES is Senior Professor and Head of Geophysics at the Dublin Institute for Advanced Studies, and has been using magnetotellurics since the early 1970s. He has undertaken magnetotellurics in Europe, southern Africa, Canada and China, for problems ranging from the near-surface (groundwater contamination) to mining, geothermal studies and tectonics of the deep mantle (to 1200 km). He has been instrumental in many developments of magnetotellurics, from processing and analysis to modeling/inversion and interpretation. He was awarded the Tuzo Wilson Medal of the Canadian Geophysical Union in 2006, appointed to *Academia Europaea* in 2009, and made a member of the Royal Irish Academy in 2010.

THE MAGNETOTELLURIC METHOD

Theory and practice

Edited by

ALAN D. CHAVE
Woods Hole Oceanographic Institution

ALAN G. JONES
Dublin Institute for Advanced Studies

CAMBRIDGE
UNIVERSITY PRESS

University Printing House, Cambridge CB2 8BS, United Kingdom

One Liberty Plaza, 20th Floor, New York, NY 10006, USA

477 Williamstown Road, Port Melbourne, VIC 3207, Australia

4843/24, 2nd Floor, Ansari Road, Daryaganj, Delhi - 110002, India

79 Anson Road, #06-04/06, Singapore 079906

Cambridge University Press is part of the University of Cambridge.

It furthers the University's mission by disseminating knowledge in the pursuit of education, learning and research at the highest international levels of excellence.

www.cambridge.org
Information on this title: www.cambridge.org/9781108446808

© Cambridge University Press 2012

First published 2012
3rd printing 2013
First paperback edition 2017

A catalogue record for this publication is available from the British Library

ISBN 978-0-521-81927-5 Hardback
ISBN 978-1-108-44680-8 Paperback

During the course of writing this book, we received the sad news of the death of Peter Weidelt while on a visit to Turkey, where he was planning to complete Chapter 4. Peter's influence on magnetotellurics was profound, as he was especially responsible for giving the method a rigorous mathematical and physical grounding that constitutes the basis for many of the developments of the past few decades. Peter was a humble and generous man whose scientific contributions and humanity are sorely missed. This book is dedicated with warmth to his memory and with respect to his legacy.

Picture taken by Bai Denghai at the Schmucker Symposium on 27 July 2009, just four days before Peter's untimely death.

Contents

The color plate section can be found between pages 302 and 303.

Preface

Just as electromagnetics was the last aspect of classical physics to be fully understood and theoretically described in the mid-nineteenth century, so was electromagnetics the last of the classical methods of physics to be utilized in geophysics, particularly exploration geophysics, to understand the Earth. Of the two basic types of electromagnetic methods, namely controlled-source and natural-source, this book describes the theory and application of natural-source electromagnetics, named "magneto-tellurics" ironically by a renowned seismologist, Louis Cagniard, in 1953, from "magneto" inferring magnetic fields and "telluric" inferring electric fields in the ground (*tellus*, Latin for "earth"). Magnetotellurics (the hyphen was dropped during the mid-1970s) has, since its early inception and embryonic years in the 1950s and 1960s, grown in stature to the extent that it is now a formidable geophysical tool for obtaining high-resolution information about the lateral and vertical variations in electrical conductivity that can be related to resources and geological processes.

This book was originally conceived as being written entirely by two people, but given the extensive breadth of the subject, they modestly decided that others needed to be invited to contribute chapters in their areas of specialty. Thus, the book comprises 10 chapters, each penned by one or more leading experts, and is organized in logical order (at least to the editors). The hope, of course, is that the whole is greater than the sum of its parts, such that the individual styles do not detract from the continuous theme.

The motivation behind the book is that there is no comprehensive and rigorous volume on modern magnetotellurics that is current, circumscribing today's thinking, approaches and methods. The extensive review papers from the biennial "EM Induction Workshops" are excellent, but are not cohesive. The recent volume *Practical Magnetotellurics* by Fiona Simpson and Karsten Bahr, also published by Cambridge University Press, serves as a useful introductory text describing practical aspects, but is not as comprehensive.

The book is aimed to educate and inform at many levels. It is intended for the whole spectrum of readers, from the established practitioner in magnetotellurics, to graduate and advanced undergraduates in geophysics, to other geophysicists and other geoscientists. It can be read continuously, or can be read in parts, as the need arises.

Chapter 1, by Alan Chave and Alan Jones, provides an introduction to the book and particularly describes the historical perspective up to around 1960. A special place is reserved for the role of the Japanese, whose investigations in the 1910s to 1940s have not been appreciated at the same level as the two papers, by Andrey Tikhonov and Louis Cagniard, that are often cited as establishing the field. Certainly, the Cagniard–Tikhonov magnetotelluric method should be renamed the Cagniard–Rikitake–Tikhonov method (the order of the names is immaterial, as the work was undertaken independently by all of them).

Chapter 2, by Chave and the late Peter Weidelt, describes the theoretical basis for magnetotellurics, starting with the Maxwell equations and working through one-, two- and three-dimensional (1D, 2D and 3D) solutions. The 1D magnetotelluric response is presented as a limiting case of a vertical magnetic dipole source, bringing out the role of a quasi-uniform source. Much of this is standard, but, for the first time in a book, a thorough treatment is presented of the electromagnetic fields produced by water motion.

Chapter 3 that follows – given the breadth of the topic, the Earth's electromagnetic environment – is split into two parts. The first part, by Rob Evans, deals with laboratory studies of the electrical conductivity of rocks and minerals. This is a field that saw significant investment through the 1970s and 1980s, but suffered from quiescence through the 1990s. Encouragingly, there are more groups now undertaking measurements on rocks and understanding the physics of electrical conduction through them. In the second part, Ari Viljanen then follows by covering the nature and influence of external source currents flowing above Earth. For the most part, magnetotelluricists can reasonably assume a plane-wave model, but in equatorial and auroral latitudes this is not the case, and consideration has to be given to the effects of non-uniform sources.

Chapter 4, by Weidelt and Chave, gives a thorough treatment of the magnetotelluric response and magnetic transfer functions. The mathematical properties of the response function and its rotational invariants are explored in 1D, 2D and 3D. Chave also wrote Chapter 5, which follows on from Chapter 4 and describes the estimation of the response function using modern robust methods.

Chapter 6, by Jones, describes the next step in the logical chain of processing and analysis, which is evaluation of the derived response functions for distortion effects and its inherent dimensionality and directionality. Older magnitude-based methods are shown to be unsuitable, and newer phase-based methods are advocated.

Chester (Chet) Weiss penned Chapter 7, and presents the forward problem in magneto-tellurics – determining the fields that would be observed given a particular conductivity distribution. Particular focus is given to the similarities and distinctions between finite differences and finite elements. Chapter 8 is concerned with the magnetotelluric inverse problem, and William (Bill) Rodi and Randall (Randy) Mackie describe various minimization algorithms in 1D and 3D.

The last two chapters that complete the book are concerned with practical aspects and the purposes of magnetotellurics, namely instrumentation and field procedures (Chapter 9) and case histories and geological applications (Chapter 10), authored by Ian Ferguson, with Jones and Chave also participating in the last chapter. It is hoped that those outside magnetotellurics will be enthralled by Chapter 10, and will appreciate what magnetotellurics can bring to addressing geological problems.

This book would not exist without the dedication, warmth and wisdom of those who were leaders of the field when the editors and chapter authors were young, aspiring students. To recognize some of these people always runs the danger of, by omission, inadvertently not recognizing others, but from a very personal perspective Alan Chave wishes to thank Chip Cox, Jean Filloux and Nigel Edwards, and Alan Jones wishes to recognize Rosemary Hutton, Ian Gough, Ulrich Schmucker and Peter Weidelt, who aided them in their training and development during their formative graduate and postgraduate years. The reviewers of the chapters of this book are all gratefully thanked for their generous advice. They were: Nestor Cuevas, Gary Egbert, Mark Everett, Colin Farquharson, Uli Matzander, Nils Olsen, Anne Pommier, Pilar Queralt, Art Richmond, Jeff Roberts, Weerachai Siripunvaraporn, David Thomson, Martyn Unsworth and John Weaver.

Contributors

Alan D. Chave
Deep Submergence Laboratory, Department of Applied Ocean Physics and Engineering Woods Hole Oceanographic Institution, Woods Hole, MA 02543, USA

Rob L. Evans
Department of Geology and Geophysics, Woods Hole Oceanographic Institution, Woods Hole, MA 02543, USA

Ian J. Ferguson
Department of Geological Sciences, University of Manitoba, Wallace Building, 125 Dysart Road, Winnipeg, R3T 2N2, Canada

Alan G. Jones
Dublin Institute for Advanced Studies, 5 Merrion Square, Dublin 2, Ireland

Randall L. Mackie
Formerly: Schlumberger EMI Technology Center, Berkeley, CA 94804, USA

William L. Rodi
Department of Earth, Atmospheric and Planetary Sciences, Massachusetts Institute of Technology, 77 Massachusetts Avenue, Cambridge, MA 02139, USA

Ari Viljanen
Finnish Meteorological Institute, Arctic Research Unit, Erik Palmenin aukio, FI-00560 Helsinki, Finland

Peter Weidelt (*deceased*)
Formerly: Institut für Geophysik und Extraterrestrische Physik, Technische Universität Braunschweig, D-38106 Braunschweig, Germany

Chester Weiss
Department of Geosciences, Virginia Tech, 4044 Derring Hall (0420), Blacksburg, VA 24061, USA

1

Introduction to the magnetotelluric method

ALAN D. CHAVE AND ALAN G. JONES

1.1 Introduction

Over the past three decades, and particularly over the intervals 1985–1995 and 2005–2010, magnetotellurics has undergone a revolution driven by four main factors: (1) the emergence of low-power, low-cost, 24-bit digital electromagnetic sensing and recording technologies, (2) dramatic improvements in the understanding of noise in electromagnetic measurements, with the concomitant evolution of data processing algorithms, (3) substantial advances in the ability to recognize and remove distortion by near-surface structure local to the measurement point that is the bane of practical magnetotellurics, and (4) the development of fast two- and three-dimensional (2D and 3D) modeling and inversion capabilities concurrent with the constantly increasing power of computers. In the 1970s, a typical magnetotelluric survey consisted of a handful of sites whose data were analyzed using ordinary least-squares methods, smoothed in the frequency domain to reduce data scatter, and interpreted using one-dimensional (1D) models "stitched" together into a 2D pseudo-section that may, or may not, be tested through 2D forward modeling. By the 1990s, surveys comprising several tens of sites along a single line were common, data were processed using robust methods, which produced substantially more reliable response estimates that were subsequently analyzed for galvanic distortion, and rapid 2D modeling and inversion were standard. By the 2010s, magnetotelluric surveys consisting of many hundreds of sites, with areal rather than linear coverage, are being carried out, data processing is semi-automatic, usually using bounded influence or multivariate approaches, multi-site distortion removal is being applied routinely, 2D interpretation, often including anisotropy, is routine, while 3D interpretation based on 3D inversion is becoming commonplace. The key purposes of this book are documenting this magnetotelluric revolution and providing an up-to-date, rigorous reference on the field that is much more than a practical guide, and that is useful to the novice, expert practitioner and interested non-magnetotelluric geoscientist alike.

As an illustration of the technological and scientific distance that magnetotellurics has come over the past three decades, two representative studies from the mid-1970s and mid-2000s will be compared and contrasted. The first is the thorough analysis and geological interpretation of 16 magnetotelluric sites from the eastern Snake River Plain–Yellowstone region of the western USA presented by Stanley *et al.* (1977). The second is a 3D magnetotelluric study utilizing 100 areally distributed sites from the Betic Cordillera on the Iberian peninsula published by Rosell *et al.* (2011).

Stanley *et al.* (1977) is a (for the era) cutting-edge magnetotelluric data collection, analysis and interpretation effort that is also notable for its rigorous geological interpretation, a practice that was not widespread in the 1970s. The strong guidance of Francis Bostick, who published infrequently but was very influential in the 1970s and 1980s, is evident in the paper. Data were collected using the magnetotelluric–audiomagnetotelluric system of the University of Texas Geomagnetics Laboratory based on coil magnetometers and electric field lines recorded in analog form on frequency-modulated (FM) tape after band-pass filtering. These were supplemented by a digital cryogenic system developed by the US Geological Survey. Little is stated about processing the data into magnetotelluric response functions, but it can be assumed that a least-squares approach without remote referencing was applied. The resulting responses were represented in a 2D sense following Sims & Bostick (1969) and Word *et al.* (1970), with an adopted geoelectric strike angle consistent with surface geology. However, the magnetotelluric responses were not smooth functions of frequency, especially at low frequencies, probably due to a lack of robustness in response function estimation, and were often not interpretable due to the presence of galvanic distortion of the electric fields. As was sometimes the practice at the time, the response functions were smoothed after the fact, in this case using the novel approach of enforcing a Hilbert transform relationship between the apparent resistivities and phases (Figure 1.1). Each site was then inverted for a 1D model, and the results at adjacent sites were "stitched" together to produce the electrical cross-section shown in Figure 1.2. The key notable feature in the cross-section is the conductive zone at a depth ranging from 7 km beneath the Raft River thermal area to 18 km beneath the central part of the Snake River Plain to 5 km beneath Yellowstone. The resistivity at these depths ranges from 1 to 10 Ω m, and was interpreted to have a primarily thermal origin.

Rosell *et al.* (2011) utilized 100 broadband magnetotelluric sites (41 of which also included long-period data) distributed across a 400 km × 200 km area of southern Spain. The data were processed into response tensors using robust methods, and covered a period range of 0.001 to 20 000 s. The dimensionality of the dataset was assessed using the method of Martí *et al.* (2009) based on rotational invariants (see Chapter 6). The predominance of 3D behavior led to 3D inversion of the off-diagonal tensor elements using the WSINV3DMT code of Siripunvaraporn *et al.* (2005). Figure 1.3 depicts three vertical slices through the model, along with earthquake hypocenters, together with a geological map showing the locations of the magnetotelluric sites. The CB2 conductive body is the upward-oriented tongue near the center of each slice that divides the structure into three domains as follows: (1) a southwest domain D1 characterized by a resistive lithosphere and hypocenters with increasing depth toward the east and south, (2) a less resistive southeast domain D2 with hypocenter density concentrated at upper crustal levels, and (3) a northern domain D3

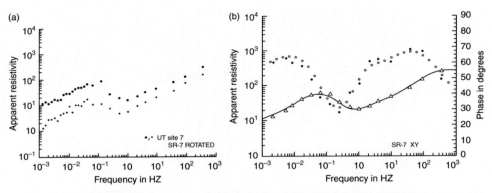

Figure 1.1. (a) The principal apparent resistivities for Site 7 from Stanley *et al.* (1977) after rotation to maximize the sum of the absolute values of the off-diagonal response tensor elements. (b) The transverse electric (TE) phase (solid circles) and the apparent resistivity computed from the phase through a Hilbert transform relationship (triangles). The open circles and solid line are the phase and apparent resistivity for a 1D model fitted to the transformed data. Taken from Stanley *et al.* (1977).

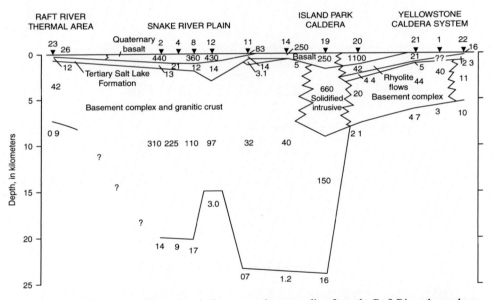

Figure 1.2. Southwest to northeast electrical cross-section extending from the Raft River thermal area to Yellowstone obtained by stitching together 1D inversions at each of the sites shown as triangles at the top. The numbers in the cross-section are the electrical resistivities obtained at that depth. The vertical exaggeration is 10:1. Taken from Stanley *et al.* (1977).

that is resistive to greater depth than in D1 and D2 and displays a paucity of hypocenters. Rosell *et al.* (2011) interpreted CB2 as the intrusion of asthenospheric material into the lithosphere caused by lateral lithospheric tearing and fracture of the eastward-directed subducting Ligurian slab beneath the Alboran region.

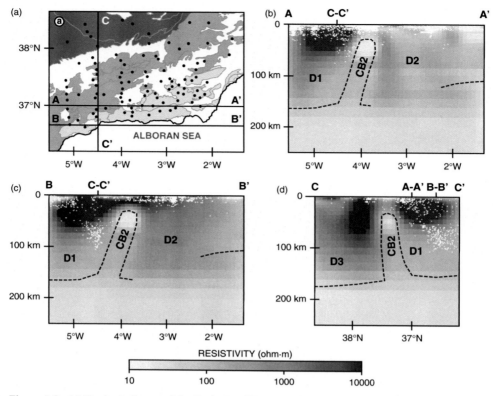

Figure 1.3. (a) Geological map of the Betic Cordillera showing the locations of the magnetotelluric sites (black dots) and (b)–(d) the slices A–A', B–B' and C–C', respectively, that are vertical slices of the 3D resistivity model crossing the CB2 body shown as the vertical tongue near the center of each slice. Earthquake hypocenter locations within 8 km of each profile recorded since 1900 are shown by white dots. The dashed black line shows the lithosphere–asthenosphere boundary inferred from the resistivity distribution. D1, D2 and D3 are the main tectonic domains described in the text. Grayscale adaption of color figure from Rosell *et al.* (2011).

1.2 A quick tour of magnetotellurics

The propagation of electromagnetic waves through a medium with homogeneous physical properties is proportional to $e^{-\gamma x}$, where the complex propagation constant γ is

$$\gamma = \sqrt{\omega\mu(i\sigma - \omega\varepsilon)} \tag{1.1}$$

and ω is the angular frequency under the assumption of $e^{i\omega t}$ time dependence, μ is the magnetic permeability, σ is the electrical conductivity and ε is the electric permittivity. The constitutive relations, μ, σ and ε in (1.1) that also link the electric and magnetic fields in the Maxwell equations parameterize the physical properties of Earth that may be sensed in an electromagnetic experiment of any design (i.e. from megahertz ground-penetrating radar to sub-millihertz deep-mantle studies using the 11 yr solar cycle).

As shown in Chapter 2, at the frequencies used and for the targets of interest in magnetotellurics, the magnetic permeability may be taken as the free-space value μ_0 in nearly all Earth materials. In addition, $\omega\varepsilon \ll \sigma$, with the difference amounting to many orders of magnitude. The second term under the square root in (1.1) is due to displacement current in the Maxwell equations (the so-called Maxwellian term), and, once its effect is neglected, the electromagnetic fields are not governed by a wave equation. Instead, they are governed by a diffusion equation that also obtains in other fields of geophysics such as heat flow, with the important distinction that in electromagnetism they are vector fields rather than scalar in form. The diffusion equation limit of the Maxwell equations yields electromagnetic induction, and the magnetotelluric method is its primary geophysical application.

In a uniformly conducting medium, the length scale for electromagnetic induction is the familiar skin depth, which is the distance over which the electromagnetic field decays by $1/e \approx 0.37$, and is given by

$$\delta(\omega) = \sqrt{\frac{2}{\omega\mu\sigma}} \tag{1.2}$$

Using SI units, (1.2) may be simplified to

$$\delta = 503\sqrt{\rho T} \tag{1.3}$$

where δ has units of meters, T is the period in seconds and ρ is the inverse of the electrical conductivity, or the electrical resistivity in Ω m. At a zeroth-order conceptual level, the magnetotelluric method comprises measurement of the skin depth as a function of period to infer resistivity as a function of position in Earth. As shown in Chapter 3, the period range over which magnetotellurics can operate is about 10^{-4}–10^5 s, and most bulk Earth materials have a resistivity range of 10^{-1}–10^5 Ω m. More typical values in Earth's crust and uppermost mantle are 10–10 000 Ω m, usually decreasing with increasing depth (due to rising temperature), yielding a skin depth δ ranging over 50 m to 500 km and beyond, covering everything from the surface into the deep upper mantle. This broad range of penetration depths is one of the appeals of the magnetotelluric method, as it can be used in studies whose focus is near-surface to upper mantle without altering the underlying principles. Only the instrumentation needs to be changed.

The magnetotelluric method uses time variations of Earth's magnetic field caused at low frequencies (<10 Hz) by the interaction of the solar plasma with the ionosphere and magnetosphere as a source, and at high frequencies (>10 Hz) by global lightning activity (Chapter 3). The fluctuating magnetic field induces an electric current within Earth whose magnitude depends on electrical conductivity, and from Ampere's law, measurements of magnetic field fluctuations at Earth's surface determine the total electric current in the subsurface. The addition of an electric field measurement at Earth's surface yields the electrical conductivity at that point, and transformation of electric and magnetic field data into the frequency domain allows the geophysicist to map the electrical conductivity as a function of depth and position (and occasionally, with time, in continuous or repeat four-dimensional experiments). Since the electromagnetic fields are vector entities, it is possible to measure three components of the magnetic field and the two horizontal components of the electric field at Earth's surface, where the vertical electric field vanishes due to the presence of the insulating atmosphere. This limitation does not exist at the seafloor, but other, non-magnetospheric source types typically dominate the vertical electric field in the ocean, as shown in Chapter 2.

Owing to the vector form of electromagnetic fields, the fundamental datum in the magnetotelluric method is a tensor relationship $\overset{\leftrightarrow}{\mathbf{Z}}$ (called the magnetotelluric response function or tensor) between the surface vector horizontal electric and magnetic fields as a function of period, so that

$$\mathbf{E} = \overset{\leftrightarrow}{\mathbf{Z}} \cdot \mathbf{B} \qquad (1.4)$$

where • denotes the inner product, along with an auxiliary equation relating the vertical and horizontal magnetic fields

$$\mathbf{B}_z = \mathbf{T} \cdot \mathbf{B}_h \qquad (1.5)$$

where $\mathbf{T}$ is a vector called the Z/B transfer function, geomagnetic depth sounding transfer function or tipper function. Owing to the tensor form of the relation between the electric and magnetic fields, $\overset{\leftrightarrow}{\mathbf{Z}}$ and $\mathbf{T}$ from a single site contain information about the electrical conductivity as a function of coordinate direction and depth. Combining $\overset{\leftrightarrow}{\mathbf{Z}}$ and $\mathbf{T}$ from sites along a line or over an area can yield a 2D or 3D understanding of Earth structure. A formal physical and mathematical treatment of the properties of the magnetotelluric response $\overset{\leftrightarrow}{\mathbf{Z}}$ for 1D, 2D and 3D media is contained in Chapters 2 and 4. Its estimation from measurements is a difficult statistical problem whose solution is described in Chapter 5.

Once reliable estimates of the magnetotelluric response $\overset{\leftrightarrow}{\mathbf{Z}}$ and Z/B transfer function $\mathbf{T}$ are obtained from a set of data, the focus of the magnetotelluric practitioner shifts to their analysis, modeling and ultimately a geological interpretation. Distortion of regional electric fields by local structures is arguably the greatest bane of the magnetotelluric method, and one that has caused poor interpretations directly leading to low acceptance of magneto-tellurics within the broad geoscience community. Consequently, the first step in under-standing estimates of $\overset{\leftrightarrow}{\mathbf{Z}}$ and $\mathbf{T}$ is assessing the extent and nature of their distortion. Approaches and methods developed over the past two decades to recognize and remove such distortion, primarily through tensor decomposition, enabled significant advances in reliable imaging of the subsurface electrical conductivity structure using magnetotellurics, as discussed in Chapter 6.

For most geologically plausible structures, numerical methods are required to solve the forward problem of predicting the magnetotelluric responses that would be observed at arbitrary locations and frequencies given a hypothetical model of Earth's resistivity structure. This is straightforward for 1D structures, as described in Chapter 2, and requires the use of sophisticated numerical finite-difference, finite-element or integral equation methods in

2D and 3D, as presented in Chapter 7. Solving the forward problem is an essential element in the second step of magnetotelluric data interpretation, but is only a means to an end.

Simply producing a model that qualitatively fits the measurements does not allow the magnetotelluric practitioner to fully assess and interpret a given set of measurements. This requires consideration of the inverse problem of inferring Earth's conductivity structure on the basis of observed magnetotelluric responses at specific locations and frequencies. Solving the magnetotelluric inverse problem involves finding one or more models of conductivity whose predicted responses fit the observed responses. The notion of "fit" must take into consideration measurement errors in the observations, computational errors in theoretical predictions, and even the appropriateness of the model itself. This is exactly the same notion as used to assess whether a given model of distortion is appropriate for a dataset. In most circumstances, a wide range of conductivity models will provide acceptable fits to a set of magnetotelluric observations. In other words, solutions to the inverse problem are non-unique, and it becomes necessary to impose additional constraints on the model or to find a concise characterization of the range of acceptable models. The revolution in magnetotelluric inversion to deal with such non-uniqueness is covered in Chapter 8.

In magnetotelluric surveys, it is essential to acquire high-quality data over a broad frequency or period range. The components of a magnetotelluric recording system can be divided into three parts: electrometers for sensing electric fields (or, more correctly, potential differences), magnetometers for sensing the magnetic field (or, more correctly, the magnetic induction) and recording/timing units for controlling the timing, digitization, filtering, and recording of the data, all of which have improved dramatically in recent decades. However, modern instrumentation alone will not yield time series of high fidelity; this also necessitates thoughtful site selection, careful installation and thorough field documentation that comprise state-of-the-art magnetotelluric field procedures. The advances in magnetotelluric data acquisition are documented in Chapter 9.

Chapter 10 concludes the book with a review of a selected set of case histories, with the emphasis placed on illustrating many of the analysis and modeling methods described in Chapters 2–8, and using magnetotelluric data either alone or with other geophysical data to make geological inferences about Earth. The selected examples span near-surface to crustal to mantle studies on both land and the seafloor, and provide an up-to-date set of examples of what can be accomplished using the magnetotelluric method.

1.3 Historical perspective

Electromagnetic induction studies have their theoretical origin in the spherical Earth treatment of Lamb (1883). Subsequently, Schuster (1889) applied Lamb's theory to data from globally distributed geomagnetic observatories to show that Earth is a conducting body. Chapman (1919) and Chapman & Price (1930) produced spherically symmetric but disparate models for the electrical resistivity of Earth, consisting of a thin surface insulating shell overlying 28 or 2.3 Ω-m conductors, respectively. The difference was attributed to a decreasing resistivity with depth because the Chapman & Price study used longer-period data than the earlier Chapman analysis. In a sequence of now classic studies, Lahiri & Price (1939) and Rikitake (1950, 1951a,b,c,e) extended these results to produce more detailed, radially varying models for the resistivity of Earth. By the early 1980s, the ability to resolve large-scale lateral variability in Earth structure emerged, as reviewed by Tarits (1994). All of these studies used the *geomagnetic depth sounding* (GDS) method based on measurements of the geomagnetic (but not the geoelectric) fields at periods of a day to a week or more. Under the assumption that the driving external source is dominantly an axisymmetric ring current flowing many Earth radii above the geomagnetic equator, the fundamental GDS datum is the ratio of the vertical to the horizontal magnetic field as a function of period, which may be used to infer the electrical resistivity as a function of depth and, in some cases, location. It differs significantly from magnetotellurics in many important respects, not least of which is that classical GDS operates at substantially longer periods, hence is primarily sensitive in the mid-mantle transition zone (400–600 km), and will not be further covered in this book. However, from a historical perspective, it is important because the theory for GDS is similar to that for magnetotellurics, and the academic rigor from GDS studies provided much of the theoretical basis for magnetotellurics.

In parallel to the evolution of GDS, electrical geophysics emerged as a viable tool in the exploration for petroleum and conductive mineral ores beginning in the 1920s with the Schlumberger brothers. In contrast to GDS, exploration electrical geophysics typically utilizes an artificial source to inject or induce electric currents in the subsurface. As a result, the physical scale of structures that can be studied is much smaller than with GDS, and in fact is typically limited to, at most, the uppermost crust. A comprehensive survey of electromagnetic exploration methods appears in the two-volume compendium edited by Misac Nabighian (1987, 1991) that remains the standard reference today. Electrical exploration methods differ substantially and practically from magnetotellurics in many ways, but represent a field that was more rapidly evolving than academic GDS. The combination of the need to span the gap lying between the exploration and global scales, and the constant drive for innovation in the exploration world, contributed to the emergence of the magnetotelluric method by the middle of the twentieth century.

In the nineteenth century, descriptions of geomagnetically induced fluctuations on telegraph cables led to the simultaneous measurement of the terrestrial electric and magnetic fields at the Royal Observatory, Greenwich (Airy, 1868). While their relationship was not studied quantitatively, Airy did note that the electric field frequently led the magnetic field, and inferred a subsurface origin for the field variations. This observation led a German research team to make measurements of the time-varying magnetic and electric fields during their contribution to the 1887 First International Polar Year in the Arctic. These data, the first ever field campaign magnetotelluric data recorded, were analyzed in the mid-1980s by Jones & Garland (1986). Subsequently, in an attempt to explain the polarization and temporal behavior of short-period (<2 h) geomagnetic fluctuations such as bay disturbances and micropulsations, Terada (1917) calculated the phase difference between the vertical and horizontal components of geomagnetic variations assuming that it was caused by electric currents in Earth. The relationship between orthogonal electric and magnetic field components in (1.4) was introduced by Hirayama (1934), and was at one time known as Terada's relation in Japan (Hirayama, 1934; Rikitake, 1951d). Hirayama (1934), and later Rikitake (1948), showed that the E/B amplitude is proportional to the square root of period with a constant phase difference of 45° for a uniform half-space. However, Hatakeyama & Hirayama (1934) reported simultaneous measurements of time-varying surface electric and magnetic fields at Toyohara, Japan (now Yuzhno-Sakhalinsk, Russia), showing that their phase difference varies with period, being smaller than 45° at long periods. The recurring observation that the measured E/B amplitude and phase changes with period was rigorously explained by Kato & Kikuchi (1950a,b) and Rikitake (1951d) through a varying resistivity with depth under the assumption of a plane-wave source outside the atmosphere. The content of Rikitake (1951d) was first presented at a Japanese national meeting in October 1949.

Geophysical tradition holds that the theory of magnetotellurics was proposed simultaneously and independently by Tikhonov (1950) in the USSR and Cagniard (1953) in France. Cagniard notes in the Acknowledgements to his paper that: "The theoretical work reported in this paper was done some time ago and has been mentioned in applications for patents which have been made in several countries to protect the new prospecting method involved. Because of the potential practical applications I have had to postpone any publication related to magneto-telluric phenomena for many years." But he goes on to say that: "Meanwhile, the Russian scientist Tikhonov, and the Japanese scientists Kato, Kikuchi and Rikitake had also recognized the existence of such an effect. To my knowledge, they have not pointed out the possibility disclosed by my work of applying these results to practical geophysical exploration. They have, however, paid attention to their possible use for investigating the electrical conductivities of very deep regions in the Earth's crust." Thus, although Cagniard deserves the credit for appreciating that magnetotellurics could become a geophysical exploration tool for resources and gives practical examples of its use, the key concepts underlying magnetotellurics originated earlier in Japan than in Europe, and had achieved an advanced level of development and application by the early 1950s. Rikitake deserves as much of the credit for its development as the customary choices of Tikhonov and Cagniard, although magnetotellurics is really the cumulative result of research by a number of investigators over the first half of the twentieth century.

It should be noted that Tikhonov (1950), Rikitake (1951d) and Cagniard (1953) are all physically wrong because they are based on the concept of a plane-wave source that cannot be responsible for electromagnetic induction, as shown in Chapters 2 and 4. However, they remain mathematically correct, and constitute the historical basis for the developments in the subsequent 60 years, with primarily those of the past 25 years comprising the subject of this book.

However, controversy over the plane-wave source model continued into the 1960s. Wait (1954, 1962) cast aspersions on the validity of the assumption of plane-wave sources, since finite ionospheric sources do not give rise to normally incident plane waves. Further, magnetotelluric results inferring conductivity distributions well into the mantle (Migaux *et al.*, 1960) did not agree with geomagnetic depth sounding analyses, and the difference was sometimes ascribed to finite source fields. This led Price (1962) to develop a general theory for the magnetotelluric method with a finite dimension for the source fields. Srivastava (1965) extended the plane-wave recursion algorithms of Wait (1954) to include the effect of finite source dimensions, and gave a curve-matching technique for estimating not only the conductivity layering parameters but also the scale of the source when both horizontal components of the magnetic field are measured.

Computer modeling studies undertaken by Madden & Nelson (1964) and Srivastava (1965) indicated that, for realistic Earth conductivity profiles, the plane-wave source field assumption is valid for periods up to 10^3 s. Swift (1967) analyzed magnetograms from two stations 1300 km apart (Dallas and Tucson) at mid-geomagnetic latitude, and concluded that, for the period band 10^3–10^5 s, the wavelength of the source field must be in excess of 10^4 km. This corresponds to a wavenumber ($k = 2\pi/\lambda$) of the order of 10^{-5} m^{-1}, and therefore from Srivastava's (1965) studies, the plane-wave mathematical assumption is valid at mid-geomagnetic latitudes for periods up to 10^5 s. This conclusion has stood the test of time, and it is only in unusual locations (e.g. beneath the equatorial electrojet or during auroral activity at high latitudes) that finite source fields seriously bias magnetotelluric data.

During the 1950s and 1960s, it also became apparent that the scalar nature of the magnetotelluric response from Tikhonov (1950), Rikitake (1951d) and Cagniard (1953) was inadequate for Earth studies. Neves (1957) appears to have been the first to recognize the tensor nature of the relationship between the electric and magnetic fields, and defined a finite-difference algorithm for solving the 2D magnetotelluric forward problem.

The authors have chosen to stop the historical perspective at this point out of deference to living colleagues who played a role after about 1960, and in the belief that the authors of the chapters in this book have provided their perspective about each specialty area that should take precedence.

1.4 Commercial use of magnetotellurics

Although almost all of the authors of the chapters in this book are academics, magnetotellurics is used in industry to a far larger extent than in academia. Large academic magnetotelluric groups have perhaps 10 or so broadband magnetotelluric (BBMT) systems, whereas there are over 500 BBMT systems from one manufacturer in continuous use in China alone for oil and gas exploration.

As far as the editors of this book are aware, the first commercial use of magnetotellurics was for geothermal exploration in the USA in the late 1950s and early 1960s. Magnetotellurics was in significant use in the Soviet Union during the 1970s, with over 100 crews actively recording data primarily for hydrocarbon exploration (Spies, 1983), although the majority used analog equipment and determined the apparent resistivity in a scalar manner by measuring the heights and periods of sinusoids on paper chart recorders.

Audiomagnetotellurics has been used for mineral exploration for both base and precious metals since the early 1970s (Strangway *et al.*, 1971; Garcia & Jones, 2000), and has seen much growth over the past 15 years, with over 25 000 sites collected in Canada alone. Audiomagnetotellurics is being used to explore depths that are beyond those easily attainable with controlled-source electromagnetic surveys (i.e. 250–2000 m). Most of the surveys were either for anomaly definition, where the original anomalies were detected by other means, or for "sterilization" (ensuring that there is not a large deposit in an area before allowing its lease to expire). Areas of particular interest for nickel exploration have been the Sudbury structure (Livelybrooks *et al.*, 1996; Stevens & McNeice, 1998; Zhang *et al.*, 1998), the Voisey's Bay deposit in Newfoundland (Balch *et al.*, 1998; Zhang *et al.*, 1998; Watts & Balch, 2000), and the Thompson Nickel Belt in Canada. Key areas for gold exploration include the Carlin Trend in Nevada (Morrison *et al.*, 1990; Wannamaker & Doerner, 2002; Nutt & Hofstra, 2003; Petrick, 2007). However, problems can occur in small-scale surveys owing to the regional nature of the governing current systems (Jones & Garcia, 2003).

The depth range of hundreds of meters to kilometers is also of interest for geothermal exploration, where the target in high-enthalpy regions is the conductive, impermeable clay cap layer (montmorillonite) that seals the geothermal (sericite) reservoir below (e.g. Jones & Dumas, 1993; Newman *et al.*, 2008). Work has also been conducted on low-enthalpy geothermal prospects (Muñoz *et al.*, 2010a,b).

Deeper still, with regard to hydrocarbon exploration, the primary uses of magnetotellurics have been in greenfield areas that are logistically challenging (e.g. heavily forested areas such as Papua New Guinea; Christopherson, 1989), and in areas with a resistive overburden (e.g. either basalt and volcanic cover or basement overthrust regions), leading to a large contrast with the underlying sediments (e.g. Anderson & Pelton, 1985; Watts & Pince, 1998; Colombo *et al.*, 2011). During the early 1980s, when oil rose to the dizzying height of US $35/barrel, many major US oil companies formed in-house magnetotelluric groups, often constituted from personnel with physics backgrounds and without any training in magnetotellurics, and there were four or five large and very busy independent contractors (e.g. AET, Phoenix, Z-axis, Zonge) in the USA alone. Those groups and companies mostly were out of business by the late 1980s as the price of oil dropped and the major oil companies no longer experimented with "exotic" technologies. Interest in magnetotellurics has resurged over the past decade, particularly in the marine realm, but there remains a boom-and-bust cycle tied to the price of oil.

Finally, the deepest mineral of commercial interest is diamonds, and magnetotellurics has been promoted primarily to define the depth to the base of the cold, cratonic lithosphere for production area selection (Jones & Craven, 2004). Results from northwestern Canada and southern Africa are characterizing the extent of existing producing zones and defining possible new areas of interest (Jones *et al.*, 2003, 2009; Muller *et al.*, 2009; Evans *et al.*, 2011).

1.5 The future of magnetotellurics

Predicting the future is always a challenge (and often wrong), but some trends are apparent in magnetotellurics. Starting at the front end (instrumentation), it is clear that, even today, most surveys produce data that are spatially aliased for the geological problems being addressed. For surveys imaging the top ~5 km of the crust, continuous profiling is necessary using instrumentation similar to Francis Bostick's EMAP (Bostick, 1986; Torres-Verdin & Bostick, 1992) or Nick Sheard's MIMDAS (Garner & Webb, 2000; Goldie, 2007), as used by Quantec in their TITAN24 system, but with a full five electromagnetic field components at each location to obtain full resolution of the anomalies and avoid the problems associated with steeply dipping targets that respond weakly in the along-profile transverse magnetic response (Jones & McNeice, 2002). Regional surveys should have sites some 1–2 km apart, instead of the more common 10–20 km. The sensors can still be improved, with the greatest weakness in magnetotellurics being electrodes: stable, robust, low-noise electrodes that can be installed quickly and last for survey after survey for many months to years are not yet available. Recorders can move to a full 32-bit resolution to enable greater signal-to-noise ratio through oversampling. The biggest challenge is cost – whereas commercial seismic reflection systems have become an order of magnitude cheaper over the past decades, resulting in substantial increases in data redundancy due to the low cost of geophones and channels, a five-component magnetotelluric system is of the order of US$50 000, and needs to come down to one-tenth of that through production manufacturing, rather than the current "cottage industry" approach, to enable greater station density.

One significant advantage of profile, or preferably array, acquisition is that not only can each single-station magnetotelluric response function be estimated, but inter-station response functions (both electric and magnetic) can also be determined. As shown by Soyer & Brasse (2001), the inter-station magnetic response functions can yield almost as much information as the conventional magnetotelluric one, and have the significant advantage that they are not affected by galvanic distortion of the electric fields. Indeed, Ledo *et al.* (2002) showed that such transfer functions can be used to define static shifts in magnetotelluric data.

Once time-series data have been recorded, they must be processed into the frequency domain for conventional magnetotellurics. Robust, remote reference techniques, developed in the mid- to late-1980s by Alan Chave, Gary Egbert, John Booker and Alan Jones, have proven their worth by producing dramatically more precise and accurate responses. Significant problems still exist with correlated (across sites) noise sources, such as the low-frequency disruption by direct-current (DC) train lines and repetitive high-frequency noise sources such as electric fences. Improved processing algorithms that deal with these noise complications need to be developed.

Having derived the most precise and accurate magnetotelluric response functions possible, the results need to be appraised for distortion (see Chapter 6). Current methods deal well with 3D galvanic distortion of the electric fields over 1D or 2D regional structures. Work needs to be undertaken to advance methods of appraisal of 3D distortion over 3D regional structures. Some work has been undertaken that highlights the issues (Garcia & Jones, 2002). Perhaps the optimum approach will be incorporating consideration of galvanic distortion as part of 3D inversion for structure, as is currently being investigated by Miensopust (2009) and Avdeeva *et al.* (2011).

Once the most optimum regional responses have been defined, they must be modeled and inverted for structure beneath an array. Advances are progressing at a significant rate, but nevertheless the computing requirements are high. Hybrid methods of decomposing a large inverse problem into many smaller ones may prove fruitful (Curtis, 2011). As noted recently by Zhdanov (2010), there is merit in recognizing that in magnetotellurics the electric field *per se* is not measured, but rather is derived from potential differences. Modeling those voltages, as discussed and implemented originally in 2D by Poll *et al.* (1989) and employed by Jones (1988) in the examination of static shift effects, is important for mining-scale problems.

However, almost all current inversion codes do not recognize that the data come from an Earth that will permit only certain ranges, given permissible temperatures, pressures, salinity, permeability, porosity and so on. Future codes should invert the data in a manner that will ensure consistency with those other constraints; an example is the 1D inversion of mantle-probing magnetotelluric data from the Kaapvaal Craton that is constrained by mineral physics in Fullea *et al.* (2011) within the LITMOD approach (Afonso *et al.*, 2008; Fullea *et al.*, 2009).

1.6 More information on magnetotellurics

Magnetotellurics is a rapidly advancing field, and some aspects of this book may be virtually out of date when its content is read. There is an Internet portal for magnetotellurics, called MTNet and located at the URL www.mtnet.info, that tries to keep the community informed of advances. MTNet is an independent resource for the electromagnetic induction community, and is intended as an international electronic forum for the free exchange of knowledge, programs and data between scientists engaged in the study of the Earth using electromagnetic methods, principally but not exclusively magnetotellurics.

On MTNet, links can be found to publications (both in press and within the last year), and also links to the extensive review papers given at the biennial "EM Induction Workshops" that the reader is strongly recommended to consult for more recent developments.

A number of books have been published in recent years, including *Practical Magnetotellurics* by Simpson & Bahr (2005) also published by Cambridge University Press, which is an entry-level text on the subject. Most of the other books treat the theoretical aspects in greater detail, including Berdichevsky & Dmitriev (2008) and Zhdanov (2009), but lack the thorough treatment of the experimental and observational aspects presented herein.

Unfortunately, formal courses on magnetotellurics or that contain it are few and far between. Sadly, magnetotellurics is not taught routinely as part of undergraduate- or graduate-level courses in applied geophysics. The Society of Exploration Geophysicists (SEG) invited Karen Christopherson to present a course on magnetotellurics at the Annual Meeting in 2002. The course, *Magnetotellurics for Natural Resources: From Acquisition Through Interpretation*, was given by Christopherson, together with Randy Mackie and Alan Jones, in 2002, and has become one of the standard Continuing Education courses offered by the SEG. The course is also presented irregularly at other locations globally by Jones (Dublin, 2008 and 2010; South Africa, 2008; Egypt, 2009; China, 2010). In 2011 a new course was presented at the SEG, *Marine Electromagnetic Methods for Hydrocarbon Exploration*, focused on marine controlled-source electromagnetics but with marine magnetotellurics as a component.

1.7 Epilogue

As the contrast between the two leading examples of magnetotellurics using the top methods of their day – namely Stanley *et al*. (1977) and Rosell *et al*. (2011) – shows, magnetotellurics has come a long way, primarily because of the advances presented in this book. Who knows where magnetotellurics will be in 2040?

Magnetotellurics is a fascinating field that demands a diverse set of skills for it to be practiced at its highest levels. In both academia and industry, it brings together a diverse array of scientists and students with varying backgrounds. All the authors hope not only that this book will both educate and enthrall, but also that the reader will keep in mind that the field can evolve so rapidly that some aspects presented will only be current for perhaps a decade, whereas others, such as the theory parts, will last forever.

References

Afonso, J. C., M. Fernàndez, G. Ranalli, W. L. Griffin & J. A. D. Connolly (2008). Combined geophysical–petrological modeling of the lithospheric–sublithospheric upper mantle: methodology and applications. *Geochem. Geophys. Geosyst.*, **9**, Q05008.

Airy, G. B. (1868). Comparison of magnetic disturbances recorded by the self-registering magnetometers at the Royal Observatory, Greenwich, with magnetic disturbances deduced from the corresponding terrestrial galvanic currents recorded by the self-registering galvanometers of the Royal Observatory. *Phil. Trans. R. Soc. Lond.*, **158**, 465–472.

Anderson, R. G. & W. H. Pelton (1985). MT exploration in volcanic cover, overthrust belts, and rift zones. *SEG Tech. Prog. Exp. Abstr.*, **4**, 274–276.

Avdeeva, A., D. B. Avdeev & M. Moorkamp (2011). Inverting Dublin secret dataset 2 with x3Di. In *MT3DINV2: Second Three-Dimensional Magnetotelluric Inversion Workshop*, Dublin, Ireland.

Balch, S. J., T. J. Crebs, A. King & M. Verbiski (1998). Geophysics of the Voisey's Bay Ni–Cu–Co deposits. *SEG Tech. Prog. Exp. Abstr.*, **17**, 784–787.

Berdichevsky, M. N. & I. V. Dmitriev (2008). *Models and Methods of Magnetotellurics*. Berlin: Springer.

Bostick, F. X. (1986). Electromagnetic array profiling (EMAP). *SEG Tech. Prog. Exp. Abstr.*, **5**, 60–61.

Cagniard, L. (1953). Basic theory of the magnetotelluric method of geophysical prospecting. *Geophysics*, **18**, 605–635.

Chapman, S. (1919). The solar and lunar diurnal variation of the earth's magnetism. *Phil. Trans. R. Soc. Lond.*, **A218**, 1–118.

Chapman, S. & A. Price (1930). The electric and magnetic state of the interior of the earth as inferred from terrestrial magnetic variations. *Phil. Trans. R. Soc. Lond.*, **A229**, 427–460.

Christopherson, K. R. (1989). Magnetotellurics in Papua New Guinea. *SEG Tech. Prog. Exp. Abstr.*, **8**, 160–164.

Colombo, D., T. Keho, E. Janoubi & W. Soyer (2011). Sub-basalt imaging with broadband magnetotellurics in NW Saudi Arabia. *SEG Tech. Prog. Exp. Abstr.*, **30**, 619–623.

Curtis, A. (2011). Fast, nonlinear, probabilistic inversion of large geophysical problems. In *Proc. 8th EGU General Assembly, Vienna, Austria. Geophys. Res. Abstr.*, **13**, EGU2011-9155-1.

Evans, R. L., A. G. Jones, X. Garcia, M. Muller, M. Hamilton, S. Evans, S. Fourie, J. Spratt, S. Webb, H. Jelsma & D. Hutchins (2011). The electrical lithosphere beneath the Kaapvaal Craton, Southern Africa. *J. Geophys. Res.*, **116**, B04105, doi:10.1029/2010JB007883.

Fullea, J., J. C. Afonso, J. A. D. Connolly, M. Fernandez, D. Garcia-Castellanos & H. Zeyen (2009). LitMod3D: an interactive 3-D software to model the thermal, compositional, density, seismological, and rheological structure of the lithosphere and sublithospheric upper mantle, *Geochem. Geophys. Geosyst.*, **10**, Q08019, doi:10.1029/2009GC002391.

Fullea, J., M. R. Muller & A. G. Jones (2011). Electrical conductivity of continental lithospheric mantle from integrated geophysical and petrological modeling: application to the Kaapvaal Craton and Rehoboth Terrane, southern Africa. *J. Geophys. Res.*, **116**, B10202, doi:10.1029/2011JB008544.

Garcia, X. & A. G. Jones (2000). Advances in aspects of the application of magnetotellurics for mineral exploration. *SEG Tech. Prog. Exp. Abstr.*, **19**, 1115–1118.

Garcia, X. & A. G. Jones (2002). Decomposition of three-dimensional magnetotelluric data. In *Three-Dimensional Electromagnetics*, ed. M. S. Zhdanov & P. E. Wannamaker. Tulsa: Society of Exploration Geophysicists, pp. 235–250.

Garner, S. & D. Webb (2000). Broadband MT and IP electrical property mapping with MIMDAS. *SEG Tech. Prog. Exp. Abstr.*, **19**, 1085–1088.

Goldie, M. (2007). A comparison between conventional and distributed acquisition induced polarization surveys for gold exploration in Nevada. *The Leading Edge*, **26**, 180–183.

Hatakeyama, H. & M. Hirayama (1934). On the phase difference between the pulsation of terrestrial magnetism and of earth current. *J. Meteorol. Soc. Japan*, **12**, 449–459 (in Japanese).

Hirayama, M. (1934). On the relations between the variations of earth potential gradient and terrestrial magnetism. *J. Meteorol. Soc. Japan*, **12**, 16–22 (in Japanese).

Jones, A. G. (1988). Static shift of magnetotelluric data and its removal in a sedimentary basin environment. *Geophysics*, **53**, 967–978.

Jones, A. G. & J. A. Craven (2004). Area selection for diamond exploration using deep-probing electromagnetic surveying. *Lithos*, **77**, 765–782.

Jones, A. G. & I. Dumas (1993). Electromagnetic images of a volcanic zone. *Phys. Earth Planet. Inter.*, **81**, 289–314.

Jones, A. G. & X. Garcia (2003). Okak Bay AMT data-set case study: lessons in dimensionality and scale. *Geophysics*, **68**, 70–91.

Jones, A. G. & G. D. Garland (1986). Preliminary interpretation of the upper crustal structure beneath Prince Edward Island. *Ann. Geophys.*, **4B**, 157–164.

Jones, A. G. & G. McNeice (2002). Audio-magnetotellurics (AMT) for steeply-dipping mineral targets: importance of multi-component measurements at each site. In *Proc. 72nd SEG Meeting*, Salt Lake City, Utah, Abstracts, pp. 496–499.

Jones, A. G., P. Lezaeta, I. J. Ferguson, A. D. Chave, R. L. Evans, X. Garcia & J. Spratt (2003). The electrical structure of the Slave Craton, *Lithos*, **71**, 505–527.

Jones, A. G., R. L. Evans, M. R. Muller, M. P. Hamilton, M. P. Miensopust, X. Garcia, P. Cole, T. Ngwisanyi, D. Hutchins, C. J. S. Fourie, H. Jelsma, S. Evans, T. Aravanis, W. Pettit, S. Webb, J. Wasborg & SAMTEX Team (2009). Area selection for diamonds using magnetotellurics: examples from southern Africa. *Lithos*, **112**, 83–92, doi:10.1016/j.lithos.2009.06.011.

Kato, Y. & T. Kikuchi (1950a). On the phase difference of earth current induced by the changes of the earth's magnetic field, part 1. *Sci. Rep. Tohoku Univ.*, 5th Ser., **2**, 139–141.

Kato, Y. & T. Kikuchi (1950b). On the phase difference of earth current induced by the changes of the earth's magnetic field, part 2. *Sci. Rep. Tohoku Univ.*, 5th Ser., **2**, 142–145.

Lahiri, B. & A. Price (1939). Electromagnetic induction in non-uniform conductors, and the determination of the conductivity of the earth from terrestrial magnetic variations. *Phil. Trans. R. Soc. Lond.*, **A237**, 509–540.

Lamb, H. (1883). On electrical motions in a spherical conductor. *Phil. Trans. R. Soc. Lond.*, **A174**, 519–549.

Ledo, J., A. Gabas & A. Marcuello (2002). Static shift leveling using geomagnetic transfer functions. *Earth Planets Space*, **54**, 493–498.

Livelybrooks, D., M. Mareschal, E. Blais & J. T. Smith (1996). Magnetotelluric delineation of the Trillabelle massive sulfide body in Sudbury, Ontario. *Geophysics*, **61**, 971–986.

Madden, T. R. & P. Nelson (1964). A defense of Cagniard's magnetotelluric method. *Geophys. Lab*. ONR NR-371–401, Final Report, MIT, Cambridge, MA.

Martí, A., P. Queralt & J. Ledo (2009). WALDIM: a code for the dimensionality analysis of magnetotelluric data using the rotational invariants of the magnetotelluric tensor. *Comput. Geosci.*, **35**, 2295–2303.

Miensopust, M. P. (2009). *Multidimensional magnetotellurics – a 2D case study and a 3D approach to simultaneously invert for resistivity structure and distortion parameters*. Ph.D. thesis, National University of Ireland, Galway.

Migaux, L., J. L. Astier & P. H. Reval (1960). Un essai de determination experimentale de la resistivite electrique des couches profondes de l'ecorce terreste. *Ann. Geophys.*, **16**, 555–560.

Morrison, H. F., E. A. Nichols, C. Torres-Verdin, J. R. Booker & S. C. Constable (1990). Comparison of magnetotelluric inversion techniques on a mineral prospect in Nevada. *SEG Tech. Prog. Exp. Abstr.*, **9**, 516–519.

Muller, M. R., A. G. Jones, R. L. Evans, H. S. Grütter, C. Hatton, X. Garcia, M. P. Hamilton, M. P. Miensopust, P. Cole, T. Ngwisany, D. Hutchins, C. J. Fourie, H. A. Jelsma, S. F. Evans, T. Aravanis, W. Pettit, S. J. Webb, J. Wasborg & SAMTEX Team (2009). Lithospheric structure, evolution and diamond prospectivity of the Rehoboth Terrane and western Kaapvaal Craton, southern Africa: constraints from broadband magnetotellurics. *Lithos*, **112**, 93–105.

Muñoz, G., O. Ritter & I. Moeck (2010a). A target-oriented magnetotelluric inversion approach for characterizing the low enthalpy Gross Schonebeck geothermal reservoir, *Geophys. J. Int.*, **183**, 1199–1215.

Muñoz, G., K. Bauer, I. Moeck, A. Schulze & O. Ritter (2010b). Exploring the Gross Schonebeck (Germany) geothermal site using a statistical joint interpretation of magnetotelluric and seismic tomography models. *Geothermics*, **39**, 35–45.

Nabighian, M. N. (1987). *Electromagnetic Methods in Applied Geophysics*, Vol. **1**, Theory. Tulsa: Society of Exploration Geophysicists.

Nabighian, M. N. (1991). *Electromagnetic Methods in Applied Geophysics*, Vol. **2**, Applications. Tulsa: Society of Exploration Geophysicists.

Neves, A. S. D. (1957). *The generalized magneto-telluric method*. Ph.D. thesis, Massachusetts Institute of Technology.

Newman, G. A., E. Gasperikova, G. M. Hoversten & P. E. Wannamaker (2008). Three-dimensional magnetotelluric characterization of the Coso geothermal field. *Geothermics*, **37**, 369–399.

Nutt, C. J. & A. H. Hofstra (2003). Alligator ridge district, east-central Nevada: Carlin-type gold mineralization at shallow depths. *Econ. Geol. Bull. Soc. Econ. Geol.*, **98**, 1225–1241.

Petrick, W. R. (2007). Practical 3D magnetotelluric inversion: finally dispensing with TE and TM. *SEG Tech. Prog. Exp. Abstr.*, **26**, 521–523.

Poll, H. E., J. T. Weaver & A. G. Jones (1989). Calculations of voltages for magnetotelluric modelling of a region with near-surface inhomogeneities. *Phys. Earth Planet. Inter.*, **53**, 287–297.

Price, A. T. (1962). The theory of magnetotelluric fields when the source field is considered. *J. Geophys. Res.*, **67**, 1907–1918.

Rikitake, T. (1948). Note on the electromagnetic induction within the earth. *Bull. Earthq. Res. Inst., Univ. Tokyo*, **24**, 1–9.

Rikitake, T. (1950). Electromagnetic induction within the earth and its relation to the electrical state of the earth's interior, part I(1). *Bull. Earthq. Res. Inst., Univ. Tokyo*, **28**, 45–100.

Rikitake, T. (1951a). Electromagnetic induction within the earth and its relation to the electrical state of the earth's interior, part I(2). *Bull. Earthq. Res. Inst., Univ. Tokyo*, **28**, 219–262.

Rikitake, T. (1951b). Electromagnetic induction within the earth and its relation to the electrical state of the earth's interior, part II. *Bull. Earthq. Res. Inst., Univ. Tokyo*, **28**, 263–283.

Rikitake, T. (1951c). Electromagnetic induction within the earth and its relation to the electrical state of the earth's interior, part III. *Bull. Earthq. Res. Inst., Univ. Tokyo*, **29**, 61–69.

Rikitake, T. (1951d). Changes in earth current and their relation to the electrical state of the earth's crust. *Bull. Earthq. Res. Inst., Univ. Tokyo*, **29**, 271–276.

Rikitake, T. (1951e). Electromagnetic induction within the earth and its relation to the electrical state of the earth's interior, part IV. *Bull. Earthq. Res. Inst., Univ. Tokyo*, **29**, 539–547.

Rosell, O., A. Martí, À. Marcuello, J. Ledo, P. Queralt, E. Roca & J. Campanyà (2011). Deep electrical resistivity structure of the northern Gibraltar Arc (western Mediterranean): evidence of lithospheric slab break-off. *Terra Nova*, **23**, 179–186.

Schuster, A. (1889). The diurnal variation of terrestrial magnetism. *Phil. Trans. R. Soc. Lond.*, **A180**, 467–518.

Simpson, F. & K. Bahr (2005). *Practical Magnetotellurics*. Cambridge: Cambridge University Press.

Sims, W. E. & F. X. Bostick, Jr. (1969). *Methods of magnetotelluric analysis*. Tech. Rep. 58, Electrical Engineering Research Laboratory, University of Texas, Austin.

Siripunvaraporn, W., G. Egbert, Y. Lenbury & M. Uyeshima (2005). Three-dimensional magnetotelluric data: data space method. *Phys. Earth Planet. Inter.*, **150**, 3–14.

Soyer, W. & H. Brasse (2001). A magneto-variation array study in the central Andes of N Chile and SW Bolivia. *Geophys. Res. Lett.*, **28**, 3023–3026.

Spies, B. R. (1983). Recent developments in the use of surface electrical methods for oil and gas exploration in the Soviet Union. *Geophysics*, **48**, 1102–1112.

Srivastava, S. P. (1965). Method of interpretation of magnetotelluric data when the source field is considered. *J. Geophys. Res.*, **70**, 945–954.

Stanley, W. D., J. E. Boehls, F. X. Bostick & H. W. Smith (1977). Geothermal significance of magnetotelluric sounding in the eastern Snake River Plain–Yellowstone region. *J. Geophys. Res.*, **82**, 2501–2514.

Stevens, K. M. & G. W. McNeice (1998). On the detection of Ni–Cu ore hosting structures in the Sudbury Igneous Complex using the magnetotelluric method. *SEG Tech. Prog. Exp. Abstr.*, **17**, 751–755.

Strangway, D. W., C. M. Swift & R. C. Holmer (1971). Application of audio frequency magnetotellurics (AMT) to mineral exploration. *Geophysics*, **36**, 1159–1175.

Swift, C. M., Jr. (1967). *A magnetotelluric investigation of an electrical conductivity anomaly in the southwestern United States*. Ph.D. thesis, Massachusetts Institute of Technology.

Tarits, P. (1994). Electromagnetic studies of global geodynamic processes. *Surv. Geophys.*, **15**, 209–238.

Terada, T. (1917). On rapid periodic variations of terrestrial magnetism. *J. Coll. Sci., Imp. Univ. Tokyo*, **37**, art. 9, 1–85.

Tikhonov, A. N. (1950). On determination of electric characteristics of deep layers of the earth's crust. *Dokl. Acad. Nauk SSSR*, **151**, 295–297.

Torres-Verdin, C. & F. X. Bostick (1992). Principles of spatial surface electric-field filtering in magnetotellurics – electromagnetic array profiling (EMAP). *Geophysics*, **57**, 603–622.

Wait, J. R. (1954). On the relation between telluric currents and the earth's magnetic field. *Geophysics*, **19**, 281–2389.

Wait, J. R. (1962). Theory of the magnetotelluric field. *J. Res. Natl Bur. Stand.*, **66D**, 590–641.

Wannamaker, P. E. & W. M. Doerner (2002). Crustal structure of the Ruby Mountains and southern Carlin trend region, Nevada, from magnetotelluric data. *Ore Geol. Rev.*, **21**, 185–210.

Watts, M. D. & S. J. Balch (2000). AEM-constrained 2D inversion of AMT data over the Voisey's Bay massive sulfide body, Labrador. *SEG Tech. Prog. Exp. Abstr.*, **19**, 1119–1121.

Watts, M. D. & A. Pince (1998). Petroleum exploration in overthrust areas using magneto-telluric and seismic data. *SEG Tech. Prog. Exp. Abstr.*, **17**, 429–431.

Word, D. R., H. W. Smith & F. X. Bostick, Jr. (1970). *An investigation of the magnetotelluric tensor impedance method*. Tech. Rep. 82, Electrical Engineering Research Laboratory, University of Texas, Austin.

Zhang, P., A. King & D. Watts (1998). Using magnetotellurics for mineral exploration. *SEG Tech. Prog. Exp. Abstr.*, **17**, 776–779.

Zhdanov, M. S. (2009), *Geophysical Electromagnetic Theory and Methods*. Amsterdam: Elsevier.

Zhdanov, M. S. (2010). Electromagnetic geophysics: notes from the past and the road ahead. *Geophysics*, **75**, A49–A66.

2

The theoretical basis for electromagnetic induction

ALAN D. CHAVE AND PETER WEIDELT

2.1 The Maxwell equations

The four Maxwell equations taken in aggregate are a complete description of the relationships between electric and magnetic fields in any medium. Because the Maxwell equations are covariant under Lorentz transformations, their differential form in either a resting or moving inertial reference frame using the MKS (SI) system is

$$\nabla \cdot \mathbf{D} = \rho_e \tag{2.1}$$

$$\nabla \cdot \mathbf{B} = 0 \tag{2.2}$$

$$\nabla \times \mathbf{E} = -\partial_t \mathbf{B} \tag{2.3}$$

$$\nabla \times \mathbf{H} = \mathbf{J} + \partial_t \mathbf{D} \tag{2.4}$$

where $\mathbf{B}$ is the magnetic induction, $\mathbf{H}$ is the magnetic field, $\mathbf{D}$ is the electric displacement, $\mathbf{E}$ is the electric field, $\mathbf{J}$ is the electric current density and ρ_e is the electric charge density. Constitutive relations are also required that connect the field components with material properties of the medium in which they occur, as further described below. Equations (2.1)–(2.4) respectively state that: (1) the electric field diverges from electric charges (Gauss's law for electricity); (2) there are no magnetic monopoles (Gauss's law for magnetism); (3) circulating electric fields are produced by time-varying magnetic fields (Faraday's law); and (4) circulating magnetic fields are produced by the vector sum of electric currents and time-varying electric fields (Ampere's law).

For a medium in which the constitutive relationships are linear in the material properties and that is moving with velocity $\mathbf{v}$, the fields are connected by (Sommerfeld, 1952, Ch. 34)

$$\mathbf{D} + \frac{1}{c^2} \mathbf{v} \times \mathbf{H} = \varepsilon(\mathbf{E} + \mathbf{v} \times \mathbf{B}) \tag{2.5}$$

$$\mathbf{B} - \frac{1}{c^2} \mathbf{v} \times \mathbf{E} = \mu(\mathbf{H} - \mathbf{v} \times \mathbf{D}) \tag{2.6}$$

$$\mathbf{J} = \sigma(\mathbf{E} + \mathbf{v} \times \mathbf{B}) + \rho_e \mathbf{v} + \mathbf{J}^0 \tag{2.7}$$

The Magnetotelluric Method: Theory and Practice, ed. Alan D. Chave and Alan G. Jones. Published by Cambridge University Press © Cambridge University Press 2012.

where ε is the electric permittivity, μ is the magnetic permeability, σ is the electrical conductivity (all of which are taken to be scalars for the present, so the medium is isotropic), c is the free-space speed of light and $\mathbf{J}^0$ is an extrinsic (i.e. applied) electric current density. Equations (2.5) and (2.6) are exact, while (2.7) excludes terms of $O(v^2/c^2)$ and smaller, where $v = |\mathbf{v}|$. On the spatial scale over which magnetotellurics operates, except within certain types of ore bodies, Earth materials are not magnetizable, and hence the magnetic permeability may be taken as that of free space μ_0.

To within terms of $O(v^2/c^2)$ and smaller, (2.5) may be simplified to

$$\mathbf{D} = \varepsilon_0 \mathbf{E} + \mathbf{P} \tag{2.8}$$

with the electric polarization given by

$$\mathbf{P} = \varepsilon_0 (\kappa - 1)(\mathbf{E} + \mathbf{v} \times \mathbf{B}) \tag{2.9}$$

where the free-space electric permittivity is ε_0 and the dimensionless quantity $\kappa = \varepsilon/\varepsilon_0$ is the dielectric constant. Substituting (2.8) into (2.6), and placing the result together with (2.7) into (2.4) yields Ampere's law in the form

$$\nabla \times \mathbf{B} = \mu_0 [\mathbf{J}_c + \rho_e \mathbf{v} + \partial_t \mathbf{D} + \nabla \times (\mathbf{P} \times \mathbf{v}) + \mathbf{J}^0] \tag{2.10}$$

where $\mathbf{J}_c = \sigma(\mathbf{E} + \mathbf{V} \times \mathbf{B})$. The five terms on the right side of (2.10) represent conduction current, advected electric charge, displacement current, polarization current and extrinsic current, respectively. Taking the divergence of (2.10) and using (2.1) gives the equation for conservation of charge

$$\nabla \cdot \mathbf{J}_c = -\partial_t \rho_e - \nabla \cdot (\rho_e \mathbf{v}) - \cdot \mathbf{J}^0 \tag{2.11}$$

The complete set of Maxwell equations (2.1)–(2.3) and (2.10) apply to many classes of geophysical problems, including (when combined with the equations of motion) magneto-hydrodynamic processes in Earth's core, many ionospheric and magnetospheric current systems, motional induction caused by moving seawater cutting the main geomagnetic field, controlled-source electromagnetics and magnetotellurics. In each instance, a suitable scaling of the Maxwell equations may be used to determine which terms on the right side of (2.10) are significant, further simplifying the set. The physics of electromagnetic fields of internal (Earth's core) and external (ionospheric and magnetospheric) origin are beyond the scope of this book, and will not be treated further. The remaining three classes will be considered in turn.

2.2 Motional electromagnetic induction

Electromagnetic fields induced by water motion in the geomagnetic field have important applications in understanding the energy and momentum balances of the ocean (i.e. in physical oceanography), and can be a significant source of noise for magnetotelluric or controlled-source electromagnetic studies within the ocean, especially in coastal areas where the water depth is under 1000 m or so. While physical oceanography applications are beyond the scope of this book, an understanding of motional induction noise will aid in interpreting seafloor magnetotelluric data, especially as applications of electromagnetic methods for hydrocarbon exploration on the continental shelves proliferate.

Sanford (1971) considered the motional induction problem beginning from (2.1)–(2.3) and (2.10). The extrinsic current $\mathbf{J}^0$ will be taken as zero. Since motionally induced magnetic fields $\mathbf{b}$ are substantially weaker than the main geomagnetic field $\mathbf{F}$ that is taken to be constant, $\mathbf{B} = \mathbf{F} + \mathbf{b}$, where $\nabla \cdot \mathbf{F} = \nabla \times \mathbf{F} = 0$ and $|\mathbf{b}| \ll |\mathbf{F}|$. Scale the velocity $\mathbf{v}$ by a characteristic value v_0, x and y by a characteristic horizontal length scale L, and z by a characteristic vertical length scale H (all of which will be defined later), so that the magnitudes of $\mathbf{J}_c$ and $\mathbf{E}$ are $\sigma F v_0$ and $F v_0$, respectively, where $F = |\mathbf{F}|$. Then, the first four terms on the right side of (2.10) are in the ratio

$$1 : \frac{\varepsilon_0 \kappa v_0}{\sigma\varsigma} : \frac{\varepsilon_0 \kappa}{\sigma T} : \frac{2\varepsilon_0(\kappa - 1)v_0}{\sigma\varsigma} \tag{2.12}$$

where ς is the smaller of (L, H) and T is a characteristic time scale. Typical SI values in seawater are $\varepsilon_0 \approx 10^{-11}$ F/m, $\kappa \approx 80$, $v_0 \approx 1$ m/s and $\sigma \approx 3$ S/m.

At periods substantially longer than the inertial time scale (half of the rotation period of Earth on a latitude-dependent tangent or f plane, or $43\,200/\sin \lambda$ in s, where λ is latitude; at mid-latitudes a typical value would be $\approx 10^5$ s), which characterizes a diverse set of open-ocean phenomena, $L \gg H$ and H is the water depth (≈ 5000 m in the open ocean). A typical size for L is 100 km or more. In this case, advected charge, displacement current and polarization current are all less than 10^{-13} times the size of the conduction current and may be neglected, so that Ampere's law may be written as

$$\nabla \times \mathbf{b} = \mu_0 \sigma(\mathbf{E} + \mathbf{v} \times \mathbf{F}) \tag{2.13}$$

Sanford (1971) and Chave & Luther (1990) have considered the theory of motional induction in the sub-inertial limit, concluding that the motional horizontal electric field is proportional to the vertically integrated, seawater conductivity-weighted horizontal water velocity

$$\langle \mathbf{v}_h \rangle^* = \frac{\int\limits_H^0 \sigma(z)\mathbf{v}(z)dz}{\int\limits_H^0 \sigma(z)dz} \tag{2.14}$$

which is very nearly the vertically averaged or barotropic water velocity in most of the world oceans. The horizontal electric field is nearly independent of location in the water column, and is given by

$$\mathbf{E}_h = C F_z \, \hat{\mathbf{z}} \times \langle \mathbf{v}_h \rangle^* \tag{2.15}$$

where C is a scalar constant and F_z is the vertical component of the geomagnetic field. A similar expression to (2.15) applies for the magnetic field, while the vertical electric field is proportional to the geomagnetic east–west water velocity at the point of measurement. The scale factor C was originally thought to reflect the influence of electric current leakage

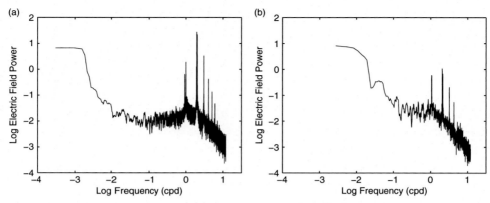

Figure 2.1. Multi-taper power spectra (see Section 5.3) in $mV^2 \, km^{-2} \, d^{-1}$ computed (a) from nine years of hourly means of the electric field from the north segment of the HAW-1 undersea cable extending from Point Arena, California to Hanauma Bay, Hawaii, and (b) for a year-long time series of the east electric field from north of the Hawaiian Islands ($28°41.7'N$, $163°44.9'W$) obtained with a horizontal electrometer that measures the electric field over a 3 m span on the seafloor. The horizontal electrometer utilizes water chopping technology (see Chapter 9) that eliminates the effect of electrode drift whose influence is also minimized when measuring over a large geographic span. The HAW-1 estimate has a resolution bandwidth of $0.0037 \, d^{-1}$ at all frequencies and possesses about 22 degrees of freedom at each frequency. The Hawaii estimate has a resolution bandwidth of $0.022 \, d^{-1}$ at all frequencies and possesses about 14 degrees of freedom at each frequency. Note that the electrometer power spectrum rises rapidly at periods longer than a few days, while the HAW-1 power spectrum does not show this effect until a period of ~100 d.

into the seafloor, but is now known also to include the effect of galvanic distortion by electric charge on the seawater–rock interface (Chave *et al.*, 2004).

The relationship of the motional electric field to (2.14) has been exploited for investigation of boundary currents such as the Gulf Stream and deep western boundary current (Larsen, 1992; Chave *et al.*, 1997; Baringer & Larsen, 2001) and to investigate the oceanic response to wind forcing in weak oceanic eddy kinetic energy regions (Chave *et al.*, 1992). The barotropic velocity field of the ocean is responsible for the very sharp rise in spectral level for oceanic electric field measurements at periods longer than a few days (Figure 2.1) that masks the much weaker electric field induced by external sources. As a consequence, very long-period magnetotelluric measurements cannot be obtained within the ocean. The use of extended (several thousand kilometers) seafloor cables averages out the motional electric field whose spatial scale is smaller than the cable length, substantially reducing the long-period increase in power (Figure 2.1), but even then does not provide usable magnetotelluric data at periods longer than a few days (Lizarralde *et al.*, 1995). In fact, terrestrial electric field measurements may be influenced by oceanic sources even at substantial distances from the coast (e.g. at Tucson, AZ; Egbert *et al.*, 1992). By contrast, motional magnetic fields are typically weak in most of the oceans, although a few measurements have been obtained in extremely energetic boundary currents (Lilley *et al.*, 1993, 2001).

While a wide range of much shorter-period phenomena may affect the motional electromagnetic field in coastal waters, wind waves and swell are ubiquitous. A typical time scale for swell is 10 s, and its length scale follows from the dispersion relation for gravity waves $\omega^2 = gk \tanh kH$ (Phillips, 1977), where ω is angular frequency, g is the acceleration of gravity, k is the wavenumber and H is the water depth. For 10 s waves, the corresponding length scale is about 160 m in water over 200 m deep. Using these values in (2.12), advected charge, displacement current and polarization currents are all less than 10^{-11} times the size of the conduction current, so that (2.13) still obtains. The electromagnetic fields induced by surface gravity waves have been treated by Weaver (1965), Larsen (1971) and Podney (1975). Cox *et al.* (1978) and Webb & Cox (1986) extended the theory to microseisms, while Webb & Cox (1982) derived expressions for the motional electric field produced by seismic waves. These share similar scales to surface gravity waves, hence the same approximations in the Maxwell equations pertain. It should be noted that the cited references pertain to the motional electromagnetic field induced by moving seawater, and not to the seismoelectric effect that is important in groundwater exploration.

Chave & Cox (1982) presented rough order-of-magnitude estimates for motional noise in the period band where marine controlled-source electromagnetics (and magnetotellurics on continental shelves) typically operate. Chave *et al.* (1991) extended this to longer periods, showing that direct forcing by surface gravity waves below 10–40 s (where their wavelength becomes comparable to or larger than the water depth) produces a rapidly rising noise spectrum even in the deep ocean. The spectral level is poorly known, and varies regionally and with the passage of storms, but can be comparable to the signal level from external sources, especially when the latter are weak. The long-period limit for the electromagnetic effects of long waves is not well characterized. Microseisms are produced by the nonlinear interference of opposing surface gravity wave trains that produce a pressure disturbance at twice the intrinsic frequency of the waves that propagates to the seafloor in all water depths (Longuet-Higgins, 1950). Their ensuing electromagnetic effects result in sharply peaked spectra that may be one to three orders of magnitude larger than the ionospheric background at 1–3 s period (Figure 2.2), varying with significant wave height. Multiple peaks can occur due to the presence of swell from distant storms and reflections from coastlines. Seafloor acoustic and seismic disturbances are also a sharply peaked source of noise at shorter periods. While it is not possible to produce a quantitative model for these phenomena, motional electromagnetic noise that is temporally and spatially non-stationary is ubiquitous at periods shorter than 100–500 s in all water depths, and extends to at least 0.1 s, especially in coastal waters. Magnetotelluric measurements at the seafloor that cover this period band may be affected by these short spatial scale sources, resulting in biased and/or erratic response functions.

There are three additional sources of motional noise that should be considered by the magnetotelluric practitioner: the tides, internal waves and turbulence. Tidally induced electromagnetic fields were first predicted by Faraday (1832). Motional induction by the ocean tides is typically comparable to that from the solar daily variation and its first harmonic, but at the lunar semidiurnal period ($\approx$12.42 h), motional effects are dominant,

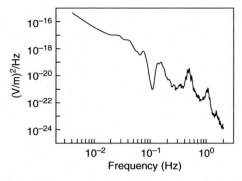

Figure 2.2. Electric field power spectrum computed from measurements 200 km southwest of San Diego in 3700 m of water. The f^{-4} slope at low frequencies is caused by ionospheric activity and forcing by long gravity waves. The peak near 0.1 Hz is due to microseism activity at the seafloor, while the higher-frequency peaks are due to Rayleigh wave-induced motion of the seafloor antenna. Taken from Chave *et al.* (1991).

and have even been detected in satellite magnetic field data (Tyler *et al.*, 2003). While the tidal forcing function does not contain periods shorter than about 12 h, hydrodynamic nonlinearity may result in tidal fluctuations at periods around 8, 6, 4.8 and 4 h, especially in coastal waters. Because the periods of the tides are well known, their effects in magneto-telluric data can be predicted, and hence they are more of a nuisance than a problem.

Internal waves are an omnipresent phenomenon in the deep ocean from periods of 1 h to the local inertial period ($\approx 10^5$ s at mid-latitudes). The theory of motional induction by internal waves was presented by Podney (1975) and Chave (1984a). The electromagnetic fields from internal waves are typically much weaker than those from external sources except in the vertical electric field, where they are dominant over the internal wave band (Chave & Filloux, 1985). In the presence of rough topography, the internal wave-induced vertical electric field may be scattered into the horizontal components, resulting in noise contamination.

The electromagnetic effect of turbulence was considered by Cox *et al.* (1971). The magnetic signature is negligible, while the electric field is of order $\mathbf{v} \times \mathbf{F}$, where $\mathbf{v}$ is the turbulent velocity. Turbulence caused by vortex shedding from instrument housings can occur during intervals of strong flow, such as the peak of the tidal cycle, and is typically broadband. In addition, vortex shedding can physically move instrument housings, resulting in substantial magnetic field noise due to sensor motion relative to the geomagnetic field. The lead author has data showing a large instance of this phenomenon in water over 1000 m deep during a fall storm off the Faroe Islands in 2005. Instrument motion noise can be removed if an independent measurement of instrument motion is available. It is easy to measure the two components of tilt about horizontal axes for this purpose, although they are not a complete description of instrument motion. Lezaeta *et al.* (2005) describe an adaptive correlation cancelation filter for instrument motion noise removal using tilt measurements.

A final source of electromagnetic noise in coastal waters is passing ships. Most vessels have a net direct current electric dipole moment due to either active hull corrosion protection or corrosion currents flowing from the hull through the propeller shaft. When the ship is in motion, this can produce a large, time-dependent electromagnetic field that is readily apparent in seafloor measurements. In addition, fluctuations in the electrical connectivity between the hull and the propeller shaft can produce an electromagnetic field at the shaft rate (typically 50–100 rpm for commercial vessels) and its first few harmonics.

2.3 Electromagnetic induction by extrinsic sources

Setting the velocity $\mathbf{v}$ to zero in (2.10) yields Ampere's law for a fixed medium. The time scale on which the ratio of the displacement to the conduction current is unity is the charge relaxation time $\tau = \varepsilon_0 \kappa \rho$, where $\rho = 1/\sigma$ is the electrical resistivity. In the ocean, τ is about 0.2 ns, and in Earth, τ typically lies between 1 and 100 ns, with a somewhat wider range (a couple of orders of magnitude on either end) encompassing virtually all Earth materials. All of these values are many orders of magnitude smaller than a typical period used in magnetotellurics. Thus, the displacement current term in (2.10) is negligible, yielding the pre-Maxwell equations with Ampere's law written

$$\nabla \times \mathbf{B} = \mu_0 (\sigma \mathbf{E} + \mathbf{J}^0) \tag{2.16}$$

Taking the divergence of (2.16), the total electric current density $\sigma \mathbf{E} + \mathbf{J}^0$ must be divergence-free, and hence the time rate of change of charge density in (2.11) is zero. However, this does not imply that charge has no role: charge appears and disappears instantaneously at material interfaces and gradients in the pre-Maxwell limit, and does not produce any magnetic field.

An alternative way to understand the pre-Maxwell limit follows from considering Ampere's law with both the displacement and conduction currents retained written in the form

$$\tau \, \partial_t \mathbf{E} + \mathbf{E} = \frac{\rho}{\mu_0} \nabla \times \mathbf{B} \tag{2.17}$$

Equation (2.17) may be regarded as a first-order differential equation for $\mathbf{E}$ whose driving term is given by the right-hand side. It is easily integrated to give

$$\mathbf{E}(\mathbf{r}, t) = \mathbf{E}(\mathbf{r}, 0) e^{-t/\tau} + \frac{1}{\tau} \frac{\rho}{\mu_0} \int_0^t ds \, e^{-(t-s)/\tau} \nabla \times \mathbf{B}(\mathbf{r}, s) \tag{2.18}$$

The electric field on the left of (2.18) is independent of the initial condition after a few times τ, and thereafter is the convolution of the exponential term with $\nabla \times \mathbf{B}$. The result is an exponentially weighted running mean of past values of $\rho(\nabla \times \mathbf{B})/\mu_0$, where the averaging time τ is very small compared to a typical period used in magnetotellurics, so that the convolving exponential is very sharply peaked at $t = 0$. Consequently, the averaging is unimportant, and the electric field $\mathbf{E}$ in (2.18) becomes the pre-Maxwell value.

Combining (2.3) and (2.16) yields a single equation for the electric field

$$\nabla \times \nabla \times \mathbf{E} + \mu_0 \sigma \, \partial_t \mathbf{E} = \mu_0 \partial_t \mathbf{J}^0 \tag{2.19}$$

Because it has only a single time derivative, (2.19) is a parabolic diffusion equation rather than the hyperbolic wave equation that appears in most elementary textbooks on electromagnetism. In the theory of partial differential equations, the characteristics are curves or surfaces along which a partial differential equation is transformed into an ordinary differential equation. For the simplest case of one spatial dimension, a parabolic diffusion equation has only a single family of characteristics (lines of constant t), while a hyperbolic wave equation has two such families (lines of constant $x \pm ct$, where c is the phase velocity). Since they are not invariant under the transformation $t \rightarrow -t$, solutions to parabolic equations evolve unidirectionally forward in time simultaneously at all points away from a source. This set of traits precludes the existence of reflection (and concomitantly, refraction) at interfaces, so that ray concepts such as angle of incidence or Brewster's angles are meaningless. Mandelis *et al.* (2001) provide a thorough elaboration of these issues.

Electromagnetic induction by artificial (or controlled) electromagnetic sources is an important topic in exploration and environmental geophysics. While some further consideration of the topic will appear in the sequel, these areas are themselves worthy of a book, and hence most details will be omitted. A thorough overview of the topic appears in two volumes edited by Nabighian (1987, 1991).

2.4 The one-dimensional approximation

2.4.1 *Modal solutions for the pre-Maxwell equations*

Define a Cartesian coordinate system with $\hat{\mathbf{z}}$ positive downward and zero at Earth's surface, and assume that the electrical conductivity σ varies only with this coordinate. Such a simple model is called one-dimensional (1D), and may be approximated as a stack of layers to represent the conductivity structure (see Figure 2.3).

Using the Helmholtz representation theorem (Backus, 1986), any vector field $\mathbf{S}$ may be written in terms of its toroidal, consoidal and vertical parts

$$\mathbf{S} = \hat{\mathbf{z}} \times \nabla_h r + \nabla_h s + t\hat{\mathbf{z}} \tag{2.20}$$

where r, s and t are scalar functions, and r and s are unique up to additive constants. The surface curl and surface gradient operators are given in Cartesian coordinates by $\hat{\mathbf{z}} \times \nabla_h = (-\hat{\mathbf{x}} \, \partial_y, \hat{\mathbf{y}} \, \partial_x)$ and $\nabla_h = (\hat{\mathbf{x}} \, \partial_x, \hat{\mathbf{y}} \, \partial_y)$, respectively. For a spherical geometry, Backus (1986) showed that the additive constants for r and s may be specified by requiring that their average value vanish on a sphere of arbitrary radius. In this section, a planar geometry is assumed, and the additive constants may be chosen by requiring that the average value of r and s on an arbitrary plane, denoted by $\langle r \rangle$ and $\langle s \rangle$, vanish where

$$\langle f(x,y) \rangle = \lim_{L \to \infty} \frac{1}{L^2} \int\limits_{-L}^{L} \int\limits_{-L}^{L} f(x,y) dx dy \qquad (2.21)$$

By virtue of (2.2), a condition on the Helmholtz representation of the magnetic induction is $\nabla_h^2 s + \partial_z t = 0$, where the surface Laplacian is $\nabla_h^2 = \partial_x^2 + \partial_y^2$. This condition may be satisfied by choosing $s = \partial_z \Psi$ and $t = -\nabla_h^2 \Psi$, and, after setting $r = \Pi$, the magnetic induction becomes

$$\mathbf{B} = \hat{\mathbf{z}} \times \nabla_h \Pi + \nabla_h \partial_z \Psi - \nabla_h^2 \Psi \hat{\mathbf{z}} \qquad (2.22)$$

This is equivalent to using the more familiar vector potential with $\mathbf{B} = \nabla \times \mathbf{A}$ and $\mathbf{A} = (\partial_y \Psi, -\partial_x \Psi, \Pi)$. The scalar functions Ψ and Π represent poloidal and toroidal magnetic (hereafter PM and TM) modes for which governing differential equations will be derived using (2.3) and (2.16) combined with a Helmholtz representation of the electric field

$$\mathbf{E} = \hat{\mathbf{z}} \times \nabla_h A + \nabla_h B + X \hat{\mathbf{z}} \qquad (2.23)$$

and the source current density

$$\mathbf{J}^0 = \hat{\mathbf{z}} \times \nabla_h \Upsilon + \nabla_h T + \Xi \hat{\mathbf{z}} \qquad (2.24)$$

The toroidal and consoidal source current scalar functions satisfy

$$\nabla_h^2 \Upsilon = (\nabla_h \times \mathbf{J}_h^0) \cdot \hat{\mathbf{z}} \qquad (2.25)$$

$$\nabla_h^2 T = \nabla_h \cdot \mathbf{J}_h^0 \qquad (2.26)$$

It can be shown that $\langle (\hat{\mathbf{z}} \times \nabla_h) \cdot \mathbf{J}_h^0 \rangle = \langle \nabla_h \cdot \mathbf{J}_h^0 \rangle = 0$, so any constants in Υ and T may be chosen by requiring that $\langle \Upsilon \rangle = \langle T \rangle = 0$.

The time dependence of all variables is assumed to be $e^{i\omega t}$ where ω is the angular frequency. Substituting (2.22) and (2.23) into Faraday's law (2.3) and writing out the three Cartesian components yields

$$\partial_x(-\partial_z A + i\omega \, \partial_z \Psi) - \partial_y(-X + \partial_z B + i\omega \, \Pi) = 0 \qquad (2.27)$$

$$\partial_y(-\partial_z A + i\omega \, \partial_z \Psi) + \partial_x(-X + \partial_z B + i\omega \, \Pi) = 0 \qquad (2.28)$$

$$\nabla_h^2(i\omega \, \Psi - A) = 0 \qquad (2.29)$$

Equations (2.27) and (2.28) are the Cauchy–Riemann conditions from the theory of complex variables. Taking

$$\begin{aligned} u_1 &= -\partial_z A + i\omega \, \partial_z \Psi \\ v_1 &= -X + \partial_z B + i\omega \, \Pi \end{aligned} \qquad (2.30)$$

requires

$$u_1 + i v_1 = f_1(x + iy) \qquad (2.31)$$

where f_1 is an analytic function of a complex variable, and u_1 and v_1 are harmonic functions.

Any transformation may be made to the modal scalars provided that the pre-Maxwell equations are not violated and the fields are unchanged. Let $\Psi = \Psi' - \upsilon$, where $\upsilon = \Psi - A/i\omega$. Substituting into (2.22) yields

$$\mathbf{B} = \hat{\mathbf{z}} \times \nabla_h \Pi + \nabla_h \partial_z \Psi' - \nabla_h \upsilon - \nabla_h^2 \Psi \hat{z} + \nabla_h^2 \upsilon \hat{\mathbf{z}} \tag{2.32}$$

where the final term vanishes because of (2.29). Consequently, writing $\Pi = \Pi' + \zeta$ and requiring $\hat{\mathbf{z}} \times \nabla_h \zeta = \nabla_h \upsilon$ yields (2.22), and hence υ could have been taken as zero from the outset. Since $u_1 = \partial_z \upsilon$ except for a scale factor, that quantity is also zero, and hence v_1 is independent of x and y in light of (2.27) and (2.28). However, a constant may be added to Π without changing (2.22), and hence v_1 may also be taken as zero. This establishes two conditions on A, B and X.

Repeating the same procedure using Ampere's law (2.16) yields

$$\partial_x(\nabla^2 \Psi - \mu_0 \sigma A - \mu_0 \Upsilon) - \partial_y(\partial_z \Pi + \mu_0 \sigma B + \mu_0 T) = 0 \tag{2.33}$$

$$\partial_y(\nabla^2 \Psi - \mu_0 \sigma A - \mu_0 \Upsilon) - \partial_x(\partial_z \Pi + \mu_0 \sigma B + \mu_0 T) = 0 \tag{2.34}$$

$$\nabla_h^2 \Pi - \mu_0 \sigma X - \mu_0 \Xi = 0 \tag{2.35}$$

The first two equations are again the Cauchy–Riemann conditions. Defining

$$\begin{aligned} u_2 &= \nabla^2 \Psi - \mu_0 \sigma A - \mu_0 \Upsilon \\ v_2 &= \partial_z \Pi + \mu_0 \sigma B + \mu_0 T \end{aligned} \tag{2.36}$$

these require

$$u_2 + iv_2 = f_2(x + iy) \tag{2.37}$$

where f_2 is an analytic function. Since (2.22) is independent of $\partial_z \Pi$, that quantity may be chosen arbitrarily (note that this is equivalent to specifying $\nabla \cdot \mathbf{A}$, or the gauge, for the ordinary vector potential), and hence v_2 may be taken as zero. This means that u_2 is independent of x and y by the Cauchy–Riemann conditions, but a constant may be added to Υ in (2.24) without changing the source current density, and hence u_2 is also zero. Combining the terms yields the electric field

$$\mathbf{E} = i\omega \hat{\mathbf{z}} \times \nabla_h \Psi - \nabla_h(\partial_z \Pi + \mu_0 T)/\mu_0 \sigma + (\nabla_h^2 \Pi - \mu_0 \Xi)/\mu_0 \sigma \hat{\mathbf{z}} \tag{2.38}$$

where the PM and TM mode scalar functions satisfy

$$\nabla^2 \Psi - i\omega\mu_0 \sigma \Psi = \mu_0 \Upsilon \tag{2.39}$$

$$\nabla_h^2 \Pi + \sigma \partial_z(\partial_z \Pi/\sigma) - i\omega\mu_0 \sigma \Pi = \mu_0 \Xi - \mu_0 \sigma \partial_z(T/\sigma) \tag{2.40}$$

under the condition that $\langle \Psi \rangle = \langle \Pi \rangle = 0$. The PM mode is associated with the toroidal part of the source current, while the TM mode is driven by its vertical and consoidal parts.

The modal form of the fields is more useful than the widely used vector potential because it reduces the physics into two relatively simple and independent components. The PM mode is characterized by electric currents flowing in horizontal planes that couple purely by induction, and has no vertical electric field component. The TM mode is associated with electric currents that flow in planes containing the vertical direction, couples both inductively and galvanically, and has no vertical magnetic field component.

The usual boundary conditions on the horizontal components of the electric and magnetic fields and the vertical electric current must be satisfied at horizontal interfaces, and require inclusion of the source current when it is located in the interface. These require continuity of $\Pi + \mu_0 T'$, $\partial_z \Psi - \mu_0 \Upsilon'$, Ψ and $(\partial_z \Pi + \mu_0 T)/\mu_0 \sigma$, where the primed variables denote surface current expressions of the corresponding terms in (2.24). Since the boundary conditions are not coupled, the PM and TM modes represented by solutions of (2.39) and (2.40) are independent, and the conditions $\langle \Psi \rangle = \langle \Pi \rangle = 0$ guarantee their uniqueness, so they may be studied separately.

2.4.2 Green's functions

It will be more convenient to work with the horizontal Fourier transforms of the modal differential equations (2.39) and (2.40) using the convention

$$
\tilde{f}(\eta, \xi) = \int\limits_{-\infty}^{\infty} \int\limits_{-\infty}^{\infty} f(x, y) e^{i(\eta x + \xi y)} \, dx \, dy
$$

$$
f(x, y) = \frac{1}{(2\pi)^2} \int\limits_{-\infty}^{\infty} \int\limits_{-\infty}^{\infty} \tilde{f}(\eta, \xi) e^{-i(\eta x + \xi y)} \, d\eta \, d\xi
$$

(2.41)

where η and ξ are horizontal wavenumbers. Applying (2.41) to (2.39) and (2.40) yields

$$
\partial_z^2 \tilde{\Psi} - \beta^2 \tilde{\Psi} = \mu_0 \tilde{\Upsilon}
$$

(2.42)

$$
\sigma \partial_z (\partial_z \tilde{\Pi} / \sigma) - \beta^2 \tilde{\Pi} = \mu_0 \tilde{\Xi} - \mu_0 \sigma \partial_z (\tilde{T} / \sigma)
$$

(2.43)

where $k = \sqrt{\eta^2 + \xi^2}$ is the magnitude of the horizontal wavenumber and $\beta = \sqrt{k^2 + i\omega \mu_0 \sigma}$ is the diffusion parameter.

Green's functions for (2.42) and (2.43) facilitate their solution for arbitrary sources. Consider a half-space of conductivity σ_0 containing a current source at a height $z' < 0$ overlying a layered structure below Earth's surface at $z = 0$. The Green's functions satisfy

$$
\partial_h^2 g_{\psi\{\pi\}}(z, z') - \beta_0^2 g_{\psi\{\pi\}}(z, z') = \delta(z - z')
$$

(2.44)

where $\beta_0 = \sqrt{k^2 + i\omega \mu_0 \sigma_0}$. The solution is subject to a finiteness condition as $z, z' \to -\infty$, and boundary conditions at Earth's surface given by

$$
g_\psi(0, z') + \tilde{\Lambda} \partial_z g_\psi(0, z') = 0
$$

(2.45)

$$
g_\pi(0, z') + \frac{\tilde{K}}{\sigma_0} \partial_z g_\pi(0, z') = 0
$$

(2.46)

The modal response functions $\tilde{\Lambda}$ and $\tilde{K}$ specify the structure beneath Earth's surface, and are given by

$$\tilde{\Lambda} = -\left.\frac{\tilde{\Psi}}{\partial_z \tilde{\Psi}}\right|_{z=0^+} \tag{2.47}$$

$$\tilde{K} = -\left.\frac{\sigma \tilde{\Pi}}{\partial_z \tilde{\Pi}}\right|_{z=0^+} \tag{2.48}$$

where $z = 0^+$ implies that $z = 0$ is approached from below. The Green's functions may easily be constructed using the method of variation of parameters (Roach, 1982), yielding

$$g_{\psi\{\pi\}}(z, z') = -\left[e^{-\beta_0|z-z'|} + \Omega_0^{\text{PM}\{\text{TM}\}} e^{\beta_0(z+z')}\right]/2\beta_0 \tag{2.49}$$

where $\Omega_0^{\text{PM}\{\text{TM}\}}$ are complex, wavenumber-dependent diffusion interaction coefficients at Earth's surface for the PM or TM mode. They are related to $\tilde{\Lambda}$ and $\tilde{K}$ through

$$\Omega_0^{\text{PM}} = \frac{\beta_0 \tilde{\Lambda}_0 - 1}{\beta_0 \tilde{\Lambda}_0 + 1} \tag{2.50}$$

$$\Omega_0^{\text{TM}} = \frac{\beta_0 \tilde{K}_0/\sigma_0 - 1}{\beta_0 \tilde{K}_0/\sigma_0 + 1} \tag{2.51}$$

Consider a stack of N layers beneath Earth's surface each having thickness h_i and conductivity σ_i (Figure 2.3) . Let $z = z_i$ denote the ith interface and terminate the stack in a half-space of conductivity σ_N at $z = z_N$. The PM mode response function (2.47) may be generalized so that $\tilde{\Lambda}_i$ denotes its value at the ith interface and $\tilde{\Lambda}_0$ is the surface value used in (2.50). In a similar way, the PM mode diffusion interaction coefficient may be generalized so that Ω_i^{PM} is its value at the ith interface. Within Earth, the PM mode scalar function $\tilde{\Psi}$ satisfies the homogeneous form of (2.42) with $\sigma(z) = \sigma_i$ within the ith layer. Using its solution with the continuity conditions on $\tilde{\Psi}$ and $\partial_z \tilde{\Psi}$ at each interface, it is easy to derive a recursive relation for $\tilde{\Lambda}_i$ in terms of $\tilde{\Lambda}_{i+1}$:

$$\tilde{\Lambda}_i = \frac{\beta_{i+1}\tilde{\Lambda}_{i+1} + \tanh(\beta_{i+1}h_{i+1})}{\beta_{i+1}\left[1 + \beta_{i+1}\tanh(\beta_{i+1}h_{i+1})\tilde{\Lambda}_{i+1}\right]} \tag{2.52}$$

where $\beta_i = \sqrt{k^2 + i\omega\mu_0\sigma_i}$. Inverting (2.50), the ith modal response function and the ith diffusion interaction coefficient are related by

$$\tilde{\Lambda}_i = \frac{1 + \Omega_i^{\text{PM}}}{\beta_i(1 - \Omega_i^{\text{PM}})} \tag{2.53}$$

Combining (2.52) and (2.53) gives an expression for the ith diffusion interaction coefficient in terms of the $(i+1)$th one:

$$\Omega_i^{\text{PM}} = \frac{\alpha_i + e^{-2\beta_{i+1}h_{i+1}}\Omega_{i+1}^{\text{PM}}}{1 + \alpha_i e^{-2\beta_{i+1}h_{i+1}}\Omega_{i+1}^{\text{PM}}} \tag{2.54}$$

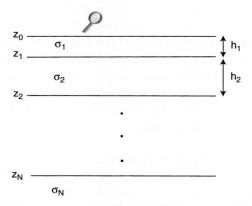

Figure 2.3. Cartoon illustrating a layered approximation to a 1D structure. A magnetotelluric measurement on the surface of the Earth at $z = z_0$ is represented by the magnifying glass. The first layer has a conductivity of σ_1 and a thickness of h_1, and terminates at a depth z_1. The stack of layers continues down to a depth of z_N, below which the conductivity is constant at σ_N.

where the interface interaction coefficients are

$$\alpha_i = \frac{\beta_i - \beta_{i+1}}{\beta_i + \beta_{i+1}} \tag{2.55}$$

The diffusion interaction coefficient in (2.49) is obtained when $i = 0$. The recursion (2.54) is initialized at $z = z_N$ with α_N for a conducting or insulating half-space or -1 for a perfect conductor.

The same process may be carried out for the TM mode, yielding the diffusion interaction coefficient relationship

$$\Omega_i^{\mathrm{TM}} = \frac{\chi_i + e^{-2\beta_{i+1}h_{i+1}}\Omega_{i+1}^{\mathrm{TM}}}{1 + \chi_i e^{-2\beta_{i+1}h_{i+1}}\Omega_{i+1}^{\mathrm{TM}}} \tag{2.56}$$

and the interface interaction coefficients

$$\chi_i = \frac{\beta_i \sigma_{i+1} - \beta_{i+1}\sigma_i}{\beta_i \sigma_{i+1} + \beta_{i+1}\sigma_i} \tag{2.57}$$

The TM mode diffusion interaction coefficient is initialized with χ_N for a conducting half-space, $+1$ for a perfect conductor or -1 for an insulator.

Using the Green's functions, solutions to the Fourier-domain versions of (2.39) and (2.40) for a general source may be written

$$\tilde{\Psi}(z) = \mu_0 \int_0^\infty g_\psi(z, z')\, \tilde{\Upsilon}(z')\, dz' \tag{2.58}$$

$$\tilde{\Pi}(z) = \mu_0 \int_0^\infty g_\pi(z, z')\tilde{\Xi}(z')dz' + \mu_0 \int_0^\infty \partial_{z'} g_\pi(z, z')\tilde{T}(z')dz' \tag{2.59}$$

where the second term in (2.59) is obtained after a single integration by parts. Solutions for the electromagnetic fields follow by taking inverse Fourier transforms using (2.41) and computing the appropriate horizontal and vertical derivatives in (2.22) and (2.38). Note that the vertical derivatives (either in z or z') of a Green's function appear in the expressions for the horizontal magnetic and vertical electric fields. Care must be taken in obtaining these, as they are discontinuous by ± 1 at $z = z'$. In addition, the TM mode horizontal electric field contains the mixed second derivative of the Green's function that possesses a delta function discontinuity at $z = z'$, and may be written $\partial_z \partial_{z'} g_\pi(z, z') = g_\pi^\dagger(z, z') - \delta(z - z')$, where $g_\pi^\dagger(z, z')$ is continuous.

There are four fundamental electromagnetic source types: the vertical and horizontal, electric and magnetic dipoles, typically abbreviated as VED, VMD, HED and HMD. The electric sources at Earth's surface must be in electrical contact with the ground to function, while the magnetic types consist of closed current loops, and may be located above the ground. In addition, the vertical sources are unimodal (the VED produces only a TM mode and the VMD produces only a PM mode), while the HED produces both modes. The HMD induces both modes when immersed in a conductive medium like the ocean, but when placed in the insulating atmosphere, it produces only a PM mode. Chave (2009) derived the electromagnetic fields for finite versions of all four source types. For magnetotellurics, only the magnetic sources have any relevance.

2.4.3 The poloidal magnetic (PM) mode

Consider a closed loop of radius a oriented in the x–y plane at a height z' above Earth that is constructed from infinitesimally thin wire and carries an electric current I (Figure 2.4). The electric current is entirely azimuthal, so that cylindrical rather than Cartesian coordinates are the natural choice. The source electric current density is given by

$$\mathbf{J}^0(\mathbf{r}) = I\delta(\rho - a)\delta(z - z')\hat{\boldsymbol{\varphi}} \qquad (2.60)$$

where the sign has been chosen so that the magnetic moment points in the $\hat{\mathbf{z}}$ direction. From (2.26), the consoidal part of the horizontal source current vanishes, and, since there is no vertical current source, there is no TM mode. Taking the horizontal Fourier transform (2.41) of (2.25), converting from Cartesian to polar coordinates, where $(x, y) = \rho(\cos\varphi, \sin\varphi)$ and $(\eta, \zeta) = k(\cos\theta, \sin\theta)$, applying Poisson's formula

$$\frac{1}{2\pi} \int_0^{2\pi} e^{\pm ik\rho \cos(\theta - \varphi)} d\theta = J_0(k\rho) \qquad (2.61)$$

where $J_0(x)$ is a Bessel function of the first kind of order zero, and performing a single integration by parts yields

$$\tilde{\Upsilon}(z) = \frac{2\pi I a}{k} J_1(ka) \, \delta(z - z') \qquad (2.62)$$

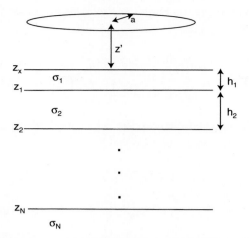

Figure 2.4. Schematic showing an idealized vertical magnetic dipole (VMD) representing the magnetotelluric source current overlying a layered half-space. The VMD is at a height z' above the surface and has radius a. The traditional magnetotelluric response that is independent of source wavenumber is obtained as a, $z' \to -\infty$ as described in the text.

Substituting into (2.58) and applying the inverse Fourier transform along with Poisson's formula yields

$$\Psi(\mathbf{r}, z') = \mu_0 I a \int_0^\infty g_\psi(z, z') \, J_1(ka) \, J_0(k\rho) \, dk \tag{2.63}$$

where it may be shown that $\langle \Upsilon \rangle = \langle \Psi \rangle = 0$, so that the result is unique. By symmetry, the radial and vertical electric and azimuthal magnetic fields vanish. The single electric field component follows from (2.38)

$$E_\varphi(\mathbf{r}, z') = i\omega \, \mu_0 I a \int_0^\infty g_\psi(z, z') k J_1(ka) J_1(k\rho) \, dk \tag{2.64}$$

while the two magnetic field components obtain from (2.22)

$$B_\rho(\mathbf{r}, z') = -\mu_0 I a \int_0^\infty \partial_z g_\psi(z, z') k J_1(ka) \, J_1(k\rho) \, dk \tag{2.65}$$

$$B_z(\mathbf{r}, z') = \mu_0 I a \int_0^\infty g_\psi(z, z') k^2 J_1(ka) \, J_0(k\rho) \, dk \tag{2.66}$$

Equations (2.64)–(2.66) are exact, and include the geometric effects of a finite size current loop through the $J_1(ka)$ term.

Letting $z = 0$ (i.e. an observer on Earth's surface), the PM mode Green's function (2.49) and its vertical derivative become

$$g_\psi(0, z') = -\frac{1 + \Omega_0^{\mathrm{PM}}}{2\beta_0} e^{-\beta_0 z'}$$

$$\partial_z g_\psi(0, z') = -\frac{1 - \Omega_0^{\mathrm{PM}}}{2} e^{-\beta_0 z'}$$

(2.67)

where σ_0 is the conductivity of the atmosphere, which is nominally zero. Taking the ratio of the electric to the horizontal magnetic field yields

$$\frac{E_\varphi(\rho, z')}{B_\rho(\rho, z')} = \frac{i\omega \int_0^\infty (1 + \Omega_0^{\mathrm{PM}}) (k/\beta_0) e^{-\beta_0 z'} \mathrm{J}_1(ka) \mathrm{J}_1(k\rho) \, dk}{\int_0^\infty (1 - \Omega_0^{\mathrm{PM}}) k e^{-\beta_0 z'} \mathrm{J}_1(ka) \mathrm{J}_1(k\rho) \, dk}$$

(2.68)

The term $k e^{-\beta_0 z'} \mathrm{J}_1(ka) \mathrm{J}_1(k\rho)$ may be regarded as a weight function applied to the remainder of the numerator and denominator that contain information about the electrical structure of the subsurface. In the limit where $a = z' \gg \rho$, the weight function is peaked at approximately $2/a$ with a halfwidth extending from about $1/(2a)$ to $3/a$, and hence has an increasingly smaller wavenumber bandwidth as the source simultaneously grows larger and moves farther from the observer (Figure 2.5). In the limit where the source becomes extremely large and is at a considerable distance above a Earth, it will produce nearly planar waves normally incident on Earth's surface. In this limit, the weight function becomes a delta function centered at a wavelength that is arbitrarily close to zero wavenumber. However, the presence of the multiplicative factor k in $k e{-}^{ka}\mathrm{J}_1(ka)\mathrm{J}_1(k\rho)$ guarantees that the delta function cannot occur exactly at zero wavenumber, and hence the incident field is not quite planar. Ignoring this for the moment and taking the zero-wavenumber limit in (2.68) yields

$$\frac{E_\varphi}{B_\rho} = \frac{i\omega (1 + \Omega_0^{\mathrm{PM}})}{\gamma_0 (1 - \Omega_0^{\mathrm{PM}})} = i\omega \tilde{\Lambda}_0$$

(2.69)

using (2.53), where $\beta_0 = \gamma_0 = \sqrt{i\omega\mu_0\sigma_0}$ when $k = 0$. From (2.47), $\tilde{\Lambda}_0$ depends only on the structure beneath the surface, and is independent of the conductivity of the upper half-space. Thus, it explicitly holds in the instance where the upper half-space is the insulating atmosphere.

However, there is an obvious problem upon examining (2.66) in the zero-wavenumber limit. It is not difficult to show that B_z decreases as the source simultaneously gets larger and farther from an observer, until it vanishes in the plane-wave limit. From Faraday's law (2.3), this would imply that E_φ must also vanish, meaning that (2.69) must be zero. This seeming contradiction occurs because the physical model for magnetotellurics requires that the source have a finite extent, as is discussed at greater length in Section 4.1. The physical requirement for a quasi-uniform source is captured in the derivation of (2.68) and (2.69), but

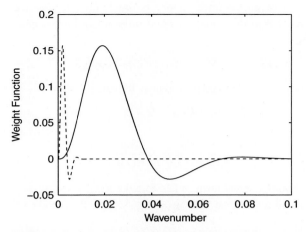

Figure 2.5. The weight function $k\alpha^2 e^{-k\alpha} J_1(k\alpha) J_1(k)$ from (2.68) as a function of the wavenumber k. The parameter α is the ratio of the current loop radius a to the radial distance ρ, and the solid and dashed lines show the weight function for $\alpha = 100$ and 1000, respectively. As the loop gets larger and farther from Earth's surface, the weight function becomes more sharply peaked at a point closer to zero wavenumber, although it cannot actually reach that value because of the leading term k.

is frequently overlooked in the magnetotelluric literature through invoking the plane wave limit from the outset by enforcing unphysical and inappropriate boundary conditions at Earth's surface instead of including the source currents in the model.

The mathematical relationship (2.69) lies at the center of the theory of magnetotelluric sounding as proposed by Rikitake (1951), Tikhonov (1950) and Cagniard (1953). The simple (and physically incorrect) model underlying these works consists of a plane electromagnetic wave vertically incident on an Earth consisting of homogeneous, horizontal, isotropic layers. Measurements are made of the time variations of the horizontal electric and magnetic fields. After transforming these time series into the frequency domain using methods described in Chapter 5, the ratio of the electric field to the orthogonal magnetic field at a given frequency is obtained. The set of such ratios over a range of frequencies is called the magnetotelluric response function $\mathbf{Z}$, with units of the electric field $\mathbf{E}$ divided by the magnetic induction, with units of velocity in m/s in SI units, and whose mathematical properties are described in Chapter 4. The magnetotelluric response function may be computed for any layered structure using (2.69) with (2.52).

As a potential source of confusion, the closely related magnetotelluric impedance is defined using the magnetic field $\mathbf{H}$ instead of the magnetic induction $\mathbf{B}$. The ratio of $\mathbf{E}$ and $\mathbf{H}$ does have units of ohms, leading to its appeal, and is equivalent to $\mu_0 \mathbf{Z}$. In the literature, the terms magnetotelluric response function and magnetotelluric impedance are often used interchangeably, although they are physically and mathematically distinct. In the remainder of the book, the context will determine whether $\mathbf{Z}$ or $\mu_0 \mathbf{Z}$ is being used.

One-dimensional magnetotelluric theory is not limited to discrete layered structures. Boyce & DiPrima (1986) show that a second-order linear homogeneous differential equation in the standard form

$$R(z)\partial_z^2 y - [\partial_z R(z) + Q(z)\,R(z)]\partial_z y + R^2(z)\,P(z)\,y = 0 \qquad (2.70)$$

may be transformed to the first-order nonlinear Riccati equation

$$\partial_z w = P(z) + Q(z)\,w + R(z)\,w^2 \qquad (2.71)$$

by the relationship

$$w = -\frac{\partial_z y}{R(z)y} \qquad (2.72)$$

Considering the homogeneous form of (2.58), take $R(z) = 1$, $Q(z) = 0$, $P(z) = -\beta^2(z) = -k^2 - i\omega\mu_0\sigma(z)$ and $y = \tilde\Psi$, so that $w = -1/\tilde\Lambda$. The PM mode response function for arbitrary $\sigma(z)$ satisfies

$$\partial_z\tilde\Lambda + \beta^2(z)\tilde\Lambda^2 = 1 \qquad (2.73)$$

Equation (2.73) may easily be solved numerically for any functional form $\sigma(z)$. Its value at $z = 0$ is $\tilde\Lambda_0$.

A remaining issue is the validity of the plane wave mathematical (but quasi-uniform physical) approximation that is required in (2.69). This was controversial in the early years of magnetotellurics, with strong differences of opinion being manifest (e.g. Wait, 1954, 1962; Price, 1962; Madden & Nelson, 1964). More recently, Dmitriev & Berdichevsky (1979) showed that the zero source field wavenumber mathematical approximation applies if the depth of investigation is less than one-third the scale over which linear variations in the source field are observed. Many years of experience indicate that finite source field wavenumbers are not a major source of bias in magnetotellurics at low to mid-latitudes, at least by comparison to other causes of interpretational complexity. The characteristics of the ionospheric and magnetospheric sources are further discussed in Chapter 3. In reality, the source field frequency–wavenumber structure is dynamic and time-dependent, and data processing methods that remove this effect have been developed, as described in Chapter 5.

2.4.4 The toroidal magnetic (TM) mode

In the theory of magnetotellurics, the TM mode does not play a role because the TM mode magnetic field vanishes at and above insulating boundaries. As a result, a source located above Earth's surface (such as in the ionosphere and magnetosphere) cannot induce a TM mode within it. Instead, the source must be in electrical contact with conducting material for a TM mode to exist. For example, a long insulated horizontal wire that is bared at its ends would produce a TM mode. However, such a model is not consistent with natural sources of electromagnetic energy.

This fact of physics is unfortunate, as it is well known that the TM mode is preferentially sensitive to low-conductivity layers in Earth that are nearly invisible to the PM mode, and that this sensitivity is substantially better than the corresponding ability of the PM mode to detect conductors. This has been demonstrated through the Fréchet derivatives that relate small changes in the conductivity as a function of depth to perturbations of the surface fields (Chave, 1984b), and is the basis for commercial use of artificial horizontal electric dipole sources in the marine environment to search for resistive hydrocarbon zones (e.g. Constable & Srnka, 2007; Chave, 2009).

Motional induction in the ocean is the only near-surface natural source that produces both PM and TM modes. However, the use of motional induction to probe the electrical structure of Earth requires an understanding of its wavenumber characteristics, as the horizontal length scales for most oceanic disturbances are comparable to the penetration depth of the electromagnetic field. A sufficiently detailed understanding of the wavenumber properties of the ocean is lacking, with the notable exception of the tides. To date, the tides have not been used as a source to probe the electrical structure of Earth, although tide models are now accurate enough for that purpose.

2.5 The two-dimensional approximation

2.5.1 The Maxwell equations in a two-dimensional medium

Two-dimensional models are characterized by a strike direction along which both the conductivity structure and the source field are constant (Figures 2.6 and 2.7). As a consequence, all field quantities remain unchanged in this orientation. Let the x axis be the strike direction. With $\partial_x \equiv 0$ and assuming $e^{i\omega t}$ time dependence, Faraday's law (2.3) and Ampere's law (2.16) yield the two decoupled sets of equations

$$\partial_y B_z - \partial_z B_y = \mu_0 (\sigma E_x + J_x^0) \tag{2.74}$$

$$\partial_z E_x = -i\omega B_y \tag{2.75}$$

$$-\partial_y E_x = -i\omega B_z \tag{2.76}$$

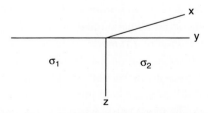

Figure 2.6. A simple 2D model consisting of two abutting quarter-spaces with different conductivities meeting at a vertical contact that extends to infinity. The magnetotelluric response will be distinct on either side of and far from the contact, and its characteristics will be different for the electric field polarized across-strike (TM mode) and along-strike (TE mode).

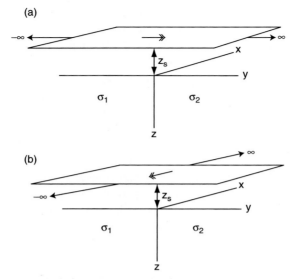

Figure 2.7. The 2D example of Figure 2.6 showing simple models for the source electric currents for (a) the TM mode and (b) the TE mode. For the former, the source consists of a sheet of current flowing in the y direction at a height z_s extending to infinity in $\pm y$. The current direction may also have a z component. For the TE mode, the source is a sheet of current flowing in the $-x$ direction at a height z_s extending to infinity in $\pm x$. The sheet must have a horizontal scale in the y (TE) and x (TM) directions that is much larger than z_s, and may be taken as infinite.

and

$$\partial_y E_z - \partial_z E_y = -i\omega B_x \qquad (2.77)$$

$$\partial_z B_x = \mu_0(\sigma E_y + J_y^0) \qquad (2.78)$$

$$-\partial_y B_x = \mu_0(\sigma E_z + J_z^0) \qquad (2.79)$$

The first set (2.74)–(2.76), in which **E** points in the strike direction and **B** lies in the plane orthogonal to strike, is called the *E-polarization* or *TE mode*. The complementary second set (2.77)–(2.79), with **B** in the strike direction and **E** in the horizontal plane orthogonal to strike, is called the *B-polarization* or *TM mode*. Note that this is *not* the same mode as the toroidal magnetic mode defined in Section 2.4. However, the physical behavior of the 2D and 1D TM modes are analogous given that both are influenced by boundary electric charge, while the 2D TE and 1D PM modes are both purely inductive, as further developed in the sequel. In addition, the 2D TM mode exists when the source currents are not in galvanic contact with the Earth, unlike the 1D TM mode.

Take the z coordinate positive downward, with $z = 0$ at the air–Earth interface, and specialize the model to the requirements of magnetotellurics by assuming that the sources lie in the insulating air half-space $z < 0$. Despite the similarity in the formal description of the two polarizations, their physical behaviors are quite different. This becomes evident when considering the TM mode magnetic field in $z < 0$. Assuming $J_z^0(z = 0^-) = 0$, meaning that the sources in $z < 0$ are coupled to the conducting half-space in $z > 0$ by induction only and not by galvanic contact, (2.79) implies $\partial_y B_x = 0$ at $z = 0$, and therefore

$$B_x(y, 0) = B_0 \tag{2.80}$$

where B_0 is a constant. In addition, (2.78) and (2.79) imply that inside the air half-space, but away from the source currents, the magnetic field $B_x(y,z)$ is a constant. It follows that in the air half-space $\partial_y J_y^0 + \partial_z J_z^0 = \nabla_2 \cdot \mathbf{J}^0 = 0$, where the two-dimensional nabla operator $\nabla_2 = \hat{\mathbf{y}} \partial_y + \hat{\mathbf{z}} \partial_z$. This solenoidal condition on the source currents in $z < 0$ can be satisfied by closed current loops, yielding a horizontal magnetic field B_x that is constant inside and zero outside the loop. Such loops are irrelevant for electromagnetic induction in the Earth. Another source current that satisfies the solenoidal condition is a sheet current with density $(B_0/\mu_0)\hat{\mathbf{t}}$, where $\hat{\mathbf{t}}$ is a function of y and z, flowing at a height $z_s < 0$ in the direction of the tangential unit vector $\hat{\mathbf{t}}$ and extending from $y = -\infty$ to $y = +\infty$. Such a sheet current exhibits a constant magnetic field $B_x = B_0$ between the sheet and Earth's surface at $z = 0$ and $B_x = 0$ above the current.

As a specific example, take the horizontal sheet current to be

$$\mathbf{J}^0(\mathbf{r}) = (B_0/\mu_0)\delta(z - z_s)\hat{\mathbf{y}} \tag{2.81}$$

The sheet current (2.81) produces $B_x = 0$ in $z < z_s$ and $B_x = B_0$ in $z_s < z \leq 0$ in the presence of the conducting substrate, as is evident physically since the sheet current in free space generates the magnetic field $B_0/2$ in $z > z_s$ and $-B_0/2$ in $z < z_s$. The magnetic field of the currents induced in the conducting Earth in $z > 0$ is the same size as that produced by the source, and hence their superposition in $z_s \leq z \leq 0$ doubles B_x and cancels it above the current sheet.

As a consequence, when modeling the TM mode electromagnetic field, Eqs. (2.77)–(2.79) are solved in the conductor $z > 0$ subject to the boundary condition (2.80) at $z = 0$. The air half-space $z < 0$ need not be further considered.

For the TE mode, a simple boundary condition such as (2.80) does not exist, and the field in the whole y–z plane has to be considered. The sources may be any set of line currents in the x direction, ranging from a single line current with density $\mathbf{J}^0(\mathbf{r}) = I\delta(\mathbf{r} - \mathbf{r}_s)\hat{\mathbf{x}}$, where I is the current and $\mathbf{r}_s$ is its position, to a uniform horizontal sheet current at $z = z_s < 0$:

$$\mathbf{J}^0(\mathbf{r}) = -(B_0/\mu_0)\delta(z - z_s)\hat{\mathbf{x}} \tag{2.82}$$

The sheet current (2.82) gives rise to a magnetic field $(B_0/2)\hat{\mathbf{y}}$ in $z_s \leq z \leq 0$, and is the source that is assumed in magnetotellurics. The field induced in Earth remains unaffected by the value of the source height z_s.

After eliminating B_y and B_z from (2.74)–(2.76) and using (2.82), the differential equation governing the magnetotelluric TE mode is

$$\partial_y^2 E_x + \partial_z^2 E_x = \nabla_2^2 E_x = i\omega \left[\mu_0 \sigma E_x - B_0 \delta(z - z_s) \right] \tag{2.83}$$

Assuming $\sigma > 0$ for $z \to \infty$, the electric field E_x vanishes in the same limit. Choosing $z_s > -\infty$, the boundary condition in the air half-space is obtained by integrating over the source level, yielding

$$\partial_x E_x(z = z_s^+) - \partial_x E_x(z = z_s^-) = -i\omega B_0 \tag{2.84}$$

where $\partial_z E_x \to 0$ as $z \to -\infty$. For the choice $z_s = -\infty$, (2.84) simplifies to

$$\partial_z E_x \to -i\omega B_0 \qquad \text{for } z \to -\infty \tag{2.85}$$

In practice, the TE mode must be modeled by explicitly including an air half-space above the Earth and applying (2.85) at the top, in contrast to the TM mode, where no air half-space is required.

Using the same approach utilized to obtain (2.83), the TM mode governing differential equation is obtained by eliminating the electric fields from (2.77)–(2.79), yielding

$$\partial_y (\partial_y B_x/\sigma) + \partial_z (\partial_z B_x/\sigma) = \nabla_2 \bullet \left(\frac{\nabla_2 B_x}{\sigma} \right) = i\omega \mu_0 B_x \tag{2.86}$$

This equation is valid for $z \geq 0$, and at $z = 0$ the boundary condition is (2.80).

2.5.2 *Inductive/galvanic coupling and the role of electric charge*

The simple boundary condition (2.80) does not conceal the fact that the physics of the TM mode is more complicated than that of the TE mode. An electric field can be created both through induction by a time-variable magnetic field and by electric charge. Whereas induction is ubiquitous, electric field generation by electric charge can occur only at places where the electric current has a component parallel to the conductivity gradient, as further described below. This happens only for the TM mode; for the TE mode, the current is always orthogonal to the conductivity gradient. Because of mutual interaction, the electromagnetic field in a particular volume element depends on the electromagnetic field in all other volume elements. In the TM mode, this mutual coupling is both inductive and galvanic, whereas only inductive coupling occurs in the TE mode. This leads to a different sensitivity of the modes to poor conductors: a thin resistive zone, for instance, is easily bridged via induction for the TE mode, whereas in the TM mode it presents a barrier for the currents trying to penetrate it. This difference is analogous to that for the PM and TM modes in 1D induction, except that the thin zones can occur in either the horizontal or along-strike vertical planes.

From Gauss's law (2.1) and assuming the absence of polarizable material so that $\mathbf{D} = \varepsilon_0 \mathbf{E}$, outside of extrinsic sources, $\nabla_2 \bullet \mathbf{J} = \nabla_2 \bullet (\sigma \mathbf{E}) = 0$ and the density of electric charge is given by

$$\rho_e = \varepsilon_0 \nabla_2 \bullet \mathbf{E} = \varepsilon_0 \nabla_2 \bullet (\mathbf{J}/\sigma) = \varepsilon_0 \mathbf{J} \bullet \nabla_2 (1/\sigma) \tag{2.87}$$

Charge accumulates at places where the electric current density **J** has a component parallel to the conductivity gradient $\nabla_2\sigma$. The physical explanation is simple. Consider two adjacent regions with different conductivities σ_1 and σ_2 (Figure 2.6), and let a current with normal current density J_n cross the boundary from region 1 to region 2. The continuity of J_n requires that $\sigma_1 E_{n1} = \sigma_2 E_{n2}$, and therefore $E_{n1} \neq E_{n2}$. To produce this difference in normal electric field, a surface charge must be set up that has a density

$$\sigma_e = \varepsilon_0(E_{n2} - E_{n1}) = \varepsilon_0 J_n \left(\frac{1}{\sigma_2} - \frac{1}{\sigma_1} \right) \tag{2.88}$$

Taking direct currents as the simplest example, the case $\sigma_1 > \sigma_2$ requires that $E_{n1} < E_{n2}$, and is achieved by a positive surface charge density that decreases E_n in region 1 and increases it in region 2.

2.5.3 Transverse magnetic mode: the electric field in the air half-space

When the magnetic field B_x is calculated in the conductive Earth in $z \geq 0$ from (2.86), then the electric fields are easily obtained from the horizontal and vertical derivatives using (2.78) and (2.79). However, since $\sigma = 0$ for $z < 0$, these equations do not allow the calculation of the electric field in the air half-space. Assuming a plane current sheet at $z = z_s < 0$, then (2.77)–(2.79) can be satisfied in the air by

$$\mathbf{E} = i\omega B_0 \max(z, z_s)\hat{\mathbf{y}} + \nabla_2\Phi \tag{2.89}$$

where the first part is solenoidal and the second part is irrotational. Apart from a constant electric field in the y direction, the scalar electric potential Φ describes the field caused by electric charge at $z = 0$. Since there is no charge in the air half-space, $\nabla_2 \cdot \mathbf{E} = 0$ implies $\nabla_2^2\Phi = 0$. Moreover, $\nabla_2\Phi$ tends to a constant for $z \to -\infty$ and assumes the boundary value $\partial_y\Phi(y,0) = E_y(y,0)$ at $z = 0$, where the continuous tangential component $E_y(y,0)$ can be obtained via (2.78) from B_x in $z \geq 0$. It may easily be verified that the solution of this boundary value problem is given by

$$\partial_y\Phi(y,z) = -\frac{z}{\pi} \int_{-\infty}^{\infty} \frac{E_y(\eta,0)}{(y-\eta)^2 + z^2} d\eta \tag{2.90}$$

$$\partial_y\Phi(y,z) = -\frac{1}{\pi} \int_{-\infty}^{\infty} \frac{(y-\eta)E_y(\eta,0)}{(y-\eta)^2 + z^2} d\eta \tag{2.91}$$

where $z < 0$. For $z \to 0^-$, the solution simplifies to

$$\partial_y\Phi(y,0^-) = E_y(y,0)$$

$$\partial_z\Phi(y,0^-) = E_z(y,0^-) = \frac{1}{\pi} \int_{-\infty}^{\infty} \frac{E_y(\eta,0)}{y-\eta} d\eta \tag{2.92}$$

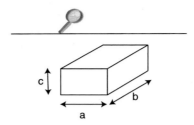

Figure 2.8. A simple 3D model consisting of a prism with sides $a \neq b \neq c$ buried at some depth below Earth's surface. The magnetotelluric response will be a function of position on the 2D surface of the Earth, and will also display different values at a given frequency depending on the horizontal coordinate direction.

where the Cauchy principal value (e.g. Mathews & Walker, 1970, p. 480) is taken in the integral in the second term. For the integrals in (2.90)–(2.92) to exist, it has to be assumed that $E_y(y,0)$ tends to the same value for $y \rightarrow \pm\infty$, meaning that the conductivity structure has to be identical in that limit.

2.6 Three-dimensional electromagnetic induction

2.6.1 The Maxwell equations for three-dimensional media

Figure 2.8 shows a simple 3D structure. The basic 3D equations under the pre-Maxwell approximation are (2.3) and (2.16). After eliminating the magnetic field **B**, they lead to the second-order equation (2.19) for **E**. Recalling that the time factor is $e^{i\omega t}$, the frequency-domain version of (2.19) is

$$\nabla \times \nabla \times \mathbf{E}(\mathbf{r}) + i\omega\mu_0\sigma(\mathbf{r})\mathbf{E}(\mathbf{r}) = -i\omega\mu_0\mathbf{J}^0(\mathbf{r}) \qquad (2.93)$$

where $\mathbf{J}^0(\mathbf{r})$ is the extrinsic source current density and $\sigma(\mathbf{r}) > 0$ is assumed for $z \geq 0$. When writing (2.93) in Cartesian components, three coupled scalar equations for the three unknowns E_x, E_y and E_z obtain. The coupling results from the divergence operator in the last term of the identity

$$\nabla \times \nabla \times \mathbf{E} = -\nabla^2\mathbf{E} + \nabla(\nabla \cdot \mathbf{E}) \qquad (2.94)$$

where, for instance, the x component of this term is $\partial_x(\partial_x E_x + \partial_y E_y + \partial_z E_z)$. For a localized source (e.g. a point dipole), all fields have to vanish at infinity. For a plane infinite horizontal current sheet at source level $z = z_s < 0$, which is the standard source current in magnetotellurics, the boundary condition is less simple: the magnetic field vanishes for $z \rightarrow \infty$ and for $z \rightarrow -\infty$; and it either vanishes if $z_s > -\infty$ or approaches a constant value $\mathbf{B}_0$ if $z_s \rightarrow -\infty$, where $\mathbf{B}_0$ is twice the horizontal source field. For $|x| \rightarrow \infty$ and $|y| \rightarrow \infty$, it is assumed that 2D conditions apply for definiteness.

The electric field **E** has a non-vanishing curl induced by the time variations of **B** from (2.3), and generally also has a non-vanishing divergence generated by charge accumulation at places where the current flows parallel to conductivity gradients (see Section 2.5.2) from

$$\rho_e = \varepsilon_0 \nabla \cdot \mathbf{E} = \varepsilon_0 \mathbf{J} \cdot \nabla(1/\sigma) \tag{2.95}$$

where $\mathbf{J}^0 = 0$ has been assumed for simplicity of notation.

In the differential equation (2.93), the air half-space with $\sigma(\mathbf{r}) = 0$ gives rise to a problem of non-uniqueness: let the volume V_0 be the air half-space $z < 0$ and let ∂V_0 be its boundary. If $\mathbf{E}$ is a solution in V_0, then $\mathbf{E} + \mathbf{E}_0$ with $\mathbf{E}_0 = \nabla U$ (implying $\nabla \times \mathbf{E}_0$) is also a solution in the same volume. With an arbitrary source term s, the potential U solves $\nabla \cdot \mathbf{E}_0 = \nabla^2 U = s \neq 0$ with $U = 0$ on the boundary ∂V_0. The term $\mathbf{E}_0$ vanishes by explicitly imposing the side condition $\nabla \cdot \mathbf{E} = 0$ on V_0. Then, $\nabla^2 U = 0$ in V_0 and $U = 0$ on ∂V_0 implies $U \equiv 0$ in V_0.

In theory, this instability will not arise in the conductor $z > 0$, where a positive conductivity is assumed. However, in numerical work, approximations to (2.93) become unstable for low frequencies when the second term on the left-hand side becomes so small compared to the first term that it drowns in the approximation errors of the latter. At that point, the solenoidal condition for the total current

$$\nabla \cdot (\sigma \mathbf{E} + \mathbf{J}^0) = 0 \tag{2.96}$$

obtained by taking the divergence of (2.93) is no longer guaranteed, and should be enforced as a side condition. The incorporation of side conditions can be achieved by representing the electromagnetic field by a source-free magnetic vector potential $\mathbf{A}$ and a scalar electric potential Φ such that

$$\begin{aligned}
\mathbf{B} &= \nabla \times \mathbf{A} \\
\mathbf{E} &= -i\omega\mathbf{A} + \nabla\Phi \\
\nabla \cdot \mathbf{A} &= 0
\end{aligned} \tag{2.97}$$

where $\mathbf{A}$ and $\nabla\Phi$ account for the curl and the divergence of $\mathbf{E}$, respectively. Inserting (2.97) into (2.93) and (2.96) yields

$$\begin{aligned}
\nabla^2 \mathbf{A} + \mu_0 \sigma(-i\omega\mathbf{A} + \nabla\Phi) &= -\mu_0 \mathbf{J}^0 \\
\nabla \cdot (\sigma \nabla\Phi) &= \nabla \cdot (i\omega\mathbf{A} - \mathbf{J}^0)
\end{aligned} \tag{2.98}$$

for $z > 0$. In the air half-space $z < 0$, enforcing the condition $\nabla \cdot \mathbf{E} = 0$,

$$\begin{aligned}
\nabla^2 \mathbf{A} &= -\mu_0 \mathbf{J}^0 \\
\nabla^2 \Phi &= 0
\end{aligned} \tag{2.99}$$

The systems (2.98) and (2.99) each consist of four scalar equations for the four unknown functions Φ and the three components of $\mathbf{A}$. However, because of the side condition $\nabla \cdot \mathbf{A} = 0$, the vector potential $\mathbf{A}$ has only two degrees of freedom. The demanding numerical implementation of these equations, or of their variants, is described in Chapter 7.

A simple example for the representation of $\mathbf{E}$ by (2.97) was encountered in (2.89), where $\mathbf{A} = A_y\hat{\mathbf{y}}$ with $A_y = -\max(z, z_s)B_0$. According to (2.99), this is a solution of

$$\partial_z^2 A_y(z) = -\mu_0 J_y^0(z) = -B_0 \delta(z - z_s) \tag{2.100}$$

where $A_y(0) = 0$.

2.6.2 The role of anisotropy

In an isotropic conductor, the electric field $\mathbf{E}$ and the electric current density $\mathbf{J} = \sigma \mathbf{E}$ are always parallel, whereas in an anisotropic conductor in which the conductivity depends on the direction of the electric field, parallelism between $\mathbf{E}$ and $\mathbf{J}$ is restricted to three mutually orthogonal directions of $\mathbf{E}$ in the general case.

The anisotropy of electrical conductivity is essentially a scale effect: even if the conductivity is isotropic on the microscale, it will become anisotropic on a larger scale if a preferred orientation (e.g. layering or lamination) exists in the averaging volume. As a simple example, consider a model in the x–y plane with two adjacent stripes of equal width in the y direction and extending to infinity in the x direction (Figure 2.9). To quantify the model, let the isotropic conductivities in the stripes be σ_1 and σ_2 with $\sigma_1 \neq \sigma_2$, and assume that the parallel vectors $\mathbf{E}_1$ and $\mathbf{J}_1$ in stripe 1 subtend an angle α_1 with the x direction. At the interface between the stripes, the tangential (x) component of $\mathbf{E}$ and the normal (y) component of $\mathbf{J}$ are continuous. Then, in stripe 2 the parallel vectors $\mathbf{E}_2$ and $\mathbf{J}_2$ subtend the angle α_2 with the x axis, where

$$\tan \alpha_2 = \frac{\sigma_1}{\sigma_2} \tan \alpha_1 \tag{2.101}$$

and the length ratio of the vectors changes according to

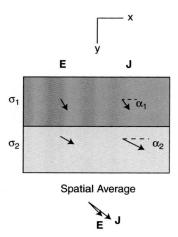

Spatial Average

Figure 2.9. A simple example showing the origin of structural anisotropy from the spatial average of $\mathbf{E}$ and $\mathbf{J}$ over isotropic conductors with a preferred orientation. The averaged current density is deflected toward the preferred direction. The conductors have values of σ_1 and σ_2, with $\sigma_1 < \sigma_2$.

$$\frac{|\mathbf{J}_2|}{|\mathbf{E}_2|} = \frac{\sigma_2 |\mathbf{J}_1|}{\sigma_1 |\mathbf{E}_2|} \tag{2.102}$$

When averaging $\mathbf{E}$ and $\mathbf{J}$ over the two stripes of equal width such that $\bar{\mathbf{E}} = (\mathbf{E}_1 + \mathbf{E}_2)/2$ and $\bar{\mathbf{J}} = (\mathbf{J}_1 + \mathbf{J}_2)/2$, the average vectors will be parallel only along the principal directions $\alpha_1 = 0$ and $\alpha_1 = \pi/2$, and for all other values of α_1 will form an acute angle $\beta > 0$ whose maximum value β_m is given by

$$\sin \beta_m = \frac{\sigma_l - \sigma_t}{\sigma_l + \sigma_t} \tag{2.103}$$

which is attained for

$$\sin^2 \alpha_1 = \frac{\sigma_2}{\sigma_1 + \sigma_2} \tag{2.104}$$

In (2.103), σ_l is the macroscopic longitudinal conductivity in the x direction (along-strike) and σ_t is the smaller macroscopic transverse conductivity in the y direction (normal to strike) given by

$$\begin{aligned} \sigma_l &= (\sigma_1 + \sigma_2)/2 \\ 1/\sigma_t &= (1/\sigma_1 + 1/\sigma_2)/2 \end{aligned} \tag{2.105}$$

where $\sigma_l > \sigma_t$. The average current density $\bar{\mathbf{J}}$ forms a smaller angle with the more conductive x direction than $\bar{\mathbf{E}}$. In the case of maximum deflection, the angle between $\bar{\mathbf{J}}$ and the x axis is the same as the angle between $\bar{\mathbf{E}}$ and the y axis. For conductivity contrasts of 3, 10, 100 and 1000, the maximum deflection β_m is, respectively, 8°, 30°, 68° and 83°. This structural anisotropy exists on the larger macroscale. It must be distinguished from the intrinsic anisotropy caused by ordered inhomogeneities that may exist on a smaller scale. Therefore, the assumption of (structural) anisotropy reflects an innate inability to adequately recognize the structure on the microscale.

In an anisotropic medium, Ohm's law is

$$\mathbf{J} = \vec{\sigma} \cdot \mathbf{E} \tag{2.106}$$

where $\vec{\sigma}$ is the second-rank conductivity tensor

$$\vec{\sigma} = \begin{bmatrix} \sigma_{xx} & \sigma_{xy} & \sigma_{xz} \\ \sigma_{xy} & \sigma_{yy} & \sigma_{yz} \\ \sigma_{xz} & \sigma_{yz} & \sigma_{zz} \end{bmatrix} \tag{2.107}$$

and, for example, $J_x = \sigma_{xx} E_x + \sigma_{xy} E_y + \sigma_{xz} E_z$. The conductivity tensor $\vec{\sigma}$ has the following basic properties.

1. **Symmetry:** $\vec{\sigma}$ is symmetric whenever the magnetic field does not play a role in the conduction process (Onsager, 1931). In the presence of Hall currents, as in a plasma, $\vec{\sigma}$ is non-symmetric. Symmetry holds for the purely ohmic conduction that occurs within the Earth.

2. **Positivity**: in a conductor, the energy dissipation per unit volume from Joule heating for direct currents is $\mathbf{E} \cdot \mathbf{J} = \mathbf{E} \cdot \vec{\sigma} \cdot \mathbf{E}$. For a time-harmonic field, the time-averaged energy dissipation is $(\mathbf{E} \cdot \vec{\sigma} \cdot \mathbf{E}^*)/2$, where the superscript * denotes the complex conjugate. Since the energy dissipation is positive, $\vec{\sigma}$ has to satisfy $\mathbf{v} \cdot \vec{\sigma} \cdot \mathbf{v} > 0$ for any $\mathbf{v}$ with $|\mathbf{v}| > 0$. Such a tensor is positive definite.

The positivity of $\vec{\sigma}$ implies that geometrically $\mathbf{E}$ and $\mathbf{J}$ always form an acute angle. For time-harmonic fields, this holds both for the real and imaginary vectors.

Since $\vec{\sigma}$ is a symmetric positive tensor, the eigenvalue problem

$$\vec{\sigma} \cdot \hat{\mathbf{e}} = \sigma \hat{\mathbf{e}} \tag{2.108}$$

has as eigensolutions three mutually orthogonal electric field directions $\hat{\mathbf{e}}_k$ with associated positive eigenvalues σ_k along which $\mathbf{E}$ is parallel to $\mathbf{J}$, so that $J_k = \sigma_k E_k$ for $k = 1$–3. For the example in Figure 2.9, if a third direction $\hat{\mathbf{z}}$ is added that is mutually orthogonal to $\hat{\mathbf{x}}$ and $\hat{\mathbf{y}}$, the principal directions and conductivities are $\hat{\mathbf{e}}_1 = \hat{\mathbf{x}}$, $\hat{\mathbf{e}}_2 = \hat{\mathbf{y}}$, $\hat{\mathbf{e}}_3 = \hat{\mathbf{z}}$, $\sigma_1 = \sigma_3 = \sigma_l$ and $\sigma_2 = \sigma_t$.

The origin of macroscopic electrical anisotropy is the result of spatially averaging the field over regions where the isotropic conductivity has preferred orientations on the microscale. For a volume with a given composition of isotropic conductors of known conductivity, homogenization theory provides estimates of the achievable anisotropy by considering all possible electrical connections of the constituents. For more details, see the monograph by Jikov *et al.* (1994, pp. 187–198) and the summary by Weidelt (1999).

The basic equation for modeling electrical anisotropy remains (2.93) through replacement of $\sigma(\mathbf{r})$ by $\vec{\sigma}(\mathbf{r})$. Although macroscopic anisotropy can be modeled by sufficiently refined isotropic models, the consideration of macro-anisotropy is useful for cryptic 3D models that sometimes can be modeled as anisotropic 2D conductors; see for instance the instructive study of Pek & Verner (1997). Consider as a simple example an isotropic 2D conductor covered by an anisotropic uniform layer consisting of vertical thin dikes on the microscale (Figure 2.10) . These dikes define a vertical plane of preferred current flow and a horizontal direction of hindered current flow. Assume an external magnetic field in the strike direction of the isotropic 2D conductor. If the strike direction of the dikes in the surface layer coincides with this direction, there is a pure TM mode configuration and no magnetic

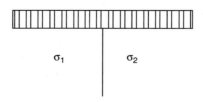

Figure 2.10. A 2D model consisting of abutting quarter-spaces of different conductivity overlain by an anisotropic layer. If the source magnetic field and the preferred direction of the anisotropic layer are aligned with the 2D strike direction, then the magnetotelluric response is purely TM mode. However, if the preferred direction differs from the strike direction, both a TE and a TM mode response will ensue as electric currents are deflected into the preferred direction in the anisotropic layer.

field disturbance at the surface would occur. There is no current flow in the preferred plane. However, if the horizontal direction of the source current sheet deviates from the 2D strike, strong induced currents may flow along the preferred plane and will produce an inducing field for the 2D conductor with a component perpendicular to the strike. This TE mode geometry produces magnetic field variations at the surface and a vertical field that would not occur for an isotropic overburden.

References

Backus, G. E. (1986). Poloidal and toroidal fields in geomagnetic modeling. *Rev. Geophys.*, **24**, 75–109.

Baringer, M. & J. C. Larsen (2001). Sixteen years of Florida Current transport at 27 degrees N. *Geophys. Res. Lett.*, **28**, 3179–3182.

Boyce, W. E. & R. C. DiPrima (1986). *Elementary Differential Equations and Boundary Value Problems*, 4th edn. New York: John Wiley.

Cagniard, L. (1953). Basic theory of the magnetotelluric method of geophysical prospecting. *Geophysics*, **18**, 605–635.

Chave, A. D. (1984a). On the electromagnetic fields induced by oceanic internal waves. *J. Geophys. Res.*, **89**, 10 519–10 528.

Chave, A. D. (1984b). The Fréchet derivatives of electromagnetic induction. *J. Geophys. Res.*, **89**, 3373–3380.

Chave, A. D. (2009). The electromagnetic fields produced by marine frequency domain controlled sources. *Geophys. J. Int.*, **179**, 1429–1457.

Chave, A. D. & C. S. Cox (1982). Controlled electromagnetic sources for measuring electrical conductivity beneath the oceans. 1. Forward problem and model study. *J. Geophys. Res.*, **87**, 5327–5338.

Chave, A. D. & J. H. Filloux (1985). Observation and interpretation of the seafloor vertical electric field in the eastern North Pacific. *Geophys. Res. Lett.*, **12**, 793–796.

Chave, A. D. & D. S. Luther (1990). Low-frequency, motionally induced electromagnetic fields in the ocean. 1. Theory. *J. Geophys. Res.*, **95**, 7185–7200.

Chave, A. D., S. C. Constable & R. N. Edwards (1991). Electrical exploration methods for the seafloor. In *Electromagnetic Methods in Applied Geophysics*, Vol. **2**, ed. M. N. Nabighian. Tulsa: Society of Exploration Geophysicists, pp. 931–966.

Chave, A. D., D. S. Luther & J. H. Filloux (1992). The barotropic electromagnetic and pressure experiment. 1. Barotropic current response to atmospheric forcing. *J. Geophys. Res.*, **97**, 9565–9593.

Chave, A. D., D. S. Luther & J. H. Filloux (1997). Observations of the boundary current system at 26.5N in the subtropical North Atlantic Ocean. *J. Phys. Oceanogr.*, **27**, 1827–1848.

Chave, A. D., D. S. Luther & C. S. Meinen (2004). Correction of motional electric field measurements for galvanic distortion. *J. Atmos. Ocean. Technol.*, **21**, 317–330.

Constable, S. C. & L. J. Srnka (2007). An introduction to marine controlled-source electromagnetic methods for hydrocarbon exploration. *Geophysics*, **72**, WA3–WA12.

Cox, C. S., J. H. Filloux & J. C. Larsen (1971). Electromagnetic studies of ocean currents and the electrical conductivity below the seafloor. In *The Sea*, Vol. 4, Pt. 2, ed. A. E. Maxwell. New York: John Wiley, pp. 637–693.

Cox, C. S., N. Kroll, P. Pistek & K. Watson (1978). Electromagnetic fluctuations induced by wind waves on the deep-sea floor. *J. Geophys. Res.*, **83**, 431–442.

Dmitriev, V. I. & M. N. Berdichevsky (1979). The fundamental model of magnetotelluric sounding. *Proc. IEEE*, **67**, 1034–1044.

Egbert, G. D., J. R. Booker & A. Schultz (1992). Very long period magnetotellurics at Tucson Observatory: estimation of impedances. *J. Geophys. Res.*, **97**, 15 113–15 128.

Faraday, M. (1832). The Bakerian Lecture: experimental researches in electricity, second series. *Phil. Trans. R. Soc. Lond.*, **122**, 163–194.

Jikov, V. V., S. M. Kozlov & O. A. Oleinik (1994). *Homogenization of Differential Operators and Integral Functionals*. Berlin: Springer.

Larsen, J. C. (1971). The electromagnetic field of long and intermediate waves. *J. Mar. Res.*, **29**, 28–45.

Larsen, J. C. (1992). Transport and heat flux of the Florida Current at 27° N derived from cross-section voltages and profiling data: theory and observations. *Phil. Trans. R. Soc. Lond.*, **A338**, 169–236.

Lezaeta, P., A. D. Chave & R. L. Evans (2005). Correction of shallow water electromagnetic data for noise induced by instrument motion. *Geophysics*, **70**, G127–G133, doi:10.1190/1.2080748.

Lilley, F. E. M., J. H. Filloux, N. L. Bindoff, I. J Ferguson & P. J. Mulhearn (1993). Magnetic signals from an ocean eddy. *J. Geomagn. Geoelectr.*, **45**, 403–422.

Lilley, F. E. M., A. White & G. S. Heinson (2001). Earth's magnetic field: ocean current contributions to vertical profiles in deep oceans. *Geophys. J. Int.*, **147**, 163–175.

Lizarralde, D., A. D. Chave, J. G. Hirth & A. Schultz (1995). Long period magnetotelluric study using Hawaii-to-California submarine cable data: implications for mantle conductivity. *J. Geophys. Res.*, **100**, 17 837–17 854.

Longuet-Higgins, M. (1950). A theory of microseisms. *Phil. Trans. R. Soc. Lond.*, **A243**, 1–35.

Madden, T. R. & P. Nelson (1964). A defense of Cagniard's magnetotelluric method. *Geophys. Lab.* ONR NR-371–401, Final Report, MIT, Cambridge, MA.

Mandelis, A., L. Nicolaides & Y. Chen (2001). Structure and the reflectionless/refractionless nature of parabolic diffusion-wave fields. *Phys. Rev. Lett.*, **87**, 020801.

Mathews, J. & R. L. Walker (1970). *Mathematical Methods for Physics*, 2nd edn. Menlo Park: Benjamin/Cummings.

Nabighian, M. N. (1987). Electromagnetic Methods in Applied Geophysics, Vol. **1**, *Theory*. Tulsa: Society of Exploration Geophysicists.

Nabighian, M. N. (1991). Electromagnetic Methods in Applied Geophysics, Vol. **2**, *Applications*. Tulsa: Society of Exploration Geophysicists.

Onsager, L. (1931). Reciprocal relations in irreversible processes. *Phys. Rev.*, **37**, 405–426.

Pek, J. & T. Verner (1997) Finite-difference modeling of magnetotelluric fields in two dimensional anisotropic media. *Geophys. J. Int.*, **128**, 505–521.

Phillips, O. M. (1977). *The Dynamics of the Upper Ocean*. Cambridge: Cambridge University Press.

Podney, W. (1975). Electromagnetic fields generated by ocean waves. *J. Geophys. Res.*, **80**, 2977–2990.

Price, A. T. (1962). Theory of the magnetotelluric field when the source field is considered. *J. Geophys. Res.*, **67**, 1907–1918.

Rikitake, T. (1951). Changes in earth current and their relation to the electrical state of the earth's crust. *Bull. Earthq. Res. Inst., Univ. Tokyo*, **29**, 271–276.

Roach, G. F. (1982). *Green's Functions*, 2nd edn. Cambridge: Cambridge University Press.

Sanford, T. B. (1971). Motionally induced electric and magnetic fields in the sea. *J. Geophys. Res.*, **76**, 3476–3492.

Sommerfeld, A. (1952). *Electrodynamics*. London: Academic Press.

Tikhonov, A. N. (1950). On determination of electric characteristics of deep layers of the earth's crust. *Dokl. Acad. Nauk SSSR*, **151**, 295–297.

Tyler, R., S. Maus & H. Luhr (2003). Satellite observations of magnetic fields due to ocean tidal flow. *Science*, **299**, 239–241.

Wait, J. R. (1954). On the relation between telluric currents and the earth's magnetic field. *Geophysics*, **19**, 281–2389.

Wait, J. R. (1962). Theory of the magnetotelluric field. *J. Res. Natl Bur. Stand.*, **66D**, 590–641.

Weaver, J. C. (1965). Magnetic variations associated with ocean waves and swell. *J. Geophys. Res.*, **70**, 1921–1929.

Webb, S. C. & C. S. Cox (1982). Electromagnetic fields induced at the seafloor by Rayleigh–Stonely waves. *J. Geophys. Res.*, **87**, 4093–4102.

Webb, S. C. & C. S. Cox (1986). Observations and modeling of seafloor microseisms. *J. Geophys. Res.*, **91**, 7343–7358.

Weidelt, P. (1999). 3-D conductivity models: implications of electrical anisotropy. In *Three-Dimensional Electromagnetics*, ed. M. Oristaglio & B. Spies. Tulsa: Society of Exploration Geophysicists.

3

Earth's electromagnetic environment

3A. Conductivity of Earth materials

ROB L. EVANS

3A.1 Introduction

The aim of a magnetotelluric survey is constraining the electrical conductivity structure of the Earth. Conductivity is a transport property: it defines how well a material is able to carry an electric current. At the atomic level, the mobility or energy state of charged particles (e.g. electrons, ions, defects) is influenced heavily by temperature, and so it is not surprising that one of the first-order controls on Earth's conductivity with increasing depth is its thermal structure. In addition to temperature and modal composition, there are other important factors, often related to tectonic activity, that can influence or even control bulk conductivity. Principal among these are partially molten materials associated with upwelling at plate boundaries or mantle plumes, aqueous fluids, and metallic, carbon or sulfidic content.

This chapter will attempt to provide an overview of the conductivity structures found within Earth. It is impossible, within the confines of a single book chapter, to cover all of the material relevant to a complete understanding of Earth conductivity. As a result, the chapter will cover issues that the author feels are the most important, and will also highlight some current controversies in the field. There have been a number of recent reviews of Earth conductivity from various angles, and in what follows there will be several areas that will not be covered in detail. Specific topic areas that will not be treated include: how laboratory measurements are made on minerals – see reviews by Roberts & Tyburczy (1994) and Nover (2005); details about conductivity mechanisms at the atomic scale – see Gueguen & Palciauskas (1994) for a review in terms of rocks and minerals, Blakemore (1985) for a solid-state physics viewpoint, and recent reviews by Tyburczy (2007) and Yoshino (2010); and mixing rules for multiphase materials will only be briefly touched upon, with more details given in the references provided.

The chapter will start with a basic overview of conduction mechanisms operative in the Earth (see Table 3A.1 for exemplars), and in each case will give an idea of their expected magnitudes. When possible, examples will be shown of the tectonic settings under discussion and where magnetotellurics has been used to provide key constraints. The strategy will be to start with the oceanic structure – partly because it is most familiar to the author, but also because it nicely demonstrates the underlying controls on conductivity away from the

The Magnetotelluric Method: Theory and Practice, ed. Alan D. Chave and Alan G. Jones. Published by Cambridge University Press © Cambridge University Press 2012.

Table 3A.1. *Important targets for magnetotelluric transects.*

	Target/Process	Example
Crust	Fluids and zones of fluid alteration	Lower crustal conductors Uranium deposits
	Crustal deformation	Faults and shear zone imaging (Grenville Front, Great Slave Lake Shear Zone)
	Crustal amalgamation	Suture zones
	Crustal melt accumulation	
	Economic minerals	Gold, copper, nickel
Mantle	Thermal structure of the mantle	Lithosphere–asthenosphere boundary Craton structure
	Partial melting	Mid-ocean ridge melting. Subduction zone volcanics
	Asthenospheric rheology	Water in the upper mantle. Subduction zone processes
	Hydration of the upper mantle	Serpentinization. Water release into subduction zones

complications of continental structure, since the oceanic crust and mantle have a more straightforward thermal and compositional makeup. What is known about oceanic mantle composition will be explored, and geothermal structure and the conductivities of constituent mantle minerals as measured in the laboratory to demonstrate the expected primary conductivity structures will be presented. On top of this background structure, key processes, such as mantle melting, are implanted, pointing to instances where magnetotelluric studies are particularly useful.

The continents, representing as they do the end-product of many stages of orogeny and volcanism, are inherently more complex. Yet, except for certain "exotic" phases (as well as melts in areas of active volcanism), composition plays second fiddle to temperature in controlling conductivity, at least in the lithospheric mantle. One of the primary exotic phases often invoked to explain high conductivities is graphite, and also metals (iron oxides) and sulfides are important, particularly in the crust.

It is important to point out that the interpretation of field magnetotelluric data in terms of rock properties is only as good as the extant laboratory data on conductivity. Although there have been significant advances in this area, there still remain some significant unknowns. For example, serpentinite is often cited as an electrical conductor, particularly in the 1970s, but this conclusion is based on only a very small number of measurements, none of which are able to characterize the important spatial distribution of magnetite through the samples. Other hydrous phases, such as phlogopite, have also been suggested as explanations for high mantle conductivity (Boerner *et al.*, 1999), yet again the laboratory data to back up the claim are extremely limited, and in any case what exists is not supportive. In fact, Olhoeft (1981) showed that the reported low resistivity of hydrated minerals may be associated with poor laboratory technique, with prior results likely measuring outgassing water.

At present, there is considerable controversy over the role of dissolved hydrogen in olivine in enhancing bulk mantle conductivity. Three sets of laboratory data show apparently conflicting results, with one in favor of enhanced conductivity at asthenospheric temperatures (Wang *et al.*, 2006) and the other two significantly less so (Yoshino *et al.*, 2006; Poe *et al.*, 2010), although all of the data show significant conductivity enhancement at lithospheric temperatures. Since water plays such an important role in altering mantle rheology and instigating deep melting, the ability of magnetotellurics to constrain water content in the mantle is of considerable interest.

One issue that has been widely discussed is the relevance of laboratory data, measured on small hand samples and often at high frequencies, to magnetotelluric field data that sense length scales of hundreds of meters to tens of kilometers deep at periods of up to 10 000 s (Duba, 1976; Roberts & Tyburczy, 1994). Certainly, measurements on small samples or on compacted powders miss structural complexity that must be accounted for in the final interpretation. It is also critical that the laboratory measurements be carried out under conditions representative of the region of Earth to which they are applied (Duba, 1976). Recent measurements have attempted to do this, although the controversy over the impact of water on the conductivity of olivine demonstrates how difficult some of these measurements can be to obtain.

3A.2 Conductivity mechanisms: electronic and semiconduction

The electrical conductivity of Earth materials is a transport property that varies over many orders of magnitude, more so than any other physical property, with the exception of viscosity. Earth materials fall into three categories defined by their conductivity that in turn are related to the type of charge carrier and its mobility:

Conductors:	10^8 to 10^5 S/m
Semiconductors:	10^5 to 10^{-7} S/m
Insulators:	$< 10^{-7}$ S/m

Conduction in the Earth occurs through the transport of several types of charge carriers, including electrons (in metals), ions (in aqueous fluids and melts) or electron "holes" or vacancies (in semiconductors). In minerals, point defects, or imperfections in the crystal lattice, give rise to excess or deficits of charge and provide the conduction mechanism. The chemistry of defects is complex and not always well understood, and concentrations of one kind of defect can influence others (Hirsch & Shankland, 1993; Hirsch *et al.*, 1993; Tyburczy, 2007), leading to complex conductivity relationships with temperature, pressure and effective oxygen partial pressure (fugacity).

Since the process that is occurring is conduction, or more correctly for most Earth materials semiconduction, it seems natural to refer to rock conductivity measured in siemens per meter (S/m), and all laboratory studies do. However, magnetotelluric models are often shown and discussed in terms of the reciprocal property, resistivity, measured in ohm meters (Ω m), as it is simply easier to discuss and compare numbers that are greater than one than those less than one. Given the wide range of variation in values, log conductivity or log resistivity is often used, which makes the conversion from one to the other more straightforward as they are simply the negative of each other.

Silicic rocks are poor electrical conductors. "Dry" basalt, for example, has a conductivity of around 10^{-12} S/m (10^{12} Ω m) at standard temperature and pressure (STP), which classifies it as an insulator. However, most rocks have semiconductor properties in which conductivity is strongly temperature-dependent, and so at mantle temperatures conductivity is elevated. Furthermore, the addition of small amounts of fluid in cracks within the crust will tend to dominate the bulk conductivity, as is discussed below in more detail.

A brief description of the most important conduction mechanisms operative within rocks and minerals found within the Earth will be presented. More details are given in Guegen & Palciauskas (1994) and Tyburczy (2007). Tools to calculate conductivity for many of the situations discussed below are available (e.g. Pommier & Le Trong, 2011).

3A.2.1 Electronic conduction

Magnetotellurics and higher-frequency audiomagnetotellurics have been used extensively on land, along with other electromagnetic sounding methods, to find economically valuable mineral and ore deposits. These deposits are typically conductive because they have sufficiently high metal content that controls the bulk conductivity.

Examples of metallic bodies include polymetalliferous sulfide mounds (e.g. chalcopyrite or pyrite). These mounds are formed through hydrothermal processes at mid-ocean ridges and in continental crust, with hot fluids precipitating a range of metals, particularly valuable platinum-group elements. Pyrrhotite seems to be a critical conductor, with a conductivity of around 1.4×10^5 S/m measured in the laboratory (Worm *et al.*, 1993). Despite the high intrinsic conductivities of the major minerals found in these settings, the effects of oxidation and dispersal amid less conductive minerals tends to reduce the *in situ* bulk conductivity. However, the *in situ* conductivities of massive sulfide mounds, as measured on the seafloor, are on the order of 10 S/m (Cairns *et al.*, 1996), which is still a significant conductor. Although these seafloor features are too small for magnetotelluric surveys to detect, audio-magnetotellurics is becoming increasingly used for detailed prospecting on land, and, in addition, the presence of such highly conductive features beneath a survey profile could distort magnetotelluric response functions at adjacent sites (see Chapter 6).

Graphite, along with other forms of carbon, is a commonly found mineral that exhibits electronic conduction, and it is frequently invoked to explain regions of high conductivity in the continental crust and lithospheric mantle, particularly in sheared zones (faults, planes, etc.). Because graphite is such an efficient conductor, only trace amounts are needed to enhance bulk conductivity. However, interconnection of the graphite through the rock matrix is essential for increased regional bulk conductivity, and evidence for such interconnection over large scales is limited. Duba & Shankland (1982) calculated that an interconnected volume fraction of graphite of 5×10^{-6} is all that is needed to enhance bulk conductivity to values around 0.1 S/m.

The observations that increased pressure can aid in connecting graphite networks (Shankland *et al.*, 1997) has led to carbon as a preferred explanation for high conductivities in crustal shear zones, where extensive deformation can be used as a mechanism for forming the graphite network. In such cases, conductivity might be expected to show anisotropy, with a preferred conduction direction related to the direction of shearing. Such a pattern was observed around the Grenville Front in Canada (Mareschal *et al.*, 1995). However, not all samples from depth show interconnected graphite (see discussion in Sections 3.3 and 3.4).

3A.2.1.1 Surface conduction

In some rocks, there are additional charge carriers in the fluid phase adjacent to the solid grain surface. Conduction can occur along this grain surface. Clay minerals (e.g. smectites, montmorillonite) are common in areas that have undergone hydrothermal alteration. They have an abundance of charged impurities, such as Al^{3+}, that replace Si^{4+}, leaving a net negative surface charge. This is countered by adsorption of cations onto the surface, forming an electric double layer. These cations form the charge carriers for surface conduction, which can be an especially important mechanism at low porosity and fluid saturation (fluid content). The presence of a conducting clay cap (usually attributed to montmorillonite and confirmed by drilling) has been observed in several volcanic geothermal settings (e.g. Martinez-Garcia, 1992; Jones & Dumas, 1993; Nurhasan *et al.*, 2006).

3A.2.2 Semiconduction: mantle conductivity

Within the mantle, the constituent minerals exhibit semiconductor behavior, with the bulk conductivity σ determined by a number of thermally activated processes described by the solid-state Arrhenius equation:

$$\sigma = \sum_{i=1}^{N} \sigma_i e^{-E_i/kT} \tag{3A.1}$$

The value i refers to the ith conduction mechanism with the zero-temperature conductivity Ω_i and the activation energy E_i for that mechanism, T is the temperature in Kelvin and k is Boltzmann's constant.

The conductivities of olivine, clinopyroxene, orthopyroxene and garnet, the primary modal minerals in the mantle, have been measured in the laboratory under controlled conditions (T, P and fO_2; e.g. Huebner *et al.*, 1979; Xu *et al.*, 2000; Romano *et al.*, 2006). Olivine (dry and hydrous) has been particularly well studied (e.g. Constable & Duba, 1990; Roberts & Tyburczy, 1991; Constable *et al.*, 1992; Xu *et al.*, 2000; Du Frane *et al.*, 2005; Wang *et al.*, 2006; Yoshino *et al.*, 2006, 2009; Poe *et al.*, 2010; Watson *et al.*, 2010; Dai *et al.*, 2010). Controlling the oxygen fugacity (fO_2) is critical to obtaining estimates appropriate for mantle conditions, as discussed in Duba (1976). Du Frane *et al.* (2005), Constable (2006) and Dai *et al.* (2010) give expressions for olivine conductivity as a function of oxygen fugacity.

The bulk conductivity of olivine is generally given as the sum of two thermally activated transport processes. Between about 800°C and 1400°C, the principal mechanism is electron holes that move from one Fe site to another, transforming Fe^{2+} to Fe^{3+}.

Oxygen fugacity (the effective partial pressure of oxygen) describes the oxidation state of the mantle, and in essence controls the ratio of variable-valence cations, in particular Fe^{2+} and Fe^{3+} ions. Differences in charge availability, resulting from differences in oxidation state, impact the bulk conductivity. Also fO_2 controls the point-defect concentrations in olivine and is suggested as a controlling factor on hydroxyl solubility in olivine (Grant *et al.*, 2007). Thus, fO_2 can further impact bulk conductivity in the mantle by limiting the amount of water that olivine can hold. High fO_2 has been shown to weaken olivine, and so understanding the oxidation state of the mantle has important implications for the rheological behavior of the planet.

Because of the importance of Fe as a charge carrier, the iron content of minerals is important in their bulk conductivity. Most mantle xenoliths show iron contents that vary between Fo_{88} and Fo_{93}, where *Fo* is a measure of the Fe content defined by $Fo = Mg/(Fe + Mg)$. This can cause about a factor of 2 change in conductivity (Hirsch & Shankland, 1993). Garnet conductivity has been shown to be particularly sensitive to iron content, with almost seven orders of magnitude difference in the conductivity (at 1000°C) of a purely Mg-bearing pyrope compared to an Fe-bearing almandine (Romano *et al.*, 2006; Dai & Karato, 2009a). However, garnet is a minor phase, stable below ~80 km in the mantle (Robinson & Wood, 1998), and is usually found as isolated grains within the mantle matrix in xenolith samples. The same is true for chromite, a conductive (metallic) but minor mantle phase.

The effect of pressure on bulk conductivity has not been extensively studied. Measurements that have been made, mostly on olivine, but also for pyroxenite (Wang *et al.*, 2008), suggest that there is a negligible pressure effect on conductivity. The major influence of pressure (ignoring the effects on grain boundaries) is to control, along with temperature, phase stabilities. The phase transition from olivine to wadsleyite has been suggested to cause a two orders of magnitude increase in conductivity across the 410 km discontinuity (Xu *et al.*, 1998, 2000). New data do not support this conclusion, but also show a high degree of variability for reasons similar to those discussed in Section 3.4.5 (Huang *et al.*, 2005; Karato, 2006; Yoshino *et al.*, 2006, 2009; Romano *et al.*, 2009) (Figure 3A.1). Magnetotelluric data that are able to image this transition are limited because deep electromagnetic probing is usually undertaken using observatory standard magnetic field data rather than the use of tellurics, which can suffer from electrode-related drift at long periods. Those that are extant for the most part do not see definitive evidence for a jump in conductivity across the 410 km discontinuity, but are consistent with it (Schultz *et al.*, 1993). Other suggestions for structure across the 410 km discontinuity include the presence of a thin melt layer at the boundary. Such a feature has been tested in several datasets, but is not a required feature of most profiles that have high-quality data at long periods (Toffelmier & Tyburczy, 2007).

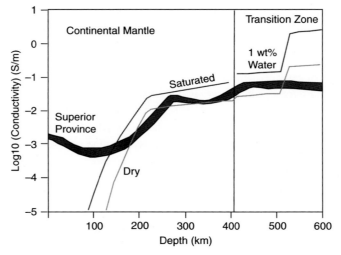

Figure 3A.1. Conductivity–depth profile based on laboratory measurements through the mantle. New laboratory data show large variability in the conductivities of both olivine and wadsleyite as a function of water content (Huang *et al.*, 2005; Karato, 2006; Wang *et al.*, 2006; Yoshino *et al.*, 2006, 2008, 2009; Romano *et al.*, 2009; Poe *et al.*, 2010), but evidence from these laboratory data and the few deep-probing MT soundings that penetrate the 410 km discontinuity lean toward a dry transition zone model. The curve for the Superior Province is the inversion result from a long-period magnetotelluric measurement (Schultz *et al.*, 1993). Curves are shown for a dry mantle, an upper mantle saturated in water, and 1 wt% water below the transition zone. Figure redrawn from Yoshino *et al.* (2009).

3A.3 Multiple phases and fluids

In Earth, the situation where a single conductive phase dominates is almost never seen. While it may be true that one phase may dominate over the others present, as for example with dry basalt and seawater, it is usually necessary to consider the bulk conductivity as a mixture of multiple phases. In such cases, the bulk conductivity depends on the conductivities of the available phases, their relative proportions and the way in which the most conductive phases are connected through the rock matrix. Since most rocks are poor conductors, the prevalent case to be considered is two-phase mixtures of a rock and a conductive fluid. The two most important cases of fluid phases are discussed next: free aqueous fluids in cracks and pore spaces within the crust; and partial melt in the crust and mantle.

3A.3.1 Aqueous fluids

The conductivity of H_2O depends on the concentration of dissolved ionic species and on the temperature through changes in fluid density and charge mobility. For the most part, and certainly in the oceans, this primarily means the concentration of NaCl (salinity), although in some continental settings KCl can also be a factor, and the two impact conductivity in essentially the same way (Nesbitt, 1993). Dissolved CO_2 can also impact conductivity by producing bicarbonates, as well as by influencing deep melting of the mantle (Dasgupta & Hirschmann, 2007b). Even relatively fresh water typically encountered within the Earth will likely have sufficient ionic concentration to have conductivities of around 0.1 S/m at low temperatures. Although there exist laboratory data for a range of compositions and conditions (pressure and temperature), the range is somewhat limited and does not extend to some of the more complex fluids that might be expected in subduction zones, for example. This is an area where more data would be welcome.

Because the conductivity of chloride solutions is a strong function of temperature and salinity, the magnetotelluric method is able to identify regions of anomalous heat and/or fluid flux. Over modest temperature ranges, the conductivity of fluids similar to seawater is very well known, as the conductivity is used in concert with temperature to ascertain salinity in oceanic conductivity–temperature–depth (CTD) instrumentation (Perkin & Lewis, 1980). Relationships between salinity, temperature and conductivity are given as empirical fits to observations. From ambient seafloor temperatures to around 350°C, the conductivity of seawater and similar chloride-bearing solutions increases nearly linearly with temperature from roughly 3 S/m to around 30 S/m (Quist & Marshall, 1968). If conditions pass the critical point, then two phases are formed, with a highly conductive brine coexisting with a very resistive vapor phase.

Within the continental crust, Nesbitt (1993) has calculated conductivity–depth profiles for typical geotherms and salinities. The conductivities of typical brines are predicted to be as high as 100 S/m, with maximum values predicted at depths of around 8 km (Figure 3A.2). Under crustal conditions, the resistivity of these brines is predicted to stay more or less steady at temperatures above 300°C.

3A.3.2 Silicate melts

Shankland & Waff (1977) showed that silicate melt is a semiconductor with conductivity dependent on temperature and weakly on pressure. Partial melts of mantle materials have conductivities in the range of 1–10 S/m, although the conductivity of a particular melt depends on composition in addition to temperature and pressure (Roberts & Tyburczy, 1999; Toffelmier & Tyburczy, 2007) (Figure 3A.3). However, for a given temperature and melt fraction, order-of-magnitude variations in electrical conductivity can result from relatively minor changes in the melt interconnection geometry or texture (Schmeling, 1986; Roberts & Tyburczy, 1999; ten Grotenhuis *et al.*, 2005). Note that the redox conditions (oxygen fugacity) have a negligible impact on the conductivity of silicate melts (Waff & Weill, 1975; Pommier *et al.*, 2010a,b).

Although melt conductivity decreases with decreasing temperature and increasing pressure, the addition of water to the melt can cause the conductivity to increase along an isobar. For a dry phonolite, for example, the conductivity at 1200°C and 200–300 MPa is around

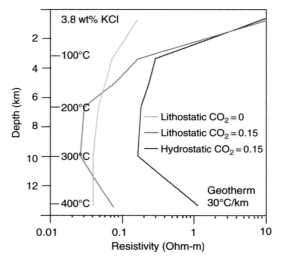

Figure 3A.2. The resistivity as a function of depth for an aqueous fluid within the continental lithosphere under a range of pressure conditions and CO_2 concentrations. The fluid considered has a KCl concentration of 3.8 wt% and the geotherm is assumed to have a constant 30°C/km gradient. Figure redrawn from Nesbitt (1993).

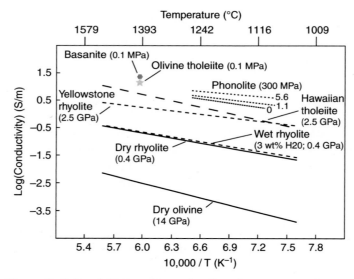

Figure 3A.3. The conductivity of silicate melts as a function of temperature and composition. Figure redrawn from Toffelmier & Tyburczy (2007). Phonolite data are from Pommier *et al.* (2008) and are labeled according to their water content (wt%).

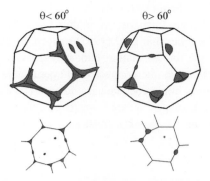

Figure 3A.4. The effect of dihedral angle on melt topology. The figure shows a 3D image of a grain and melt pockets in black and (below) a cross-section through the grain, for dihedral angles θ less than and greater than 60°, as labeled. Values of dihedral angle less than 60° lead to connected melt distributions, while larger values lead to melt sitting in isolated pockets. At large dihedral angles, even large melt fractions will not impact bulk conductivity. Melt pockets on grain faces are metastable. Figure redrawn from Watson & Brennan (1987) and Von Bargen & Waff (1986).

3 S/m. With the addition of 1.1 wt% water, the phonolite has a conductivity of around 4 S/m, and at 5.6 wt% water, the conductivity is ~5.6 S/m (Pommier *et al*., 2008). This change in conductivity (~0.27 log units) is smaller than the change from temperature variations (see Figure 3A.3).

In the mantle, under conditions of textural equilibrium that occurs when the key interfaces between melt and solid are controlled by interfacial energies, the key parameter controlling melt distribution (aside from the obvious volume of melt present), and hence the bulk conductivity, in a partial melt-bearing region is the dihedral angle, which describes the angle made by melt pockets at grain boundaries (Figure 3A.4). Values of dihedral angle less than about 60° lead to interconnected melt topologies. Since the dihedral angle is controlled by the surface energy, the composition of the melt and the surrounding matrix, the temperature and pressure conditions as well as the time scales of deformation are important in determining whether textural equilibrium will be a valid assumption (Cheadle *et al*., 2004). Under crustal (brittle) conditions, bulk conductivity will be controlled by the topology of the crack network.

Evidence that the partial melt in the mantle, even at low melt fractions, is well interconnected has come from laboratory simulations (Toramaru & Fujii, 1986; Riley *et al*., 1990). Most of the evidence suggests that melt has at least some degree of connectivity at even very low melt fractions. Drury & FitzGerald (1996) observed that very thin (1.0–1.5 nm) films are found on grain boundaries that can only be observed using high-resolution electron microscopy, and are not observable by light or scanning electron microscope observations. These films constitute 0.02% of the fraction of the rock and are highly interconnected. This is an important observation, as the vast majority of laboratory studies do not undertake such high-resolution imaging and thus are unable to see these very low-order connections. Faul (1997) also saw evidence for tubules that connect melt along grain edges at all melt fractions. Very thin films are sufficient to enhance the conductivity, although the tubes that Faul (1997) imaged do not have a simple geometry, and so simple relationships for calculating bulk conductivity may not be valid. Further evidence that melt must form an interconnected and permeable network, escaping rapidly from its source region, comes from geochemical studies of U/Th disequilibria in young lavas at mid-ocean ridges. These suggest a rapid escape of deep melts at small melt fractions (<0.1%). A full review of the geochemical constraints on melting is given in Kelemen *et al.* (1997).

3A.3.3 Carbonatite melts

Recent work has shown that carbonatite melts are significantly more conductive (by at least two orders of magnitude) than the silicate melts described in Section 3.2.2 (Gaillard *et al.*, 2008; Yoshino *et al.*, 2010). Carbonatite melts can form in trace amounts deep in the mantle (Dasgupta & Hirschmann, 2006). Because of their high conductivity, volume fractions as small as 0.01% have the potential to impact bulk conductivity at measurable levels, and carbonatite melt is interconnected to fractions as low as 0.05 wt% (Minarik & Watson, 1995). Thus, magnetotellurics offers the possibility of imaging the very early stages of deep melting beneath mid-ocean ridges, and there is good evidence that it has already done so (e.g. Baba *et al.*, 2006).

Carbonatite melts are quite rare in surface samples, and so it is unclear how common they are within the mantle and, if formed, what their fate is. Hirschmann (2010) has suggested that, at depths shallower than ~100 km in the oceanic mantle, carbonatites are probably restricted to older regions of the seafloor ($\gtrsim 40$ Ma). Carbonatites are expected to be quite buoyant, based on their density and low viscosity (Dobson *et al.*, 1996). It is not clear how a rising carbonatite melt would mix with silicate melt, but this mixing is critical to determining the bulk conductivity of a predominately silicate melt–carbonatite melt mixture. There is some evidence that the two are immiscible (Lee & Wyllie, 1997), in which case the carbonatite would likely sit in isolated inclusions within the silicate melt, and the bulk conductivity of the melt would match that of the silicate phase.

3A.3.4 Sulfidic melts

Ducea & Park (2000) suggested that the content of the metals (Fe, Cu, Ni) in sulfidic melts could further enhance conductivity. Xenoliths from the Sierra Nevada in the USA show sulfide minerals along grain boundaries. When melt is present, the additional conductivity afforded by the metals reduces the melt fractions needed to explain a given conductivity anomaly. Once the melt has crystallized, the sulfide minerals could potentially continue to act as a conductor, provided they form a well-connected network. Xenoliths from a variety of settings show evidence for sulfide inclusions, suggesting that these may be a widespread phenomenon (Szabo & Bodnar, 1995), and potentially a source of elevated conductivity in continental lithosphere. Watson *et al.* (2010) showed that sulfide contents of 1 vol% have a significant impact on bulk conductivity, even though a very small component of the sulfide was interconnected in their samples.

3A.3.5 Mixing relationships and interconnectivity

When multiple phases coexist, it is important to understand how the distribution of each phase impacts the conductivity. There are a number of theoretical and empirical approaches to addressing this issue, and they will only be summarized here. In general, there are relationships that deal with situations where one phase dominates the conduction and others that more formally account for the network topology of each phase in calculating bulk conductivity.

Under crustal (brittle) conditions, it is typical for the conductivity to depend on the volume fraction of pore fluid and its degree of connectivity (or tortuosity). In the oceanic crust and sediments, for example, electrical resistivity has been used as a proxy for porosity in numerous studies. An original study on well log data allowed the development of the empirical Archie's law (Archie, 1942). Many further studies have added to this relationship, some of them of sufficiently small scale that samples collected could be analyzed in the laboratory (e.g. Andrews & Bennett, 1984; Wheatcroft *et al.*, 1996; Jackson *et al.*, 2002). Archie's law can be written as

$$\rho_m = A\rho_f \theta^{-m} \tag{3A.2}$$

where ρ_m is the measured or observed resistivity, ρ_f is that of the pore fluid and θ is the porosity. The term A is commonly used to describe the degree of saturation, and is often treated as an empirical fitting factor when samples are recovered and both bulk porosity and resistivity can be measured. Although the saturation is written here as a constant, it is not, as the behavior of bulk conduction on the degree of saturation can be quite complex, with surface conduction mechanisms dominating at low degrees of saturation before the pore fluid is able to dominate (Roberts, 2002). Glover (2010) recently presented a formulation of Archie's law for multiple phases.

Archie's law does not contain any information on the resistivity of the host rock itself, which is assumed to be sufficiently higher than that of the pore fluid that it does not contribute to the bulk conductivity. What controls the bulk conductivity is the degree of interconnection of the pore fluid, described by the exponent m. Typical values of m range from 1.5 to 1.8 for marine sands (Jackson *et al.*, 1978, 2002). Higher values of m in Archie's law reflect less well-connected, lower-permeability fluid distributions, with values of 2–3 not uncommon for crustal rocks (e.g. Evans, 1994). Modeling work has shown the importance of interconnection, and variables describing the degree of interconnection have been suggested as a more reliable indicator of bulk conductivity (Hautot & Tarits, 2002).

Although Archie's law was developed based on empirical observations, there have been numerous numerical and theoretical studies that show how a power-law relationship between resistivity and porosity naturally arises in fluid-bearing materials (e.g. Shankland & Waff, 1974; Madden, 1976; Wong et al., 1984; Roberts & Schwartz, 1985; Schwartz & Kimminau, 1987; Evans, 1994; ten Grotenhuis et al., 2004). A formal range on the conductivity of an isotropic two-phase material is given by the Hashin–Shtrikman bounds, which consider a suspension of isolated particles in a medium (Hashin & Shtrikman, 1962).

The simplest geometrical conduction model is that of a stack of layers with two conductivities σ_1 and σ_2 with respective volume fractions ϕ and $(1 - \phi)$. Conduction parallel and perpendicular to the layers gives the familiar parallel and series conduction relations:

$$\sigma_{parallel} = \phi\sigma_1 + (1 - \phi)\sigma_2$$

$$\sigma_{series} = \left(\frac{\phi}{\sigma_1} - \frac{1 - \phi}{\sigma_2}\right)^{-1} \tag{3A.3}$$

These relationships were expanded for a more general and anisotropic case of conductive plates embedded in a uniform medium by Yu et al. (1997).

Berryman (1995, 2006) presents extended Hashin–Shtrikman bounds that allow for multiphase materials. His formulas for the lower σ_{HS}^- and upper σ_{HS}^+ bounds on electrical conductivity are:

$$\sigma_{HS}^- = \left(\sum_{i=1}^{N} \frac{x_i}{\sigma_i + 2\sigma_{min}}\right)^{-1} - 2\sigma_{min}$$

$$\sigma_{HS}^+ = \left(\sum_{i=1}^{N} \frac{x_i}{\sigma_i + 2\sigma_{max}}\right)^{-1} - 2\sigma_{max} \tag{3A.4}$$

where x_i denotes the volume fraction of the ith component and σ_i is its conductivity. The values s_{min} and s_{max} are those of the least and most conductive of the N phases being mixed.

In some cases, the behavior of the fluid can determine the geometry and, despite the likely complexity of networks, simple analogs might appropriately be used to estimate the bulk conductivity. For melt networks in the mantle, for example, it has been suggested that a description of electrical transport based on a cubic array of cylindrical tubes may be appropriate for a simple olivine–basalt system that has dihedral angles of ~45° (e.g. Cheadle, 1989; Dullien, 1992). In this case, wetting is along grain edges only, an expression for the bulk conductivity is given as (e.g. Schmeling, 1986; Dullien, 1992)

$$\sigma = \tfrac{1}{3}\phi\sigma_{melt} + (1 - \phi)\sigma_{rock} \tag{3A.5}$$

where σ_{melt} and σ_{rock} are the conductivities of the melt and host rock, respectively, and φ is the melt fraction. Schmeling (1986) also discusses a formalism for fluids in films based on an earlier model by Waff (1974).

Madden (1976) argues that the geometric mean of the individual conductivities is the appropriate metric for a multiphase medium. While this does not bound the likely range, Madden argues on the basis of network theory that it is the most likely representation of a multiphase distribution. However, as can be seen in Figure 3A.5, for a melt network, either

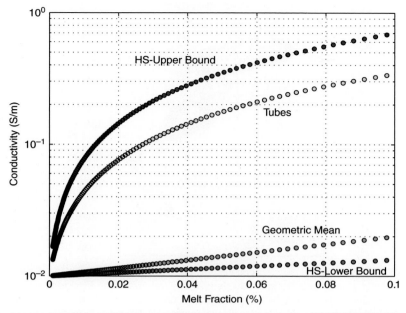

Figure 3A.5. Conductivity as a function of melt fraction for the various relationships discussed in the text (upper and lower Hashin–Shtrikman (HS) bounds; melt in tubes; and the geometric mean of conductivities). The background conductivity is set at 0.01 S/m, essentially that of oceanic mantle on an adiabat. The melt is assigned a conductivity of 10 S/m, consistent with a hydrous basalt at high temperature (>1200°C; Pommier *et al.*, 2010a,b). For both the HS upper bound and melt in tubes, both of which assume an interconnected melt distribution, the bulk conductivity is greatly enhanced even at low melt fractions. Typical melt fractions envisioned in the mantle beneath active spreading centers is of the order of 1–2%, but with such melt fractions an order-of-magnitude increase in bulk conductivity is possible. (See plate section for color version.)

the upper Hashin–Shtrikman bound or the expression for a network of tubes predict significantly enhanced conductivity at very low melt fractions. In contrast, the geometric mean only enhances conductivity at higher melt fractions, requiring a melt fraction of ~30% to obtain an order-of-magnitude increase in bulk conductivity (a conductivity of ~0.1 S/m in Figure 3A.5), while the lower Hashin–Shtrikman bound predicts that conductivities are enhanced by less than an order of magnitude (<0.1 S/m in Figure 3A.5) for melt fractions less than ~80%.

3A.4 Conductivity structure beneath the oceans

The discussion of Earth's conductivity structure will start in the oceans for the simple reason that the controls on conductivity in the oceans are more straightforward to understand. A picture will be built up of the structure through key regions of oceanic crust and mantle, and

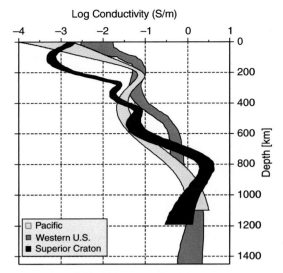

Figure 3A.6. Three deep-probing, 1D conductivity profiles from different tectonic settings. All of the profiles show a general increase in conductivity with depth, reflecting increasing temperature with depth and reinforcing the notion that temperature has a first-order control on mantle conductivity. The differences between the profiles, at least in the uppermost ~300 km, relate to tectonic environment. The Pacific profile (Lizarralde *et al.*, 1995) shows a peak in conductivity at 200 km depth that has been interpreted to be due to the presence of water in the form of dissolved hydrogen in olivine. The western USA profile (Egbert & Booker, 1992) comes from a region of thin lithosphere and a partial melt-bearing mantle. The Superior Craton profile (Schultz *et al.*, 1993) reflects a more resistive and thicker lithospheric root. Redrawn from Lizarralde *et al.* (1995)

factors that might impact conductivity in active tectonic regions will be discussed. Some first-order differences between the mantle in different settings can be seen in Figure 3A.6. A comprehensive picture of the structure of the oceanic crust and mantle, including more details of some of the constraints discussed below, appears in Sinha & Evans (2004).

3A.4.1 *Oceanic crust and sediments*

Until very recently, the oceanic crust has been out of the reach of seafloor magnetotelluric experiments owing to an inability to measure weak short-period magnetic fields. As a result of increased interest in marine electromagnetic methods for oil exploration, new instrumentation has made imaging the crust more feasible, particularly in the shallow waters of the continental shelf. In deeper waters, controlled-source methods have constrained the shallow crust and sediments, and these constraints will be important to deep-ocean seafloor magnetotelluric surveys. Efforts using magnetotelluric techniques for oil exploration are focused in sedimentary basins and continental margins. For these reasons, there follows a brief description of the conductivity structure of the oceanic crust and overlying sediment column.

In marine sediments, Archie's law seems to be a reasonable predictor of bulk resistivity from porosity – see the discussion in Evans (2001). Typical resistivities in marine sediments are between 1 and 10 Ω m. In constructing a typical resistivity–depth profile through a sedimentary sequence, there are several structural and compositional considerations that may make the conductivity at depth higher than expected. Most importantly, the geothermal gradient is critical for interpreting the conductivity through a sedimentary basin as the pore-fluid conductivity will rise with increasing temperatures at depth, and this can offset some, if not all, of the impact of decreasing porosity. There may also be areas of fluid overpressure at depth that maintain high porosity and fluid permeability, and hence electrical conductivity (Dugan *et al.*, 2003).

3A.4.2 Compaction and diagenesis

Once compaction, consolidation or diagenesis occurs, reducing porosity–conduction relationships become more complicated. The shutting down of conduction paths through precipitation and compaction has a dramatic impact on electrical resistivity. Studies on the change in conductivity and permeability resulting from the closing of pore spaces have been carried out for sedimentary rocks as well as for materials that act as analogs of compacting sediments (e.g. Bernabé *et al.*, 1982; Zhang *et al.*, 1994; Zhu *et al.*, 1995). Models have also been developed to quantify the loss of connectivity upon compaction (e.g. Zhu *et al.*, 1995). Both permeability and conductivity vary with changing porosity according to a power law. There is typically a critical cross-over porosity at which the behavior of the sample changes, typically somewhere between 10 and 20%. At porosities greater than the cross-over porosity, the power law typically has a cubic relationship, reflecting the tortuous nature of the pore space. Below the cross-over point, the conductivity starts to decrease much more rapidly with decreasing porosity. The cross-over porosity represents the point at which connectivity starts to be reduced within the rock network. In this case, the throats between adjacent pores become the critical transport property. As these throats become pinched or closed, connectivity is lost and conductivity rapidly drops.

3A.4.3 Basaltic crust

Although magnetotellurics has been used in sedimentary basins for oil exploration, its use in deeper water over bare oceanic crust remains somewhat band-limited owing to the much deeper water in these settings ($\gtrsim 2500$ m). Thus, magnetotellurics to date has provided constraints on the lower oceanic crust, but the uppermost section of crust requires controlled-source electromagnetic methods. For completeness, however, a brief description of the oceanic crustal electrical structure will be provided.

Just as for the sedimentary column, the conductivity of basaltic crust is controlled by the nature of the seawater network that infiltrates cracks and pores. The uppermost section of extrusive lavas tends to be heavily fractured, and in some places is simply a rubble pile. At depth, repeat eruptions and dike intrusions reduce the porosity, and the dike complex that forms the mid-part of the crust has a very low porosity (~1%; Evans, 1994). Beneath this, the intrusive gabbroic section has essentially no porosity. Archie's law relationships have also been used to estimate porosity from crustal resistivity measurements, although more attention needs to be paid to the exponent in the Archie's law formulation, as heavily fractured rubble near the seafloor and cracked sections of crust may require low values. At depth, porosity will decrease dramatically and resistivities will accordingly increase by one or two orders of magnitude.

Typical resistivities for upper oceanic crust run between 1 and 10 Ω m, and increase sharply with depth to around 1000 Ω m or greater at the base of the crust, depending on the crustal age and thermal structure (Evans *et al.*, 1991, 1994). Compositional variations may, in some cases, enhance conductivity of the oceanic lower crust. Within Ocean Drilling Program (ODP) hole 735B on the Southwest Indian Ridge, low resistivities (around 5 Ω m) were observed in a band of Fe–Ti oxide-rich gabbros (Pezard *et al.*, 1991). Fe–Ti oxide gabbros may be related to localized intrusions of evolved magmas (Ozawa *et al.*, 1991). There is, as yet, no evidence that such features impact the bulk conductivity on a scale measurable by magnetotellurics. In fact, most evidence of the electrical structure of oceanic crust (*in situ*) comes from controlled-source electromagnetic measurements (e.g. Cox *et al.*, 1986; Evans *et al.*, 1991, 1994).

The pore fluid controlling bulk resistivity is obviously critical in determining the bulk resistivity and so, once again, the geothermal gradient must be understood if good porosity–conductivity relationships are to be established.

At mid-ocean ridges, mid-crustal conductors associated with melt, and further melt ponding at the base of the crust and through the upper mantle, may constitute electrically conductive targets that can be constrained by magnetotellurics using modern instrumentation (Key & Constable, 2002).

3A.4.4 Clays and surface conduction

Laboratory data suggest that, under normal seafloor salinity conditions, the impact of clay minerals on bulk conductivity is minimized, with the pore fluid dominating (Wildenschild *et al.*, 2000). The same can be true in the oceanic crust in areas of hydrothermal alteration where deposits such as smectites that are electrically conductive can be present (Drury & Hyndman, 1979; Pezard *et al.*, 1989; Pezard, 1990). Drury & Hyndman (1979) examined a suite of basalts from various oceanic boreholes: Deep Sea Drilling Project (DSDP), Ocean Drilling Program (ODP) and Integrated Ocean Drilling Program (IODP). They found evidence for the modification of conductivity by the presence of clay minerals, but also found that the conduction was dominated by the pore fluid. However, samples from DSDP hole 504B are characterized by higher values of cationic exchange capacities in the pillow section than in the dike complex. With increasing depth, clay content decreases and there is also an apparent increase in the abundance of resistive minerals such as talc and quartz

(Pezard *et al.*, 1989). In this case, estimates of porosity based on Archie's law will under-estimate the total porosity, most of which is filled by resistive minerals.

3A.4.5 Oceanic mantle

Thermal structure provides a first-order control toward delineating oceanic lithosphere from the underlying asthenospheric mantle, where asthenospheric mantle is defined as that part of the mantle that can participate in mantle convection. However, compositional controls have also been shown to impact oceanic mantle near mid-ocean ridges (R. L. Evans *et al.*, 2005).

The thermal structure of oceanic mantle is reasonably well understood, although there are disagreements over the adiabatic temperature. Similarly, the bulk composition of oceanic mantle is also reasonably well understood, and it will be shown that variations in modal composition, at the levels argued over by petrologists, do not greatly influence bulk conductivity. This is also true in the continents.

Using an average pyrolite composition for the oceanic upper mantle above the transition zone at 410 km depth, and a thermal model that accounts for plate cooling and an assumed adiabat (Stein & Stein, 1992, 1994), a conductivity–age section across an ocean basin can be calculated. Two important results come from this exercise:

1. The thermal lithosphere can be seen as a resistive layer of increasing thickness with plate age (Figures 3A.7 and 3A.8).
2. The underlying adiabatic mantle has a conductivity that is around 0.01 S/m (resistivity of 100 Ω m) (Figures 3A.7 and 3A.8).

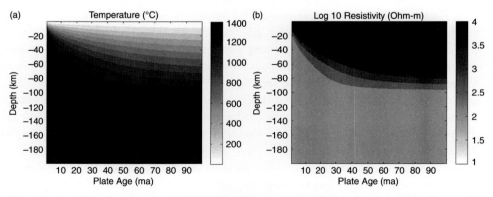

Figure 3A.7. (a) A temperature model (GDH1) of the oceanic mantle based on thermal cooling models (Stein & Stein, 1992, 1994). (b) The conversion of that thermal model to log resistivity using the data reported by Xu *et al.* (2000) and assuming a pyrolite mantle composition. The iron content (Fo_{90}) of the samples used by Xu *et al.* (1998) is in the middle of the expected variation in olivine Fe content. Note that the thermal lithosphere (regions with log resistivity greater than 4) increases in thickness with age. Resistivities in the lithosphere are predicted to be much higher than the values shown, which are capped at 4, and this can be seen in Figure 3A.8.

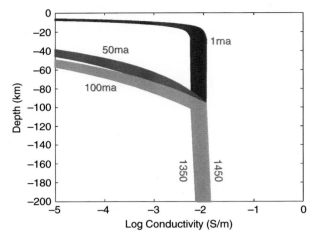

Figure 3A.8. Conductivity–depth profiles from the GDH1 thermal model calculated as in Figure 3A.7, but shown as depth profiles for three different plate ages. In the oceans below ~100 km depth, the conductivity is essentially independent of age if only thermal factors influence the structure. Variations in adiabatic temperature from 1350 to 1450°C bracket the curves as labeled. The uppermost lithosphere is predicted to be highly resistive. (See plate section for color version.)

Water is believed by some to have a substantial impact on mantle conductivity at very low concentrations, and has been invoked to explain the observed increase of measured mantle conductivity over that predicted for a dry olivine mantle. Karato (1990) suggested that hydrogen diffusion could be a significant contributor to electrical conductivity in the mantle, and that the magnetotelluric method could be used to determine the hydrogen distribution (e.g. Karato, 2006). However, at present the laboratory data confirming this belief are conflicting (Figure 3A.9). Reasons for this relate to the difficulty of the measurements, particularly in controlling and constraining water content in the samples at high temperatures. The three extensive published datasets on the subject are extrapolations of lower-temperature measurements to mantle conditions: Wang *et al.* (2006) measured to 1273 K, and Yoshino *et al.* (2006) and Poe *et al.* (2010) to around 1000 K. There seems little doubt that hydrous olivine is more electrically conductive than is dry olivine, although there are differences in the measured activation energies between the different datasets resulting in different magnitudes of enhancement.

The Wang *et al.* (2006) measurements made on olivine aggregates show significantly enhanced conductivities in wet samples, although their activation energy is different from that predicted by Karato's (1990) original hydrogen diffusion model. This original model is a Nernst–Einstein relationship in which conductivity is proportional to the concentration of the charge-carrying species multiplied by the square of its charge. Thus, for conduction accommodated by diffusion of protons, conductivity is directly proportional to hydrogen (water) concentration. Laboratory data on the diffusivity (Mackwell & Kohlstedt, 1990; Kohlstedt & Mackwell, 1998) and solubility (Kohlstedt & Mackwell, 1999) of hydrogen in

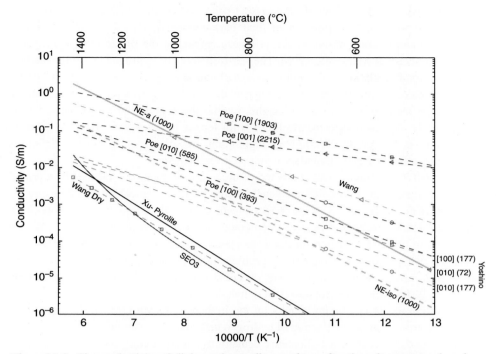

Figure 3A.9. The conductivity of olivine and a pyrolite mantle as a function of temperature based on a variety of laboratory measurements and theoretical models. The curves are as follows: 'SEO3' is olivine conductivity accounting for fO_2 state of the mantle (Constable, 2006); 'Xu-Pyrolite' uses the modal conductivities of Xu *et al.* (2000) for a pyrolite composition; 'Wang Dry' are the conductivities of dry samples reported by Wang *et al.* (2006) and 'Wang' are their wet samples; 'Yoshino' are hydrous samples of Yoshino *et al.* (2006) along the *a*-[100], *b*-[010] and *c*-[001] axes (water contents in ppm are in parentheses); 'Poe' are the hydrous measurements of Poe *et al.* (2010) as labeled with crystallographic direction and water contents in ppm; 'Ne-iso' and 'Ne-a' show the predicted conductivity caused by hydrogen diffusion in olivine using the Nernst–Einstein relationship for 1000 ppm water. Note that, for the hydrous measurements, symbols shown on the lines show the approximate limit in temperature of the laboratory measurements.

olivine are used to estimate conductivity for a given water concentration. The approach adopted for high-temperature measurements on olivine aggregates is to make measurements quickly and to measure water contents before and after measurement (Wang *et al.*, 2006). Samples without substantial water loss are used to calculate activation energies. However, since the activation energies are dependent on water content, uncertainties in water content at each data point could greatly influence the interpretation. Because the high-temperature measurements have been made on aggregates, they do not shed any light on conductivity anisotropy.

Yoshino *et al.* (2006) and Poe *et al.* (2010) observed lower conductivities (but still enhanced relative to dry compositions) in olivine single crystals than Wang *et al.* (2006)

at a given water content. They do not predict enhanced conductivity along the *a* axis at mantle conditions (an extrapolation of their measurements), something that the Karato diffusion model predicts based on earlier measurements of the diffusion of hydrogen in olivine (Kohlstedt & Mackwell, 1998). The self-diffusivity of hydrogen in olivine has been seen to be significantly faster along the *a* axis of olivine at temperatures up to about 1100°C. However, more recent measurements suggest that different diffusivity mechanisms are operative at high temperatures, and in this regime the *a* axis (the axis expected to be preferentially aligned due to mantle flow) is no longer the direction of fastest diffusion (Demouchy & Mackwell, 2006). The single-grain measurements do not, by definition, include any effects of grain-boundary conduction mechanisms.

Poe *et al.* (2010) suggest that, at low temperature, conductivity is controlled by a proton-bearing charge carrier, with a different small polaron conduction mechanism taking over at a temperature that is dependent on the water concentration. This interpretation is more or less consistent with the recent diffusivity measurements of Demouchy & Mackwell (2006). The possibility of different transport mechanisms operating at these higher temperatures could make the extrapolation of conductivity problematic, but, in any event, the *a* axis is not predicted to be the most conductive one, except at very high water contents (~2000 ppm; Poe *et al.*, 2010), which are not observed in mantle xenoliths.

In addition to the conflicting results on the influence of hydrogen on conductivity of olivine, determination of the fraction of hydrogen that contributes to the conductivity is also uncertain. In the typical Nernst–Einstein calculations, it is commonly assumed that all of the hydrogen contributes, thus maximizing the influence of hydrogen on conductivity. The lower activation energy (compared to the Nernst–Einstein relation) seen in the Wang *et al.* (2006) results suggests that free protons are the charge mechanism and that less of the hydrogen contributes to the overall conduction than in the original model. Discrepancies between the temperature dependence of conductivity, as measured and as predicted from diffusion measurements, may be inherent to the way that the two types of measurements are made (Hirth, 2006).

Recent data suggest that water can also influence the conductivities of both orthopyroxene and garnet (Dai & Karato, 2009a,b). The enhancement in conductivity is pronounced for both minerals at low temperatures, where conductivities are several orders of magnitude greater than for equivalent dry mineral compositions. Mierdel *et al.* (2007) show that aluminous orthopyroxene can hold large quantities of water under lithospheric mantle conditions. Given that conductivities of hydrous olivine, orthopyroxene and garnet have all been shown to be significantly enhanced at lithospheric temperatures with fairly small water contents, there perhaps is an explanation for observations of lithospheric resistivities that are lower than expected for dry compositions on typical conductive geotherms (e.g. Bagdassarov *et al.*, 2007).

The Nernst–Einstein relationship predicts that a water concentration of 10^3 H/10^6 Si associated with 0.13 wt% water, or more accurately dissolved water-derived hydrogen at 200 km depth, will result in a ~1.5 order-of-magnitude increase in mantle conductivity over that for dry olivine. Water may also impact conductivity by lowering the solidus and initiating melting. New data suggest that the water-saturated solidus may be as low as 800°C (Grove *et al.*, 2006). Although experimental data on wet solidi at non-saturated conditions are limited, it seems reasonable to expect small amounts of melt to be present when water is available. Water behaves incompatibly (i.e. it travels with the melt), and has a partition coefficient of 0.01 (e.g. 1% melting will remove about half of the water from the mantle and put it into the melt), and so, as melting progresses, the residual mantle becomes increasingly dehydrated. In contrast, as the melt holds more and more water, the conductivity of the melt can be further enhanced. The addition of water enhances conductivity by increasing sodium mobility for rhyolites (Gaillard, 2004) and for less polymerized melts such as tephrite, phonotephrite and phonolite (Pommier *et al.*, 2008). The enhancement in conductivity is more pronounced at lower temperatures.

Areas of the oceanic mantle that have conductivities significantly elevated above 0.01 S/m (Figure 3A.8) most likely reflect the effects of either water, partial melt or their combination. There is growing evidence for enhanced conductivity in the oceanic mantle that has been interpreted in terms of these mechanisms. For example, a profile of conductivity measured on a submarine cable from Hawaii to Oregon showed enhanced conductivity at a depth of around 200 km (Lizarralde *et al.*, 1995). This conductivity profile is consistent with an olivine water content of approximately 1000–3000 H/10^6 Si, within the bounds of the water content of olivine in the oceanic mantle estimated by Hirth & Kohlstedt (1996) based on independent laboratory (e.g. Kohlstedt & Mackwell, 1999) and petrological (e.g. Bell & Rossman, 1992) constraints. The conductivity around 200 km depth is consistent with conduction in an anisotropic mantle, although this could not be confirmed by the single-component data.

Data from the MELT experiment (R. L. Evans *et al.*, 1999, 2005; Baba *et al.*, 2006) show features that can be interpreted in terms of melt and water. Away from the ridge, there is a region of anisotropic conductivity, with conductivity higher in the direction of plate motion, which R. L. Evans *et al.* (2005) interpret as a signal of high water content. They also suggest that some melt may be present, although Yoshino *et al.* (2006), based on an inferred lack of anisotropy in their laboratory data extrapolated to mantle conditions, suggest that the result must reflect partial melt. Baba *et al.* (2006) show that the melt distribution beneath the ridge axis is a broad region (>100 km wide) and is also asymmetric.

Deep-probing magnetotelluric data have been used to test a controversial model in which the 410 km discontinuity is marked by an accumulation of dense melt (Bercovici & Karato, 2003). Although data from the tectonically active area of the southwestern USA are consistent with a melt body at the transition zone, several other deep-probing datasets are not (Toffelmier & Tyburczy, 2007).

3A.4.6 Serpentinization

Serpentinite is an important alteration mineral in the oceanic upper mantle, and potentially is a major source of fluids into subduction zones initiating deep melting of the mantle wedge. It may also play an important role in controlling the rheology in areas of alteration such as oceanic transform faults, as serpentinite has been shown to be considerably weaker than mantle peridotite (Escartin *et al.*, 2001). Serpentinite is also the heat source for the Lost City vent field (Kelley *et al.*, 2001), and so the presence of widespread serpentinization may further enhance hydrothermal heat exchange with the oceans. Seismic constraints on serpentinization are weak, as there is little difference in velocity between serpentinized peridotite and gabbro at modest levels of serpentinization. So here is an area where magnetotellurics may potentially provide valuable constraints.

Laboratory measurements of electrical resistivity of serpentinite samples span seven orders of magnitude at low temperature (e.g. Stesky & Brace, 1973; Popp & Kern, 1993; Xie *et al.*, 2002), although measurements made on seawater-saturated samples have a smaller spread (typically 3–1000 Ω m), and show generally lower resistivities than those reported for serpentinite-free peridotite, gabbro or basalt (Stesky & Brace, 1973; Gillis *et al.*, 1993; Cannat *et al.*, 1995). A discussion of the existing data on serpentinite at low-temperature and low-pressure conditions is given in Evans *et al.* (2010). In summary, seafloor samples of serpentinites generally appear to have higher conductivities largely as a result of having higher porosities.

Despite the role of porosity, it has been suggested that the conductivity of serpentinized peridotite is controlled by the presence, oxidation state and interconnectivity of magnetite (Stesky & Brace, 1973). Magnetite has a resistivity of 4×10^{-5} Ω m (Lefever, 1980), and so a small amount of interconnected magnetite can dramatically impact bulk conductivity. Despite this, there is little evidence to support the interconnection model, at least on the scale of laboratory samples and in the field (Evans *et al.*, 2010). Whether magnetite is a viable conduction mechanism on the larger scale and at depths as porosity is reduced is uncertain. It is possible that serpentinized regions of the mantle have higher conductivities as a result of higher fracture porosity.

3A.5 Continents

The larger range of compositions and tectonic regimes found in the continents makes the task of summarizing their conductivity structure more difficult. Yet, even in the continents, compositional variations are usually swamped by the impact of thermal structure, the presence of fluids or other exotic phases. However, there are first-order differences in the lithosphere beneath continents and oceans that relate to differences in thermal history and also to composition that can be inferred from conductivity profiles (e.g. Hirth *et al.*, 2000).

3A.5.1 Continental crust

It has been a common observation that the continental lower crust is electrically conductive, and indeed it has been argued that it is conductive almost everywhere, with resistivity values about 100 times more than predicted by laboratory measurements (e.g. Hyndman *et al.*,

1993). There does seem to be an age progression in the conductance (the product of conductivity and thickness) of the lower crust (Jones, 1981; Shankland & Ander, 1983; Haak & Hutton, 1986; Jones, 1992). Recent data demonstrate that Mesoarchean terranes, southern Slave Craton (Jones & Ferguson, 2001), the Rae terrane (Jones *et al.*, 2002; S. Evans *et al.*, 2005) and the Kaapvaal Craton (Muller *et al.*, 2009) exhibit conductance values of <2 S, which is broadly consistent with dry mafic rocks at low temperatures. In contrast, Phanerozoic regions generally exhibit conductances in excess of 200–2000 S, with Neoarchean and Proterozoic regions lying in between.

Areas of high conductivity in the continental mid- to lower crust can usually be ascribed to three mechanisms: melt, which will only be relevant in certain locations and cannot explain consistently high conductivities; aqueous fluids with potentially high chlorinity; and metallic conductors, carbon, sulfides and iron oxides (Jones, 1992). Wannamaker (2000) presents a summary discussion highlighting the roles that fluids and graphite may play in the conductivity increase within the continental mid-crust. Importantly, he points out that prior attempts to ascribe a single mechanism likely oversimplified the situation, as concluded by Duba *et al.* (1994) in their examination of rocks exhumed from the mid-crust by the German deep drilling experiment (KTB), and that the tectonic history of a survey area needs to be taken into consideration in interpreting crustal conductors. The thermal structure of continental crust will vary significantly between different regions of similar age, depending on the timing of the most recent orogeny or tectonic event. Some areas of lower crust may be in the ductile regime, while others will be able to sustain crack networks; these differences alone likely rule out a single conduction mechanism.

3A.5.1.1 Sedimentary basins

The relatively high porosities of sedimentary basins result in resistivities on the order of 1–100 Ω m (e.g. Williston Basin, central North America; Jones, 1988), depending on the degree of compaction and the salinity of the pore fluid. If the basins are filled with carbon-rich organic deposits that may have been converted to coal, then resistivity will drop dramatically at their horizons. Some of these coal horizons may contain graphite, others pyrite. Branch *et al.* (2007) show conductivities in excess of 1 S/m in a pyrite-rich band in South Africa. Petroleum deposits, which are found in sedimentary basin settings, are resistive horizons (e.g. Lucius *et al.*, 1989) and, although controlled-source electromagnetic methods have been used to search for such deposits offshore, magnetotellurics is not sensitive to the kinds of thin resistive layers presented by these targets. In some settings, shallow marine basins that have subsequently been uplifted may have salt horizons that are highly conductive, especially if groundwater is able to dissolve the salt (e.g. Sainato *et al.*, 2000).

3A.5.1.2 Metamorphic rocks

By their nature, metamorphic rocks contain a high degree of complexity in terms of both fabric and composition, and it is difficult to know how representative samples of a given rock type from a specific region are likely to be. Granulites, for example, have wide variations in the amount of carbon they contain, which, in turn, impacts their bulk conductivity. Most laboratory measurements of metamorphic rocks (granulites, gneisses, amphibolites, serpentinite and eclogite) show fairly high resistivities under conditions throughout the crust (e.g. Glover & Vine, 1995; and references below). The strong foliation that is commonly found in deformed metamorphic rocks can also give rise to high degrees of electrical anisotropy, especially when carbon is present. Such anisotropy has been seen in gneisses and in granulites (Glover & Vine, 1992; Lastovickova *et al.*, 1993; Fuji-ta *et al.*, 2007; see Section 3A.6 for more discussion).

Measurements have been made on granulites from the Hidaka metamorphic belt in Japan under conditions representative of the mid- to lower crust (Fuji-ta *et al.*, 2004). In contrast to arguments (see Section 3A.5.1) for conductive phases to explain highly conductive lower crust, the granulites have resistivities (~700 Ω m at 400°C and ~170 Ω m at 600°C) consistent with those measured by magnetotelluric soundings in the area (e.g. Ogawa *et al.*, 1994). Granulites that contain connected carbon show about an order-of-magnitude anisotropy between directions perpendicular and parallel to the foliation, but resistivities are in the range of 10 000 Ω m and higher, even at high pressure when carbon interconnection is enhanced (Glover & Vine, 1992). Saturating these samples (carbon- and non-carbon-bearing) further enhances conduction (with resistivities around 2000–5000 Ω m), but in this case conduction decreases with increasing pressure as fluid pathways are closed.

There are very few laboratory data on the conductivity of eclogites, but what do exist suggest that they are resistive, but also highly variable, with at least two orders of magnitude spread in values seen at a given temperature (Lastovickova, 1975; Lastovickova & Parchomenko, 1976). However, high-pressure subduction zone eclogites commonly consist of high volume fractions (around 50%) of electrically conductive Fe-rich garnet (Konrad-Schmolke *et al.*, 2008; Romano *et al.*, 2006). If the garnets are deformed to the point where they interconnect, or are connected through metasomatic events, then perhaps these rocks could be highly conducting, although laboratory measurements of conductivity have not been made on samples of this rock type.

Serpentinite was discussed in detail in Section 3A.4.6 and the situation in the continental crust is much the same except for the absence of seawater, which appears to control the conduction of shallow serpentinite bodies on the seafloor. Although serpentinite is primarily thought of in an oceanic context, forming as a result of interaction between seawater and the oceanic upper mantle, serpentinite bodies do exist on the continents, either at their point of formation, for example in the cool part of a mantle wedge in a subduction zone setting (e.g. Bostock *et al.*, 2002), or as a result of exhumation (e.g. Pilchin, 2005) or obduction processes (e.g. Dick, 1977). Unless the hypothesis that magnetite enhances conduction in serpentinite holds true (Stesky & Brace, 1973) – and, as discussed earlier, the evidence for this is slight – then serpentinite is expected to be electrically resistive in a continental setting.

With the exception of the granulites in the Hidaka metamorphic belt (Fuji-ta *et al.*, 2004), most laboratory measurements of rocks representative of the mid- to lower continental crust, including high-grade metamorphic rocks, require some additional conductive phase to explain widespread observations of a conducting lowermost crust. In the next sections, several possibilities for enhancing crustal conductivity are discussed.

3A.5.1.3 Melts

Where melt occurs in the crust, there are several important distinctions to the previous discussion on mantle melts. First, equilibrium conditions probably do not hold in the crust, and the bulk conductivity will likely be controlled by crack geometries except in areas of melt ponding such as magma chambers. Second, Wannamaker (1986, 2000) discusses important differences in the conductivity of melts at the lower temperatures and pressures of the lower crust. Chief among these is that the conductivity appears, on the basis of laboratory measurements, to depend on the water content of the melt (see discussion in Sections 3A.3.2–3A.3.3). As discussed above, Gaillard (2004) suggested that the addition of water enhances sodium mobility, resulting in an increase in conductivity.

3A.5.1.4 Aqueous fluids

As discussed above, the conductivity of aqueous fluids depends on temperature and chloride content (salinity for seawater). While in oceanic sediments, and even the upper oceanic crust, a clear case can be made that interconnected and conductive fluids control bulk conductivity, the situation is less clear for the continental crust. Data from the German deep KTB borehole suggest that cracks are closed at depth and generally do not permit interconnected brines to form effective conductors (ELEKTB Group, 1997), although a small percentage of conductivity anomalies are ascribed to ionic conduction by fluids. Indeed, it was surprising that the upper crust was seen to contain so much brine and yet be so resistive, perhaps supporting the hypothesis that the stress regime plays an important role in fluid interconnectivity in the continental crust (e.g. Gough, 1986).

Nesbitt (1993) points out that, although crustal fluids could have resistivities as low as $0.01 \, \Omega$ m, these values are typically reached at depths shallower than 8 km, particularly in volcanic contexts where hydrothermal brines have been observed (e.g. Pommier *et al.*, 2010c). Mid- and lower crustal conductors most likely result from mechanisms other than aqueous fluids, unless there are dramatic changes in pore-space geometry local to an observed conductor (Hyndman *et al.*, 1993).

In areas of lower crust that are in the ductile regime, connectivity of fluids may be enhanced, requiring porosities of less than 0.5% in order to achieve bulk resistivities of less than $10 \, \Omega$ m (Wannamaker *et al.*, 2008), and so in these situations fluids become a viable mechanism. In this case (assuming equilibrium conditions), the connectivity of the fluids depends on the dihedral angle in much the same way as for melts in the mantle (see Figure 3A.4). Dihedral angles in quartzite rocks are seen to have values as low as 40° and are dependent on concentrations of solutes (particularly NaCl and KCl), permitting interconnection (although pure H_2O does not interconnect). Dihedral angles are always above 60° (unconnected) in dunite (Watson & Brennan, 1987). Low dihedral angles, essential for connectivity, are usually also taken to suggest a permeable fluid network. If fluids are a viable explanation, then there must also be some form of impermeable barrier in the mid-crust (Etheridge *et al.*, 1983, 1984) trapping fluids below (Jones 1987, 1992). Although strongly opposed by some petrologists – for example, Yardley & Valley (1997), and see discussion by Wannamaker (2000) and reply by Yardley & Valley (2000) – the rapid dewatering of extensive fluids in the Juneau Gold field during plate reorganization (Goldfarb *et al.*, 1991) is evidence that fluids can be retained on time scales of tens to hundreds of million years.

3A.5.1.5 Carbon

Carbon is often called upon to explain areas of high conductivity in the lower continental crust. There is significant controversy in the literature over the forms of carbon that are deposited – it is common in the magnetotelluric community to refer more or less exclusively to graphite, but other forms of carbon are possible. As discussed above, graphite is extremely electrically conductive, and only trace amounts are needed to raise bulk conductivity, provided it is interconnected. Because conduction is electronic, areas of conductivity enhanced by graphite do not have measurable temperature dependence. Carbon is thought to be deposited either as a result of fluxing of CO_2 in subduction zone settings or in settings with abundant deep fluids carrying volatiles, or as a result of fracturing, including during earthquake activity (Mathez *et al.*, 2008).

Estimates for the concentration of CO_2 in the mantle beneath oceanic ridges range from 50 to 2000 ppm (Saal *et al.*, 2002). However, as pointed out by Dasgupta & Hirschmann (2007a, and references therein), mantle xenoliths frequently lose carbon during ascent and magmas also degas upon eruption, leading to significant uncertainty about concentrations of carbon in mantle source regions. The form of carbon held in the mantle may be either reduced carbon (diamond or graphite) or crystalline carbonate, depending on depth and on the oxygen fugacity (Dasgupta & Hirschmann, 2006).

Shear strain activity, such as that concentrated along faults, appears to enhance graphite interconnection, and consequently graphite in crustal settings is often associated with former suture zones or regions of major deformation (e.g. Mareschal *et al.*, 1995). Because of the links to hydrothermal fluid flow, graphite-bearing zones may commonly contain economic minerals, particularly uranium (e.g. Heinson *et al.*, 2006; Tuncer *et al.*, 2006; Farquharson & Craven, 2009).

Samples from the KTB deep borehole in Germany show evidence for interconnected graphite in the middle continental crust, and downhole electrical measurements suggest that the carbon must be interconnected over distances on the order of hundreds of meters (e.g. ELEKTB Group, 1997). Yet, not all of the graphite present is interconnected, and so much of it does not impact the bulk conductivity. Interestingly, the resistivity of samples from the borehole shows no correlation with lithology, again demonstrating that other factors (e.g. cracking, fluids, graphite) dominate the bulk conduction mechanism. Where the graphite is highly connected, the bulk resistivity seen in logging is as low as 0.1 Ω m compared to several thousand ohm meters in the surrounding rock.

Electrical responses associated with earthquake activity, including magnetotelluric signals (e.g. Fraser-Smith *et al.*, 1990), are thought to be related to changes in bulk conductivity. Mechanisms suggested for these changes include redistribution of brines and the deposition of carbon as films on fracture surfaces (Roberts *et al.*, 1999). Mathez *et al.* (2008) ran a series of experiments in which they showed rapid carbon deposition on freshly fractured surfaces. Magnitudes of conductivity enhancement seen in the laboratory were relatively modest, although in a real fault network carbon accumulation over repeat events could result in high bulk conductivities.

3A.5.1.6 Electric moho?

Unlike seismic reflection profiling, for which there is a marked change in seismic impedance at the crust–mantle interface (the Mohorovičić discontinuity or moho), most magnetotelluric surveys do not reveal a significant difference in resistivity across the crust–mantle interface. The reasons for this are that the compositional change across the boundary does not usually lie within the range that leads to large changes in bulk conductivity. Laboratory studies suggest that the resistivity of ultramafic rocks is about half an order of magnitude to an order of magnitude (at most) greater than mafic rocks (see compilation by Haak, 1982). Also, the conductance of the conducting lower crust can be represented as a thin highly conducting layer or a thicker less conducting layer (Jones, 1992), and has a shielding effect on the resistivity of the uppermost mantle (Jones, 1999). The first solid evidence for an electrical moho comes from the Slave Craton (Jones & Ferguson, 2001), in which conductivity jumps by about an order of magnitude. This observation was made possible because of anomalously resistive lower crust in the Slave that enhances the contrast with the underlying mantle. Other studies since also present evidence for a resistivity change at the moho (Bhattacharya & Shalivahan, 2002; Jones *et al.*, 2002; Bouzid *et al.*, 2008), but the pervasive conducting lower crustal layer makes resolution of an electrical moho difficult.

3A.5.2 Continental lithospheric mantle

3A.5.2.1 Composition

Xenolith suites, which span a broad range of compositions, can be used to estimate the range of conductivity expected from compositional variations in the lithospheric mantle. In general, these differences are not large, and are typically below the level of uncertainty in both magnetotelluric inversion models and estimates of the regional temperature that have a first-order effect on the bulk conductivity.

Jones *et al.* (2009) demonstrated the small impact that composition has on bulk conductivity of the continental lithospheric mantle compared to temperature. For example, the conductivity contrast from a depleted harzburgitic region (Mg# = 92.0) to the less-depleted, more iron-rich, lherzolite (Mg# = 90.2) would be ~0.1 log units (assuming representative temperatures for the mid-lithosphere). This change is far smaller than would be caused by uncertainty in temperature.

Several observations show that the uppermost lithospheric mantle beneath Archean terranes is more conductive than predicted on the basis of laboratory measurements and the best estimates of continental geotherms for the regions (Jones & Ferguson, 2001; Bagdassarov *et al.*, 2007; Spratt *et al.*, 2009; Evans *et al.*, 2011). Typically, the shielding effect of the conducting lower crust makes determination of the actual resistivity of the uppermost mantle impossible – only a minimum bound can be set (Jones, 1999). However, there are regions where this layer is weak to absent. The usual explanation is that a connected conducting phase is required, but it is unclear what that phase might be and why it should be so ubiquitous. Graphite is the most commonly invoked explanation, particularly for the very low-resistivity regions (<100 Ω m), but other possibilities exist.

3A.5.2.2 *Water*

Geochemical evidence does not support the idea of a wet lithospheric mantle beneath continents, largely due to the notion that extensive melting should have depleted the mantle of most of its water. That does not rule out late-stage metasomatic introduction of water, but geochemists seem content with a dry lithospheric mantle, particularly beneath the older cratons. Bell *et al.* (2004) describe OH concentrations in cratonic kimberlite samples, and concentrations of <200 ppm OH (by weight) is the upper limit seen in cratonic settings. As discussed in Section 3A.4.5, recent data show significant enhancement of conductivity for olivine, orthopyroxene and garnet at low temperatures and for fairly modest water contents consistent with the xenolith data. Under conditions expected for the uppermost ~150 km of lithospheric mantle, conductivities could be several orders of magnitude greater than for equivalent dry mineral compositions, potentially providing an explanation for observed low resistivities in cratonic settings (e.g. Bagdassarov *et al.*, 2007; Evans *et al.*, 2011). Deeper into the lithospheric mantle, as the temperatures increase, this effect is less important, except at high water contents, and so, unless the kimberlite samples have failed to accurately account for water content, it is unreasonable to expect that water will have a significant impact on the lowermost reaches of the continental lithospheric mantle.

3A.5.2.3 *Carbon*

There is only limited literature on the presence and distribution of carbon in xenolith samples that might improve understanding of the likely abundance of carbon in the continental lithospheric mantle. As discussed above, the dominant mechanism by which carbon is believed to be introduced into the mantle is by the release of fluids from subducting oceanic crust, although the possibility of an abiotic origin for carbon, particularly in Archean rocks, has been suggested by some (e.g. Fedo & Whitehouse, 2002; Whitehouse *et al.*, 2009). For a biotic origin, proposed models of continent formation that involve accretion of arc terranes would potentially introduce carbon throughout the lithospheric mantle. Pearson *et al.* (1994) examined a suite of xenoliths from the Kaapvaal and elsewhere, and documented the graphite distribution in 26 samples. Interestingly, graphite is only documented in kimberlites found within the boundaries of cratons; no graphite is recorded in off-craton kimberlites. Graphite is found in two morphologies, single isolated flakes and multicrystal-line stacks, but in both cases is dispersed and does not form the interconnected network necessary to enhance conductivity. Watson *et al.* (2010) measured the conductivity of San Carlos olivine containing carbon impurities. In contrast to the notion that only very small volumes of carbon are needed to enhance bulk conductivity, they found that ~0.16 vol% of carbon had no measurable effect on bulk conductivity. In this case, the carbon was thought to be unevenly distributed on grain boundaries. It further suggests a percolation threshold, or a critical volume of carbon needed to result in interconnection. In contrast, 1 vol% of iron-rich sulfide did result in a conductivity enhancement, even though only ~0.004 vol% was estimated to be interconnected, the rest sitting as isolated inclusions.

In the crust, it appears likely that widespread strain or fracturing is needed to interconnect the graphite. While there may be areas of lithospheric mantle that have undergone such strain, and Mareschal *et al.*'s (1995) example is excellent, it is not clear that graphite represents a reasonable explanation for the higher than expected conductivity of lithospheric mantle beneath some cratons. Graphite has been used as an explanation for a strong conductor seen beneath the central Slave Craton (Jones *et al.*, 2001) that is coincident with a seismic discontinuity (Chen *et al.*, 2009), strong geochemical layering (juxtaposition of an ultra-depleted harzburgite with a lherzolite) and the location of a cluster of hundreds of Eocene-aged diamondiferous kimberlites.

3A.5.2.4 Hydrous phases

In some cases, hydrous phases, such as phlogopite, have been invoked to explain high mantle conductivities (Boerner *et al.*, 1999), but laboratory data confirming this assertion have been limited. Recent work suggests that phlogopite is not highly conductive, with resistivity values of around 1000 Ω m at 1000°C, increasing substantially with falling temperatures (Guseinov *et al.*, 2005). The conduction mechanisms in phlogopite are complex, and at least three different conduction regimes are seen as a function of temperature; however, none are sufficient to appreciably raise the bulk conductivity of the mantle. Laboratory data on other hydrous phases such as chlorite, amphibole and antigorite, which play a critical role in subduction zones, are similarly lacking. Issues with measuring conductivity of hydrous olivine and its high-pressure polymorphs (wadsleyite and ringwoodite) at the high temperatures of the mid-mantle were discussed earlier in Section 3A.4.5.

3A.6 Anisotropy

Conductivity is a second-rank tensor that is usually reduced to a scalar value in practice. However, under certain conditions it is possible to describe conductivity as a diagonal tensor rotated to a specified coordinate system, with a distinct value for each of the three axial directions (e.g. Yu *et al.*, 1997). Such conditions are usually the most complicated that one is able (or willing) to consider. Transverse anisotropy is a special case, where the vertical conductivity differs from the horizontal one, which is independent of azimuth.

In magnetotellurics, not everything that is presented as evidence for anisotropy is actually representative of anisotropic fabric, and there is ambiguity between fabric anisotropy and structural anisotropy, i.e. 2D or 3D heterogeneity in conductivity structure. This subject is discussed in extensive detail by Yin & Weidelt (2009) and Wannamaker (2005), with the latter giving an in-depth discussion of where anisotropy has been seen in surveys. Details about how anisotropy appears in magnetotelluric data are presented in Chapter 2, and its incorporation in forward and inverse models is described in Chapters 7 and Chapters 8, respectively.

The differences between the phases of the TE and TM modes are frequently referred to as evidence for anisotropy. However, such differences are to be expected over an Earth with non-1D structure under many circumstances (see Section 2.6.2). In some cases, separate inversion of the TE and TM modes may be used to demonstrate that higher conductivities are required (or at least preferred) in one of the principal directions (e.g. Wannamaker *et al.*, 2008). Evans *et al.* (2005) and Baba *et al.* (2006) present results from a formal inversion in which anisotropy is explicitly accounted for in addition to 2D structure.

Fluid-filled cracks, faults and dikes are features that have a clear strike direction and hence can cause conductivity to exhibit directionality (see Section 2.6.2). Thus, such crustal features might be expected to give rise to anisotropic responses. Areas that are likely to contain such features, either through collisional or extensional tectonics, or volcanism, are candidates for apparent anisotropic conductivity. Examples include collisional belts such as New Zealand (Wannamaker *et al.*, 2002), orogenic belts and suture zones, and areas of crustal and lithospheric shearing (Mareschal *et al.*, 1995; Ji *et al.*, 1999; Eaton *et al.*, 2004) or extension (Wannamaker *et al.*, 2008). In some of these cases, magnetotelluric anisotropy has been compared to seismic anisotropy, most commonly from polarization of shear-wave velocities. In such cases, magnetotellurics offers a significant advantage over the SKS splitting approach (SKS is a seismic wave that is converted to a shear wave on leaving the core), as it is better able to determine the depths at which anisotropic structure occurs. A note of caution though is given by Jones (2006) – SKS-like displays should not be automatically attempted with magnetotelluric data.

Laboratory measurements on metamorphic rocks from the continental crust that have a high degree of foliation show evidence for electrical anisotropy. Gneisses from Japan have conductivities that are an order of magnitude higher in the foliation direction (Fuji-ta *et al.*, 2007). Similarly, carbon-bearing granulites show higher conductivities in the foliation direction (Glover & Vine, 1992), with the magnitude of the conductivity and the anisotropy dependent on pressure, with increasing pressure enhancing the connectivity of the carbon.

Data from the Grenville Front (Mareschal *et al.*, 1995; Ji *et al.*, 1999) show a compelling case for graphite, aligned and connected through extensive shearing, causing anisotropic electrical anisotropy. An obliquity between the seismic and electric anisotropic directions is interpreted to show that seismic velocities respond to the lattice-preferred orientation of olivine, while the electrical conductivity records the lineation of the olivine, but with graphite aligned in the lineation direction and enhancing conductivity. Obliquity between the dominant slip direction (shear plane) and the foliation most likely reflects the limited degree of accumulated shear strain.

A similar comparison between seismic and magnetotelluric anisotropies in continental crust and lithospheric mantle was carried out by Hamilton *et al.* (2006). Electrical anisotropy in the crust appears to correlate with large-scale deformation patterns seen in the surface geology. In the mantle, there is greater ambiguity between electrical and seismic anisotropy, and it seems likely that what appears as electrical anisotropy has a significant component arising from structural heterogeneity.

Anisotropy can also be intrinsic to the minerals making up the host rock. In the mantle, dry olivine is only modestly anisotropic – a factor of ~2–3 between the *c* axis (most conductive) and the other two axes (Constable & Duba, 1990) – but the diffusion of hydrogen is significantly faster along the *a* axis of olivine. In areas of *a* axis alignment, such as is observed seismically in the oceanic asthenospheric mantle, a wet mantle might be expected to show electrical anisotropy, and this has been observed on the southern East Pacific (R. L. Evans *et al.*, 2005). Others have also discussed water in cratonic mantle as a possible source of electrical anisotropy (Simpson, 2002). It is not clear either that geo-chemical evidence supports the levels of water in the cratonic lithospheric mantle that would be required to cause anisotropic conductivities, or that the data are showing true anisotropy and not the result of regional heterogeneity. Simpson & Tommasi (2005) show that the anisotropy arising from hydrous olivine is heavily dependent on the degree of accumulated shear strain that aligns the olivine crystals. Depending on the mantle composition, aniso-tropic factors on the order of 3–4 are possible, similar to that seen by R. L. Evans *et al.* (2005) in the oceanic mantle, but larger anisotropies require additional mechanisms. Others have argued that water is unlikely to cause electrical anisotropy at all, and that melt (either silicate or carbonatite) is a likely explanation (Yoshino *et al.*, 2006; Gaillard *et al.*, 2008; Poe *et al.*, 2010). While there may be instances where melt causes electrical anisotropy, particularly at shallower depths, where it may occur in dunite channels or in cracks, the controlling factors of melt in the mantle are unclear. In particular, whether melt topology is controlled entirely by surface energy and buoyancy, or by the regional stress regime, and the impact that melt may have on mantle fabric itself, are matters of considerable debate (e.g. Wark & Watson, 2002; Holtzman *et al.*, 2003; Caricchi *et al.*, 2011).

3A.7 Comments on permeability and conductivity

Since electrical conductivity is a transport property, it would appear obvious that it should somehow relate to fluid permeability in the crust and to melt permeability in the mantle. For example, one of the reasons given against fluids in the mid-continental crust is that the connectivity of the fluids that explain the high conductivity would also require high permeability, which would allow the fluid to escape. Despite this notional link, actual data tying conductivity to permeability have been hard to acquire, and efforts to do so require assumptions about the host rock properties (e.g. Evans, 1994; Roberts & Tyburczy, 1999; Wright *et al.*, 2009).

Laboratory measurements on permeability and conductivity at low porosities typically show evidence for threshold porosities that are necessary for fluids to flow, but not for conductivity to be enhanced. The reasons are that conductivity at low porosity is often controlled by secondary porosity (small cracks and fissures), which does not allow fluid flow, but is perfectly able to carry electrical current. In many cases, these small cracks are too small to image effectively (e.g. Doyen, 1988; Friedrich *et al.*, 1993). This difference, as well as the scaling distinction between conductivity and permeability and the length scale of the pore space, make it difficult to relate one to the other, even though both are transport properties (e.g. David, 1993). Admittedly, many laboratory studies are done on sandstones, which will have very different crack networks to the rocks within the continental mid-crust, for example. Further, in the laboratory, changes in porosity are realized by adding pressure, but closure of pore throats and networks can also occur through metamorphic reaction and deposition of minerals by fluid flow, although it is not clear that the two processes result in the same behavior at very low porosity in the Earth (Roberts & Schwartz, 1985).

In a partial melt network in the mantle, the geometry of melt is determined by interfacial energies between the melt and the solid matrix. Numerous articles have presented power-law relationships between bulk permeability and porosity, with permeability, $\kappa \propto \phi^{\,n}$, where n is usually between 2 and 3 (e.g. Doyen, 1988; Cheadle, 1989; Dullien, 1992; Faul, 1997; Wark & Watson, 1998; Cheadle *et al.*, 2004). One example is the relationship given by Wark *et al.* (2003), who estimate the Darcy porous flow permeability k of a melt network as

$$k = \frac{\phi^3 d^2}{270} \tag{3A.6}$$

where d is the grain size and φ is the porosity. Recent numerical work details the contributions of effective and ineffective porosity to the overall permeability for a simple sphere and tube network. In this case, the permeability estimated by a tube model is an upper bound on the permeability with an exponent of ~2 (Simpson *et al.*, 2010).

In theory, porosity estimates from conductivity allow melt permeability to be further estimated. Although typical bounds on melt porosity are wide, because of the uncertainties in melt connectivity, the lower bound on porosity probably provides the best estimate of permeability, as contributions from melt in isolated pockets do not impact the bulk permeability.

3A.8 Summary

The past several years have seen exciting developments in our understanding of the conductivity of a variety of materials. Yet, as discussed above, there remains a good deal that is either contested, uncertain or has yet to be studied. A summary of what is known, along with some of the uncertainties, is shown in Figure 3A.10. The role of water on the conductivity of olivine and other mantle phases is perhaps the murkiest of the recent stories, but the ability of magnetotellurics to provide constraints on mantle water contents, in both the asthenosphere and lithosphere, is a strong selling point for the technique, and work in

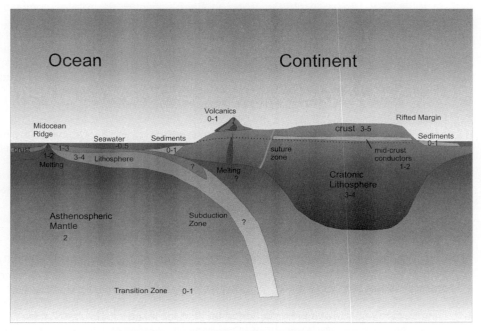

Figure 3A.10. A cartoon of the continents and oceans showing key features and typical ranges of resistivity in each; numbers shown are $\log_{10}$(resistivity/Ω m).

this area continues. The understanding that carbonatite melts are so highly conductive has allowed identification of incipient melting deep in the mantle beneath mid-ocean ridges at melt fractions that are beyond seismic sensitivity.

Some of the most exciting studies that have been done recently relate to the generation and transport of melt at subduction zones. The fluids released from the downgoing slab are an important part of this process, yet their conductivities are almost entirely unknown – representing a large area for future work.

References

Andrews, D. & A. Bennett (1984). Measurements of diffusivity near the sediment–water interface with a fine scale resistivity probe. *Geochim. Cosmochim. Acta*, **45**, 2169–2175.

Archie, G. E. (1942). The electrical resistivity log as an aid in determining some reservoir characteristics. *J. Petr. Tech.*, **5**, 1–8.

Baba, K., A. D. Chave, R. L. Evans, G. Hirth & R. Mackie (2006). Mantle dynamics beneath the East Pacific Rise at 17 S: insights from the Mantle Electromagnetic and Tomography (MELT) experiment. *J. Geophys. Res.*, **111**, B02101, doi:10.1029/2004JB003598.

Bagdassarov, N. S., M. G. Kopylova & S. Eichert (2007). Laboratory derived constraints on electrical conductivity beneath Slave Craton. *Phys. Earth Planet. Inter.*, **161**, 126–133.

Bell, D. R. & G. R. Rossman (1992). Water in Earth's mantle: the role of nominally anhydrous minerals. *Science*, **255**, 1391–1397.

Bell, D. R., G. R. Rossman & R. O. Moore (2004). Abundance and partitioning of OH in a high-pressure magmatic system: megacrysts from the Monastery kimberlite, South Africa. *J. Petrol.*, **45**, 1539–1564.

Bercovici, D. & S.-I. Karato (2003). Whole mantle convection and the transition zone water filter. *Nature*, **425**, 39–44.

Bernabé, Y., W. F. Brace & B. Evans (1982). Permeability, porosity and pore geometry of hot-pressed calcite. *Mech. Mater.*, **1**, 173–183.

Berryman, J. G. (1995). Mixture theories for rock properties. In *Rock Physics and Phase Relations: A Handbook of Physical Constants*, ed. T. J. Ahrens. Washington: American Geophysical Union.

Berryman, J. G. (2006). Measures of microstructure to improve estimates and bounds on elastic constants and transport coefficients in heterogeneous media. *Mech. Mater.*, **38**, 732–747.

Bhattacharya, B. B. & Shalivahan (2002). The electric moho underneath Eastern Indian Craton. *Geophys. Res. Lett.*, **29**, 1376, doi:10.1029/2001GL014062.

Blakemore, J. S. (1985). *Solid State Physics*. Cambridge: Cambridge University Press.

Boerner, D., R. D. Kurtz, J. A. Craven, G. M. Ross, F. W. Jones & W. J. Davis (1999). Electrical conductivity in the Precambrian lithosphere of western Canada. *Science*, **283**, 668–670.

Bostock, M. G., R. D. Hyndman, S. Rondenay & S. M. Peacock (2002). An inverted continental moho and serepentinization of the fore-arc mantle. *Nature*, **417**, 536–538.

Bouzid, A., N. Akacem, M. Hamoudi, K. Ouzegane, A. Abtout & J. R. Kienast (2008). Magnetotelluric modeling of the deep geologic structure of In Ouzzal Granulitic Unit (western Hoggar). *Comptes Rendus Geosci.*, **340**, 711–722.

Branch, T., O. Ritter, U. Weckmann & F. Schilling (2007). The Whitehill Formation – a high conductivity marker horizon in the Karoo Basin. *S. Afr. J. Geol.*, **110**(2–3), 465–476.

Cairns, G., R. L. Evans & R. N. Edwards (1996). A time domain electromagnetic survey of the TAG hydrothermal mound. *Geophys. Res. Lett.*, **23**, 3455–3458.

Cannat, M., J. A. Karson, D. J. Miller, *et al.* (1995). *Proc. Ocean Drilling Program*, Initial Report 153, doi:10.2973/odp.proc.ir.153.1995.

Caricchi, L., F. Gaillard, J. Mecklenburgh & E. Le Trong (2011). Experimental determination of electrical conductivity during deformation of melt-bearing olivine aggregates and electrical anisotropy in the oceanic low velocity zone. *Earth Planet. Sci. Lett.*, **302**, 81–94.

Cheadle, M. J. (1989). *Properties of texturally equilibrated two-phase aggregates*. Ph.D. thesis, University of Cambridge.

Cheadle, M. J., M. T. Elliott & D. McKenzie (2004). Percolation threshold and permeability of crystallizing igneous rocks: the importance of textural equilibrium. *Geology*, **32**, 757–760.

Chen,, C.-W., S. Rondenay, R. L. Evans & D. B. Snyder (2009). Geophysical detection of relict metasomatism from an Archaean subduction zone. *Science*, **326**, 1089–1091.

Constable, S. C. (2006). SEO3: a new model of olivine electrical conductivity. *Geophys. J. Int.*, **166**, 435–437.

Constable, S. C. & A. Duba (1990). The electrical conductivity of olivine, a dunite and the mantle. *J. Geophys. Res.*, **95**, 6967–6978.

Constable, S., T. J. Shankland & A. Duba (1992). The electrical conductivity of an isotropic olivine mantle. *J. Geophys. Res.*, **97**, 3397–3404.

Cox, C. S., C. S. Constable, A. D. Chave & S. C. Webb (1986). Controlled source electro-magnetic sounding of the oceanic lithosphere. *Nature*, **320**, 52–54.

Dai, L. & S.-I. Karato (2009a). Electrical conductivity of pyrope-rich garnet at high temperature and high pressure. *Phys. Earth Planet. Inter.*, **176**, 83–88.

Dai, L. & S.-I. Karato (2009b). Electrical conductivity of orthopyroxene: implications for the water content of the asthenosphere. *Proc. Japan. Acad., Ser. B*, **85**, 466–475.

Dai, L., H. Li, C. Li, H. Hu & S. Shan (2010). The electrical conductivity of dry polycrystalline olivine compacts at high temperatures and pressures. *Mineral. Mag.*, **74**, 849–857.

Dasgupta, R. & M. M. Hirschmann (2006). Melting in the Earth's deep upper mantle caused by carbon dioxide. *Nature*, **440**, 659–662.

Dasgupta, R. & M. M. Hirschmann (2007a). A modified iterative sandwich method for determination of near-solidus partial melt compositions. II. Application to determination of near-solidus melt compositions of carbonated peridotite. *Contrib. Mineral. Petrol.*, **154**, 647–661.

Dasgupta, R. & M. M. Hirschmann (2007b). Effect of variable carbonate concentration on the solidus of mantle peridotite. *Am. Mineral.*, **92**, 370–379.

David, C. (1993). Geometry of flow paths for fluid transport in rocks. *J. Geophys. Res.*, **98**, 12 267–12 278.

Demouchy, S. & S. Mackwell (2006). Mechanisms of hydrogen incorporation and diffusion in iron-bearing olivine. *Phys. Chem. Minerals*, **33**(5), 347–355.

Dick, H. J. B. (1977). Partial melting in the Josephine peridotite: I, the effects on mineral composition and its consequence for geobarometry and geothermometry. *Am. J. Sci.*, **277**, 801–832.

Dobson, D. P., A. P. Jones, R. Rabe, T. Sekine, K. Kurita, T. Taniguchi, T. Kondo, T. Kato, O. Shimomura & S. Urakawa (1996). In-situ measurement of viscosity and density of carbonate melts at high pressure. *Earth Planet. Sci. Lett.*, **143**, 207–215.

Doyen, P. M. (1988). Permeability, conductivity and pore geometry of sandstone. *J. Geophys. Res.*, **93**, 7729–7740.

Drury, M. J. & R. D. Hyndman (1979). The electrical resistivity of oceanic basalts. *J. Geophys. Res.*, **84**, 4537–4545.

Drury, M. R. & J. D. FitzGerald (1996). Grain boundary melt films in an experimentally deformed olivine–orthopyroxene rock; implications for melt distribution in upper mantle rocks. *Geophys. Res. Lett.*, **23**, 701–704.

Duba, A. (1976). Are laboratory electrical conductivity data relevant to the Earth? *Acta Geodaet. Geophys. Montanist. Acad. Sci. Hung.*, **11**, 485–495.

Duba, A. & T. J. Shankland (1982). Free carbon and electrical conductivity in the Earth's mantle. *Geophys. Res. Lett.*, **9**, 1271–1274.

Duba, A., S. Heikamp, W. Meurer, G. Nover & G. Will (1994). Evidence from borehole samples for the role of accessory minerals in lower-crustal conductivity. *Nature*, **367**, 59–61.

Ducea, M. N. & S. K. Park (2000). Enhanced mantle conductivity from sulphide minerals, southern Sierra Nevada, California. *Geophys. Res. Lett.*, **27**, 2405–2408.

Du Frane, W. L., J. J. Roberts, D. A. Toffelmier & J. A. Tyburczy (2005). Anisotropy of electrical conductivity in dry olivine. *J. Geophys. Res.*, **32**, L24315, doi:10.1029/2005GL023879.

Dugan, B., P. B. Flemings, D. L. Olgaard & M. J. Gooch (2003). Consolidation, effective stress, and fluid pressure of sediments from ODP Site 1073, US mid-Atlantic continental slope. *Earth Planet. Sci. Lett.*, **215**, 13–26.

Dullien, F. A. L. (1992). *Porous Media; Fluid Transport and Pore Structure*. San Diego: Academic Press.

Eaton, D. W., A. G. Jones & I. J. Ferguson (2004). Lithospheric anisotropy structure inferred from collocated teleseismic and magnetotelluric observations: Great Slave Lake shear zone, northern Canada. *Geophys. Res. Lett.*, **31**, L19614, doi:10.1029/2004GL020939.

Egbert, G. & J. R. Booker (1992). Very long period magnetotellurics at Tucson observatory: implications for mantle conductivity. *J. Geophys. Res.*, **98**, 15 099–15 112.

ELEKTB Group (1997). KTB and the electrical conductivity of the crust. *J. Geophys. Res.*, **102**, 18 289–18 305.

Escartin, J., G. Hirth & B. Evans (2001). Strength of slightly serpentinized peridotites: implications for the tectonics of oceanic lithosphere. *Geology*, **29**, 1023–1026.

Etheridge, M. A., V. J. Wall & R. H. Vernon (1983). The role of the fluid phase during regional metamorphism and deformation. *J. Metamorph. Geol.*, **1**, 205–226.

Etheridge, M. A., V. J. Wall, S. F. Cox & R. H. Vernon (1984). High fluid pressures during regional metamorphism and deformation – implications for mass-transport and deformation mechanisms. *J. Geophys. Res.*, **89**, 4344–4358.

Evans, R. L. (1994). Constraints on the large scale porosity of young oceanic crust from seismic and resistivity data. *Geophys. J. Int.*, **119**, 869–879.

Evans, R. L. (2001). Measuring the shallow porosity structure of sediments on the continental shelf: a comparison of an electromagnetic approach with cores and acoustic backscatter. *J. Geophys. Res.*, **106**, 27 047–27 060.

Evans, R. L., S. C. Constable, M. C. Sinha, C. S. Cox & M. J. Unsworth (1991). Upper crustal resistivity structure of the East Pacific Rise near 13N. *Geophys. Res. Lett.*, **18**, 1917–1920.

Evans, R. L., M. C. Sinha, S. C. Constable & M. J. Unsworth (1994). On the electrical nature of the axial melt zone on the East Pacific Rise near 13N. *J. Geophys. Res.*, **99**, 577–589.

Evans, R. L., P. Tarits, A. D. Chave, A. White, G. Heinson, J. H. Filloux, H. Toh, N. Seama, H. Utada, J. R. Booker & M. Unsworth (1999). Asymmetric electrical structure in the mantle beneath the East Pacific Rise at 17S. *Science*, **286**, 756–759.

Evans, R. L., G. Hirth, K. Baba, D. Forsyth, A. Chave & R. Mackie (2005). Geophysical controls from the MELT area for compositional controls on oceanic plates. *Nature*, **437**, 249–252.

Evans, R. L., J. Escartin & M. Cannat (2010). A short electromagnetic profile across the Kane Oceanic Core Complex. *Geophys. Res. Lett.*, **37**, L15309, doi:10.1029/2010GL043813.

Evans, R. L., A. G. Jones, X. Garcia, M. Muller, M. Hamilton, S. Evans, S. Fourie, J. Spratt, S. Webb, H. Jelsma & D. Hutchins (2011). The electrical lithosphere beneath the Kaapvaal Craton, Southern Africa. *J. Geophys. Res.*, **116**, B04105, doi:10.1029/2010JB007883.

Evans, S., A. G. Jones, J. Spratt & J. Katsube (2005). Central Baffin Electromagnetic Experiment (CBEX) maps the NACP in the Canadian arctic. *Phys. Earth Planet. Inter.*, **150**, 107–122.

Farquharson, C. & J. Craven (2009). Three-dimensional inversion of magnetotelluric data for mineral exploration; an example from the McArthur River uranium deposit, Saskatchewan, Canada. *J. Appl. Geophys.*, **68**, 450–458.

Faul, U. H. (1997). Permeability of partially molten upper mantle rocks from experiments and percolation theory. *J. Geophys. Res.*, **102**, 10 299–10 311.

Fedo, C. M. & M. J. Whitehouse (2002). Metasomatic origin of quartz–pyroxene rock, Akilia, Greenland, and implications for Earth's earliest life. *Science*, **296**, 1448–1452.

Fraser-Smith, A. C., A. Bernardi, P. R. McGill, M. E. Ladd, R. A. Helliwell & O. G. Villard Jr. (1990). Low-frequency magnetic field measurements near the epicenter of the Ms 7.1 Loma Prieta earthquake. *Geophys. Res. Lett.*, **17**, 1465–1468.

Friedrich, J. T., K. H. Greaves & J. W. Martin (1993). Pore geometry and transport properties of Fontainbleau sandstone. *Int. J. Rock Mech. Min. Sci. Geomech. Abstr.*, **30**, 691–697.

Fuji-ta, K., T. Katsura & Y. Tainosho (2004). Electrical conductivity measurement of granulite under mid- to lower crustal pressure–temperature conditions. *Geophys. J. Int.*, **157**, 79–86.

Fuji-ta, K., T. Katsura, T. Matsuzaki, M. Ichiki & T. Kobayashi (2007). Electrical conductivity measurement of gneiss under mid- to lower crustal *P–T* conditions. *Tectonophysics*, **434**, 93–101.

Gaillard, F. (2004). Laboratory measurements of electrical conductivity of hydrous and dry silicic melts under pressure. *Earth Planet. Sci. Lett.*, **218**, 215–228.

Gaillard, F., M. Malki, G. Iacono-Marziano, M. Pichavant & B. Scaillet (2008). Carbonatite melts and electrical conductivity in the asthenosphere. *Science*, **28**, 1363–1365.

Gillis, K., C. Mével, J. Allan, *et al.* (1993). *Proc. Ocean Drilling Program*, Initial Report 147.

Glover, P. W. J. (2010). A generalized Archie's law for *n* phases. *Geophysics*, **75**, E247–E265, doi:10.1190/1.3509781.

Glover, P. W. J. & F. J. Vine (1992). Electrical conductivity of carbon-bearing granulite at raised temperatures and pressures. *Nature*, **360**, 723–726.

Glover, P. W. J. & F. J. Vine (1995). Beyond KTB – electrical conductivity of the deep continental crust. *Surv. Geophys.*, **16**, 5–36.

Goldfarb, R. J., L. W. Snee, L. D. Miller & R. J. Newberry (1991). Rapid dewatering of the crust deduced from ages of mesothermal gold deposits. *Nature*, **354**, 296–298.

Gough, D. I. (1986). Seismic reflectors, conductivity, water and stress in the continental crust. *Nature*, **323**, 143–144.

Grant, K. J., R. A. Brooker, S. C. Kohn & B. J. Wood (2007). The effect of oxygen fugacity on hydroxyl concentrations and speciation in olivine: implications for water solubility in the upper mantle. *Earth Planet. Sci. Lett.*, **261**, 217–229.

Grove, T. L., N. Chaterjee, S. W. Parman & E. Medard (2006). The influence of H_2O on mantle wedge melting. *Earth Planet. Sci. Lett.*, **249**, 74–89.

Gueguen, Y. & V. Palciauskas (1994). *Introduction to the Physics of Rocks*. Princeton, NJ: Princeton University Press.

Guseinov, A. A., I. O. Gargatsev & R. U. Gabitova (2005). Electrical conductivity of phlogopites at high temperatures. *Izv., Phys. Solid Earth*, **41**, 670–679.

Haak, V. (1982). *Numerical Data and Functional Relationships in Science and Technology*, Vol. **1**, Subvol. b, Ch. 5, Sect. 4, ed. G. Angenheister. Berlin: Springer, pp. 291–307.

Haak, V. & V. R. S. Hutton (1986). Electrical resistivity in continental lower crust. In *The Nature of the Lower Continental Crust*, ed. J. B. Dawson, D. A. Carswell, J. Hall & K. H. Wedepohl, Spec. Publ. 24. London: Geological Society, pp. 35–49.

Hamilton, M. P., A. G. Jones, R. L. Evans, S. Evans, C. J. S. Fourie, X. Garcia, A. Mountford, J. Spratt & SAMTEX Team (2006). Electrical anisotropy of South African lithosphere compared with seismic anisotropy from shear-wave splitting analyses. *Phys. Earth Planet. Inter.*, **158**, 226–239.

Hashin, Z. & S. Shtrikman (1962). A variational approach to the theory of the effective magnetic permeability of multiphase materials. *J. Appl. Phys.*, **33**, 3125–3131.

Hautot, S. & P. Tarits (2002). Effective electrical conductivity of 3-D heterogeneous porous media. *Geophys. Res. Lett.*, **29**, 1669, doi:10.1029/2002GL014907.

Heinson, G. S., N. G. Direen & R. M. Gill (2006). Magnetotelluric evidence for a deep-crustal mineralizing system beneath the Olympic Dam iron oxide copper–gold deposit, southern Australia. *Geology*, **34**, 573–576.

Hirsch, L. M. & T. J. Shankland (1993). Quantitative olivine-defect chemical model: insights on electrical conduction, diffusion and the role of Fe content. *Geophys. J. Int.*, **114**, 21–35.

Hirsch, L. M., T. J. Shankland & A. G. Duba (1993) Electrical conduction and polaron mobility in Fe-bearing olivine. *Geophys. J. Int.*, **114**, 36–44.

Hirschmann, M. M. (2010). Partial melt in the oceanic low velocity zone. *Phys. Earth Planet. Inter.*, **179**, 60–61.

Hirth, G. (2006). Protons lead the charge. *Nature*, **443**, 927–928.

Hirth, G. & D. L. Kohlstedt (1996). Water in the oceanic upper mantle: implications for rheology, melt extraction and the evolution of the lithosphere. *Earth Planet. Sci. Lett.*, **144**, 93–108.

Hirth, G., R. L. Evans & A. D. Chave (2000). Comparison of continental and oceanic mantle electrical conductivity: Is the Archean lithosphere dry? *Geochem. Geophys. Geosyst.*, **1** (12), 1030, doi:10.1029/2000GC000048.

Holtzman, B. K., D. L. Kohlstedt, M. E. Zimmerman, F. Heidelbach, T. Hiraga & J. Hustoft (2003). Melt segregation and strain partitioning: implications for seismic anisotropy and mantle flow. *Science*, **301**, 1227–1230.

Huang, X., Y. Xu & S. Karato (2005). Water content in the transition zone from electrical conductivity of wadsleyite and ringwoodite. *Nature*, **434**, 746–749.

Huebner, J. S., A. Duba & L. B. Wiggins (1979). Electrical conductivity of pyroxene which contains trivalent cations: laboratory measurements and the lunar temperature profile. *J. Geophys. Res.*, **84**, 4652–2656.

Hyndman, R. D., L. L. Vanyan, G. Marquis & L. K. Law (1993). The origin of electrically conductive lower continental crust: saline water or graphite? *Phys. Earth Planet. Inter.*, **81**, 325–344.

Jackson, P. D., D. Taylor-Smith & P. N. Stanford (1978). Resistivity–porosity–particle shape relationships for marine sands. *Geophysics*, **43**, 1250–1268.

Jackson, P. D., K. B. Briggs, R. C. Flint, R. J. Holyer & J. C. Sandidge (2002). Two- and three-dimensional heterogeneity in carbonate sediments using resistivity imaging. *Mar. Geol.*, **182**, 55–76.

Ji, S., S. Rondenay, M. Mareschal & G. Senechal (1999). Obliquity between seismic and electrical anisotropies as a potential indicator of movement sense for ductile shear aones in the upper mantle. *Geology*, **24**, 1033–1036.

Jones, A. G. (1981). On a type classification of lower crustal layers under Precambrian regions. *J. Geophys. (Z. Geophys.)*, **49**, 226–233.

Jones, A. G. (1987). MT and reflection: an essential combination. *Geophys. J. R. Astron. Soc.*, **89**, 7–18

Jones, A. G. (1988). Static shift of magnetotelluric data and its removal in a sedimentary basin environment. *Geophysics*, **53**, 967–978.

Jones, A. G. (1992). Electrical conductivity of the continental lower crust. In *Continental Lower Crust*, ed. D. M. Fountain, R. J. Arculus & R. W. Kay. Amsterdam: Elsevier, ch. 3, pp. 81–143.

Jones, A. G. (1999). Imaging the continental upper mantle using electromagnetic methods. *Lithos*, **48**, 57–80.

Jones, A. G. (2006). Electromagnetic interrogation of the anisotropic Earth: looking into the Earth with polarized spectacles. *Phys. Earth Planet. Inter.*, **158**, 281–291

Jones, A. G. & I. Dumas (1993). Electromagnetic images of a volcanic zone. *Phys. Earth Planet. Inter.*, **81**, 289–314.

Jones, A. G. & I. J. Ferguson (2001). The electric moho. *Nature*, **409**, 331–333.

Jones, A. G., I. J. Ferguson, A. D. Chave, R. L. Evans & G. W. McNeice (2001). The electric lithosphere of the Slave Craton. *Geology*, **29**, 423–426.

Jones, A. G., D. Snyder, S. Hanmer, I. Asudeh, D. White, D. Eaton & G. Clarke (2002). Magnetotelluric and teleseismic study across the Snowbird Tectonic Zone, Canadian shield: a Neoarchean mantle suture? *Geophys. Res. Lett.*, **29**, 1829, doi:10.1029/2002GL015359.

Jones, A. G., R. L. Evans & D. W. Eaton (2009). Velocity–conductivity relationships for mantle mineral assemblages in Archean cratonic lithosphere based on extremal bounds. *Lithos*, **109**, 131–143.

Karato, S. (1990). The role of hydrogen in the electrical conductivity of the upper-mantle. *Nature*, **347**, 272–273.

Karato, S. (2006). Remote sensing of hydrogen in Earth's mantle. *Rev. Min. Geochem.*, **62**, 343–375.

Kelemen, P. B., G. Hirth, N. Shimizu, M. Spiegelman & H. J. B. Dick (1997). A review of melt migration processes in the adiabatically upwelling mantle beneath oceanic spreading ridges. *Phil. Trans. R. Soc. Lond.*, **A355**, 1–35.

Kelley, D. S., J. A. Karson, D. K. Blackman, G. L. Früh-Green, D. A. Butterfield, M. D. Lilley, E. J. Olson, M. O. Schrenk, K. K. Roe, G. T. Lebon, P. A. Rivizzigno & AT3-60 Shipboard Party (2001). An off-axis hydrothermal vent field near the Mid-Atlantic Ridge at 30N. *Nature*, **412**, 145–149.

Key, K. & S. Constable (2002). Broadband marine MT exploration of the East Pacific Rise at 9°50′N. *Geophys. Res. Lett.*, **29**, 2054.

Kohlstedt, D. L. & S. J. Mackwell (1998). Diffusion of hydrogen and intrinsic point defects in olivine. *Z. Phys.Chem.*, **207**, 147–162.

Kohlstedt, D. & S. J. Mackwell (1999). Solubility and diffusion of "water" in silicate minerals. In *Microscopic Properties and Processes in Minerals*, ed. K. Wright & R. Catlow. Amsterdam: Kluwer, pp. 539–559.

Konrad-Schmolke, M., P. J. O'Brien, C. de Capitani & D. A. Carswell (2008). Garnet growth at high- and ultra-high-pressure conditions and the effect of element fractionation on mineral modes and composition, *Lithos*, **103**, 309–332.

Lastovickova, M. (1975). The electrical conductivity of eclogites measured by two methods. *Studia Geophys. Geod.*, **19**, 394–398.

Lastovickova, M. & E. I. Parchomenko (1976). The eletrical properties of eclogites from the Bohemian Massif under high temperatures and pressures. *Pure Appl. Geophys.*, **114**, 451–460.

Lastovickova, M., G. Losito & A. Trova (1993). Anisotropy of electrical conductivity of dry and saturated KTB samples. *Phys. Earth Planet. Inter.*, **81**(1–4), 315–324.

Lee, W. & P. Wyllie (1997). Liquid immiscibility between nephelinite and carbonatite from 1.0 to 2.5 GPa compared with mantle melt compositions. *Contrib. Mineral. Petrol.*, **127**, 1–16.

Lefever, R. A. (1980). Fe^{2+}–Fe^{3+} spinels and Fe^{2+}–Fe^{3+} spinels with substitutions. In *Magnetic and Other Properties of Oxides and Related Compounds*, Landolt-Bornstein, Vol. **12**, Part B, ed. K.-H. Hellwege. Berlin: Springer, pp. 55–87.

Lizarralde, D., A. D. Chave, G. Hirth & A. Schultz (1995). Northeastern Pacific mantle conductivity profile from long-period magnetotelluric sounding using Hawaii to California submarine cable data. *J. Geophys. Res.*, **100**, 17 837–17 854.

Lucius, J. E., G. R. Olhoeft, P. L. Hill & S. K. Duke (1989). *Properties and hazards of 108 selected substances*. USGS Open File Report, 89–491.

Mackwell, S. J. & D. L. Kohlstedt (1990). Diffusion of hydrogen in olivine: implications for water in the mantle. *J. Geophys. Res.*, **95**, 5079–5088.

Madden, T. R. (1976). Random networks and mixing laws. *Geophysics*, **41**, 1104–1125.

Mareschal, M., R. L. Kellett, R. D. Kurtz, J. N. Ludden, S. Ji & R. C. Bailey (1995). Archaean cratonic roots, mantle shear zones, and deep electrical anisotropy. *Nature*, **375**, 134–137.

Martinez-Garcia, M. (1992). Electromagnetic induction in geothermal fields and volcanic belts. *Surv. Geophys.*, **13**, 409–434.

Mathez, E. A., J. J. Roberts, A. G. Duba, A. K. Kronenberg & S. L. Karner (2008). Carbon deposition during brittle rock deformation: changes in electrical properties of fault zones and potential geoelectric phenomena during earthquakes. *J. Geophys. Res.*, **113**, B12201.

Mierdel, K., H. Keppler, J. R. Smyth & F. Langenhorst (2007). Water solubility in aluminous orthopyroxene and the origin of Earth's asthenosphere. *Science*, **315**, 364–368.

Minarik, W. G. & E. B. Watson (1995). Interconnectivity of carbonate melt at low melt fraction. *Earth Planet. Sci. Lett.*, **133**, 423–437.

Muller, M. R., A. G. Jones, R. L. Evans, H. S. Grütter, C. Hatton, X. Garcia, M. P. Hamilton, M. P. Miensopust, P. Cole, T. Ngwisany, D. Hutchins, C. J. Fourie, H. A. Jelsma, S. F. Evans, T. Aravanis, W. Pettit, S. J. Webb, J. Wasborg & SAMTEX Team (2009). Lithospheric structure, evolution and diamond prospectivity of the Rehoboth Terrane and western Kaapvaal Craton, southern Africa: constraints from broadband magnetotellurics. *Lithos*, **112S**, 93–105.

Nesbitt, B. E. (1993). Electrical resistivities of crustal fluids. *J. Geophys. Res.*, **98**, 4301–4310.

Nover, G. (2005). Electrical properties of crustal and mantle rocks – a review of laboratory measurements and their explanation. *Surv. Geophys.*, **26**, 593–651.

Nurhasan,, Y. Ogawa, N. Ujihara, S. B. Tank, Y. Honkura, S. Onizawa, T. Mori & M. Makino (2006). Two electrical conductors beneath Kusatsu-Shirane volcano, Japan, imaged by audiomagnetotellurics, and their implications for the hydrothermal system. *Earth Planets Space*, **58**, 1053–1059.

Ogawa, Y., Y. Nishida & M. Makino (1994). A collision boundary imaged by magnetotellurics, Hidaka mountains, central Hokkaido, Japan. *J. Geophys. Res.*, **99**, 22 373–22 388.

Olhoeft, G. R. (1981). Electrical properties of granite with implications for the lower crust. *J. Geophys. Res.*, **86**, 931–936.

Ozawa, K., P. S. Meyer & S. H. Bloomer (1991). Mineralogy and textures of iron–titanium oxide gabbros and associated olivine gabbros from Hole 735B. *Proc. Ocean Drilling Program, Sci. Results*, **118**, 41–73, doi:10.2973/odp.proc.sr.118.125.1991.

Pearson, D. G., F. R. Boyd, S. E. Haggerty, J. D. Pasteris, S. W. Field, P. H. Nixon & N. P. Pokhilenko (1994). The characterization and origin of graphite in cratonic lithospheric mantle: a petrological carbon isotope and Raman spectroscopic study. *Contrib. Mineral. Petrol.*. **115**, 449–466.

Perkin, R. G. & E. L. Lewis (1980). The practical salinity scale 1978: fitting the data. *IEEE J. Oceanic Eng.*, **5**, 9–16.

Pezard, P. A. (1990). Electrical properties of MOR basalt and implications for the structure of the upper oceanic crust in hole 504B. *J. Geophys. Res.*, **95**, 9237–9264.

Pezard, P. A., J. J. Howard & M. A. Lovell (1989). Clay conduction and pore structure of oceanic basalts from DSDP hole 504B. *Proc. Ocean Drilling Program, Sci. Results*, **111**, 133–146.

Pezard, P. A., J. J. Howard & D. Goldberg (1991). Electrical conduction in oceanic gabbros, Hole 735B, Southwest Indian Ridge. *Proc. Ocean Drilling Program, Sci. Results*, **118**, 323–331, doi:10.2973/odp.proc.sr.118.161.1991.

Pilchin, A. (2005). The role of serpentinisation in exhumation of high- to ultra-high-pressure metamorphic rocks. *Earth Planet. Sci. Lett.*, **237**, 815–828.

Poe, B. T., C. Romano, F. Nestola & J. R. Smyth (2010). Electrical conductivity anisotropy of dry and hydrous olivine at 8 GPa. *Phys. Earth Planet. Inter.*, **181**, 103–111.

Pommier, A. & E. Le Trong (2011). SIGMELTS: a web portal for electrical conductivity calculations in geosciences. *Comput. Geosci.*, **37**(9), 1450–1459, doi:10.1016/j.cageo.2011.01.002.

Pommier, A., F. Gaillard, M. Pichavant & B. Scaillet (2008). Laboratory measurements of electrical conductivities of hydrous and dry Mount Vesuvius melts under pressure. *J. Geophys. Res.*, **113**, B05205, doi:10.1029/2007JB005269.

Pommier, A., F. Gaillard & M. Pichavant (2010a). Time-dependent changes of the electrical conductivity of basaltic melts with redox state. *Geochim. Cosmochim. Acta*, **74**, 1653–1671.

Pommier, A., F. Gaillard, M. Malki & M. Pichavant (2010b). . Methodological re-evaluation of the electrical conductivity of silicate melts. *Am. Mineral.*, **95**, 284–291.

Pommier, A., P. Tarits, S. Hautot, M. Pichavant, B. Scaillet & F. Gaillard (2010c). A new petrological and geophysical investigation of the present-day plumbing system of Mount Vesuvius. *Geochem. Geophys. Geosyst.*, **11**, Q07013, doi:10.1029/2010GC003059.

Popp, T. & H. Kern (1993). Thermal dehydration reactions characterized by combined measurements of electrical conductivity and elastic wave velocities. *Earth Planet. Sci. Lett.*, **120**, 43–57.

Quist, A. S. & W. L. Marshall (1968). Electrical conductances of aqueous sodium chloride solutions from 0 to 800 C and at pressures to 4000 bars. *J. Phys. Chem.*, **71**, 684–703.

Riley, G. N., Jr., D. L. Kohlstedt & F. M. Richter (1990). Melt migration in a silicate liquid–olivine system: an experimental test of compaction theory. *Geophys. Res. Lett.*, **17**, 2101–2104.

Roberts, J. J. (2002). Electrical properties of a microporous rock as a function of saturation and temperature. *J. Appl. Phys.*, **91**, 1687–1694.

Roberts, J. J. & J. A. Tyburczy (1991). Frequency dependent electrical properties of polycrystalline olivine compacts. *J. Geophys. Res.*, **96**, 16 205–16 222.

Roberts, J. J. & J. A. Tyburczy (1994). Frequency dependent electrical properties of minerals and partial-melts. *Surv. Geophys.*, **15**, 239–262.

Roberts, J. J. & J. A. Tyburczy (1999). Partial-melt electrical conductivity: influence of melt composition. *J. Geophys. Res.*, **104**, 7055–7065.

Roberts, J. J., A. G. Duba, E. A. Mathez, T. J. Shankland & R. Kinzler (1999). Carbon-enhanced electrical conductivity during fracture of rocks. *J. Geophys. Res.*, **104**, 737–747.

Roberts, J. N. & L. M. Schwartz (1985). *Grain consolidation and electrical conductivity in porous media. Phys. Rev.*, **B31**, 5990–5996.

Robinson, J. A. C. & B. J. Wood (1998). The depth of the spinel to garnet transition at the peridotite solidus. *Earth Planet. Sci. Lett.*, **164**, 277–284.

Romano, C., B. T. Poe, N. Kreidie & C. A. McCammon (2006). Electrical conductivities of pyrope–almandine garnets up to 19 GPa and 1700 C. *Am. Mineral.*, **91**, 1371–1377.

Romano, C., B. T. Poe, J. Tyburczy & F. Nestola (2009). Electrical conductivity of hydrous wadsleyite. *Eur. J. Mineral.*, **21**, 615–622.

Saal, A. E., E. Hauri, C. H. Langmuir & M. R. Perfit (2002). Vapour undersaturation in primitive mid-ocean-ridge basalt and the volatile content of Earth's upper mantle. *Nature*, **419**, 451–455.

Sainato, C., M. C. Pomposiello, A. Landini, G. Galindo & H. Malleville (2000). Hydrogeological sections of the Pergamino basin (Buenos Aires province,

Argentina): audio magnetotelluric and geochemical results. *Rev. Bras. Geofis.*, **18**, 187–200, doi:10.1590/S0102-261X2000000200007.

Schmeling, H. (1986). Numerical models on the influence of partial melt on elastic, anelastic and electrical properties of rocks. Part II: Electrical conductivity. *Phys. Earth Planet. Inter.*, **43**, 123–136.

Schultz, A., R. D. Kurtz, A. D. Chave & A. G. Jones (1993). Conductivity discontinuities in the upper mantle beneath a stable craton. *Geophys. Res. Lett.*, **20**, 2941–2944.

Schwartz, L. M. & S. Kimminau (1987). Analysis of electrical conduction in the grain consolidation model. *Geophysics*, **52**, 1402–1411.

Shankland, T. J. & M. E. Ander (1983). Electrical conductivity, temperatures, and fluids in the lower crust. *J. Geophys. Res.*, **88**, 9475–9484.

Shankland, T. J. & H. S. Waff (1974). Conductivity in fluid bearing rocks. *J. Geophys. Res.*, **79**, 4863–4868.

Shankland, T. J. & H. S. Waff (1977). Partial melting and electrical conductivity anomalies in the upper mantle. *J. Geophys. Res.*, **82**, 5409–5417.

Shankland, T. J., A. G. Duba, E. A. Mathez & C. L. Peach (1997). Increase of electrical conductivity with pressure as an indicator of conduction through a solid phase in midcrustal rocks. *J. Geophys. Res.*, **102**, 14 741–14 750.

Simpson, F. (2002). Intensity and direction of lattice preferred orientation of olivine: Are electrical and seismic anisotropies of the Australian mantle reconcilable? *Earth Planet. Sci. Lett.*, **203**, 535–547.

Simpson, F. & A. Tommassi (2005). Hydrogen diffusivity and electrical anisotropy of a peridotite mantle. *Geophys. J. Int.*, **160**, 1092–1102.

Simpson, G., M. Spiegelman & M. I. Weinstein (2010). A multi-scale model of partial melts: 2. Numerical results. *J. Geophys. Res.*, **115**, B04411, doi:10.1029/2009JB006376.

Sinha, M. C. & R. L. Evans (2004). Geophysical constraints upon the thermal regime of the oceanic crust. In *Mid-Ocean Ridges; Hydrothermal Interactions Between the Lithosphere and Oceans*, ed. C. R. German, J. Lin & L. M. Parsons, *Geophys. Monogr. 148. Washington: AGU Washington*, pp. 19–62.

Spratt, J. E., A. G. Jones, V. Jackson, L. Collins & A. Avdeeva (2009). Lithospheric geometry of the Wopmay Orogen from a Slave Craton to Bear Province magneto-telluric transect. *J. Geophys. Res.*, **114**, B01101, doi:10.1029/2007JB005326.

Stein, C. A. & S. Stein (1992). A model for the global variation in oceanic depth and heat flow with lithospheric age. *Nature*, **359**, 123–129.

Stein, C. A. & S. Stein (1994). Comparison of plate and asthenospheric flow models for the thermal evolution of oceanic lithosphere. *Geophys. Res. Lett.*, **21**, 709–712.

Stesky, R. M. & W. F. Brace (1973). Electrical conductivity of serpentinized rocks to 6 kilobars. *J. Geophys. Res.*, **78**, 7614–7620.

Szabo, C. & R. J. Bodnar (1995). Chemistry and origin of mantle sulfides in spinel peridotitc xenoliths from alkaline basaltic lavas, Nograd-Gomor Volcanic field. *Geochim. Cosmochim. Acta*, **59**, 3917–3972.

ten Grotenhuis, S. M., M. R. Drury, C. J. Peach & C. J. Spiers (2004). Electrical properties of fine-grained olivine: evidence for grain boundary transport. *J. Geophys. Res.*, **109**, B06203, doi:10.1029/2003JB002799.

ten Grotenhuis, S. M., M. R. Drury, C. J. Spiers & C. J. Peach (2005). Melt distribution in olivine rocks based on electrical conductivity measurements. *J. Geophys. Res.*, **110**, B12201, doi:10.1029/2004JB003462.

Toffelmier, D. A. & J. A. Tyburczy (2007). Electromagnetic detection of a 410-km-deep melt layer in the southwestern United States. *Nature*, **447**, 991–994.

Toramaru, A. & N. Fujii (1986). Connectivity of melt phase in a partially molten peridotite. *J. Geophys. Res.*, **91**, 9239–9252.

Tuncer, V., M. Unsworth, W. Siripunvaraporn & J. A. Craven (2006). Exploration for unconformity-type uranium deposits with audiomagnetotelluric data: a case study from the McArthur River mine, Saskatchewan, Canada. *Geophysics*, **71**, 201–209.

Tyburczy, J. A. (2007). Properties of rocks and minerals – the electrical conductivity of rocks, minerals and the Earth. In *Treatise on Geophysics*, Vol. **2**, *Mineral Physics*, ed. G. D. Price. Oxford: Elsevier, pp. 631–642.

Von Bargen, N. & H. S. Waff (1986). Permeabilities, interfacial areas and curvatures of partially molten systems, results of numerical computations of equilibrium micro-structures. *J. Geophys. Res.*, **91**, 9261–9276.

Waff, H. S. (1974). Theoretical considerations of electrical conductivity in a partially molten mantle and implications for geothermometry. *J. Geophys. Res.*, **79**, 4003–4010.

Waff, H. S. & D. F. Weill (1975). Electrical conductivity of magmatic liquids – effects of temperature, oxygen fugacity and composition. *Earth Planet. Sci. Lett.*, **28**, 254–260.

Wang, D., M. Mookherjee, Y. Xu & S.-I. Karato (2006). The effect of water on the electrical conductivity of olivine. *Nature*, **443**, 977–980, doi:10.1038/nature05256.

Wang, D. J., H. P. Li, L. Yi & B. P. Shi (2008). The electrical conductivity of upper-mantle rocks: water content in the upper mantle. *Phys. Chem. Minerals*, **35**, 157–162.

Wannamaker, P. E. (1986). Electrical conductivity of water-undersaturated crustal melting. *J. Geophys. Res.*, **91**, 6321–6327.

Wannamaker, P. E. (2000). Comment on "The petrologic case for a dry lower crust" by B. W. D. Yardley and J. W. Valley. *J. Geophys. Res.*, **105**, 6057–6064.

Wannamaker, P. E. (2005). Anisotropy versus heterogeneity in continental solid earth electromagnetic studies: fundamental response characteristics and implications for physiochemical state. *Surv. Geophys.*, **26**, 733–765.

Wannamaker, P. E., G. R. Jiracek, J. A. Stodt, T. G. Caldwell, A. D. Porter, V. M. Gonzalez & J. D. McKnight (2002). Fluid generation and movement beneath an active compressional orogen, the New Zealand Southern Alps, inferred from magnetotelluric (MT) data. *J. Geophys. Res.*, **107**, 1–22.

Wannamaker, P. E., D. P. Hasterok, J. M. Johnston, J. A. Stodt, D. B. Hall, T. L. Sodergren, L. Pellerin, V. Maris, W. M. Doerner, K. A. Groenewold & M. J. Unsworth (2008). Lithospheric dismemberment and magmatic processes of the Great Basin–Colorado Plateau transition, Utah, implied from magnetotellurics. *Geochem. Geophys. Geosyst.*, **9**, Q05019, doi:10.1029/2007GC001886.

Wark, D. A. & E. B. Watson (1998). Grain-scale permeabilities of texturally equilibrated, monomineralic rocks. *Earth Planet. Sci. Lett.*, **164**, 591–605.

Wark, D. A. & E. B. Watson (2002). Grain scale channelisation of pores due to gradients in temperature or composition of intergranular fluid or melt. *J. Geophys. Res.*, **107**, 2040, doi:10.1029/2001JB000365.

Wark, D. A., C. A. Williams, E. B. Watson & J. D. Price (2003). Reassessment of pore shapes in microstructurally equilibrated rocks, with implications for permeability of the upper mantle. *J. Geophys. Res.*, **108**, 2050, doi:10.1029/2001JB001575.

Watson, E. B. & J. M. Brennan (1987). Fluids in the lithosphere, 1. Experimentally-determined wetting characteristics of CO_2–H_2O fluids and their implications for fluid transport, host-rock physical properties, and fluid inclusion formation. *Earth Planet. Sci. Lett.*, **85**, 497–515.

Watson, H. C., J. J. Roberts & J. A. Tyburczy (2010). The effect of conductive impurities on electrical conductivity in polycrystalline olivine. *Geophys. Res. Lett.*, **37**, L02302, doi:10.1029/2009GL041566.

Wheatcroft, R. A., J. C. Borgeld, R. S. Born, D. E. Drake, E. L. Leithold, C. A. Nittrouer & C. K. Sommerfield (1996). The anatomy of an oceanic flood deposit. *Oceanography*, **9**, 158–162.

Whitehouse, M. J., J. S. Myers & C. M. Fedo (2009). The Akilia controversy: field, structural and geochronological evidence questions interpretations of > 3.8 Ga life in SW Greenland. *J. Geol. Soc.*, **166**, 335–348.

Wildenschild, D., J. J. Roberts & E. D. Carlberg (2000). On the relationship between microstructure and electrical and hydraulic properties of sand–clay mixtures. *Geophys. Res. Lett.*, **27**, 3085–3088.

Wong, P. Z., J. Koeplik & J. P. Tomanic (1984). Conductivity and permeability of rocks. *Phys. Rev.*, **B30**, 6606–6614.

Worm, H.-U., D. Clark & M. J. Dekkers (1993). Magnetic susceptibility of pyrrhotite: grain size, field and frequency dependence. *Geophys. J. Int.*, **114**, 127–137.

Wright, H. M. N., K. V. Cashman, E. H. Gottesfeld & J. J. Roberts (2009). Pore structure of volcanic clasts: measurements of permeability and electrical conductivity. *Earth Planet. Sci. Lett.*, **280**, 93–104.

Xie, H.-S., W.-G. Zhou, M.-X. Zhu, Y.-G. Liu, Z.-D. Zhao & J. Guo (2002). Elastic and electrical properties of serpentinite dehydration at high temperature and pressure. *J. Phys. Condens. Matter*, **14**, 11359–11363.

Xu, Y., B. T. Poe, T. J. Shankland & D. C. Rubie (1998). Electrical conductivity of olivine, wadsleyite and ringwoodite under upper mantle conditions. *Science*, **280**, 1415–1418.

Xu, Y. S., T. J. Shankland & B. T. Poe (2000). Laboratory-based electrical conductivity of the Earth's mantle. *J. Geophys. Res.*, **105**, 27 865–27 875.

Yardley, B. W. D. & J. W. Valley (1997). The petrologic case for a dry lower crust. *J. Geophys. Res.*, **102**, 12 173–12 185.

Yardley, B. W. D. & J. W. Valley (2000). Comment on "The petrologic case for a dry lower crust" by Bruce W. D. Yardley and John W. Valley – Reply. *J. Geophys. Res.*, **105**, 6065–6068.

Yin, C. & P. Weidelt (2009). Geoelectrical fields in a layered earth with arbitrary anisotropy. *Geophysics*, **64**, 426–434, doi:10.1190/1.1444547.

Yoshino, T. (2010). Laboratory electrical conductivity measurement of mantle materials. *Surv. Geophys.*, **31**, 163–206.

Yoshino, T., T. Matsuzaki, S. Yamashita & T. Katsura (2006). Hydrous olivine unable to account for conductivity anomaly at the top of the asthenosphere. *Nature*, **443**, 973–976, doi:10.1038/nature05223.

Yoshino, T., G. Manthilake, T. Matsuzaki & T. Katsura (2008). Dry mantle transition zone inferred from electrical conductivity of wadsleyite and ringwoodite. *Nature*, **451**, 326–329.

Yoshino, T., T. Matsuzaki, A. Shatskiy & T. Katsura (2009). The effect of water on the electrical conductivity of olivine aggregates and its implications for the electrical structure of the upper mantle. *Earth Planet. Sci. Lett.*, **288**, 291–300.

Yoshino, T., M. Laumonier, E. McIsaac & T. Katsura (2010). Electrical conductivity of basaltic and carbonatite melt-bearing peridotites at high pressures: implications for melt distribution and melt fraction in the upper mantle. *Earth Planet. Sci. Lett.*, **295**, 593–602.

Yu, L., R. L. Evans & R. N. Edwards (1997). Transient electromagnetic responses in seafloor with tri-axial anisotropy. *Geophys. J. Int.*, **129**, 292–304.

Zhang, S., M. S. Paterson & S. F. Cox (1994). Porosity and permeability evolution during hot isostatic pressing of calcite aggregates. *J. Geophys. Res.*, **99**, 15 741–15 760.

Zhu, W., C. David & T.-F. Wong (1995). Network modeling of permeability evolution during cementation and hot isostatic pressing. *J. Geophys. Res.*, **100**, 15 451–15 464.

3B. Description of the magnetospheric/ionospheric sources

3B.1 Overview of interaction of Earth with solar wind

Interaction between the solar wind plasma and the geomagnetic field, and ionizing effects of solar radiation on the upper atmosphere, produce the electromagnetic environment that is utilized in magnetotelluric studies. Plasma can be considered to be a distinct state of matter, apart from neutral gases, because of its unique properties. Plasma is essentially neutral, with nearly equal numbers of singly charged positive ions and negative electrons in most space applications. As an ionized medium, it contains free electric charges that make the plasma electrically conductive so that it responds strongly to electromagnetic fields. This gives rise to electric currents in near-Earth space.

Only a brief introduction to the near-Earth space environment will be presented. For further details, see space plasma textbooks such as Baumjohann & Treumann (1996), Kivelson & Russell (1995), Parks (2004) and Treumann & Baumjohann (1997).

3B.1.1 Major regions

The solar wind is a plasma stream ejected from the upper atmosphere of the Sun. It consists mostly of high-energy electrons and protons. The average values of the most important solar wind parameters close to the Earth are: velocity 440 km/s, density 10 particles/cm^3, and magnetic field 6 nT (values based on the OMNI data from the GSFC/SPDF OMNIWeb interface at http://omniweb.gsfc.nasa.gov/).

Earth has a strong internal magnetic field, whose magnitude near the surface is about 30 000 nT at the equator and 60 000 nT at the poles. In the absence of any external drivers, the geomagnetic field can be roughly approximated by a dipole field with an axis tilted about 11 degrees from the spin axis. Forcing by the solar wind modifies the field, creating a cavity called the magnetosphere (Figure 3B.1). The outer boundary of the magnetosphere is called the magnetopause. In front of the dayside magnetopause, another boundary called the bow shock is formed because the solar wind is supersonic. The region between the bow shock and the magnetopause is the magnetosheath. The solar wind compresses the sunward side to

The Magnetotelluric Method: Theory and Practice, ed. Alan D. Chave and Alan G. Jones. Published by Cambridge University Press © Cambridge University Press 2012.

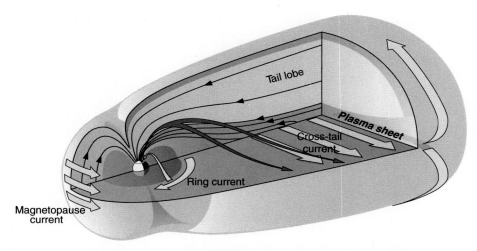

Figure 3B.1. Sketch of the magnetosphere. Illustration by Teemu Mäkinen (Finnish Meteorological Institute).

a distance of typically 10 Earth radii (RE), and it drags the nightside magnetosphere out to some 1000 RE (beyond the Moon's orbit).

At low altitudes, the magnetosphere ends at the ionosphere at about 70–1500 km height without any sharp upper limit (Figure 3B.2). Solar ultraviolet radiation is the main cause of ionization there. At high latitudes, particle precipitation from the magnetosphere also causes significant ionization. No more than about on the order of 1/1000th of atmospheric molecules are ionized. This means that the mass of neutrals is much larger than that of ions and electrons. Consequently, neutrals also have a major effect on the motion of the ionized particles. Despite a low fraction of ionization, the ionosphere still behaves like a plasma.

While at low latitudes the ionospheric plasma is co-rotating with the Earth, at higher latitudes it is convecting under the influence of the large-scale magnetospheric electric field created by the solar wind–magnetosphere interaction. Ionospheric currents flow along the auroral oval, so that two convection electrojets are formed at an altitude of about 100 km, with an eastward electrojet on the duskside and a westward electrojet on the dawnside.

3B.1.2 Key physical concepts

A dynamic description of plasma requires theory at different levels. The simplest single-particle approach studies the motion of charged particles in a specified electromagnetic field. This is a suitable approximation if interactions between particles are negligible, as they are in Earth's radiation belts in the magnetosphere.

The most accurate modeling is based on statistical physics that describes the behavior of large particle populations. Instead of the equation of motion of single particles, the evolution of distribution functions in the space–velocity or space–momentum domains are studied.

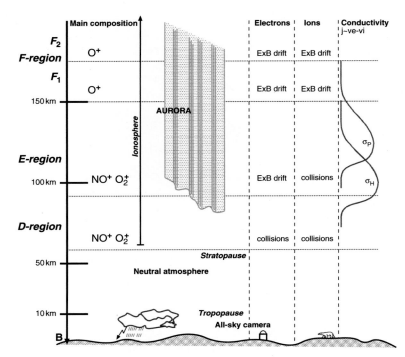

Figure 3B.2. Regions and the main ion constituents of the ionosphere, the typical altitudes of the aurora, as well as a sketch of the ionospheric Hall and Pedersen conductivity profiles. The main drivers of the ion and electron motion are also presented. Illustration by Mikko Syrjäsuo (Finnish Meteorological Institute).

Statistical plasma theory is able to explain wave–particle interactions, for example, which are beyond the single-particle approach.

If each particle population is in a local equilibrium state, then a hydrodynamic model is suitable. This deals with macroscopic parameters such as density, velocity, temperature and pressure. A simplified version is a one-fluid model in which the whole plasma is treated as an electrically conducting fluid. The best-known example is magnetohydrodynamics (MHD), which has been applied quite successfully in Earth's environment.

The basic mechanism forcing charged particles to move horizontally in the ionosphere is the $\mathbf{E} \times \mathbf{B}$ drift due to the electric ($\mathbf{E}$) and magnetic ($\mathbf{B}$) fields. As such, it does not produce any net current. However, collisions between charged and neutral particles cause deviations from the pure drift motion, especially at altitudes of 90–130 km, so there is a net horizontal electric current. The ionospheric Ohm's law takes a tensor form, which means that the conductivities parallel to and transverse to the geomagnetic field are substantially different. At high latitudes, with a nearly vertical background magnetic field, Pedersen (parallel to $\mathbf{E}$) and Hall (perpendicular to $\mathbf{E}$) conductivities drive horizontal currents parallel to the electric field and opposite to the $\mathbf{E} \times \mathbf{B}$ direction, respectively. Along the magnetic field lines, particles are in principle free to move, and they carry field-aligned currents between the ionosphere and magnetosphere. Close to the equator, the geomagnetic field is nearly horizontal, so that the current density in the direction perpendicular to the geomagnetic field is determined by the Cowling conductivity, which is larger than the Pedersen and Hall conductivities. The Cowling mechanism can be explained as follows: ionospheric currents build up electric charges and thus create a secondary electric field, which in turn causes secondary currents. The associated effective conductivity is called the Cowling conductivity.

In the magnetosphere, solar wind energy is dissipated into various sinks, especially during magnetic storms and substorms. The main energy dissipation channels are the ring current encircling Earth at the equatorial plane, ionospheric Joule heating, auroral precipitation, plasma sheet heating in the nightside magnetosphere, and release of compact volumes of plasma and magnetic fields with closed field lines (plasmoids) from the magnetospheric tail. About half, perhaps even a larger part, of the energy ends up in the ionosphere, and the Joule heating consumes the main part. However, a vast majority of the incident energy at the magnetopause passes by Earth, and only a small fraction maintains the magnetosphere and further dissipates in the magnetosphere and ionosphere.

3B.2 General description of Earth's external field sources

3B.2.1 Observation of external current systems

Ionospheric and magnetospheric current systems create disturbances in the geomagnetic field. All currents above Earth's surface will be called primary or external sources of the (geo)electromagnetic field. Earth itself acts as a secondary or internal source due to currents induced in the conducting ground. The total field measured at the surface is the vector sum of the primary and secondary fields. The geomagnetic variation field at Earth's surface can reach a few thousand nanoteslas at most.

By measuring the variation field with ground-based or satellite magnetometers, one can continuously monitor the magnetospheric activity. The advantage of ground-based instruments is good temporal coverage over decades, but the presence of the oceans limits possible measurement sites. Spacecraft, in turn, provide good spatial coverage, but missions last at most a few years.

Modern standard magnetometers record field variations with a time resolution of 1 s or better and with an accuracy of at least 0.1 nT. Pulsation magnetometers that are used to study geomagnetic variations in the ultra-low-frequency (ULF) range have a resolution of 0.1 s or less.

The global INTERMAGNET network contains about 110 magnetic observatories fulfilling certain quality conditions (http://www.intermagnet.org/). There are also more than 100 continuously operating variometer stations where the absolute field is not recorded. Especially at northern auroral and subauroral latitudes, there are several such networks: the IMAGE, SAMNET and Greenland networks in Europe, and the Carisma network in Canada. Furthermore, temporary arrays are often run, for example, in magnetospheric or magnetotelluric campaigns such as the Scandinavian Magnetometer Array (SMA) in 1976 to 1979 (Küppers *et al.*, 1979) or the Baltic Electromagnetic Array Research (BEAR) in 1998 (Varentsov *et al.*, 2002).

A fundamental limitation of ground-based magnetic recordings is that it is impossible to uniquely derive the true horizontal ionospheric current distribution (Fukushima, 1976). It is only possible to deduce an equivalent horizontal current system in the ionospheric plane that completely explains the ground variations. The equivalent currents give sufficient information for studying the source effect problem in magnetotellurics. For a deeper understanding of ionosphere–magnetosphere coupling, the true 3D current system should be known. It is possible to derive this with additional ground or satellite observations of other ionospheric parameters and with some assumptions, as discussed by Kamide *et al.* (1981), Richmond & Kamide (1988), Inhester *et al.* (1992) and Amm (1995, 1998, 2001).

Satellite-borne magnetometers provide, in principle, global magnetic mapping. However, the high velocity of the spacecraft, combined with the small scale size of the systems of interest, enforce sampling windows of the order of 1–10 s for tens to hundreds of kilometer scale sizes, thus providing nothing more than snapshots of the *in situ* magnetic field. The most important low-orbit spacecraft for observing the geomagnetic field have been MAGSAT (Langel & Estes, 1985), Ørsted (Olsen *et al.*, 2000) and CHAMP (Reigber *et al.*, 2002). The upcoming SWARM mission with a three-spacecraft constellation opens unique possibilities for low-orbit studies by allowing the measurement of gradients of the magnetic field (Kuvshinov *et al.*, 2006; Olsen *et al.*, 2006a,b).

3B.2.2 Magnetic storms; Dst and the ring current

Geomagnetic activity can be divided into two main categories: storms and substorms. Storms start when enhanced energy transfer from the interplanetary magnetic field (IMF) of the solar wind into the magnetosphere leads to intensification of the magnetospheric ring current at a distance of 3–8 RE. The ring current development can be monitored with the *Dst* index. The following storm definition has been proposed by Gonzales *et al.* (1994): a storm is an interval of time when a sufficiently intense and long-lasting interplanetary convection electric field leads, through a substantial energization in the magnetosphere–ionosphere system, to an intensified ring current strong enough to exceed some key threshold of the quantifying storm time *Dst* index.

The *Dst* index is obtained from magnetometer stations near the equator, but not in the immediate vicinity of the equatorial electrojet (see Section 3B.2.4). At such latitudes, the north component of the magnetic field variation (after removal of *Sq*, the regular daily variation) is dominated by the ring current. *Dst* is a measure of the hourly average of this perturbation. Storms are typically divided into three distinct phases according to the signatures in *Dst*. Figure 3B.3 shows August 2003 as an example, with a major storm on August 17–18. The initial phase lasts from minutes to hours, and *Dst* increases to positive values of up to tens of nanoteslas. The main phase lasts from half an hour to several hours, and *Dst* can then reach negative values of hundreds of nanoteslas. During the recovery phase, lasting from tens of hours to a week, *Dst* gradually returns to the quiet-time level.

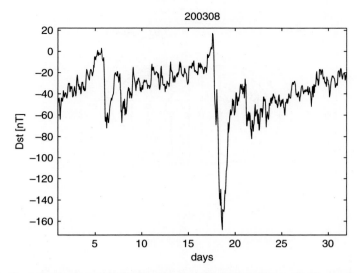

Figure 3B.3. *Dst* index in August 2003. Data source: World Data Center for Geomagnetism, Kyoto (http://swdcwww.kugi.kyoto-u.ac.jp/dstdir/index.html).

3B.2.3 Polar substorms; auroral electrojet; field-aligned currents

The magnetosphere and currents in the polar ionosphere are not stationary. The magnetosphere is activated when the IMF turns southward. Dynamic changes in the form of magnetospheric substorms then take place in the magnetosphere. In the ionosphere, they are observed as auroral substorms and variations in the size of the polar cap. The energy released in the ionosphere during substorms originates from the solar wind, but the magnetosphere has an active role in storing the energy and releasing it. Although known for nearly 50 years (Akasofu, 1964), substorms are still an active research topic raising controversial issues (e.g. Akasofu, 2004; Cheng, 2004; Meng & Liou, 2004).

Close to its earthward edge, the magnetospheric cross-tail current is blocked and connected to field-aligned currents (cf. Figures 3B.1 and 3B.4). In the dawnside of the tail, the current flows to the ionosphere, where it is connected to a westward current called the auroral electrojet (cf. Figure 3B.4). In the pre-midnight sector, this current is connected to an upward field-aligned current that is again connected into a cross-tail current in the magnetosphere. This happens in both the northern and southern hemispheres. At altitudes of about 100 km, the currents flow horizontally as auroral electrojets and turn back to the magnetosphere, where they merge to re-create the cross-tail current. Untiedt & Baumjohann (1993) give an extensive review of polar current systems.

A substorm consists of three different phases: a growth phase, an expansion phase and a recovery phase (Akasofu, 1968; McPherron, 1970) (Figure 3B.5). The substorm starts with a growth phase when the energy transfer from the solar wind to the magnetosphere is enhanced. This happens when the IMF B_z component turns negative (toward the south). Part

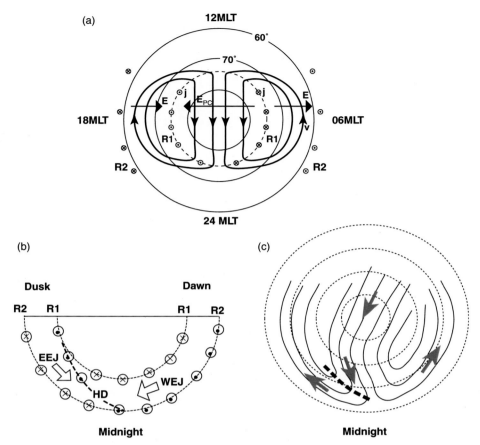

Figure 3B.4. (a) An ideal view of the ionospheric convection pattern from above the northern geomagnetic pole shows the average locations of the plasma flow **v**, electric field **E** and field-aligned currents **j** between the ionosphere and magnetosphere. The inner and outer field-aligned current regions are marked by R1 and R2. $\mathbf{E}_{PC}$ stands for the polar cap electric field, and the latitudes refer to the geomagnetic latitudes. The dashed circle indicates the open–closed field line boundary, i.e. the boundary of the polar cap. Illustration by Mikko Syrjäsuo (Finnish Meteorological Institute), after Heppner & Maynard (1987). (b) Schematic picture of eastward (EEJ) and westward (WEJ) auroral electrojets, the Harang discontinuity (HD), and field-aligned currents (cross, downward; dot, upward). (c) The shaded arrows indicate the plasma flow. After Koskinen & Pulkkinen (1995).

of the solar wind energy flows directly through the magnetosphere and part of it is stored in the form of magnetic energy in the lobes of the magnetotail. The substorm expansion phase starts with an onset, where the westward electrojet intensifies rapidly and the whole auroral oval expands poleward and equatorward. At the end of the expansion phase, the westward electrojet reaches its maximum and begins to recover toward the quiet-time level. During the recovery phase, the magnetosphere returns back to its quiet-time state.

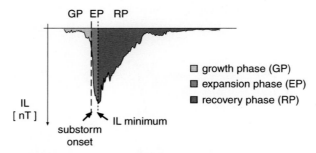

GP EP RP

□ growth phase (GP)
■ expansion phase (EP)
■ recovery phase (RP)

IL
[nT]

IL minimum

substorm
onset

Figure 3B.5. A sketch of substorm phases (Tanskanen, 2002). The *IL* indicator is the lower envelope of the northward component of the magnetic field variation at the IMAGE magnetometer stations in northern Europe.

Figure 3B.6 shows observations from a chain of magnetometers from Hankasalmi (HAN) in central Finland through northern Norway up to Bear Island (BJN) in the Arctic Ocean. The geographic north (*X*) component of the magnetic field shows several typical features of high-latitude variations. The first half of August 17 is relatively quiet, with only the regular daily variation (*Sq*) visible. At about 1330 UT, there is a sudden impulse indicating the arrival of a solar wind shock. At the same time, *Dst* increases as seen in Figure 3B.3. The impulse is followed by an increase of *X* that is caused by an eastward electrojet. During the following night, there are several negative excursions of *X* related to intensification of a westward electrojet during auroral substorms. This example shows an intense event, since *X* variations are also large at subauroral latitudes. The total amplitude of the electrojet currents can be as high as millions of amps, or, as current sheet densities, up to several thousand amps per kilometer.

A different presentation of high-latitude ionospheric currents during another event is shown in Figure 3B.7. Panel (a) shows the interpolated and rotated horizontal ground magnetic field in northern Europe on October 30, 2003, at around 2007 UT. This pattern mimics ionospheric equivalent currents. The two other panels present the interpolated time derivative of the horizontal field. The equivalent current pattern typically looks smooth and changes slowly in time, whereas the spatial structure of $d\mathbf{B}/dt$ is much more dynamic. According to an extensive study by Pulkkinen *et al.* (2006), there is a significant change in the dynamics of high-latitude field fluctuations in the temporal scale of 80–100 s, where $d\mathbf{B}/dt$ undergoes a transition from correlated to uncorrelated temporal behavior. The spatio-temporal behavior of $d\mathbf{B}/dt$ above temporal scales of 100 s resembles that of uncorrelated white noise. The spatial symmetry of field fluctuations increases during substorms, in that the difference between the scaling behavior of the *X* and *Y* components decreases during substorm events. This indicates the presence of spatially less ordered ionospheric equivalent currents. In other words, it can also be stated that $d\mathbf{B}/dt$ is randomly drawn from a pre-defined distribution without direct correlation to driving conditions (ultimately the solar wind forcing). Concerning the increasing spatial symmetry of field fluctuations during

IMAGE magnetometer network 2003-08-17

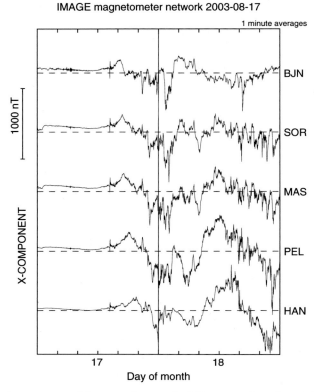

Figure 3B.6. North component of the ground magnetic field measured on August 17–18, 2003. Five sites in Finland and Norway at about geomagnetic latitudes from 59°N (HAN) to 71°N (BJN) are shown.

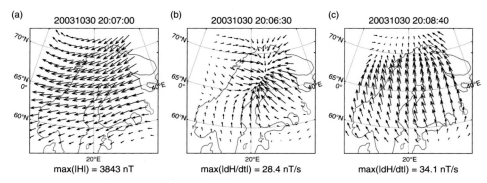

Figure 3B.7. (a) Equivalent currents in northern Europe formed by rotating the ground magnetic field 90 degrees clockwise at 20:07:00 UT; and the time derivative of the horizontal ground magnetic field at (b) 20:06:30 UT and (c) 20:08:40 UT.

substorms, the dominance of the westward electrojet decreases then, for example, due to vortex-type currents affecting equally B_x and B_y components. (A strictly eastward or westward current causes only B_x.)

3B.2.4 Sq *and ionospheric tides; equatorial electrojet*

Even during quiet solar wind conditions, the geomagnetic field has regular diurnal variations with dependence on season and solar cycle. The currents are caused by the atmospheric dynamo and variations in the ionospheric conductivity. The neutral winds are controlled by the atmospheric tides, which are mainly due to solar heating. Therefore, the global pattern is called the *Sq* (solar quiet) current system. The seasonal changes in the *Sq* system are connected to the fact that the relative solar illumination of the two hemispheres varies with season. The Moon also produces gravitational tides in the atmosphere, so it is possible to divide the magnetic variations and ionospheric currents into solar and lunar parts. The magnitude of the lunar variation is much smaller than the *Sq* variation, and it is semidiurnal, but is suppressed – like solar variations – during local nighttime due to vanishing ionospheric *E*-layer conductivity. A sketch of the *Sq* current system is shown in Figure 3B.8. The direction of the current flow close to the equator is eastward in the daytime and westward at night.

Close to the magnetic equator, the high Cowling conductivity results in quite a strong equatorial electrojet (EEJ). Its direction is the same as the direction of the electric field or eastward in the daytime, and during the nighttime it nearly vanishes due to a very low

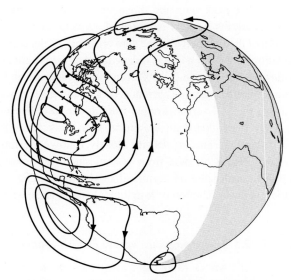

Figure 3B.8. Sketch of the *Sq* current system in the dayside hemisphere. The current direction is indicated by arrows. (Credit: US Geological Survey.)

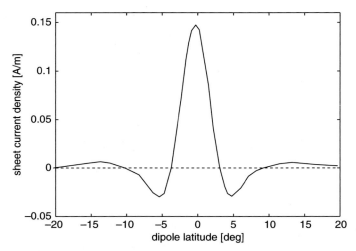

Figure 3B.9. Average current density profile of the equatorial electrojet around noon obtained from magnetic recordings of the CHAMP satellite. A positive value refers to an eastward current. Replotted from Lühr *et al.* (2004).

conductivity. The EEJ flows at low altitudes between 90 and 130 km. There is also a counter- or return electrojet that is a narrow band of westward current frequently appearing on either sides of the dip equator in the morning and afternoon.

The profile of the current density of the EEJ at its maximum at a time around noon is shown in Figure 3B.9 (Lühr *et al.*, 2004). A prominent feature of the mean EEJ is the precise coincidence of the current peak with the dip equator without seasonal or longitudinal variations. The return currents peak at latitudes of ±5 degrees on either side of the dip equator. The full width of the EEJ between the zero crossings of the current density profile is about 7 degrees in latitude, and the width is independent of the longitude or amplitude. The peak current densities are 0.15 A/m and −0.03 A/m for the eastward and westward currents, respectively. The total average eastward current is 65 kA, and the westward return current is 21 kA. The intensity of the EEJ varies strongly from day to day. The average peak current density exhibits a clear dependence on longitude, with peaks over South America and South-East Asia. The average current density closely follows the monthly mean of the solar flux index $F10.7$. Based on data from several satellites crossing the dip equator, Alken & Maus (2007) constructed an empirical model of the EEJ climatological mean and day-to-day variability as a function of longitude, local time, season and solar flux.

3B.2.5 Hydromagnetic waves; Pc disturbances

Ultra-low-frequency (ULF) waves (300 Hz–3 kHz) incident on Earth are produced by processes in the magnetosphere and the solar wind. They produce a wide variety of ULF hydromagnetic wave types that are classified on the ground as either Pc or Pi pulsations

Table 3B.1. *Pulsation classes. Pc refers to continuous pulsations, Pi to irregular ones, and* T *is the period.*

	Pc1	Pc2	Pc3	Pc4	Pc5	Pi1	Pi2
T (s)	0.2–5	5–10	10–45	45–150	150–600	1–40	40–150

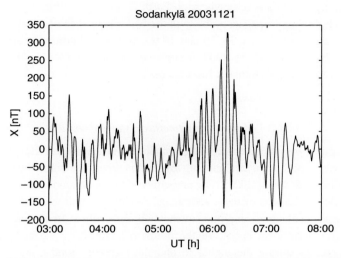

Figure 3B.10. Pc5 pulsations at Sodankylä, Finland (magnetic latitude (MLAT) about 64°) on November 21, 2003. The unfiltered geographic north component of the variation field is shown. For a detailed study of this event, see Kleimenova *et al.* (2005).

(continuous or irregular, see Table 3B.1). The magnetosphere is a resonant cavity and waveguide for waves that either originate within or propagate through the system. This cavity responds to broadband sources by resonating at discrete frequencies. The cavity modes couple to geomagnetic field line resonances that drive currents in the ionosphere. These currents re-radiate the energy as electromagnetic waves that propagate to the ground. Because these ionospheric currents are localized in latitude, there are very rapid variations in the wave phase at Earth's surface. The occurrence of various waves depend on conditions in the solar wind and in the magnetosphere. At Earth's surface, pulsations are seen as more or less sinusoidal variations of the magnetic field (Figure 3B.10). For extensive reviews of pulsations, see Kangas *et al.* (1998) and McPherron (2005).

3B.2.6 *Change of mode at about 1 Hz and dead band; Schumann resonances; lightning*

The long-period variations (> 1 s) discussed above are produced by dynamic variations of space plasma. At shorter periods, the origin of geoelectromagnetic fluctuations lies in the meteorological activity of the neutral atmosphere. The power spectrum has a minimum around a period of 1 s, so it is called a "dead band" since it results in reduced quality of the estimated magnetotelluric transfer functions (e.g. Simpson & Bahr, 2005). This band is thus also especially sensitive to industrial electromagnetic noise.

The atmospheric region between the highly conducting ground and ionosphere forms a resonance cavity. Schumann resonances are made up of the background signal resulting from global lightning activity and extremely low-frequency (ELF) transients resulting from intense lightning discharges somewhere on Earth. Transients within the Earth–ionosphere cavity due to lightning propagate globally in the ELF range. The nominal average frequencies observed are 8, 14, 20, 26, 33, 39 and 45 Hz, with a slight diurnal variation. A special section in *Radio Science*, **42**(2), 2007, introduced earlier by Pasko (2006), reviews recent advances in the study of Schumann resonances.

3B.2.7 Audiomagnetotelluric sources to 10 kHz

The energy source for natural-source magnetotellurics in the range 1 Hz–10 kHz is electromagnetic waves caused by distant lightning storms that propagate within the Earth–ionosphere waveguide (Garcia & Jones, 2002). The properties of this waveguide cause diurnal, seasonal and solar-cycle fluctuations of these audiomagnetotelluric variations. The temporal fluctuations cause significant signal amplitude attenuation especially at frequencies in the 1–5 kHz "audiomagnetotelluric dead band". In the northern hemisphere, the audiomagnetotelluric amplitude increases during the nighttime and the summer months, and correspondingly decreases during the daytime and the winter months.

3B.2.8 Annual and solar cycle variations of geomagnetic activity

Annual and longer-term variations of geomagnetic activity have been known since the nineteenth century, but they are still a subject of active research (Ahn *et al.*, 2000; Pulkkinen *et al.*, 2001; Richardson *et al.*, 2002; Usoskin & Mursula, 2003; Clilverd *et al.*, 2006), for which global climate change has given it a new aspect (e.g. special issue of *Advances in Space Research*, **34**(2); Labitzke *et al.*, 2004). New activity indices are being suggested to supplement or replace the standard ones (Svalgaard *et al.*, 2004; Karinen & Mursula, 2005; Mursula & Martini, 2007).

Figure 3B.11 gives an overview of magnetic activity during about the five latest sunspot cycles. Magnetic activity has a maximum around the equinoxes (spring and fall) and a minimum around the solstices (summer and winter). During a solar cycle, activity is at its minimum approximately when the number of sunspot numbers is smallest. There are typically two maxima of geomagnetic activity during a sunspot cycle: during the ascending and descending phases. At the time of the largest sunspot number, there is often a decrease in magnetic activity. There are also prominent differences between sunspot cycles.

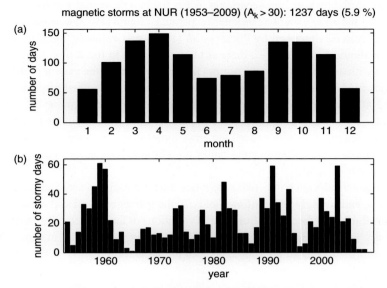

Figure 3B.11. (a) Monthly number of magnetically stormy days when the daily activity index A_k exceeded 30 at the Nurmijärvi observatory, Finland (about 57 MLAT). (b) Number of stormy days in 1953–2009 at Nurmijärvi.

3B.3 Description of the variation of source field characteristics

3B.3.1 Separation of external and internal contributions

When the spatial and temporal characteristics of the source field are considered, it is in principle necessary first to separate the primary and secondary fields. For the measured magnetic variation field at Earth's surface, mathematical separation utilizes potential theory, with the reasonable assumption of quasi-static variations (Siebert & Kertz, 1957; Weaver, 1964). Implementation as a regularized inverse problem, where the unknowns are the internal and external scalar magnetic potentials, is a natural extension (Richmond & Baumjohann, 1984). With this approach, interpolation of the data on a regular grid and field separation are accomplished in a single step. Pulkkinen *et al.* (2003) used a set of spherical elementary current systems (Amm & Viljanen, 1999) to parameterize the internal and external magnetic fields in terms of equivalent currents.

In practice, a successful separation requires a dense chain or network of magneto-meters. The use of a chain requires the additional assumption of the independence of the field on one horizontal coordinate. There have been only a few attempts to separate the magnetic field with 2D magnetometer networks (e.g. Porath *et al.*, 1970; Gough, 1973; Richmond & Baumjohann, 1984; Pulkkinen *et al.*, 2003). Other examples deal with long magnetometer chains and electrojet situations that are dominantly 1D cases (Tanskanen *et al.*, 2001).

A major difficulty with any separation is that fields with spatial wavelengths exceeding the extent of the array cannot be uniquely separated. Gough (1973) concluded that "… The separation exercise described by Porath *et al.* (1970) was valuable, but it may be doubted whether it should be repeated … there is little advantage in the formal field separation as compared with an estimation of the normal field by smoothing the map or profile, and of the anomalous field as the difference between the original and smoothed maps or profiles." As Egbert (2002) commented, Gough's conclusions are based on experience with regional-scale arrays mostly at mid-latitudes, and at periods from a few minutes to an hour, where source wavelengths typically exceed the array dimensions. In other circumstances (higher latitudes, arrays covering larger areas, and over a broader frequency range), formal separation can be more useful.

From a practical viewpoint, two categories may be roughly considered: nearly plane-wave fields, and strongly non-uniform fields. In the former case, all spatial variations of the field at the Earth's surface result from lateral variations of the conductivity. In the latter case, the primary field is clearly larger than the induced part, so the total magnetic field gives relevant information about the source without the separation.

3B.3.2 Spatial characteristics

Rostoker & Phan (1986) showed that, during periods of moderate activity (auroral electrojet index *AE* between 50 and 500 nT), the width of the auroral electrojet remains at about 6 degrees. With increasing activity, both the eastward and westward electrojets expand equatorward. This shift is more pronounced for the eastward current. Ahn *et al.* (2005) noted that there seems to be a lower limit for the equatorward boundary, and particularly for the westward electrojet, of about 60 degrees in geomagnetic latitude, regardless of magnetic activity level.

3B.3.3 Spectral characteristics

Campbell (1973) considered geomagnetic recordings of 16 observatories in the North American quadrant in 1965, which was relatively low in magnetic activity. Field amplitudes were found to increase almost linearly with period in the 5 min to 2 h range. The horizontal components are about equal at all latitudes, with slightly larger values of the north component in the auroral and equatorial electrojet region. The vertical component is smaller than the horizontal ones except in the auroral region. This is due to the fact that the electrojets tend to flow parallel to geomagnetic latitudes, so they contribute mostly to the north and vertical components of the field. Seasonal changes in the amplitude appear everywhere, but they are more pronounced in the polar cap. The largest day-to-day variations occur at the equatorial and auroral zones. The lowest field values occur between geomagnetic latitudes 10–35 degrees, with a minimum near 20–25 degrees. Maximum values of the field are found at 65–75 degrees.

3B.3.4 Consequences for the magnetotelluric method

Experience has shown that magnetotelluric observations recorded over a long enough time period can compensate for source field effects. This is best achieved at mid-latitudes far from non-uniform ionospheric currents, whereas at high latitudes this approach has to be used with caution (e.g. Mareschal, 1986; Lezaeta *et al.*, 2007). After a short overview of theoretical results, the characteristics of geomagnetic transfer functions (or induction arrows) will be discussed. They do not involve electric field measurements, but source effects cause similar problems with telluric recordings. Finally, the source effect in magneto-tellurics will be considered.

As a general theoretical result, Dmitriev & Berdichevsky (1979) showed for 1D conductivity models that, in addition to plane waves (zero wavenumber), induction is independent of the source morphology for linearly varying and odd-order wavenumber sources when their wavelengths are comparatively long. The assumption of a uniform source is most valid at mid-latitudes, where the distance to the equatorial and auroral electrojets is largest. Short-wavelength sources are associated with energetic, transient events during active periods in areas close to the equatorial and auroral electrojets (e.g. Mareschal, 1986). This can lead to considerable difficulty in estimating the magnetotelluric or tipper responses even using modern processing methods (Garcia *et al.*, 1997).

Osipova *et al.* (1989) considered a simplified line source as a model for auroral currents. They found that, for periods longer than 15 min, the magnetic field depends only weakly on Earth's conductivity. Consequently, the resolution of apparent resistivity curves obtained by the horizontal spatial gradient technique becomes poor at longer periods close to the source region.

Pirjola (1992) used a more complex, yet still simplified, model of the auroral electrojet with field-aligned currents. In the case of a uniform Earth, he found that the apparent resistivity can differ remarkably from the plane-wave value starting from a period of about 10 s. The location of the observation point relative to the source current also affects how this deviation behaves as a function of period. Although the Earth model was simplified, similar conclusions were obtained for a layered-Earth model (Pirjola & Häkkinen, 1991).

Illustrative examples are shown in Figure 3B.12. The model ionospheric current is infinitely long and has a Gaussian transverse distribution. The Earth has a layered structure, so the fields at the surface vary only as a function of the transverse distance. For a fixed period, deviation from the plane-wave result decreases when the observation point is farther away from the current. For a fixed observation site, the deviation generally increases as a function of period. However, even in this simplified case, the source effect produces some complex features. For example, for a fixed period, there is a point where the apparent resistivity for the complex source equals the value for a plane wave.

More realistic models of auroral currents make the situation even more complicated. Viljanen *et al.* (1999) considered dynamic magnetic variations at high latitudes and

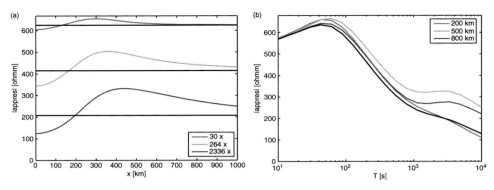

Figure 3B.12. (a) Apparent resistivity in the case of a time-harmonic Gaussian electrojet at a height of 110 km, with halfwidth 100 km, and centered at $x = 0$. (b) Apparent resistivity as a function of period at three fixed locations. The black lines show the value due to a plane-wave source. (See plate section for color version.)

compared their magnetotelluric responses to the plane-wave result. The strong spatial variation of the electromagnetic field produces spiky magnetotelluric curves as a function of both period and observation site. An interesting exception are pulsations caused by traveling ionospheric current vortices, for which the apparent resistivity is nearly undistorted, although the fields are still spatially strongly non-uniform.

Pulkkinen *et al.* (2006) showed that, below periods of 10 000 s for auroral and subauroral latitudes, the spatial variations of the horizontal magnetic field are symmetric, producing a planar average field. With the additional assumption of a uniform Earth, the magnetotelluric relation will then hold. The average planar behavior can be obtained only if the averaging is carried out over a long enough time period (tens of days). In addition, the conventional requirement for planar or linear spatial behavior of the source field can possibly be relaxed to that of average power-law-type behavior.

As a general conclusion based on theoretical models, the magnetotelluric source effect can cause deviations from the plane-wave result starting at periods of only a few seconds. The effects are more prominent for resistive than conductive Earth models. When the distance from the source increases, the magnetotelluric responses approach their plane-wave values. However, the detailed behavior depends strongly on the spatial and temporal structure of the source current. Crude models of stationary line or sheet currents may give an oversimplified impression. On the other hand, long-term recordings seem to provide a practical way to compensate source effects.

3B.3.4.1 Geomagnetic transfer functions

Transfer functions between the vertical and horizontal components of the magnetic field are used to define induction arrows that qualitatively indicate the location of induced currents.

However, source effects may contaminate this interpretation. As a simple example, consider a 1D conductivity structure and a non-uniform ionospheric current. There are no conductivity anomalies, but still the induction arrows would point toward the ionospheric sources (or away from them, depending on the sign convention). If the Earth is not 1D, then induction arrows are affected by both conductivity anomalies and ionospheric sources. Again, the use of long time series may help in extracting the desired Earth response.

Anderson *et al.* (1976, 1978) studied a 12-day sequence of induction arrows (periods of 37–1390 s) at subauroral latitudes, and found significant variations as a function of the local time. They also showed that averaging over long intervals does not completely remove the source effect.

Viljanen (1996) calculated induction arrows (at periods of 64–1280 s) in Fennoscandia using hundreds of 4 h events of quiet-time variations, and of high magnetic activity due to eastward and westward electrojets. Quiet-time arrows were very similar to those calculated previously from a set of carefully selected plane-wave events. Owing to the source effect during electrojet events, the average direction of the induction vectors may rotate tens of degrees compared to the quiet time. The length generally increased tens of percent, but a decrease was observed at a couple of stations near the electrojet and highly conductive anomalies. If the quiet-time induction vector is longer than about 0.50, then the source effect does not influence the direction of the vector even if its length undergoes considerable change. If the conductivity anomaly is weaker (quiet-time vector shorter than about 0.30), then the source effect dominates near the electrojet, but is less important at subauroral latitudes. If a large number of magnetically quiet events are available in the local morning hours, no special effort to reduce the source effect is needed.

From a one-year geomagnetic depth sounding survey in Greenland, Engels (1997) concluded that the induction arrows at periods below 2000 s represent the response to a quasi-uniform source. At longer periods, non-uniform effects became evident in the dataset, but could be compensated by processing the time series robustly over a year. Engels (1997) further asserted that the induction arrows were more sensitive to conductivity variations than to source geometry.

Varentsov *et al.* (2003) used simultaneous array soundings on the Fennoscandian (Baltic) Shield in June–July 1998. The study region extended from subauroral to auroral and partly into polar cap latitudes. The main tool used for diagnosis of the distortion was comparison of the time dependence of the transfer operators with variations in the intensity and inhomogeneity of the primary magnetic field, as well as comparative analysis of estimates obtained from various data samples. Suppressing the distortion was based on the application of coherence criteria to the selection of partial estimates of transfer operators during robust averaging (e.g. Varentsov, 2007). A more elaborate analysis with various signal-to-noise ratios in the ensemble of data demonstrated that non-stationary effects can be reliably suppressed on hourly and daily scales. The local transfer operators (impedance and tipper) were shown to be generally consistent with model results at periods of up to 3 h; in the case of the impedance, this agreement was valid up to diurnal harmonics for values at some individual sites and for the array as a whole.

Lezaeta *et al.* (2007) showed in a study from the auroral region in Canada that power spectra of the magnetic fields over monthly and daily intervals have maxima during summertime at periods over 1000 s. In contrast, the maximum power at short periods in the year 2000 were seen in winter. At long periods, the geomagnetic north component was dominant by one or two orders of magnitude with respect to the east component, as is generally expected at high latitudes. Short-period power spectra have in contrast east components that are as large as the north ones. This may reflect the 2D nature of ionospheric (equivalent) currents related to geomagnetic pulsations and other rapid variations, where no preferred direction of the horizontal current flow prevails (cf. Pulkkinen *et al.*, 2006).

The nature of the source is thus different at 100–1000 s than at > 1000 s period. It seems that the source also changes in morphology with season at periods below half a day, suggesting dynamic behavior over time. At long periods (>1000 s), the year 2000 displayed geomagnetic activity with low energy in March and a storm event in July. This is not typical because spring (and fall) close to the equinox are usually more active than near the solstice (Russell & McPherron, 1973).

Source field effects are most prominent at high latitudes. However, the equatorial electrojet also produces a strongly non-uniform primary field. Problems due to the equatorial electrojet can evidently be avoided quite simply by using data sequences recorded during the local nighttime (e.g. Arora *et al.*, 1998).

As a global study, Kuvshinov *et al.* (2007) investigated the spatiotemporal behavior of the vertical magnetic field (Z) from the daily ionospheric current system: the equatorial electrojet (EEJ) and solar quiet (Sq) variations. The focus was on induction in the mantle and oceans. The EEJ and Sq current systems were provided by the model of Sabaka *et al.* (2004). A 3D conductivity model of the Earth included oceans of laterally variable conductance and a radially symmetric conductor (1D) underneath. The results demonstrated that induction effects in Z due to the EEJ were negligible everywhere inland for all local times. At the altitude of the CHAMP satellite (400 km), the magnetic signal induced by the EEJ above the oceans did not exceed 2–5% of the external field at local noon. This means that consideration of induction effects is not necessary when modeling the EEJ from inland ground magnetic measurements and/or satellite data. As expected, induction in the oceans strongly affected the Sq field. The model studies showed that the anomalous induction effect (the difference between results obtained with 1D and 3D conductivity models) of Sq is substantial at CHAMP altitude, comprising 50% of the total field. It is therefore necessary to consider induction in the oceans when modeling Sq variations for both ground-based and satellite data.

3B.3.4.2 Magnetotelluric impedance

Padilha *et al.* (1997) studied the source effect due to the EEJ in a broadband (0.0006–2048 s) magnetotelluric survey of eight stations across the magnetic equator in Brazil.

Measurements were carried out both during the daytime, under the presence of the EEJ, and at night, when the EEJ almost vanishes (cf. Arora *et al.*, 1998). There were no significant differences between daytime and nighttime soundings, contrary to the expectation from theoretical calculations. Thus, the traditional plane-wave formula can provide reliable subsurface conductivity structures even under the influence of the EEJ, at least in this specific region up to periods of 2000 s. Model calculations indicate that a line current model generates departures from the plane-wave result that are not observed in practice, and that a Gaussian model only agrees with the experimental data if the geoelectric section approaches its most conductive limit defined from a 1D inversion.

As a supplementary study, Padilha (1999) used new magnetic recordings obtained simultaneously with the magnetotelluric experiment at a chain of equatorial and mid-latitude stations. An attempt was made to explain the magnetotelluric results in terms of the behavior of the primary inducing field. Most of the analysis was performed by considering a single period (1130 s). It was observed that the amplitude of the geo-magnetic variations is horizontally uniform within the study area (from −3 to +3 degrees of geomagnetic latitude), indicating that the primary field in the analyzed frequency range may be considered to be sufficiently uniform in the horizontal direction to satisfy the quasi-uniform criterion. The same geomagnetic data also showed that, if any EEJ source effect exists, it would be restricted to the transition zone (between 3–5 degrees, at both sides of the dip equator). The Dmitriev & Berdichevsky (1979) constraints calculated at two different frequencies and a modeling exercise using EEJ parameters derived from a magnetometer array were able to explain the magnetotelluric observations, and showed that source effects would just appear at periods longer than 1000 s (resistive regions) and 10 000 s (conductive regions). Considering the characteristics of propagation and the increase of geomagnetic variations in the equatorial zone, it was concluded that EEJ currents could be used as a source for lithospheric magnetotelluric studies in this region.

Viljanen *et al.* (1993) considered a line current close to a 2D conductivity structure, and a 3D ionospheric current system above a layered Earth. The location and geometric dimension of the currents were varied to simulate different geomagnetic events at high latitudes. Several simple averaging methods of the modeled electromagnetic field amplitudes were applied to calculate magnetotelluric responses and geomagnetic induction vectors. The results showed that averaging cannot remove source field effects on the magnetotelluric response, but can reduce them.

Garcia *et al.* (1997) considered magnetotelluric data from northern Canada collected in September–October, 1994, which were strongly influenced by auroral currents. The largest effect on the impedance tensor occurred during intervals of the highest magnetic activity during the local nighttime. In the daytime, the effect on the normal impedance tensor response was usually small. A robust processing algorithm was applied to these data in an attempt to extract the stable uniform field estimates of the impedance (Chave & Thomson, 1989). The non-robust estimate using all of the data was controlled by the nocturnal data, which are dominated by non-uniform source effects. Non-robust process-ing of only the daytime data failed to recover a useful result. There was little difference between the robust response for the entire and daytime data provided that the fraction of auroral activity was not large. The strongest bias was observed during the initial 25% of auroral events.

Jones & Spratt (2002) used the vertical component of the magnetic field time series to eliminate highly disturbed intervals of auroral origin, successfully reducing source effects in the magnetotelluric response under weak lateral conductivity contrasts. Times were chosen where the variations stay within prescribed limits defined on the basis of a histogram of the variations for the whole recording interval.

Since array data provide information about the actual source fields, both coherent noise and complications in the geometry of natural sources can be much more easily diagnosed than with single-site data, as discussed in the extensive review by Egbert (2002). In an ideal case with dense arrays (like BEAR; see Korja et al., 2002), it is possible to directly map polar ionospheric current systems. Information about source complexity determined from the array can then be used to select data with the best source characteristics for magnetotelluric processing. However, if even a single remote site is separated from the local site by many skin depths, remote reference processing might be expected to eliminate serious source bias in magnetotelluric impedances since source fields with wavelengths short enough to cause bias would not generally be coherent between distant sites.

Simple robust remote reference processing may be less effective for vertical field transfer functions at lower frequencies, or for inter-station transfer functions with large site separations. For both of these situations, systematic effects due to the finite spatial scale of the source will be more severe, and selecting data segments based on source diagnostics can offer significant advantages. For example, for an array of 29 sites under the equatorial electrojet in Brazil, Arora et al. (1998) found from examination of simultaneous time series that the source had a much larger spatial scale in the evening hours. Inter-station vertical field transfer functions estimated from nighttime data were found to be significantly less affected by the short spatial scales associated with the electrojet, and thus more indicative of internal structure.

Clearly the best solution to the problem of coherent noise is to maintain one or more remote sites a number of skin depths distant from the local sites. One true remote is almost certainly more useful for cancelation of coherent noise than a large array of nearby sites.

References

Ahn, B.-H., H. W. Kroehl, Y. Kamide & E. A. Kihn (2000). Seasonal and solar cycle variations of the auroral electrojet indices. *J. Atmos. Sol.-Terr. Phys.*, **62**, 1301–1310.
Ahn, B.-I., G. X. Chen, W. Sun, J. W. Gjerloev, Y. Kamide, J. B. Sigwarth & L. A. Frank (2005). Equatorward expansion of the westward electrojet during magnetically disturbed periods. *J. Geophys. Res.*, **110**, A01305, doi:10.1029/2004JA010553.
Akasofu, S.-I. (1964). The development of the auroral substorm. *Planet. Space Sci.*, **12**, 273.

Akasofu, S.-I. (1968). *Polar and Magnetospheric Substorms*. Astrophysics and Space Science Library, Vol. **11**, ed. S.-I. Akasofu. Dordrecht: Reidel.

Akasofu, S.-I. (2004). Several "controversial" issues on substorms. *Space Sci. Rev.*, **113**, 1–40.

Alken, P. & S. Maus, (2007). Spatio-temporal characterization of the equatorial electrojet from CHAMP, Ørsted, and SAC-C satellite magnetic measurements. *J. Geophys. Res.*, **112**, A09305, doi:10.1029/2007JA012524.

Amm, O. (1995). Direct determination of the local ionospheric Hall conductance distribution from two-dimensional electric and magnetic field data: application of the method using models of typical ionospheric electrodynamic situations. *J. Geophys. Res.*, **100**, 21473–21488.

Amm, O. (1998). Method of characteristics in spherical geometry applied to a Harang-discontinuity situation. *Ann. Geophys.*, **16**, 413–424.

Amm, O. (2001). The elementary current method for calculating ionospheric current systems from multisatellite and ground magnetometer data. *J. Geophys. Res.*, **106**, 24 843–24 855.

Amm, O. & A. Viljanen, (1999). Ionospheric disturbance magnetic field continuation from the ground to the ionosphere using spherical elementary current systems. *Earth Planets Space*, **51**, 431–440.

Anderson, C., L. Lanzerotti & C. Maclennan, (1976). Local time variation of induction vectors as indicators of internal and external current systems. *Geophys. Res. Lett.*, **3**, 495–498.

Anderson, C., L. Lanzerotti & C. Maclennan, (1978). Local time variation of geomagnetic induction vectors. *J. Geophys. Res.*, **83**, 3469–3484.

Arora, B. R., A. Rigoti, I. Vitorello, A. L. Padilha, N. B. Trivedi & F. H. Chamalaun, (1998). Magnetometer array study in north-northeast Brazil: conductive image building and functional induction modes. *Pure Appl. Geophys.*, **152**, 349–375.

Baumjohann, W. & R. Treumann, (1996). *Basic Space Plasma Physics*. London: Imperial College Press.

Campbell, W. H. (1973). Spectral composition of geomagnetic field variations in the period range of 5 min to 2 hr as observed at the earth's surface. *Radio Sci.*, **8**, 929–932.

Chave, A. D. & D. J. Thomson, (1989). Some comments on magnetotelluric response function estimation. *J. Geophys. Res.*, **94**, 14 215–14 225.

Cheng, C. Z. (2004). Physics of substorm growth phase, onset, and dipolarization. *Space Sci. Rev.*, **113**, 1–40.

Clilverd, M. A., E. Clarke, T. Ulich, H. Rishbeth & M. J. Jarvis, (2006). Predicting solar cycle 24 and beyond. *Space Weather*, **4**, S09005, doi:10.1029/2005SW000207.

Dmitriev, V. I. & M. N. Berdichevsky, (1979). The fundamental model of magnetotelluric sounding. *Proc. IEEE*, **67**, 1034–1044.

Egbert, G. D. (2002). Processing and interpretation of electromagnetic induction array data. *Surv. Geophys.*, **23**, 207–249.

Engels, M. (1997). *Untersuchungen zur elektromagnetischen Induktion in Grönland*. Ph.D. thesis, Georg-August Universität zu Göttingen.

Fukushima, N. (1976). Generalized theorem for no ground magnetic effect of vertical currents connected with Pedersen currents in the uniform-conductivity ionosphere. *Rep. Ionosph. Space Res. Japan*, **30**, 35–40.

Garcia, X. & A. G. Jones, (2002). Atmospheric sources for audio-magnetotelluric (AMT) sounding. *Geophysics*, **67**, 448–458.

Garcia, X., A. D. Chave & A. G. Jones, (1997). Robust processing of magnetotelluric data from the auroral zone. *J. Geomagn. Geoelectr.*, **49**, 1451–1468.

Gonzales, W. D., J. A. Joselyn, Y. Kamide, H. W. Kroehl, G. Rostoker, B. T. Tsurutani & V. M. Vasyliunas, (1994). What is a geomagnetic storm? *J. Geophys. Res.*, **99**, 5771–5792.

Gough, D. I. (1973). The interpretation of magnetometer array studies. *Geophys. J. R. Astron. Soc.*, **35**, 83–98.

Heppner, J. & N. Maynard, (1987). Empirical high-latitude electric field model. *J. Geophys. Res.*, **92**, 4467–4489.

Inhester, B., J. Untiedt, M. Segatz & M. Kürschner, (1992). Direct determination of the local ionospheric Hall conductance distribution from two-dimensional electric and magnetic field data. *J. Geophys. Res.*, **97**, 4073–4083.

Jones, A. G. & J. Spratt, (2002). A simple method for deriving the uniform field MT responses in auroral zones. *Earth Planets Space*, **54**, 443–450.

Kamide, Y., A. D. Richmond & S. Matsushita, (1981). Estimation of ionospheric electric fields, ionospheric currents, and field-aligned currents from ground magnetic records. *J. Geophys. Res.*, **86**, 801–813.

Kangas, J., A. Guglielmi & O. Pokhotelov, (2004). Morphology and physics of shortperiod magnetic pulsations (a review). *Space Sci. Rev.*, **83**, 435–512.

Karinen, A. & K. Mursula, (2005). A new reconstruction of the D_{st} index for 1932–2002. *Ann. Geophys.*, **23**, 475–485.

Kivelson, M. G. & C. T. Russell, (ed.) (1995). *Introduction to Space Physics*. Cambridge: Cambridge University Press.

Kleimenova, N. G., O. V. Kozyreva, J. Manninen & A. Ranta, (2005). Unusual strong quasi-monochromatic ground Pc5 geomagnetic pulsations in the recovery phase of November 2003 superstorm. *Ann. Geophys.*, **23**, 2621–2634.

Korja, T., M. Engels, A. A. Zhamaletdinov, A. A. Kovtun, N. A. Palshin, M. Yu. Smirnov, A. D. Tokarev, V. E. Asming, L. L. Vanyan, I. L. Vardaniants & BEAR Working Group (2002). Crustal conductivity in Fennoscandia – a compilation of a database on crustal conductance in the Fennoscandian Shield. *Earth Planets Space*, **54**, 535–558.

Koskinen, H. E. J. & T. I. Pulkkinen, (1995). Midnight velocity shear zone and the concept of Harang discontinuity. *J. Geophys. Res.*, **100**, 9539–9547.

Küppers, F., J. Untiedt, W. Baumjohann, K. Lange & A. G. Jones, (1979). A two-dimensional magnetometer array for ground-based observations of auroral zone electric currents during the International Magnetospheric Study (IMS). *J. Geophys.*, **46**, 429–450.

Kuvshinov, A., T. Sabaka & N. Olsen, (2006). 3-D electromagnetic induction studies using the Swarm constellation: mapping conductivity anomalies in the earth's mantle. *Earth Planets Space*, **58**, 417–427.

Kuvshinov, A., C. Manoj, N. Olsen & T. Sabaka, (2007). On induction effects of geomagnetic daily variations from equatorial electrojet and solar quiet sources at low and middle latitudes. *J. Geophys. Res.*, **112**, B10102, doi:10.1029/2007JB004955.

Labitzke, K., J. Pap, J. R. Kuhn & M. A. Shea, (2004). Solar variability and climate changes. *Adv. Space Res.*, **34**, 224.

Langel, R. A. & R. H. Estes, (1985). The near-earth magnetic field at 1980 determined from MAGSAT data. *J. Geophys. Res.*, **90**, 2495–2509.

Lezaeta, P., A. D. Chave, A. G. Jones & R. Evans, (2007). Source field effects in the auroral zone: evidence from the Slave Craton (NW Canada). *Phys. Earth Planet. Inter.*, **164**, 323–239.

Lühr, H., S. Maus & M. Rother, (2004). Noon-time equatorial electrojet: its spatial features as determined by the CHAMP satellite. *J. Geophys. Res.*, **109**, A01306, doi:10.1029/ 2002JA009656.

Mareschal, M. (1986). Modeling of natural sources of magnetospheric origin in the interpretation of regional induction studies: a review. *Surv. Geophys.*, **8**, 261–300.

McPherron, R. L. (1970). Growth phase of magnetospheric substorms. *J. Geophys. Res.*, **75**, 5592.

McPherron, R. L. (2005). Magnetic pulsations: their sources and relation to solar wind and geomagnetic activity. *Surv. Geophys.*, **26**, 545–592.

Meng, C.-I. & K. Liou, (2004). Substorm timings and timescales: a new aspect. *Space Sci. Rev.*, **113**, 41–75.

Mursula, K. & D. Martini, (2007). New indices of geomagnetic activity at test: comparing the correlation of the analogue ak index with the digital Ah and IHV indices at the Sodankylä station. *Adv. Space Res.*, **40**, 1105–1111.

Olsen, N. *et al.* (2000). Ørsted initial field model. *Geophys. Res. Lett.*, **27**, 3607–3610.

Olsen, N., R. Haagmans, T. J. Sabaka, A. Kuvshinov, S. Maus, M. E. Purucker, M. Rother, V. Lesur & M. Mandea, (2006a). The Swarm end-to-end mission simulator study: a demonstration of separating the various contributions to Earth's magnetic field using synthetic data. *Earth Planets Space*, **58**, 359–370.

Olsen, N., H. Lühr, T. J. Sabaka, M. Mandea, M. Rother, L. Tøffner-Clausen & S. Choi, (2006b). CHAOS a model of the Earth's magnetic field derived from CHAMP, Ørsted, and SAC-C magnetic satellite data. *Geophys. J. Int.*, **166**, 67–75, doi:10.1111/ j.1365-246X.2006.029.

Osipova, I. L., S. E. Hjelt & L. L. Vanyan, (1989). Source field problems in northern parts of the Baltic Shield. *Phys. Earth Planet. Inter.*, **53**, 337–342.

Padilha, A. L. (1999). Behavior of magnetotelluric source fields within the equatorial zone. *Earth Planets Space*, **51**, 1119–1125.

Padilha, A. L., I. Vitorello & L. Rijo, (1997). Effects of the equatorial electrojet on magnetotelluric surveys: field results from Northwest Brazil. *Geophys. Res. Lett.*, **24**, 89–92.

Parks, G. K. (2004). *Physics of Space Plasmas. An Introduction*. Boulder: Westview Press.

Pasko, V. P. (2006). Introduction to special section: recent advances in studies of Schumann resonances on Earth and other planets of the solar system. *Radio Sci.*, **41**, RS2S01, doi:10.1029/2006RS003576.

Pirjola, R. (1992). On magnetotelluric source effects caused by an auroral electrojet system. *Radio Sci.*, **27**, 463–468.

Pirjola, R. J. & L. T. V. Häkkinen, (1991). Electromagnetic field caused by an auroral electrojet current system model. In *Environmental and Space Electromagnetics*, ed. H. Kikuchi. Berlin: Springer, pp. 288–298.

Porath, H., D. W. Oldenburg & D. I. Gough, (1970). Separation of magnetic variation fields and conductive structures in the western United States. *Geophys. J. R. Astron. Soc.*, **19**, 237–260.

Pulkkinen, T. I., H. Nevanlinna, P. J. Pulkkinen & M. Lockwood, (2001). The Sun–Earth connection in time scales from years to decades and centuries. *Space Sci. Rev.*, **95**, 625–637.

Pulkkinen, A., O. Amm, A. Viljanen & BEAR Working Group (2003). Separation of the geomagnetic variation field into external and internal parts using the spherical elementary current system method. *Earth Planets Space*, **55**, 117–129.

Pulkkinen, A., A. Klimas, D. Vassiliadis, V. Uritsky & E. Tanskanen, (2006). Spatiotemporal scaling properties of the ground geomagnetic field variations. *J. Geophys. Res.*, **111**(A3), A03305, doi:10.1029/2005JA011294.

Reigber, C., H. Lühr & P. Schwintzer, (2002). CHAMP mission status. *Adv. Space Res.*, **30**, 129–134.

Richardson, I. G., H. V. Cane & E. W. Cliver, (2002). Sources of geomagnetic activity during nearly three solar cycles (1972–2000). *J. Geophys. Res.*, **107**, 1187, doi:10.1029/2001JA000504.

Richmond, A. D. & W. Baumjohann, (1984). Three-dimensional analysis of magnetometer array data. *J. Geophys.*, **54**, 138–156.

Richmond, A. D. & Y. Kamide (1988). Mapping electrodynamic features of the high-latitude ionosphere from localized observations: technique. *J. Geophys. Res.*, **93**, 5741–5759.

Rostoker, G. & T. D. Phan, (1986). Variation of auroral electrojet spatial location as a function of the level of magnetospheric activity. *J. Geophys. Res.*, **91**, 1716–1722.

Russell, C. T. & R. L. McPherron, (1973). Semiannual variation of geomagnetic activity. *J. Geophys. Res.*, **78**, 92–108.

Sabaka, T. J., N. Olsen & M. Purucker, (2004). Extending comprehensive models of the earth's magnetic field with Ørsted and CHAMP data. *Geophys. J. Int.*, **159**, 521–547, doi:10.1111/j.1365-246X.2004.02421.x.

Siebert, M. & W. Kertz, (1957). Zur Zerlegung eines lokalen erdmagnetischen Feldes in äusseren und inneren Anteil. *Nachr. Akad. Wiss. Göttingen, Math.-Physik. Kl. IIa*, 89–112.

Simpson, F. & K. Bahr (2005). *Practical Magnetotellurics*. Cambridge: Cambridge University Press.

Svalgaard, L., E. W. Cliver & P. Le Sager, (2004). IHV: a new long-term geomagnetic index. *Adv. Space Res.*, **34**, 436–439.

Tanskanen, E. (2002). Terrestrial Substorms as a Part of Global Energy Flow. *Finnish Meteorological Institute Contributions*, 36.

Tanskanen, E. I., A. Viljanen, T. I. Pulkkinen, R. Pirjola, L. Häkkinen, A. Pulkkinen & O. Amm, (2001). At substorm onset, 40% of AL comes from underground. *J. Geophys. Res.*, **106**, 13 119–13 134.

Treumann, R. & W. Baumjohann, (1997). *Advanced Space Plasma Physics*. London: Imperial College Press.

Untiedt, J. & W. Baumjohann, (1993). Studies of polar current systems using the IMS Scandinavian magnetometer array. *Space Sci. Rev.*, **63**, 245–390.

Usoskin, I. G. & K. Mursula, (2003). Long-term solar cycle evolution: review of recent developments. *Solar Physics*, **218**, 319–343

Varentsov, I. M. (2007). Arrays of simultaneous electromagnetic soundings: design, data processing and analysis. In *Electromagnetic Sounding of the Earth's Interior*, ed. V. V. Spichak. Amsterdam: Elsevier, pp. 259–273.

Varentsov, I. M., M. Engels, T. Korja, M. Yu. Smirnov & BEAR Working Group (2002). A generalized geoelectric model of Fennoscandia: a challenging database for long-period 3D modeling studies within the Baltic Electromagnetic Array Research (BEAR) Project. *Izv., Phys. Solid Earth*, **38**, 855–892.

Varentsov, I. M., E. Y. Sokolova & BEAR Working Group (2003). Diagnostics and suppression of auroral distortions in the transfer operators of the electromagnetic field in the BEAR experiment. *Izv., Phys. Solid Earth*, **39**, 283–307.

Viljanen, A. (1996). Source effect on geomagnetic induction vectors in the Fennoscandian auroral region. *J. Geomagn. Geoelectr.*, **48**, 1001–1009.

Viljanen, A., R. Pirjola & L. Häkkinen (1993). An attempt to reduce induction source effects at high latitudes. *J. Geomagn. Geoelectr.*, **45**, 817–831.

Viljanen, A., R. Pirjola & O. Amm, (1999). Magnetotelluric source effect due to 3D ionospheric current systems using the complex image method for 1D conductivity structures. *Earth Planets Space*, **51**, 933–945.

Weaver, J. T. (1964). On the separation of local geomagnetic fields into external and internal parts. *Z. Geophys.*, **30**, 29–36.

4

The magnetotelluric response function

PETER WEIDELT AND ALAN D. CHAVE

4.1 General concepts

4.1.1 *Quasi-uniform source fields*

The basic magnetotelluric model is based on the assumption of a time-varying uniform horizontal magnetic field from sources above Earth in $z < 0$ that induce an electric field $\mathbf{E}$ within Earth in $z \geq 0$. For simplicity, the conductor in $z \geq 0$ is assumed to be layered. Consider an arbitrary area A on the surface $z = 0$ with boundary ∂A. Assuming harmonic time dependence ($e^{i\omega t}$), the integral formulation of Faraday's law (2.3) yields

$$\oint_{\partial A} \mathbf{E} \cdot d\mathbf{r} = -i\omega \int_A B_z \, da \tag{4.1}$$

where ∂A is followed in the clockwise direction. By assumption in this simplified model, the vertical component of the magnetic induction B_z vanishes, and therefore no electric field is induced. This contradicts the experimental finding that there is a horizontal electric field at Earth's surface that is in fact essential for the magnetotelluric method to function.

The obvious reason for this contradiction is that the basic model incorporates such a high degree of simplification that it becomes meaningless physically (although not mathematically; see Section 2.4.3). An inducing magnetic field component normal to the surface can be produced either by retaining the assumption of a half-space, but replacing the strictly uniform inducing field by a slightly non-uniform field (as was done in Section 2.4.3), or by retaining the uniform field assumption and replacing the half-space by a sphere of very large radius. In the sequel, the first remedy will be utilized.

Assume that the magnetic source field in the insulating air half-space but below a source level $z_s < 0$ is derived via $\mathbf{B}^0 = -\nabla W^0$ from the scalar potential

$$W^0(x, y) = -\frac{B_0}{2k} \sin(kx) e^{-kz} \tag{4.2}$$

where $z > z_s$ and the wavenumber $k > 0$, corresponding to a wavelength $2\pi/k$, characterizes the heterogeneity of the field. This potential gives rise to the source field

The Magnetotelluric Method: Theory and Practice, ed. Alan D. Chave and Alan G. Jones. Published by Cambridge University Press © Cambridge University Press 2012.

$$B_x^0(x, z) = \frac{B_0}{2} \cos(kx) e^{-kz} \tag{4.3}$$

$$B_z^0(x, z) = -\frac{B_0}{2} \sin(kx) e^{-kz} \tag{4.4}$$

$$E_y^0(x, z) = -\frac{i\omega B_0}{2k} \cos(kx) e^{-kz} \tag{4.5}$$

The vertical magnetic field B_z^0 is positive in $-\pi/k < x < 0$ and negative in $0 < x < \pi/k$, and produces two oppositely directed electric field vortices centered at $x = \pm\pi/(2k)$. By constructive superposition, a strong induced horizontal electric field arises in the neighborhood of $x = 0$. For $k \to 0$, this neighborhood can be quite large. For simplicity, the Earth model is assumed to be a uniform half-space of conductivity σ. With $\beta^2 = k^2 + \gamma^2$, where $\gamma^2 = i\omega\mu_0\sigma$, the electromagnetic field at $z = 0$ is given by

$$B_x(x, 0) = \frac{B_0 \beta}{\beta + k} \cos(kx) \tag{4.6}$$

$$B_z(x, 0) = -\frac{B_0 k}{\beta + k} \sin(kx) \tag{4.7}$$

$$E_y(x, 0) = -\frac{i\omega B_0}{\beta + k} \cos(kx) \tag{4.8}$$

The horizontal magnetic and electric field components are continuous for $k \to 0$ and tend to the limits $B_x = B_0$ and $E_y = -i\omega B_0/\gamma$. The magnetotelluric response function $Z = -E_y/B_x$ that is the quantity of interest in magnetotellurics tends to the continuous limit $Z = i\omega/\gamma$.

In contrast to the horizontal magnetic field components, the magnitude of B_z becomes arbitrarily small as $k \to 0$, but the region where B_z retains its sign increases concomitantly such that

$$\lim_{k \to 0} \int_0^{\pi/k} B_z(x, 0)dx = -\lim_{k \to 0} \frac{2B_0}{\beta + k} = -\frac{2B_0}{\gamma} \tag{4.9}$$

which attains a definite but finite limit. When assuming $k = 0$ from the outset, this limit would be undefined. The finite limit (4.9) is a reminder that a large-scale vertical magnetic field is present that is responsible for the induced electric field in the domain of measurement near $x = 0$. Typically, the domain of measurement will have a width of several penetration depths $1/|\gamma|$, or, in the case of a layered Earth, a width of several inductive scale lengths $|\tilde{\Lambda}_0|$ as defined via the recursion (2.52). By decreasing k, the scale length $2\pi/k$ of the external field can be made much larger than this width. Then, the source field in the domain of measurement can be considered as *quasi-uniform*. This assumption underlies the magnetotelluric method and will be adopted for the rest of this chapter.

In summary, for magnetotellurics, the assumption of a small but finite wavenumber k of the source field is required physically, but not mathematically. The plane-wave limit $k = 0$

can be taken from the outset because the response, along with the tangential components of **B** and **E**, attains a well-defined plane-wave limit for $k \to 0$. The same is not true when the vertical magnetic field is considered, as was seen in Section 2.4.3.

Alternatively, when a uniform source field is retained and a large conductor of finite size L_σ is considered, where L_σ is much larger than the size of the domain of measurements, then for $L_\sigma \to \infty$, the tangential electric and magnetic components tend to limits that, for a given source field, may be different from those obtained for a plane Earth and $k \to 0$. The magnetotelluric response still attains the same plane-wave limit as $k \to 0$; see the exposition in Weaver (1994, p. 68).

4.1.2 Tensor relation between the electric and magnetic fields

Assume a quasi-uniform, horizontal, external magnetic field $\mathbf{B}_0/2$ caused by the source current density $\mathbf{J}^0$ such that

$$\mu_0 \mathbf{J}^0(\mathbf{r}) = \hat{\mathbf{z}} \times \mathbf{B}_0 \delta(z - z_s) \tag{4.10}$$

where $z_s < 0$ is the source level. The time dependence of $\mathbf{J}^0$ is harmonic according to $e^{i\omega t}$. Let $\mathbf{E}_h(\mathbf{r}_0, \omega)$ and $\mathbf{B}_h(\mathbf{r}_0, \omega)$ be the horizontal electric and magnetic fields measured at a point $\mathbf{r}_0$ on the surface of or inside the Earth (i.e. $z_0 \geq 0$). $\mathbf{E}_h$ and $\mathbf{B}_h$ are linear functions of $\mathbf{B}_0$, as will be justified below, so that

$$\begin{aligned} \mathbf{E}_h(\mathbf{r}_0, \omega) &= \vec{\vec{K}}^e(\mathbf{r}_0, \omega) \bullet \mathbf{B}_0(\omega) \\ \mathbf{B}_h(\mathbf{r}_0, \omega) &= \vec{\vec{K}}^m(\mathbf{r}_0, \omega) \bullet \mathbf{B}_0(\omega) \end{aligned} \tag{4.11}$$

where $\vec{\vec{K}}^e$ and $\vec{\vec{K}}^m$ are complex second-rank tensors and $\bullet$ denotes the inner product. In the absence of a rigorous proof, the reasonable assumption will be made that the x component of the source field $B_0\hat{\mathbf{x}}$ will create a magnetic field $\mathbf{B}_h$ at $\mathbf{r}_0$ that is not parallel to the field generated at the same site by the orthogonal component of the source field $B_0\hat{\mathbf{y}}$. Mathematically, this means that the two columns of $\vec{\vec{K}}^m$ are not parallel. Then, $\det\left(\vec{\vec{K}}^m\right) \neq 0$, and therefore the inverse of $\vec{\vec{K}}^m$ exists such that

$$\mathbf{B}_0 = \left(\vec{\vec{K}}^m\right)^{-1} \bullet \mathbf{B}_h \tag{4.12}$$

After suppressing the dependence on $\mathbf{r}_0$ and ω, substitution of (4.12) into the first equation of (4.11) yields

$$\mathbf{E}_h = \vec{\vec{K}}^e \bullet \left(\vec{\vec{K}}^m\right)^{-1} \bullet \mathbf{B}_h = \vec{\vec{Z}} \bullet \mathbf{B}_h \tag{4.13}$$

where $\vec{\vec{Z}}$ is the second-rank *magnetotelluric response tensor*. The *magnetotelluric impedance tensor* is $\mu_0 \vec{\vec{Z}}$, and is obtained by replacing the magnetic induction **B** with the magnetic field **H**. The tensor $\vec{\vec{Z}}$ is the fundamental measured parameter of magnetotellurics, and is the point where experiment and theory are connected. It is a function of frequency and, when the underlying conductivity structure is two- or three-dimensional (2D or 3D), of position. The tensor elements are estimated from data and are compared with the tensor elements resulting from numerical modeling to infer the electrical structure of the subsurface.

The relation (4.13) maps the horizontal vector $\mathbf{B}_h$ onto the horizontal vector $\mathbf{E}_h$. This is a geometric procedure that does not require the specification of a horizontal coordinate system. However, in practical applications, individual elements of $\bar{\mathbf{Z}}$ have to be addressed in a specified horizontal coordinate system, and the coordinate representation of $\bar{\mathbf{Z}}$ depends on the selection of this system. If the geophysical situation does not favor a particular system, one may select geographical coordinates with the x and y directions defining, respectively, the directions of north and east (and the z axis pointing down). The response tensor has the representation

$$\bar{\mathbf{Z}} = \begin{bmatrix} Z_{xx} & Z_{xy} \\ Z_{yx} & Z_{yy} \end{bmatrix} \tag{4.14}$$

The representation in other coordinate systems will be considered in Section 4.1.4.

The proof for the linear relations (4.11) will now be sketched; see Berdichevsky & Dmitriev (2002, pp. 125–127) for details. The Maxwell equations yield the differential equation (2.93)

$$\nabla \times \nabla \times \mathbf{E}(\mathbf{r}) + i\omega\mu_0\sigma(\mathbf{r})\mathbf{E}(\mathbf{r}) = -i\omega\mu_0\mathbf{J}^0(\mathbf{r}) \tag{4.15}$$

with $\mathbf{J}^0$ given by (4.10). Let $\mathbf{G}_x^e(\mathbf{r}|\mathbf{r}_0)$ be the electric field vector observed at $\mathbf{r}$ from a point horizontal electric dipole with unit current oriented in the x direction placed at $\mathbf{r}_0$ on or inside the Earth. It is assumed that $\sigma(\mathbf{r}) > 0$ in the vicinity of $\mathbf{r}_0$ such that the dipole is grounded. Then $\mathbf{G}_x^e(\mathbf{r}|\mathbf{r}_0)$ satisfies

$$\nabla \times \nabla \times \mathbf{G}_x^e(\mathbf{r}|\mathbf{r}_0) + i\omega\mu_0\sigma(\mathbf{r})\mathbf{G}_x^e(\mathbf{r}|\mathbf{r}_0) = -i\omega\mu_0\delta(\mathbf{r} - \mathbf{r}_0)\hat{\mathbf{x}} \tag{4.16}$$

Taking the inner product of (4.15) with $\mathbf{G}_x^e(\mathbf{r}|\mathbf{r}_0)$ and of (4.16) with $\mathbf{E}$, subsequent subtraction and integration over the full space R^3 with respect to $\mathbf{r}$ yields

$$E_x(\mathbf{r}_0) = \int_{R^3} \mathbf{J}^0(\mathbf{r}) \cdot \mathbf{G}_x^e(\mathbf{r}|\mathbf{r}_0)dv \tag{4.17}$$

where the identity

$$\int_{R^3} \left(\mathbf{G}_x^e \cdot \nabla \times \nabla \times \mathbf{E} - \mathbf{E} \cdot \nabla \times \nabla \times \mathbf{G}_x^e \right)dv = 0 \tag{4.18}$$

has been used. Equation (4.18) is a special case of the divergence theorem that depends on a surface integral term vanishing at infinity (see Morse & Feshbach, 1953, p. 1768).

Inserting (4.10) into (4.17) and using

$$(\hat{\mathbf{z}} \times \mathbf{B}_0) \cdot \mathbf{G}_x^e = (\mathbf{G}_x^e \times \hat{\mathbf{z}}) \cdot \mathbf{B}_0 = -(\hat{\mathbf{z}} \times \mathbf{G}_x^e) \cdot \mathbf{B}_0 \tag{4.19}$$

yields

$$E_x(\mathbf{r}_0) = -\frac{1}{\mu_0} \left[\hat{\mathbf{z}} \times \int \int_{-\infty}^{\infty} \mathbf{G}_x^e(x, y, z_s|\mathbf{r}_0)dx\,dy \right] \cdot \mathbf{B}_0 = (K_{xx}^e\hat{\mathbf{x}} + K_{xx}^e\hat{\mathbf{y}}) \cdot \mathbf{B}_0 \tag{4.20}$$

and generates the upper row of the tensor $\vec{\mathbf{K}}^e$ in (4.11). The integral in (4.20) exists since $\mathbf{G}_x^e$ decays in the air half-space for $(x, y) \to \pm\infty$ as $(x^2 + y^2)^{-3/2}$. In a similar vein, consideration of the field $\mathbf{G}_y^e$ of an electric dipole in the y direction leads to K_{yx}^e and K_{yy}^e.

The tensor $\vec{\mathbf{K}}^m$ is obtained by using magnetic rather than electric dipoles. The electric field $\mathbf{G}_x^m(\mathbf{r}|\mathbf{r}_0)$ observed at $\mathbf{r}$ of a magnetic dipole of unit moment oriented in the x direction placed at $\mathbf{r}_0$ satisfies

$$\nabla \times \nabla \times \mathbf{G}_x^m(\mathbf{r}|\mathbf{r}_0) + i\omega\mu_0\sigma(\mathbf{r})\mathbf{G}_x^m(\mathbf{r}|\mathbf{r}_0) = -i\omega\mu_0\nabla \times (\delta(r - r_0)\hat{\mathbf{x}}) \quad (4.21)$$

Proceeding as for (4.17)

$$\int_{R^3} \mathbf{E}(\mathbf{r}) \cdot \nabla \times [\delta(\mathbf{r} - \mathbf{r}_0)\hat{\mathbf{x}}] \, dv = \int_{R^3} [\nabla \times \mathbf{E}(\mathbf{r})] \cdot [\delta(\mathbf{r} - \mathbf{r}_0)\hat{\mathbf{x}}] \, dv = \hat{\mathbf{x}} \cdot \nabla \times \mathbf{E}(\mathbf{r}_0) \quad (4.22)$$

and using Faraday's law (2.3) yields

$$B_x(\mathbf{r}_0) = -\frac{1}{i\omega} \int_{R^3} \mathbf{J}^0(\mathbf{r}) \cdot \mathbf{G}_x^m(\mathbf{r}|\mathbf{r}_0) \, dv \quad (4.23)$$

Insertion of (4.10) into (4.23) then gives

$$B_x(\mathbf{r}_0) = \frac{1}{i\omega\mu_0} \left[\hat{\mathbf{z}} \times \int\int_{-\infty}^{\infty} \mathbf{G}_x^m(x, y, z_s|\mathbf{r}_0)dx \, dy \right] \cdot \mathbf{B}_0 = (K_{xx}^m\hat{\mathbf{x}} + K_{xy}^m\hat{\mathbf{y}}) \cdot \mathbf{B}_0 \quad (4.24)$$

In this case, the integral also exists since $\left|\mathbf{G}_x^m\right| = O\left[(x^2 + y^2)^{-3/2}\right]$ for $(x, y) \to \pm\infty$. With a similar derivation for $B_y(\mathbf{r}_0)$, the proof of (4.11) is complete.

Experimentally, the response tensor $\vec{\mathbf{Z}}$ can be determined from two datasets $\mathbf{E}_h^{(i)}$ and $\mathbf{B}_h^{(i)}$ for $i = 1, 2$ generated by two linearly independent polarizations of $\mathbf{B}_0$ with amplitudes $B_0^{(i)}$ and azimuths $\alpha_{(i)}$. Then Z_{xx} and Z_{xy} are obtained from the two equations

$$E_x^{(i)} = Z_{xx}B_x^{(i)} + Z_{xy}B_y^{(i)}, \quad i = 1, 2 \quad (4.25)$$

and Z_{yx} and Z_{yy} from

$$E_y^{(i)} = Z_{yx}B_x^{(i)} + Z_{yy}B_y^{(i)}, \quad i = 1, 2 \quad (4.26)$$

Using (4.11), it is apparent that a solution exists in both cases since the determinant

$$B_x^{(1)}B_y^{(2)} - B_x^{(2)}B_y^{(1)} = B_0^{(1)}B_0^{(2)} \sin(\alpha^{(2)} - \alpha^{(1)})\det\left(\vec{\mathbf{K}}^m\right) \quad (4.27)$$

is non-zero because of the linear independence of the polarizations and the assumed invertibility of $\vec{\mathbf{K}}^m$, i.e. $\det\left(\vec{\mathbf{K}}^m\right) \neq 0$.

4.1.3 Magnetic transfer functions

It is often advantageous to complement the information contained in the response tensor through transfer functions between the magnetic field components, most notably between B_z and B_h. Let the mapping of $\mathbf{B}_0$ onto the three-component magnetic field $\mathbf{B}(\mathbf{r}_0)$ be

$$\mathbf{B} = \begin{pmatrix} \vec{\mathbf{K}}^m & \\ K_{zx}^m & K_{zy}^m \end{pmatrix} \cdot \mathbf{B}_0 \tag{4.28}$$

where the components of $\vec{\mathbf{K}}^m$ are defined in (4.24). By analogy, the elements K_{zx}^m and K_{zy}^m are obtained by placing a unit vertical magnetic dipole at $\mathbf{r}_0$, yielding

$$B_z = (K_{zx}^m \hat{\mathbf{x}} + K_{zy}^m \hat{\mathbf{y}}) \cdot \mathbf{B}_0 = (K_{zx}^m \hat{\mathbf{x}} + K_{zy}^m \hat{\mathbf{y}}) \cdot (\vec{\mathbf{K}}^m)^1 \cdot \mathbf{B}_h = \mathbf{T} \cdot \mathbf{B}_h \tag{4.29}$$

where

$$\mathbf{T} = \left(K_{zx}^m L_{xx} + K_{zy}^m L_{yx} \right) \hat{\mathbf{x}} + \left(K_{zx}^m L_{xy} + K_{zy}^m L_{yy} \right) \hat{\mathbf{y}} \tag{4.30}$$

and $\vec{\mathbf{L}} = (\vec{\mathbf{K}}^m)^{-1}$. The vector $\mathbf{T}$ is called the *tipper*, and is a helpful dimensionality indicator since $\mathbf{T} = 0$ in a 1D situation, and $\mathbf{T}$ is directed orthogonal to strike in 2D. Further, the tensor $\vec{\mathbf{K}}^m$ defined in (4.11) is of use if $\mathbf{B}_0$ is identified with the horizontal magnetic field at some "normal" site assumed to be only marginally affected by lateral conductivity variations.

The tipper vector $\mathbf{T} = T_x \hat{\mathbf{x}} + T_y \hat{\mathbf{y}}$ is determined experimentally by again assuming that B_z and $\mathbf{B}_h^{(i)}$ for $i = 1, 2$ are generated by two linearly independent polarizations of $\mathbf{B}_0$ such that T_x and T_y can be determined from the non-singular system

$$B_z^{(i)} = T_x B_x^{(i)} + T_y B_y^{(i)} \tag{4.31}$$

4.1.4 Rotational invariants

The magnetotelluric response tensor $\vec{\mathbf{Z}}$ maps $\mathbf{B}_h$ onto $\mathbf{E}_h$, and can be defined without any reference to a horizontal coordinate system. However, when individual tensor elements have to be addressed, the specification of a coordinate system is required, and the coordinate representation of $\vec{\mathbf{Z}}$ depends on the orientation of this system. Let (4.14) be the representation of $\vec{\mathbf{Z}}$ in a specified x, y coordinate system. In an x', y' coordinate system rotated in a clockwise direction through an angle α, the representations in $x'y'$ and x,y coordinates are related by

$$\begin{aligned} \mathbf{E}_h' &= \vec{\mathbf{R}}(\alpha) \cdot \mathbf{E}_h \\ \mathbf{B}_h' &= \vec{\mathbf{R}}(\alpha) \cdot \mathbf{B}_h \end{aligned} \tag{4.32}$$

where

$$\vec{\mathbf{R}}(\alpha) = \begin{bmatrix} \cos\alpha & \sin\alpha \\ -\sin\alpha & \cos\alpha \end{bmatrix} \tag{4.33}$$

and $\vec{\mathbf{R}}(-\alpha) = (\vec{\mathbf{R}}(\alpha))^{\mathrm{T}} = (\vec{\mathbf{R}}(\alpha))^{-1}$, so that $\vec{\mathbf{R}}$ is unitary. Substituting into (4.13) yields

$$\begin{aligned} \mathbf{E}_h' &= \vec{\mathbf{R}}(\alpha) \cdot \mathbf{E}_h = \vec{\mathbf{R}}(\alpha) \cdot \vec{\mathbf{Z}} \cdot \mathbf{B}_h \\ &= \vec{\mathbf{R}}(\alpha) \cdot \vec{\mathbf{Z}} \cdot \vec{\mathbf{R}}(-\alpha) \cdot \mathbf{B}_h' = \vec{\mathbf{Z}}'(\alpha) \cdot \mathbf{B}_h' \end{aligned} \tag{4.34}$$

where

$$\vec{\mathbf{Z}}'(\alpha) = \vec{\mathbf{R}}(\alpha) \cdot \vec{\mathbf{Z}}(\alpha) \cdot \vec{\mathbf{R}}(-\alpha) \tag{4.35}$$

Writing out the terms explicitly, using the simplified notation $Z'_{xx}(\alpha)$ instead of $Z_{x'x'}(\alpha)$, etc., gives

$$Z'_{xx}(\alpha) = Z_{xx}\cos^2\alpha + (Z_{xy} + Z_{yx})\sin\alpha\cos\alpha + Z_{yy}\sin^2\alpha \tag{4.36}$$

$$Z'_{xy}(\alpha) = Z_{xy}\cos^2\alpha - (Z_{xx} - Z_{yy})\sin\alpha\cos\alpha - Z_{yx}\sin^2\alpha \tag{4.37}$$

$$Z'_{yx}(\alpha) = Z_{yx}\cos^2\alpha - (Z_{xx} - Z_{yy})\sin\alpha\cos\alpha - Z_{xy}\sin^2\alpha \tag{4.38}$$

$$Z'_{yy}(\alpha) = Z_{yy}\cos^2\alpha - (Z_{xy} + Z_{yx})\sin\alpha\cos\alpha + Z_{xx}\sin^2\alpha \tag{4.39}$$

Since $\vec{Z}'(\alpha + \pi) = \vec{Z}'(\alpha)$, the angle can be restricted to $0 \le \alpha \le \pi$. Moreover, from (4.36) through (4.39), it is easy to show that

$$\begin{aligned} Z'_{xx}(\alpha + \pi/2) &= Z'_{yy}(\alpha) \\ Z'_{xy}(\alpha + \pi/2) &= -Z'_{yx}(\alpha) \\ Z'_{yx}(\alpha + \pi/2) &= -Z'_{xy}(\alpha) \\ Z'_{yy}(\alpha + \pi/2) &= Z'_{xx}(\alpha) \end{aligned} \tag{4.40}$$

The response tensor has a number of properties that hold for any orientation of the horizontal coordinate system. As a simple example, it may be deduced from (4.36) and (4.39) that the trace of $\vec{Z}$ is invariant under rotation, so that

$$Z'_{xx}(\alpha) + Z'_{yy}(\alpha) = Z_{xx} + Z_{yy} \tag{4.41}$$

These so-called *rotational invariants* are of interest if they can serve as dimensionality indicators (e.g. the skew derived from $Z_{xy}-Z_{yx}$) or as a first guess for the underlying conductivity structure, e.g. the effective response derived from $\det(\vec{Z})$. Rotational invariants have been used in magnetotellurics since its early beginnings, but their systematic investigation did not occur until recently, when Szarka & Menvielle (1997) presented a very thorough exposition on the subject. These authors showed that the complex magnetotelluric response tensor $\vec{Z}$ that is described by eight real parameters can be characterized by seven real rotational invariants. Since any function of rotational invariants is itself rotationally invariant, the selection of the seven invariants is not unique. The formally simplest set from Szarka & Menvielle (1997) is

$$\begin{aligned} I_1 &= \mathrm{Re}(Z_{xx} + Z_{yy}) \\ I_2 &= \mathrm{Im}(Z_{xx} + Z_{yy}) \end{aligned} \tag{4.42}$$

$$\begin{aligned} I_3 &= \mathrm{Re}(Z_{xy} - Z_{yx}) \\ I_4 &= \mathrm{Im}(Z_{xy} - Z_{yx}) \end{aligned} \tag{4.43}$$

$$\begin{aligned} I_5 &= \det[\mathrm{Re}(\vec{Z})] \\ I_6 &= \det[\mathrm{Im}(\vec{Z})] \\ I_7 &= \mathrm{Im}[\det(\vec{Z})] \end{aligned} \tag{4.44}$$

It must be noted that the complex rotational invariant $\det\left(\vec{\vec{Z}}\right)$ is equivalent to the three real invariants given in (4.44), where I_7 is the only invariant that connects the real and imaginary parts of $\vec{\vec{Z}}$. All possible invariants can be expressed in terms of I_1 to I_7. Examples include

(a) the real part of $\det\left(\vec{\vec{Z}}\right)$

$$\mathrm{Re}\left(\det \vec{\vec{Z}}\right) = I_5 - I_6 \tag{4.45}$$

(b) the square of the Frobenius norm of $\vec{\vec{Z}}$

$$|Z_{xx}|^2 + |Z_{xy}|^2 + |Z_{yx}|^2 + |Z_{yy}|^2 = I_1^2 + I_2^2 + I_3^2 + I_4^2 - 2(I_5 + I_6) \tag{4.46}$$

(c) the skew

$$S = \frac{|Z_{xx} + Z_{yy}|}{|Z_{xy} - Z_{yx}|} = \frac{\sqrt{I_1^2 + I_2^2}}{\sqrt{I_3^2 + I_4^2}} \tag{4.47}$$

(d) and the rotational invariant C using the definition of Bahr (1988) that serves as a dimensionality indicator in Section 4.1.5

$$C = 2\mathrm{Im}(Z_{xx}Z_{yx}^* - Z_{yy}Z_{xy}^*)$$

$$= I_1 I_4 - I_2 I_3 - \sqrt{(I_1^2 + I_3^2 - 4I_5)(I_2^2 + I_4^2 - 4I_6) - (I_1 I_2 + I_3 I_4 - 2I_7)^2}$$

The selection of the invariants (4.42)–(4.44) was governed by their formal simplicity. However, the last example underlines impressively that other selections of rotational invariants could be more useful for diagnostic purposes. This issue was carefully pursued by Weaver *et al.* (2000), who introduced additional invariants, in particular for geometric visualization of $\vec{\vec{Z}}$ by Mohr circles.

By definition, rotational invariants contain no angular information. Therefore, in addition to the seven real values I_1 to I_7, the specification of an angle as the eighth parameter is required to recover the four complex elements of $\vec{\vec{Z}}$. For example, this angle can be the azimuth at which the maximum of $\left|Z_{xy}'(\alpha)\right|$ is attained. Apart from the fundamental indeterminacy of the orientation, the specification of the invariants leaves an additional ambiguity in the symmetry of $\vec{\vec{Z}}$. From the definition of the invariants in (4.42)–(4.44), it can be inferred that the invariants remain unchanged when exchanging Z_{xx} with Z_{yy} and/or Z_{xy} with $-Z_{yx}$. According to (4.40), the performance of both exchanges simply rotates the axes of $\vec{\vec{Z}}$ by $\pi/2$; however, the performance of a single exchange generally produces a new tensor.

In principle, the reconstruction of $\vec{\vec{Z}}$ from the invariants can be performed as in the following example. From the second equation in (4.37), determine the extrema of $\mathrm{Re}\left[Z_{xy}'(\alpha)\right]$, yielding that this quantity varies in the range

$$\left|I_3 - 2\mathrm{Re}\left[Z_{xy}'(\alpha)\right]\right|^2 \le I_1^2 + I_3^2 - 4I_5 \tag{4.49}$$

Take for $\mathrm{Re}\left[Z_{xy}'(\alpha)\right]$ any value inside this feasible range, such as $I_3/2$, and determine the remaining seven real parameters of $\vec{\vec{Z}}$ with the seven invariants (4.42)–(4.44). The nonlinearity in the invariants I_5 to I_7 gives rise to four different reconstructions of $\vec{\vec{Z}}$. This can be handled by assigning the angle $\alpha = 0$ to each reconstruction, then taking the first reconstruction as a reference and rotating the other three reconstructions via (4.35) through appropriate angles until all four reconstructions agree in the complex element $Z'_{xy}(\alpha)$. At this point, the set of rotated tensors $\vec{\vec{Z}}'(\alpha)$ disintegrates into two subsets. Each subset consists of two identical tensors, and the difference between the subsets is the interchange of the diagonal elements $Z'_{xx}(\alpha)$ and $Z'_{yy}(\alpha)$. After rotating the tensors in one subset by $\pi/2$, the disparity is shifted from the diagonal elements to the off-diagonal elements (see (4.40)). Thus, taking only the seven invariants as data, both subsets are valid reconstructions.

4.1.5 Electric field distortion: first encounter

This section provides some basic facts about the subject of electromagnetic distortion. An in-depth presentation taking into account realistic geophysical situations is postponed to Chapter 6.

Interpretation of the magnetotelluric response tensor $\vec{\vec{Z}}$ faces a great problem due to the distortion of $\mathbf{E}_h$ at the measurement site $\mathbf{r}$ by a (possibly small) local conductivity anomaly. As a result, the measured local field $\mathbf{E}_h$ may not be representative of the undisturbed regional electric field $\mathbf{E}_{Rh}$ that would be measured in the absence of the anomaly, and that remains the quantity of interest for inferring the conductivity distribution at depth. The regional field $\mathbf{E}_{Rh}$ is mapped onto the local field $\mathbf{E}_h$ by the 2×2 *distortion tensor* $\vec{\vec{D}}$

$$\mathbf{E}_h(\mathbf{r}, \omega) = \vec{\vec{D}}(\mathbf{r}) \cdot \mathbf{E}_{Rh}(\mathbf{r}, \omega) \tag{4.50}$$

This is a common model based on the assumption that distortion is caused by galvanic deflection of the electric field rather than by induction. Therefore, $\vec{\vec{D}}$ is assumed to be real and frequency-independent, implying that $\mathbf{E}_h$ and $\mathbf{E}_{Rh}$ may differ in direction and magnitude, but not in phase. The distortion model is a low-frequency approximation that requires (roughly) that $\omega\mu_0|\delta\sigma|L^2 \ll 1$, where $\delta\sigma$ and L are a typical conductivity contrast and length scale characterizing the conductivity anomaly. Chave & Smith (1994) give a complete first principles derivation of (4.50) beginning from the integral equation governing scattering by the distorting inhomogeneity.

Consider as the simplest example a uniform thin sheet of conductance τ_R in which an elliptical local conductivity anomaly of conductance τ is inserted (Figure 4.1). The major and minor axes of the ellipse are a and b, respectively, where axis a has an azimuth γ measured from the x axis to the y axis. The distortion tensor at the observation point $\mathbf{r}$ is given in Morse & Feshbach (1953, p. 1199)

$$\vec{\vec{D}} = \vec{\vec{R}}(-\gamma) \cdot \begin{pmatrix} \frac{a+b}{a+\kappa b} & 0 \\ 0 & \frac{a+b}{\kappa a+b} \end{pmatrix} \cdot \vec{\vec{R}}(\gamma) \tag{4.51}$$

where $\kappa = \tau/\tau_R$ and the rotation tensor $\vec{\vec{R}}$ is defined in (4.33). If $\mathbf{E}_{Rh}$ is parallel to the minor or major axis of the conductivity anomaly, the uniform fields $\mathbf{E}_{Rh}$ and $\mathbf{E}_h$ differ only in

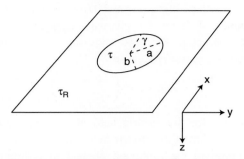

Figure 4.1. Simple distortion model described in the text consisting of a thin sheet of conductance τ_R with an embedded elliptical region of conductance τ. The elliptical region has a major axis of length a and a minor axis of length b, and the major axis makes an angle γ with respect to the x axis.

magnitude, and otherwise they also differ in direction. The distortion tensor depends only on b/a and not on the absolute size of the ellipse, which may be quite small. The low-frequency approximation requires that $\omega\mu_0|\tau - \tau_R|a \ll 1$ in this model.

Another assumption underlying the distortion model is that distortion in the local magnetic field $\mathbf{B}_h$ can be neglected, implying $\mathbf{B}_h = \mathbf{B}_{Rh}$. The validity of this condition is discussed by Singer (1992) and Chave & Smith (1994), and in Chapter 6.

From the response tensor relationship for the regional field

$$\mathbf{E}_{Rh} = \vec{\mathbf{Z}}_R \cdot \mathbf{B}_{Rh} \tag{4.52}$$

and

$$\mathbf{E}_h = \vec{\mathbf{D}} \cdot \mathbf{E}_{Rh} = \vec{\mathbf{D}} \cdot \vec{\mathbf{Z}}_R \cdot \mathbf{B}_{Rh} = \vec{\mathbf{D}} \cdot \vec{\mathbf{Z}}_R \cdot \mathbf{B}_h \tag{4.53}$$

it follows by comparison with (4.13) that

$$\vec{\mathbf{Z}} = \vec{\mathbf{D}} \cdot \vec{\mathbf{Z}}_R \tag{4.54}$$

In a 1D conductivity structure, the magnetotelluric response Z_{1D} is simply the ratio of the two orthogonal horizontal components of $\mathbf{E}$ and $\mathbf{B}$,

$$\hat{\mathbf{e}} \cdot \mathbf{E} = (\hat{\mathbf{z}} \times \hat{\mathbf{e}}) \cdot \mathbf{B}\, Z_{1D} \tag{4.55}$$

where $\hat{\mathbf{e}}$ is a horizontal unit vector. With $\hat{\mathbf{e}} = \hat{\mathbf{x}}$ and $\hat{\mathbf{e}} = \hat{\mathbf{y}}$ in turn, this yields

$$\vec{\mathbf{Z}}_R = \begin{pmatrix} 0 & Z_{1D} \\ -Z_{1D} & 0 \end{pmatrix} = \vec{\mathbf{Z}}'_R(\alpha) \tag{4.56}$$

The second identity in (4.56), which can be verified directly either with horizontal isotropy (4.55) or by applying (4.36) to (4.39), confirms the expected result that no horizontal coordinate system is privileged. The measured response is

$$\vec{\mathbf{Z}} = \vec{\mathbf{D}} \cdot \vec{\mathbf{Z}}_R = \begin{pmatrix} -D_{xy}Z_{1D} & D_{xx}Z_{1D} \\ -D_{yy}Z_{1D} & D_{yx}Z_{1D} \end{pmatrix} = \begin{pmatrix} -D_{xy} & D_{xx} \\ -D_{yy} & D_{yx} \end{pmatrix} Z_{1D} \tag{4.57}$$

This allows the determination of $Z_{1D}(\omega)$ and $\bar{\bar{D}}$ up to a frequency-independent multiplicative scalar, the static shift factor. The phase of Z_{1D} is uniquely determined.

In a 2D conductivity structure, the strike direction defines a privileged azimuth α_s in which the regional response tensor reduces to

$$\vec{Z}'_R(\alpha_s) = \begin{pmatrix} 0 & Z_1 \\ -Z_2 & 0 \end{pmatrix} \tag{4.58}$$

where Z_1 is the response for the electric field in the strike direction (TE mode; see Section 2.5.1) and Z_2 results from the electric field orthogonal to the strike (TM mode).

Bahr (1988) has proposed the following method to determine the unknown strike direction α_s from data (see also Berdichevsky & Dmitriev, 2002, pp. 133–135). The distorted tensor in strike coordinates is

$$\vec{Z}'(\alpha_s) = \bar{\bar{D}}'(\alpha_s) \bullet \vec{Z}'(\alpha_s) = \begin{pmatrix} -D'_{xy}(\alpha_s)Z_2 & D'_{xx}(\alpha_s)Z_1 \\ -D'_{yy}(\alpha_s)Z_2 & D'_{yx}(\alpha_s)Z_1 \end{pmatrix} \tag{4.59}$$

Because the elements of $\bar{\bar{D}}$ are real, the two elements in each of the two columns of $\vec{Z}'(\alpha_s)$ in (4.59) have the same phase, meaning that their ratio is real, so that

$$\mathrm{Im}\left[\frac{Z'_{xx}(\alpha_s)}{Z'_{yx}(\alpha_s)}\right] = 0$$
$$\mathrm{Im}\left[\frac{Z'_{xy}(\alpha_s)}{Z'_{yy}(\alpha_s)}\right] = 0 \tag{4.60}$$

or equivalently

$$\mathrm{Im}\left[Z'_{xx}(\alpha_s)\,Z'_{yx}(\alpha_s)^*\right] = 0$$
$$\mathrm{Im}\left[Z'_{xy}(\alpha_s)\,Z'_{yy}(\alpha_s)^*\right] = 0 \tag{4.61}$$

By inserting (4.36)–(4.39) into (4.61), equations (4.60) and (4.61) can be formulated in terms of the measured tensor elements Z_{ik} for $i, k = x, y$,

$$A\sin 2\alpha_s - B\cos 2\alpha_s + C = 0$$
$$A\sin 2\alpha_s - B\cos 2\alpha_s - C = 0 \tag{4.62}$$

with

$$A = \mathrm{Im}\left(Z_{xx}Z^*_{yy} + Z_{xy}Z^*_{yx}\right) \tag{4.63}$$

$$B = \mathrm{Im}\left(Z_{xy}Z^*_{yy} + Z_{xx}Z^*_{yx}\right) \tag{4.64}$$

$$C = \mathrm{Im}\left(Z_{xx}Z^*_{yx} - Z_{yy}Z^*_{xy}\right) \tag{4.65}$$

where C has previously occurred in (4.48) except for a scale factor of 2. A single strike angle α_s satisfying both equations in (4.62) exists if the data satisfy the compatibility condition

$$C = 0 \tag{4.66}$$

By inserting (4.36)–(4.39) into (4.65), or by referring to (4.48), it can easily be verified that C is a rotational invariant. As a result, the validity of (4.66) can be checked in an arbitrary coordinate system. Under the condition (4.66), an undistorted estimate of the strike angle α_s is determined via (4.62) from

$$\tan 2\alpha_s = \frac{B}{A} = \frac{\mathrm{Im}\left(Z_{xy}Z_{yy}^* - Z_{xx}Z_{yx}^*\right)}{\mathrm{Im}\left(Z_{xx}Z_{yy}^* + Z_{xy}Z_{yx}^*\right)} = \frac{2\,\mathrm{Im}\left(Z_{xy}Z_{yy}^*\right)}{\mathrm{Im}\left(Z_{xx}Z_{yy}^* + Z_{xy}Z_{yx}^*\right)} \tag{4.67}$$

Since the tangent has a period of π, (4.67) allows the determination of $2\alpha_s$ only with an ambiguity of that angle. The strike angle α_s then has $\pi/2$ ambiguity such that one cannot distinguish between the strike direction and the direction orthogonal to it without additional information. The phases of the responses for the electric field aligned along the principal directions can be determined uniquely as

$$\phi(\alpha_s) = \arg\left[Z'_{xy}(\alpha_s)\right] = \arg\left[Z'_{yy}(\alpha_s)\right] \tag{4.68}$$

$$\phi(\alpha_s \pm \pi/2) = \arg[Z'_{xx}(\alpha_s)] = \arg\left[Z'_{yx}(\alpha_s)\right] \tag{4.69}$$

where the phase determination from the diagonal elements may become indefinite. If the phases of the principal responses agree such that $\arg(Z_1) = \arg(Z_2)$, but $|Z_1| = |Z_2|$ does not hold, then all elements of $\overline{\mathbf{Z}}$ have the same phase such that the products in (4.63) and (4.64) are real. Consequently, $A = B = 0$, and the computation of α_s via (4.67) is indeterminate (Berdichevsky & Dmitriev, 2002, p. 135).

The validity of the condition (4.66) signals that the regional structure is strictly 2D, although this rarely occurs in practice. As a measure of the departure from a 2D structure, Bahr (1988) introduced a non-dimensional rotational invariant, the *phase-sensitive skew*

$$\eta = \frac{\sqrt{|C|}}{|Z_{xy} - Z_{yx}|} \tag{4.70}$$

where C is given by (4.48), and complements the traditional skew given by (4.47) as a measure of the deviation from a 2D structure. The skew (4.47) is also a non-dimensional rotational invariant, with $S = 0$ in a 2D situation, since the rotationally invariant numerator vanishes in that instance according to (4.58).

The problem in the practical application of (4.70) and (4.47) is defining suitable thresholds for η and S that mark the transition from 2D to 3D. Moreover, it should be noted that the definitions of η and S are based on the measured distorted magnetotelluric responses. This is particularly serious for S, whereas it will be shown in Section 4.1.6 that an undistorted version $\bar{\eta}$ of η is easily obtained by changing the normalization of C to

$$\bar{\eta}^2 = \frac{|C|}{\left|\det[\mathrm{Re}(\overline{\mathbf{Z}})]\right|} = \frac{|C|}{|I_5|} \tag{4.71}$$

which is again a non-dimensional rotational invariant.

4.1.6 The phase tensor

Galvanic distortion is one of the key problems in magnetotellurics. A major breakthrough in distortion analysis was achieved by Caldwell *et al.* (2004) and Bibby *et al.* (2005) through introduction of the phase tensor. This novel concept brought great clarity and unity into the problem, and shed new light onto previous efforts at distortion removal.

Consider the two tensors $\vec{\mathbf{P}} = \mathrm{Re}(\vec{\mathbf{Z}})$ and $\vec{\mathbf{Q}} = \mathrm{Im}(\vec{\mathbf{Z}})$. With similar tensors for $\vec{\mathbf{Z}}_R$, because of (4.54)

$$\vec{\mathbf{Z}} = \vec{\mathbf{P}} + i\vec{\mathbf{Q}} = \vec{\mathbf{D}} \cdot \vec{\mathbf{P}}_R + i\vec{\mathbf{D}} \cdot \vec{\mathbf{Q}}_R \tag{4.72}$$

The phase tensor $\vec{\mathbf{\Phi}}$ is defined as

$$\vec{\mathbf{\Phi}} = \begin{pmatrix} \Phi_{xx} & \Phi_{xy} \\ \Phi_{yx} & \Phi_{yy} \end{pmatrix} = \vec{\mathbf{P}}^{-1} \cdot \vec{\mathbf{Q}}. \tag{4.73}$$

Writing the terms out explicitly

$$\vec{\mathbf{\Phi}} = \frac{1}{\det(\vec{\mathbf{P}})} \begin{pmatrix} P_{yy}Q_{xx} - P_{xy}Q_{yx} & P_{yy}Q_{xy} - P_{xy}Q_{yy} \\ P_{xx}Q_{yx} - P_{yx}Q_{xx} & P_{xx}Q_{yy} - P_{yx}Q_{xy} \end{pmatrix} \tag{4.74}$$

The phase tensor has real elements with the fundamental property

$$\vec{\mathbf{\Phi}} = \left(\vec{\mathbf{D}} \cdot \vec{\mathbf{P}}_R\right)^{-1} \cdot \vec{\mathbf{D}} \cdot \vec{\mathbf{Q}}_R = \vec{\mathbf{P}}_R^{-1} \cdot \vec{\mathbf{D}}^{-1} \cdot \vec{\mathbf{D}} \cdot \vec{\mathbf{Q}}_R = \vec{\mathbf{P}}_R^{-1} \cdot \vec{\mathbf{Q}}_R = \vec{\mathbf{\Phi}}_R \tag{4.75}$$

The distortion tensor $\vec{\mathbf{D}}$ has dropped out, so that the locally determined phase tensor $\vec{\mathbf{\Phi}}$ equals the undistorted regional phase tensor $\vec{\mathbf{\Phi}}_R$, and hence the conclusions drawn from $\vec{\mathbf{\Phi}}$ immediately apply to the regional conductivity structure. With the alternative definition of the phase tensor as $\vec{\mathbf{Q}} \cdot \vec{\mathbf{P}}^{-1}$, which has the same trace and determinant as $\vec{\mathbf{\Phi}}$, the distortion does not drop out.

The real tensor $\vec{\mathbf{\Phi}}$ is characterized by three rotational invariants. In accordance with Weaver *et al.* (2006), these are

$$J_1 = \frac{\Phi_{xx} + \Phi_{yy}}{2} \tag{4.76}$$

$$J_3 = \frac{\Phi_{xy} - \Phi_{yx}}{2} \tag{4.77}$$

$$J_2^2 = J_1^2 + J_3^2 - \det\vec{\mathbf{\Phi}} = \frac{\left(\Phi_{xx} - \Phi_{yy}\right)^2 + \left(\Phi_{xy} + \Phi_{yx}\right)^2}{4} \tag{4.78}$$

with the dependent invariant

$$J_0^2 = J_1^2 + J_3^2 \tag{4.79}$$

The parameters J_1 and J_3 are well-known invariants of a 2×2 tensor. The determinant with the third invariant replaced by J_2 turns out to be more useful in the following discussion.

The phase tensor $\overset{\leftrightarrow}{\Phi}$ admits the two remarkable representations

(a) Weaver *et al.* (2006)

$$\overset{\leftrightarrow}{\Phi} = \begin{pmatrix} J_1 & 0 \\ 0 & J_1 \end{pmatrix} + J_2 \begin{pmatrix} \cos 2\alpha & \sin 2\alpha \\ \sin 2\alpha & -\cos 2\alpha \end{pmatrix} + \begin{pmatrix} 0 & J_3 \\ -J_3 & 0 \end{pmatrix} \tag{4.80}$$

(b) Caldwell *et al.* (2004)

$$\overset{\leftrightarrow}{\Phi} = \overset{\leftrightarrow}{\mathbf{R}}(\beta - \alpha) \cdot \begin{pmatrix} J_0 + J_2 & 0 \\ 0 & J_0 - J_2 \end{pmatrix} \cdot \overset{\leftrightarrow}{\mathbf{R}}(\beta + \alpha) \tag{4.81}$$

with

$$\tan 2\alpha = \frac{\Phi_{xy} + \Phi_{yx}}{\Phi_{xx} - \Phi_{yy}}$$

$$\tan 2\beta = \frac{J_3}{J_1} = \frac{\Phi_{xy} - \Phi_{yx}}{\Phi_{xx} + \Phi_{yy}} \tag{4.82}$$

implying, with reference to the definitions of J_2 and J_0 in (4.78) and (4.79),

$$2J_2 \sin 2\alpha = \Phi_{xy} + \Phi_{yx}$$
$$2J_2 \cos 2\alpha = \Phi_{xx} - \Phi_{yy}$$
$$2J_0 \sin 2\beta = \Phi_{xy} - \Phi_{yx} = 2J_3$$
$$2J_0 \cos 2\beta = \Phi_{xx} + \Phi_{yy} = 2J_1 \tag{4.83}$$

Since the angle β, with $tan2\beta = J_3/J_1$, is also an invariant, $\overset{\leftrightarrow}{\Phi}$ may be expressed in both representations in terms of three invariants and an angle α fixing the orientation of $\overset{\leftrightarrow}{\Phi}$ in the horizontal plane. For $J_0 > J_2$, i.e. for $\det(\overset{\leftrightarrow}{\Phi}) > 0$, the representation (4.81) is the well-known singular value decomposition of a 2×2 matrix. In the sequel, it will be assumed that $\det(\overset{\leftrightarrow}{\Phi}) > 0$. Caldwell *et al.* (2004, p. 469) discuss the anomalous case $\det(\overset{\leftrightarrow}{\Phi}) \le 0$.

The identity of (4.80) with (4.73) can immediately be verified by combining (4.76) and (4.77) with (4.83). To prove the identity of (4.81) with (4.73) and (4.80), the matrix multiplications in (4.81) must first be performed, yielding

$$\overset{\leftrightarrow}{\Phi} = J_0 \begin{pmatrix} \cos 2\beta & \sin 2\beta \\ -\sin 2\beta & \cos 2\beta \end{pmatrix} + J_2 \begin{pmatrix} \cos 2\alpha & \sin 2\alpha \\ \sin 2\alpha & -\cos 2\alpha \end{pmatrix} \tag{4.84}$$

and then arriving at (4.73) and (4.80) using (4.83).

The graphical depiction of the phase tensor is an ellipse (Bibby *et al.*, 2005, p. 930). For a demonstration, it will be assumed that the magnetic field is a linearly polarized unit vector with azimuth γ given by

$$\hat{\mathbf{b}} = \hat{\mathbf{x}} \cos \gamma + \hat{\mathbf{y}} \sin \gamma \tag{4.85}$$

where $0 \le \gamma < 2\pi$. Consider the mapping

$$\mathbf{e}(\gamma) = \overset{\leftrightarrow}{\Phi} \cdot \hat{\mathbf{b}}(\gamma) \tag{4.86}$$

or in components using (4.84)

$$
\begin{aligned}
e_x(\gamma) &= J_0 \cos(\gamma - 2\beta) + J_2 \cos(\gamma - 2\alpha) \\
e_y(\gamma) &= J_0 \sin(\gamma - 2\beta) - J_2 \sin(\gamma - 2\alpha)
\end{aligned}
\tag{4.87}
$$

The geometrical meaning of (4.87) becomes more obvious after eliminating the parameter γ. This is easily achieved in a system rotated in a clockwise direction through an angle $\delta = \alpha - \beta$ against the (e_x, e_y) system, yielding

$$
\begin{aligned}
e_x(\gamma) \cos \delta + e_y(\gamma) \sin \delta &= (J_0 + J_2) \cos(\gamma - \alpha - \beta) \\
e_y(\gamma) \cos \delta - e_x(\gamma) \sin \delta &= (J_0 - J_2) \sin(\gamma - \alpha - \beta)
\end{aligned}
\tag{4.88}
$$

The parameter γ is readily eliminated, giving

$$
\left(\frac{e_x \cos \delta + e_y \sin \delta}{J_0 + J_2} \right)^2 + \left(\frac{e_y \cos \delta - e_x \sin \delta}{J_0 - J_2} \right)^2 = 1
\tag{4.89}
$$

Equation (4.89) describes an ellipse with semi-major axis $J_0 + J_2$ and semi-minor axis $J_0 - J_2$ (Figure 4.2). The major axis has the azimuth $\delta = \alpha - \beta$, and is generated by a magnetic field of azimuth $\gamma = \alpha - \beta$. This is also obvious from (4.87), which in this case yields

$$
\frac{e_y(\alpha + \beta)}{e_x(\alpha + \beta)} = \tan(\alpha - \beta)
\tag{4.90}
$$

The representation (4.80) of Weaver *et al.* (2006) neatly describes the dimensionality of the regional structure: in 1D, only J_1 is non-zero; while in 2D, both J_1 and J_2 contribute, whereas the skew J_3 vanishes; and in 3D, all three invariants are non-zero.

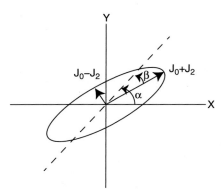

Figure 4.2. Graphical representation of the phase tensor as an ellipse. The lengths of the semi-major and semi-minor axes of the ellipse are defined by the invariants $J_0 + J_2$ and $J_0 - J_2$, and represent the principle axes of the phase tensor. For a non-symmetric phase tensor, a third invariant is needed, the skew angle β, which gives the rotation of the major axis of the ellipse away from an identically shaped ellipse represented by a symmetric tensor. The direction of the semi-major axis is defined by the angle $\delta = \alpha - \beta$, where α is the angle between the semi-major axis and the reference x axis.

Since $\vec{\Phi} = \vec{\Phi}_R$, it is sufficient to check the invariants for the latter in an arbitrary coordinate system to deduce the necessary conditions for a tensor $\vec{\Phi}$ to be interpretable by a regional 1D or 2D conductivity model as follows:

1. The regional structure is 1D $\Rightarrow J_2 = J_3 = 0$. From the representation of $\vec{Z}_R$ in (4.49), with $Z_{1D} = P + iQ$, it follows that

$$
\begin{aligned}
\vec{\Phi}_R &= \begin{pmatrix} 0 & P \\ -P & 0 \end{pmatrix}^{-1} \cdot \begin{pmatrix} 0 & Q \\ -Q & 0 \end{pmatrix} \\
&= \begin{pmatrix} 0 & -1/P \\ 1/P & 0 \end{pmatrix} \cdot \begin{pmatrix} 0 & Q \\ -Q & 0 \end{pmatrix} \\
&= \begin{pmatrix} Q/P & 0 \\ 0 & Q/P \end{pmatrix} = \begin{pmatrix} 1 & 0 \\ 0 & 1 \end{pmatrix} Q/P
\end{aligned}
\tag{4.91}
$$

Equations (4.76)–(4.78) then yield $J_2 = J_3 = 0$. Conversely, if these conditions are satisfied, then $\Phi_{xy} = \Phi_{yx} = 0$, $\Phi_{xx} = \Phi_{yy} = \tan\phi$ and $\vec{\Phi}$ is necessarily represented by

$$
\vec{\Phi} = \begin{pmatrix} 1 & 0 \\ 0 & 1 \end{pmatrix} \tan\phi
\tag{4.92}
$$

Only the phase of the regional response Z_{1D} is determinable from distorted measurements, in agreement with the conclusion from (4.57). According to (4.89), the graphical representation of $\vec{\Phi}$ is a circle of radius $J_0 = \tan\phi$.

2. The regional structure is 2D $\Rightarrow J_3 = 0$. The regional response in the strike direction is given by (4.58). Let $Z_k = P_k + iQ_k$ for $k = 1,2$. Then the regional phase tensor in strike coordinates is

$$
\begin{aligned}
\vec{\Phi}'_R &= \begin{pmatrix} 0 & P_1 \\ P_2 & 0 \end{pmatrix}^{-1} \cdot \begin{pmatrix} 0 & Q_1 \\ Q_2 & 0 \end{pmatrix} \\
&= \begin{pmatrix} 0 & 1/P_2 \\ 1/P_1 & 0 \end{pmatrix} \cdot \begin{pmatrix} 0 & Q_1 \\ Q_2 & 0 \end{pmatrix} \\
&= \begin{pmatrix} Q_2/P_2 & 0 \\ 0 & Q_1/P_1 \end{pmatrix}
\end{aligned}
\tag{4.93}
$$

such that the symmetry $J_3 = 0$ is the necessary condition for the interpretation of $\vec{\Phi}$ by a regional 2D structure. According to (4.82), this condition is equivalent to $\beta = 0$. Consider the general structure of $\vec{\Phi}$ that is compatible with $J_3 = 0$. The assumption of the normal case $\det(\vec{\Phi}) = \det(\vec{\Phi}_R) = (Q_1/P_1)(Q_2/P_2) > 0$ requires that the principal phases lie in the first and/or third quadrant such that $Q_k/P_k > 0$ for $k = 1, 2$. Then, (4.76)–(4.78) yield

$$
\begin{aligned}
J_0 = J_1 &= \frac{Q_1/P_1 + Q_2/P_2}{2} \\
J_2 &= \frac{|Q_1/P_1 - Q_2/P_2|}{2}
\end{aligned}
\tag{4.94}
$$

and therefore

$$J_0 + J_2 = \max(Q_1/P_1, Q_2/P_2) = \tan\phi_{max}$$
$$J_0 - J_2 = \min(Q_1/P_1, Q_2/P_2) = \tan\phi_{min} \tag{4.95}$$

The representation (4.81) of Caldwell *et al.* (2004) reduces to

$$\vec{\Phi} = \vec{R}(-\alpha) \cdot \begin{pmatrix} \tan\phi_{max} & 0 \\ 0 & \tan\phi_{min} \end{pmatrix} \cdot \vec{R}(\alpha) \tag{4.96}$$

which is the principal axis representation of a symmetric tensor. The corresponding ellipse has the semi-axes $\tan\phi_{max}$ and $\tan\phi_{min}$. Since $\beta = 0$, the azimuth α of the major axis agrees with the azimuth of the magnetic field that generates the greater principal phase. This azimuth is either parallel or orthogonal to the strike. According to (4.82) on using (4.73), the angle α satisfies

$$\tan 2\alpha = \frac{\Phi_{xy} + \Phi_{yx}}{\Phi_{xx} - \Phi_{yy}} = \frac{2\Phi_{xy}}{\Phi_{xx} - \Phi_{yy}}$$
$$= \frac{2\mathrm{Im}(Z_{xy}Z_{yy}^*)}{\mathrm{Im}(Z_{xx}Z_{yy}^* + Z_{xy}Z_{yx}^*)} \tag{4.97}$$

and agrees with Bahr's undistorted estimate α_s given in (4.67). The direction α of the major axis is uniquely determined (apart from a trivial uncertainty of π) since both $\sin 2\alpha$ and $\cos 2\alpha$ are known from (4.83), whereas the azimuth α_s determined from (4.67) has a $\pi/2$ ambiguity. Nevertheless, the $\pi/2$ ambiguity in the identification of the strike direction remains unresolved in both approaches.

In the 2D case, three undistorted parameters of $\vec{Z}_R$ can be estimated uniquely, namely the three parameters occurring in the representation (4.96) of $\vec{\Phi}$: the phases ϕ_{max} and ϕ_{min} of the principal responses as rotational invariants and the azimuth α of the magnetic field that gives rise to ϕ_{max}. The strike direction itself retains a $\pi/2$ ambiguity.

For an arbitrary value of J_3, from (4.48), (4.73) and (4.77) it can be inferred that

$$2J_3 = \Phi_{xy} - \Phi_{yx}$$
$$= \frac{\mathrm{Im}\left(Z_{xx}Z_{yx}^* - Z_{yy}Z_{xy}^*\right)}{\det(\vec{P})}$$
$$= \frac{C}{\det(\vec{P})} \tag{4.98}$$

Therefore, $J_3 = 0$ is equivalent to $C = 0$, which is just the 2D compatibility condition (4.66). An undistorted estimate of the skew, a measure for the deviation from 2D, should be a rotational invariant derived from $\vec{\Phi}$. Such a measure $\bar{\eta}$ is defined in (4.71). According to (4.98), $\bar{\eta}$ is related to $\vec{\Phi}$ by $\bar{\eta}^2 = 2|\Phi_{xy} - \Phi_{yx}| = 4|J_3|$.

3. The regional structure is 3D ⇒ all three invariants contribute. The phase tensor admits the determination of four undistorted parameters, namely α and three rotational invariants. According to the representations (4.80) or (4.81), the latter are either J_1, J_2 and J_3, or

J_0, J_2 and β. Unfortunately, none of these four undistorted parameters (or a combination of them) can be assigned a clear geophysical meaning. However, it can qualitatively be inferred from "high" values of $|\beta|$ or $|J_3|$ that the regional conductivity distribution is essentially 3D, although threshold values have to be determined experimentally. The 3D phase tensor is displayed graphically by an ellipse with semi-minor axis $J_0 - J_2$ and semi-major axis $J_0 + J_2$, where the latter has the azimuth $\alpha - \beta$. The fourth parameter, the 3D indicator β, may also be plotted (see e.g. Caldwell *et al.*, 2004, p. 463).

The phase tensor does not eliminate the task of attacking the (strictly unsolvable) problem of distortion removal prior to inversion by introducing additional assumptions. The strength of a phase-tensor analysis lies in identifying and displaying those parts of the response tensor that are unaffected by local distortion, and in the ability to impose additional assumptions in a more controlled manner.

4.2 Properties of the magnetotelluric response function in one dimension

4.2.1 Definitions

In Sections 4.2 and 4.3, the analytical properties of the magnetotelluric response function for 1D and 2D conductivity distributions will be discussed in detail. Particular emphasis will be placed on its frequency dependence. For conciseness of presentation, only the results are presented, and the reader will have to address the specified references for details.

Assume a 1D conductivity structure with the electric field given by $\mathbf{E} = E_x \hat{\mathbf{x}}$ and the magnetic field specified by $\mathbf{B} = B_y \hat{\mathbf{y}}$ with E_x and B_y interrelated through Faraday's law (2.3) and Ampere's law (2.16) in σ, so that

$$\partial_z E_x(z, \omega) = -i\omega B_y(z, \omega)$$
$$-\partial_z B_x(z, \omega) = \mu_0 \sigma(z) E_x(z, \omega) \tag{4.99}$$

Therefore, $E_x(z)$ is the solution of

$$\partial_z^2 E_x(z) = i\omega \mu_0 \sigma(z) E_x(z) \tag{4.100}$$

with the boundary condition $\partial_z E_x(z) \to 0$ for $z \to \infty$. A boundary condition on $\partial_z E_x$ rather than on E_x has to be imposed because the solution E_x of (4.100) tends to a finite non-zero value for $z \to \infty$ if $\sigma(z)$ decays faster than $1/z^2$.

The source is taken into account by assigning an arbitrary non-zero value B_0 to $B_y(0, \omega)$ that drops out when considering 1D response functions. Candidate parameters of interest include the 1D response function

$$Z_{1D}(\omega) = \frac{E_x(0, \omega)}{B_y(0, \omega)} \tag{4.101}$$

and the Schmucker (1987) *c*-response (see also (2.69))

$$c(\omega) = \tilde{\Lambda}_0(\omega) = \frac{Z_{1D}(\omega)}{i\omega} = -\frac{E_x(0, \omega)}{\partial_z E_x(0, \omega)} \tag{4.102}$$

Response functions are usually converted and displayed in terms of *apparent resistivity*. This quantity is generally defined so that the apparent resistivity agrees with the true resistivity when the Earth is a uniform half-space. According to (2.69) with (2.52), the response of a uniform half-space of resistivity ρ is

$$Z = \sqrt{\frac{i\omega\rho}{\mu_0}} \qquad (4.103)$$

Using (4.102), the apparent resistivity ρ_a is related to c and Z_{1D} by

$$\rho_a(\omega) = \omega\mu_0\,|c(\omega)|^2 = \mu_0\,|Z_{1D}(\omega)|^2\,/\omega \qquad (4.104)$$

One advantage of $c(\omega)$ over $Z_{1D}(\omega)$ is that $\mathrm{Re}[c(\omega)]$ or $|c(\omega)|$ has the dimension of length, and gives the approximate depth of investigation for a given frequency (see Section 4.2.5).

4.2.2 Analytical properties in the complex frequency plane

The frequency dependence of $c(\omega)$ is defined by the following theorem (Weidelt, 1972, 2005; Parker, 1980; Yee & Paulson, 1988a):

The Schmucker response $c(\omega)$ is an analytic function in the whole complex ω-plane, except on the positive imaginary axis $\omega = i\lambda$ for $\lambda \geq 0$, where it has poles and branch cuts (Figure 4.3). Outside the positive imaginary axis, $c(\omega)$ admits the spectral representation

$$c(\omega) = a_0 + \int_0^\infty \frac{a(\lambda)d\lambda}{\lambda + i\omega} \qquad (4.105)$$

where $a_0 \geq 0$ and $a(\lambda) \geq 0$, and a_0 is the depth to the shallowest conductor.

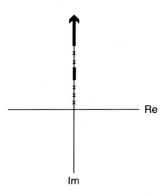

Figure 4.3. Complex plane representation of $c(\omega)$ that is analytic everywhere except on the positive imaginary axis, where it displays alternating sets of poles ($\times$) and branch cuts (heavy lines), with the last branch cut extending to $i\infty$. There may be more than the two sets of poles and branch cuts that are shown, which are intended to be representative. Data collected at a discrete set of frequencies will plot on the real axis.

The integral (4.105) has to be considered in a generalized sense (as a Riemann–Stieltjes integral) to include both the continuous part (branch cuts) and the discontinuous part (poles). The points λ with $a(\lambda) > 0$ correspond to the decay constants of current systems in the conductor. However, this aspect will not be pursued any further at this point; see Parker (1980) and Weidelt (2005) for details. The existence of the integral requires the following limits:

$$
\begin{array}{lll}
\lambda \to 0: & a(\lambda) = O(\lambda^{-\varepsilon_1}) & \text{for } \varepsilon_1 < 1 \\
\lambda \to \infty: & a(\lambda) = O(\lambda^{-\varepsilon_2}) & \text{for } \varepsilon_2 > 0
\end{array}
\tag{4.106}
$$

In particular, the power law $a(\lambda) \sim \lambda^{-\varepsilon}$ for $0 < \varepsilon < 1$ is associated with $c(\omega) \sim (i\omega)^{-\varepsilon}$ and $\sigma(z) \sim z^{-2+1/\varepsilon}$ (Weidelt, 2005). The function $a(\lambda)$ is related to $c(\omega)$ on the positive imaginary semi-axis by (e.g. Yee & Paulson, 1988a)

$$
a(\lambda) = -\frac{1}{\pi} \lim_{\varepsilon \to 0^+} \text{Im}[c(i\lambda + \varepsilon)]
\tag{4.107}
$$

Two simple examples of the representation of $c(\omega)$ by (4.105) are as follows.

(a) Uniform half-space with $\sigma(z) = \sigma_0$:

$$
c(\omega) = \frac{1}{\sqrt{i\omega\mu_0\sigma_0}} \quad \Rightarrow \quad a_0 = 0, \; a(\lambda) = \frac{1}{\pi\sqrt{\lambda\mu_0\sigma_0}}
\tag{4.108}
$$

The singularity of $c(\omega)$ is a branch cut from $\omega = 0$ to $\omega = i\infty$.

(b) Thin sheet of conductance τ at depth $z = d_1$ and a perfect conductor at $z = d_1 + d_2$:

$$
c(\omega) = d_1 + \frac{d_2}{1 + i\omega\mu_0\tau d_2} \quad \Rightarrow \quad a_0 = d_1, \; a(\lambda) = \frac{1}{\mu_0\tau}\delta\left(\lambda - \frac{1}{\mu_0\tau d_2}\right)
\tag{4.109}
$$

The singularity of $c(\omega)$ is a pole at $\omega = i/(\mu_0\tau d_2)$.

From (4.106) with (4.102), it immediately follows that

$$
Z_{1D} = |Z_{1D}|e^{i\phi} = i\omega c = i\omega|c|e^{i(\phi - \pi/2)} = i\omega a_0 + \int_0^\infty \frac{\omega(\omega + i\lambda)a(\lambda)}{\lambda^2 + \omega^2}\,d\lambda
\tag{4.110}
$$

Assuming that $\omega > 0$, Z_{1D} lies in the first quadrant with $0 \le \phi \le \pi/2$ and c lies in the fourth quadrant with phase $\phi - \pi/2$.

4.2.3 Existence conditions for a set of discrete frequencies

For a set of $M \ge 1$ Schmucker responses $c(\omega_k) = c_k$ measured at the discrete frequencies $\omega_k > 0$ for $k = 1,\ldots,M$, it can be deduced whether the data allow the integral representation (4.105) that is the necessary condition for the existence of a 1D conductivity model $\sigma(z)$ simply by checking $2M$ signs. This problem was first addressed by Weidelt (1986) and later more satisfactorily examined by Yee & Paulson (1988b).

Let

$$d_{mn} = \frac{i(c_m - c_n^*)}{\omega_m + \omega_n}$$

$$\bar{d}_{mn} = \frac{\omega_m c_m + \omega_n c_n^*}{\omega_m + \omega_n}$$

(4.111)

for $m, n = 1, \ldots, M$, so that the conditions $d_{mn}^* = d_{nm}$, $\bar{d}_{mn}^* = \bar{d}_{nm}$ are satisfied, and form the $2M$ real-valued Hermitian determinants

$$D_k = \begin{vmatrix} d_{11} & d_{12} & \cdots & d_{1k} \\ d_{21} & d_{22} & \cdots & d_{2k} \\ \cdots & \cdots & \cdots & \cdots \\ d_{k1} & d_{k2} & \cdots & d_{kk} \end{vmatrix}$$

$$\bar{D}_k = \begin{vmatrix} \bar{d}_{11} & \bar{d}_{12} & \cdots & \bar{d}_{1k} \\ \bar{d}_{21} & \bar{d}_{22} & \cdots & \bar{d}_{2k} \\ \cdots & \cdots & \cdots & \cdots \\ \bar{d}_{k1} & \bar{d}_{k2} & \cdots & \bar{d}_{kk} \end{vmatrix}$$

(4.112)

for $k = 1, \ldots, M$. One must distinguish between the regular case, where a finite dataset is compatible with an infinite number of models, and the singular case, where the dataset admits only one model. The necessary and sufficient conditions that the M-frequency dataset admits a 1D interpretation are

regular case:

$$D_k > 0$$
$$\bar{D}_k > 0 \quad k = 1, \ldots, M$$

(4.113)

singular case:

$$\begin{aligned} D_k &> 0 \\ \bar{D}_k &> 0 \qquad k = 1, \ldots, m < M \\ D_{m+1} \bar{D}_{m+1} &= 0 \\ D_k &= 0 \\ \bar{D}_k &= 0 \qquad k = m + 2, \ldots, M \end{aligned}$$

(4.114)

where in the singular case the first set of conditions is ignored for $m = 0$ and the last set for $m = M - 1$. Necessity and sufficiency of the conditions means, respectively, that the violation of only one constraint excludes the existence of a 1D model and that satisfaction of all $2M$ constraints grants the existence of at least one 1D model.

The regular case will be considered first. Starting with $k = 1$, the inequalities $D_1 > 0$ and $\bar{D}_1 > 0$ give the necessary conditions for c_1 to be 1D interpretable. After assigning a feasible value to c_1, the inequalities $D_2 > 0$ and $\bar{D}_2 > 0$ define the compatibility region for c_2 such that both c_1 and c_2 are 1D interpretable. If a feasible value for c_2 is fixed, the inequalities $D_3 > 0$ and $\bar{D}_3 > 0$ give the conditions under which c_3 is compatible with the assigned values c_1 and c_2, and so on. Yee & Paulson (1988b) show that, for $k \geq 2$, both $D_k > 0$ and $\bar{D}_k > 0$ define the interior of a circle in the complex c_k plane (Figure 4.4). A feasible value of c_k that is compatible with c_j for $j = 1, \ldots, k - 1$ lies inside the lens-shaped intersection of these circles.

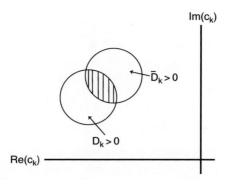

Figure 4.4. Compatibility condition for a 1D model for $k \geq 2$ as described in the text. The determinants D_k and $\bar{D}_k$ each define the interiors of circles in the complex c_k plane when they are strictly positive. For the regular case, an admissible value of c_k that is compatible with c_j for $j = 1, ..., k-1$ lies within the intersection of the two circles. For the singular case, c_k lies on the boundary of the intersection, hence the compatibility region shrinks to a point for subsequent frequencies.

With increasing k, the radii of the circles decrease such that the compatibility region shrinks, and little freedom is left for a feasible assignment of c_k. The singular case arises if c_k lies on the boundary of the lens in the c_k plane, so that the compatibility regions for subsequent frequencies shrink to a point.

Consider the cases $k = 1$ and $k = 2$ in more detail. With $c_k = g_k - ih_k$, after some rearrangement, (4.111)–(4.113) yield

$k = 1$:

$$h_1 > 0 \tag{4.115}$$
$$g_1 > 0$$

$k = 2$:

$$\left| \frac{c_2 - c_1^*}{\omega_2 - \omega_1} \right|^2 < \frac{h_1 h_2}{\omega_1 \omega_2} \tag{4.116}$$
$$\left| \frac{\omega_2 c_2 - \omega_1 c_1^*}{\omega_2 - \omega_1} \right|^2 < g_1 g_2$$

From (4.115), it can be inferred that c_1 has to lie inside the fourth quadrant of the complex c_1 plane (see also (4.110)), and from (4.116) it can be deduced (after fixing c_1) that c_2 has to lie in the c_2 plane inside the lens-shaped intersection of the two circles,

$$(g_2 - g_1)^2 + \left(h_2 - \frac{\omega_1^2 + \omega_2^2}{2\omega_1 \omega_2} h_1 \right)^2 = \left(\frac{\omega_1^2 - \omega_2^2}{2\omega_1 \omega_2} \right)^2 h_1^2 \tag{4.117}$$

and

$$\left(g_2 - \frac{\omega_1^2 + \omega_2^2}{2\omega_2^2} g_1 \right)^2 + \left(h_2 - \frac{\omega_1}{\omega_2} h_1 \right)^2 = \left(\frac{\omega_1^2 - \omega_2^2}{2\omega_2^2} \right)^2 g_1^2 \tag{4.118}$$

The two points of intersection A and B are

A:

$$g_2 = g_1$$
$$h_2 = (\omega_1/\omega_2)h_1$$

(4.119)

B:

$$g_2/|c_2|^2 = g_1/|c_1|^2$$
$$h_2/|c_2|^2 = (\omega_2/\omega_1)h_1/|c_1|^2$$

(4.120)

Depending on the actual position of c_2 either inside the lens, on the boundary of the lens or outside of it, there exist an infinite number of models, exactly one model or no model compatible with c_1 and c_2. The singular case occurs if c_2 is assigned to a boundary point. The resulting unique conductivity model is a thin-sheet model: a two-parameter model at A and B and a three-parameter model at boundary points outside the vertices. In particular, at vertex A the model is a thin sheet of conductance τ at $z = d_1$ and at vertex B the model consists of a thin sheet of conductance τ at $z = 0$ and a perfect conductor at $z = d_2$. The resulting response functions are (see also (4.109))

$$c_A(\omega) = d_1 + \frac{1}{i\omega\mu_0\,\tau}$$
$$\frac{1}{c_B(\omega)} = \frac{1}{d_2} + i\omega\mu_0\,\tau$$

(4.121)

No freedom is left for the subsequent response functions c_k for $k \geq 3$, where the compatibility regions shrink to points. More generally, if c_{m+1} is assigned to one of the vertices of the lens-shaped compatibility region in the c_{m+1} plane, the corresponding unique conductivity models are thin-sheet models with $2m$ parameters. The models are compatible with c_1 to c_m, and either maximize the depth to the first conductor and at the same time minimize the total conductance (vertex A) or minimize the depth to a perfect conductor and at the same time maximize the conductance of the surface sheet (vertex B).

The content of this section is mainly theoretical: the intimate relationship between 1D responses for M different frequencies may be examined for 1D interpretability without modeling on the basis of $2M$ necessary and sufficient data constraints. In practice, data errors limit the applicability to real data. In particular, it cannot be decided whether the conditions (4.114) are satisfied for a singular dataset.

4.2.4 Dispersion relations

The fact that $c(\omega)$ is an analytic function in the lower ω half-plane implies that the real and imaginary parts of $c(\omega) = g(\omega) - ih(\omega)$ are dependent functions related by dispersion relations (e.g. Toll, 1956), and also implies causality of the underlying physical system. From (4.105), it follows that, for real ω, the symmetry relation $c(-\omega) = c^*(\omega)$ holds, and therefore $g(-\omega) = g(\omega)$, $h(-\omega) = -h(\omega)$. This leads to the two dispersion relations (e.g. Mathews & Walker, 1970, p. 129)

$$g(\omega) = \frac{1}{\pi} \int\limits_{-\infty}^{\infty} \frac{h(x)dx}{x-\omega} = \frac{2}{\pi} \int\limits_{0}^{\infty} \frac{xh(x)dx}{x^2-\omega^2} \tag{4.122}$$

$$h(\omega) = -\frac{1}{\pi} \int\limits_{-\infty}^{\infty} \frac{g(x)dx}{x-\omega} = -\frac{2\omega}{\pi} \int\limits_{0}^{\infty} \frac{g(x)dx}{x^2-\omega^2} \tag{4.123}$$

where Cauchy principal values are taken in the integrals (e.g. Mathews & Walker, 1970, p. 480). More relevant for 1D magnetotellurics is that $\log c(\omega)$ is also analytic in the lower ω half-plane since $c(\omega)$ is free of zeros in that region. The latter follows from (4.105) with $\omega = \omega_r - ip$, where ω_r is real and $p \geq 0$, so that

$$\mathrm{Re}[c(\omega_r - ip)] = a_0 + \int\limits_{0}^{\infty} \frac{(\lambda+p)a(\lambda)d\lambda}{(\lambda+p)^2 + \omega_r^2} > 0 \tag{4.124}$$

The analyticity of $\log c(\omega)$ for $\mathrm{Im}(\omega) \leq 0$ allows the formulation of dispersion relations between $\mathrm{Re}(\log c) = \log|c|$ and $\mathrm{Im}(\log c) = \phi - \pi/2$, or between $\log \rho_a$ and ϕ. Transfer functions that, in one ω half-plane, are analytic, free of zeros and vanish at infinity are called *minimum phase*.

For ease of notation, attention is confined to $\omega < 0$; for $\omega > 0$ all phases change sign. From (4.102)–(4.104), it follows that

$$c = \sqrt{\frac{\rho_a}{\omega\mu_0}} e^{i(\phi-\pi/2)} \tag{4.125}$$

Assuming that $c(\omega)$ tends to the response $c_0 = \sqrt{\rho_0/(\omega\mu_0)} e^{-i\pi/4}$ of a uniform half-space of resistivity $\rho_0 = 1/\sigma(z=0)$ for $\omega \to \infty$, c is normalized with c_0 and the logarithm is taken, yielding

$$c/c_0 = \sqrt{\rho_a/\rho_0}\, e^{i(\phi-\pi/4)}$$
$$\log(c/c_0) = \frac{1}{2}\log(\rho_a/\rho_0) + i(\phi - \pi/4) \tag{4.126}$$

The function $\log(c/c_0)$ is analytic for $\mathrm{Im}(\omega) \leq 0$ and vanishes for $\omega \to \infty$. This is sufficient for the existence of dispersion relations between the amplitude and phase of c/c_0. With the replacements $g \to \log(\rho_a/\rho_0)/2$ and $h \to \pi/4 - \phi$, the second version of (4.122) and (4.123) immediately produces

$$\log[\rho_a(\omega)/\rho_0] = \frac{4}{\pi} \int\limits_{0}^{\infty} \frac{x[\pi/4 - \phi(x)]}{x^2-\omega^2} dx \tag{4.127}$$

$$\phi(\omega) = \frac{\pi}{4} + \frac{\omega}{\pi} \int\limits_{0}^{\infty} \frac{\log[\rho_a(x)/\rho_0]}{x^2-\omega^2} dx \tag{4.128}$$

where Cauchy principal values are again taken in the integrals. In principle, an arbitrary positive value can be assigned to the resistivity ρ_0: when the apparent resistivity is computed from the phase via (4.127), ρ_0 fixes the non-determinable high-frequency limit of $\rho_a(\omega)$. On the other hand, in view of the identity

$$\int_0^\infty \frac{dx}{x^2 - \omega^2} = 0 \tag{4.129}$$

the computation of the phase from the apparent resistivity via (4.128) is unaffected by the choice of ρ_0. However, to accelerate the convergence of the integral when using numerical data, the best choice of ρ_0 is the known high-frequency limit of $\rho_a(\omega)$.

The approximate version of (4.128) given by Weidelt (1972) and Weaver (1994, p. 106)

$$\phi(\omega) \approx \frac{\pi}{4} \left[1 + \frac{d\log \rho_a(\omega)}{d\log \omega} \right]$$
$$\phi(T) \approx \frac{\pi}{4} \left[1 - \frac{d\log \rho_a(T)}{d\log T} \right] \tag{4.130}$$

is useful for visually checking the consistency of experimental data when $\log \rho_a$ is plotted against $\log T$, where $T = 2\pi/\omega$ is the period. Phases smaller than $\pi/4$ occur where $\rho_a(T)$ is ascending, whereas phases greater than $\pi/4$ coincide with a descending branch. Since

$$|\partial_T \log \rho_a(T)| \leq 1 \tag{4.131}$$

the approximate phase lies in the correct range $0 \leq \phi(T) \leq \pi/2$; see (4.110), (4.148) and (4.149).

4.2.5 Simple approximate mappings of the true resistivity

Several simple mapping methods are available that allow the immediate estimation of a first guess for the underlying conductivity profile from $c(\omega)$, or from $\rho_a(\omega)$ and $\phi(\omega)$. In contrast to formal inversion methods, only a single frequency is processed at a time. The result is a parametric representation of an approximate depth profile with frequency or period as the independent parameter.

Schmucker's $\rho^*(z^*)$ method will first be considered. The physical interpretation of this method is based on the identification of $c(\omega)$ as the normalized first moment of the induced electric currents. With the current density $J_x(z, \omega) = \sigma(z)E_x(z, \omega)$, (4.99) yields

$$E_x(0, \omega) = -i\omega\mu_0 \int_0^\infty z J_x(z, \omega)dz$$
$$\partial_z E_x(0, \omega) = i\omega\mu_0 \int_0^\infty J_x(z, \omega)dz \tag{4.132}$$

where the first equation has been obtained after integrating by parts. Therefore, (4.102) gives

$$c(\omega) = \frac{\int\limits_0^\infty z J_x(z, \omega)dz}{\int\limits_0^\infty J_x(z, \omega)dz} \tag{4.133}$$

Integration of (4.99) shows that the denominator of (4.133) is in phase with $B_y(0)$, so that it is real, hence

$$g(\omega) = \frac{\int\limits_0^\infty z \operatorname{Re}[J_x(z, \omega)]dz}{\int\limits_0^\infty J_x(z, \omega)dz} \tag{4.134}$$

is the representative depth for the induced in-phase current. Owing to the skin effect, this depth decreases as the frequency increases. This is corroborated by

$$g(\omega) = a_0 + \int\limits_0^\infty \frac{\lambda a(\lambda)d\lambda}{\lambda^2 + \omega^2} \tag{4.135}$$

which follows from (4.105). To a given $c(\omega)$, Schmucker (1987, p. 348) assigns the depth $z^*(\omega) = g(\omega) = \operatorname{Re}[c(\omega)]$ and estimates the resistivity ρ^* at depth z^* in two different ways, depending on the phase $\phi(\omega)$.

(a) $0 \leq \phi \leq \pi/4$: According to (4.130), for this range of phase, $c(\omega)$ senses an increase of resistivity with depth. This can be taken into account by a two-parameter model with a thin surface sheet of conductance τ over a uniform half-space of resistivity ρ^*, yielding

$$1/c = i\omega\mu_0 \tau + \sqrt{i\omega\mu_0/\rho*} \tag{4.136}$$

Applying (4.125) gives

$$\sqrt{\omega\mu_0/\rho_a} \sin\phi = \operatorname{Re}(1/c) = \sqrt{\omega\mu_0/(2\rho*)} \tag{4.137}$$

(b) $\pi/4 \leq \phi \leq \pi/2$: In this instance, $c(\omega)$ senses a decrease of resistivity with depth, and the appropriate two-parameter model is an insulator of thickness d over a uniform half-space of resistivity ρ^*. This yields

$$c = d + \sqrt{\rho* /(i\omega\mu_0)} \tag{4.138}$$

and on using (4.125)

$$-\sqrt{\rho_a/(i\omega\mu_0)} = \operatorname{Im}(c) = -\sqrt{\rho * /(2\omega\mu_0)} \tag{4.139}$$

Solving (4.137) and (4.139) for ρ^*, the $\rho^*(z^*)$ method yields

$$z^* = \sqrt{\rho_a/(\omega\mu_0)} \sin\phi \tag{4.140}$$

$$\rho^* = \begin{cases} \rho_a/(2\sin^2\phi) & 0 \le \phi \le \pi/4 \\ 2\rho_a \cos^2\phi & \pi/4 \le \phi \le \pi/2 \end{cases} \tag{4.141}$$

The second parameters of the models

$$\begin{aligned} \tau &= \sqrt{2/(\omega\mu_0\rho_a)} \sin(\pi/4 - \phi) \\ d &= \sqrt{2\rho_a/(\omega\mu_0)} \sin(\phi - \pi/4) \end{aligned} \tag{4.142}$$

are meaningful only in the appropriate phase range.

The $\rho^*(z^*)$ method performs well if the conductivity increases with depth ($\phi > \pi/4$). On the other hand, in the case of a thick poor conductor below a good conductor ($\phi < \pi/4$), the resistivity starts to increase significantly within the conductor well before reaching the resistor, meaning that $z^*(\omega)$ increases too slowly with decreasing frequency. In particular, in the extreme case of a uniform layer of thickness d over an insulator, $\rho^* = \infty$ is assigned to the depth $z^*_{max} = z^*(\omega = 0) = d/3$.

A second classical mapping method is the Niblett–Bostick transform (Niblett & Sayn-Wittgenstein, 1960; Bostick, 1977), which is traditionally formulated in terms of the period T. As data it uses $\rho_a(T)$ and (instead of the phase) the slope $m(T)$ of $\rho_a(T)$ in a double-logarithmic plot

$$m(T) = \partial_{\log T} \log \rho_a(T) \tag{4.143}$$

where $-1 \le m(T) \le 1$; see also (4.130) and (4.131). The bounds of $m(T)$ are attained in the following two extremal models.

(a) Insulator of thickness D over a perfect conductor:

$$\begin{aligned} c &= D, & \rho_a &= \frac{2\pi\mu_0 D^2}{T} \\ m &= -1, & D^2 &= \frac{T\rho_a}{2\pi\mu_0} \end{aligned} \tag{4.144}$$

(b) Thin surface sheet of conductance τ over an insulator:

$$\begin{aligned} c &= \frac{T}{i2\pi\mu_0 \tau}, & \rho_a &= \frac{T}{2\pi\mu_0 \tau^2} \\ m &= 1, & \tau^2 &= \frac{T}{2\pi\mu_0\rho_a} \end{aligned} \tag{4.145}$$

To a measured pair (T, ρ_a) with $|m| < 1$, one can formally assign a pair (D, τ) by the last identities in (4.144) and (4.145). In the Niblett–Bostick transform, the depth $D(T) = |c(T)|$

is considered to be the depth of penetration at the period T and $\tau(T)$ is interpreted to be the integrated conductivity up to this depth. This concept is supported by the fact that both τ and D increase with T. Hence, the assumption is that

$$\tau(T) = \int_0^{D(T)} \frac{dz}{\rho_b(z)} \qquad \text{or} \qquad \partial_T \tau = \frac{1}{\rho_b(D)} \partial_T D \tag{4.146}$$

where $\rho_b(z)$ is an approximation to $\rho_a(z)$. Then,

$$\rho_b(D) = \partial_\tau D = \frac{D}{\tau} \left(\frac{2dD}{D} \right) \Big/ \left(\frac{2dt}{\tau} \right) = \rho_a \partial_{\log \tau^2} \log D^2$$

$$= \rho_a \frac{d \log T + d \log \rho_a}{d \log T - d \log \rho_a} = \rho_a \frac{1+m}{1-m} \tag{4.147}$$

where the fact that (4.144) and (4.145) imply $D/\tau = \rho_a$ has been utilized. Therefore, the Niblett–Bostick transform $\rho_b(D)$ reads

$$D = \sqrt{\frac{T\rho_a}{2\pi\mu_0}}$$

$$\rho_b = \rho_a \frac{1+m}{1-m} \tag{4.148}$$

The variation of T then yields a parametric representation of $\rho_b(D)$. The sounding curve ρ_a that inherently smooths true resistivity contrasts is roughened by forming ρ_b.

The transform $\rho_b(D)$ performs well for an Earth displaying increasing conductivity. It alleviates the problem encountered when using $\rho^*(z^*)$ for a thick resistor below a conductor, but tends generally to stronger overshooting. Comparisons of the two methods are given by Schmucker (1987).

4.2.6 Interpretation of a(λ) in terms of thin sheets: the D+ model

The inverse problem in 1D magnetotellurics can be approached from two different directions. The standard way is solution of the nonlinear inverse problem by fitting the free parameters of an *a priori* 1D conductivity structure (e.g. a set of layer parameters) using least-squares principles starting from an initial guess. The nonlinear problem is solved iteratively using an algorithm such as the Levenberg–Marquardt method. Chapter 8 discusses this approach in detail.

The second approach is initial extraction of that part of the data that can be represented by the non-negative generalized function $a(\lambda)$ in the fundamental integral representation (4.105) by solving a linear inverse problem with positivity constraints, and subsequent interpretation of $a(\lambda)$ in terms of the conductivity structure. By fully extracting the interpretable part of the data in the first step, the conductivity model gives the best possible fit to the data. A stable algorithm (the D+ model) was presented by Parker (1980) and Parker & Whaler (1981) (Figure 4.5). This method will be discussed beginning with the relationship

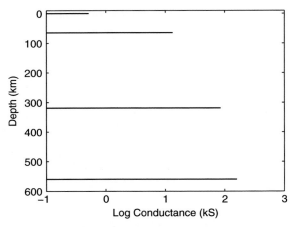

Figure 4.5. Best-fitting (D+) model consisting of thin sheet conductors at four depths obtained from estimates of $c(\omega)$ at seven periods ranging from 3.7 h to 48 h on the Hawaiian Islands as given in Larsen (1975). The normalized chi-squared misfit statistic is 7.412. Any smoother model than the optimal one will have a higher misfit statistic. Model results taken from Parker & Whaler (1981, their table 3).

between $a(\lambda)$ and $\sigma(z)$. For simplicity, it will be assumed that $a_0 = 0$ in (4.109), so that all conductivity models must have a conductor at $z = 0$.

In general, $a(\lambda)$ consists of a continuous part $a_c(\lambda)$ resulting from branch cuts and a discrete part resulting from N poles at $\lambda = \lambda_n \geq 0$ with residual $a_n > 0$,

$$a(\lambda) = a_c(\lambda) + \sum_{n=1}^{N} a_n \delta(\lambda - \lambda_n) \tag{4.149}$$

Let $f(\lambda)$ and $g(\lambda)$ be two real-valued square-integrable functions in the interval $0 \leq \lambda \leq \infty$ with $a(\lambda)$ as weight function, and let

$$
\begin{aligned}
(f, g) &= \int_0^\infty a(\lambda) f(\lambda) g(\lambda) d\lambda \\
&= \int_0^\infty a_c(\lambda) f(\lambda) g(\lambda) d\lambda + \sum_{n=1}^{N} a_n f(\lambda_n) g(\lambda_n)
\end{aligned}
\tag{4.150}
$$

Moreover, let m_k denote the kth non-central moment of $a(\lambda)$,

$$m_k = \int_0^\infty a(\lambda) \lambda^k d\lambda, \qquad k = 0, 1, 2, \dots \tag{4.151}$$

which is $(1, \lambda^k)$ using the notation in (4.150). The classification of the pertinent conductivity distributions depends on the existence (meaning finiteness) of the moments m_k. There are three cases (Weidelt, 2005).

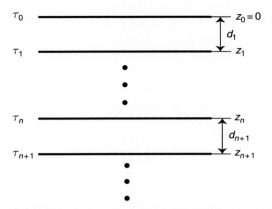

Figure 4.6. Infinite thin sheet model with conductances τ_n and inter-sheet separations d_n.

(a) All moments exist: $\sigma(z)$ is a finite or infinite sequence of thin sheets. An example is (4.109) and is shown in Figure 4.6.
(b) A few moments exist: $\sigma(z)$ is a continuous conductor below a set of thin sheets.
(c) No moment exists: $\sigma(z)$ is a continuous conductor, possibly with a few embedded thin sheets. An example is (4.108).

In the sequel, attention is concentrated on case (a) that, for example, applies if $a(\lambda) = 0$ for $\lambda > b$ (with $b < \infty$), or if $a(\lambda)$ decays very rapidly as $\lambda \to \infty$. The first condition is satisfied for all numerical determinations of $a(\lambda)$ that define its discrete part with $\max(\lambda_n) < \infty$; see the discussion at the end of this section.

The existence of all moments allows the construction of a set of n-degree polynomials $p_n(\lambda)$, for $n = 0, 1, 2, \ldots$, that are mutually orthogonal with respect to the weight function $a(\lambda)$, so that

$$(p_m, p_n) = h_n \delta_{mn} \tag{4.152}$$

where δ_{mn} is the Kronecker symbol and $h_n = (p_n, p_n)$ is the L_2-norm of $p_n(\lambda)$ in the sense of (4.150). The polynomials are uniquely determined up to an amplitude factor that is a free parameter. Let this factor be the coefficient $k_n \neq 0$ of the leading power λ^n. Then, $a(\lambda)$ can be interpreted in terms of a stack of thin sheets with conductances τ_n for $n = 0, 1, 2, \ldots$ and inter-sheet separations d_n for $n = 1, 2, 3, \ldots$ (Figure 4.6) that can be expressed in terms of the orthogonal polynomials evaluated at $\lambda = 0$,

$$\mu_0 \tau_n = \frac{p_n^2(0)}{h_n}$$

$$d_{n+1} = -\frac{k_{n+1} h_n}{k_n p_n(0) p_{n+1}(0)} \tag{4.153}$$

for $n \geq 0$ (Weidelt, 2005). The number of thin sheets is finite only if $a_c = 0$ and $N < \infty$ in (4.149). So far, no standardization (e.g. $h_n = 1$ or $k_n = 1$) has been applied, and the two dependent parameters h_n and k_n have been retained for flexibility.

Before turning to the algorithm for the recursive computation of τ_n and d_n from $a(\lambda)$ via (4.153), an example of the construction of the response function $c(\omega)$ and the sequence of thin sheets for a given set of orthogonal polynomials $p_n(\lambda)$ will be presented. Select the Chebyshev polynomials $T_n(x)$ shifted from the standard interval $-1 < x < 1$ to the interval $0 \leq e < \lambda < f < \infty$,

$$p_n(\lambda) = T_n\left(\frac{2\lambda - e - f}{f - e}\right) \tag{4.154}$$

with the weight function

$$a(\lambda) = a_c(\lambda) = \frac{m_0}{\pi\sqrt{(\lambda - e)(f - \lambda)}}, \qquad e < \lambda < f \tag{4.155}$$

and $a(\lambda) = 0$ elsewhere. The zero-order moment $m_0 > 0$ is a free parameter. According to (4.105), the weight function is associated with the response function (Gradshteyn & Ryzhik, 1980, formula 3.121.2)

$$c(\omega) = \frac{m_0}{\sqrt{(e + i\omega)(f + i\omega)}} \tag{4.156}$$

with

$$p_n(0) = T_n(-u) = (-1)^n T_n(u)$$
$$u = \frac{f + e}{f - e} \geq 1 \tag{4.157}$$

and

$$h_0 = m_0$$
$$h_n = \frac{m_0}{2}$$
$$\frac{k_{n+1} h_n}{k_n} = \frac{2m_0}{f - e} \tag{4.158}$$

for non-negative n. The thin-sheet parameters creating $c(\omega)$ are given by

$$\mu_0 \tau_0 = \frac{1}{m_0}$$
$$\mu_0 \tau_n = \frac{2[T_n(u)]^2}{m_0} \tag{4.159}$$
$$d_n = \frac{2m_0}{(f - e)T_{n-1}(u)T_n(u)}$$

for non-negative n according to (4.153), where the first sheet with conductance τ_0 lies at $z = 0$. For $e > 0$, implying $T_{n+1}(u) > T_n(u) \geq 1$, (4.159) represents an infinite sequence of thin sheets with increasing conductance τ_n and decreasing separation d_n clustering at depth

$c(0) = m_0/\sqrt{ef}$. For $e = 0$, implying $T_n(u) = 1$, the nearly uniform sequence $\mu_0 \tau_0 = 1/m_0$, $\mu_0 \tau_n = 2/m_0$ and $d_n = 2m_0/f$ for $n \geq 1$ is obtained. All of the properties of the Chebyshev polynomials used in this example can be found in Abramowitz & Stegun (1972, chapter 22).

The recursive computation of the thin-sheet parameters from $a(\lambda)$ must now be addressed. If $a(\lambda)$ is discrete (i.e. is the second term in (4.149)), the Rutishauser algorithm proposed by Parker & Whaler (1981) offers an efficient method. The present approach, based on orthogonal polynomials, is an equivalent alternative that in addition allows consideration of the continuous part of $a(\lambda)$. For numerical simplicity, the orthogonal polynomials $p_n(\lambda)$ are now assumed to be normalized, so that

$$h_n = (p_n, p_n) = 1 \\ k_n > 0 \tag{4.160}$$

Then, three consecutive orthogonal polynomials are related by (Abramowitz & Stegun, 1972, formula 22.1.4)

$$\beta_{n+1}p_{n+1}(\lambda) = (\lambda - \alpha_n)p_n(\lambda) - \beta_n \, p_{n-1}(\lambda) \tag{4.161}$$

where $\beta_{n+1} = k_n/k_{n+1} > 0$ and $\alpha_n = (\lambda p_n, p_n)$. Equation (4.161) provides the following simple way for the recursive computation of the polynomials and the thin-sheet parameters for a given $a(\lambda)$:

Starting with

$$p_{-1}(\lambda) = 0 \\ p_0(\lambda) = 1/\sqrt{m_0} = 1/\sqrt{(1,1)} \\ \beta_0 = 0 \tag{4.162}$$

then for $n = 0, 1, 2, \dots$

$$\begin{aligned} (1) \qquad & \alpha_n = (\lambda p_n, p_n) \\ (2) \qquad & \bar{p}_{n+1}(\lambda) = (\lambda - \alpha_n)p_n(\lambda) - \beta_n p_{n-1}(\lambda) \\ (3) \qquad & \beta_{n+1} = \sqrt{(\bar{p}_{n+1}, \bar{p}_{n+1})} \\ (4) \qquad & p_{n+1}(\lambda) = \bar{p}_{n+1}(\lambda)/\beta_{n+1} \end{aligned} \tag{4.163}$$

Taking into account that $h_n = 1$ and $k_n/k_{n+1} = \beta_{n+1}$, the resulting thin-sheet parameters obtained via (4.153) are simply

$$\mu_0 \tau_n = p_n^2(0) \\ d_{n+1} = -\frac{1}{p_n(0)\bar{p}_{n+1}(0)} \tag{4.164}$$

In steps (2) and (4) of the algorithm (4.163), $\bar{p}_{n+1}(0)$ and $p_{n+1}(0)$ are updated along with $\bar{p}_{n+1}(\lambda)$ and $p_{n+1}(\lambda)$. The weight function $a(\lambda)$ enters in steps (1) and (3).

In the case of a discrete weight function $a(\lambda)$ with $N < \infty$ (see (4.149)), the algorithm reaches an end: if $\min(\lambda_n) = 0$, the last determined parameter is τ_{N-1} (and $d_N = \infty$), while if $\min(\lambda_n) > 0$, the algorithm stops with d_N (and $\tau_N = \infty$). The algorithm is stable even for a large number of closely spaced sheets.

A method for the extraction of $a(\lambda)$ from a given set of measured response functions is given by Parker (1980). For M frequencies ω_m, the dataset consists of response functions

$c_m = c(\omega_m)$ with standard deviations s_m for $1 \leq m \leq M$. After a dense sampling of the λ axis yielding $J \gg M$ samples $\bar{\lambda}_j$ ranging from $\bar{\lambda}_1 \ll \min(\omega_m)$ to $\bar{\lambda}_1 \gg \max(\omega_m)$, the $J+1$ non-negative constants $\bar{a}_j$ for $0 \leq j \leq J$ are determined by minimizing the quadratic functional resulting from (4.105),

$$Q(\bar{a}_0, ..., \bar{a}_J) = \sum_{m=1}^{M} \frac{1}{s_m^2} \left| c_m - \left(\bar{a}_0 + \sum_{j=1}^{J} \frac{\bar{a}_j}{\bar{\lambda}_j + i\omega_m} \right) \right|^2 \qquad (4.165)$$

subject to the non-negativity constraints

$$\bar{a}_j \geq 0, \qquad 0 \leq j \leq J \qquad (4.166)$$

This minimization is efficiently achieved by quadratic programming – for example, by using the NNLS program of Lawson & Hanson (1974). In general, only a small number $N \ll J$ of positive constants $\bar{a}_j = a_n$ evolve. They are collected, along with $\bar{\lambda}_j = \lambda_n$, in the discrete weight function

$$a(\lambda) = \sum_{n=1}^{N} a_n \delta(\lambda - \lambda_n) \qquad (4.167)$$

where $a_n > 0$, and are interpreted as a sequence of thin sheets. The shallowest thin sheet lies at $z = \bar{a}_0$. Then, the cleaned response estimates

$$\bar{c}_m = \bar{a}_0 + \sum_{n=1}^{N} \frac{a_n}{\lambda_n + i\omega_m} \qquad (4.168)$$

where $\bar{a}_0\,0$, $a_n > 0$ for $1 \leq n \leq N$, give the best possible fit to the data c_m.

4.2.7 The rho+ model

Parker & Booker (1996) recast the D+ problem from the real and imaginary parts of $c(\omega)$ to the more commonly used experimental variables, apparent resistivity and phase. The formalism follows from considering

$$\log c(\omega) = \frac{1}{2} \log[\rho_a(\omega)/(\mu_0\omega)] + i[\phi(\omega) - \pi/2] \qquad (4.169)$$

Equation (4.105) for a finite set of thin conducting sheets has $a(\lambda)$ represented by a finite set of delta functions in λ space as in (4.167), and can also be written as the ratio of two polynomials of degree N,

$$c(\omega) = a_0 \prod_{n=1}^{N} \frac{(1 + v_n/i\omega)}{(1 + \lambda_n/i\omega)} \qquad (4.170)$$

where $\omega = iv_n$ are the zeros of $c(\omega)$. Taking the logarithm of (4.170)

$$\log c(\omega) = \log a_0 + \sum_{n=1}^{N} \log(\omega - iv_n) - \sum_{n=1}^{N} \log(\omega - i\lambda_n) \qquad (4.171)$$

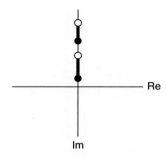

Figure 4.7. Complex plane representation of $\log c(\omega)$ for the instance of two poles. The result is analytic everywhere except on the positive imaginary axis, where it displays logarithmic branch points at interlacing sets of poles and zeros of equal numbers. Branch cuts (heavy lines) are chosen to join the successive poles (filled circles) and zeros (open circles).

so that $\log c(\omega)$ is analytic except for a set of logarithmic branch points at the location of the poles and zeros on the positive imaginary axis. Branch cuts join the successive poles and zeros of $c(\omega)$ as shown in Figure 4.7; since the numbers of poles and zeros are identical, there is no branch cut extending to infinity, in contrast to the situation for the real and imaginary parts of $c(\omega)$. Parker & Booker (1996) constructed a contour in the complex plane that encloses all of the branch points and a single real frequency point, showing that

$$\log c(\omega) = \log a_0 + \int_0^\infty \frac{u(\lambda)}{\lambda + i\omega} d\lambda \tag{4.172}$$

where

$$
\begin{aligned}
u(\lambda) &= 1 & \lambda_m \leq \lambda \leq \nu_m \\
&= 0 & \text{otherwise}
\end{aligned}
\tag{4.173}
$$

Parker & Booker (1996) considered the problem of matching measurements of apparent resistivity and phase with their uncertainties to a D+ model using an optimization technique, and also treated bounds on linear functionals of (4.172). The latter was applied to the apparent resistivities and phases themselves, yielding a consistency test of one variable with the other within their respective uncertainties under the assumption of a 1D structure. Figure 4.8 shows an example based on the COPROD dataset (Jones & Hutton, 1979). This approach can be useful for detecting outlying values of the apparent resistivity or phase, or for directly estimating one quantity from the other.

4.3 Properties of the magnetotelluric response function in two dimensions

Whereas a wealth of knowledge exists about 1D transfer functions, information about the analytical properties of 2D responses is relatively sparse (and nearly non-existent for 3D responses). Assuming in the 2D situation that the x axis is in the strike direction, the

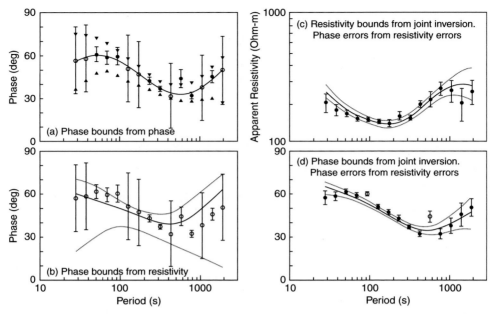

Figure 4.8. COPROD data with errors that are either used (solid circles) or not used (open circles) shown with the D+ response (solid line) and bounds computed in four ways. (a) Phase bounds computed from inversion of the phase, shown as solid triangles with the apex pointing up or down, respectively. (b) Phase bounds computed from the apparent resistivity, shown as dotted lines. For both panels, the errors are those given by Jones & Hutton (1979). (c) Apparent resistivity bounds computed from joint inversion of the apparent resistivity and phase, with the stated apparent resistivity errors. (d) Phase bounds computed from joint inversion of the apparent resistivity and phase, with the phase errors predicted from the apparent resistivity errors. Two phase outliers are especially apparent in (d). Taken from Parker & Booker (1996).

pertinent transfer functions at the surface $z = 0$ are (see also (4.101) and (4.102) and Section 2.5.1):

TE mode

$$Z_e(y, \omega) = \frac{E_x(y, z = 0, \omega)}{B_y(y, z = 0, \omega)}$$

$$c_e(y, \omega) = \frac{Z_e(y, \omega)}{i\omega}$$

(4.174)

TM mode

$$Z_m(y, \omega) = -\frac{E_y(y, z = 0, \omega)}{B_0(\omega)}$$

$$c_m(y, \omega) = \frac{Z_m(y, \omega)}{i\omega}$$

(4.175)

For the TM mode, $B_0(\omega) = B_x(y, z = 0, \omega)$ is the uniform surface field.

In Section 2.5.2, it was pointed out that the occurrence of both electromagnetic induction and galvanic current flow renders the physics of the TM mode more complex than that of the TE mode, where only electromagnetic induction is involved. However, mathematically the opposite is true: the TM mode boundary condition $B_x = B_0$ at $z = 0$ leads to significant simplifications, such that at present general results can be formulated only for that mode. Moreover, the TM mode response Z_m is the disturbed horizontal electric field, normalized with the constant B_0, whereas the TE mode response Z_e is the ratio of two disturbed quantities. Let

$$Z_m = |Z_m|e^{i\phi_m}$$
$$\rho_m = \mu_0 |Z_m|^2 /\omega \tag{4.176}$$

Assuming $\omega > 0$ and $\sigma(y, z) > 0$ in $z = 0$, it can be shown that the following hold:

(a) The response function $Z_m(y, \omega)$ lies in the first quadrant, so that $0 < \phi_m(y, \omega) < \pi/2$.
(b) Dispersion relations (4.127) and (4.128) with $\rho_a = \rho_m$ and $\phi = \phi_m$ exist between the TM mode apparent resistivity $\log[\rho_m(y, \omega)]$ and the phase $\phi = \phi_m(y, \omega)$.

These results mean that data assessment using the D+ or rho+ models applies to 2D TM mode response functions in addition to 1D ones.

The proof of these two statements is given by Weidelt & Kaikkonen (1994). For that purpose, the notation is simplified by switching from conductivity σ to resistivity $\rho = 1/\sigma$. Let

$$B_x(y, z)/B_0 = b(y, z)e^{i\psi(y,z)} \tag{4.177}$$

with $b(y, 0) = 1$ and $\psi(y, 0) = 0$. Then, (4.175), (4.177) and (2.78) yield

$$Z_m(y) = -E_y(y, 0)/B_0 = -\frac{\rho(y, 0^+)}{\mu_0 B_0} \partial_z B_x(y, z)\Big|_{z=0^+}$$
$$= -\frac{\rho(y, 0^+)}{\mu_0} [\partial_z b(y, z) + i\partial_z \psi(y, z)]\Big|_{z=0^+} \tag{4.178}$$

It will be shown that the amplitude b and phase ψ attain their maxima $b = 1$ and $\psi = 0$ at each point of the boundary $z = 0$. Therefore, b and ψ will not increase when moving from $z = 0$ downward, so that $\partial_z b|_{z=0^+} \leq 0$ and $\partial_z \psi|_{z=0^+} \leq 0$, implying according to (4.178) that Z_m lies in the first quadrant. The differential equation (2.86) for B_x reads

$$\nabla_2 \bullet (\rho\nabla_2 B_x) = i\omega\mu_0 B_x \tag{4.179}$$

After inserting (4.177) and separating the real and imaginary parts, (4.179) disintegrates into the system of nonlinear equations

$$\nabla_2 \bullet (\rho\nabla_2 b) = b\rho (\nabla_2\psi)^2$$
$$\nabla_2 \bullet (b^2\rho\nabla_2\psi) = \omega\mu_0 b^2 \tag{4.180}$$

Let $f = b$ or ψ. Then, the left-hand sides of (4.180) have the form

$$\nabla_2 \bullet (\alpha\nabla_2 f) = \nabla_2\alpha \bullet \nabla_2 f + \alpha\nabla_2^2 f \tag{4.181}$$

where $\alpha > 0$. With $\nabla_2 f = 0$ as the necessary condition for a maximum, it follows from (4.180) and (4.181) that $\nabla_2^2 f \geq 0$ at a possible maximum. If under this condition the function f attains a maximum M at any point in $z > 0$, then $f \equiv M$ at all points in $z > 0$ according to the maximum principle (e.g. Protter & Weinberger, 1967, p. 53). In reality, f is not a constant. Therefore, neither b nor ψ can have a local maximum at an interior point of the half-space $z > 0$. Assuming that $\rho(y,z)$ tends to a 1D distribution $\rho(z)$ for $y \rightarrow \pm\infty$ that is not necessarily the same on both sides, it follows from (4.180) that

$$\partial_z b(z) = -\frac{1}{\rho(z)} \int_z^\infty b(t)\rho(t) \left[\partial_t \psi(t)\right]^2 dt < 0 \qquad (4.182)$$

$$\partial_z \psi(z) = -\frac{\omega\mu_0}{b^2(z)\rho(z)} \int_z^\infty b^2(t)dt < 0 \qquad (4.183)$$

such that b and ψ decrease with increasing depth on the left and right boundary. On the lower boundary ($z \rightarrow \infty$) it follows that $b \rightarrow 0$ and $\psi \rightarrow -\infty$. Therefore, b and ψ attain their maxima at $z = 0$ such that $\partial_z b|_{z=0^+} \leq 0$ and $\partial_z \psi|_{z=0^+} \leq 0$. In fact, these derivatives are strictly negative: the converse assumption $\partial_z \psi|_{z=0^+} = 0$ along with $\partial_y \psi = 0$ at $z = 0$ would lead to $\partial_z^2 \psi|_{z=0^+} = \omega\mu_0/\rho(y, 0^+) > 0$ via the second equation of (4.180), implying an increase of ψ when moving into the conductor. Therefore, $\partial_z \psi|_{z=0^+} < 0$. With this result, it follows from the first equation of (4.180), again by contradiction, that $\partial_z b|_{z=0^+} < 0$. This completes the proof that $0 < \phi_m(y, \omega) < \pi/2$. It can be shown (Weidelt & Kaikkonen, 1994) that $c_m(y, \omega)$ defined in (4.175) admits the representation

$$c_m(y, \omega) = \int_0^\infty \frac{a_m(y, \lambda)\, d\lambda}{\lambda + i\omega} \qquad (4.184)$$

where, contrary to the 1D representation (4.105), $a_m(y, \omega)$ can have short λ sections with $a_m(y, \omega) < 0$, although non-negative values always prevail. Because of this sporadic violation of the 1D non-negativity constraint, a strict local 1D interpretation of $c_m(y,\omega)$ is possible only in exceptional cases (e.g. for two quarter-spaces). In particular, $c_m(y,\omega)$ responses sufficiently averaged in the y direction will satisfy the non-negativity constraint.

From (4.184), it follows that $c_m(y, \omega)$ is an analytic function of ω in the whole complex ω plane except on the positive imaginary axis $\omega = i\lambda$ for $\lambda \geq 0$. The analyticity for $\text{Im}(\omega) \leq 0$ grants the existence of dispersion relations like (4.122) and (4.123). Dispersion relations between apparent resistivity and phase require in addition that the lower ω half-plane be free of zeros such that $\log c_m$ is analytic there. For this reason, consider (4.179) with (4.177) and the complex frequency $\omega = \omega_r - ip$, $\omega_r > 0$ and $p \geq 0$. After separating the real and imaginary parts, the first equation in (4.180) becomes

$$\nabla_2 \cdot (\rho \nabla_2 b) = b\left[p\mu_0 + \rho \left(\nabla_2 \psi\right)^2\right] \qquad (4.185)$$

and must be non-negative. For $p \geq 0$, this equation has the same properties as the first equation of (4.180) for real frequencies. Therefore, it can be concluded that

$$\text{Re}[Z_m(y, \omega_r - ip)] > 0 \qquad (4.186)$$

for $p \geq 0$ such that $Z_m(y, \omega) \neq 0$ and $c_m(y, \omega) \neq 0$ in m. This grants the existence of the dispersion relations (4.127) and (4.128).

The phase constraint $0 < \phi_m(y, \omega) < \pi/2$ and the dispersion relations also hold for a conductivity model that is generalized in two directions (Weidelt & Kaikkonen, 1994; Weidelt, 1999):

(a) The flat air–Earth interface with downward normal vector $\hat{\mathbf{n}} = \hat{\mathbf{z}}$ is replaced by smooth topography with normal vector $\hat{\mathbf{n}}(y)$ such that, at a boundary point $\mathbf{r}_0$ (with $\mathbf{r}_0^+$ immediately below $\mathbf{r}_0$),

$$Z_m(\mathbf{r}_0) = -\frac{\rho(\mathbf{r}_0^+)}{\mu_0 B_0} \hat{\mathbf{n}} \cdot \nabla_2 B_x(\mathbf{r}_0^+) \qquad (4.187)$$

(b) The isotropic resistivity ρ in Ohm's law $\mathbf{E} = \rho\mathbf{J}$ is replaced by an anisotropic resistivity $\vec{\rho}$ such that $\mathbf{E} = \vec{\rho} \cdot \mathbf{J}$, or explicitly

$$\begin{aligned} E_y &= \rho_{yy}J_y + \rho_{yz}J_z \\ E_z &= \rho_{zy}J_y + \rho_{zz}J_z \end{aligned} \qquad (4.188)$$

where $\vec{\rho}$ is a symmetric positive definite tensor such that $\rho_{yz} = \rho_{zy}$ and $\mathbf{v} \cdot \vec{\rho} \cdot \mathbf{v} > 0$ for all non-trivial vectors $\mathbf{v}$. This result is not surprising since a macroscale anisotropic resistor can be considered as an isotropic non-uniform resistor with preferred orientations (e.g. laminations) on the microscale (see Section 2.6.2).

The methods that have been so powerful for the TM mode cannot be adapted to the TE mode. Although the application of the maximum principle to (2.83) will show that the maxima of modulus and phase of E_x are attained at $z = 0$, these maxima do not necessarily coincide with the point where $Z_e(y, \omega)$ has been determined. Moreover, a representation of the type in (4.184) exists only separately for the numerator and denominator of $Z_e(y, \omega)$ or $c_e(y, \omega)$,

$$\begin{aligned} \frac{E_x(y, 0, \omega)}{i\omega B_0(\omega)} &= \int_0^\infty \frac{a_E(y, \lambda)d\lambda}{\lambda + i\omega} \\ \frac{B_y(y, 0, \omega)}{i\omega B_0(\omega)} &= 1 + \int_0^y \frac{a_B(y, \lambda)d\lambda}{\lambda + i\omega} \end{aligned} \qquad (4.189)$$

where $B_0(\omega)/2$ is the horizontal source field such that $B_y(\pm\infty, 0, \omega) = B_0(\omega)/2$ (see Section 2.5.1). The functions $a_E(y, \lambda)$ and $a_B(y, \lambda)$ have no definite sign.

Although so far no strict proof is available, numerical modeling has provided strong empirical evidence that the TE mode phase never leaves its quadrant, and that dispersion relations exist between apparent resistivity and phase. However, Parker (2010) has shown by counter-example that a model exists where the TE mode response cannot be represented as a 1D response, although the model consists of a variable-conductance surface thin sheet underlain by an insulator and terminated by a perfect conductor at depth, hence is not geologically plausible.

These results hold when measurements are made at Earth's surface, where induced currents are always beneath the observer, but do not necessarily apply at the seafloor, where induced currents may flow both above and below the observation point. In this instance, out-of-quadrant phases can and do occur for 2D structures.

4.4 Properties of the magnetotelluric response function in three dimensions

For a 3D structure, the magnetotelluric fields are characterized by elliptical polarization and non-orthogonality of the electric and magnetic fields, as is also the case for a 1D structure excited by a 3D source that occurs in controlled-source electromagnetics. As a result, the elements of the magnetotelluric response tensor $\overline{\overline{\mathbf{Z}}}$ are dependent on the orientation of the measurement coordinate system in a complicated manner, and the simple 2D symmetries present in (4.36)–(4.39) do not pertain. Exceptions occur when the conductive structure has a vertical plane of symmetry and the observations are collected along it, in which case $\overline{\overline{\mathbf{Z}}}$ is antidiagonal, or when a conductive body is axisymmetric with a vertical symmetry axis and the observations are collected at its center, in which case $\overline{\overline{\mathbf{Z}}}$ will appear to be 1D. More generally, interpretation of $\overline{\overline{\mathbf{Z}}}$ is facilitated if it is cast in its simplest possible form, which is rarely obtained in the measurement coordinates unless there is *a priori* knowledge about the shape and orientation of the subsurface structure.

The conventional approach to magnetotelluric interpretation requires finding the rotation angle α in (4.35) by some criterion, such as minimizing the squared sum of the diagonal elements or maximizing the squared sum of the off-diagonal elements. For a 3D structure, such an approach is equivalent to approximating a 3D with a 2D structure. While this may work well under some circumstances – for example, at the middle of an elongate 3D body (see Wannamaker *et al.*, 1984; Ledo, 2005) – it is not generally applicable, as no real rotation angle that antidiagonalizes $\overline{\overline{\mathbf{Z}}}$ into (4.58) may exist. Perhaps more importantly, the representation of the magnetotelluric response function is incomplete. For a 3D structure, there are eight degrees of freedom in $\overline{\overline{\mathbf{Z}}}$, while the antidiagonal form (4.58) plus a real rotation angle comprise only five degrees of freedom. This is sometimes addressed by adding semi-quantitative parameters such as the skew (4.47) and the ellipticity,

$$\beta = \frac{\left| Z_{xx} - Z_{yy} \right|}{\left| Z_{xy} + Z_{yx} \right|} \tag{4.190}$$

Alternatively, the magnetotelluric response tensor can be cast in terms of rotational invariants as described in Section 4.1.4.

An alternative that does not cast an *a priori* interpretation structure onto the observations is available. It is possible to let the measured response tensor define the best coordinate system through an eigenvalue–eigenvector decomposition, as is commonly practiced in statistics to reveal the simplest possible set of independent parameters that describe a complicated dataset – the principal components (see Jolliffe, 2002). This concept was introduced into magnetotellurics by Eggers (1982), although he imposed the constraint $\mathbf{E} \cdot \mathbf{B} = 0$, limiting the formulation to those states that are inherently orthogonal. This condition need not apply in 3D induction. Subsequently, Spitz (1985) applied a Cayley factorization of $\overset{\leftrightarrow}{\mathbf{Z}}$ to determine two independent coordinate systems that describe the data. However, there remained ambiguity about which coordinate system serves best to describe $\overset{\leftrightarrow}{\mathbf{Z}}$ with either of these approaches.

La Torraca *et al.* (1986) generalized the eigenvector analysis through use of the singular value decomposition (SVD), which imposes no constraints on the solution, and instead expresses $\overset{\leftrightarrow}{\mathbf{Z}}$ through the maximum and minimum possible values of $|\mathbf{E}|/|\mathbf{B}|$. Yee & Paulson (1987) used a different approach to get the same result, and provide a thorough comparison with other response tensor representations.

The SVD of any $N \times p$ matrix $\mathbf{A}$ may be written

$$\mathbf{A} = \mathbf{U} \cdot \boldsymbol{\Sigma} \cdot \mathbf{V}^H \qquad (4.191)$$

where the superscript H denotes the Hermitian (complex conjugate) transpose, $\mathbf{U}$ contains the eigenvectors of $\mathbf{A} \cdot \mathbf{A}^H$, $\mathbf{V}$ contains the eigenvectors of $\mathbf{A}^H \cdot \mathbf{A}$, and $\boldsymbol{\Sigma}$ is zero except on the p-diagonal and contains the square roots of the real eigenvalues of $\mathbf{A} \cdot \mathbf{A}^H$ and $\mathbf{A}^H \cdot \mathbf{A}$. La Torraca *et al.* (1986) recognized that, when $\mathbf{A} = \overset{\leftrightarrow}{\mathbf{Z}}$, so that $N = p = 2$, the SVD becomes

$$\overset{\leftrightarrow}{\mathbf{Z}} = \mathbf{U}_E \cdot \boldsymbol{\Sigma} \cdot \mathbf{U}_B \qquad (4.192)$$

where $\mathbf{U}_E$ and $\mathbf{U}_B$ are unitary matrices that contain the eigenvectors of $\overset{\leftrightarrow}{\mathbf{Z}} \cdot \overset{\leftrightarrow}{\mathbf{Z}}^H$ and $\overset{\leftrightarrow}{\mathbf{Z}}^H \cdot \overset{\leftrightarrow}{\mathbf{Z}}$, respectively. Because $\boldsymbol{\Sigma}$ is real and $\mathbf{U}_E$ and $\mathbf{U}_B$ are complex, the right-hand side of (4.192) contains 10 parameters while $\overset{\leftrightarrow}{\mathbf{Z}}$ contains only eight. This ambiguity can be resolved because the two columns of $\mathbf{U}_E$ and $\mathbf{U}_B$ are simultaneously uncertain by a phase factor $e^{i\theta}$, which may be chosen by making the column vector inner products real and positive, resulting in $\boldsymbol{\Sigma}$ becoming complex. This reduces the parameter count to eight because the constrained versions of $\mathbf{U}_E$ and $\mathbf{U}_B$ contain two parameters each.

La Torraca *et al.* (1986) then presented a physical interpretation of the result in terms of elliptically polarized fields. The elements of $\boldsymbol{\Sigma}^H \cdot \boldsymbol{\Sigma} = \boldsymbol{\Sigma} \cdot \boldsymbol{\Sigma}^H$ are the extreme values of the magnetotelluric response with both an amplitude and a phase. The columns of $\mathbf{U}_E$ and $\mathbf{U}_B$ are the corresponding electric and magnetic field elliptical polarization states, which may be parameterized in terms of the orientation γ of each principal axis with respect to the original coordinate system. The result is a complete description of the eight elements in $\overset{\leftrightarrow}{\mathbf{Z}}$ as the four parameters γ_{Ej}, γ_{Bj}, the amplitudes r_j and the phases ϕ_j for each

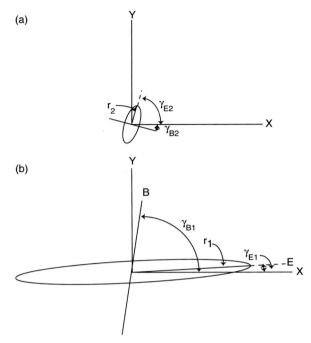

Figure 4.9. The state representation of a single element of the magnetotelluric response tensor: shown are the states corresponding to (a) the minimum value and (b) the maximum value of $|E|/|B|$. Each state is represented by the amplitude of the response r_j, the orientation of the electric γ_{Ej} and magnetic γ_{Bj} field ellipses, and the phase difference ϕ_j between the electric and magnetic fields (not shown).

of the two polarization states indicated by the subscript j. Figure 4.9 shows an example. Note that the electric and magnetic polarization ellipses are not orthogonal, as would be expected for a 3D medium.

References

Abramowitz, M. & I. A. Stegun (1972). *Handbook of Mathematical Functions*. New York: Dover.

Bahr, K. (1988). Interpretation of the magnetotelluric impedance tensor: regional induction and local telluric distortion. *J. Geophys.*, **62**, 119–127.

Berdichevsky, M. N. & V. I. Dmitriev (2002). *Magnetotellurics in the Context of the Theory of Ill-Posed Problems*. Tulsa: Society of Exploration Geophysicists.

Bibby, H. M., T. G. Caldwell & C. Brown (2005). Determinable and non-determinable parameters of galvanic distortion in magnetotellurics. *Geophys. J. Int.*, **163**, 915–930.

Bostick, F. X. (1977). A simple almost exact method of MT-analysis. In *Workshop on Electrical Methods in Geothermal Exploration*, U.S. Geological Survey Contract No. 14080001–8–359.

Caldwell, T. G., H. M. Bibby & C. Brown (2004). The magnetotelluric phase tensor. *Geophys. J. Int.*, **158**, 457–469.

Chave, A. D. & J. T. Smith (1994). On electric and magnetic galvanic distortion tensor decompositions. *J. Geophys. Res.*, **99**, 4669–4682.

Eggers, D. (1982). An eigenstate formulation of the magnetotelluric impedance tensor. *Geophysics*, **47**, 1204–1214.

Gradshteyn, I. S. & I. M. Ryzhik (1980). *Table of Integrals, Series, and Products*. New York: Academic Press.

Jolliffe, I. T. (2002). *Principal Component Analysis*, 2nd edn. New York: Springer.

Jones, A. G. & V. R. S. Hutton (1979). A multi-station magnetotelluric study in southern Scotland, I. Fieldwork, data analysis and results. *Geophys. J. R. Astron. Soc.*, **56**, 329–349.

Larsen, J. C. (1975). Low frequency (0.1–0.6 cpd) electromagnetic study of deep mantle electrical conductivity beneath the Hawaiian Islands. *Geophys. J. R. Astron. Soc.*, **43**, 17–46.

La Torraca, G. A., T. R. Madden & J. Korringa (1986). An analysis of the magnetotelluric impedance for three-dimensional conductivity structure. *Geophysics*, **51**, 1819–1829.

Lawson, C. L. & R. J. Hanson (1974). *Solving Least Squares Problems*. Englewood Cliffs, NJ: Prentice-Hall.

Ledo, J. (2005). 2-D versus 3-D magnetotelluric data interpretation. *Surv. Geophys.*, **26**, 511–543.

Mathews, J. & R. L. Walker (1970). *Mathematical Methods for Physics*, 2nd edn. Menlo Park: Benjamin/Cummings.

Morse, P. M. & H. Feshbach (1953). *Methods of Theoretical Physics*. New York: McGraw-Hill.

Niblett, E. R. & C. Sayn-Wittgenstein (1960). Variation of electrical conductivity with depth by the magnetotelluric method. *Geophysics*, **25**, 998–1008.

Parker, R. L. (1980). The inverse problem of electromagnetic induction: existence and construction of solutions based on incomplete data. *J. Geophys. Res.*, **85**, 4421–4428.

Parker, R. L. (2010). Can a 2D MT frequency response always be interpreted as a 1D response? *Geophys. J. Int.*, **181**, 269–274.

Parker, R. L. & J. R. Booker (1996). Optimal one-dimensional inversion and bounding of magnetotelluric apparent resistivity and phase measurements. *Phys. Earth Planet. Inter.*, **98**, 269–282.

Parker, R. L. & K. A. Whaler (1981). Numerical methods for establishing solutions to the inverse problem of electromagnetic induction. *J. Geophys. Res.*, **86**, 9574–9584.

Protter, H. M. & H. F. Weinberger (1967). *Maximum Principles in Differential Equations*. Englewood Cliffs, NJ: Prentice-Hall.

Schmucker U. (1987). Substitute conductors for electromagnetic response estimates. *Pure Appl. Geophys.*, **125**, 341–367.

Singer, B. S. (1992). Correction for distortion of magnetotelluric fields: limits of validity of the static approach. *Surv. Geophys.*, **13**, 309–340.

Spitz, S. (1985). The magnetotelluric impedance tensor properties with respect to rotations. *Geophysics*, **50**, 1610–1617.

Szarka, L. & M. Menvielle (1997). Analysis of rotational invariants of the magnetotelluric impedance tensor. *Geophys. J. Int.*, **129**, 133–142.

Toll, J. (1956). Causality and the dispersion relation: logical foundations. *Phys. Rev.*, **104**, 1760–1770.

Wannamaker, P. E., G. W. Hohmann & S. H. Ward (1984). Magnetotelluric responses of three-dimensional bodies in layered earths. *Geophysics*, **49**, 1517–1533.

Weaver, J. T. (1994). *Mathematical Methods for Geo-electromagnetic Induction*. Taunton, UK: Research Studies Press.

Weaver, J. T., A. K. Agarwal & F. E. M. Lilley (2000). Characterization of the magneto-telluric tensor in terms of its invariants. *Geophys. J. Int.*, **141**, 321–336.

Weaver, J. T., A. K. Agarwal & F. E. M. Lilley (2006). The relationship between the magnetotelluric tensor invariants and the phase tensor of Caldwell, Bibby, and Brown. *Explor. Geophys.*, **37**, 261–267.

Weidelt, P. (1972). The inverse problem of geomagnetic induction. *Z. Geophys.*, **38**, 257–289.

Weidelt, P. (1986). Discrete frequency inequalities for magnetotelluric impedances of one-dimensional conductors. *J. Geophys.*, **59**, 171–176.

Weidelt, P. (1999). 3-D conductivity models: implications of electrical anisotropy. In *Three-Dimensional Electromagnetics*, ed. M. Oristaglio & B. Spies. Tulsa: Society of Exploration Geophysicists.

Weidelt, P. (2005). The relationship between the spectral function and the underlying conductivity structure in 1-D magnetotellurics. *Geophys. J. Int.*, **161**, 566–590.

Weidelt, P. & P. Kaikkonen (1994). Local 1-D interpretation of magnetotelluric B-polarization impedances. *Geophys. J. Int.*, **117**, 733–748.

Yee, E. & K. V. Paulson (1987). The canonical decomposition and its relationship to other forms of magnetotelluric impedance tensor analysis. *J. Geophys.*, **61**, 273–289.

Yee, E. & K. V. Paulson (1988a). Properties of the c-response function for conductivity distributions of class S+. *Geophys. J.*, **93**, 265–278.

Yee, E. & K. V. Paulson (1988b). Necessary and sufficient conditions for the existence of a solution to the one-dimensional magnetotelluric inverse problem. *Geophys. J.*, **93**, 279–293.

5

Estimation of the magnetotelluric response function

ALAN D. CHAVE

5.1 The statistical problem

As described throughout Chapters 2 and 4, the fundamental datum in magnetotellurics is a location-specific, frequency-dependent tensor linearly connecting the horizontal electric and magnetic fields measured at Earth's surface (or at the seafloor). Under general conditions specified in Section 4.1.2, in the absence of noise, with precise data and independent of the dimensionality of the underlying electrical structure, the relationship may be written

$$\mathbf{E} = \vec{\mathbf{Z}} \cdot \mathbf{B} \qquad (5.1)$$

where $\mathbf{E}$ and $\mathbf{B}$ are 2-vectors of the horizontal electric and magnetic field components at a specific site and frequency, $\vec{\mathbf{Z}}$ is the second-rank, 2×2 magnetotelluric response tensor connecting them, and $\cdot$ denotes the inner product. For the remainder of this chapter, vectors will be shown as bold, and matrices or tensors as bold with a double arrow above them.

The solution to (5.1) is

$$\vec{\mathbf{Z}} = (\mathbf{E} \otimes \mathbf{B}^*) \cdot (\mathbf{B} \otimes \mathbf{B}^*)^{-1} \qquad (5.2)$$

where $\otimes$ is the outer product, and the terms in parentheses are the exact cross- and auto-power spectra. Section 4.1.3 introduces additional transfer functions between the vertical and horizontal magnetic fields, or between the horizontal magnetic fields at different sites, for which the estimation methodologies are similar, and hence only the magnetotelluric response will be considered further.

When $\mathbf{E}$ and $\mathbf{B}$ are actual measurements, (5.1) and (5.2) do not hold exactly due to the finite size of the data sample and the presence of noise, and it becomes necessary to estimate both $\vec{\mathbf{Z}}$ and its uncertainty $\delta\vec{\mathbf{Z}}$ in a statistical manner. The classical approach in magnetotellurics is the application of least-squares principles (e.g. Sims *et al.*, 1971; Vozoff, 1972). In the standard linear regression model for the row-by-row solution of (5.1), the equivalent set of matrix equations is

$$\mathbf{e} = \vec{\mathbf{b}} \cdot \hat{\mathbf{z}} + \boldsymbol{\varepsilon} \qquad (5.3)$$

The Magnetotelluric Method: Theory and Practice, ed. Alan D. Chave and Alan G. Jones. Published by Cambridge University Press © Cambridge University Press 2012.

where there are N observations (i.e. N Fourier transforms of N independent data sections at a given frequency, as described in Section 5.2), so that $\mathbf{e}$ is the *response* N-vector (i.e. N rows and 1 column), $\bar{\bar{\mathbf{b}}}$ is the $N \times p$ *predictor* matrix (where $p = 2$ for magnetotellurics), $\hat{\mathbf{z}}$ is an estimate of the *parameter* p-vector, and $\boldsymbol{\varepsilon}$ is an N-vector of unobservable *random errors*.

In elementary statistics texts, $\bar{\bar{\mathbf{b}}}$ is usually assumed to contain non-stochastic (i.e. fixed) variables, in which case the statistical model corresponding to (5.3) becomes

$$
\begin{aligned}
\mathrm{E}(\mathbf{e}) &= \bar{\bar{\mathbf{b}}} \cdot \mathbf{z} \\
\mathrm{Cov}(\mathbf{e}) &= \sigma^2 \bar{\mathbf{I}}
\end{aligned}
\tag{5.4}
$$

where $\mathrm{E}(.)$ and $\mathrm{Cov}(.)$ denote the expected value and covariance operators, respectively, $\mathbf{z}$ and σ^2 are the population rather than estimated values of the parameters and variance, and $\bar{\mathbf{I}}$ is the $p \times p$ identity matrix.

In magnetotellurics, the predictor variables in $\bar{\bar{\mathbf{b}}}$ are themselves random variables. This can be accommodated in the statistical model by using the conditional expected value and covariance

$$
\begin{aligned}
\mathrm{E}(\mathbf{e}|\bar{\bar{\mathbf{b}}}) &= \bar{\bar{\mathbf{b}}} \cdot \mathbf{z} \\
\mathrm{Cov}(\mathbf{e}|\bar{\bar{\mathbf{b}}}) &= \sigma^2 \bar{\mathbf{I}}
\end{aligned}
\tag{5.5}
$$

where $\mathbf{e} \mid \bar{\bar{\mathbf{b}}}$ means $\mathbf{e}$ conditional on $\bar{\bar{\mathbf{b}}}$, and

$$
\mathrm{E}(x|y) = \int_{\Omega} x f(x|y) dx
\tag{5.6}
$$

where the conditional probability distribution $f(x|y)$ has support (or domain of existence) Ω, and is just the joint distribution of x and y divided by the x-marginal distribution. $\mathrm{E}(\mathbf{e}|\bar{\bar{\mathbf{b}}})$ is itself a random variable; a different result will ensue for each instance of $\bar{\bar{\mathbf{b}}}$ that is utilized. The theory of least squares applies equally well when $\bar{\bar{\mathbf{b}}}$ contains random variables under very general conditions (Shaffer, 1991). It will further be assumed that the statistical model is linear in the parameters, a condition that holds for magnetotellurics based on physics, and that $\hat{\mathbf{z}}$ is not affected by linear equalities among the predictor variables, so that $\mathrm{rank}(\bar{\bar{\mathbf{b}}}) = p$. Finally, the response and predictor variables in magnetotellurics are complex rather than real. This requires only minor changes to standard theory, as shown by Miller (1973).

The least-squares estimator for the statistical model (5.5) is

$$
\hat{\mathbf{z}} = \left(\bar{\bar{\mathbf{b}}}^{H} \cdot \bar{\bar{\mathbf{b}}}\right)^{-1} \cdot \left(\bar{\bar{\mathbf{b}}}^{H} \cdot \mathbf{e}\right)
\tag{5.7}
$$

where the superscript H denotes the Hermitian (complex conjugate) transpose, the elements of $\bar{\bar{\mathbf{b}}}^{H} \cdot \bar{\bar{\mathbf{b}}}$ and $\bar{\bar{\mathbf{b}}}^{H} \cdot \mathbf{e}$ are the averaged estimates of the auto- and cross-power spectra based on the available data, and $\hat{\mathbf{z}}$ is a random variable. The following properties and definitions apply to the least-squares solution.

1. The estimator is unconditionally unbiased

$$E(\hat{z}) = E\left[E\left(\hat{z}|\vec{b}\right)\right] = z \tag{5.8}$$

when either $(e, \vec{b})$ are jointly multivariate Gaussian with unknown parameters, or else the distribution of $\vec{b}$ is completely unknown and that for e is arbitrary, in both cases subject to what is implied by (5.5) (Shaffer, 1991). These conditions are sufficiently broad that they can be taken as general.

2. The conditional covariance matrix for the parameters is

$$Cov\left(\hat{z}|\vec{b}\right) = \sigma^2 \left(\vec{b}^H \cdot \vec{b}\right)^{-1} \tag{5.9}$$

The unconditional covariance matrix is $\sigma^2 E\left[\left(\vec{b}^H \cdot \vec{b}\right)^{-1}\right]$. However, Shaffer (1991) showed that if $E\left[\left(\vec{b}^H \cdot \vec{b}\right)^{-1}\right]$ is known, then an optimal least-squares estimator does not exist, and so the conditional form will be retained.

3. The N-vector of *predicted values* for the response variable is

$$\hat{e} = \vec{b} \cdot \hat{z} = \vec{b} \cdot \left(\vec{b}^H \cdot \vec{b}\right)^{-1} \cdot \vec{b}^H \cdot e \equiv \vec{\vec{H}} \cdot e \tag{5.10}$$

where $\vec{\vec{H}}$ is the $N \times N$ predictor or *hat matrix*. The predicted values have the following statistical properties:

(a) $E\left(\hat{e}|\vec{b}\right) = \vec{b} \cdot z$

(b) $Cov\left(\hat{e}|\vec{b}\right) = \sigma^2 \vec{\vec{H}}$

The hat matrix has the following properties:

(c) Because of the relation between the predicted and observed response variables, $\vec{\vec{H}}$ is a projection matrix that maps e onto $\hat{e}$. As a result, $\vec{\vec{H}}$ must be Hermitian, idempotent ($\vec{\vec{H}} \cdot \vec{\vec{H}} = \vec{\vec{H}}$), and have eigenvalues of either 0 or 1.

(d) The entries on the main diagonal (denoted by h_{ii}) satisfy $0 \leq h_{ii} \leq 1$.

(e) If $h_{ii} = 0$, then the corresponding entry in $\hat{e} = 0$, and hence is not affected by that entry in e. If $h_{ii} = 1$, then the corresponding entries in $\hat{e}$ and e are identical and the model fits the datum exactly. In this instance, only the ith row of $\vec{b}$ has any influence on the regression problem, leading to the term high (or extreme) *leverage* to describe that point.

(f) Chave & Thomson (2003) showed that, when the rows of $\vec{b}$ are complex p-variate normal, the distribution of the hat matrix diagonal is the beta distribution $\beta(h_{ii}, p, N - p)$. From the properties of the beta distribution, the expected value of h_{ii} is p/N.

4. The regression *residuals* are given by

$$r = e - \vec{b} \cdot \hat{z} = \left(\vec{\vec{I}} - \vec{\vec{H}}\right) \cdot e \tag{5.11}$$

5. The *sum of squared residuals* is given by

$$\mathbf{r}^H \cdot \mathbf{r} = \mathbf{e}^H \cdot \left(\vec{\mathbf{I}} - \vec{\mathbf{H}} \right) \cdot \mathbf{e} \tag{5.12}$$

6. The unbiased estimate for σ^2 conditional on $\mathbf{b}$ is

$$\hat{\sigma}^2 = \frac{\mathbf{r}^H \cdot \mathbf{r}}{N - p} \tag{5.13}$$

7. The residuals $\mathbf{r}$ are uncorrelated with the predicted values $\hat{\mathbf{e}}$.
8. The parameter vector $\hat{\mathbf{z}}$ is the best linear unbiased estimator for $\mathbf{z}$. This statement is the Gauss–Markov theorem that underlies least squares, and holds for random regressors under the conditions listed in 1.
9. While statements 1–8 do not depend on distributional assumptions, if $\boldsymbol{\varepsilon} \sim C_N \left(\mathbf{0}, \sigma^2 \right)$ (i.e. the random errors are N-variate complex normal with zero mean and common variance σ^2; this implies that they are identically distributed), then $\hat{\mathbf{z}}$ is also the maximum likelihood estimate, and the regression quantities are distributed according to

$$\hat{\mathbf{z}} \sim C_p \left[\mathbf{z}, \sigma^2 \left(\vec{\mathbf{b}}^H \cdot \vec{\mathbf{b}} \right)^{-1} \right]$$
$$\hat{\mathbf{e}} \sim C_N \left(\vec{\mathbf{b}} \cdot \mathbf{z}, \sigma^2 \vec{\mathbf{H}} \right) \tag{5.14}$$
$$\frac{(N - p)\hat{\sigma}^2}{\sigma^2} \sim \chi^2_{2(N-p)}$$

where χ^2_v is the chi-squared distribution with v degrees of freedom. Further, $\hat{\mathbf{z}}$ and $\hat{\sigma}^2$ are independent, and serve as sufficient statistics for estimating $\mathbf{z}$ and σ^2. The term "sufficient statistics" means that no other statistic for a parameter calculated from a given data sample can provide additional information about the value of the parameter. The properties in (5.14) apply in the asymptotic (large-sample) limit, and much greater complexity ensues in the small-sample case. Sawa (1969) provides the exact sampling distribution for the least-squares estimator with a finite sample size.

The theory outlined in (5.5) and (5.7)–(5.14) strictly applies when $\vec{\mathbf{b}}$, while a random variable, is measured without error. A more realistic model incorporates measurement error into both $\mathbf{e}$ and $\vec{\mathbf{b}}$, and is called an *errors-in-variables* model. It has been known since at least the time of Adcock (1878) that the presence of errors in all of the measured variables makes the ordinary least-squares estimator (5.7) biased (i.e. it fails to converge to the population value as the number of data rises). Additional analysis steps are required to minimize bias when an errors-in-variables model obtains, as discussed in Section 5.4.

With natural-source electromagnetic data, the conditions on the least-squares estimate are rarely tenable even when $\vec{\mathbf{b}}$ is measured without error, for at least six reasons:

1. The variance of the residuals **r** is often dependent on that of the data, especially when energetic intervals coincide with source field complexity, as is the case for many classes of geomagnetic disturbances, or in the presence of correlated cultural noise.

2. The finite duration of many geomagnetic or cultural events results in data anomalies that occur in patches, violating the independent residual requirement.

3. Owing to marked *non-stationarity*, extreme residuals are much more common with magnetotelluric data than would be expected for a Gaussian model, and hence the residual distribution is typically very long-tailed with a Gaussian center. The data corresponding to large residuals are called *influential*, and can result in serious bias to $\hat{z}$ and even larger bias in parametric (i.e. those based on assumptions about distributions) estimates of its uncertainty. In addition, correlation of estimates at distinct frequencies may become appreciable in the presence of non-stationarity.

4. Because the Fourier transform is the sum of many terms, it is often assumed that its distribution will tend to multivariate Gaussian through loose central limit theorem reasoning. However, this argument fails in the presence of non-stationarity or influential data, and the non-Gaussian nature of the data must be accommodated during analysis.

5. The presence of a deterministic component at a given frequency introduces *non-centrality* into the statistical model. Failure to accommodate non-centrality will result in bias to $\hat{z}$ and parametric estimates of its uncertainty. Non-centrality has long been considered unimportant for magnetotelluric data except at the periods of the solar daily variation and its harmonics, but the recent discovery by Thomson *et al.* (2007) that the geomagnetic field contains the high-Q signature of normal solar modes over a wide frequency range requires re-examination of this assumption.

6. Leakage from high-power portions of the spectrum into lower-power ones due to poorly chosen data tapers (see Section 5.2) can induce non-centrality, with a highly correlated but random non-centrality parameter, biasing estimates of $\hat{z}$ and its uncertainty.

Any one of these issues can seriously impact the least-squares solution (5.7) in unpredictable and sometimes insidious ways; in the presence of more than one, difficulty is guaranteed.

These problems have led to the introduction into statistics and hence magnetotellurics of procedures that are robust, in the sense that they are relatively insensitive to a moderate amount of bad data or to inadequacies in the model, and that they react gradually rather than abruptly to perturbations of either. Stigler (2010) provides a historical perspective on the evolution of robust statistics. Robust spectral analysis and its variants have revolutionized the practice of magnetotellurics over the past two decades, making it possible to nearly automatically obtain good response estimates under most conditions. Robust methods and their extensions will be described in the following sections, after an initial set of comments on the art of spectral analysis.

5.2 Discourse on spectral analysis in magnetotellurics

A detailed elaboration of the principles and practice of the analysis of finite time sequences is beyond the scope of this book, and the reader is referred to the excellent treatment of Percival & Walden (1993) for the univariate case and to Koopmans (1995) for multivariate extensions. An overview will be presented here, with an emphasis on approaches that lend themselves to a robust treatment.

It will be assumed that finite time sequences of the horizontal electric and magnetic field variations from one or more sites over one or more contemporaneous time intervals are available. It is also presumed that the time series have been digitized without aliasing. The initial step in magnetotelluric processing is inspection of the data in the time domain to identify and note any unusual segments, and to plan manual or automated editing of them. While it is best to remove gross errors like boxcar shifts or large spikes, well-designed robust processing algorithms can usually accommodate such problems. Finally, serious long-term trends should be removed by high-pass filtering or least-squares spline fitting.

An essential step in any spectral analysis is the point-by-point multiplication of a data sequence by one or more data windows or tapers prior to taking the Fourier transform to control spectral leakage or bias. Historically, these have been chosen on the basis of computational simplicity or other *ad hoc* criteria (e.g. Harris, 1978). However, an optimal class of data tapers can be obtained by considering those finite time sequences (characterized by a length N) that have the largest possible concentration of energy in the frequency interval $(-W, W)$, where W is a free parameter that has units of $1/N$ or Nyquists, ranging over $[0.0, 0.5)$. Nyquists may be converted to dimensional frequency through division by the sample interval. The discrete time version of the optimal taper problem was solved by Slepian (1978), and the result is an orthogonal family of discrete prolate spheroidal or Slepian sequences $v_n^{(k)}(N, W)$, where n is the time index and k is the order number. Numerical solutions for the Slepian sequences are obtained by solving an eigenvalue problem; a numerically robust tridiagonal form is given by Slepian (1978), and a more accurate variant is described in Thomson (1990, appendix B). For given choices of N and W, there are $2NW$ (the Shannon number from information theory) Slepian sequences that reduce bias due to spectral leakage. The eigenvalue λ_k for the eigenfunction $v_n^{(k)}(N, W)$ gives the fractional energy concentration inside $(-W, W)$. Because of their orthogonality, spectra obtained using successive Slepian sequences are approximately independent, but as the order number increases, the bias properties become increasingly suboptimal.

Of all possible functions, the zeroth-order Slepian sequence $v_n^{(0)}(N, W)$ has the highest energy concentration in $(-W, W)$, making it the best choice as a single data taper for spectral analysis (Thomson, 1977, 1982; Percival & Walden, 1993). The time–bandwidth product $\tau = NW$ determines both the amount of bias protection outside the main lobe of the window and the main lobe halfwidth (Figure 5.1). A useful range of τ for magnetotelluric data processing is 1 to 4. A $\tau = 1$ window provides limited (about 23 dB) bias protection, but yields raw estimates that are independent on the standard discrete Fourier transform grid with frequencies separated by $1/N$. In this instance, pre-whitening to reduce the dynamic

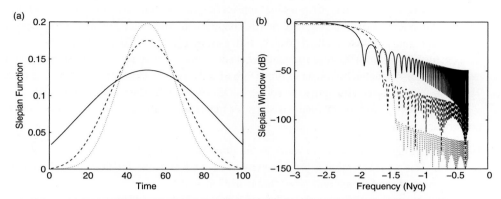

Figure 5.1. (a) Three 100-point Slepian sequence data windows with a time–bandwidth product τ of 1 (solid), 2.5 (dashed) and 4 (dotted). Note the increase in time concentration as the time × bandwidth product rises. (b) Spectral windows for the Slepian sequences shown in (a). For a 100-point sequence, the fundamental (Rayleigh resolution) frequency increment is 1/100 or −2 on a log scale. The $\tau = 1$ spectral window very nearly has this resolution, but its first sidelobe is only ~23 dB down from the peak. By contrast, the $\tau = 4$ window has a spectral resolution of 4/100 or −1.4 on a log scale, while the first sidelobe is more than 100 dB down from the peak.

range of the spectrum is essential to avoid bias. A $\tau = 4$ window provides over 100 dB of bias protection (Figure 5.1), nearly negating the requirement for pre-whitening, but yields raw estimates that are highly correlated over $(f - W, f + W)$, where f is the frequency of interest. As a rule of thumb, frequencies that are spaced τ apart on the discrete Fourier transform grid may be taken as independent. It is also important to avoid frequencies within W of zero frequency, as these are often contaminated by unresolved low-frequency components.

Pre-whitening may be achieved by time-domain filtering with a short autoregressive (AR) sequence fit to the time series. Standard time-domain AR estimators are based on least-squares principles; for example, the Yule–Walker equations (Yule, 1927; Walker, 1931) may be solved for the autocovariance sequence (acvs), from which the AR filter can be obtained using the Levinson–Durbin recursion (Levinson, 1947; Durbin, 1960). However, in the presence of extreme data such as are common in magnetotellurics, least-squares estimators for the acvs are often seriously biased, yielding AR filters that may be either unstable (i.e. the zeros of the filter z-transform lie outside the unit circle) or highly persistent (i.e. the zeros of the filter z-transform lie inside but very close to the unit circle). In such instances, the filter may actually enhance spectral leakage by increasing the dynamic range of the data spectrum. A robust approach to acvs estimation is essential to avoid these difficulties. A simple robust AR estimator may be devised using acvs estimates obtained from the Fourier transform of a robust estimate of the power spectrum of the time series. The robust power spectrum may be computed using the WOSA method (see next paragraph) with a low-bias data taper and taking the frequency-by-frequency median rather than mean of the raw section estimates. The AR filter follows from the robust acvs using the Levinson–Durbin recursion.

While there are a number of possible approaches to spectral analysis, magnetotelluric response function estimation coverage in this chapter will emphasize the Welch overlapped section averaging (WOSA) method (Welch, 1967; see Percival & Walden, 1993, section 6.17, for a detailed discussion) because anomalous data detection is facilitated by comparing spectra estimated over distinct subsections in time (Egbert & Booker, 1986; Banks, 1998; Chave & Thomson, 2004). Some alternatives to robust WOSA will be discussed in Section 5.10.

At a high level, the spectral analysis steps for magnetotelluric response function estimation are as follows:

1. Pre-whitening of the raw time series in the time domain to reduce the spectral dynamic range. This step is optional but recommended, and depends on the spectral leakage properties of the Slepian data taper that is used at step 3.
2. Starting at the lowest frequency (or longest period) of interest, a subset length N that is of order a few over the frequency of interest is selected. N is typically much smaller than the length of the time series.
3. The resolution bandwidth W is chosen and each subset is tapered with the zeroth-order Slepian sequence. Data sections may be overlapped to improve statistical efficiency. The amount that data sections can be overlapped without affecting their independence varies, ranging from 50% to 71% as τ ranges from 1 to 4 (Percival & Walden, 1993, section 6.17).
4. Discrete Fourier transforms of each tapered data section are taken, pre-whitening is corrected for if it was used and the instrument response is removed as applicable. In the context of this chapter, the data become the Fourier transforms of the windowed data from each section at the frequencies of interest.
5. The least-squares solution (5.7) is obtained using the section Fourier transforms of the electric and magnetic fields at a single frequency. This is the step at which robust or bounded influence weighting may also be applied, as described in Sections 5.5 and 5.6.
6. An estimate of the covariance is obtained, either parametrically using (5.9) with (5.13) or non-parametrically using the jackknife, as described in Section 5.9. Confidence interval construction requires further distributional assumptions, either using (5.14) or a non-parametric approach.
7. The section length is then repetitively reduced as higher frequencies are addressed. A variable section length WOSA analysis of this type is philosophically akin to wavelet analysis in that the time scale of the basis functions is changed along with the frequency scale to optimize resolution.
8. The quality of the response estimate is assessed through some combination of coherence analysis, residual analysis and hat matrix diagonal analysis (Section 5.8).

5.3 Data assessment

For the purposes of illustration, three exemplar long-period magnetotelluric datasets collected in November 2003 in the Kaapvaal province of South Africa will be utilized. These are Site 127 (28°48′04″S, 23°47′14″E), Site 145 (26°19′13″S, 26°05′18″E) and Site 172

(22°37′49″S, 29°30′40″E). They lie on a roughly southwest to northeast line separated by 357 km (Sites 127–145), 537 km (Sites 145–172) and 829 km (Sites 127–172), respectively. Each time series is sampled at 5 s intervals, and consists of two horizontal electric and magnetic field components at Sites 127 and 172 and two horizontal magnetic field components at Site 145. In all instances, the x and y components are aligned with local geomagnetic north and east, respectively. Each time series has been anti-alias filtered with an analog six-pole Bessel filter with -3 dB point at 5 Hz and a digital multi-stage Chebyshev finite impulse response filter with the final stage at twice the sampling rate. The telluric channels have also been high-pass filtered with an analog single-pole Butterworth filter having the -3 dB point at 30 000 s.

The present section will define a set of useful summary statistics and their application to the exemplar dataset. This begins with time-domain inspection of the data. The horizontal magnetic field data are fairly clean and visually coherent. There are no obvious problems for Sites 127 and 145, while Site 172 displays occasional large spikes and a single boxcar offset in B_y late in the record. As expected, the electric field data show larger inter-site variations. The Site 127 and 172 data show clear evidence for the solar daily variation, with a few obvious high-amplitude spikes. In the spirit of providing a realistic test for the methods discussed in this chapter, no correction for these data problems will be applied.

It is appropriate to assess the data in the frequency domain using both univariate and multivariate approaches. Two tools based on the multi-taper method of Thomson (1982) will be described. This method applies the family of orthogonal Slepian sequences $v_n^{(k)}(N, W)$ to data sequences, as described by Thomson (1982) and Percival & Walden (1993, chapter 7). To be specific, let the raw Fourier transform at each order number k be

$$a_k(f) = \sum_{n=0}^{N-1} v_n^{(k)}(N, W)x_n e^{-i2\pi fn} \tag{5.15}$$

where x_n is the data sequence. The absolute square of (5.15) is called the eigenspectrum $\hat{S}_k(f)$. Because of the orthogonality of the Slepian sequences, the set $\{\hat{S}_k(f)\}$ is approximately independent for distinct choices of order number k; this is exact for white noise, emphasizing the importance of pre-whitening.

The raw estimates may be combined to yield the multi-taper power spectrum through adaptive weighting. The multi-taper estimate is

$$\hat{S}(f) = \frac{\sum_{k=0}^{K-1} \lambda_k d_k^2(f)\, \hat{S}_k(f)}{\sum_{k=0}^{K-1} \lambda_k d_k^2(f)} \tag{5.16}$$

where $K \leq 2NW$ is a free parameter and the real adaptive weights are

$$d_k(f) = \frac{\sqrt{\lambda_k} S(f)}{\lambda_k S(f) + \sigma^2 (1 - \lambda_k)} \tag{5.17}$$

$S(f)$ and σ^2 are the population values of the power spectrum and data variance, respectively. The adaptive weights reduce the influence of the raw estimates at frequencies and for order numbers where the broadband bias $\sigma^2(1 - \lambda_k)$ becomes comparable to the spectrum, eliminating spectral leakage as a source of bias. In the absence of broadband bias, the estimate (5.16) has $2K$ degrees of freedom per frequency.

The population parameters in (5.17) must be replaced with estimates to get the adaptive weights and power spectrum estimate. Substituting (5.17) into (5.16) and rearranging terms gives the nonlinear equation

$$\sum_{k=0}^{K-1} \frac{\lambda_k^2 [\hat{S}(f) - \hat{S}_k(f)]}{[\lambda_k \hat{S}(f) + s^2(1 - \lambda_k)]^2} = 0 \tag{5.18}$$

where s^2 is the sample variance for the data sequence. Equation (5.18) may be solved numerically for $\hat{S}(f)$ at each frequency, after which the weights follow from (5.17).

It is quite useful to compute the data and noise power spectral density of the time series to characterize their behavior as a function of frequency and their evolution over time. The data spectra of the M time sequences (i.e. the electromagnetic field components at a set of stations) are obtained from the multi-taper algorithm using a section length that is a few times the inverse of the lowest frequency of interest. The time evolution follows by computing the spectrum over a succession of data sections.

An approximation to the noise spectrum may be obtained from the residual variance obtained after regressing the section Fourier transforms for a given time series against those for all of the remaining ones. The adaptively weighted Fourier transform of the mth data sequence is given by

$$c_k^m(f) = \frac{\sqrt{\lambda_k} d_k^m(f) a_k^m(f)}{\sqrt{\sum_{k=0}^{K-1} \lambda_k [d_k^m(f)]^2}} \tag{5.19}$$

Let $\mathbf{c}_{(m)}$ be the K-vector obtained by stacking (5.19) for the mth time sequence and $\mathbf{C}_{(m)}$ be the $K \times M - 1$ matrix comprising $\mathbf{c}_{(i)}$ for the remaining time series; K must be larger than $M - 1$. The noise power spectral density is given by

$$N^m(f) = \mathbf{c}_m^H \cdot \left(\bar{\bar{\mathbf{I}}} - \bar{\bar{\mathbf{H}}}_m \right) \cdot \mathbf{c}_m \tag{5.20}$$

where $\bar{\bar{\mathbf{I}}}$ is the $K \times K$ identity matrix and

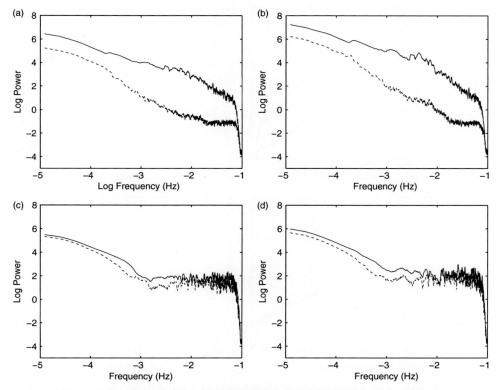

Figure 5.2. Data (solid) and noise (dashed) spectra for (a,c) E_x and (b,d) E_y at Site 172 for two distinct time intervals. (a,b) Spectra for the highly energetic initial ~12 h, as reflected in a signal-to-noise ratio of 100 or more over most of the frequency range. (c,d) By contrast, an interval about 6 d later where the signal-to-noise ratio is only ~1, and the data spectrum is white at frequencies above 10^{-3} Hz. The spectra have a time–bandwidth product of 13 and 25 Slepian tapers, yielding about 50 degrees of freedom at each frequency. The ordinate scale is identical for all plots. The sharp roll-off at high frequencies is caused by anti-alias filtering that is identical for all time series.

$$\vec{\bar{H}}_m = \vec{\bar{C}}_m \cdot \left(\vec{\bar{C}}_m^H \cdot \vec{\bar{C}}_m \right)^{-1} \cdot \vec{\bar{C}}_m^H \qquad (5.21)$$

is the $K \times K$ hat matrix.

For the purposes of illustration, Site 172 will be regarded as the local one, where electric and magnetic fields are measured, while Sites 127 and 145 will be treated as reference ones, where only the magnetic field is measured; consequently, $M = 8$. Data sections 8192 points long (or about 11.4 h) will be used with a time bandwidth of 13 (or a resolution bandwidth of 6×10^{-4} Hz) and 25 Slepian tapers, yielding about 46 degrees of freedom per frequency after adaptive weighting. Figure 5.2 shows multi-taper spectra for the Site 172 electric field in two sections of data that illustrate the range of behavior. Panels (a) and (b)

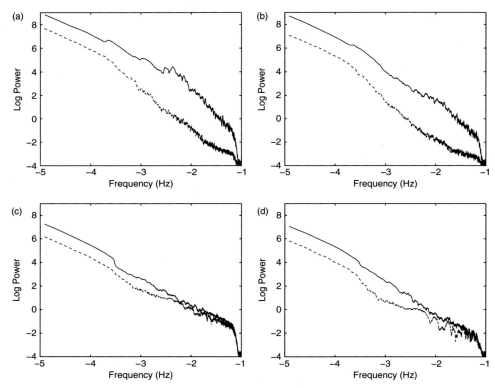

Figure 5.3. Data (solid) and noise (dashed) spectra for (a,c) B_x and (b,d) B_y at Site 172 for two distinct time intervals. (a,b) Spectra for the highly energetic initial ~12 h, as reflected in a signal-to-noise ratio of 100 or more over most of the frequency range. (c,d) By contrast, an interval about 4 d later where the signal-to-noise ratio is only slightly better than 1 and the noise spectra contain a variety of narrowband features, especially for B_y. Inspection of the data shows that this is due to natural variations in source field energy level. See Figure 5.2 caption for further details.

show the E_x and E_y spectra for an interval characterized by an energetic source field. The signal-to-noise ratio is above 100 over the entire frequency range. Panels (c) and (d) show the same components for an interval that is extremely quiet except for a short ~1000 point burst of high-frequency activity that is not seen in any of the other data series, and hence must be due to a disturbance local to the Site 172 electric field measurements. These two sections represent extremes, and examination of spectra for each of the 50 data sections shows that the signal-to-noise ratio is at least 10 for all except two of them, indicating that the data quality is high.

Figure 5.3 shows analogous results for the B_x and B_y components at Site 172. Panels (a) and (b) are for the same section as in Figure 5.2, while panels (c) and (d) are for a different one. These two examples again show the extremes in behavior, with energetic time intervals displaying a signal-to-noise ratio of well over 100 and quiet intervals where it approaches 1.

About 10% of the data sections are comparable to panels (c) and (d). An additional 10% show the effect of data spikes through a nearly white noise data spectrum. Similar results are observed for the reference site magnetic fields.

The power spectrum characterizes the variability of a given dataset, but does not provide any information about its relationship to other time series. While a wide range of statistical tools exist to characterize the frequency-domain covariance of time series, a simple approach is based on examination of the eigenvalues of the matrix comprising all possible cross-spectra between the data series. This is effectively the first step in the multivariate approach to magnetotelluric response function estimation of Egbert (1997), as described in Section 5.10. The input data are the adaptively weighted Fourier trans-forms of the data sections (5.19) used to obtain power spectra. These constitute the columns of the $K \times M$ spectral matrix $\vec{\mathbf{A}}$ at a given frequency; K must be larger than M, and typically is chosen so that $2(K-M) \geq 20$. The spectral matrix is then decomposed using the singular value decomposition

$$\vec{\mathbf{A}} \cdot \vec{\mathbf{D}} = \vec{\mathbf{U}} \cdot \vec{\Sigma} \cdot \vec{\mathbf{V}}^H \tag{5.22}$$

where $\vec{\mathbf{D}}$ is a diagonal weight matrix specified below, $\vec{\mathbf{U}}$ and $\vec{\mathbf{V}}$ are $K \times K$ and $M \times M$ matrices containing the eigenvectors of $\vec{\mathbf{A}} \cdot \vec{\mathbf{D}} \cdot \vec{\mathbf{D}} \cdot \vec{\mathbf{A}}^H$ and $\vec{\mathbf{D}} \cdot \vec{\mathbf{A}}^H \cdot \vec{\mathbf{A}} \cdot \vec{\mathbf{D}}$, respectively, and $\vec{\Sigma}$ is a $K \times M$ matrix containing M singular values on the diagonal. Only the singular values are of interest for the present purpose, and algorithms that obtain these without computing the eigenvectors are available. The weights in $\vec{\mathbf{D}}$ are the inverse square roots of the noise spectrum (5.20); this is equivalent to modeling the noise covariance of $\vec{\mathbf{A}}$ as a diagonal matrix, meaning that there is presumed to be no correlated noise in the data. The square of the singular values in $\vec{\Sigma}$ plotted against frequency characterize the dimensionality of the data. Because of the normalization, the squared singular values are dimensionless and serve as an approximation to the signal-to-noise ratio. The number of squared singular values significantly greater than unity is the effective coherence dimension of the dataset. For the quasi-uniform source magnetotelluric model, this should be two; larger values indicate the presence of additional, non-magnetotelluric sources. This allows the coherence dimension to be visualized as a function of frequency. As a simple way to obtain a robust result, the frequency-by-frequency order statistics of the squared singular values obtained by ranking them will be used.

Figure 5.4 shows the frequency-by-frequency median squared singular values for all of the available data (10 channels) compared to four channels (horizontal electric and magnetic fields) at various sites with two horizontal magnetic channels at others as defined in the caption. The result for all of the data (panel (a)) indicates a coherence dimension of three or four across the entire frequency band, with the spread between the largest and next two squared singular values increasing at high frequencies. This suggests the presence of correlated noise in some data sections, or else the presence of a source field gradient across the three-station array, or a combination. By contrast, the result using four channels at Site 172 and two channels at Site 145 (panel (b)) has a coherence dimension of two. A nearly

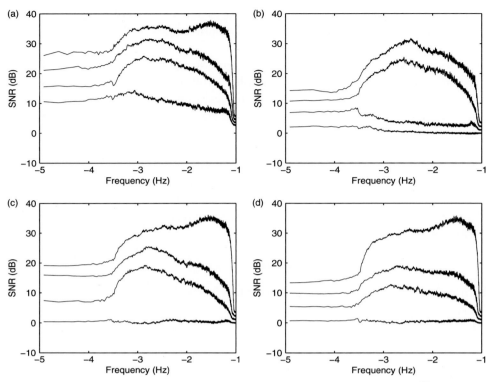

Figure 5.4. The frequency-by-frequency median four largest squared singular values in $\bar{\bar{\mathbf{A}}}$ for: (a) Site 127 (four), Site 145 (two) and Site 172 (four); (b) Site 172 (four) and Site 145 (two); (c) Site 127 (four) and Site 145 (two); and (d) Site 127 (four) and Site (172) (two). The number in parentheses denotes the number of data channels, where "four" means horizontal electric and magnetic, and "two" means horizontal magnetic. The estimates have a time–bandwidth product of 13 and 25 Slepian tapers, yielding about 34 degrees of freedom at each frequency, except for panel (a), where there are about 30 degrees of freedom. The ordinate scale is identical for all plots. The sharp roll-off at high frequencies is caused by anti-alias filtering.

identical result is obtained if the magnetic channels at Site 127 are substituted for those at Site 145. Panels (c) and (d) show the median squared singular values for four channels at Site 127 and two channels at (c) Site 145 and (d) Site 172. In both cases, the coherence dimension is three. Further, the shape of the two largest squared singular values as a function of frequency when four channels are used at Site 172 (panel (b)) is mimicked by the second and third squared singular values when four channels from Site 127 are processed. Taken in aggregate, this result suggests that the Site 172 data are dominated by a magnetotelluric source, that the magnetic fields at Sites 127 and 145 are relatively clean, but the Site 127 electric field is strongly contaminated by non-geomagnetic coherent noise. Given the proximity of Site 127 to train tracks and diamond mines, the cause is most likely DC train noise and/or mining equipment.

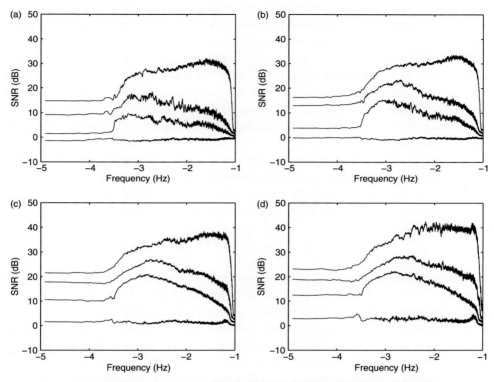

Figure 5.5. The frequency-by-frequency (a) 10th, (b) 25th, (c) 75th and (d) 90th percentiles for the four largest squared eigenvalues of $\ddot{\mathbf{A}}$ using the horizontal electric and magnetic fields from Site 127 and the horizontal magnetic fields from Site 145. See Figure 5.4 caption for further details.

Figure 5.5 shows the frequency-by-frequency 10th, 25th, 75th and 90th percentiles for four channels at Site 127 and two channels at Site 145. These results make it clear that the electric field contamination is pervasive over most, if not all, of the time series.

5.4 The remote reference method

Unless the local magnetic field variable $\vec{\mathbf{b}}$ is noise-free, the least-squares estimator (5.7) will always be downward-biased. To see this, let $\vec{\mathbf{b}} = \vec{\mathbf{b}}_0 + \vec{\mathbf{n}}_b$, where $\vec{\mathbf{b}}_0$ is noise-free and uncorrelated with the noise term $\vec{\mathbf{n}}_b$. It is easy to show that $\vec{\mathbf{b}}^H \cdot \vec{\mathbf{b}} = \vec{\mathbf{b}}_0^H \cdot \vec{\mathbf{b}}_0 + \vec{\mathbf{n}}_b^H \cdot \vec{\mathbf{n}}_b$, and hence $\left(\vec{\mathbf{b}}^H \cdot \vec{\mathbf{b}} \right)^{-1}$ in (5.7) will be systematically smaller than $\left(\vec{\mathbf{b}}_0^H \cdot \vec{\mathbf{b}}_0 \right)^{-1}$ as the noise power increases. This issue has long been recognized by statisticians and magneto-telluric practitioners, and a number of mitigation strategies have been proposed. One approach is solving (5.7) by interchanging $\mathbf{e}$ and $\mathbf{b}$, and then average its inverse with the original response. This will work if the relative signal-to-noise ratio is the same in both the electric and magnetic fields, but otherwise the average should be weighted according to the true signal-to-noise ratio, which is rarely well characterized. For single-station magneto-tellurics, a variety of other solutions to the bias problem have been proposed. For example, Sims *et al.* (1971) applied six permutations of the response and predictor variables to characterize the bias. Kao & Rankin (1977) introduced an iterative version of (5.7) weighted by the coherence in an attempt to minimize the bias. Pedersen (1982) provides a thorough analysis of bias effects on single-station magnetotelluric responses. Müller (2000) derived expressions for the bias under specific models for the uncorrelated noise and the structure of the response tensor.

The shortcomings of existing methods led Goubau *et al.* (1978) and Gamble *et al.* (1979) to introduce the now-standard approach to bias reduction, the remote reference method, in which an auxiliary set of magnetic field variables $\bar{\mathbf{b}}_r$ are obtained that contain the same signal that is in $\bar{\mathbf{b}}$ but different noise. The remote reference estimator for (5.5) is

$$\hat{\mathbf{z}}^r = \left(\bar{\mathbf{b}}_r \cdot \bar{\mathbf{b}}\right)^{-1} \cdot \left(\bar{\mathbf{b}}_r \cdot \mathbf{e}\right) \tag{5.23}$$

in which the cross-powers between the remote and local variables replace the local auto- and cross-powers. If the noise in the local and remote variables is uncorrelated, and the coherence between $\bar{\mathbf{b}}$ and $\bar{\mathbf{b}}_r$ is very high, then (5.23) is nearly unbiased. While there is an almost universal tendency to use remote magnetic sites as references, Pomposiello *et al.* (2009) have shown that continuing improvement in electric field data quality allows their use in some circumstances. In this instance, $\bar{\mathbf{b}}_r$ is replaced with $\bar{\mathbf{e}}_r$ in (5.23).

The remote reference method is actually a rediscovery of the method of instrumental variables (the statistical term for the remote reference variables) to solve errors-in-variables problems in econometrics, and was introduced by Wald (1940), Reiersol (1945) and Geary (1949). While it is widely believed by magnetotelluric practitioners that the remote reference method is unbiased independent of the degree of correlation between $\bar{\mathbf{b}}$ and $\bar{\mathbf{b}}_r$, the statistical literature shows this to be incorrect. The remote reference solution will be biased in unpredictable ways by correlated noise between the local and reference variables. In addition, weak correlation of the local and reference variables results in downward bias for two reasons: in this instance, even a slight correlation between $\bar{\mathbf{b}}_r$ and the regression residuals $\mathbf{r}$ can lead to large inconsistency, and for finite samples, the remote reference estimate (5.23) is always biased in the direction of ordinary least squares, with comparable bias as the correlation approaches zero (Bound *et al.*, 1995). Further, the remote reference solution is not minimum variance, and hence will always yield a solution with larger confidence intervals compared to the least-squares one.

The necessary separation of the two measurements to ensure lack of correlation of the noise in $\bar{\mathbf{b}}$ and $\bar{\mathbf{b}}_r$ depends on the local geology, the source field behavior, instrument noise and cultural noise. For example, if the noise source is purely instrumental, then it might be expected that paired, co-located magnetic field measurements will be sufficient. If the noise source is instrument motion induced by the wind, then separation by more than its correlation scale (typically, a few tens of meters) will work. If the noise source is DC trains, then a separation of hundreds of kilometers may be required to obtain a noise-free reference. In practice, the separation must be determined empirically – for three relevant studies, see Goubau *et al.* (1984), Larsen *et al.* (1996) and Shalivahan & Bhattacharya (2002).

Another problem with the remote reference method is that there has historically been no obvious choice for either the residuals or the hat matrix. Chave & Thomson (1989) defined the residuals using the local magnetic field

$$\mathbf{r} = \mathbf{e} - \vec{\mathbf{b}} \cdot \hat{\mathbf{z}}^r \tag{5.24}$$

and introduced a pseudo-hat matrix based on the standard definition (5.10)

$$\vec{\mathbf{H}}_r^0 = \vec{\mathbf{b}} \cdot \left(\vec{\mathbf{b}}_r^H \cdot \vec{\mathbf{b}} \right)^{-1} \cdot \vec{\mathbf{b}}_r \tag{5.25}$$

However, (5.25) is neither Hermitian nor a projection matrix, so does not share the standard properties of the hat matrix.

While the ordinary remote reference method utilized the same number of references as local magnetic channels, Chave & Thomson (2004) introduced a generalization that encompasses many reference channels. Assume that an auxiliary set of remote reference measurements $\vec{\mathbf{Q}}$ are available, where for a given frequency, $\vec{\mathbf{Q}}$ is $N \times q$ with $q \geq p$. For example, if more than one remote reference site is available, the q may be a multiple of 2 and $\vec{\mathbf{Q}}$ contains the horizontal magnetic fields from all of the sites. The local magnetic field $\vec{\mathbf{b}}$ may be regressed on $\vec{\mathbf{Q}}$ in the usual way by solving

$$\vec{\mathbf{b}} = \vec{\mathbf{Q}} \cdot \hat{\vec{\mathbf{t}}} + \vec{\boldsymbol{\varepsilon}} \tag{5.26}$$

whose solution $\hat{\vec{\mathbf{t}}}$ is a $q \times 2$ vector of inter-site transfer functions between the local and remote horizontal magnetic fields. Equation (5.26) may be solved by least squares or by any of the robust/bounded influence methods described in subsequent sections. By analogy to (5.10), the predicted local magnetic field is

$$\hat{\vec{\mathbf{b}}} = \vec{\mathbf{Q}} \cdot \left(\vec{\mathbf{Q}}^H \cdot \vec{\mathbf{Q}} \right)^{-1} \cdot \vec{\mathbf{Q}}^H \cdot \vec{\mathbf{b}} \tag{5.27}$$

and is just a projection of $\vec{\mathbf{b}}$. Substituting $\hat{\vec{\mathbf{b}}}$ for $\vec{\mathbf{b}}$ in (5.7) yields the generalized remote reference estimator

$$\hat{\mathbf{z}}^q = \left(\hat{\vec{\mathbf{b}}}^H \cdot \hat{\vec{\mathbf{b}}} \right)^{-1} \cdot \left(\hat{\vec{\mathbf{b}}}^H \cdot \mathbf{e} \right) \tag{5.28}$$

It is easy to show that $\hat{\vec{\mathbf{b}}}^H \cdot \hat{\vec{\mathbf{b}}} = \hat{\vec{\mathbf{b}}}^H \cdot \vec{\mathbf{b}}$. Expanding (5.28) using (5.27) yields

$$\hat{\mathbf{z}}^{q} = \left[\vec{\mathbf{b}}^{H} \cdot \vec{\mathbf{Q}} \cdot \left(\vec{\mathbf{Q}}^{H} \cdot \vec{\mathbf{Q}} \right)^{-1} \cdot \vec{\mathbf{Q}}^{H} \cdot \vec{\mathbf{b}} \right]^{-1} \cdot \left[\vec{\mathbf{b}}^{H} \cdot \vec{\mathbf{Q}} \cdot \left(\vec{\mathbf{Q}}^{H} \cdot \vec{\mathbf{Q}} \right)^{-1} \cdot \vec{\mathbf{Q}}^{H} \cdot \mathbf{e} \right] \qquad (5.29)$$

Since $\left(\vec{\mathbf{Q}}^{H} \cdot \vec{\mathbf{Q}} \right)^{-1}$ appears in both the numerator and denominator of (5.29), and only cross-products of $\vec{\mathbf{b}}$ and $\vec{\mathbf{Q}}$ appear in the remaining inverse terms, downward bias by uncorrelated noise in the reference fields is eliminated. When $q = 2$, (5.29) reduces to (5.23). Just as the ordinary remote reference method is identical to a previously discovered technique, the generalized remote reference method is equivalent to the two-stage least-squares solution of econometrics (Anderson & Rubin, 1949, 1950). The use of multiple references can improve magnetotelluric response estimates if different channels become noisy at different times, as shown by example in Chave & Thomson (2004). However, as for the remote reference estimator (5.23), (5.29) will be biased by correlated noise between $\vec{\mathbf{b}}$ and $\vec{\mathbf{Q}}$.

Using (5.28), the hat matrix (5.10) becomes

$$\vec{\mathbf{H}}_{q} = \hat{\mathbf{b}} \cdot \left(\hat{\mathbf{b}}^{H} \cdot \hat{\mathbf{b}} \right)^{-1} \cdot \hat{\mathbf{b}}^{H} \qquad (5.30)$$

and for a single reference site reduces to

$$\vec{\mathbf{H}}_{r} = \vec{\mathbf{b}}_{r} \cdot \left(\vec{\mathbf{b}}_{r}^{H} \cdot \vec{\mathbf{b}}_{r} \right)^{-1} \cdot \vec{\mathbf{b}}_{r}^{H} \qquad (5.31)$$

Equation (5.31) has the main properties of a hat matrix. Chave & Thomson (2004) advocate replacing (5.25) with (5.31) for ordinary remote reference processing.

Figure 5.6 compares the off-diagonal elements of the magnetotelluric response at Site 172 from ordinary least squares using (5.7) with that from (5.23) using the Site 145 magnetic fields as a reference. Nearly identical results are obtained using Site 127 as the reference. In this and succeeding examples, a Slepian sequence with a time bandwidth of 1 and 50% overlap between sections was applied, and the data were pre-whitened with a robust AR filter containing 11 terms. While no attempt has been made to determine the regional strike of the response tensor (nor is that necessary for the purposes of this chapter), the diagonal elements are both much smaller than the off-diagonal ones across the frequency band. The least-squares apparent resistivity for the Z_{xy} element is seriously downward-biased over most of the period range as compared to the remote reference estimate, and the least-squares phase is statistically meaningless given the size of the error bars. The bias in the Z_{yx} apparent resistivity (panels (c) and (d)) is less severe, being substantial only at the longest periods, while the phase is affected more seriously. It is not obvious why the least-squares estimate for Z_{xy} is more biased than that for Z_{yx}. From Figure 5.3, there is not a large difference in power between the components of either the electric or magnetic fields at Site 172, so it is unlikely to be due to a higher relative noise level in some of the data.

Figure 5.7 compares the ordinary remote reference result from Figure 5.6 with the generalized remote reference estimate from (5.28) using both Sites 145 and 127 as references.

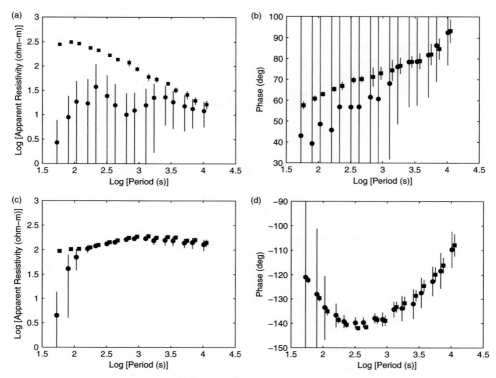

Figure 5.6. (a,c) Apparent resistivity and (b,d) phase as a function of period for the off-diagonal elements of the magnetotelluric response function at Site 172: (a,b) Z_{xy} and (c,d) Z_{yx}; see text for details. The vertical lines are the double-sided jackknife Bonferroni confidence intervals using the delta method, as described in Section 5.8. The circles denote the least-squares estimate without a remote reference, while the squares denote the ordinary remote reference estimate using Site 145. The latter have been offset to the right on the *x* axis for clarity. The effective degrees of freedom varies from 150 to over 39 000 as period decreases.

Unsurprisingly given that the ordinary remote reference estimates using Sites 145 and 127 separately are essentially identical, there is not much change from using two reference sites. However, the confidence limits are typically wider for the generalized estimate, reflecting the increasing departure from a minimum-variance result as the number of reference variables rises.

The value of the generalized remote reference estimator can be illustrated with an artificial example. The first one-eighth and middle one-eighth of the Site 145 reference data are replaced by Gaussian random numbers having the mean and one-tenth of the standard deviation of the original data series. Figure 5.8 shows the Z_{xy} apparent resistivity computed using the ordinary remote reference estimator with the contaminated Site 145 data and the generalized remote reference result using both the contaminated data and the original Site 127 data as references. As expected, the ordinary remote reference result is badly biased, displaying erratic apparent resistivity and especially phase as a function of period. However, the generalized estimate is

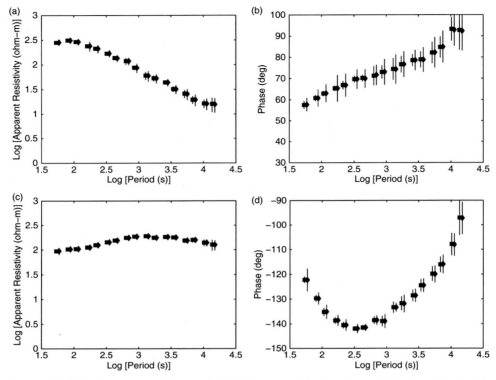

Figure 5.7. (a,c) Apparent resistivity and (b,d) phase as a function of period for the off-diagonal elements of the magnetotelluric response function at Site 172. The squares denote the remote reference estimate with Site 145, as shown in Figure 5.6, while the diamonds denote the two-stage estimate with both Sites 145 and 127. See Figure 5.6 caption for further details.

nearly unaffected, as can be seen by comparing Figure 5.7a with Figure 5.8. The only change is somewhat larger error bars when the contaminated Site 145 data are used.

The remote reference estimators (5.23) and (5.28) are usually quite effective at removing downward bias due to random noise in the local magnetic field, presuming that the coherence of those two entities is close to one, and remain an essential adjunct to successful robust processing methods. However, like the ordinary least-squares estimator (5.7), they remain susceptible to influential data in either the electric or magnetic fields. They will also break down in the presence of correlated noise in the local and reference fields.

5.5 Robust magnetotelluric processing 1: M-estimators

The six issues listed at the end of Section 5.1 have led to the development of magnetotelluric processing procedures that are robust. Chave *et al.* (1987) presented a tutorial review of robust methods in the context of geophysical data processing. Chave & Thomson (1989,

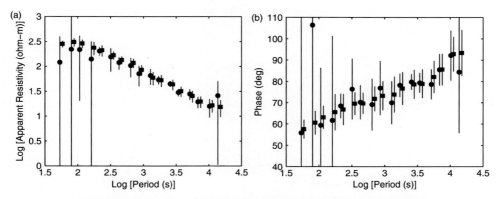

Figure 5.8. (a) Apparent resistivity and (b) phase for the Z_{xy} element of the magnetotelluric response function at Site 172. The circles denote the remote reference estimate with the contaminated version of the Site 145 data described in the text, while the squares denote the two-stage estimate using both the contaminated Site 145 and original Site 127 data. See Figure 5.6 caption for further details.

2004) describe the robust principles for magnetotelluric data analysis that constitute the core of this section. Egbert & Booker (1986) adapted robust methods to estimation of the geomagnetic transfer functions, yielding an algorithm that differs from the one described here only in a few implementation details. Jones *et al.* (1989) provide a comparison of magnetotelluric processing methods, showing the general advantages of robust algorithms. Alternative robust methods will be described in Section 5.10.

The most suitable type of robust estimator for magnetotelluric data is the *M-estimator* that is motivated by the maximum likelihood method of statistical inference (Stuart *et al.*, 1999, chapter 18). M-estimation is similar to least squares in that it minimizes a norm of the random errors in (5.3), but the misfit measure is chosen so that a few extreme values (called *outliers*) cannot dominate the result. The M-estimate is obtained computationally by minimizing $\mathbf{R}^H \cdot \mathbf{R}$ instead of $\mathbf{r}^H \cdot \mathbf{r}$, where $\mathbf{R}$ is an N-vector whose ith entry is $\sqrt{\rho(r_i/d)}$, d is a scale factor that makes the argument of ρ independent of the size of the entries in $\mathbf{r}$ and $\rho(x)$ is a *loss function*. The term "loss function" originates in statistical decision theory (Stuart & Ord, 1994, section 26.52) and, loosely speaking, is a measure of the distance between the true and estimated values of a statistical parameter. For standard least squares, $\rho(x) = x^2/2$, and (5.7) is immediately obtained (Figure 5.9). More generally, if $\rho(x)$ is chosen to be $-\log[f(x)]$, where $f(x)$ is the true probability density function (p.d.f.) of the regression residuals $\mathbf{r}$ that includes extreme values, then the M-estimate is maximum likelihood. In practice, $f(x)$ cannot be determined reliably from finite samples, and the loss function must be chosen on theoretical or empirical grounds.

Minimizing $\mathbf{R}^H \cdot \mathbf{R}$ gives the M-estimator analog to the normal equations

$$\vec{\mathbf{b}}^H \cdot \mathbf{\Psi} = 0 \qquad (5.32)$$

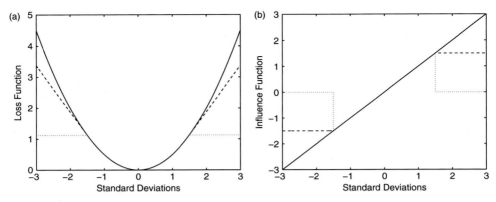

Figure 5.9. (a) The loss function and (b) the influence function for least squares (solid), the Huber estimator (5.36) or (5.37) (dashed) and a trimmed estimator (dotted). The Huber parameter α is set to 1.5. The trimmed estimator discards data larger than 1.5 standard deviations from the middle of the distribution, leading to an influence function that descends to zero and a loss function that becomes constant.

where $\boldsymbol{\Psi}$ is an N-vector whose ith entry is the *influence function* $\psi(x) = \partial_x \rho(x)$ evaluated at $x = r_i/d$, where d is a scale factor. For ordinary least squares, $\psi(x) = x$ (Figure 5.9), reducing (5.32) to (5.7).

Unless the influence function has a very simple form, equations (5.32) are nonlinear. They may be linearized and solved using iteratively re-weighted least squares. The solution is initialized with the ordinary least-squares estimate, or, for badly contaminated data, with a least absolute values (L_1) estimate obtained from a linear programming algorithm. At the kth subsequent iteration, $\boldsymbol{\Psi}$ is replaced by $\vec{\mathbf{v}}^{[k]} \bullet \mathbf{r}^{[k]}$, so that (5.32) becomes

$$\vec{\mathbf{b}}^{H} \bullet \vec{\mathbf{v}}^{[k]} \bullet \mathbf{r}^{[k]} = 0 \tag{5.33}$$

where $\vec{\mathbf{v}}^{[k]}$ is an $N \times N$ diagonal weight matrix whose ith entry is

$$v_{ii}^{[k]} = \frac{\psi\left(r_i^{[k-1]}/d^{[k-1]}\right)}{r_i^{[k-1]}/d^{[k-1]}} \tag{5.34}$$

The problem is linearized through use of the residuals and scale from the immediate prior iteration to compute the weights in (5.34). The weighted least-squares estimator for (5.32) at the kth iteration is

$$\hat{\mathbf{z}}_*^{[k]} = \left(\vec{\mathbf{b}}^{H} \bullet \vec{\mathbf{v}}^{[k]} \bullet \vec{\mathbf{b}}\right)^{-1} \bullet \left(\vec{\mathbf{b}}^{H} \bullet \vec{\mathbf{v}}^{[k]} \bullet \mathbf{e}\right) \tag{5.35}$$

The sequence of estimators terminates when the weighted sum of squared residuals $\mathbf{r}^{H} \bullet \vec{\mathbf{v}} \bullet \mathbf{r}$ does not change appreciably (for example, by 1% between iterations), typically after 3–5 steps. Since the weights are chosen to minimize the influence of data corresponding to large residuals in a well-defined sense, the M-estimator (5.35) is inherently data-adaptive.

A simple form for the loss and influence functions is due to Huber (1964), as shown in Figure 5.9. The conceptual model behind the Huber estimator is a Gaussian in the center of the data distribution and a double exponential (the natural distribution for the L_1 estimator, just as the Gaussian is the natural distribution for least squares) in the tails. The Huber loss function is

$$\rho(x) = \begin{cases} x^2/2 & (|x| \leq \alpha) \\ \alpha|x| - \alpha^2/2 & (|x| > \alpha) \end{cases} \tag{5.36}$$

and the Huber influence function is

$$\psi(x) = \begin{cases} x & (|x| \leq \alpha) \\ \alpha \, \mathrm{sgn}(x) & (|x| > \alpha) \end{cases} \tag{5.37}$$

The corresponding weights have diagonal elements

$$v_{ii} = \begin{cases} 1 & (|x_i| \leq \alpha) \\ \alpha/|x_i| & (|x_i| > \alpha) \end{cases} \tag{5.38}$$

The choice $\alpha = 1.5$ gives better than 95% efficiency with outlier-free Gaussian data. Downweighting with (5.38) begins when $|x_i| = |r_i/d| = \alpha$, so that the scale factor d determines which of the residuals are regarded as large. It is necessary because the weighted least-squares problem is not scale invariant without it, in the sense that multiplication of the data by a constant will not produce a comparably affine change in the solution.

The scale estimate d must be robust, and is chosen to be the ratio of the sample and theoretical values of some statistic based on a target distribution for the residuals. The data in a magnetotelluric context are Fourier transforms and therefore complex, but the complex Gaussian is not necessarily the optimal choice for the residual distribution. It is more appropriate to measure residual size using the magnitude since that quantity is rotationally (phase) invariant, and the appropriate distribution for the magnitude of a complex number is Rayleigh. The properties of the Rayleigh distribution are summarized in Johnson *et al.* (1994, chapter 18). The Rayleigh p.d.f. in standard form is

$$f(x) = xe^{-x^2/2} \tag{5.39}$$

from which standard statistical parameters such as the mean or median can be derived by integration.

A number of choices for d exist. Owing to its sensitivity to outliers, the ratio of the sample and theoretical standard deviations is not suitable. Better solutions can be obtained from the order statistics of the residuals obtained by sorting them into ascending ranks. Let $r_{(i)}$ be the ith order statistic, so that the middle order statistic or median is $r_{(\lfloor N/2 \rfloor + 1)}$, where the floor function $\lfloor x \rfloor$ denotes the integer part of x rounded down. The interquartile range is defined as the difference between the 75% and 25% points of the distribution. The sample value can be obtained from the order statistics as $r_{(\lfloor 3N/4 \rfloor + 1)} - r_{(\lfloor N/4 \rfloor + 1)}$, and compared to the theoretical value of $\sqrt{2\log 4} - \sqrt{2\log(4/3)} \approx 0.906\,58$ for the Rayleigh distribution. However, the interquartile range breaks down when the outlier fraction approaches 50%, and half of that value if it is only large residuals that are of concern. An extremely robust scale statistic is the median absolute deviation (MAD) from the median. The sample value is

$$S_{MAD} = \left| \mathbf{r} - r_{(\lfloor N/2 \rfloor + 1)} \right|_{\lfloor N/2 \rfloor + 1} \tag{5.40}$$

and can be obtained through two sort operations. The theoretical MAD is the solution to

$$F(\tilde{\mu} + \sigma_{MAD}) - F(\tilde{\mu} - \sigma_{MAD}) = 1/2 \tag{5.41}$$

where $\tilde{\mu}$ is the theoretical median of $\sqrt{2\log 2}$ and $F(x)$ denotes the Rayleigh cumulative distribution function (c.d.f.). This yields the transcendental equation

$$2\sinh(\tilde{\mu}\sigma_{MAD}) = e^{\sigma_{MAD}^2/2} \tag{5.42}$$

whose solution is $\sigma_{MAD} \approx 0.448\,45$.

The weights (5.38) fall off slowly for large residuals and never fully descend to zero, so that they provide inadequate protection against severe outliers. However, the loss function corresponding to (5.38) is convex (Figure 5.9), and hence convergence to a global minimum is assured, rendering it safe for the initial iterations of a robust procedure to obtain a reliable estimate of the scale. Motivated by the form of the Gumbel extreme value distribution, Thomson (1977) and Chave & Thomson (1989) suggested the more severe weight function

$$v_{ii} = \exp\left[e^{-\xi^2} \right] \exp\left[-e^{-\xi(|x_i| - \xi)} \right] \tag{5.43}$$

for the last one or two iterations, where the parameter ξ determines the residual size at which downweighting begins and the first term in (5.43) forces $v_i = 1$ when $x_i = 0$. The weight (5.43) is essentially a hard limit at $x_i = \xi$ except that the truncation function is continuous and continuously differentiable. Chave *et al.* (1987) advocated setting ξ to the Nth quantile $\sqrt{2\log(2N)}$ of the Rayleigh distribution to automatically increase the allowed residual magnitude as the sample size rises.

The M-estimator algorithm can be summarized as follows:

1. Given N estimates of the Fourier transforms of the horizontal electric and horizontal magnetic fields at the frequency of interest obtained as outlined in Section 5.2, compute the least-squares estimate (5.7).
2. Compute the residuals $\mathbf{r} = \mathbf{e} - \vec{\mathbf{b}} \cdot \hat{\mathbf{z}}$, the scale $d = s_{MAD}/0.448\,45$ using (5.40) and the initial residual sum of squares $\mathbf{r}^H \cdot \mathbf{r}$.

3. Compute the Huber weights (5.38) and the M-estimate using (5.35).
4. Compute the residuals $\mathbf{r} = \mathbf{e} - \vec{\mathbf{b}} \cdot \hat{\mathbf{z}}_*$, the scale $d = s_{MAD}/0.448\,45$ and the weighted residual sum of squares $\mathbf{r}^H \cdot \vec{\mathbf{v}} \cdot \mathbf{r}$.
5. Repeat steps 3 and 4 until the weighted residual sum of squares does not change by more than 1%.
6. Compute the Thomson weights (5.43), eliminate those data for which the weight is zero and compute the M-estimate using (5.35).
7. Compute the residuals $\mathbf{r} = \mathbf{e} - \vec{\mathbf{b}} \cdot \hat{\mathbf{z}}_*$, the scale $d = s_{MAD}/0.448\,45$ (fixed after the first iteration) and the weighted residual sum of squares $\mathbf{r}^H \cdot \vec{\mathbf{v}} \cdot \mathbf{r}$.
8. Repeat steps 6 and 7 with the scale d fixed until the residual sum of squares does not change by more than 1%.
9. Assess the fit (see Section 5.8), and compute the covariance matrix and confidence intervals (see Section 5.9).

The methodology described here can easily be adapted to use one or more remote references as described in Section 5.4. This is essential in practice, as M-estimators are not capable of eliminating random noise in the predictor magnetic variables, and downward bias will ensue. The remote reference or generalized remote reference M-estimator will be designated $\hat{\mathbf{z}}_*^r$ or $\hat{\mathbf{z}}_*^q$. Finally, it must be emphasized that M-estimators are sensitive only to extreme residuals that are usually caused by outliers in the electric field response variables. They are typically not able to handle extreme values in the predictor variables.

Figure 5.10 compares the remote reference estimate shown in Figures 5.6 and 5.7 with the robust estimate $\hat{\mathbf{z}}_*^r$, in both cases using Site 145 as the magnetic field reference. The robust estimate was computed using the Nth quantile of the Rayleigh distribution to define ξ in (5.43). In contrast to the differences between the ordinary least-squares and remote reference results in Figure 5.6, the distinctions between the remote reference and robust estimates are small, being manifest especially in the error estimates. However, the robust estimate is in general a smoother function of frequency, as is especially evident in the phase curves. The M-estimator weights discard about 10% of the data, suggesting the presence of a limited fraction of outliers.

Direct comparison of the apparent resistivity and phase, as in Figure 5.10, does not illustrate the actual differences between estimates very effectively. Following Chave & Jones (1997), Figure 5.11 shows the Studentized differences between the remote reference and robust response function estimates obtained by differencing them and normalizing by the jackknife estimate of standard error for the robust result. Studentization yields units of standard deviations. Figure 5.11 also shows the simultaneous bound for all of the response tensor elements obtained from the Bonferroni critical value of the t-distribution with very large degrees of freedom. The Bonferroni bound of 2.24 is larger than the traditional value of 1.95 because it spreads the tail probability among all of the response elements. Figure 5.11 shows the real and imaginary parts of all elements of the magnetotelluric response tensor; there are 136 values, so that, at the 95% level for Gaussian data, one would expect that about seven will lie outside the Bonferroni bounds. In fact, there are 64 values (or nearly half of the

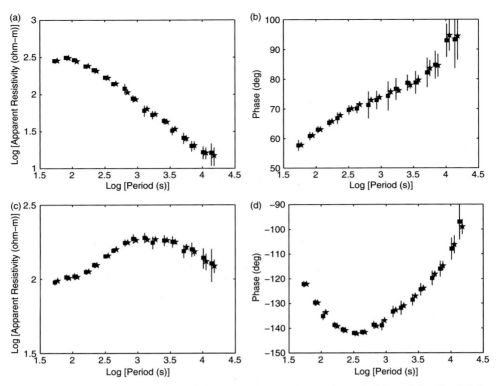

Figure 5.10. The remote reference (solid squares) and the robust estimate (solid stars) for Site 172, in both cases using Site 145 as the reference. The remote reference estimate is identical to that shown in Figures 5.6 and 5.7; note that the ordinate scale has changed for some of the panels. The robust results have been offset slightly to the right for clarity. See Figure 5.6 caption for further details.

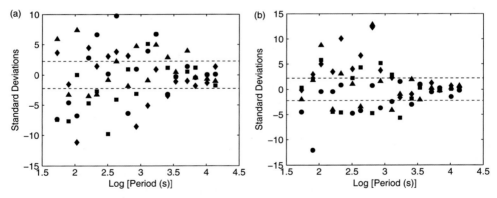

Figure 5.11. The real and imaginary parts of the Studentized differences between (a) Z_{xx} and Z_{xy}, and (b) Z_{yx} and Z_{yy}, for the remote reference and robust estimates shown in Figure 5.10. In each instance, the difference in the response is normalized by the jackknife standard error for the robust estimate, so the ordinate has units of standard deviations. The symbols denote diagonal element real and imaginary parts (circles and squares) and off-diagonal element real and imaginary parts (diamonds and triangles). The dashed lines show the Bonferroni simultaneous confidence bounds for all estimates.

total) outside the bounds, with some excursions in excess of 10 standard deviations. This graphically illustrates the substantial difference between the remote reference and robust estimates and the marked effect of non-Gaussianity on the former.

A major problem in magnetotelluric analysis is the presence of correlated noise that appears in all local electromagnetic field components and, in some circumstances, the remote magnetic field. Correlated noise can be due to ground currents from electric fences and railways, mining equipment and electric trains, and is reviewed by Szarka (1988) and Junge (1996). Lowes (2009) provides a detailed discussion of the electromagnetic characteristics of DC trains. Under these circumstances, remote reference and/or robust estimators will produce biased response function estimates. However, if one or more clean reference sites are available, a robust extension of the generalized remote reference algorithm can produce good magnetotelluric responses. At the first stage, the clean reference sites result in a correlated noise-free estimate for $\hat{\vec{b}}$ in (5.27), and at the second stage, correlated noise in the local electric field is eliminated. Note that complete two-stage least squares is required even when only a single reference site is available, as the remote reference estimator (5.23) does not eliminate correlated local noise even when robustified. This approach works well when the correlated noise is intermittent in time, and breaks down when the correlated noise is sufficiently pervasive that clean data sections become rare events.

Varentsov *et al.* (2003) proposed an alternative method that takes advantage of multiple reference sites in an array. Separate robust remote reference estimates of the local responses were computed using different single reference sites, and then robustly averaged. Experiments with magnetotelluric data from the Baltic Electromagnetic Array Research (BEAR) project yielded some improvement in response function estimates with 3–10 separate remote sites. However, this approach does not reduce the estimator error because the same local electromagnetic fields are used for each remote reference estimate.

Larsen *et al.* (1996) and Oettinger *et al.* (2001) proposed multiple transfer function methods to remove cultural noise using distant, uncontaminated sites as references. Both of these algorithms are special cases of robust two-stage processing using clean reference sites. For example, Larsen *et al.* (1996) dealt with the correlated electric and magnetic field noise from DC electric trains by explicitly separating the magnetotelluric and noise responses. A tensor relating the local and remote magnetic fields equivalent to $\vec{t}$ in (5.26) is first computed to give a correlated-noise-free $\hat{\vec{b}}$, so that the correlated noise in $\vec{b}$ is given by $\vec{b} - \hat{\vec{b}}$. Both the magnetotelluric response and the correlated noise tensor are obtained from a robust solution to

$$\mathbf{e} = \hat{\vec{b}} \cdot \hat{\mathbf{z}} + \left(\vec{b} - \hat{\vec{b}} \right) \cdot \hat{\mathbf{z}}^0 + \boldsymbol{\varepsilon} \tag{5.44}$$

that explicitly separates $\mathbf{e}$ into magnetotelluric signal and correlated noise parts plus an uncorrelated random error. Equation (5.44) is easily solved using the methods of this section by recasting the problem so that $\vec{b}$ and $\vec{b} - \hat{\vec{b}}$ comprise a four-column $\vec{b}$ in (5.35), and the four-parameter solution contains robust versions of $\hat{\mathbf{z}}$ and $\hat{\mathbf{z}}^0$. However, uncorrelated noise in

$\bar{\mathbf{b}}$ is combined with the correlated noise in $\bar{\mathbf{b}} - \hat{\bar{\mathbf{b}}}$ and hence the estimates of $\hat{\mathbf{z}}$ and $\hat{\mathbf{z}}^0$ will always be biased to some degree. The bias will rise as the relative size of the correlated noise to signal increases. In contrast, the two-stage estimate $\hat{\mathbf{z}}_*^q$ is not biased in this way because the correlated noise tensor $\hat{\mathbf{z}}^0$ is never explicitly computed.

5.6 Robust magnetotelluric processing 2: bounded influence estimators

Two statistical concepts that facilitate the understanding of problems with least squares and M-estimators are the breakdown point and the influence function in the sense of Hampel (1974). Loosely speaking, the breakdown point is the smallest fraction of bad data that can carry an estimate beyond all statistical bounds that can be placed on it. Ordinary least squares has a breakdown point of $1/N$, as a single outlying data point can inflict unlimited damage on the estimate. The Hampel influence function measures the effect of an additional observation on the estimate of a statistic given a large sample. Influence functions may be unbounded or bounded, as the estimator is or is not sensitive to an additional bad datum. Further details may be found in Hampel *et al.* (1986, section 1.3).

The procedures described in Sections 5.4 and 5.5 are capable of eliminating bias due to uncorrelated noise in the local predictor fields and contamination from data that produce outlying residuals in a semi-automatic fashion. However, they do require that the reference channels in $\bar{\mathbf{Q}}$ (5.27) or $\bar{\mathbf{b}}_r$ (5.23) and the local magnetic field $\bar{\mathbf{b}}$ be reasonably clear of extreme values caused by instrument malfunctions, impulsive events in the natural-source fields, or man-made sources such as electric fences, mines and trains. Extreme data in the predictor or reference channels are called *leverage points* because they have undue influence on the outcome. When leverage points are present, any conventional robust M-estimator, such as described in the last section, can be severely biased. As for least-squares estimators, M-estimators possess a breakdown point of only $1/N$, so that a single leverage point can completely dominate the estimate, and their Hampel influence functions are unbounded.

The hat matrix defined in (5.10) is an important auxiliary quantity in regression theory, and has long been used to detect unusually high leverage data (e.g. Hoaglin & Welsch, 1978; Belsley *et al.*, 1980; Chatterjee & Hadi, 1988; Chave & Thomson 2003). As described in Section 5.1, the diagonal elements of the hat matrix h_{ii} are a measure of the leverage exerted by a row in $\bar{\mathbf{b}}$ or $\bar{\mathbf{Q}}$. The factor by which h_{ii} must exceed the expected value p/N to be considered a leverage point is not well defined, but statistical lore suggests that values more than 2–3 times p/N are a concern (Hoaglin & Welsch, 1978).

Chave & Thomson (2003) derived the exact distribution of h_{ii} for complex Gaussian predictors, as might be expected for Fourier transforms in the absence of influential data. The resulting p.d.f. is the beta distribution $\beta(h_{ii}, p, N - p)$, and the c.d.f. is the incomplete beta function ratio $I_x(p, N - p)$. For magnetotellurics, where $p = 2$ and when $N \gg p$, Chave & Thomson (2004) presented some critical values for the probability that a given h_{ii} will exceed the expected value of $2/N$ by a factor of η. They showed that data for which h_{ii} is only twice the expected value will occur with probability 0.091. Rejecting values at this level will typically remove a significant number of valid data in an uncontaminated population. Data for which h_{ii} is 4.6 times the expected value occur with a probability of only 0.001. Thus, a typical rejection threshold lies between 2 and 4 times $2/N$. To introduce automatic dependence on N, so that the threshold will automatically adjust upward for large datasets, Chave & Thomson (2003) advocated rejecting data corresponding to h_{ii} values larger than the inverse of the beta distribution at a reasonable probability level. For example, selecting the 0.95 threshold will carry a 5% penalty for Gaussian data, yet effectively protect against large leverage points. In datasets for which h_{ii} is longer-tailed than beta (a common occurrence in magnetotellurics), larger cutoff values should be used.

This discussion indicates that the hat matrix diagonal elements serve as indicators of leverage, and Chave & Thomson (2004) show that the hat matrix diagonal differs from simpler leverage metrics such as the power in a given data section, as it depends on the correlation between the magnetic field channels as well as their amplitude. Extensions to robust M-estimators that can detect both outliers based on the regression residuals and leverage points based on the hat matrix diagonal elements can be devised (e.g. Mallows, 1975; Chave & Thomson, 2003). When properly implemented, these algorithms have bounded Hampel influence functions, and hence are called bounded influence (BI) estimators. The bounded influence form of the normal equations analogous to (5.32) is

$$\vec{\mathbf{b}}^{H} \cdot \mathbf{\Psi} \cdot \vec{\mathbf{w}} = 0 \qquad (5.45)$$

where $\vec{\mathbf{w}}$ is an $N \times N$ diagonal leverage weight matrix. Mallows (1975) suggested setting $w_{ii} = \sqrt{1 - h_{ii}}$ and $\mathbf{\Psi} = \{\psi(r_i/d)\}$ as in (5.32), so that residual and leverage weighting are decoupled and leverage points are gently downweighted according to the size of the hat matrix diagonal elements. However, the Mallows estimator does not perform satisfactorily with magnetotelluric data because downweighting of high-leverage data is mild and limited, and its breakdown point is only slightly better than that of conventional M-estimators.

More capable, high-breakdown-point (up to 50% bad data) bounded influence estimators have been proposed (Siegel, 1982; Rousseeuw, 1984; Coakley & Hettmansperger, 1993), but these entail a major increase in computational overhead that limits their suitability for the large, complex datasets that occur in magnetotellurics. They also typically lose the advantage of being formulated in terms of Gaussian statistics for uncontaminated data, which is a limitation for interval and hypothesis estimation. An alternative bounded influence estimator that combines high asymptotic efficiency for Gaussian data, high-breakdown-point performance with contaminated data and computational simplicity that is suitable for large datasets was proposed by Chave & Thomson (2003), and adapted to magnetotelluric applications by Chave & Thomson (2004). Based on simulations, this estimator has a breakdown point of $1/p$, hence will function in the presence of 50% bad data in magnetotellurics.

The bounded influence estimator is implemented by replacing the weight matrix $\mathbf{v}$ in (5.35) with the product of two diagonal matrices $\vec{\mathbf{u}} = \vec{\mathbf{v}} \cdot \vec{\mathbf{w}}$, where $\vec{\mathbf{v}}$ is the M-estimator weight matrix defined in (5.34) whose diagonal entries are based on the size of the regression residuals, and $\vec{\vec{w}}$ is a leverage weight matrix whose elements depend on the size of the hat matrix diagonal elements. The bounded influence estimator at the kth iteration obtained using $\vec{\vec{u}}$ in place of $\vec{\vec{v}}$ in (5.35) is

$$\hat{\mathbf{z}}_{\#}^{[k]} = \left(\vec{\mathbf{b}}^{H} \cdot \vec{\vec{w}}^{[k]} \cdot \vec{\vec{v}}^{[k]} \cdot \vec{\mathbf{b}}\right)^{-1} \cdot \left(\vec{\mathbf{b}}^{H} \cdot \vec{\vec{w}}^{[k]} \cdot \vec{\vec{v}}^{[k]} \cdot \mathbf{e}\right) \tag{5.46}$$

As with the M-estimator $\hat{\mathbf{z}}_{*}$, one or more remote references may easily be incorporated to yield $\hat{\mathbf{z}}_{\#}^{r}$ or $\hat{\mathbf{z}}_{\#}^{q}$.

Based on simulations and trials with actual data, the robust weights in (5.35) are frequently seen both to increase the amount of leverage exerted by existing influential predictor data and to create entirely new ones. This occurs because M-estimators do not have bounded influence, and probably accounts for situations where robust procedures do not perform as well as might be expected. Even with Mallows weights in (5.46), $\vec{\vec{v}}$ and $\vec{\vec{w}}$ can interact to produce an unstable solution to the robust regression problem unless care is used in applying the leverage weights. This is especially true when the leverage weights are recomputed at each iteration step, which often results in oscillation between two incorrect solutions, each of which is dominated by a few extreme leverage points. Instead, the leverage weights in the Chave & Thomson (2003) estimator are initialized with unity on the main diagonal, and, at the kth iteration, the ith diagonal element is given by

$$w_{ii}^{[k]} = w_{ii}^{[k-1]} \exp[e^{-\chi^2}] \exp[-e^{\chi(y_i - \chi)}] \tag{5.47}$$

where the driving statistic is

$$y_i = \Xi h_{ii}^{[k]}/p \tag{5.48}$$

The parameter Ξ is the trace of $\vec{\vec{u}}^{[k-1]}$ that is initially the number of data N, and the hat matrix diagonal must include the effect of bounded influence weighting, so that

$$\vec{\vec{H}}^{[k]} = \sqrt{\vec{\vec{u}}^{[k-1]}} \cdot \vec{\mathbf{b}} \cdot \left(\vec{\mathbf{b}}^{H} \cdot \vec{\vec{u}}^{[k-1]} \cdot \vec{\mathbf{b}}\right)^{-1} \cdot \vec{\mathbf{b}}^{H} \cdot \sqrt{\vec{\vec{u}}^{[k-1]}} \tag{5.49}$$

yielding

$$h_{ii}^{[k]} = u_{ii}^{[k-1]} \mathbf{b}_i \cdot \left(\vec{\mathbf{b}}^{H} \cdot \vec{\vec{u}}^{[k-1]} \cdot \vec{\mathbf{b}}\right)^{-1} \cdot \mathbf{b}_i^{H} \tag{5.50}$$

where $\mathbf{b}_i$ is the ith row of $\vec{\mathbf{b}}$. The parameter χ in (5.47) determines the level at which leverage point downweighting begins, and is chosen to be $\beta^{-1}(\alpha, p, \Xi - p)$ with $\alpha = 0.95$ to 0.9999 as previously discussed. To further ensure a stable solution, downweighting of leverage points is applied in half-decade steps beginning with χ set just below the largest normalized hat matrix diagonal element and ending with the final value $\beta^{-1}(\alpha, p, \Xi - p)$, rather than all at once.

The bounded influence algorithm can be summarized as follows:

1. Given N estimates of the Fourier transforms of the horizontal electric and magnetic fields at the frequency of interest computed as outlined in Section 5.2, obtain the least-squares estimator (5.7).
2. Compute the residuals $\mathbf{r} = \mathbf{e} - \vec{\mathbf{b}} \bullet \hat{\mathbf{z}}$, the scale using (5.40), the initial residual sum of squares $\mathbf{r}^H \bullet \mathbf{r}$ and the hat matrix diagonal using (5.50) with $\vec{\mathbf{u}}^{[0]} = \vec{\mathbf{I}}$.
3. Begin an outer loop that starts with the most extreme value for χ minus one half-decade.
4. Compute the Huber weights (5.38), hat matrix weights (5.47), eliminate those data for which the weight is zero, and compute the BI estimate (5.46).
5. Compute the residuals $\mathbf{r} = \mathbf{e} - \vec{\mathbf{b}} \bullet \hat{\mathbf{z}}_{\#}$, the scale $d = s_{MAD}/0.44845$, and the weighted residual sum of squares $\mathbf{r}^H \bullet \vec{\mathbf{u}} \bullet \mathbf{r}$.
6. Repeat steps 4 and 5 until the residual sum of squares does not change by more than 1%.
7. Decrement the hat matrix parameter χ by a half-decade step and repeat, starting at step 3, stopping when the minimum value specified by $\beta^{-1}(\alpha, p, \Xi - p)$ is reached.
8. Compute the Thomson weights (5.43) and hat matrix weights (5.47), eliminate those data for which the weight is zero and compute the BI estimate (5.46).
9. Compute the residuals $\mathbf{r} = \mathbf{e} - \vec{\mathbf{u}} \bullet \hat{\mathbf{z}}_{\#}$, the scale $d = s_{MAD}/0.448\,45$ (at the first iteration only) and the weighted residual sum of squares $\mathbf{r}^H \bullet \vec{\mathbf{u}} \bullet \mathbf{r}$.
10. Repeat steps 8 and 9 until the residual sum of squares does not change by more than 1%.
11. Assess the fit (see Section 5.8), and compute the covariance matrix and confidence intervals (see Section 5.9).

In the presence of correlated noise in the local variables, instead of a bounded influence remote reference estimator, a two-stage version of the bounded influence estimator of this section is advocated. At the first stage, a bounded influence estimator is applied to (5.26), yielding an estimate for the projected local magnetic field

$$\hat{\vec{\mathbf{b}}}_{\#} = \sqrt{\vec{\mathbf{u}}_1} \bullet \vec{\mathbf{Q}} \bullet (\vec{\mathbf{Q}}^H \bullet \vec{\mathbf{u}}_1 \bullet \vec{\mathbf{Q}})^{-1} \bullet \vec{\mathbf{Q}}^H \bullet \sqrt{\vec{\mathbf{u}}_1} \bullet \mathbf{b} \tag{5.51}$$

where $\vec{\mathbf{u}}_1$ is the final first-stage bounded influence weight matrix computed as described above. At the second stage, a bounded influence estimator version of (5.28) using (5.51) is solved to yield

$$^{\#}\hat{\mathbf{z}}_{\#}^q = \left(\hat{\vec{\mathbf{b}}}_{\#} \bullet \vec{\mathbf{u}}_2 \bullet \vec{\mathbf{b}}\right)^{-1} \bullet \left(\hat{\mathbf{b}}_{\#}^H \bullet \vec{\mathbf{u}}_2 \bullet \mathbf{e}\right) \tag{5.52}$$

where $\vec{\mathbf{u}}_2$ is the product of the second-stage bounded influence weight matrix and $\vec{\mathbf{u}}_1$.

Smirnov (2003) presented an alternative bounded influence estimator based on the repeated median algorithm of Siegel (1982). This approach is more computationally intensive than that described in this section, and lacks optimality in the event that the data are Gaussian, but does show promise with synthetic data.

Figure 5.12 shows the Z_{xy} and Z_{yx} response tensor robust and bounded influence elements for Site 172, in both cases using Site 145 as a reference. The two estimates are statistically similar at all periods. The bounded influence weights show that 5–7% of the data have been discarded due to leverage, in addition to the approximately 10% removed by the M-estimator weights. Decreasing χ in (5.47) does not change the appearance of Figure 5.12, except that the error estimates for the bounded influence result become larger as more data are removed. This indicates that leverage effects in the dataset are weak.

Figure 5.13 shows the Studentized differences between the robust and bounded influence estimates obtained by differencing them and normalizing by the jackknife estimate of standard error for the bounded influence result. It also shows the simultaneous bound for all of the response function elements obtained from the Bonferroni critical value of the t-distribution with very large degrees of freedom. There are 10 values (or only slightly more

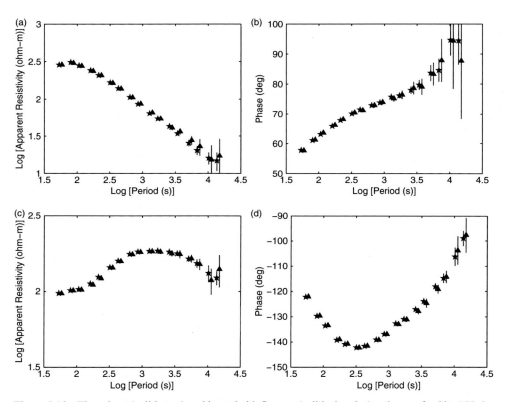

Figure 5.12. The robust (solid stars) and bounded influence (solid triangles) estimates for Site 172, in both cases using Site 145 as the reference. The robust estimate is identical to that shown in Figure 5.10. The bounded influence estimate was computed by setting χ in (5.47) to 0.999. See Figure 5.6 caption for further details.

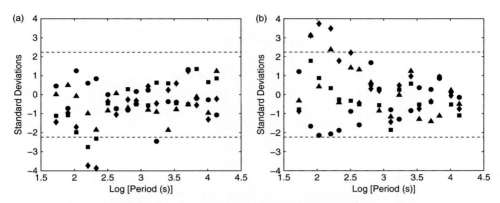

Figure 5.13. The real and imaginary parts of the Studentized differences between (a) Z_{xx} and Z_{xy}, and (b) Z_{yx} and Z_{yy}, for the robust and bounded influence estimates shown in Figure 5.12. In each instance, the difference in the impedance is normalized by the jackknife standard error for the bounded influence estimate, so the ordinate has units of standard deviations. The symbols denote diagonal element real and imaginary part (circles and squares) and off-diagonal element real and imaginary parts (diamonds and triangles). The dashed lines show the Bonferroni simultaneous confidence bounds for all estimates.

than the seven values expected for Gaussian data) outside the bounds, with all excursions under four standard deviations. This shows that the difference between the robust and bounded influence estimates is small for this dataset, in contrast to the dramatic difference between the remote reference and robust results in Figure 5.11.

Figure 5.14 provides a graphical display of what bounded influence weighting is doing to the time series by illustrating those parts of the original data (solid trace) that are removed by robust (dark gray trace) and bounded influence (light gray trace) weights for E_x and B_y at a period of 106.6 s. Robust weighting has eliminated most of the magnetic storm events near the beginning of the record and starting at about 500 h, as well as a series of impulsive spikes in the Site 172 electric and magnetic fields. Recalling that the hat matrix for a remote reference estimate involves only the reference magnetic field at Site 145, the bounded influence weights remove a spike event at about 480 h along with elements of the storm time variations and other scattered intervals throughout.

Figure 5.15 compares the Z_{xy} and Z_{yx} response tensor robust (solid squares) and two-stage bounded influence (solid triangles) elements for Site 172, using either Site 145 as a reference (robust) or both Sites 145 and 127 as references (two-stage). The two estimates are statistically indistinguishable, as can be demonstrated by a comparison like that in Figure 5.13 (not shown), which gives only six estimates outside of the Bonferroni bounds. However, at the lowest frequencies the two-stage estimator yields results that are physically reasonable for all tensor quantities. This example illustrates the value of using multiple remotes whenever possible.

Figure 5.16 shows the effect of two-stage weighting on the E_y and B_x time series at a period of 3414 s. As for Figure 5.14, the solid trace is the original data, while the dark gray and light

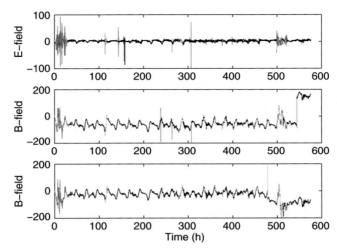

Figure 5.14. The E_x and B_y components at Site 172 and the B_y component at Site 145 (from top to bottom). The solid trace is the part of the time series used to estimate the response, and, for a frequency of 0.009 375 Hz (period of 106.6 s), the dark gray trace is the part of the data eliminated by robust weighting, and the light gray trace is the part of the data eliminated by bounded influence weighting, both for the bounded influence estimate of Figure 5.12.

gray traces show those intervals eliminated at the first and second stages, respectively. The preponderance of dark gray and paucity of light gray illustrate that the combination of robust and bounded influence weighting at the first stage eliminates most of the anomalous data.

5.7 Preprocessing

The procedures described in Sections 5.4–5.6 work well unless a large fraction of the available data are quite noisy, as can occur when the natural signal level is frequently comparable to or below the instrument noise level with infrequent excursions to a higher signal-to-noise ratio, or in the presence of some types of cultural or mechanical (e.g. magnetometer motion) noise that contaminates most of the data. The solution to problems of this class is often the addition of an initial screening step that pre-selects those data whose signal-to-noise ratio is adequate, prior to application of a robust or bounded influence estimator, or that corrects for noise based on additional, non-electromagnetic data.

The predominance of a weak signal is typical in the so-called dead bands between approximately 0.5 and 5 Hz for magnetotellurics and between 1000 and 5000 Hz for audiomagnetotellurics, where natural-source electromagnetic power spectra display relative minima. It is also frequently observed in the magnetic field at frequencies above 0.1–1 Hz with fluxgate sensors and at frequencies below 0.001 Hz with coil sensors due to instrument noise characteristics. In these instances, robust or bounded influence estimators can introduce bias into the response tensor beyond that which would ensue from the use of ordinary

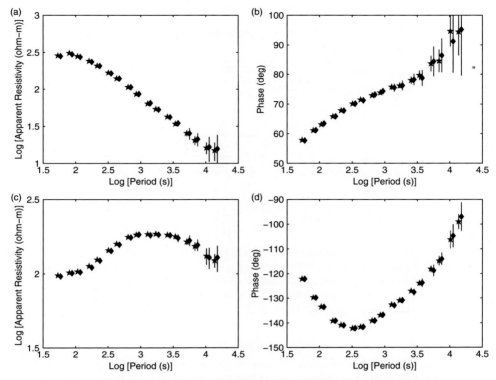

Figure 5.15. The robust (solid stars) and two-stage bounded influence (solid diamonds) estimates for Site 172, using Site 145 as the reference for the former and both Sites 145 and 127 as the references for the latter. The robust estimate is identical to that shown in Figures 5.10 and 5.12. See Figure 5.6 caption for further details.

least squares because most of the data are noisy, and the unusual values detected and removed by data-adaptive weighting are actually those rare ones containing useful information. A number of methods have been suggested to ameliorate this type of problem, and others will undoubtedly be introduced as future situations require them, since magnetotelluric data processing algorithms are strongly data-driven.

Jones & Jödicke (1984), Travassos & Beamish (1988) and Lamarque (1999) introduced data screening based on the coherence between the electric and magnetic fields in discrete data segments, rejecting those displaying coherence below a threshold value, although not in the context of preprocessing to a robust algorithm. Egbert & Livelybrooks (1996) extended these results by first applying coherence thresholding to single-station dead-band data and subsequently using robust processing. However, Chave & Jones (1997) give examples where this approach fails, and hence it must be applied with care.

Garcia & Jones (2008) combined the use of a wavelet transform procedure to pre-select energetic audiomagnetotelluric dead-band events, then applied two coherence thresholds

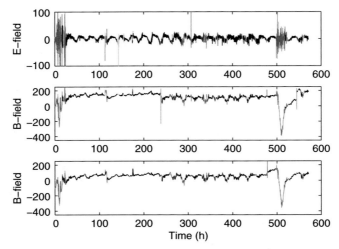

Figure 5.16. The E_y and B_x components at Site 172 and the B_x component at Site 145 (from top to bottom). The solid trace is the part of the time series used to estimate the response, and, for a frequency of 0.000 293 Hz (period of 3413 s), the dark gray trace is the part of the data eliminated in the first stage, and the light gray trace is the part of the data eliminated in the second stage, both for the two-stage bounded influence estimate shown in Figure 5.15.

and bounded influence processing. The results were superior to ordinary bounded influence processing for these data, but produced inferior results for magnetotellurics, probably because the source field transitions from being impulsive (due to an origin in lightning strikes) to broadband below the audiomagnetotelluric band. In a reversal of this philosophy, Jones & Spratt (2002) pre-selected data segments whose vertical magnetic field power was below a threshold value to minimize auroral source field bias in high-latitude data.

Weckmann *et al.* (2005) combined a number of these steps for single-site magnetotelluric processing, introducing a mixture of frequency-domain thresholding based on the section-by-section power spectral density in the local electromagnetic fields, the multiple and partial coherences between the electric and magnetic fields, the polarization directions of the electromagnetic fields, and the responses and their uncertainties. Through a series of examples encompassing a variety of incoherent and coherent local noise sources, they showed substantial improvement over straight application of robust processing.

Lezaeta *et al.* (2005) describe a time-domain adaptive correlation canceler to remove instrument motion noise on electric and magnetic field data collected in shallow lakes during high-wind intervals. An independent measure of instrument motion (tilt sensors) was used as the noise reference, and the adaptive correlation canceler removed the noise from the electromagnetic data prior to estimation of the magnetotelluric response using a bounded influence estimator. The data were not interpretable without this stage.

All of these studies demonstrated substantially better performance for robust weighting schemes after pre-sorting. Similar preprocessing stages can be applied to data with single or multiple reference sites, although the number of datasets and hence the number of candidate selection criteria is increased.

5.8 Statistical verification

Statistical tests for the presence of outliers and leverage points do not exist for realistic situations where anomalous data occur in clusters and are intercorrelated. In addition, the difficulty of devising such tests rises rapidly with the number of channels or data, and when the population distribution of the outliers and leverage points is unknown. This means that *a priori* examination of a dataset for bad values is not feasible, and hence it is necessary to construct *a posteriori* tests to validate the result of a bounded influence analysis. Further, owing to the inherently nonlinear form of robust or bounded influence regression solutions like (5.35) or (5.46), along with their remote reference or two-stage extensions, and because such estimators implicitly involve the elimination of a significant fraction of the original dataset, it is prudent to devise methods for assessing the statistical veracity of the result. These can take many forms, as described in Belsley *et al.* (1980), Randles (1984) and Chatterjee & Hadi (1988). In a magnetotelluric context, the most useful have proven to be the multiple coherence suitably modified to allow for remote reference variables and adaptive weights, and quantile–quantile plots of the regression residuals and hat matrix diagonal elements against their respective target distributions for Gaussian data.

In classical spectral analysis, the multiple coherence is defined as the proportion of the power in a response variable at a given frequency that can be explained through a linear regression with all of the predictor variables (Koopmans, 1995, p. 155). This quantity is usually expressed as the magnitude squared coherence (MSC), which is a measure of the similarity of the corresponding variables. The MSC is just the squared coherence between the observed electric field $\mathbf{e}$ and its prediction $\hat{\mathbf{e}}$, and can easily be derived for any of the estimators previously discussed. For example, for the two-stage bounded influence estimator (5.52), the MSC is

$$\vec{\gamma}_{\mathbf{e}\hat{\mathbf{e}}}^{\,2} = \left[\vec{\mathbf{S}}_{\mathbf{ee}} \bullet \left({}^{\#} \hat{\mathbf{z}}_{\#}^{q} \right)^{H} \bullet \vec{\mathbf{S}}_{\hat{\mathbf{b}}_{\#}\hat{\mathbf{b}}_{\#}} \bullet {}^{\#} \hat{\mathbf{z}}_{\#}^{q} \right]^{-1} \bullet \left({}^{\#} \hat{\mathbf{z}}_{\#}^{q} \right)^{H} \bullet \vec{\mathbf{S}}_{\mathbf{e}\hat{\mathbf{b}}_{\#}}^{H} \bullet \vec{\mathbf{S}}_{\mathbf{e}\hat{\mathbf{b}}_{\#}} \bullet {}^{\#} \hat{\mathbf{z}}_{\#}^{q} \tag{5.53}$$

where $\vec{\mathbf{S}}_{\mathbf{xy}}$ is the spectral density matrix between $\mathbf{x}$ and $\mathbf{y}$. Equation (5.53) defines four frequency-dependent quantities: the diagonal elements are the MSC for two measured orientations of the electric field. Ideally, the MSC should lie close to unity. Departures from this condition indicate the presence of noise and/or decorrelation between the local and remote variables due to a rise in source wavenumber bandwidth.

A second and essential tool for analyzing the results of a bounded influence magneto-telluric analysis is comparison of the original and final quantile–quantile plots of the residuals and the hat matrix diagonal. Quantile–quantile (q–q) plots compare the statistical distribution of an observed quantity with a corresponding theoretical entity, and hence provide a qualitative means for assessing the statistical outcome of robust processing. The N quantiles $\{q_j\}$ of a distribution divide the area under the p.d.f. into $N+1$ equal-area pieces, and hence define equal probability intervals. They are easily obtained by solving

$$F(q_j) = \frac{j - \frac{1}{2}}{N} \tag{5.54}$$

for $j = 1, \ldots, N$, where $F(x)$ is the target c.d.f. A q–q plot is a comparison of the quantiles with the order statistics obtained by ranking and sorting the data along with a suitable scaling to make them independent of data units. The advantage of q–q plots over some alternatives is that they emphasize the distribution tails; most of a q–q plot covers only the last few percent of the distribution range.

For residual q–q plots, the N quantiles of the Rayleigh distribution (5.39) are plotted against the residual absolute value order statistics $\mathbf{r}_{(i)}$ scaled so that their second moment is 2, as would be expected for a true Rayleigh variate. The quantiles of the Rayleigh distribution may easily be derived from (5.39), yielding

$$q_j = \sqrt{2\log\left(\frac{N}{N - j + \frac{1}{2}}\right)} \tag{5.55}$$

for $j = 1, \ldots, N$.

For hat matrix diagonal q–q plots, the quantiles of the $\beta(p, N - p)$ distribution may be compared to the order statistics with both quantities divided by p/N, so that a value of 1 corresponds to the expected value. The quantiles are the solutions to

$$I_{qj}(p, N - p) = \frac{j - \frac{1}{2}}{N} \tag{5.56}$$

Chave & Thomson (2003, Appendix B) give a closed-form expression for the incomplete beta function ratio $I_x(m, n)$ in (5.56).

The use of data-adaptive weighting that eliminates a fraction of the data requires that the quantiles be obtained from the truncated form of the original target distribution, or else the result will inevitably appear to be short-tailed. The truncated distribution is easily obtained from the original one. Suppose data are censored in the process of robust or bounded influence weighting using an estimator such as those described in Sections 5.5 and 5.6. Let $f_X(x)$ and $F_X(x)$ be the p.d.f. and c.d.f. of a random variable $\mathbf{X}$ prior to censoring; this may be the Rayleigh distribution for the residuals or the beta distribution for the hat matrix diagonal. After truncation, the p.d.f. of the censored random variable $\mathbf{X}'$ is

$$f_{\mathbf{X}'}(x) = \frac{f_{\mathbf{X}}(x')}{F_{\mathbf{X}}(b) - F_{\mathbf{X}}(a)} \tag{5.57}$$

where $a \leq x' \leq b$. Let N be the original and M be the final number of data, so that $M = N - m_1 - m_2$, where m_1 and m_2 are the number of data censored from the bottom and top of the distribution, respectively. Suitable choices for a and b are the m_1th and $(N - m_2)$th quantiles of the original distribution $f_{\mathbf{X}}(x)$. The M quantiles of the truncated distribution can then be computed from that of the original one using

$$F_{\mathbf{X}}(q_j) = [F_{\mathbf{X}}(b) - F_{\mathbf{X}}(a)] \frac{j - \frac{1}{2}}{M} + F_{\mathbf{X}}(a) \qquad (5.58)$$

for $j = 1, \ldots, M$.

Whether the data have been censored or not, a straight-line q–q plot indicates that the residual or hat matrix diagonal elements are drawn from the target distribution. Data that are inconsistent with the target distribution appear as departures from linearity, and are usually manifest as sharp upward shifts in the order statistics at the distribution ends. With magneto-telluric data, this is often extreme, and is diagnostic of long-tailed behavior where a small fraction of the data occur at improbable (for Gaussian variates) distances from the distribution center, and hence will exert undue influence on the regression solution. The residual q–q plots from the output of a robust or bounded influence estimator should be approximately linear or slightly short-tailed to be consistent with the optimality requirements of least-squares theory.

There is not a theoretical requirement for the hat matrix diagonal distribution to be $\beta(p, N - p)$ unless the magnetic field predictors actually are Gaussian. Nevertheless, hat matrix diagonal q–q plots are useful for detecting extreme values. Extremely long-tailed hat matrix diagonal distributions are especially prevalent in the auroral oval, where substorms can produce spectacular leverage – see Garcia *et al.* (1997) for representative examples. In these instances, the non-bounded influence hat matrix diagonal distribution is sometimes closer to log beta than beta. It is not difficult to rationalize the underlying log beta form. Ionospheric and magnetospheric processes are extremely nonlinear, and more so at auroral than lower latitudes. As a result, their electromagnetic effects are the result of many multi-plicative steps. This means that the statistical distribution of the magnetic variations will tend toward log normal rather than normal (Lanzerotti *et al.*, 1991), and hence the resulting hat matrix diagonal will tend toward log beta rather than beta.

Figure 5.17 shows q–q plots for various response estimates at a period of 1280 s. Panels (a) and (b) compare the ordinary least-squares and robust results whose response are shown in Figures 5.6 and 5.10, while panels (c) and (d) compare the robust and bounded influence results whose response are shown in Figure 5.12. The ordinary least-squares q–q plots for the residuals and hat matrix diagonal are both long-tailed owing to the presence of a fraction of outliers and leverage points. The application of the robust estimator removes the outliers (panel (a)) and yields a straight-line q–q plot, but in the process the effect of leverage has actually become much worse (panel (b)). This is not unexpected: robust M-estimators are insensitive to leverage, and it is not unusual to observe increased leverage due to the adaptive weights. In some cases, this can be spectacular, as shown by Chave & Thomson

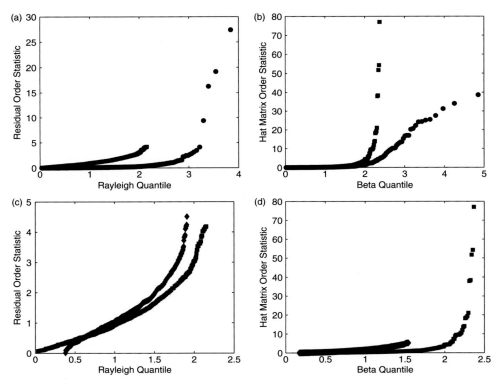

Figure 5.17. Quantile–quantile plots for (a,c) the residual order statistics against the Rayleigh distribution and (b,d) the hat matrix diagonal order statistics against the beta distribution at a period of 1280 s for the E_x response function element. Comparison of (a,b) ordinary least-squares (circles) and robust (squares) results shown in Figures 5.6 and 5.10, and (c,d) robust (squares) and bounded influence (diamonds) estimates shown in Figure 5.12. The ordinary least-squares result contains 612 data, while robust and bounded influence weighting reduces this quantity. In that instance, the truncated version of the Rayleigh and beta distributions are used, as described in the text. The residual order statistic has been normalized to a second moment of 2, while the hat matrix diagonal order statistic has been normalized to its expected value.

(2004). The bounded influence estimator (panels (c) and (d)) maintains control of both outliers and leverage points, minimizing the effect of influential data. However, the final residual and hat matrix diagonal distributions are only approximately Rayleigh or beta, respectively, and display a tendency to be slightly longer-tailed than expected. This effect is limited and unlikely to bias the result, and in fact is present in all of the q–q plots to varying degrees. The q–q plots shown in Figure 5.17 are typical rather than extreme.

  Quantile–quantile plots are useful qualitative tools for assessing the results from robust or bounded influence magnetotelluric analyses, but they do not provide a quantitative statement about the residual distribution. In situations where specificity is needed, suitable non-parametric tests for the goodness of fit of a dataset to a target distribution are available, of which the best known is the Kolmogorov–Smirnov test described in elementary statistics texts. However, this test has the disadvantage that it is most sensitive at the distribution median and least sensitive at the tails, yet it is in the tails where statistical differences in magnetotelluric data are most strongly manifest.

A more powerful non-parametric test for goodness of fit, under the null hypothesis that a set of data are drawn from a specified distribution against the alternative hypothesis that they are not, is that of Anderson & Darling (1952,1954). The Anderson–Darling test is most sensitive in the distribution tails, but has the disadvantage that the critical values depend on the target distribution. However, random variables from any distribution may be transformed to those from the uniform distribution by

$$u_j = F^*(x_j) \tag{5.59}$$

where $F^*(x)$ is the target c.d.f. As a result, the test statistic

$$A^2 = -N - \frac{1}{N} \sum_{i=1}^{N} (2i-1) \left\{ \log F^*\left(x_{(i)}\right) + \log\left[1 - F^*\left(x_{(N+i-1)}\right)\right] \right\} \tag{5.60}$$

may be assessed against critical values of the Anderson–Darling statistic. The null hypothesis rejects if A^2 is greater than 0.751 or 1.029 at the 95% and 99% confidence levels, respectively.

5.9 Uncertainty estimates for the magnetotelluric response

An essential feature of any method for computing magnetotelluric responses is the provision of both an estimate and a measure of its accuracy. Estimates of uncertainty may be obtained using parametric approaches that depend on explicit statistical models ultimately based on a Gaussian distribution for the data, or non-parametric approaches where the statistical assumptions can be relaxed but the computational overhead is higher. These will be introduced in turn.

Huber (1967) first proved that M-estimators are asymptotically normal under certain conditions on the influence function that was introduced in Section 5.5. Specifically, under the three conditions:

1. the influence function ψ is continuously differentiable with respect to the elements of $\mathbf{z}$,
2. $(1/N)\,\mathbf{i}^T \bullet \partial_{z_{ij}} \mathbf{\Psi}|_{z_{ij}=\hat{z}_{ij}}$ converges in probability to $a(z_{ij})$, where $\mathbf{i}$ is a column vector of ones, and
3. $(1/\sqrt{N})\,\mathbf{i}^T \bullet \mathbf{\Psi}$ converges in distribution to $N\left[0, b(z_{ij})\right]$,

then $\sqrt{N}(z_{ij} - \hat{z}_{ij})$ converges in distribution to $N\left[0, b(z_{ij})/a^2(z_{ij})\right]$. The asymptotic model can be implemented with sample estimates for a and b using the influence function and residuals from the last iteration of the M-estimator (5.34). Both the influence function and its derivative are readily obtained for the Thomson weight function (5.42). The weakness of this approach is that it is a large-sample result, and may break down for finite samples. This is apt to be an issue at low frequencies, where the number of data sections is inevitably limited. In addition, it strictly applies to M-estimators, and an equivalent result for the bounded influence estimator (5.45) has not been derived.

Exact expressions for the distributions of the magnetotelluric response amplitude and phase, or the apparent resistivity, are available when the data are Gaussian and correlation between the elements in a given row of the response tensor is neglected. Chave & Lezaeta (2007) derived these three distributions from first principles. The use of parametric estimates based on these distributions will be reasonably accurate if the least-squares model in Section 5.1 holds, as would be the case for robust/bounded influence estimates that have been verified using the approaches of Section 5.8. In that case, individual response function elements are approximately Gaussian according to (5.14). However, parametric estimates will be inaccurate (and frequently wildly so) when the least-squares conditions are violated.

The magnitude of an element of the magnetotelluric response $z_a = |\hat{z}_{ij}|$ follows the Rice distribution with p.d.f.

$$f_m(z_a|\lambda_0, \sigma) = \frac{z_a}{\sigma^2} e^{-(z_a^2 + \lambda_0^2)/2\sigma^2} I_0\left(\frac{\lambda_0 z_a}{\sigma^2}\right) \tag{5.61}$$

where the support is $[0, \infty)$, λ_0 is the population value for the magnitude of the response function, σ^2 is the population variance and $I_0(x)$ is a modified Bessel function of the first kind of order zero. Equation (5.61) reduces to the Rayleigh distribution when $\lambda_0 = 0$. The Rice distribution is skewed to the right, with the distribution peak moving to the right and the distribution becoming more symmetric as λ_0 increases. The expected value of (5.61) is

$$E(z_a) = \sqrt{\frac{\pi}{2}} \sigma L_{1/2}\left(\frac{-\lambda_0^2}{2\sigma^2}\right) \tag{5.62}$$

where $L_{1/2}(x)$ is a generalized Laguerre function given by

$$L_{1/2}(x) = e^{x/2}[(1 - x)I_0(-x/2) - xI_1(-x/2)] \tag{5.63}$$

As a result, the expected value is a biased estimator for the mean.

The squared magnitude of the magnetotelluric response $z_b = |z_{ij}|^2$, from which the distribution of the apparent resistivity is a simple scaling by μ_o/ω, follows the non-central chi-squared distribution with two degrees of freedom

$$f_\rho(z_b|\lambda, \sigma^2) = \frac{1}{2\sigma^2} e^{-(z_b + \lambda)/2\sigma^2} I_0\left(\frac{\lambda z_b}{\sigma^2}\right) \tag{5.64}$$

whose support is $[0, \infty)$ with properties as described in Johnson *et al.* (1995, chapter 29). The parameter λ is the population value for the squared response function. Equation (5.64) reduces to the exponential distribution when $\lambda = 0$. Like the Rice distribution (5.61), (5.64) is

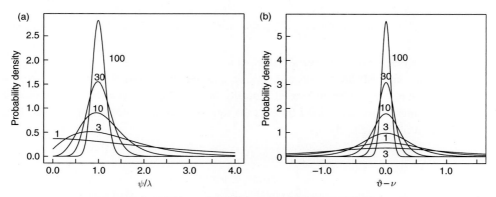

Figure 5.18. (a) The non-dimensional magnitude squared impedance p.d.f. (5.65), where $\psi = z_b$ in the notation of this chapter, for values of $\kappa = \lambda/(2\sigma^2)$ of 1, 3, 10, 30 and 100. The distribution is increasingly skewed and non-Gaussian for $\kappa < 30$, and lacks an obvious mode as κ gets small. (b) The non-dimensional phase distribution (5.68), where $\vartheta - \nu = \phi - \varphi$ in the notation of this chapter, for $\kappa = \lambda/(2\sigma^2)$ of 0.3, 1, 3, 10, 30 and 100. The distribution is nearly uniform for small κ and becomes increasingly concentrated as it rises with an increasing base value. Taken from Chave & Lezeata (2007).

asymmetric, shifting to the right and becoming more symmetric as λ increases. This is most easily visualized by transforming into non-dimensional variables $\kappa = \lambda/(2\sigma^2)$ and $\eta = z_b/\lambda$, so that (5.64) becomes

$$g(\eta|\kappa) = \kappa e^{-\kappa(\eta+1)} I_0\left(2\kappa\sqrt{\eta}\right) \qquad (5.65)$$

Figure 5.18 shows (5.65) for several values of κ.

The expected value of (5.64) is

$$E(z_b) = \lambda + 2\sigma^2 \qquad (5.66)$$

Consequently, the expected value is an upward-biased estimator for the mean, with the bias increasing with the variance of the estimate.

The p.d.f. of the phase of the magnetotelluric response is

$$f_\varphi(\vartheta|\lambda, \sigma^2) = \frac{e^{-\lambda/(2\sigma^2)}}{2\pi}\left[1 + \sqrt{\frac{\pi\lambda}{2\sigma^2}} \cos\vartheta\, e^{\lambda\cos^2\vartheta/(2\sigma^2)} \mathrm{erfc}\left(-\sqrt{\frac{\lambda}{2\sigma^2}}\cos\vartheta\right)\right] \qquad (5.67)$$

where $\vartheta = \phi - \varphi$, φ is the population value for the phase ϕ and $\mathrm{erfc}(x)$ is the complementary error function. The support of (5.67) is $[0, 2\pi)$, and it reduces to the uniform distribution when $\lambda = 0$. The expected value of ϑ is zero, hence the expected value of ϕ is the population value ϕ and is an unbiased estimator for the phase. The phase distribution has a mixture-like appearance, with a base at the level $\exp(-\kappa)/(2\pi)$ and a Gaussian-like center. The non-dimensional form of (5.67) is

$$h(\vartheta|\kappa) = \frac{e^{-\kappa}}{2\pi}\left[1 + \sqrt{\pi\kappa}\,\cos\vartheta\,e^{\kappa\,\cos^2\vartheta}\,\mathrm{erfc}\left(-\sqrt{\kappa}\cos\vartheta\right)\right] \qquad (5.68)$$

Figure 5.18 shows (5.68) for several values of κ.

Confidence intervals are always non-unique, and may be central or non-central about a specified value, but minimum size is typically achieved in the central case. A central confidence interval about a value $\tilde{\lambda}$ for λ for the squared magnitude of the response function may be derived through solving

$$\int_{\lceil 0,\tilde{\lambda}-c\rceil}^{\tilde{\lambda}+c} f_\rho(z|\lambda,\sigma^2)\,dz = 1-\alpha \qquad (5.69)$$

for c, where α is the probability allocated to the tails and $\lceil x,y\rceil$ denotes the larger of x and y. A similar equation holds for the magnitude of the response function (5.61). For instance, to compute the 95% confidence interval on the expected value for the squared magnitude of the response function, $\alpha = 0.05$ and $\tilde{\lambda}$ is the expected value of λ. A confidence interval for the phase requires solution of

$$\int_{\tilde{\phi}-c}^{\tilde{\phi}+c} f_\phi(\phi|\lambda,\sigma^2)\,d\phi = 1-\alpha \qquad (5.70)$$

Equations (5.69) and (5.70) are easily solved using numerical quadrature combined with a nonlinear equation solver. In either instance, estimation of the confidence interval requires replacement of the population values for λ and σ^2 with sample estimates. A naïve but adequate method-of-moments approach sets $\tilde{\lambda}$ and λ to the solution to (5.23), (5.28), (5.35), (5.46) or (5.52), $\tilde{\phi}$ to the corresponding value for phase and σ^2 to (5.13) modified to include robust and/or bounded influence weights. A better, although computationally more intensive, approach would use the maximum likelihood estimates for the distribution parameters, requiring the solution of three simultaneous nonlinear equations. In either instance, solutions to (5.69) and (5.70) are typically more accurate than the approximate forms given in texts on multivariate spectral analysis, such as Bendat & Piersol (1986, chapter 9).

In practical applications, information about the data distribution is sparse and/or simplifying assumptions may be untenable, leading to an unknown level of accuracy (or inaccuracy) for parametric estimates based on (5.61), (5.64), (5.67), (5.69) and (5.70). The complications include:

- the presence of mixtures of distributions rather than a single one
- influential data
- non-centrality

- unknown degrees of freedom due to correlation and heteroscedasticity
- multivariate relationships between data.

These are especially likely when a standard least-squares estimator is used, a robust/ bounded influence estimator is incorrectly configured, or with data where contamination by correlated noise is especially severe. As a consequence, magnetotelluric practitioners have adopted variance and confidence limit estimators that are inherently non-parametric. Instead of assuming a particular form for the distribution and estimating its parameters, the empirical distribution based on the available sample plays the central role in non-parametric methods.

The simplest resampling estimator is the jackknife, which is a linear approximation to the more general bootstrap estimator, and is computationally simple and efficient, making it suitable for large datasets. Thomson & Chave (1991) give a detailed description of its implementation and performance in spectral analysis, while Chave & Thomson (1989) describe its use for estimating confidence intervals on the magnetotelluric response tensor.

The jackknife will first be introduced in the context of estimating a parameter from a statistical sample, and then extended to regression to cover magnetotelluric applications. Let $\{x_i\}$ be an independent and not necessarily identically distributed sample of size N drawn from an unknown distribution or mixture of distributions, and let ζ be a statistical parameter to be estimated from the sample using the estimator Z. Let $\hat{\zeta} = Z(x_1, \ldots, x_N)$ be the estimate of ζ computed using all of the data (e.g. if ζ is the variance, then Z is the unbiased sample estimator for variance). Subdivide the data into N groups of size $N-1$ by sampling with replacement. Denote the estimate of ζ obtained from the ith subset

$$\hat{\zeta}_{\backslash i} = Z(x_1, \ldots, x_{i-1}, x_{i+1}, \ldots, x_N) \tag{5.71}$$

where the subscript $\backslash i$ denotes the delete the ith datum, or delete-one, estimate obtained by applying the all-data estimator to the delete-one sample. The most important application of the jackknife occurs in estimating the variance of $\hat{\zeta}$ (e.g. the variance of the variance if ζ is the variance). The jackknife estimator for the variance is

$$\hat{\sigma}_J^2 = \frac{N-1}{N} \sum_{i=1}^{N} \left(\hat{\zeta}_{\backslash i} - \bar{\zeta} \right)^2 \tag{5.72}$$

where $\bar{\zeta}$ is the sample mean of the delete-one estimates (5.71). The jackknife variance estimate will generally be different from the standard estimate, and has been shown to be accurate over a wide range of distributions through small-sample simulations.

A significant property of the jackknife is (Efron & Stein, 1981)

$$E\left(\hat{\sigma}_J^2\right) > \frac{\sigma^2}{N} \tag{5.73}$$

This holds even for data that are not identically distributed, such as mixtures of distributions. As a consequence, the jackknife is a conservative estimator for the variance, which is an important attribute in complicated statistical situations.

A major advantage of the jackknife is computational simplicity. If the variance of a statistic using a parametric approach is desired, it is necessary first to know the sampling distribution of the statistic and then to apply an estimator like maximum likelihood to get values for its parameters. For complicated statistics (e.g. the variance of the variance) or multivariate distributions, this quickly gets hopelessly complex. With the jackknife, all that is needed is an estimator for the statistic of interest. Its variance follows from the jackknife formula (5.72).

Under general conditions, it can be shown that $(\hat{\zeta} - \zeta)/\hat{\sigma}_J \sim t_{N-1}$, where t_v is Student's t-distribution with v degrees of freedom. This allows the construction of approximate double-sided (i.e. allocating one half of the tail probability α to the lower and upper extremes of the distribution) confidence intervals using the jackknife in the usual way (Hinkley, 1977a)

$$\Pr\left[\hat{\zeta} - t_{N-1}(1 - \alpha/2)\,\hat{\sigma}_J \leq \zeta \leq \hat{\zeta} + t_{N-1}(1 - \alpha/2)\,\hat{\sigma}_J\right] = 1 - \alpha \qquad (5.74)$$

where $\Pr[x]$ denotes the probability of x.

As a cautionary note, the jackknife should not be used blindly on statistics that are markedly non-Gaussian. Based on simulations, the Student's t model breaks down unless a variance-stabilizing transformation (Stuart *et al.*, 1999, section 32.38) is applied to make the statistic more Gaussian. The solution for the apparent resistivity (as well as the variance) is jackknifing of logs rather than raw estimates, and then converting the result back to the apparent resistivity domain at the end. The details are covered in Thomson & Chave (1991). For the phase, a transformation needs to be applied to ensure that circularity is preserved (i.e. a phase of 359° is not 358° away from a phase of 1°). Thomson & Chave (1991) show that the N delete-one phase factors

$$e_{\setminus k} = \frac{(z_{ij})_{\setminus k}}{\left|(z_{ij})_{\setminus k}\right|} \qquad (5.75)$$

can be used to compute the average of the delete-one phase factors $\bar{e}$ for use with the jackknife formula.

Jackknifing for the variance of the magnetotelluric response introduces a new twist because the problem is unbalanced: there is one response variable but there are $p = 2$ predictor variables. This has led to the suggestion based on simulations (Hinkley, 1977b; Wu, 1986) to weight the jackknife estimate of the variance according to the size of the hat matrix diagonal. The jackknifed covariance matrix for a row **z** of the magnetotelluric response tensor is given by

$$\hat{\sigma}_J^2 = \frac{N}{N-p} \sum_{i=1}^{N} (1 - h_{ii})^2 \, (\hat{\mathbf{z}} - \hat{\mathbf{z}}_{\backslash i}) \cdot (\hat{\mathbf{z}} - \hat{\mathbf{z}}_{\backslash i})^H \tag{5.76}$$

where $\hat{\mathbf{z}}$ is the result from applying one of the estimators (5.61), (5.64), (5.67), (5.69) or (5.70) to all of the data and $\hat{\mathbf{z}}_{\backslash i}$ is the result from applying the same estimator with the ith row of data (i.e. the ith data section) deleted. For robust or bounded influence estimators incorporating weights, the delete-one estimate $\hat{\mathbf{z}}_{\backslash i}$ may be approximated using the weights from the solution for $\hat{\mathbf{z}}$, or else the complete solution can be recomputed for each delete-one estimate at a substantial increase in computational overhead. In either case, the standard errors on the elements of $\hat{\mathbf{z}}$ are given by the square root of its diagonal elements. The off-diagonal elements are the covariances between the two response estimates.

Confidence intervals on the elements of $\hat{\mathbf{z}}$ follow by applying the definition (5.74). However, caution is in order since applying (5.74) separately to each element of $\hat{\mathbf{z}}$ allocates all of the uncertainty independently to that estimate, resulting in confidence limits on the entire row that are underestimated. Simultaneous confidence intervals on all of the elements of $\hat{\mathbf{z}}$ can be obtained using the Bonferroni method in which the tail probability is allocated evenly by replacing α with α/p in (5.74). Simultaneous double-sided confidence intervals on all of the elements of the 2×2 response tensor at the 95% level would utilize the 98.75% critical value of Student's t-distribution. Alternatively, critical values from the multivariate analog to Student's t, Hotelling's T^2-distribution, will give a more precise result.

However, the approach advocated in this book is not direct jackknifing of the apparent resistivity or phase. Instead, the magnetotelluric response estimates may be jackknifed directly to yield an estimate for its variance that is the same for the real and imaginary parts of the response function, and then used to get uncertainties on the apparent resistivity and phase using the delta method (Stuart & Ord, 1994, section 10.5)

$$\delta\rho_{ij} = 2\mu_o |z_{ij}| \delta z_{ij}/\omega \tag{5.77}$$

$$\delta\phi_{ij} = \sin^{-1}\left(\delta z_{ij}/|z_{ij}|\right) \tag{5.78}$$

where δz_{ij} is replaced with the jackknife estimate of the standard error for the term. The quantities $\delta\rho_{ij}$ and $\delta\phi_{ij}$ serve as estimates for the standard error of the apparent resistivity and phase, respectively. Confidence intervals on these quantities then follow in the standard way. Chave & Lezaeta (2007) showed that this was quite accurate compared to parametric estimates for actual data, and it offers computational simplicity as an advantage. This quantity is plotted in all of the magnetotelluric response figures in this chapter.

Like the jackknife, the non-parametric bootstrap is based on the principle that the empirical distribution can substitute for the underlying but unknown population distribution. The underlying principle is resampling with replacement repeatedly from the empirical distribution to obtain a series of pseudo-datasets that are used to compute the statistic of interest. Unlike for the jackknife, it is possible that some data will appear either more than once or not at all, as the resampling is random. The bootstrap does not appear to have been used in magnetotelluric data processing. A description of its use in robust regression is given by Salibian-Barrera & Zamar (2002).

Using data from a continuously operating magnetotelluric array in central California, Eisel & Egbert (2001) compared the performance of parametric and jackknife estimators for the confidence limits on the response function. They argued that both the fixed-weight jackknife, as advocated here, and a computationally more intensive subset deletion jackknife overestimated the errors, as is consistent with (5.73). This led to their favoring a parametric approach. However, it appears that the residual distribution of magnetotelluric data from locations near or within the auroral oval (at a minimum) are systematically longer-tailed than Gaussian even after bounded influence estimators are applied (Chave & Thomson, 2004). In this instance, parametric approaches will consistently underestimate the true confidence limits, which could lead to interpretation errors. Whether the residual distribution is systematically long-tailed for other situations is unclear, although data from quiet intervals at mid-latitudes appear to more consistently yield nearly Gaussian residuals. Because of this uncertainty, as well as due to the other advantages of the jackknife, the use of the jackknife for magneto-telluric response estimation continues to be favored. In essence, this emphasizes confidence-limit conservatism over a difficult-to-quantify sacrifice in reliability.

5.10 Alternative magnetotelluric processing methods

The processing algorithms described in Sections 5.4–5.6 are widely used by magnetotelluric practitioners, but there are a few robust alternatives that bear mention. Prominent among these are the evolving robust algorithm of Larsen (1975, 1980, 1989) and Larsen *et al.* (1996), and the robust principal components method of Egbert (1997). Neither of these approaches bound the influence of bad data, but both offer a useful alternative and additional information under some circumstances.

The Larsen algorithm has been developed explicitly to process geomagnetic data through careful and extensive testing, and as such is primarily driven by experience rather than statistical theory. Its advantage is that a number of features are incorporated to deal with problems that occur in geomagnetic data. Its disadvantages are that it has become quite complex as attributes have been added over time, and the use of multiple types of weights produces a complicated statistical framework, making error estimation problematical.

The algorithm was robust and relatively mature in Larsen (1989), with later improvements being largely refinements, and the method as of that date will be briefly described. In contrast to Egbert & Booker (1986) and Chave & Thomson (2004), Larsen (1989) does not divide the time series into short sections prior to Fourier transformation, and instead utilizes a classic band-averaging approach on the Fourier transform of the entire dataset. He applies three types of frequency-domain and one type of time-domain weights to the data in addition to a standard data taper. The frequency-domain weights are pre-whitening continuum $W(f)$, frequency rejection $F(f)$ and post-whitening $V(f)$, while $g(t)$ is the time-domain weight function. These are applied in the order W, F, g and V, producing a relationship between a single electric and magnetic field component

$$E_{WFgV}(f) = V(f)\{G(f - f') \oplus [B_{WF}(f')Z(f')]\} + R_{WFgV}(f) \qquad (5.79)$$

where $\oplus$ denotes the convolution operator, $G(f)$ is the Fourier transform of $g(t)$, $Z(f)$ is the response function estimate, $R(f)$ is the residual and

$$B_{WF}(f) = F(f)W(f)B(f) \qquad (5.80)$$

with a similar relationship applying to the electric field.

The pre-whitening continuum weights $V(f)$ are standard (i.e. non-robust) autoregressive ones that reduce the dynamic range of the spectra. The frequency rejection weights $F(f)$ are designed to reject frequency-domain outliers. The time-domain rejection weights $g(t)$ reject time-domain outliers. The post-whitening weights $V(f)$ correct the pre-whitening, and are only necessary if different pre-whitening operators are used on the electric and magnetic field data. The entire relationship (5.79) is applied iteratively, and in addition missing data and large time-domain outliers are replaced with predicted values during the process. Finally, the response function is smoothed through parameterization by

$$Z(f) = D(f)\bar{Z}(f) \qquad (5.81)$$

where $\bar{Z}(f)$ is a 1D model response and $D(f)$ is a distortion function represented as a low-order expansion in appropriate basis functions. Additional approximations are used, as described in detail in Larsen (1989).

Larsen *et al.* (1996) extended Larsen (1989) by allowing for subdivided time series, incorporating jackknife estimates for the variance and adding stabilization through damping factors at various stages. It also allowed for explicit separation of the local fields into magnetotelluric and correlated noise responses using a clean reference site, as described at the end of Section 5.5. Overall, the result is a satisfactory algorithm that is, however, sufficiently complicated that it would be a major undertaking for a magnetotelluric practitioner to implement or extend it.

The robust and bounded influence estimators discussed in Sections 5.5 and 5.6 are fundamentally univariate in the response variables because only a single electric field component is analyzed at a time. This is not necessarily true for the predictor variables, as a two-stage least-squares approach can in principle utilize as many magnetic field measurements as are available. The dichotomy between response and predictor variables is physics-based rather than artificial. However, an alternative, multivariate errors-in-variables approach that treats all of the available data in a common manner was introduced to magnetotellurics by Egbert & Booker (1989) and Egbert (1989). It was made robust by Egbert (1997, 2002). The eigenvector analysis introduced in Section 5.3 is an initial element of this approach.

Define the $K \times N$ magnetotelluric data array $\bar{\bar{\mathbf{X}}}$ at a given frequency that contains the electric and magnetic measurements at J sites with N sections of data for each (in the sense that was introduced in Section 5.2). If there are five data channels at each site, then $K = 5J$, although other cases can easily be accommodated. Egbert (1997) defined the model

$$\bar{\bar{\mathbf{X}}} = \bar{\bar{\mathbf{U}}} \cdot \bar{\bar{\alpha}} + \bar{\bar{\mathbf{V}}} \cdot \bar{\bar{\beta}} + \bar{\bar{\varepsilon}} \tag{5.82}$$

where $\bar{\bar{\mathbf{U}}}$ is a $K \times M$ signal matrix describing the part of the electric and magnetic fields at the J sites that are consistent with a magnetotelluric model, $\bar{\bar{\alpha}}$ is an $M \times N$ matrix of weights that define the particular linear combination of signal sources for a given data section, $\bar{\bar{\mathbf{V}}}$ is a $K \times L$ coherent noise matrix describing the part of the electric and magnetic fields at the J sites that are coherent but inconsistent with the magnetotelluric model, $\bar{\bar{\beta}}$ is an $L \times N$ matrix that defines the particular combinations of noise sources for a given data section and $\bar{\bar{\varepsilon}}$ is a $K \times N$ matrix of incoherent noise that has a diagonal $N \times N$ covariance matrix $\bar{\bar{\Sigma}}$. Under the quasi-uniform source assumption for magnetotellurics, $M = 2$, so that there are two independent source polarizations for the electromagnetic fields at J sites, while the number of coherent noise sources L has to be estimated from the data.

The elements of $\bar{\bar{\Sigma}}$ may be estimated from the residual variance after regressing a given data channel against all of the remaining channels as in Section 5.3, although a robust estimator should be utilized. Once that is obtained, the coherence dimension $M + L$ of the array may be estimated from an eigenvector decomposition of the spectral matrix of the entire dataset. This can be obtained from (5.22) utilizing all of the data sections rather than a set of orthogonal Fourier estimates for a single one, and can be computed from the SVD of an $N \times K$ version of $\mathbf{A}$. If the entire dataset can be described in terms of quasi-uniform magnetotelluric sources, then there should be two eigenvalues significantly larger than unity. If there are additional sources of noise or source field complexity, then the coherence dimension will be larger than two, as in Figure 5.4.

The eigenvectors associated with the $M + L$ significant eigenvalues define the space spanned by the columns of the signal and coherent noise matrices $\bar{\bar{\mathbf{U}}}$ and $\bar{\bar{\mathbf{V}}}$. It is not possible to uniquely separate $\bar{\bar{\mathbf{U}}}$ and $\bar{\bar{\mathbf{V}}}$ without additional information or assumptions (Egbert, 1997). However, if $L = 0$ and $M = 2$, then the magnetotelluric responses and all possible inter-station transfer functions may be computed from the two eigenvectors associated with the two largest eigenvalues, as shown by Egbert & Booker (1989). This result will be unbiased by incoherent noise in the data channels.

This discussion contains the basic ideas behind the multivariate errors-in-variables method. However, Egbert (1997) describes additional algorithmic steps to make the method robust that are essential toward its use in practice. Constructing a robust strategy that deals with multivariate outliers is not a simple task, especially in the practical instance where noise is unlikely to be isotropic. In the absence of coherent noise, the resulting algorithm offers comparable performance to univariate bounded influence estimators.

When incoherent noise is present in array data, the robust errors-in-variables method offers a good chance to understand the characteristics of the noise source and, under some circumstances, the ability to separate the magnetotelluric and coherent components of the data, producing usable responses for geophysical purposes. Doing so requires that the magnetotelluric practitioner have an understanding of the multivariate nature of the algorithm, as there is no automatic approach that will yield good results.

References

Adcock, R. J. (1878). A problem in least squares. *The Analyst*, 5, 53–54.

Anderson, T. W. & D. A. Darling (1952). Asymptotic theory of certain "goodness of fit" criteria based on stochastic processes. *Ann. Math. Stat.*, 23, 193–212.

Anderson, T. W. & D. A. Darling (1954). A test of goodness of fit. *J. Am. Stat. Assoc.*, 49, 765–769.

Anderson, T. W. & H. Rubin (1949). Estimation of the parameters of a single equation in a complete system of stochastic equations. *Ann. Math. Stat.*, 20, 46–63.

Anderson, T. W. & H. Rubin (1950). The asymptotic properties of estimates of the parameters of a single equation in a complete system of stochastic equations. *Ann. Math. Stat.*, 21, 570–582.

Banks, R. J. (1998). The effects of non-stationary noise on electromagnetic response estimates. *Geophys. J. Int.*, 135, 553–563.

Belsley, D. A., E. Kuh & R. E. Welsch (1980). *Regression Diagnostics: Identifying Influential Data and Sources of Collinearity.* New York: John Wiley.

Bendat, J. S. & A. G. Piersol (1986). *Random Data: Analysis and Measurement Procedures.* New York: John Wiley.

Bound, J., D. A. Jaeger & R. M. Baker (1995). Problems with instrumental variables estimation when the correlation between the instruments and the endogenous explanatory variable is weak. *J. Am. Stat. Assoc.*, 90, 443–450.

Chatterjee, S. & A. S. Hadi (1988). *Sensitivity Analysis in Linear Regression.* New York: John Wiley.

Chave, A. D. & A. G. Jones (1997). Electric and magnetic field galvanic distortion decomposition of BC87 data. *J. Geomagn. Geoelectr.*, 49, 767–789.

Chave, A. D. & P. Lezaeta (2007). The statistical distribution of magnetotelluric apparent resistivity and phase estimates. *Geophys. J. Int.*, 171, 127–132.

Chave, A. D. & D. J. Thomson (1989). Some comments on magnetotelluric response function estimation. *J. Geophys. Res.*, 94, 14 215–14 225.

Chave, A. D. & D. J. Thomson (2003). A bounded influence regression estimator based on the statistics of the hat matrix. *J. R. Stat. Soc., Ser. C*, 52, 307–322.

Chave, A. D. & D. J. Thomson (2004). Bounded influence estimation of magnetotelluric response functions. *Geophys. J. Int.*, 157, 988–1006.

Chave, A. D., D. J. Thomson & M. E. Ander (1987). On the robust estimation of power spectra, coherences, and transfer functions. *J. Geophys. Res.*, 92, 633–648.

Coakley, C. W. & T. P. Hettmansperger (1993). A bounded influence, high breakdown, efficient regression estimator. *J. Am. Stat. Assoc.*, 88, 872–880.

Durbin, J. (1960). The fitting of time series models. *Rev. Int. Stat. Inst.*, 28, 233–244.

Efron, B. & C. Stein (1981). The jackknife estimate of variance. *Ann. Stat.*, 9, 586–596.

Egbert, G. D. (1989). Multivariate analysis of geomagnetic array data 2. Random source models. *J. Geophys. Res.*, 94, 14 249–14 265.

Egbert, G. D. (1997). Robust multiple-station magnetotelluric data processing. *Geophys. J. Int.*, 130, 475–496.

Egbert, G. D. (2002). Processing and interpretation of electromagnetic induction array data. *Surv. Geophys.*, 23, 207–249.

Egbert, G. D. & J. R. Booker (1986). Robust estimation of geomagnetic transfer functions. *Geophys. J. R. Astron. Soc.*, 87, 173–194.

Egbert, G. D. & J. R. Booker (1989). Multivariate analysis of geomagnetic array data 1. The response space. *J. Geophys. Res.*, 94, 14 227–14 247.

Egbert, G. D. & D. Livelybrooks (1996). Single station magnetotelluric impedance estimation: coherence weighting and the regression M-estimate. *Geophysics*, 61, 964–970.

Eisel, M. & G. D. Egbert (2001). On the stability of magnetotelluric transfer function estimates and the reliability of their variances. *Geophys. J. Int.*, 144, 65–82.

Gamble, T. D., W. M. Goubau & J. Clarke (1979). Magnetotellurics with a remote reference. *Geophysics*, 44, 53–68.

Garcia, X. & A. G. Jones (2008). Robust processing of magnetotelluric data in the AMT dead-band using the continuous wavelet transform. *Geophysics*, 73, F223–F234.

Garcia, X., A. D. Chave & A. G. Jones (1997). Robust processing of magnetotelluric data from the auroral zone. *J. Geomagn. Geoelectr.*, 49, 1451–1468.

Geary, R. C. (1949). Determination of linear relationships between systematic parts of variables with errors of observation the variances of which are unknown. *Econometrica*, 17, 30–58.

Goubau, W. M., T. D. Gamble & J. Clarke (1978). Magnetotelluric data analysis: removal of bias. *Geophysics*, 43, 1157–1162.

Goubau, W. M., P. M. Maxton, R. H. Koch & J. Clarke (1984). Noise correlation lengths in remote reference magnetotellurics. *Geophysics*, 49, 433–438.

Hampel, F. R. (1974). The influence curve and its role in robust estimation. *J. Am. Stat. Assoc.*, 69, 383–393.

Hampel, F. R., E. M. Ronchetti, P. J. Rousseeuw & W. A. Stahel (1986). *Robust Statistics*. New York: John Wiley.

Harris, F. J. (1978). On the use of windows for harmonic analysis with the discrete Fourier transform. *Proc. IEEE*, 66, 51–83.

Hinkley, D. V. (1977a). Jackknife confidence limits using Student t approximations. *Biometrika*, 64, 21–28.

Hinkley, D. V. (1977b). Jackknifing in unbalanced situations. *Technometrics*, 19, 285–292.

Hoaglin, D. C. & R. E. Welsch (1978). The hat matrix in regression and ANOVA. *Am. Stat.*, 32, 17–22.

Huber, P. (1964). Robust estimation of a location parameter. *Ann. Math. Stat.*, 35, 73–101.

Huber, P. (1967). The behavior of maximum likelihood estimates under nonstandard conditions. In *Proc. 5th Berkeley Symp. on Mathematical Statistics and Probability*, ed. J. M. Le Cam & J. Neyman. Berkeley: University of California Press.

Johnson, N. L., S. Kotz & N. Balakrishnan (1994). *Continuous Univariate Distributions*, Vol. 1. New York: John Wiley.

Johnson, N. L., S. Kotz & N. Balakrishnan (1995). *Continuous Univariate Distributions*, Vol. 2. New York: John Wiley.

Jones, A. G. & H. Jödicke (1984). Magnetotelluric transfer function estimation improvement by a coherence-based rejection technique. In *Proc. 54th SEG Meeting*, Atlanta, GA, Expanded Abstracts, pp. 51–55.

Jones, A. G. & J. Spratt (2002). A simple method for deriving the uniform field MT responses in auroral zones. *Earth Planets Space*, 54, 443–450.

Jones, A. G., A. D. Chave, G. D. Egbert, D. Auld & K. Bahr (1989). A comparison of techniques for magnetotelluric response function estimation. *J. Geophys. Res.*, 94, 14 201–14 214.

Junge, A. (1996). Characterization and correction for cultural noise. *Surv. Geophys.*, 17, 361–391.

Kao, D. W. & D. Rankin (1977). Enhancement of signal to noise ratio in magnetotelluric data. *Geophysics*, 42, 103–110.

Koopmans, L. H. (1995). *The Spectral Analysis of Time Series*. San Diego: Academic Press.

Lamarque, G. (1999). Improvement of MT data processing using stationary and coherence tests. *Geophys. Prosp.*, 47, 819–840.

Lanzerotti, L. J., R. E. Gold, D. J. Thomson, R. E. Decker, C. G. Maclennan & S. M. Krimigis (1991). Statistical properties of shock-accelerated ions in the outer heliosphere. *Astrophys. J.*, 380, L93–L96.

Larsen, J. C. (1975). Low frequency (0.1–6.0 cpd) electromagnetic study of deep mantle electrical conductivity beneath the Hawaiian Islands. *Geophys. J. R. Astron. Soc.*, 43, 17–46.

Larsen, J. C. (1980). Electromagnetic response functions from interrupted and noisy data. *J. Geomagn. Geoelectr.*, 32(SI), 89–103.

Larsen, J. C. (1989). Transfer functions: smooth robust estimates by least squares and remote reference methods. *Geophys. J. Int.*, 99, 655–663.

Larsen, J. C., R. L. Mackie, A. Manzella, A. Fiordelisi & S. Rieven (1996). Robust smooth magnetotelluric transfer functions. *Geophys. J. Int.*, 124, 801–819.

Levinson, N. (1947). The Wiener RMS (root-mean-square) error criterion in filter design and prediction. *J. Math. Phys.*, 25, 261–278.

Lezaeta, P., A. D. Chave & R. L. Evans (2005). Correction of shallow water electromagnetic data for noise induced by instrument motion. *Geophysics*, 70, G127–G133.

Lowes, F. J. (2009). DC railways and the magnetic fields they produce – the geomagnetic context. *Earth Planets Space*, 61, i–xv.

Mallows, C. L. (1975). *On some topics in robustness*. Tech. Mem., Bell Telephone Lab.

Miller, K. S. (1973). Complex least squares. *SIAM Rev.*, 15, 706–726.

Müller, A. (2000). A new method to compensate for bias in magnetotellurics. *Geophys. J. Int.*, 142, 257–269.

Oettinger, G., V. Haak & J. C. Larsen (2001). Noise reduction in magnetotelluric time-series with a new signal–noise separation method and its application to a field experiment in the Saxonian Granulite Massif. *Geophys. J. Int.*, 146, 659–669.

Pedersen, L. B. (1982). The magnetotelluric impedance tensor – its random and bias errors. *Geophys. Prosp.*, 30, 188–210.

Percival, D. & A. Walden (1993). *Spectral Analysis for Physical Applications*. Cambridge: Cambridge University Press.

Pomposiello, M. C., J. R. Booker & A. Favetto (2009). A discussion of bias in magnetotelluric responses. *Geophysics*, 74, F59–F65.

Randles, R. H. (1984). On tests applied to residuals. *J. Am. Stat. Assoc.*, 79, 349–354.

Reiersøl, O. (1945). Confluence analysis by means of instrumental sets of variables. *Ark. Mat. Astron. Fys.*, 32, 1–119.

Rousseeuw, P. J. (1984). Least median of squares regression. *J. Am. Stat. Assoc.*, 79, 871–880.

Salibian-Barrera, M. & R. H. Zamar (2002). Bootstrapping robust estimates of regression. *Ann. Stat.*, 30, 556–582.

Sawa, T. (1969). The exact sampling distribution of ordinary least squares and two-stage least squares estimators. *J. Am. Stat. Assoc.*, 64, 923–937.

Shaffer, J. P. (1991). The Gauss–Markov theorem and random regressors. *Am. Stat.*, 45, 269–273.

Shalivahan & B. B. Bhattacharaya (2002). How remote can the far remote reference site for magnetotelluric measurements be? *J. Geophys. Res.*, 107, 2105, doi:10.1029/2000JB000119.

Siegel, A. F. (1982). Robust regression using repeated medians. *Biometrika*, 69, 242–244.

Sims, W. E., F. X. Bostick & H. W. Smith (1971). The estimation of magnetotelluric impedance tensor elements from measured data. *Geophysics*, 36, 938–942.

Slepian, D. (1978). Prolate spheroidal wave functions, Fourier analysis, and uncertainty – V: the discrete case. *Bell. Syst. Tech. J.*, 57, 1371–1430.

Smirnov, M. Yu. (2003). Magnetotelluric data processing with a robust statistical procedure having a high breakdown point. *Geophys. J. Int.*, 152, 1–7.

Stigler, S. M. (2010). The changing history of robustness. *Am. Stat.*, 64, 277–281.

Stuart, A. & J. K. Ord (1994). *Kendall's Advanced Theory of Statistics*, Vol. 1, *Distribution Theory*. London: Edward Arnold.

Stuart, A., J. K. Ord & S. Arnold (1999). *Kendall's Advanced Theory of Statistics*, Vol. 2A, *Classical Inference and the Linear Model*. London: Edward Arnold.

Szarka, L. (1988). Geophysical aspects of man made electromagnetic noise in the earth – a review. *Surv. Geophys.*, 9, 287–318.

Thomson, D. J. (1977). Spectrum estimation techniques for characterization and development of WT4 waveguide, I. *Bell Syst. Tech. J.*, 56, 1769–1815.

Thomson, D. J. (1982). Spectrum estimation and harmonic analysis. *Proc. IEEE*, 70, 1055–1096.

Thomson, D. J. (1990). Quadratic-inverse spectrum estimates: applications to paleoclimatology. *Phil. Trans. R. Soc. Lond.*, A332, 539–597.

Thomson, D. J. & A. D. Chave (1991). Jackknifed error estimates for spectra, coherences, and transfer functions. In *Advances in Spectrum Analysis and Array Processing*, Vol. 1, ed. S. Haykin. Englewood Cliffs, NJ: Prentice Hall, pp. 58–113.

Thomson, D. J., L. J. Lanzerotti, F. L. Vernon, M. R. Lessard & L. T. P. Smith (2007). Solar modal structure of the engineering environment. *Proc. IEEE*, 95, 1085–1132.

Travassos, J. M. & D. Beamish (1988). Magnetotelluric data processing – a case study. *Geophys. J.*, 93, 377–391.

Varentsov, I. M., E. Y. Sokolova, E. R. Martanus & K. V. Nalivaiko (2003). System of electromagnetic field transfer operators for the BEAR array of simultaneous soundings: methods and results. *Izv., Phys. Solid Earth*, 39, 118–148.

Vozoff, K. (1972). The magnetotelluric method in the exploration of sedimentary basins. *Geophysics*, 37, 98–141.

Wald, A. (1940). The fitting of straight lines if both variables are subject to error. *Ann. Math. Stat.*, 11, 284–300.

Walker, G. (1931). On periodicity in series of related terms. *Proc. R. Soc.*, A131, 518–532.

Weckmann, U., A. Magunia & O. Ritter (2005). Effective noise separation for magnetotelluric single site data processing using a frequency domain selection scheme. *Geophys. J. Int.*, 161, 635–652.

Welch, P. D. (1967). The use of the fast Fourier transform for the estimation of power spectra: a method based on time averaging over short, modified periodograms. *IEEE Trans. Audio Electroacoust.*, 15, 70–73.

Wu, C. F. J. (1986). Jackknife, bootstrap and other resampling methods in regression analysis. *Ann. Stat.*, 14, 1261–1295.

Yule, G. U. (1927). On a method of investigating periodicities in disturbed series, with special reference to Wölfer's sunspot numbers. *Phil. Trans. R. Soc. Lond.*, A226, 267–298.

6

Distortion of magnetotelluric data: its identification and removal

ALAN G. JONES

6.1 Introduction

Distortion of regional electric fields by local structures is arguably the greatest bane in the application of the magnetotelluric method (MT), and one that has caused poor interpretations leading directly, within the broad geoscience community, to lower acceptance of MT than should be expected. Approaches and methods developed over the past two decades to recognize, analyze and remove these distortions have enabled significant advances in reliable imaging of the subsurface electrical conductivity structure using magnetotellurics. Without the advanced methods described in this chapter, MT would still yield unreliable models and interpretations. Accordingly, it is fitting that this book contains a significant chapter on distortion.

Precise, robust interpretation of magnetotelluric data observed at a local site from electromagnetic fields induced regionally is fraught with difficulties, primarily because of the inability to model structure at the scale of the Earth's resistivity variations, from small-scale (smaller than the experimental design, i.e. frequencies too low or electrode line length too large) to large-scale (far larger than the experimental design). This problem is usually characterized as *distortion* of the regional fields by local effects (or, perversely, distortion of local fields by regional effects, as in the case of current channeling problems). Distortion of magnetotelluric data has no formal, precise definition, as it is subjective – one person's distortion can be another person's anomaly – leading to a lot of confusion in the literature. Essentially, it means that the observed electromagnetic fields include effects due to features that are outside the scope of the experiment or the interpretation – either too small or too large or of higher dimensionality – that *distort* the regional electromagnetic fields that one wishes to observe caused by the structure(s) of interest.

Dimensionality distortion occurs when MT data (more correctly, responses) have an intrinsically higher dimension than is being used in their interpretation, such as one-dimensional (1D) interpretation of two- or three-dimensional (2D or 3D) structures, or 2D interpretation of 3D structures – see, for example, the excellent review paper by Ledo *et al.* (2005) and also Ledo (2002). One-dimensional interpretation of higher-dimension structures was studied extensively in the 1970s, as the main interpretation tool at that time was 1D forward modeling and, toward the end of the 1970s, inversion. Concepts such as *false conducting layers* (discussed in

The Magnetotelluric Method: Theory and Practice, ed. Alan D. Chave and Alan G. Jones. Published by Cambridge University Press © Cambridge University Press 2012.

Section 6.6.1.1) that arise when one interprets the TE mode data off to the side of a 2D or 3D conductor in a 1D manner, or *DC shifting* (discussed in Section 6.6.1.2) when one interprets the long-period part of the TM mode data on a conductor in 1D, were well known, particularly by the Russian school under Berdichevsky (e.g. Berdichevsky & Dmitriev, 1976a), and were studied by many groups well into the 1990s (e.g. Agarwal *et al.*, 1993). The rule of thumb in the 1970s and early 1980s, until the advent of widespread 2D inversion software late in the decade, was to model/invert TM data off a conductor and TE data on a conductor in 1D, and then stitch the 1D models together to give a pseudo-2D model. The challenge was knowing which was the *correct* mode to model at each site, depending on whether the site was on or off a conductor. This pseudo-2D model was then typically tested by 2D trial-and-error forward modeling.

Rules of thumb were similarly developed in the 1980s for 2D modeling/inversion of 3D data, including using the TE mode responses at high frequencies down to those where the ends of the conductor are one host skin depth away, but utilizing the TM mode ones to much lower frequencies, typically one-tenth of a host skin depth (Jones, 1983; Wannamaker *et al.*, 1984). Particularly the TM mode has been dominantly modeled when data are thought to be from a 3D Earth, but problems with this approach were noted early on for particular geometries (e.g. Hermance, 1982; Park & Mackie, 1997). Further, some structures (particularly narrow vertical conductors) have geometries that can be virtually invisible in the TM data and only seen in the TE data (Jones, 2006), such as the North American Central Plains (NACP) conductivity anomaly (Jones, 1993a) and sub-vertical mineralized zones (Jones & McNeice, 2002), thus requiring all of the responses to be considered.

An example of small-scale distortion, which is the primary focus of this chapter, is *static shift* (e.g. Jones, 1988b; Sternberg *et al.*, 1988), which is a subclass of the local galvanic distortion phenomena introduced by Weidelt and Chave in Chapter 4 (Section 4.1.5).

On the other hand, large-scale (tens to hundreds of kilometers) regional current systems may introduce effects that are beyond those of local induction, and their effects may not be fully appreciated (see e.g. Jones, 1983; Jones & Garcia, 2003). The usual assumption of a flat Earth also brings its own set of problems, such as over- and underestimating the resistivity of the subsurface depending on whether one is in the valley or on the mountain top, again related to an inadequate model of the Earth (Jiracek, 1990).

There would not be any problem with *distortion effects* if it were possible to model what is actually measured. If it were feasible to model the Earth in three dimensions, accounting for variations at the centimeter scale around the electrodes to the hundreds or thousands of kilometers scale of the induced regional fields, and have data capable of resolving all of these scales (and it would be good to have realistic sources rather than always adopting the plane-wave assumption), then there would be no concerns about distortion. Also, the forward code used to mimic the observations must model the experiment and recognize that, whereas virtually point measurements of magnetic fields are made, in MT, estimates of electric fields are made by determining potential differences between two locations some finite distance apart. This latter point was studied in two dimensions and incorporated into a forward code by Poll *et al.* (1989), and was used in Jones's (1988b) analysis of static shift effects, but has, as yet, not been studied in 3D nor incorporated into any forward code, even those used to model mining-scale or environmental-scale problems, where the effects are most certainly significant.

Accordingly, *distortion identification and removal techniques* must be appreciated for what they are: methods that attempt to make the modeling of observed MT data more tractable. This remains a problem today even with the latest 3D inversion codes, and in fact arguably is exacerbated in 3D as the effects of distortion are not frequency-independent (Jones, 2011), and, particularly on the seafloor, frequency-dependent magnetic distortion must be taken into account even in 2D. Currently, although such codes exist, their limitations in terms of speed and mesh size mean that the most sophisticated interpretation tool available to the practicing magnetotelluricist is still 2D inversion, increasingly with anisotropy (e.g. Baba *et al.*, 2006; Evans *et al.*, 2011; Miensopust & Jones, 2011). Therefore, the majority of distortion techniques attempt to identify those parts of a complex dataset that can validly be interpreted to within experimental error as due to a regional 2D structure.

The primary task set forth in MT is observing the time variations of *regional* electric and magnetic fields, and then interpreting from them the regional electrical structure in whatever dimension is appropriate. Of course, the scale size of "regional" is problem-dependent – for those studying mineral deposits, the region may have a dimension of a few tens to hundreds of meters, whereas for those studying continental tectonics, the region is hundreds of kilometers. In either case, these regional fields, and particularly the electric field, are perturbed by site-local features that generate *local* electric and magnetic fields. Here again the concept of *local* is scale-size dependent – regional fields for ore body imaging are local fields for lithosphere imaging.

The second problem is that of the nature of the regional fields themselves – understanding of, and assumptions about, their geometry and strength may be incorrect or inappropriate. Of course, in magnetotellurics (MT) and geomagnetic depth sounding (GDS) the total field components are measured, which are the vector sum of the regional and local fields, and it must then be considered how the local fields have affected the observations. Consequently, the problem becomes one of the separation of regional from local fields (e.g. Berdichevsky & Dmitriev, 2008, chapter 3), remembering that "regional" and "local" have no absolute definition, but are both relative to the scale of the experiment.

The literature is resplendent with innovative, evocative terms that various authors have used in attempts to describe the physics of the phenomena that is not being, in one way or another, considered or mathematically modeled, but rather treated as an unwanted effect. Essentially, to first order these can be characterized as falling into two types – those due to *currents* and those due to *charges*. Given their inter-relationship through the Maxwell equations, it is of course a simplistic conceptualization to think of the physics of induction operating in this independent manner for currents and charges, but, for the purposes of describing the distortions of magnetotelluric data, it suffices.

The integrating and continuous nature of magnetic fields is such that they are directly affected by distortion due to currents. The indirect magnetic effects of charge, caused by the deflection of current, falls off rapidly with increasing period and can, for most circumstances on land at least, be neglected (e.g. Groom & Bailey, 1991; Jones & Groom, 1993; Chave & Smith, 1994; Chave & Jones, 1997; Agarwal & Weaver, 2000). In contrast, the very local nature of electric fields, with their inherent discontinuity at resistivity contrasts due to Ohm's law current continuity, leads to their strong distortion from charges on boundaries, and charge effects are, in the inductive case, frequency-independent and persist to the longest periods.

Distortion due to deviations in regional current flow, the so-called *current channeling problem*, was examined and reviewed in detail by Jones (1983). The current channeling problem was discussed extensively during the 1970s, as many induction studies of the Earth used GDS arrays rather than MT, so concern and appreciation of regional current flow dominated the thinking. Essentially the "problem" was one of the validity of 2D interpretation in the case of data from 3D structure. This has been examined for MT by a number of others over the past 30 years, notably in the last decade by Ledo *et al.* (2002), in Ledo's excellent review (Ledo, 2005), Queralt *et al.* (2007) and Becken *et al.* (2008).

This chapter will discuss distortion effects on MT data, and present and appraise various tools for identifying and correcting for its effects. Particular attention will be given to those parts of distortion that can be determined, compared to those that are indeterminable, and also to the statistical framework within which each approach is based. Essentially, as in many aspects of science, whether distortion can be identified and removed is a function of the quality of the data, hence becomes a statistical problem. A uniform half-space will fit the poorest-quality data, regardless of distortion. For the highest-quality data, a highly complex, fine-gridded 3D model is required to fit the data to within their errors, as the Earth is 3D at high resolution.

Thus, distortion identification and removal is not only a mathematical or physical problem, but also a statistical one. Various approaches are tested statistically for their robustness by generating thousands of realizations from given 1D, 2D, 3D and 3D/2D models, and determining parameters from those realizations. As will be shown, virtually all phase-based tools will give correct results in the absence of noise, but of crucial importance is how each tool performs in the presence of noise. The effects of bias and random noise are, in most cases, not analytically derivable or predictable, especially of estimators that are given by ratios of combinations of the magnetotelluric impedance tensor (Chave, 2011) such as *skew* and other invariants and the phase tensor, so a bootstrap method is used to examine the statistical properties of them.

6.2 Theoretical considerations

6.2.1 General

The theory of the physics of distortion of both the electric and magnetic fields from charges on resistivity boundaries has been understood for over a century, and in magnetotellurics is well described by Ross Groom in his Ph.D. thesis (Groom, 1988), in the two Groom and Bailey papers (Groom & Bailey, 1989, 1991), in Groom and Bahr's excellent tutorial paper

(Groom & Bahr, 1992), and especially well by Chave & Smith (1994) in their landmark paper, to which the reader is referred for a full exposition. Using the notation from Chave & Smith (1994), under the quasi-static pre-Maxwellian approximation described in Chapter 2 (i.e. neglecting electric permittivity) in non-magnetic media (i.e. free-space magnetic permeability), in the absence of explicit sources (i.e. source-free medium), and with exp($i\omega t$) dependence suppressed, the electric field equation is

$$\nabla \times \nabla \times \mathbf{E}(\mathbf{r}) + i\omega\mu_0\sigma_0\, \mathbf{E}(\mathbf{r}) = -i\omega\mu_0\, \delta\sigma(r)\mathbf{E}(\mathbf{r}) \tag{6.1}$$

where $\delta\sigma(r) = \sigma(\mathbf{r}) - \sigma_0$, and σ_0 refers to the background (regional) conductivity structure. The scattering of the electric field at points both inside and outside an inhomogeneity is described by an integral equation (eq. (6.2) in Chave & Smith, 1994), and from it, using Ampere's law, the analogous equation for the magnetic induction can be derived (eq. (6.3) in Chave & Smith, 1994), namely

$$\begin{aligned} \mathbf{E}(\mathbf{r}) =& \mathbf{E}^0(\mathbf{r}) - i\omega\mu_0 \int_{V_s} d\mathbf{r}'\, g(\mathbf{r},\, \mathbf{r}')\delta\sigma(\mathbf{r}')\mathbf{E}(\mathbf{r}') \\ &+ \frac{1}{\sigma_0}\nabla\nabla\bullet \int_{V_s} d\mathbf{r}'\, g(\mathbf{r},\, \mathbf{r}')\delta\sigma(\mathbf{r}')\mathbf{E}(\mathbf{r}') \end{aligned} \tag{6.2}$$

and

$$\mathbf{B}(\mathbf{r}) = \mathbf{B}^0(\mathbf{r}) + \mu_0\nabla \times \int_{V_s} d\mathbf{r}'g(\mathbf{r},\mathbf{r}')\delta\sigma(\mathbf{r}')\mathbf{E}(\mathbf{r}') \tag{6.3}$$

where the Green's function $g(\mathbf{r}, \mathbf{r}')$ is given by

$$g(\mathbf{r},\mathbf{r}') = \frac{\exp(i\gamma_0|\mathbf{r} - \mathbf{r}'|)}{4\pi|\mathbf{r} - \mathbf{r}'|} \tag{6.4}$$

and γ_0 is the propagation constant in the host medium given by $\sqrt{i\omega\mu_0\sigma_0}$. Note that the second term on the right-hand side in the equation for the electric field is the inductive component, and the third term is the galvanic component.

The solution of the integral equations used by Chave & Smith (1994) takes advantage of the extended Born approximation, called the localized nonlinear (LN) approximation, advocated by Habashy *et al.* (1993). Following this approximation, and assuming that the regional field is homogeneous across the inhomogeneity, the locally observed horizontal electric ($\mathbf{E}_h$) and magnetic ($\mathbf{B}_h$) components of the electromagnetic field can be related to the regional ones ($\mathbf{E}_h{}^0$ and $\mathbf{B}_h{}^0$) by

$$\begin{aligned} \mathbf{E}_h(\mathbf{r}) &= \mathbf{C}(\mathbf{r}) \bullet \mathbf{E}_h^0(\mathbf{r}) \\ \mathbf{B}_h(\mathbf{r}) &= \mathbf{B_h}^0(\mathbf{r}) + \mathbf{D}(\mathbf{r}) \bullet \mathbf{E_h}^0(\mathbf{r}) \end{aligned} \tag{6.5}$$

(Chave & Smith, 1994, eqs. (14) and (18)), where **C** and **D** are second-rank tensors describing the electric and magnetic effects of the galvanic charges, respectively. The elements of **C** and **D** are real-valued and frequency-independent when induction can be ignored. [Note that there is no consistency in the literature about the labeling of these two tensors, and various letters have been assigned to **C**. Unfortunately, some authors used **D** for the electric distortion tensor (e.g. Smith, 1995; Caldwell *et al.*, 2004), and further compound the potential for confusion by additionally using **C** for the magnetic distortion tensor (e.g. Agarwal & Weaver, 2000). The reader is urged to use caution when comparing various expositions on distortion.]

Thus, for a regional antidiagonal MT response tensor, $\mathbf{Z}_2$, i.e. for an isotropic 1D or 2D structure, then the observed response $\mathbf{Z}_{obs}$ in strike coordinates is given by

$$\mathbf{Z}_{obs}(0) = \mathbf{C} \cdot \mathbf{Z}_2 \cdot (\mathbf{I} + \mathbf{D} \cdot \mathbf{Z}_2)^{-1} \qquad (6.6)$$

where **I** is the 2×2 identity matrix. (Note that an explicit coordinate system has been omitted, meaning that the coordinate system of observation is assumed to be that of the 2D regional structure.)

It is important to note the major approximations in this development that could lead to a breakdown in the underlying assumptions. Essentially, the distorting body must be at the galvanic limit, meaning there is little induction in it (i.e. a small induction number). This requires that the inductive scale length within the host and of the inducing source field be far greater than the size of the distorting body. As detailed by Groom & Bahr (1992) and Chave & Smith (1994), these approximations are as follows:

1. The regional electric field at the observing point and at the body have to be the same.
2. There cannot be large source field gradients over the scatterer.
3. The frequency must be low enough that the induction within the inhomogeneity is negligible.
4. The scatterer must have a conductivity that is not many orders of magnitude more conducting than the background.

6.2.2 Groom–Bailey distortion decomposition

Much of the discussion in this chapter will be framed within the context of the distortion decomposition advocated by Ross Groom and Richard (Dick) Bailey developed in the mid-1980s. Although it might logically make chronological sense to discuss the Groom–Bailey approach later in this chapter, its role is so overarching and insightful that it is introduced at this point.

As a historical note, the work was initially submitted to SEG's journal *Geophysics* (Bailey & Groom, September 1986, pers. comm.) and was first presented at the Society of Exploration Geophysicists Annual Meeting in 1987 (Bailey & Groom, 1987). Inexplicably, the paper was judged not suitable for publication, and was subsequently submitted to AGU's

Journal of Geophysical Research, where it was rapidly published (Groom & Bailey, 1989). It has since become one of the top cited publications in the whole field of magneto-tellurics. SEG realized its error, and invited Groom and Bailey to submit a paper describing extensions and further details, which was published in 1991 (Groom & Bailey, 1991). Groom developed a systematic methodological approach to the analysis of MT data (Groom *et al.*, 1993), and together with K. Bahr gave an excellent tutorial on decomposition at the EM Induction Workshop held in Ensenada, Mexico, in 1990, (Groom & Bahr, 1992).

The insight shown by the Groom–Bailey decomposition approach, which is absent in all other decomposition-based approaches except those that follow the same methodology (such as Chave & Smith, 1994; McNeice & Jones, 2001; and recently Cerv *et al.*, 2010), is that it formally separates out distortion elements that are *indeterminable* from those that are *determinable*. The former comprise effects that solely result in a change in the magnitude of the electric field, with *static shifts* being the simplest manifestation of these. The latter comprise effects that result in changes to both the amplitude and the phase of the electric field. Thus, it is a *physically based* decomposition, rather than a mathematically based one that does not recognize the nature of the problem to be solved. The phase-tensor approach of Caldwell *et al.* (2004), discussed below (Section 6.3.6) and in Chapter 4 (Section 4.1.6), uses only phases, so also derives determinable parameters (Bibby *et al.*, 2005). However, it has other weaknesses that are discussed in Section 6.3.6.

The second advantage of the approach, compared to other distortion approaches, is that it undertakes the process within a firm statistical framework. Essentially, the procedure is one of hypothesis testing and asking the question: *Can a statistically acceptable model be found of frequency-independent galvanic distortion of regional electric fields resulting from a 2D Earth that fits the observed data to within their limits?*

The third major difference between the Groom–Bailey approach and all others is that all other methods algebraically manipulate and rearrange the eight numbers in the MT response tensor (four complex numbers), whereas Groom–Bailey fits a model of distortion of the seven determinable parameters (single site, single frequency) or less (multiple sites, multiple frequencies). In this manner, all approaches, except for Groom–Bailey, are mathematically related to one another, and algebraic expressions can be found between them.

Finally, the fourth major difference is the existence of the uniqueness property for the decomposition (Groom, 1988; Groom & Bailey, 1989). A non-unique decomposition, no matter how neatly it makes the diagonal components of the regional tensor vanish, is of little value. Thus, Bailey and Groom's the first step was to derive a decomposition that removed indeterminable parts from determinable parts (see Section 6.4), because uniqueness could only be proved for entirely determinable parts. "Ross did the hard work on the uniqueness theorem, which, like uniqueness theorems in most fields, had to be proved once before it could be safely forgotten" (R. C. Bailey, 2011, pers. comm.).

Although the effects on the magnetic fields of the galvanic charges are discussed by Groom and Bailey, the Groom–Bailey distortion decomposition approach in routine use addresses purely the electric effect, and assumes that the magnetic effects of the charges, as represented by **D** in Eq. (6.6), can be ignored. This occurs at the galvanic limit when the frequencies are sufficiently low that the product **D** • $\mathbf{Z}_2$ is much smaller than the identity matrix, given the data errors, and can be ignored. Then, in an arbitrary reference frame θ,

$$\mathbf{Z}_{obs}(\theta) = \mathbf{R}(\theta) \bullet \mathbf{C} \bullet \mathbf{Z}_2 \bullet \mathbf{R}^{\mathrm{T}}(\theta) \tag{6.7}$$

where θ is the angle between the observation coordinate system and the strike of the 2D regional structure, and **R** is the 2 × 2 unitary Cartesian rotation matrix with superscript "T" denoting transpose. Groom and Bailey factored **C**, a 2 × 2 real matrix, in terms of three matrices and one scalar, and chose to use matrices that have a Pauli spin matrix basis set. The factorization is

$$\mathbf{C} = g\,\mathbf{T} \bullet \mathbf{S} \bullet \mathbf{A} \tag{6.8}$$

where g is called *site gain*, **T** is called the *twist*, **S** is *shear* and **A** is *anisotropy*. These three matrices are

$$
\begin{aligned}
\mathbf{T} &= \frac{1}{\sqrt{1+t^2}}\begin{pmatrix} 1 & -t \\ t & 1 \end{pmatrix} \\[4pt]
\mathbf{S} &= \frac{1}{\sqrt{1+s^2}}\begin{pmatrix} 1 & s \\ s & 1 \end{pmatrix} \\[4pt]
\mathbf{A} &= \frac{1}{\sqrt{1+a^2}}\begin{pmatrix} 1+a & 0 \\ 0 & 1-a \end{pmatrix}
\end{aligned}
\tag{6.9}
$$

(Note that, in Groom & Bailey (1989), the coefficients of the shear and anisotropy matrices are e, for ellipticity, and s, for split, but here s (shear) and a (anisotropy) are used for simplicity.) The leading square root terms are irrelevant, as they can be absorbed into g.

There is nothing inherently "right" about this factorization of the MT response tensor that makes any others "wrong" – any factorization that separates out the determinable from the indeterminable parts may be equally as attractive. However, any other factorization must also exhibit the uniqueness property shown by this one.

Attempts to describe the effects of the individual matrices is really a purely academic exercise, as the physics of distortion at any one site does not partition in the manner described by Eq. (6.8). However, it is useful to try to understand the effects that each have individually.

The *twist* tensor functions as a rotation operator. The local electric fields are rotated clockwise, through an angle of $\theta_{tw} = \arctan(t)$. The result of this operation relaxes the orthogonality between the electric and magnetic fields. Such an approach was the basis of the coherence-based MINMAX analysis of Volker Haak in the early 1970s (Haak, 1972), which also permitted a relaxation in the orthogonality of the magnetic field and the corresponding electric field. Clearly, this rotation cannot be greater than ±90°, so there are physical limits to t of ±∞, but these limits are not particularly useful. Distortion that can be characterized by twist alone would lead to the observed horizontal electric fields being given by

$$E_x = \frac{1}{\sqrt{1+t^2}}(E_x^0 - tE_y^0)$$
$$E_y = \frac{1}{\sqrt{1+t^2}}(tE_y^0 + E_y^0) \tag{6.10}$$

The *shear* tensor, named through analogy to deformation theory, has the form of an elastic strain tensor and produces a linear combination of the two horizontal electric field components. Distortion characterized by shear alone would lead to observed fields of

$$E_x = \frac{1}{\sqrt{1+s^2}}(E_x^0 + sE_y^0)$$
$$E_y = \frac{1}{\sqrt{1+s^2}}(sE_y^0 + E_y^0) \tag{6.11}$$

For the shear parameter s, there is a physical limit of unity on $|s|$. Distortion can never be so severe that it will cause the local fields to have a component in the reverse direction to the regional fields. When expressed as an angle, $\theta_{sh} = \arctan(s)$, this is $\pm 45°$. As this limit is approached, which can be found in highly heterogeneous regions, then the electric field essentially propagates in one direction, regardless of the direction of the inducing magnetic field. Such strong distortion occurs, for example, at long periods when the MT site is located in a narrow valley filled with conducting sediments.

The *anisotropy* tensor causes a decrease in one electric field component by the same amount that the other is increased, viz.

$$E_x = \frac{1}{\sqrt{1+a^2}}(1+a)E_x^0$$
$$E_y = \frac{1}{\sqrt{1+a^2}}(1-a)E_y^0 \tag{6.12}$$

The obvious physical limit on $|a|$ is that it must be less than unity – an anisotropy $|a| > 1$ would yield negative resistivity in one of the directions.

Figure 6.1 illustrates the effects of these three tensors operating on a family of unit vectors.

The local electric field channeling azimuth is given by the sum of three parameters, namely the regional 2D strike plus the two galvanic distortion parameters *shear* and *twist* described as angles, i.e.

$$\theta_{ch} = \theta_{st} + \theta_{sh} + \theta_{tw} \tag{6.13}$$

A problem to be solved is that the decomposition, as posed in Eqs. (6.7) and (6.8), has nine unknowns: namely the two complex regional responses Z_{xy} and Z_{yx}, the geoelectric strike direction θ_{st}, and the four parameters describing the galvanic distortion, g, t, s and a. However, there are only eight knowns: the four complex observed responses. Groom and Bailey recognized that two of the distortion parameters, g and a, operate only on the

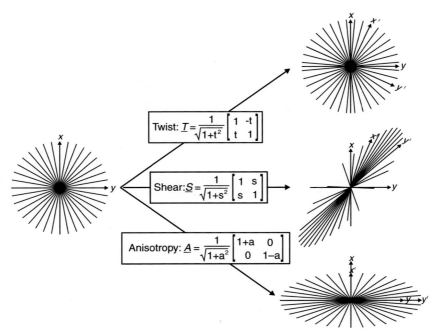

Figure 6.1. Groom–Bailey twist, shear and anisotropy distortion of a unit vector. Figure redrawn from Simpson & Bahr (2005, their figure 5.5).

magnitudes of the impedances, yielding magnitude-only *static shifts* of the regional electric fields (and thus apparent resistivities), so substituted a scaled value for the regional 2D impedance tensor $\mathbf{Z}'_2 = g\,\mathbf{A}\,\mathbf{Z}_2$ in Eq. (6.7) to give

$$\mathbf{Z}_{obs}(\theta_{st}) = \mathbf{R}(\theta_{st})\,\mathbf{T}\,\mathbf{S}\,\mathbf{Z}'_2\,\mathbf{R}^{\mathrm{T}}(\theta_{st}) \tag{6.14}$$

This is an equation of seven unknowns (θ_{st}, θ_{sh}, θ_{tw} and the two scaled complex impedances of the regional 2D Earth) and eight knowns (four complex values of the impedance tensor) at one single frequency at one site. For reasons of numerical stability, the unknown parameters are not solved directly, but combinations of them are estimated (see Groom & Bailey, 1989, eqs. (33), (34) and (35)). Note that once the seven parameters have been estimated, then, provided the phases of the regional Z_{xy} and Z_{yx} estimates are the same (modulo π) at high frequencies, the anisotropy coefficient a can be determined by merging the high-frequency asymptotes of the apparent resistivity curves to their geometric mean. Thus, the only indeterminable parameter is the *site gain* g, which multiplies the magnitudes of both regional impedance elements by the same amount.

The Groom–Bailey approach was extended by McNeice & Jones (1996, 2001) to solve simultaneously for the best-fitting strike direction and values of twists and shears at multiple sites and for multiple frequencies. Chave & Smith (1994) also undertook a multifrequency decomposition approach and found increased consistency when doing so. Recently, Cerv *et al.* (2010) proposed a multi-site, multifrequency Groom–Bailey decomposition that uses Bayesian statistics to determine the range of acceptable parameters.

6.3 Brief historical review

6.3.1 Berdichevsky's galvanic distortion effects

Marc Berdichevsky and his students were studying distortion on MT data numerically in the early 1970s, and produced albums of effects for type structures, but almost all of the early literature was published in Russian and was little known outside the Soviet Union at that time. The first papers in English came out in 1976 (Berdichevsky & Dmitriev, 1976a,b), particularly with the presentation of their excellent review paper at the EM Induction Workshop held in Sopron, Hungary, in June 1976 (Berdichevsky & Dmitriev, 1976a), which was the first EM induction workshop with significant attendance by Russian scientists.

The results in the 1976 papers have been repeated in a number of subsequent papers and publications, most recently the 2008 book by Berdichevsky & Dmitriev (2008). Owing to space restrictions, those type examples will not be repeated here, and the interested reader is referred to their 2008 book.

6.3.2 Larsen's galvanic distortion of a one-dimensional regional Earth

Recognition by the MT community of galvanic electric field distortion effects on MT data and a method of addressing them became prominently known in the Western literature following two now-classic papers by James (Jimmy) Larsen (1975, 1977), who wanted to interpret very long-period data from the island of Hawaii in a 1D manner, but knew that the coast effect, or more correctly island effect, resulted in an increase in the electric fields across the island from their regional levels. To compensate for this, Larsen (1975) represented the relationship between the horizontal seafloor field, $\mathbf{E}_0$, and the local horizontal island field, $\mathbf{E}_I$, by a 2×2 real distortion matrix of the form

$$\mathbf{E}_0 = D \begin{pmatrix} 1 - B & C + A \\ C - A & 1 + B \end{pmatrix} \mathbf{E}_I \tag{6.15}$$

(Larsen, 1975, eq. (4.15)), where D is an arbitrary scaling factor (*site gain* in Groom–Bailey parlance) and A, B and C unwrap the distortion of the island electric field to give the regional electric field. This leads to the relationship between the electric and magnetic fields on the island of

$$\mathbf{E}_I = \mathbf{Z}_I \begin{pmatrix} A + C & 1 + B \\ 1 - B & A - C \end{pmatrix} \mathbf{H}_I \tag{6.16}$$

(Larsen, 1975, eq. (4.18)). Best-fitting parameters found were $A = -0.100 \pm 0.005$, $B = -0.197 \pm 0.004$ and $C = -0.735 \pm 0.006$.

Electric field distortion is now determined in terms of the distortion of the regional electric field to the distorted field (Eq. (6.5)), which is the inverse of the unwrapping distortion matrix in Eq. (6.15), so the galvanic electric field distortion tensor $\mathbf{C}$ on Hawaii is given by

$$
\begin{aligned}
\mathbf{C}_{Hawaii} &= \left(D \begin{pmatrix} 1-B & C+A \\ C-A & 1+B \end{pmatrix} \right)^{-1} \\
&= \frac{D^{-1}}{(1 - B^2 - C^2 + A^2)} \begin{pmatrix} 1+B & -C-A \\ -C+A & 1-B \end{pmatrix} \\
&= \frac{D^{-1}}{0.431} \begin{pmatrix} 0.803 & 0.835 \\ 0.635 & 1.197 \end{pmatrix}
\end{aligned}
\tag{6.17}
$$

which is relatively strong distortion. In terms of Groom–Bailey parameters (Section 6.2.2), it has a *twist* of only 1.7° but a rather strong *shear* of 36.6° and a fairly minor *anisotropy* of 0.175. The *site gain* is $0.996/(0.431D)$, so cannot be resolved without other information.

The distortion on the island of Hawaii can now be understood as a circular resistive "distorting" body embedded in a much more conducting layer, the sea, exactly as described in Morse & Feshbach (1953, chapter 4, figure 4.1, p. 1199). From Eq. (6.31) below it can be seen that, for a perfectly circular island, the internal electric field on the island is, for $\sigma_1 \gg \sigma_2$, approximately 1.5 times the normal electric field, so that the apparent resistivities are increased by 225%. This is only an approximate figure, however, as the resistive island is clearly not floating in a conductive seawater half-space. If the embedded hemisphere model were a correct representation of Hawaii, then the diagonal elements would be equal and the off-diagonal elements zero (see Eq. (6.34) below), which is clearly not the case.

6.3.3 Schmucker's extension for a two-dimensional regional Earth

This 1D electric field distortion approach was then advanced to galvanic electric field distortion of 2D regional electric fields by Ulrich Schmucker and colleagues and presented in 1982 by M. L. Richards, U. Schmucker and E. Steveling (Richards *et al.*, 1982) at one of the German biennial EM induction workshops. The local electric fields at a site were considered to be a distorted form of the regional fields by

$$
E_x = a_{11}E_{nx} + a_{12}E_{ny} \qquad \text{and} \qquad E_y = a_{21}E_{nx} + a_{22}E_{ny}
\tag{6.18}
$$

where the subscripts n denote the regional, or "normal", field, and the terms a_{ij} are real and frequency-independent. Unfortunately, the paper was written in German and never presented at an international conference nor published, so Schmucker's role in advancing our recognition and understanding of distortion has not been appropriately recognized or appreciated.

6.3.4 Bahr's equal phases

The galvanic electric field distortion approach in 2D was taken up by Schmucker's student, Karsten Bahr, who recognized that, when a distorted regional 2D response is rotated into correct strike coordinates, i.e. ($\theta = 0$), the phases of the two elements in each column of the response tensor are equal, i.e. $\varphi_{xx}(0) = \varphi_{yx}(0)$ and $\varphi_{xy}(0) = \varphi_{yy}(0)$. In a general coordinate system at an angle of θ relative to the strike direction of the 2D structures, the distorted and rotated response that is observed is given by Eq. (6.7) as

$$\mathbf{Z}_{obs}(\theta) = \mathbf{R}(\theta) \cdot \mathbf{C} \cdot \mathbf{Z}_2 \cdot \mathbf{R}^{\mathrm{T}}(\theta)$$

When the observed data are rotated into strike coordinates such that θ is zero, then

$$\begin{aligned}
\mathbf{Z}_{obs}(0) &= \mathbf{R}(0) \cdot \mathbf{C} \cdot \mathbf{Z}_2 \cdot \mathbf{R}^{\mathrm{T}}(0) \\
&= \mathbf{C} \cdot \mathbf{Z}_2 \\
&= \begin{bmatrix} c_{11} & c_{12} \\ c_{21} & c_{22} \end{bmatrix} \begin{bmatrix} 0 & Z_{xy} \\ Z_{yx} & 0 \end{bmatrix} \\
&= \begin{bmatrix} c_{12}Z_{yx} & c_{12}Z_{xy} \\ c_{12}Z_{yx} & c_{22}Z_{xy} \end{bmatrix}
\end{aligned} \tag{6.19}$$

and, as the distortion values c_{ij} are all real, the phases of the two elements in the first column are the same as each other, and similarly the phases of the elements in the second column are the same as each other. Thus, Bahr's approach was to rotate the MT tensor until the condition in Eq. (6.19) was met as closely as possible.

Bahr presented the kernel of the concept in 1984 at the EM Workshop in Nigeria (Bahr, 1984), and it was, without question, one of the highlights of the workshop. Bahr published his thesis research in 1988 (Bahr, 1988) and extended the approach in 1991 (Bahr, 1991) for cases that are *almost* 2D (whatever that means), but no statistical measure was proposed to test the appropriateness of the 2D assumption.

6.3.5 Distortion decomposition

Laust Pedersen and Bailey in the mid-1980s independently saw the advantages of fitting a model of galvanic distortion of a 2D regional Earth of the form described by Eq. (6.14), and initiated research efforts using decomposition of the distorted MT response tensor. Pedersen's approach was 2D galvanic distortion of a 2D regional Earth (2D/2D) to explain 2D regional responses being affected by long conducting faults, and the work was led by his student, Ping Zhang, and was published in Zhang *et al.* (1987).

Bailey's approach, discussed in detail in Section 6.2.2, was of 3D distortion of a 2D regional Earth, and together with Groom advanced to what has become the most powerful tool currently available for analysis of directionality and dimensionality inherent within the observed MT data (see Section 6.2.2).

Others, particularly Alan Jones, Alan Chave and Torquil Smith (Chave & Smith, 1994; Smith, 1995, 1997; Chave & Jones, 1997) considered the problem through the 1990s. The multi-site, multifrequency *strike* code of Gary McNeice and Jones, first presented in 1996 and published in 2001 (McNeice & Jones, 1996, 2001), is based on the Groom–Bailey approach and is in use by over 150 scientists and groups worldwide. (The code is freely available to academics from Alan Jones upon request.)

Although it is not discussed in detail in this chapter, the magnetic field distortion problem can be significant, especially on the seafloor. This distortion can be characterized by a tensor **D** (Eq. (6.6)), and examination of the effects of magnetic distortion was considered by Groom & Bailey (1991), Jones & Groom (1993), Chave & Smith (1994), Smith (1997) and Agarwal & Weaver (2000), among others. However, it is not routine practice to address the possible effects of magnetic distortion, primarily as the effects drop off rapidly with increasing period and are only important for typically half a decade at most.

Smith (1995) presented a different parameterization approach to Groom–Bailey that he considered to be simpler and more understandable, in which **C** is defined in terms of two scaling factors and two angles, namely

$$\mathbf{C} = \left[g_x \begin{pmatrix} \cos\beta_x \\ \sin\beta_x \end{pmatrix} \quad g_y \begin{pmatrix} -\sin\beta_y \\ \cos\beta_y \end{pmatrix} \right] \tag{6.20}$$

The problem with this parameterization is that it does not explicitly separate determinable and indeterminable parts. Also, no model fitting is applied, so no statistical inferences can be taken about whether the model is appropriate or not.

Finally, Lezaeta & Haak (2003) describe an extension to Smith's (1997) approach, which includes magnetic effects, in the presence of strong current channeling, where standard distortion decomposition techniques may be unable to recover the correct 2D regional strike angle. Primarily, their approach is used in a qualitative sense to identify the location of 3D regional structures.

6.3.6 *Phase tensor*

One of the more intriguing approaches to have been proposed in the past decade is the phase-tensor method of Grant Caldwell, Hugh Grant and Colin Brown – see Caldwell *et al.* (2004) and Bibby *et al.* (2005), the comment by Moorkamp (2007) and the reply by Caldwell *et al.* (2007). This is discussed in detail in Chapter 4 (Section 4.1.5) and is included here for completeness. The attraction of this approach is that it makes no assumptions about the dimensionality of the regional response tensor but derives the phase tensor correctly in the presence of electric field galvanic distortion (but not when magnetic field distortion is present). Assuming there is electric-field-only galvanic distortion, as described by Eq. (6.5) above, then the local horizontal electric field is given by

$$\mathbf{E} = \mathbf{C} \cdot \mathbf{E}_R = \mathbf{C} \cdot (\mathbf{Z}_R \cdot \mathbf{B}_R) = (\mathbf{C} \cdot \mathbf{Z}_R) \cdot \mathbf{B}_R \qquad (6.21)$$

where now the subscript R represents the regional field. The relationship between the observed and regional MT response, of any dimensionality, is

$$\mathbf{Z} = \mathbf{C} \cdot \mathbf{Z}_R \qquad (6.22)$$

Separating the impedance tensors into their real and imaginary parts, such that $\mathbf{Z} = \mathbf{X} + i\mathbf{Y}$ and $\mathbf{Z}_R = \mathbf{X}_R + i\mathbf{Y}_R$, then

$$\mathbf{X} = \mathbf{C} \cdot \mathbf{X}_R \quad \text{and} \quad \mathbf{Y} = \mathbf{C} \cdot \mathbf{Y}_R \qquad (6.23)$$

From $\mathbf{X}$ and $\mathbf{Y}$ can be defined a rank-two phase tensor comprising four real elements, given by

$$\begin{aligned} \mathbf{\Phi} &= \mathbf{X}^{-1}\mathbf{Y} = (\mathbf{C}\,\mathbf{X}_R)^{-1}(\mathbf{C}\,\mathbf{Y}_R) \\ &= \mathbf{X}_R^{-1}\,\mathbf{C}^{-1}\,\mathbf{C}\,\mathbf{Y}_R = \mathbf{X}_R^{-1}\,\mathbf{Y}_R \\ &= \mathbf{\Phi}_R \quad . \end{aligned} \qquad (6.24)$$

Thus, the phase tensor derived from the observed, distorted MT response is the phase tensor of the true regional response, regardless of the dimensionality of the subsurface. Note that, of the eight parameters associated with the impedance tensor, four have been discarded and four retained. There is thus inherent information loss when using the phase tensor for modeling and\or interpretation.

The phase-tensor elements are given by

$$\begin{bmatrix} \Phi_{xx} & \Phi_{xy} \\ \Phi_{yx} & \Phi_{yy} \end{bmatrix} = \frac{1}{\det(\mathbf{X})} \begin{bmatrix} X_{yy}Y_{xx} - X_{xy}Y_{yx} & X_{yy}Y_{xy} - X_{xy}Y_{yy} \\ X_{xx}Y_{yx} - X_{yx}Y_{xx} & X_{xx}Y_{yy} - X_{yx}Y_{xy} \end{bmatrix} \qquad (6.25)$$

The four elements of the phase tensor are represented by Caldwell *et al.* (2004) in terms of the parameters of an ellipse, following Bibby's (1986) earlier work on dipole–dipole resistivity, and are the maximum (Φ_{max}) and minimum (Φ_{min}) tensor values, the skew angle β and the non-invariant angle α (Figure 4.2 in Chapter 4).

Caldwell *et al.* (2004) give analytical expressions for deriving these four parameters, with Moorkamp (2007) providing superior expressions that avoid the problem of negative determinants. The expressions are given in terms of invariants, namely the trace, determinant and skew (anti-trace) of the phase tensor, viz.

$$\begin{aligned} \Phi_1 &= \mathrm{tr}(\mathbf{\Phi})/2 = \left(\Phi_{xx} + \Phi_{yy}\right)/2 \\ \det(\mathbf{\Phi}) &= \det(\mathbf{Y})/\det(\mathbf{X}) = \Phi_{xx}\Phi_{yy} - \Phi_{xy}\Phi_{yx} \\ \Phi_3 &= \mathrm{sk}(\mathbf{\Phi})/2 = \left(\Phi_{xy} - \Phi_{yx}\right)/2 \end{aligned} \qquad (6.26)$$

and the four parameters are given by

$$\Phi_{max} = \left(\Phi_1^2 + \Phi_3^2\right)^{1/2} + \left(\Phi_1^2 + \Phi_3^2 - \det(\boldsymbol{\Phi})\right)^{1/2},$$

$$\Phi_{min} = \left(\Phi_1^2 + \Phi_3^2\right)^{1/2} - \left(\Phi_1^2 + \Phi_3^2 - \det(\boldsymbol{\Phi})\right)^{1/2},$$

$$\beta = \frac{1}{2}\tan^{-1}\left(\frac{\Phi_3}{\Phi_1}\right) = \frac{1}{2}\tan^{-1}\left(\frac{\Phi_{xy} - \Phi_{yx}}{\Phi_{xx} + \Phi_{yy}}\right), \tag{6.27}$$

$$\alpha = \frac{1}{2}\tan^{-1}\left(\frac{\Phi_{xy} + \Phi_{yx}}{\Phi_{xx} - \Phi_{yy}}\right).$$

The strike of the major axis of the ellipse is given by $\alpha - \beta$, and in the case of a 2D or 3D/2D Earth, β is zero and the 2D strike direction is given by α.

Bibby *et al.* (2005, eq. (22)) define a dimensionless fifth parameter, λ, which is a combination of the above four that is indicative of dimensionality and is given by

$$\lambda = \frac{\left[\left(\Phi_{xx} - \Phi_{yy}\right)^2 + \left(\Phi_{xy} + \Phi_{yx}\right)^2\right]^{1/2}}{\left[\left(\Phi_{xx} + \Phi_{yy}\right)^2 + \left(\Phi_{xy} - \Phi_{yx}\right)^2\right]^{1/2}} \tag{6.28}$$

In 1D, then both β and λ are zero. In 2D (and 3D/2D), then β is zero and λ is non-zero. In 3D then both β and λ are non-zero. Using threshold values for whether either can be considered to be zero is recommended by Bibby *et al.* (2005), but none are given in that publication.

One problem with the phase-tensor method is that one loses completely the impedance amplitudes. Thus modeling them requires assumptions to be made about the average resistivity or the incorporation of some structural information, such as depth to a particular layer.

Weaver *et al.* (2006) considered the algebraic relationships between MT tensor invariants (Weaver *et al.*, 2000), discussed below in Sections 6.7.2.5 and 6.7.3.4, and the phase-tensor parameters of Caldwell *et al.* (2004). They show that the two approaches agree in almost every detail. Weaver *et al.* (2006) go further, and describe how the phase tensor can be simply constructed as the sum of three tensors representative of 1D, 2D and 3D regional conductivity structures, rather than the more complex and less intuitive singular value decomposition approach of Caldwell *et al.*

Finally, it is somewhat puzzling that Bibby *et al.* (2005) should state about the Groom–Bailey type distortion decomposition techniques that "The basis of these approaches is not intuitively obvious and their physical basis remains controversial", without going into the nature of the controversy. There is no requirement in science that the correct or most optimum method is necessarily the most intuitively obvious and must be without controversy, and examples abound in science of the struggles that controversial ideas have had overcoming staunchly held belief. Indeed, that the Earth is not flat was (and arguably is today in some places!) not intuitively obvious, and was certainly controversial for thousands of years.

6.3.7 Other approaches

Other approaches have been proposed to address distortion of the MT response tensor, some of which are reviewed in Groom & Bailey (1991) and in the excellent tutorial paper of

Groom and Bahr presented at the Ensenada EM Induction Workshop in 1990 (Groom & Bahr, 1992).

All approaches essentially fall into one of two types, and are either mathematically based or physically based. The former has the advantage of implicit mathematical rigor, whereas the latter has the distinct advantage of recognizing and utilizing the nature of the MT impedance tensor.

Perhaps best known among the earlier mathematically based approaches is Eggers' eigenstate decomposition (Eggers, 1982), and others include Spitz (1985), LaTorraca *et al.* (1986) and Yee & Paulson (1987); the Eggers, Spitz and LaTorraca parameterizations are discussed in Chapter 4. Also, Lilley's Mohr circle approach (Lilley, 1993a, 1998a,b), which is a continuation of earlier graphical display presentations proposed by Lilley (1976), and the phase tensor (Section 6.3.6) can be described as mathematical in nature.

In contrast, the approaches of Bahr, Zhang *et al.*, Groom and Bailey, Chave, Smith, and McNeice and Jones all utilize the inherent properties of the response tensor, so are physically based approaches.

Another approach to distortion removal is to directly model and remove the effects. This was done on seafloor data by Baba & Chave (2005) and also Matsuno *et al.* (2010), who modeled the bathymetry, and recently for 3D distortion of 3D data by Patro & Egbert (2011), who modeled galvanic distortion in terms of a laterally heterogeneous top layer.

6.3.8 Extension for a three-dimensional regional Earth

For a 3D regional Earth, the galvanic distortion problem becomes far more complex. The observed response tensor is given by the galvanic distortion of a full 2×2 complex tensor describing the regional 3D structure, viz.

$$
\begin{aligned}
\mathbf{Z}_{obs} &= \begin{pmatrix} c_{11} & c_{12} \\ c_{21} & c_{22} \end{pmatrix} \begin{pmatrix} Z_{xx} & Z_{xy} \\ Z_{yx} & Z_{yy} \end{pmatrix} \\
&= \begin{pmatrix} c_{11}Z_{xx} + c_{12}Z_{yx} & c_{11}Z_{xy} + c_{12}Z_{yy} \\ c_{21}Z_{xx} + c_{22}Z_{yx} & c_{21}Z_{xy} + c_{22}Z_{yy} \end{pmatrix}
\end{aligned}
\tag{6.29}
$$

Note that, in the 1D and 2D (in correct strike angle) cases, the phases of the antidiagonal terms are correct, and the effect of distortion is a frequency-independent multiplicative shift of the amplitudes. In 3D, however, not only are the amplitudes affected, but also the phases – there is *phase mixing* in all elements of the phases in the respective columns. The off-diagonal elements are combinations of their true scaled elements with the scaled elements of their respective diagonal column impedance. This phase mixing inherently is not frequency-independent, as the phases themselves are not, and was studied by Ledo *et al.* (1998).

As discussed in the Introduction (Section 6.1), analysis, identification and removal of distortion effects from regional impedances will be a problem even when there are good, fast 3D inversion codes available that can model the Earth with the complexity that current 2D ones do, i.e. typically 200 horizontal cells and 100 vertical cells. Two different approaches have been proposed for dealing with 3D distortion of 3D regional structures, one by Utada & Munekane (2000) that utilizes the vertical magnetic field data, and the second by Garcia & Jones (1999, 2002) that assumes that two inductively close neighboring sites see the same regional structure but are distorted differently. Garcia & Jones (1999, 2002) adapted Groom and Bailey's parameterization, but tests revealed that the approach was not sufficiently stable.

As discussed by Jones (2011), only in two cases is it appropriate to consider the observed impedances as amplitude scaled estimates of the true impedances without any phase mixing. One of these is when the galvanic distortion is such that the distortion tensor $\mathbf{C}$ becomes diagonal only, i.e. $c_{12} = c_{21} = 0$, which is highly implausible in the real Earth at one site on its own, let alone at all sites in a survey. Such would be the case in 2D if the Groom–Bailey twist and shear distortion parameters were both zero, which is never the case for real data.

The other case where this occurs is when the regional impedance tensor has no diagonal terms, i.e. $Z_{xx} = Z_{yy} = 0$ and the regional structures are 2D, which begs the question as to why 3D inversion is being invoked in the first place.

Sasaki & Meju (2006), in an update on previous work by Sasaki (2004), describe their approach to 3D inversion of MT data taking static shifts into account. The assumption made by Sasaki & Meju (2006) is that the observed impedance tensor can be described as geometrically shifted versions of the individual impedances, viz.

$$\mathbf{Z}_{obs} = \begin{pmatrix} c_{xx}Z_{xx} & c_{xy}Z_{xy} \\ c_{yx}Z_{yx} & c_{yy}Z_{yy} \end{pmatrix} \tag{6.30}$$

For this to hold, the distortion tensor $\mathbf{C}$ must be diagonal, i.e. $c_{12} = c_{21} = 0$, and $c_{xx} = c_{xy} = c_{12}$ and $c_{yx} = c_{yy} = c_{21}$ must be true. Alternatively, given that Sasaki & Meju (2006) only fitted the off-diagonal terms in the observed impedance tensor, i.e. $c_{xy}Z_{xy}$ and $c_{yx}Z_{yx}$, then this approach is valid also when the structures are 2D.

It should be appreciated that, as a consequence of the phase mixing in Eq. (6.29), another technique routinely used in 2D inversion to deal with static shifts, namely to assign high error floors to the apparent resistivity data, is also inappropriate for galvanic distortion of 3D data. Zhdanov *et al.* (2011) adopted this 3D/2D approach to deal with 3D/3D galvanic distortion: "*We have reduced the static shift effect by normalizing the observed MT impedances with their absolute values, which effectively resulted in the phase inversion of the impedances. It is well known that the phases are less sensitive to the galvanic distortions, caused by near-surface inhomogeneities.*" This is a correct approach and correct statement only in the 2D regional case. It is incorrect in the 3D case.

Just how important and significant is this phase mixing problem? Clearly, distortion tensors that are only diagonal in form will be extremely rare – there will almost always be some component of distortion, however small, that has to be addressed. As is obvious in Eq. (6.29), the phase mixing caused by distortion is important for the off-diagonal impedances only when the distorted diagonal terms become as large as the distorted off-diagonal terms, i.e. the magnitude of $c_{12}Z_{yy}$ becomes of order $c_{11}Z_{xy}$ and/or the magnitude of $c_{21}Z_{xx}$ becomes of order $c_{22}Z_{yx}$. This will occur either when the distortion is severe, or when the diagonal terms are large, i.e. for significant 3D structure.

However, the typically smaller diagonal terms will be overwhelmed by the much larger distorted off-diagonal terms even in the case of weak distortion, i.e. the magnitude of $c_{12}Z_{yx}$ overwhelms $c_{11}Z_{xx}$ and/or the magnitude of $c_{21}Z_{xy}$ overwhelms $c_{22}Z_{yy}$. It is precisely these diagonal terms that offer the greatest increase in resolution of 3D geometries afforded by moving from 2D inversion to 3D inversion, and it is precisely these terms that are most affected by distortion.

Three-dimensional inversion of MT data is becoming more commonplace, especially as now there is a publicly available code thanks to Weerachai Siripunvaraporn (Siripunvaraporn *et al.*, 2005). The situation now is comparable to that which existed in the late 1980s and early 1990s, when freely available 2D inversion codes, spearheaded by Stephen Constable (de Groot-Hedlin & Constable, 1990), had a huge impact on MT interpretations and caused a quantum leap in MT resolution of Earth structures by allowing MT data to be inverted in 2D rather than in 1D. There are issues, though, related to the size of the models, given the very high memory requirements and long computing times, even with parallelized codes. Typical 2D models of 200 horizontal cells and 100 vertical cells still offer superior resolution where it can be demonstrated that 2D is appropriate. However, 3D inversions will become routine as faster codes and faster clusters become available. However, as was the case with 2D inversion, which took a significant leap forward when distortion effects were recognized and appropriate techniques developed, so the 3D inversion of MT data requires such attention. Given the difficulties of applying an equivalent approach in 3D as used in 2D of identifying and removing distortion effects shown by Garcia & Jones (1999, 2002), a pre-inversion analysis step may not be the most fruitful. The most optimum approach is probably to consider the frequency-independent galvanic distortion factors as four more unknowns at each site that have to be solved for, as was done in 2D by de Groot-Hedlin (1995). Such an approach is being pursued by Miensopust *et al.* (2011) and Avdeeva *et al.* (2011) independently, and both are showing promise.

Ledo *et al.* (1998) propose a hybrid method that, for valid application, relies on the highest frequencies only responding to 3D distortion of a 2D regional Earth. The distortion parameters can be estimated from those highest frequencies, then removed from longer periods by applying their inverses.

6.4 Determinable and indeterminable parts of the distortion tensor

What parts of distortion can be determined, what parts cannot be determined, and how is it possible to separate between the two? As recognized right at the very beginning by Larsen (1975), any purely amplitude scaling factor is indeterminable. Essentially, this is stating that static shifts of the apparent resistivity curves cannot even be recognized implicitly, never mind removed. Thus, any effects that are amplitude-only are, by their nature, indeterminable.

In contrast, any distortion effects that influence response phases, with or without simultaneously affecting their amplitudes, can be recognized, and potentially removed. These phase properties were first exploited by Bahr (Section 6.3.4), and consideration of phases has formed the basis of all modern distortion approaches culminating with the phase-tensor method (Section 6.3.6), which *only* considers phases and resolves four determinable parameters from the eight data at each frequency.

The great intuitive leap made by Groom and Bailey is to undertake a decomposition that naturally separates the distortion tensor into determinable and indeterminable parts, with the former being both amplitude and phase-related and the latter being solely amplitude-related. McNeice and Jones extended the work of Groom and Bailey by imposing a strict, consistent model of 3D distortion on multiple sites at multiple frequencies, then testing if the distortion model is a valid fit to the data. Other distortion decompositions, such as that of Smith (1995), while they may be thought by some to be more intuitive and more easily understandable, are simply an algebraic re-representation of the observed impedances and lead to mixed determinable and indeterminable distortion parameters.

6.5 Statistical considerations

While algebraic manipulation of ideal MT response tensors into other forms may be intellectually satisfying, it is meaningless without consideration of the nature of the errors propagating into the new forms. The primary task is not only to find the model that fits our data to within prescribed errors, but to appraise that model for resolution through statistics. This tenet of the scientific method is well appreciated when determining an electrical conductivity model that fits a dataset, but appears to be far less appreciated when fitting a model of distortion.

This lack of consideration of errors is particularly true when MT responses are reconstituted and portrayed in diagram form, such as Mohr circles (Section 6.3.7) proposed in Lilley (1976), as phase-tensor parameters (Section 6.3.6), or using the propagation number analysis of Weckmann *et al.* (2003). It is also true when conclusions are being drawn from invariants, such as Swift's skew (Swift, 1967; reproduced in part in Swift, 1986), and the invariants of Szarka & Menvielle (1997) or Weaver *et al.* (2000). To address this issue, at the suggestion of one of the reviewers of the original manuscript, Weaver *et al.* (2000) undertook a statistical examination of the so-called WAL (for Weaver–Agarwal–Lilley) invariants (Section 6.7.2.5), and proposed thresholds below which the invariants can be considered to be zero. Marti *et al.* (2005) adopted a similar bootstrap approach to test the appropriateness of Bahr's parameters, compared them to Weaver *et al.* (2000), and extended Bahr's approach to make it consistent. (Note that the analysis code (WALDIM) developed for this is published (Marti *et al.*, 2009) and is available for download.)

Again, note the warning recently declared by Chave (2011) about caution that must be exercised regarding the statistical complexities of these estimators. Chave (2011) showed that, for the skew estimator, all moments of order two and higher, including variance, are undefined, and the distributions are asymmetric, with long upper tails, leading to them exhibiting far more extreme values than would be expected for a Gaussian variate. The inference one can draw is that this result for skew applies to all estimators that are ratios of response tensor elements.

Within this statistical context, Groom & Bailey (1989) tested whether the model adopted, namely of electric field galvanic distortion of a 2D regional Earth response, fitted the data statistically on an individual frequency basis. Chave & Smith (1994) fitted a Groom–Bailey model of distortion over multiple frequencies, and also used the jackknife method to estimate the statistics. McNeice & Jones (2001) extended Groom–Bailey to determine the distortion model and strike direction that best fits a range of sites and a range of frequencies simultaneously, again within a statistical framework. In the case of McNeice & Jones (2001), this is done through resampling and bootstrapping by generating realizations, based on the errors associated with each response tensor element at each site and frequency, and performing multiple distortion decompositions to give the range of resulting distortion model parameters. The latest version of the *strike* code has the functionality to work in pseudo-depths rather than periods, where the depths are given by the Niblett–Bostick transformation. As shown in Hamilton *et al.* (2006) and Miensopust *et al.* (2011), penetration can vary significantly along a profile, and a depth-based analysis approach is required.

In contrast to bootstrapping the response estimates, which, in the absence of the sample distribution, requires an assumption to be made about the error distributions of the estimates, Chave's jackknife approach (Chave & Smith, 1994; Chave & Jones, 1997) is superior, but requires retaining all of the delete-one jackknife estimates resulting from prior time-series analysis, and using these to generate the ensemble of distortion model parameters. If one has only the response estimates, with estimates of their errors, then one must employ some sort of bootstrap approach – error propagation methods fail in these cases, as shown spectacularly by Chave (2011) for the simplest dimensionality estimator available, namely Swift's skew (Section 6.7.2.1).

6.6 Influence of distortion on the MT response

6.6.1 A simple but instructive two-dimensional model – the Rhine Graben

An instructive example that illustrates aspects of distortion effects due to inappropriate dimensionality assumptions and experiment scale size is that of the Rhine Graben, which can be simply electrically characterized as shown in Figure 6.2. The Rhine Graben geomagnetic response was a feature in the current channeling controversy of the late 1970s and early 1980s, and the evidence was examined in detail by Jones (1983) in his review of that topic. That peripheral diversion will not be explored further though, but the local MT responses at sites above and outside the graben will be studied.

The graben is modeled as 40 km wide and filled uniformly to a depth of 2.5 km with sediments of uniform isotropic resistivity of 10 Ω m, embedded in a lithosphere of 1000 Ω m, below which is an asthenosphere of 100 Ω m at a depth of 117 km and a basal layer of 10 Ω m at 350 km depth. These values are not correctly representative of the Earth beneath the

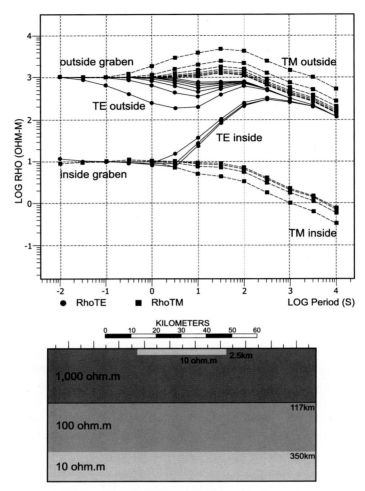

Figure 6.2. Theoretical apparent resistivity curves from 0.01 s to 10 000 s that would be observed on a profile crossing a graben structure based grossly on the Rhine Graben. The graben is 65 km wide, 2.5 km deep, with an infill of material of 10 Ω m, in a 117 km thick uniform crust and upper mantle (lithosphere) of 1000 Ω m, above a mantle layer of 100 Ω m (asthenosphere) to 350 km, with a basal layer (transition zone) of 10 Ω m. TE mode data are shown as circles connected by full lines, and TM mode data by squares connected by dashed lines. Stations are 5 km apart, and only the responses from the 11 stations on the left half of the model are shown, owing to the symmetry. The four stations inside the graben asymptote to 10 Ω m at short periods (< 1 s), whereas the seven stations outside the graben asymptote to 1000 Ω m at the shortest periods (0.01 s). The stations closest to the graben boundary are the first to respond as period is increased, and either increase (TM outside graben, TE inside graben) or decrease (TE outside graben, TM inside graben) from their short-period asymptotes.

Rhine Graben, as in reality (a) the lithosphere is likely not isotropic but there is evidence of anisotropy (Leibecker *et al.*, 2002; Gatzemeier & Moorkamp, 2005; Roux *et al.*, 2011), and (b) the basal layer should be at the 410 km phase change boundary (Xu *et al.*, 1998). However, the parameters chosen suffice to demonstrate the distortion issues that are evident at periods shorter than those probing the mantle lithosphere.

Also plotted on Figure 6.2 are the TE and TM apparent resistivity responses that would be observed at a profile of regularly spaced sites crossing the Rhine Graben every 5 km. The TE responses are shown as circles with connecting solid lines, and the TM responses as squares with connecting dashed lines. The responses are in two groups, depending on whether the site was located inside or outside the graben. The inside-graben ones have short-period apparent resistivities of 10 Ω m, with the sites closest to the edge of the graben departing at highest frequency (shortest period) from that value. The outside-graben sites have short-period apparent resistivities of 1000 Ω m, with the sites closest to the edge of the graben departing at highest frequency from that value.

6.6.1.1 False conducting layers and one-dimensional interpretation strategy

At high frequencies for sites located outside the graben, note how the TE responses at sites close to the graben are sensitive to its presence. The apparent resistivity at the closest site, which is 2.5 km from the edge of the graben, starts to decrease from the background resistivity of 1000 Ω m at around 40 Hz, which is one skin depth (in the host of 1000 Ω m) away. The observable minimum in the TE response at around 3 s is due to the current gathering effects of the sediments in the graben that are off to the side of the site, and this response is a problem when interpreting the data from that site with inappropriate dimensionality, i.e. using 1D methods when 2D is clearly required. The problem, akin to *side-swipe* in reflection seismics, has been known since the early 1970s and is named *false conducting layers* (Berdichevsky & Dmitriev, 1976b; Jones, 1988a; Jiracek, 1990; Agarwal *et al.*, 1993). The TM responses do not exhibit a local minimum, but indeed the opposite, so one mistakenly would interpret the presence of *false resistive layers*, albeit a far weaker error than the TE error.

In contrast, the TE mode data at sites within the graben can be interpreted in 1D, at least approximately. The apparent resistivities increase from 10 Ω m initially at approximately 10 s, which is the skin depth period in the sediments for the thickness of 2.5 km.

During the 1970s when the interpretation tool was 1D modeling, the standard practice was to determine whether you were located *on* or *off* the conductor. If you were on the conductor, one took the TE data for interpretation in 1D, and off the conductor one took the TM data.

6.6.1.2 Static shift correction method

The long-period responses clearly demonstrate the problem of *static shifts* (Jones, 1988b). Imagine if one had only long-period MT data from 100 s on – one would have TE responses that are close to one another, but TM responses that have almost exactly the same shape at every site, but which are different in amplitude by almost three orders of magnitude. One would not know what the true regional apparent resistivity level is.

Two points are worthy of note. First, as period increases the TE mode responses from inside the graben rise to the correct regional levels, and all TE responses, both inside and outside the graben, asymptote to the same apparent resistivity at long periods, in this case 100 Ω m at 10 000 s. Second, the high-frequency asymptotes are all at the correct levels.

These two aspects are the basis of the two-step static shift correction method suggested by Jones & Dumas (1993):

- Step 1 is to shift the TE mode apparent resistivity curves from all sites such that they all have the same long-period asymptote.
- Step 2 is shifting the TM mode apparent resistivity curves so that each has the same high-frequency asymptote as the TE mode from the same site.

The assumptions implicit in this approach are rather obvious, and are that (1) at the longest periods the Earth is laterally uniform, and (2) the highest frequencies are high enough that structures of interest are not spatially aliased. The requirement for the *same* long-period apparent resistivity was relaxed by Jones *et al.* (1992), who fitted a low-order polynomial to the long-period asymptotes, thereby permitting long-wavelength lateral variation at depth.

Shifting MT apparent resistivity curves such that they had a long-period asymptote that fitted the defined "global reference curve" of Vanyan *et al.* (1977) at periods greater than around 1000 s (34^2 s) was standard practice in the former Soviet Union through the 1970s, 1980s and 1990s (Berdichevsky *et al.*, 1989). The problem with this approach is that the modes were not defined separately, and in some cases the arithmetic averaged apparent resistivity curves of the two off-diagonal elements were used (e.g. Vanyan *et al.*, 1989). For the Rhine Graben example, this approach would be reasonable for sites outside the graben, but the arithmetic average of the TE and TM mode responses at long periods (10^4 s) inside the graben would be around 10 Ω m, which is an order of magnitude below the correct regional level of just over 100 Ω m. Shifting the averaged curves by an order of magnitude would lead to errors at short periods, with the short-period asymptotes becoming 100 Ω m instead of the correct value of 10 Ω m. Thus, the resistivity of the graben sediments would not be correctly determined.

6.6.2 *A simple but instructive three-dimensional distorting body – the embedded hemisphere*

One problem that exists in magnetotellurics is that, whereas in 1D and 2D there are analytical or pseudo-analytical solutions for certain geometries, such as the dike model of Weaver (Weaver *et al.*, 1985, 1986), and recently the finite vertical thin conductor of Parker (2011), in 3D there are no such formal analytical solutions. However, at the long-period galvanic limit, where the induction number is small, one only needs to be concerned about the galvanic charges on the boundaries of the body and not induction in the body, and there are useful solutions for some shapes such as the embedded hemisphere (Figure 6.3). The electric scattering for such a body was derived first by Ward (1967) and later Honkura

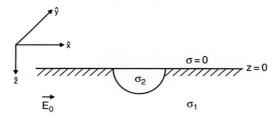

Figure 6.3. Surficial embedded hemisphere geometry. A hemisphere of conductivity σ_2 exposed in the surface ($z = 0$) is embedded within a uniform half-space of conductivity σ_1. Figure redrawn from Groom & Bailey (1991, their figure 1).

(1975), and is reviewed by Groom & Bailey (1991). The magnetic effects of such a body were originally studied by Edwards *et al.* (1978) for a grounded-dipole source, and for a uniform source field by Groom & Bailey (1991).

The electric field inside the body is given by

$$\mathbf{E}_i(x, y, z) = E_0 \left[\frac{3\sigma_1}{\sigma_2 + 2\sigma_1}, \ 0, \ 0 \right] \tag{6.31}$$

where E_0 is the external electric field, $\mathbf{E}_i$ is the electric field at point (x, y, z) in the subsurface ($z \geq 0$), σ_1 is the conductivity of the half-space and σ_2 is the conductivity of the body. Note that the electric field inside the body is independent of both location (x, y, z) and body radius (R).

Outside the body, the electric field is given by

$$\mathbf{E}_e(x, y, z) = E_0 \left[1 + K \frac{(2x^2 - y^2 - z^2)}{r^5}, \ \frac{3Kxy}{r^5}, \ \frac{3Kxz}{r^5} \right] \tag{6.32}$$

where $\mathbf{E}_e$ is the external electric field at point (x, y, z) in the subsurface ($z \geq 0$), E_0 is the external static electric field in the x direction, r is the distance of the point (x, y, z) from the center of the body, and K is related to the electric dipole moment and is given by

$$K = \frac{\sigma_2 - \sigma_1}{\sigma_2 + 2\sigma_1} R^3 \tag{6.33}$$

where R is the radius of the distorting body and is assumed to be much less than both the skin depth and variations in the electric field E_0 (reorganized from eqs. (4) and (6) in Groom & Bailey (1991)).

For highly conducting distorting bodies in a resistive medium, i.e. $\sigma_2 \gg \sigma_1$, which is often the case, then $K \approx R^3$. The secondary electric field ($\mathbf{E}_e - E_0$) drops off as $(R/r)^3$, so at a distance of R outside the body the field is one-quarter of the primary external field, and at a distance of $2R$ outside the body is one-ninth of the primary external field.

For highly resistive distorting bodies in a more conductive medium, i.e. $\sigma_2 \ll \sigma_1$, then $K \approx -R^3/2$.

The galvanic scattering tensor inside the inhomogeneity is

$$
\mathbf{C}_i(x, y, 0) = \begin{bmatrix} \dfrac{3\sigma_1}{\sigma_2 + 2\sigma_1} & 0 \\ 0 & \dfrac{3\sigma_1}{\sigma_2 + 2\sigma_1} \end{bmatrix} \tag{6.34}
$$

and outside the inhomogeneity is

$$
\mathbf{C}_e(x, y, 0) = \begin{bmatrix} 1 + K\dfrac{(2x^2 - y^2)}{r^5} & K\dfrac{3xy}{r^5} \\ K\dfrac{3xy}{r^5} & 1 + K\dfrac{(2y^2 - x^2)}{r^5} \end{bmatrix} \tag{6.35}
$$

(eqs. (11) and (12) in Groom & Bailey (1991)). For $\sigma_2 \gg \sigma_1$, $K \approx R^3$ and the scattering tensor inside the body approaches the null matrix as σ_2 becomes large. Note that the symmetry of the inhomogeneity results in a scattering tensor that is symmetric, i.e. $C_{12} = C_{21}$ both inside and outside the inhomogeneity.

This embedded hemisphere distorting body was used by Groom & Bailey (1991) and Groom *et al.* (1993) in their demonstration of the problems that can be caused by galvanic charges. A conducting body of 10 Ω m and radius 50 m was embedded in a medium of 300 Ω m that was located 6 km distant from a 2D fault striking at +30° juxtaposing a highly resistive upper layer (40 000 Ω m) (Figure 6.4). At the frequencies utilized (<3000 Hz), the galvanic response of the hemisphere dominates its inductive response. The channeling number $\sigma_{body}/\sigma_{host}$ is 30, while at 1000 Hz, the induction number $\sigma_{body}\mu_0\omega R^2$ is 8, and decreases linearly with decreasing frequency such that at 10 Hz it is only 0.08.

The numerical experimental design is shown in Figure 6.4. The data are observed in a coordinate system rotated 30° counter-clockwise to the strike of the 2D fault, and the site is at an angle of 22.5° clockwise to strike, i.e. 52.5° clockwise to the adopted coordinate system. Thus, the site is 58 m from the center of the distorting body (50 m radius body plus 8 m outside) with observational [x, y] locations of [35.3, 46.0] (x directed north and y directed east). (Note that in Groom *et al.* (1993) it states that the site is 16 m outside the inhomogeneity, but this is an error.) The site is almost 6 km away from the regional 2D structure (the fault), and in the absence of distortion the MT responses will appear 1D until approximately 0.5 s, which is the period corresponding to a skin depth of 6 km for a half-space of 300 Ω m.

However, the distortion tensor is defined in strike coordinates, where the site is located at [53.6, 22.2]. At this location, the galvanic distortion – assuming point measurement of the electric field – from Eq. (6.35) is

$$
\mathbf{C} = \begin{bmatrix} 1.91 & 0.62 \\ 0.62 & 0.67 \end{bmatrix} \tag{6.36}
$$

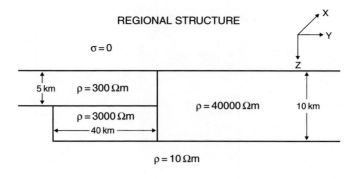

REGIONAL STRUCTURE

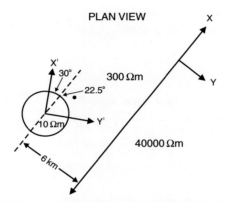

PLAN VIEW

Figure 6.4. Regional fault geometry and plan view of embedded hemisphere (not to scale). Figure redrawn from Groom & Bailey (1991, their figures 3 and 4).

In terms of Groom–Bailey parameters (Section 6.2.2), this distortion tensor has a *twist* of −12.2°, a *shear* of +30.2°, an *anisotropy* of 0.37 and a scaling (*gain*) of 1.23. This is moderate distortion, and well within the capabilities of current distortion techniques to address, *provided that the MT data are sufficiently precise!*

Note that the current channeling azimuth (Eq. (6.13)), given by the sum of the regional strike (+30.0°), the *shear* value (+30.2°) and the *twist* value (−12.2°), is +48.0°.

The data that would be observed at the site 8 m outside the body, named *far* in Groom *et al.* (1993), for relatively moderate levels of noise of 2% added to the response elements (equal to 4% in apparent resistivity and just over 1° in phase), are shown in Figure 6.5 and are named *far-hi*. This distortion tensor, and the data for site far-hi, will be analyzed below using a number of different approaches.

Consider though the effect of the assumption of point electric field measurements compared to the actual measurements of potential gradients. This can be replicated for the above location by placing an electrode array centered on location [35.3, 46.0]. For a geographically oriented

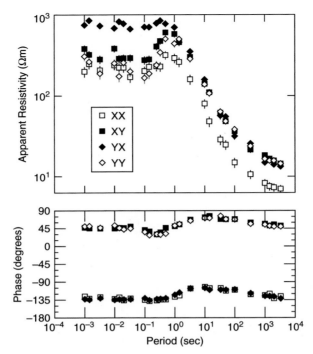

Figure 6.5. The MT data that would be observed at the site 8 m outside the embedded hemispherical body shown in Figure 6.4.

cross-configuration with 50 m arms, the electrodes would be at geographic locations North = [85.3, 46.0], East = [35.3, 96.0], South = [−14.7, 46.0] and West = [35.3, −4.0] (note that both the South and West electrodes lie within the inhomogeneity), and in strike coordinates these locations are North' = [96.9, −2.8], East' = [78.6, 65.5], South' = [10.3, 47.2] and West' = [28.6, −21.1].

The distortion tensors for these locations are:

$$\mathbf{C}_{North'} = \begin{bmatrix} 1.25 & -0.011 \\ -0.011 & 0.86 \end{bmatrix}$$

$$\mathbf{C}_{West'} = \begin{bmatrix} 0.094 & 0 \\ 0 & 0.094 \end{bmatrix}, \qquad \mathbf{C}_{East} = \begin{bmatrix} 1.08 & 0.156 \\ 0.156 & 1.02 \end{bmatrix} \qquad (6.37)$$

$$\mathbf{C}_{South'} = \begin{bmatrix} 0.094 & 0 \\ 0 & 0.094 \end{bmatrix}$$

The distortion tensor that would affect the MT data is given – to first order – by averaging these individual tensors. Averaging the top rows of $\mathbf{C}_{North}$ and $\mathbf{C}_{South}$ gives the E_x field distortion, and averaging the bottom rows of $\mathbf{C}_{West}$ and $\mathbf{C}_{East}$ the E_y field distortion, so the distortion tensor for a "real" MT site centered on [35.3, 46.0] with 50 m electrode arms is

$$\tilde{\mathbf{C}}_{[35, 46]} = \begin{bmatrix} 0.672 & -0.056 \\ 0.078 & 0.557 \end{bmatrix} \tag{6.38}$$

which is much lower phase distortion, with *twist* = +6.2° and *shear* = +0.44°, but much higher amplitude distortion, with *anisotropy* = 0.09 and scaling (*gain*) = 0.61 due to two electrodes (South and West) being within the distorting inhomogeneity.

Finally, imagine now that the distorting body is small and centered on [0, 0] with a radius *R* of 1 m, and is still of 10 Ω m embedded in a 300 Ω m half-space. Imagine further that, purely by bad luck, the site location, which is the center of the electrode array, is at [50, 0], so that with 50 m electrode array line lengths the West electrode is placed in this small body. The other three electrodes are well outside at locations North = [50, 50], East = [0, 100] and South = [−50, 50]. The distortion tensor $\mathbf{C}_{West}$ is given in Eq. (6.37), as the distortion of the electric field inside the inhomogeneity is radius- and location-independent (Eq. (6.34)). The distortion tensors at the other three electrode sites, $\mathbf{C}_{North}$, $\mathbf{C}_{East}$ and $\mathbf{C}_{South}$, are very close to the identity matrix. This results in a distortion matrix at site [0, 50] of

$$\tilde{\mathbf{C}}_{[50, 0]} = \begin{bmatrix} 1 & 0 \\ 0 & 0.75 \end{bmatrix} \tag{6.39}$$

which will result in an amplitude-only effect, decreasing $E_{y,regional}$ by 25% and not affecting $E_{x,regional}$. The result is that the effect of this small body is not insignificant at all, but is substantial; the ρ_{yx} apparent resistivity is reduced by almost a factor of 2 (9/16ths to be precise, i.e. 3/4 squared).

Such a phenomenon was observed by Jones & Garland (1986) on Prince Edward Island, Canada, in data from a survey conducted in 1984. A pair of MT sites, labeled *pei019* and *pei020*, was configured so that they shared one electrode in a "nose-to-nose" X electrode array with 50 m arms; the north electrode of pei019 was the south electrode of pei020. The sites were in an open flat farmer's field, and there were no obvious signs of local inhomogeneities. The off-diagonal MT responses (*XY* and *YX*) from pei019 and pei020 are shown in Figure 6.6, with pei019 in filled symbols and pei020 in open ones, and with the ϕ_{yx} data rotated into the first quadrant. Prince Edward Island is covered by a thick sedimentary basin, and the responses at all sites are closely 1D until basement is sensed, then become anisotropic. For pei019 and pei020, the data are 1D to a period of around 15 s in apparent resistivity, and 5 s in phase (phase always visibly responds about a half-decade before apparent resistivity). Whereas for site pei020 the ρ_{xy} and ρ_{yx} curves lie on top of each other, for pei019 the ρ_{xy} curve lies on top of both apparent resistivity curves for pei020, but the ρ_{yx} curve for pei019 lies approximately half an order of magnitude below the other three. Clearly, either the east or west electrode at pei019 was placed inside a local small-scale conductive distorting body. This is static shift in its simplest form, as discussed by Jones (1988b) and numerically described in terms of 2D distortion, but here described in terms of 3D distortion. The high-frequency asymptotes of the undistorted apparent resistivity curves are around 300 Ω m. A small-scale distorting body of 1 m radius and

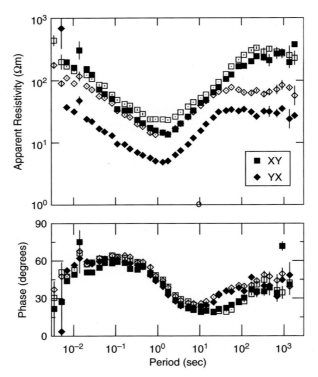

Figure 6.6. Off-diagonal MT data (*XY* as squares, and *YX* as diamonds) from sites pei019 (solid symbols) and pei020 (open symbols) located within 200 m of each other. The ϕ_{yx} data have been rotated through 180° and are plotted in the first quadrant.

resistivity 3 Ω m would result in a scattering tensor of 0.03**I** (where **I** denotes the unity tensor) for either the East or West electrode, resulting in an averaged scattering tensor with a lower row of [0, 0.515], i.e. the electric field is reduced by a factor of almost 2, and the ρ_{yx} data would therefore be reduced by a factor of almost 4, which is approximately what is observed.

The maximum distortion from a small-scale conducting body is a reduction of the electric field by a factor of 2. The electric field inside the body asymptotes to zero as the conductivity of the body increases, so the average electric field is simply half of that observed at the other electrode. Thus, the maximum effect is a downshifting of the apparent resistivities by a factor of 4. In the case of a small-scale resistive body, i.e. $\sigma_2 \ll \sigma_1$, the electric fields are enhanced in the body by 50% (3/2, see Eq. (6.34)), so the averaged electric field is enhanced by 25%, leading to an upward shift of the apparent resistivity by just over one-half (56%). Thus, whereas a conducting inhomogeneity can reduce resistivities by a factor of up to 4, a resistive inhomogeneity can increase resistivities by a factor of only 1/2 at most, explaining why most static shifted MT data are more obviously shifted downward than upward.

6.6.3 *Distorted North American Central Plains (NACP) impedance*

The dimensionality and directionality tests discussed herein will be applied to synthetic and observed MT tensors in order to evaluate the efficacy of the procedures. For synthetic data, the basic undistorted 1D impedance tensor that will be examined is

$$\mathbf{Z}_{ex}^{1D} = \begin{pmatrix} 0 & (4.72, 4.05) \\ -(4.72, 4.05) & 0 \end{pmatrix} \times 10^{-4} (\Omega \text{ m}) \tag{6.40}$$

and for 2D is

$$\mathbf{Z}_{ex}^{2D} = \begin{pmatrix} 0 & (4.72, 4.05) \\ -(8.25, 3.10) & 0 \end{pmatrix} \times 10^{-4} (\Omega \text{ m}) \tag{6.41}$$

(Note that these tensors are the MT *impedance* tensors, defined by E/H, in units of Ω m, and not the MT *response* tensors, defined by E/B, in units of m/s.) This 2D impedance tensor in Eq. (6.41) is the actual modeled impedance at 10 s derived by Jones & Craven (1990) to fit the data from a site right on top of the North American Central Plains (NACP) anomaly, discovered by Ian Gough and colleagues using magnetometer arrays in the late 1960s and early 1970s (Alabi *et al.*, 1975), which is possibly the longest conductivity anomaly in the world (Jones, 1993b). One profile of MT data crossing the NACP in southern Saskatchewan was used as a standard to assess 2D inversion methods – the so-called COPROD2 dataset (Jones, 1993a).

A 3D impedance tensor is created that is artificial but based on the 2D one above, viz.

$$\mathbf{Z}_{ex}^{3D} = \begin{pmatrix} (1.42, -0.67) & (4.72, 4.05) \\ -(8.25, 3.10) & (-2.52, -3.09) \end{pmatrix} \times 10^{-4} (\Omega \text{ m}) \tag{6.42}$$

Finally, to the 2D impedance tensor given by Eq. (6.41), distortion is added using a distortion tensor taken from Chakridi *et al.* (1992), namely

$$\mathbf{Z}_{ex}^{3D/2D} = \begin{pmatrix} 1.26 & 0.44 \\ 0.53 & 0.86 \end{pmatrix} \begin{pmatrix} 0 & (4.72, 4.05) \\ (-8.25, -3.10) & 0 \end{pmatrix} \times 10^{-4} (\Omega \text{ m})$$

$$= \begin{pmatrix} (-3.63, -1.36) & (5.95, 5.10) \\ (-7.10, -2.67) & (2.50, 2.15) \end{pmatrix} \times 10^{-4} (\Omega \text{ m}) \tag{6.43}$$

This distortion tensor has Groom–Bailey parameters of *twist* = $-2.1°$, *shear* = $25.0°$, *anisotropy* = 0.17 and scaling (*gain*) of 1.06, which is a moderate level of distortion.

Noise and scatter are added to all of the elements of all of these tensors independently to try to mimic real data as closely as possible. Noise and scatter are added at three levels, 1% (of maximum impedance magnitude value), 3% and 10%. The lowest value of 1% can be acquired using the most modern MT equipment in high-signal, low-noise conditions, with sufficient recording time and with application of modern robust time-series processing methods. The medium value of 3% is routinely achievable by most modern MT equipment in most conditions, again with sufficient recording time and use of appropriate processing codes. The high value of 10% will result with one or more of (i) poor equipment, (ii) low signal, (iii) high noise, (iv) short recording duration and/or (v) application of older non-robust codes. Even higher values of noise are possible, but then one really should not do MT, as it is being done badly and inappropriately.

6.6.4 BC87 dataset – lit007 and lit008

One particular dataset that exhibits significant and unusual distortion effects is that acquired in 1987 in southeast British Columbia over the Nelson Batholith as part of the Lithoprobe project (Clowes, 2010). The dataset, named BC87 (Jones, 1993c), was made available to attendees at the first and second MT Data Interpretation Workshops (Jones, 1993d; Jones & Schultz, 1997) for analysis and modeling. A distortion decomposition analysis is presented in Jones *et al.* (1993), and for both magnetic and electric distortion effects in Chave & Jones (1997). Lilley (1993b) analyzed the data from selected sites in terms of Mohr circles, and Eisel & Bahr (1993) performed a Bahr analysis (Section 6.3.4).

The MT responses from two sites in particular will be examined in this chapter, namely sites *lit007* and *lit008* of the BC87 dataset. The two sites provided the remote reference for each other through a hard-wire link, and were not more than 2 km apart. The four apparent resistivity and phase curves from the MT response tensor elements for the two sites are shown overlaid in Figure 6.7. Note how "well behaved" lit008 is (full symbols) compared to the obvious distortion at lit007 (open symbols). Particularly, the *XY* component at lit007 (open squares) is clearly distorted, with phase rotating through 360° and the apparent resistivity being orders of magnitude lower than any other element.

6.7 Recognizing distortion in magnetotelluric responses

6.7.1 Forms of the magnetotelluric response tensor

To recognize that the MT responses from a 1D or 2D Earth are distorted, essentially one needs to demonstrate that the MT response tensor does not conform to either of the principal shapes, either 1D or 2D, *to within experimental error*. This latter point is important – if the data are extremely poor in quality, then a uniform half-space might be the simplest valid Earth model that can be fitted to them. Conversely, even over highly homogeneous environments like sedimentary basins, if the data are sufficiently precise, then a multi-scale 3D model is demanded. Thus, how well the derived MT observations fit a particular dimensional description of the subsurface must be posed as a statistical problem in exactly the same manner as fitting a resistivity model to data. Finding a model is only the first – and

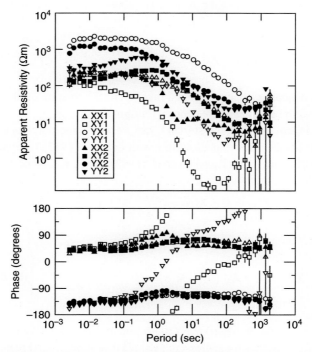

Figure 6.7. MT data from sites lit007 (open symbols) and lit008 (full symbols) of the BC87 dataset (Jones, 1993a; Jones *et al.*, 1993). All four elements are shown from both sites, with the components being: *XX*, circles; *XY*, squares; *YX*, diamonds; *YY*, circles. (Phases not rotated.)

usually easiest – step; the more important, and far more difficult, step is appraising that model against the data.

The general MT response tensor is

$$\mathbf{Z}_{obs} = \begin{bmatrix} Z_{xx} & Z_{xy} \\ Z_{yx} & Z_{yy} \end{bmatrix} \qquad (6.44)$$

and the problem is testing the observations against a variety of potential configurations.

For a 1D layered Earth, the observed MT response tensor, in the absence of noise, adopts the form

$$\mathbf{Z}_{obs} = \mathbf{Z}_{1D} = \begin{bmatrix} 0 & Z_{xy} \\ -Z_{xy} & 0 \end{bmatrix} \qquad (6.45)$$

where the minus sign on the lower off-diagonal element is to ensure compliance with the right-hand rule. This tensor is rotationally invariant.

In the case of a 2D Earth, then the observed response tensor, again in the absence of noise, is given by

$$\mathbf{Z}_{obs} = \mathbf{Z}_{2D}(\theta) = \mathbf{R}(\theta) \begin{bmatrix} 0 & Z_{xy} \\ Z_{yx} & 0 \end{bmatrix} \mathbf{R}^{\mathrm{T}}(\theta) \qquad (6.46)$$

where $\mathbf{R}$ is the standard Cartesian rotation matrix, θ is the rotation of the strike coordinates from the observation axes, and "T" denotes transpose.

In addition to these standard undistorted forms, versions incorporating galvanic distortion of the electric fields were developed, starting with Larsen's (1977) 1D approach, which was developed into 2D initially by Richards *et al.* (1982). These are, respectively,

$$\mathbf{Z}_{obs} = \mathbf{C} \cdot \mathbf{Z}_{1D} = \begin{bmatrix} c_{xx} & c_{xy} \\ c_{yx} & c_{yy} \end{bmatrix} \begin{bmatrix} 0 & Z_{xy} \\ -Z_{xy} & 0 \end{bmatrix} \qquad (6.47)$$

for the 1D case, and

$$\mathbf{Z}_{obs} = \mathbf{C} \cdot \mathbf{Z}_{2D}(\theta) = \mathbf{R}(\theta) \begin{bmatrix} c_{xx} & c_{xy} \\ c_{yx} & c_{yy} \end{bmatrix} \begin{bmatrix} 0 & Z_{xy} \\ Z_{yx} & 0 \end{bmatrix} \mathbf{R}^{\mathrm{T}}(\theta) \qquad (6.48)$$

for the 2D case.

More recently, Utada & Munekane (2000) and independently Garcia & Jones (1999, 2002) presented approaches for addressing galvanic distortion of a 3D regional structure, viz.

$$\mathbf{Z}_{obs} = \mathbf{C} \cdot \mathbf{Z}_{3D} = \begin{bmatrix} c_{xx} & c_{xy} \\ c_{yx} & c_{yy} \end{bmatrix} \begin{bmatrix} Z_{xx} & Z_{xy} \\ Z_{yx} & Z_{yy} \end{bmatrix} \qquad (6.49)$$

The rationale for such a parameterization may not be immediately obvious – if one is characterizing the subsurface as 3D, why is it necessary to consider distortion effects? Will they not be included in the 3D representation of the subsurface? It is simply that, until such time as it is possible to describe the subsurface from the centimeter scale to the hundreds of kilometers scale with sufficient spatial sampling on the surface, there will *always* be effects in the MT responses that cannot be modeled but must be treated in terms of some approximation to deal with the galvanic distortion part.

Finally, there is one other form that is little appreciated, and that is over an Earth comprising 1D anisotropic layers. This form is given by

$$\underline{\underline{\mathbf{Z}}}_{1Da} = \begin{bmatrix} Z_{xx} & Z_{xy} \\ Z_{yx} & -Z_{xx} \end{bmatrix} \qquad (6.50)$$

and is discussed in more detail in Section 6.9.

6.7.2 *Dimensionality tools*

One of the main analysis tasks of the MT data interpreter is determining the appropriate dimensionality of the data. Are the data 1D or 2D or 3D, to within experimental error? At which periods? For which sites? If 2D, are the strike directions consistent over a band of frequencies for a group of neighboring sites, as they should be if the inductive scale lengths are such that the sites sense the same regional Earth?

As discussed in the Introduction (Section 6.1), such questions must be posed within a statistical framework – if the data are very poor, with large error estimates, then the simplest model fitting them may be 1D, or even a uniform half-space. Conversely, if the data are of excellent quality, then a 3D model is likely required to adequately explain the observations (not that with current codes it could be found). The early tools developed in MT during the 1960s and 1970s were based on the rotational properties of the magnitudes of the response tensor elements, whereas modern tools developed since the mid-1980s essentially use the rotational properties of their phases.

6.7.2.1 *Swift skew*

The main dimensionality tool used for 20 years or more beginning in the late 1960s, and unfortunately still employed by some despite the many warnings about its use, is the skew of the MT response tensor. This is a mathematical quantity that is the ratio of the trace divided by the anti-trace, and was initially proposed by Charles W. Swift Jr. in his Ph.D. thesis (Swift, 1967; published in part in Swift, 1986), which is the most referenced Ph.D. thesis in magnetotellurics. This quantity has become known as *Swift skew* and was given by Swift as

$$\text{skew} \; = \; \frac{\left| Z_{xx} + Z_{yy} \right|}{\left| Z_{xy} - Z_{yx} \right|} \tag{6.51}$$

(Swift, 1986, eq. (6.14)). There are other definitions that have been called *skew*, but are different from this, such as the modulus of the complex ratio of the trace divided by the anti-trace. (Actually, Swift (1967) defined skew using both definitions in his thesis, but used the ratios of the moduli in his analysis of data.)

Note that this measure of dimensionality is based on the *amplitudes* of the response tensor. Both the trace and anti-trace of a tensor are rotationally invariant, so their ratio, skew, is also rotationally invariant. Many workers in the 1970s and 1980s expressed the view that, if skew was below 0.2, then the data were thought to be validly interpreted as either 1D or 2D. The unreliability of skew in the presence of bias noise was noted almost 30 years ago by Pedersen & Svennekjaer (1984), but their caution has often not been heeded. Recently, Chave (2011) considered the statistics of skew in the presence of random noise, and demonstrated that skew moment of order 2 is undefined and unconditionally biased, and that skew distribution is asymmetric with long upper tails, leading to skew exhibiting far more extreme values than would be expected for a Gaussian variate. These have the consequence that random fluctuations will result in large frequency-by-frequency variability of skew. Chave (2011) goes further, and expresses concern about the use of any estimators that are ratios of combinations of the MT impedance tensor elements owing to their statistical complexities, and suggests that caution be exercised before using them for quantitative purposes.

How well does this measure perform in the presence of noise and distortion? Taking Chave's (2011) concerns into account, this can be assessed by using a bootstrap approach rather than using error propagation analysis, which requires assumptions to be taken about distributions. Taking the four impedance tensors of Section 6.6.3, Equations (6.40), (6.41), (6.42) and (6.43), and creating 25 000 realizations of each from a Gaussian model of distribution of the real and imaginary response tensor elements, one can examine the mean and median values, the ranges at one (68%) and two (95%) standard errors, and the ranges for low noise (1% of maximum impedance tensor magnitude), medium noise (3%) and high noise (10%). The results are tabulated in Table 6.1. What conclusions can one draw from these results?

- Skew is longer-tailed than expected for a Gaussian distributed variate, as expected from the expression given in Chave (2011).
- Skew is highly sensitive to noise.
- In the presence of relatively high noise, there is 5% likelihood that data over a 1D or 2D Earth would be judged 3D, as the skew value would lie above 0.2.
- In the presence of the 3D structure, even at low noise levels, one could be misled into believing that a 1D or 2D model was valid for 95% of the low-noise cases, and over half the high-noise cases lie below 0.2.
- Skew cannot recognize distortion of regional 2D structures. For the example taken, even for high noise levels, over two-thirds of the estimates lie below 0.2.

6.7.2.2 Ellipticity

A second dimensionality measure, introduced in the late 1960s (Word *et al.*, 1970), is the ellipticity of the MT response tensor, given by

$$\text{ellipticity}(\theta) = \frac{|Z_{xx}(\theta) - Z_{yy}(\theta)|}{|Z_{xy}(\theta) + Z_{yx}(\theta)|} \tag{6.52}$$

which is the ratio of the minor axis of the impedance ellipse divided by the major axis – the original publication (Word *et al.*, 1970) was in an obscure report that is no longer obtainable; the reader is referred instead to Ranganayaki (1984). Ellipticity has the same distribution form as skew, so the same cautions expressed by Chave (2011) for skew will also apply. Note that ellipticity is a function of rotation angle, and that for a pure error-free 1D response, it is undefined, as both the numerator and denominator are zero, regardless of angle. Also note that ellipticity is another magnitude-based measure of dimensionality. Ellipticity will be zero for a 2D Earth when θ is in strike coordinates. Thus, ellipticity was used as both a dimensionality and directionality indicator. It was usually inspected in the strike direction, given often by Swift strike (see Section 6.7.3.1).

Table 6.1. *Swift skew.*

Model	True	Low noise (1%)					Medium noise (3%)					High noise (10%)				
		Mean	Median	68%	95%	Range	Mean	Median	68%	95%	Range	Mean	Median	68%	95%	Range
1D	0.000	0.01	0.01	0.00	0.00	0.00	0.03	0.03	0.01	0.00	0.00	0.01	0.01	0.04	0.00	0.00
				0.01	0.02	0.03			0.04	0.05	0.09			0.14	0.18	0.39
2D	0.000	0.01	0.010	0.01	0.00	0.00	0.03	0.03	0.02	0.00	0.00	0.11	0.10	0.01	0.00	0.00
				0.02	0.02	0.04			0.05	0.07	0.11			0.16	0.22	0.48
3D	0.179	0.18	0.18	0.17	0.17	0.15	0.18	0.18	0.16	0.13	0.07	0.20	0.20	0.12	0.04	0.00
				0.19	0.20	0.22			0.21	0.23	0.29			0.29	0.36	0.58
3D/2D	0.090	0.09	0.09	0.08	0.08	0.06	0.09	0.09	0.7	0.05	0.01	0.12	0.12	0.06	0.00	0.00
				0.10	0.11	0.12			0.11	0.14	0.18			0.19	0.25	0.40

How well does this measure perform in the presence of noise and distortion? Adopting the same approach as above, the results are tabulated in Table 6.2 for the strike direction of 0° in the 2D and 3D/2D cases. Note that, exactly as for skew, the appropriate distribution of ellipticity is not Gaussian, but that as given in Chave (2011), and is long-tailed. What conclusions can one draw from these results?

• Ellipticity is also highly sensitive to noise.
• In particular, in the presence of relatively high noise, the ellipticity distribution can exhibit many extreme values, regardless of the dimensionality.

6.7.2.3 Polar diagrams

The third conventional dimensionality (and directionality) tool that was heavily used prior to around 1990, and is still (unfortunately) in use by some today, is consideration of the rotational shapes of the magnitudes of the diagonal and off-diagonal tensor elements through plotting $|Z_{xx}(\theta)|$ and $|Z_{xy}(\theta)|$ as θ is rotated through 360° (of course, it is sufficient to rotate through only 90° due to the symmetry). These are named *polar diagrams*, or more colloquially *peanut diagrams*, and have been in use for almost 50 years. Note that these also are magnitude-based estimators of dimensionality and directionality.

For an error-free response over a purely 1D Earth, such as represented by Eq. (6.40), then $|Z_{xx}(\theta)|$ is zero for all angles, and $|Z_{xy}(\theta)|$ describes a perfect circle (Figure 6.8a).

For a purely 2D Earth, such as Eq. (6.41), $|Z_{xy}(\theta)|$ describes a shape that is elliptical for low to moderate anisotropy between Z_{xy} and Z_{yx}, but adopts a "peanut" form for higher anisotropy (hence the name). The diagonal term $|Z_{xx}(\theta)|$ displays a four-leaf clover pattern, with zeros at the direction of strike and perpendicular to strike (Figure 6.8b), where $|Z_{xy}(\theta)|$ reaches maxima and minima.

For a 3D Earth (Eq. (6.42)), symmetry is lost (except for data from a site on a point of geometric symmetry) and particularly the diagonal element, $|Z_{xx}(\theta)|$, does not display zeros at four cardinal points 90° apart at the angles where the off-diagonal $|Z_{xy}(\theta)|$ reaches maxima and minima (Figure 6.8c). Given noise contamination, however, an interpreter may view Figure 6.8c and decide that it is representative of a 2D Earth with a geoelectric strike of approximately −20° (or +70°).

How does distortion affect polar diagrams? Very badly, it turns out, even for low to moderate levels of distortion. Figure 6.8d displays the polar diagrams for the 3D/2D response tensor in Eq. (6.43). The polar diagram is very close to 2D, with an obvious peanut shape to the off-diagonal element and a diagonal element that is close to a four-leaf clover, with its minima at the maxima and minima of the off-diagonal element. Thus, one might erroneously conclude that the tensor was representative of a 2D Earth, with a strike of 45°.

In conclusion, all three of the cited dimensionality tools that are amplitude-based are seriously affected by distortion; hence they are all unreliable and should not be used.

Table 6.2. *Ellipticity.*

Model	True	Low noise (1%)					Medium noise (3%)					High noise (10%)				
		Mean	Median	68%	95%	Range	Mean	Median	68%	95%	Range	Mean	Median	68%	95%	Range
1D	Undef.	1.52	1.00	0.00	0.00	0.00	1.65	1.00	0.00	0.00	0.00	1.54	1.00	0.00	0.00	0.00
				3.98	6.33	86.1			12.6	23.6	1644			4.01	6.47	123
2D	0.000	0.04	0.04	0.02	0.00	0.00	0.13	0.12	0.06	0.00	0.00	0.46	0.38	0.02	0.00	0.00
				0.06	0.09	0.15			0.20	0.27	0.49			0.90	1.57	57.1
3D	1.490	1.49	1.49	1.43	1.37	1.23	1.50	1.48	1.31	1.12	0.89	1.66	1.44	0.24	0.24	0.24
				1.55	1.61	1.74			1.68	1.87	2.60			3.48	5.31	220
3D/2D	2.623	2.62	2.62	2.51	2.39	2.24	2.65	2.61	2.28	1.91	1.64	3.01	2.45	0.68	0.68	0.68
				2.74	2.86	3.14			3.01	3.37	5.50			6.80	10.6	10.6

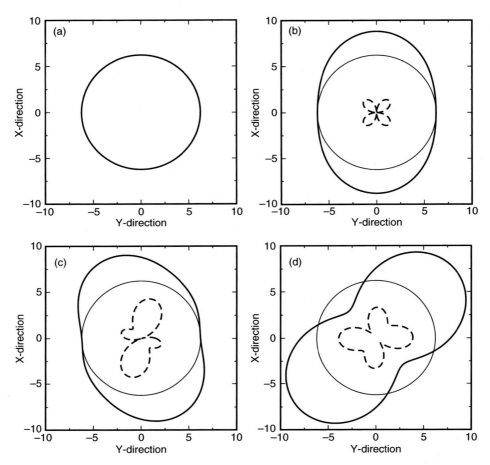

Figure 6.8. Impedance magnitude rotation diagrams ("peanut diagrams") for the synthetic (a) 1D, (b) 2D, (c) 3D and (d) 3D/2D cases. In all cases, the thin circle represents the unit circle for the 1D case, the thick full curve represents the off-diagonal impedance element, and the thick dashed curve represents the diagonal impedance.

6.7.2.4 *Bahr's phase-sensitive skew, η*

The first modern dimensionality tool to be proposed was the *phase-sensitive skew*, η, of Bahr (1988). This is a measure of the skew of the phases of the impedance tensor, so is unaffected by amplitude distortion effects, and is given by

$$\eta = \frac{\left|[D_1, S_2] - [S_1, D_2]\right|^{1/2}}{|D_2|} \tag{6.53}$$

where the S (sum) and D (difference) impedances are the so-called *modified impedances* (Bahr, 1988) given by

$$S_1 = Z_{xx} + Z_{yy}, \qquad S_2 = Z_{xy} + Z_{yx}$$
$$D_1 = Z_{xx} - Z_{yy}, \qquad D_2 = Z_{xy} - Z_{yx} \tag{6.54}$$

Bahr (1988, 1991) gave criteria for interpreting η based on its value. Values of $\eta < 0.1$ are considered to be 1D, 2D or distorted 2D (3D/2D). Values in the range $0.1 < \eta < 0.3$ are considered to be indicative of a modified 3D/2D form called the *delta* (δ) *technique*. Values of $\eta > 0.3$ are considered to represent 3D data. Marti *et al.* (2005) undertook a statistical exercise and concluded that the threshold value for η is far lower than proposed by Bahr, suggesting that values of $\eta > 0.12$ should be considered to represent 3D data.

Lezaeta (2002) developed estimates of confidence limits for phase-sensitive skew by deriving its conditional probability function in terms of the tensor elements density function, assuming the tensor elements to be normally distributed random variables.

Applying the same statistical bootstrap approach as in prior sections yields Table 6.3 and Figure 6.9 for Bahr's η.

- The η estimator is biased – its expectation value for the 1D, 2D and 3D/2D cases is not zero but increases as the noise level increases.
- The η estimator performs reasonably well when noise is low. Almost all realizations for the 1D, 2D and 3D/2D cases have $\eta < 0.12$. The only problem may be the misinterpretation of the 3D data as resulting from a 1D, 2D or 3D/2D Earth when the noise level is medium (3%), as 5% of the estimates have $\eta < 0.3$ (Bahr threshold). Note that none of the 3D data, except for some 4% of the high-noise case, yield $\eta < 0.12$ (Marti *et al.* (2005) threshold).
- The η estimator does not though perform well when the noise is high:
  - in 30% of the realizations from the 3D case, $\eta < 0.3$, and a 2D regional Earth would be concluded;
  - in 27% of the realizations from the 3D/2D case, $\eta > 0.3$, and a 3D regional Earth would be concluded; and
  - even 20% of the 1D cases and 30% of the 2D cases would be inferred to come from a 3D regional Earth, as $\eta > 0.3$.

6.7.2.5 WAL invariants

There are a family of mathematical invariants under rotation that can be constructed from the rank-two response tensor that are covered in Chapter 4 (Section 4.1.4). The obvious ones are the trace of the tensor $\text{tr}(\mathbf{Z}) = (Z_{xx} + Z_{yy})$, the anti-trace $(Z_{xy} - Z_{yx})$ and the determinant $\det(\mathbf{Z}) = \left| (Z_{xy} - Z_{yx})(Z_{xx} + Z_{yy}) \right|$, which is the ratio of the two. Weaver, Agarwal and Lilley (Weaver *et al.*, 2000), following on from the work of Szarka & Menvielle (1997), presented the most complete treatment of invariants of the MT response tensor, with a set of seven independent invariants, I_1 to I_7. Using the notation

$$\xi_1 + i\eta_1 = \tfrac{1}{2}(Z_{xx} + Z_{yy}), \qquad \xi_2 + i\eta_2 = \tfrac{1}{2}(Z_{xy} + Z_{yx})$$
$$\xi_3 + i\eta_3 = \tfrac{1}{2}(Z_{xx} + Z_{yy}), \qquad \xi_4 + i\eta_4 = \tfrac{1}{2}(Z_{xy} + Z_{yx}) \tag{6.55}$$

Table 6.3. *Bahr's phase-sensitive skew* η.

Model	True	Low noise (1%)					Medium noise (3%)					High noise (10%)				
		Mean	Median	68%	95%	Range	Mean	Median	68%	95%	Range	Mean	Median	68%	95%	Range
1D	0.000	0.07	0.07	0.03	0.01	0.00	0.12	0.12	0.07	0.02	0.00	0.22	0.22	0.13	0.03	0.00
				0.10	0.13	0.18			0.17	0.22	0.22			0.31	0.41	0.56
2D	0.000	0.08	0.08	0.04	0.01	0.00	0.13	0.13	0.06	0.02	0.00	0.24	0.24	0.14	0.04	0.00
				0.11	0.14	0.19			0.19	0.25	0.33			0.35	0.35	0.62
3D	0.368	0.37	0.37	0.35	0.34	0.31	0.37	0.37	0.32	0.28	0.16	0.35	0.37	0.23	0.11	0.002
				0.38	0.39	0.41			0.41	0.44	0.50			0.48	0.60	0.60
3D/2D	0.000	0.07	0.07	0.04	0.01	0.00	0.13	0.13	0.07	0.02	0.00	0.24	0.23	0.13	0.03	0.00
				0.11	0.14	0.18			0.18	0.24	0.32			0.34	0.44	0.65

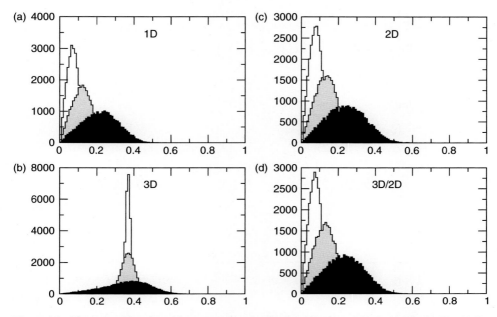

Figure 6.9. Phase-sensitive skew histograms from realizations for the synthetic (a) 1D, (b) 2D, (c) 3D and (d) 3D/2D cases, with three levels of noise added: white, low noise; gray, medium noise; black, high noise.

and the sum and difference dimensionless variables

$$
d_{ij} = \frac{\xi_i \eta_j - \xi_j \eta_i}{\left(\xi_4^2 - \xi_1^2\right)^{1/2} \left(\eta_4^2 + \eta_1^2\right)^{1/2}}
$$

$$
s_{ij} = \frac{\xi_i \eta_j + \xi_j \eta_i}{\left(\xi_4^2 + \xi_1^2\right)^{1/2} \left(\eta_4^2 + \eta_1^2\right)^{1/2}}
$$

(6.56)

the WAL invariants are given by

$$
\begin{aligned}
I_1 &= \left(\xi_4^2 + \xi_1^2\right)^{1/2} \\
&= \frac{1}{2}\left((\mathrm{Re}(Z_{xy} - Z_{yx}))^2 + (\mathrm{Re}(Z_{xx} + Z_{yy}))^2\right)^{1/2} \\
I_2 &= \left(\eta_4^2 + \eta_1^2\right)^{1/2} \\
&= \frac{1}{2}\left((\mathrm{Im}(Z_{xy} - Z_{yx}))^2 + (\mathrm{Im}(Z_{xx} + Z_{yy}))^2\right)^{1/2}
\end{aligned}
$$

$$I_3 = \left(\xi_2^2 + \xi_3^2\right)^{1/2}/I_1$$

$$= \frac{1}{2}\left((\mathrm{Re}(Z_{xy} + Z_{yx}))^2 + (\mathrm{Re}(Z_{xx} - Z_{yy}))^2\right)^{1/2}/I_1$$

$$I_4 = \left(\eta_2^2 + \eta_3^2\right)^{1/2}/I_2$$

$$= \frac{1}{2}\left((\mathrm{Im}(Z_{xy} + Z_{yx}))^2 + (\mathrm{Im}(Z_{xx} - Z_{yy}))^2\right)^{1/2}/I_2$$

$$I_5 = s_{41} = (\xi_4\eta_1 + \xi_1\eta_4)/I_1 I_2$$

$$= \left((\mathrm{Re}(Z_{xy} - Z_{yx})\mathrm{Im}(Z_{xx} + Z_{yy})) + (\mathrm{Re}(Z_{xx} + Z_{yy})\mathrm{Im}(Z_{xy} - Z_{yx}))\right)/I_1 I_2$$

$$I_6 = d_{41} = (\xi_4\eta_1 - \xi_1\eta_4)/I_1 I_2$$

$$= \left((\mathrm{Re}(Z_{xy} - Z_{yx})\mathrm{Im}(Z_{xx} + Z_{yy})) - (\mathrm{Re}(Z_{xx} + Z_{yy})\mathrm{Im}(Z_{xy} - Z_{yx}))\right)/I_1 I_2$$

$$I_7 = \frac{d_{41} - d_{23}}{\left((d_{12} - d_{34})^2 + (d_{13} + d_{24})^2\right)^{1/2}}$$

$$= \frac{((\xi_4\eta_1 - \xi_1\eta_4) - (\xi_2\eta_3 - \xi_3\eta_2))}{\left(((\xi_1\eta_2 - \xi_2\eta_1) - (\xi_3\eta_4 - \xi_4\eta_3))^2 + ((\xi_1\eta_3 - \xi_3\eta_1) + (\xi_2\eta_4 - \xi_4\eta_2))^2\right)^{1/2}}$$

$$(6.57)$$

In addition, there is an eighth invariant, Q, given by

$$Q = \left((d_{12} - d_{34})^2 + (d_{13} + d_{24})^2\right)^{1/2} \qquad (6.58)$$

but Q is not an independent invariant, as, by algebraic manipulation, it can be derived from the first four.

The first two invariants, I_1 and I_2, are defined as the fundamental pair by Weaver *et al.* (2000) and clearly are combinations of the real and imaginary parts of the trace and anti-trace of the response tensor. They do not contain any intrinsic information about dimensionality. The next two invariants, I_3 and I_4, contain information about the strength of the 2D anisotropy, ranging from 0 in a 1D case to 1 for infinite anisotropy, i.e. $|Z_{xy}|/|Z_{yx}| \to 0$ or ∞. The other three invariants, I_5, I_6 and I_7, are related to galvanic distortion. For perfect noise-free data from a 1D Earth, I_3 to I_6 are all zero and I_7 is undefined. For data from a 2D Earth, then I_5 and I_6 are zero, and I_7 is zero or undefined.

Of the WAL invariants, the most useful for discerning dimensionality are I_7 and Q: I_7 is non-zero for the 3D case and zero for the others (ignoring the undefined cases for perfect 1D data), and Q is zero for 1D and non-zero for the others. Weaver *et al.* (2000) define thresholds of 0.1 for the parameters to be meaningfully different from zero.

Undertaking the same statistical examination as above yields the values for I_7 and Q in Tables 6.4a and 6.4b. The WAL defined threshold of 0.1 for being statistically meaningful is clearly appropriate in the low-noise case – the correct dimensionality would be concluded in 95% of the cases based on examination of I_7 and Q. This threshold fails though for half of the realizations in the medium-noise case, and for most of the realizations in the high-noise case.

Table 6.4a. *WAL* I_7.

Model	True	Low noise (1%)					Medium noise (3%)					High noise (10%)				
		Mean	Median	68%	95%	Range	Mean	Median	68%	95%	Range	Mean	Median	68%	95%	Range
1D	undef.	0.01	0.00	−2.22	−4.45	−105	0.01	0.00	−2.05	−4.11	−60	0.00	0.00	−2.89	−5.76	−308
				2.22	4.45	101			2.07	−4.13	59			2.89	5.77	55.5
2D	0.000	0.00	0.00	−0.05	−0.10	−0.19	0.00	0.00	−0.16	−0.31	−0.68	−0.01	0.00	−0.85	−1.68	−26.0
				0.05	0.10	0.21			0.15	0.31	0.68			0.82	1.66	32.3
3D	0.364	0.36	0.36	0.33	0.31	0.25	0.36	0.36	0.28	0.20	0.07	0.38	0.35	0.08	−0.22	−0.55
				0.39	0.42	0.46			0.45	0.53	0.73			0.68	0.98	5.36
3D/2D	0.000	0.00	0.00	−0.05	−0.11	−0.24	0.02	0.00	−0.17	−0.36	−0.48	0.28	0.00	−1.15	−2.59	−11.8
				0.06	0.12	0.29			0.22	0.41	1.75			1.72	3.16	72.5

Table 6.4b. *WAL* Q.

Model	True	Low noise (1%)					Medium noise (3%)					High noise (10%)				
		Mean	Median	68%	95%	Range	Mean	Median	68%	95%	Range	Mean	Median	68%	95%	Range
1D	0.000	0.02	0.02	0.00	0.00	0.00	0.05	0.05	0.02	0.00	0.00	0.18	0.17	0.08	0.00	0.00
				0.03	0.04	0.06			0.08	0.11	0.19			0.28	0.37	0.68
2D	0.405	0.41	0.41	0.38	0.36	0.32	0.41	0.41	0.35	0.29	0.18	0.46	0.45	0.27	0.07	0.00
				0.43	0.45	0.48			0.47	0.53	0.69			0.66	0.85	1.61
3D	0.831	0.83	0.83	0.81	0.79	0.73	0.83	0.83	0.77	0.70	0.54	0.86	0.85	0.64	0.42	0.10
				0.85	0.87	0.92			0.90	0.96	1.11			1.08	1.30	2.06
3D/2D	0.312	0.31	0.31	0.29	0.27	0.23	0.32	0.31	0.26	0.21	0.10	0.38	0.37	0.20	0.03	0.03
				0.33	0.35	0.40			0.37	0.43	0.43			0.55	0.72	1.38

Given that there are eight independent parameters in a response tensor, then there must exist algebraic inter-relationships between the Bahr parameters and the WAL ones. Weaver *et al.* (2000) recognized this, and state that Bahr's phase-sensitive skew, η, in their notation is given by $\eta = (I_1 I_2 |d_{41} - d_{23}|)^{1/2} |\xi_4|$. Marti *et al.* (2005) showed how all of the WAL invariants relate to those of Bahr (1991), with Pracser & Szarka's (1999) modifications, and called this the *Bahr-Q* method. The approach of Marti *et al.* (2005) was coded and provided by Marti *et al.* (2009) in their WALDIM program.

6.7.2.6 Phase tensor β and λ

The phase tensor of Caldwell *et al.* (2004), described in Section 6.3.6, measures dimensionality through its four phase parameters combined into β and λ. How well do they perform in the statistical tests?

As shown in Tables 6.5a (β) and 6.5b (λ), the estimators perform well in low-noise conditions. Setting the threshold values at the 95% levels of $|\beta_c| = 1.32$ and $\lambda_c = 0.04$, one can discriminate between 1D, 2D (with 3D/2D) or 3D data. For medium-noise conditions, these threshold values would be $|\beta_c| = 4.00$ and $\lambda_c = 0.11$, but still there is very good discrimination, with less than 5% pathological cases. For high-noise conditions, however, the 95% thresholds would rise to $|\beta_c| = 13.60$ and $\lambda_c = 0.37$, for which the majority of the 3D realizations would incorrectly be judged not to be 3D.

6.7.2.7 Groom–Bailey dimensionality analysis

Dimensionality is judged in the Groom–Bailey (GB) analysis by whether the model of local 3D distortion of responses from a 1D or 2D regional Earth fits the data, to within their errors, or not. Table 6.6 lists the root mean square (RMS) error distributions for the various models assuming a single-frequency model fit of seven parameters to the eight data for the first 1000 realizations in the datasets, without an error floor being set.

For low and medium noise levels, the GB approach has identified that the 3D data are indeed 3D and cannot be represented by distortion of 1D or 2D data. At high noise levels and with the single-frequency approach, individual models can be found for each 3D realization that fits a 3D/2D distortion model.

However, the greatest strength of the GB approach is when a model is sought that is consistent with the primary assumptions about distortion, namely that the strike and distortion parameters are frequency-independent. Applying the multifrequency approach of McNeice & Jones (2001) to the first 100 realizations of these data, to mimic an MT dataset at a single site, yields the RMS, distortion parameters and strikes listed in Table 6.7

For the 3D/2D case, the McNeice–Jones multifrequency decomposition correctly identifies the strike angle and the distortion parameters very precisely. The low- and medium-noise 3D data are identified as invalid for interpretation as 3D/2D distortion. The large errors on the high-noise 3D data preclude such a conclusion, and one would instead incorrectly

Table 6.5a. *Phase tensor* β.

Model	True	Low noise (1%)					Medium noise (3%)					High noise (10%)				
		Mean	Median	68%	95%	Range	Mean	Median	68%	95%	Range	Mean	Median	68%	95%	Range
1D	0.000	0.00	0.00	-0.41	-0.81	-1.53	0.00	0.00	-1.24	-2.47	-5.02	0.00	-0.02	-4.14	-8.29	-15.3
				0.41	0.81	1.77			1.23	2.47	5.90			4.14	8.29	19.3
2D	0.000	0.00	0.00	-0.56	-1.12	-2.41	0.00	0.00	-1.71	-3.40	-7.05	-0.04	-0.04	-5.80	-11.6	-27.3
				0.57	1.13	2.06			1.69	3.39	6.69			5.71	11.5	28.4
3D	-7.68	-7.68	-7.68	-8.20	-8.72	-9.79	-7.68	-7.69	-9.25	-10.8	-14.0	-7.65	-7.62	-13.0	-18.3	-30.2
				-7.16	-6.64	-5.26			-6.11	-4.54	-0.66			-2.33	2.98	12.9
3D/2D	0.000	0.00	0.00	-0.66	-1.32	-2.42	-0.02	-0.02	-2.02	-4.01	-9.12	0.03	0.02	-6.78	-13.6	-41.9
				0.66	1.31	2.63			1.96	3.95	7.95			6.83	13.6	36.9

Table 6.5b. *Phase tensor* λ.

Model	True	Low noise (1%)					Medium noise (3%)					High noise (10%)				
		Mean	Median	68%	95%	Range	Mean	Median	68%	95%	Range	Mean	Median	68%	95%	Range
1D	0.000	0.02	0.02	0.01	0.00	0.00	0.05	0.05	0.02	0.00	0.00	0.18	0.17	0.09	0.00	0.00
				0.03	0.04	0.07			0.08	0.11	0.26			0.27	0.37	0.64
2D	0.391	0.39	0.39	0.37	0.36	0.32	0.39	0.39	0.34	0.29	0.18	0.44	0.44	0.28	0.12	0.00
				0.41	0.43	0.46			0.45	0.50	0.61			0.60	0.76	1.37
3D	0.727	0.73	0.73	0.71	0.70	0.67	0.73	0.73	0.69	0.65	0.55	0.74	0.75	0.61	0.47	0.02
				0.74	0.75	0.78			0.77	0.81	0.90			0.87	1.01	1.50
3D/2D	0.391	0.39	0.39	0.37	0.35	0.31	0.39	0.40	0.33	0.26	0.14	0.47	0.46	0.26	0.05	0.00
				0.41	0.44	0.48			0.46	0.53	0.69			0.68	0.89	2.57

Table 6.6. *Groom–Bailey root mean square.*

Model	Low noise (1%)					Medium noise (3%)					High noise (10%)				
	Mean	Median	68%	95%	Range	Mean	Median	68%	95%	Range	Mean	Median	68%	95%	Range
1D	0.28	0.23	0.06	0.00	0.00	0.29	0.25	0.08	0.00	0.00	0.28	0.25	0.08	0.00	0.00
			0.50	0.71	1.00			0.50	0.71	1.07			0.49	0.70	1.22
2D	0.29	0.24	0.07	0.00	0.00	0.29	0.25	0.08	0.00	0.00	0.29	0.24	0.07	0.00	0.00
			0.50	0.71	1.23			0.50	0.71	1.17			0.51	0.71	1.14
3D	5.17	5.17	4.80	4.43	3.97	1.73	1.75	1.37	1.00	0.69	0.55	0.52	0.21	0.00	0.00
			5.54	5.91	6.35			2.10	2.46	2.83			0.89	1.22	2.08
3D/2D	0.29	0.24	0.07	0.00	0.00	0.30	0.26	0.07	0.00	0.00	0.28	0.24	0.07	0.00	0.00
			0.50	0.72	1.18			0.52	0.75	1.17			0.48	0.69	1.19

Table 6.7 *Multifrequency McNeice–Jones analysis.*

Model	Low noise (1%)					Medium noise (3%)					High noise (10%)				
	shear	twist	strike	RMS	95% limits	shear	twist	strike	RMS	95% limits	shear	twist	strike	RMS	95% limits
1D	0.02	0.03	−12.4	0.34	0.07 / 0.64	0.00	0.08	9.8	0.65	0.16 / 1.14	0.67	−0.33	34.5	0.62	0.17 / 1.05
2D strike = 0°	0.09	−0.02	0.22	0.35	0.06 / 0.66	−0.23	0.18	−0.6	0.68	0.16 / 1.20	−1.53	−0.85	−1.04	0.65	0.25 / 1.05
3D	9.35	2.41	32.0	2.96	2.58 / 3.35	9.67	2.68	32.4	1.84	1.12 / 2.57	8.8	2.5	32.1	0.84	0.21 / 1.53
3D/2D shear = 25.0° twist = −2.1° strike = 0°	24.9	−2.04	−0.02	0.38	0.10 / 0.67	25.1	−2.30	0.36	0.65	0.25 / 1.05	24.4	−1.89	−1.52	0.55	0.17 / 0.93

conclude that the data are from a 3D/2D Earth with a strike of 32° and low distortion (*shear* = 8.8° and *twist* = 2.5°).

6.7.3 Directionality tools

Once the appropriate dimensionality has been ascertained, if the data are deemed to be representative of 2D or 3D/2D structures, then the appropriate geoelectric strike direction must be determined. A number of directionality testing tools have been developed virtually since the MT method was proposed. Again, it is critically important that the tool be statistically effective – *all* phase-based tools considered herein will give the correct result in the absence of noise, but equally imperative is how well the tool performs in the presence of noise.

The statistics performed in this section were undertaken using a circular von Mises distribution with a 90° repetition frequency, following Mardia (1972), not linear statistics.

6.7.3.1 Swift strike

As with the dimensionality tools, the early directional ones relied on the magnitudes of the response tensor, and attempted to antidiagonalize the response tensor in some optimum manner. Swift, in his Ph.D. thesis (Swift, 1967), laid out the objective of rotating the response tensor to determine the angle at which it most closely adopts the antidiagonal 2D form of Eq. (6.46). This was accomplished by finding the angle that maximizes the sum of squares of the off-diagonal terms, i.e. $|Z'_{xy}(\theta)|^2 + |Z'_{yx}(\theta)|^2$, which is given algebraically by

$$\tan(4\theta_s) = \frac{(Z_{xx} - Z_{yy})(Z_{xy} + Z_{yx})^* + (Z_{xx} + Z_{yy})^*(Z_{xy} - Z_{yx})}{|Z_{xx} - Z_{yy}|^2 - |Z_{xy} + Z_{yx}|^2} \tag{6.59}$$

(Swift, 1967; Simpson & Bahr, 2005, eq. (5.13)). This angle either minimizes or maximizes the sum of squares of the diagonal terms, i.e. $|Z'_{xx}(\theta_s)|^2 + |Z'_{yy}(\theta_s)|^2$, and in the latter case then is 45° from the strike angle, so a simple test must be performed of the angle to determine if it yields the correct one. (The formula is given in Vozoff's classic paper (Vozoff, 1972, eq. (26)), but unfortunately was printed with a sign error in the last bracket of the numerator. This was corrected in Jupp & Vozoff (1976).)

The angle determined from Eq. (6.59) has been termed "Swift strike" in the literature, and will be denoted by θ_S. It was used extensively in the 1970s (e.g. Sims & Bostick, 1969; Vozoff, 1972; Cochrane & Hyndman, 1974; Kurtz & Garland, 1976) and since, and indeed still is part of modern packages, such as the Geotools and WinGLink packages.

The values and ranges of θ_S for the 2D and 3D/2D realizations above are listed in Table 6.8. Not surprisingly, Swift strike is completely wrong in the presence of distortion (3D/2D), yielding an angle even in the low-noise case that is consistent with the inference from the polar diagram (Section 6.7.2.3) shown in Figure 6.8d. In the 2D case, the ranges of distributions of the estimates from the realizations are large, even for moderate noise levels.

Table 6.8. *Swift strike* θ_S.

Model	Low noise (1%)					Medium noise (3%)					High noise (10%)				
	Mean	Median	68%	95%	Range	Mean	Median	68%	95%	Range	Mean	Median	68%	95%	Range
2D	0.00	0.00	±1.89	±3.78	−8.35 8.00	0.00	0.00	±6.61	±13.2	−39.3 40.9	0.29	0.24	±20.4	±41.6	±45.0
3D/2D	44.1	41.7	41.0 42.3	40.4 42.9	38.5 44.1	41.7	41.5	39.8 43.5	37.9 45.3	34.6 50.4	42.2	36.2	34.4 49.9	26.7 57.6	2.1 90

6.7.3.2 Other amplitude-based approaches

Given that data are never noise-free, and that the Earth is never perfectly 2D, a number of other techniques were used through the 1960s to 1980s to try to determine the best 2D "strike" direction to adopt for interpretation. These included the following:

- rotating numerically to maximize one of the off-diagonal terms of the response tensor $|Z_{xy}(\theta)|$ (Everett & Hyndman, 1967) or of the admittance tensor $|Y_{xy}(\theta)|$ (Bostick & Smith, 1962);
- rotating numerically to maximize the sum of squares of the off-diagonal terms $|Z_{xy}(\theta)|^2 + |Z_{yx}(\theta)|^2$ (Reddy & Rankin, 1971);
- rotating numerically to minimize one of the diagonal terms, $|Z'_{xx}(\theta)|$; and
- rotating numerically to maximize a measure of the coherence between the electric field and its orthogonal magnetic field (Reddy & Rankin, 1974; Jones & Hutton, 1979).

The last is not a magnitude-based approach, but a signal-based one.

6.7.3.3 Bahr phase-sensitive strike

Bahr (1988) was the first to propose a phase-based estimator. He recognized that when the response tensor for a 2D regional structure is rotated into strike coordinates the phases of the columns are equal (see Section 6.3.4). This condition is fulfilled at angle θ_B given by

$$\tan(2\theta_B) = \frac{[S_1, S_2] - [D_1, D_2]}{[S_1, D_1] + [S_2, D_2]} \tag{6.60}$$

where S and D are given by Eq. (6.54) and the function $[x, y]$ is taken to mean $[x, y] = \mathrm{Re}\, x\, \mathrm{Im}\, y - \mathrm{Im}\, x\, \mathrm{Re}\, y$ (Simpson & Bahr, 2005, eq. (5.27)).

A conservative parametric error estimate for θ_B, derived from a propagation of errors approach implicitly assuming Gaussian statistical behavior for θ_B, was given in the appendix to Simpson (2001),

$$\Delta\theta_B = \left(\frac{1}{2}\right)\left(\frac{1}{1 + (\tan 2\theta_B)^2}\right)\Delta(\tan 2\theta_B) \tag{6.61}$$

where the error in $\tan 2\theta_B$ is given by

$$\Delta(\tan 2\theta_B) = \sqrt{\left(\frac{\Delta(|S_1, S_2| - |D_1, D_2|)}{|S_1, D_1| - |S_2, D_2|}\right)^2 + \left(\frac{\Delta\{(|S_1, D_1| + |S_2, D_2|)(|S_1, S_2| + |D_1, D_2|)\}}{(|S_1, D_1| + |S_2, D_2|)^2}\right)^2} \tag{6.62}$$

where

$$\{\Delta(|S_1, S_2| - |D_1, D_2|)\}^2 = \Delta S_1^2 S_2^2 + S_1^2 \Delta S_2^2 + \Delta D_1^2 D_2^2 + D_1^2 \Delta S_2^2$$

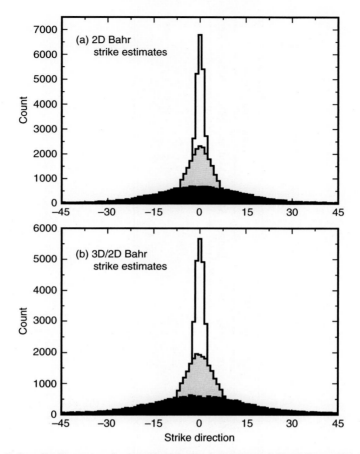

Figure 6.10. Bahr strike histograms from realizations for the (a) 2D and (b) 3D/2D models of the NACP synthetic data with three levels of noise added: white, low noise; gray, medium noise; black, high noise.

and

$$\left\{ \Delta\left(\left| S_1, D_1 \right| + \left| D_1, D_2 \right| \right) \right\}^2 = \Delta S_1^2 D_1^2 + S_1^2 \Delta D_1^2 + \Delta S_2^2 D_2^2 + S_2^2 \Delta D_2^2$$

This error estimator is included here for completeness, but is not used, as the distribution of θ_B will be as complex as that for skew (Chave, 2011).

The values and ranges of θ_B for the 2D and 3D/2D realizations are listed in Table 6.9 and are shown in histogram form in Figure 6.10.

The Bahr strike estimates are unbiased, with their mean, median and modal values all being the same (to within experimental error), but the distributions become long-tailed with high levels of noise.

Table 6.9. *Bahr phase-sensitive strike* θ_B.

Model	Low noise (1%)					Medium noise (3%)					High noise (10%)				
	Mean	Median	68%	95%	Range	Mean	Median	68%	95%	Range	Mean	Median	68%	95%	Range
2D	0.00	0.00	±1.45	±2.90	−5.84 5.65	0.00	0.00	±4.39	±8.78	−17.6 16.2	0.07	0.04	±15.5	±30.9	±45
3D/2D	0.00	0.01	±1.68	±3.37	−6.69 7.12	−0.03	−0.03	±5.17	±10.3	−25.5 20.1	0.03	0.05	±17.8	±36.5	±45

6.7.3.4 WAL strike

Weaver *et al.* (2000) define two ways of determining the strike of 2D structures in their publication, namely

$$\tan(2\theta_{W1}) = -\xi_3/\xi_2 = \frac{-\mathrm{Re}(Z_{xx} - Z_{yy})}{\mathrm{Re}(Z_{xy} - Z_{yx})} \tag{6.63}$$

(eq. (18)), which maximizes $\mathrm{Re}(Z_{xy})$, and

$$\tan(2\theta_{W2}) = \frac{d_{12} - d_{34}}{d_{13} + d_{24}} = \frac{(\xi_1\eta_2 - \xi_2\eta_1) - (\xi_3\eta_4 - \xi_4\eta_3)}{(\xi_1\eta_3 - \xi_3\eta_1) + (\xi_2\eta_4 - \xi_4\eta_2)} \tag{6.64}$$

(eq. (44)). The first of these, θ_{W1}, is clearly a magnitude-based estimator and suffers from the same issues as Swift strike (Section 6.7.3.1) in the presence of distortion; it will not be considered further. The second one, θ_{W2}, combines the real and imaginary parts of the response tensor estimates and forms a ratio, thus is equivalent to a phase-based estimator. The ability of θ_{W2} to correctly determine strike for the synthetic data is shown in Table 6.10.

As will be shown below in Section 6.7.3.7, the values in Table 6.10 are exactly the same as those in Table 6.9, as the two estimators are exactly the same.

6.7.3.5 Phase-tensor strike

The phase-tensor strike, θ_P, is given by $\alpha - \beta$ in Eq. (6.27), and for the synthetic data yields the values listed in Table 6.11.

The strike directions for the two models and three noise cases are shown as histograms in Figure 6.11. For the 2D case, the phase-tensor strike estimator is unbiased and has a distribution that appears to be Gaussian. For the 3D/2D case, as the noise increases, the strike estimator exhibits a bias, with a modal value of around $-10°$, and the distribution becomes asymmetric. The distribution of α is exactly the same as for Bahr's phase-sensitive strike η, as the two estimators are equal (Section 6.7.3.7), so the bias is introduced through β.

6.7.3.6 Groom–Bailey strike

The Groom–Bailey strike, θ_{GB}, for the test datasets is listed in Table 6.12. The GB strike estimator θ_{GB} is unbiased in the presence of noise, with tighter distributions (as expressed by the one- and two-sigma limits) compared to the other estimators.

However, as discussed in Section 6.7.2.7, the McNeice–Jones multifrequency extension to the original single-frequency approach of Groom and Bailey assures a far more stable estimate of strike. These are listed in Table 6.7 and even for the high-noise cases are within $2°$ of the correct strike value.

6.7.3.7 Comparison of Bahr, WAL and phase-tensor strike directions

The Bahr strike direction θ_B is given by Eq. (6.60). Expanding out Eq. (6.60) and collecting terms, using the Caldwell *et al.* (2004) notation $X_{xx} = \mathrm{Re}(Z_{xx})$ and $Y_{xx} = \mathrm{Im}(Z_{xx})$, etc. (Eq. (6.23)), gives

Table 6.10. *WAL strike* θ_{W2}.

Model	Low noise (1%)					Medium noise (3%)					High noise (10%)				
	Mean	Median	68%	95%	Range	Mean	Median	68%	95%	Range	Mean	Median	68%	95%	Range
2D	0.00	0.00	±1.45	±2.90	−5.84 / 5.65	0.00	0.00	±4.39	±8.78	−17.6 / 16.2	0.06	0.06	±15.4	±30.9	±45
3D/2D	0.00	0.00	±1.69	±3.37	−6.69 / 7.12	0.00	0.00	±5.17	±10.4	−25.5 / 20.1	0.03	0.03	±17.8	±35.6	±45

Table 6.11. *Phase-tensor strike* $\theta_P = \alpha - \beta$.

Model	Low noise (1%)					Medium noise (3%)					High noise (10%)				
	Mean	Median	68%	95%	Range	Mean	Median	68%	95%	Range	Mean	Median	68%	95%	Range
2D	0.00	0.00	±1.36	±2.72	−5.26 / 5.13	0.00	−0.04	±4.14	±8.29	−17.2 / 17.1	0.05	−0.05	±14.7	±29.5	−58 / 61
3D/2D	0.00	0.00	±1.82	±3.65	−6.78 / 8.05	−0.04	−0.28	−5.57 / 5.49	−11.1 / 11.0	−20.0 / 24.8	−3.65	−2.16	−21.9 / 14.6	−40.2 / 32.9	−47.9 / 32.9

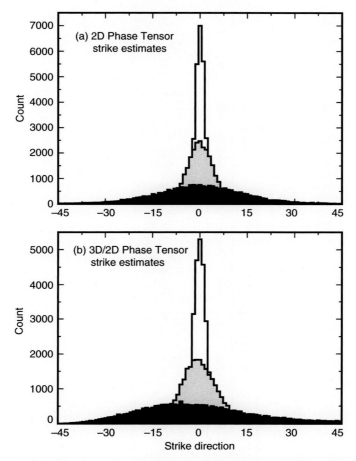

Figure 6.11. Phase-tensor strike histograms from realizations for the (a) 2D and (b) 3D/2D models of the NACP synthetic data with three levels of noise added: white, low noise; gray, medium noise; black, high noise.

$$
\tan(2\theta_B) = \frac{[S_1, S_2] - [D_1, D_2]}{[S_1, D_1] + [S_2, D_2]}
$$

$$
= \frac{\begin{aligned}&\big(\mathrm{Re}(Z_{xx} + Z_{yy})\mathrm{Im}(Z_{xy} + Z_{yx}) - \mathrm{Im}(Z_{xx} + Z_{yy})\mathrm{Re}(Z_{xy} + Z_{yx})\big)\\ &-\big(\mathrm{Re}(Z_{xx} - Z_{yy})\mathrm{Im}(Z_{xy} - Z_{yx}) - \mathrm{Im}(Z_{xx} - Z_{yy})\mathrm{Re}(Z_{xy} - Z_{yx})\big)\end{aligned}}{\begin{aligned}&\big(\mathrm{Re}(Z_{xx} + Z_{yy})\mathrm{Im}(Z_{xx} - Z_{yy}) - \mathrm{Im}(Z_{xx} + Z_{yy})\mathrm{Re}(Z_{xx} - Z_{xx})\big)\\ &+\big(\mathrm{Re}(Z_{xy} + Z_{yx})\mathrm{Im}(Z_{xy} - Z_{yx}) - \mathrm{Im}(Z_{xy} + Z_{yx})\mathrm{Re}(Z_{xy} - Z_{yx})\big)\end{aligned}}
$$

Table 6.12. *Groom–Bailey strike* θ_{GB}.

Model	Low noise (1%)					Medium noise (3%)					High noise (10%)				
	Mean	Median	68%	95%	Range	Mean	Median	68%	95%	Range	Mean	Median	68%	95%	Range
2D	0.04	0.03	-1.32	-2.70	-4.17	0.00	0.07	±4.12	±8.24		-0.03	3.51	-15.0	±29.9	
			1.42	2.79	4.23								14.9		
3D/2D	0.04	0.00	-1.57	-3.19	-4.23	0.16	0.28	-4.88	-9.92	-15.7	-0.19	3.16	±16.7	±33.1	-39.1
			1.66	3.27	5.87			5.21	10.2	14.8					+45.0

$$\big((X_{xx} + X_{yy})(Y_{xy} + Y_{yx}) - (Y_{xx} + Y_{yy})(X_{xy} + X_{yx})\big)$$

$$= \frac{-\big((X_{xx} - X_{yy})(Y_{xy} - Y_{yx}) - (Y_{xx} - Y_{yy})(X_{xy} - X_{yx})\big)}{\big((X_{xx} + X_{yy})(Y_{xx} - Y_{yy}) - (Y_{xx} + Y_{yy})(X_{xx} - X_{xx})\big) +}$$

$$+ \big((X_{xy} + X_{yx})(Y_{xy} - Y_{yx}) - (Y_{xy} + Y_{yx})(X_{xy} - X_{yx})\big)$$

$$= \frac{2(X_{xx}Y_{yx} - Y_{xx}X_{yx}) + 2(X_{yy}Y_{xy} - Y_{yy}X_{xy})}{2(X_{yy}Y_{xx} - Y_{yy}X_{xx}) + 2(X_{yx}Y_{xy} - Y_{yx}X_{xy})}$$

$$= \frac{[Z_{xx}, Z_{yx}] + [Z_{yy}, Z_{xy}]}{[Z_{yy}, Z_{xx}] + [Z_{yx}, Z_{xy}]} \tag{6.65}$$

The WAL strike direction θ_{W2} is given by Eq. (64), viz.

$$\tan(2\theta_{W2}) = \frac{d_{12} - d_{34}}{d_{13} + d_{24}} = \frac{(\xi_1\eta_2 - \xi_2\eta_1) - (\xi_3\eta_4 - \xi_4\eta_3)}{(\xi_1\eta_3 - \xi_3\eta_1) + (\xi_2\eta_4 - \xi_4\eta_2)}$$

where, in terms of Bahr's modified impedances,

$$\xi_1 = \tfrac{1}{2}\mathrm{Re}(Z_{xx} + Z_{yy}) = \tfrac{1}{2}\mathrm{Re}(S_1), \qquad \eta_1 = \tfrac{1}{2}\mathrm{Im}(Z_{xx} + Z_{yy}) = \tfrac{1}{2}\mathrm{Im}(S_1)$$

and $\xi_2 = \tfrac{1}{2}\mathrm{Re}(Z_{xy} + Z_{yx}) = \tfrac{1}{2}\mathrm{Re}(S_1), \xi_3 = \tfrac{1}{2}\mathrm{Re}(Z_{xx} - Z_{yy}) = \tfrac{1}{2}\mathrm{Re}(D_1), \xi_4 = \tfrac{1}{2}\mathrm{Re}(Z_{xy} - Z_{yx}) = \tfrac{1}{2}\mathrm{Re}(D_2)$, and similarly for η taking the imaginary parts. Thus, Eq. (64) becomes

$$\tan(2\theta_{W2}) = \frac{(\xi_1\eta_2 - \xi_2\eta_1) - (\xi_3\eta_4 - \xi_4\eta_3)}{(\xi_1\eta_3 - \xi_3\eta_1) + (\xi_2\eta_4 - \xi_4\eta_2)}$$

$$= \frac{(\mathrm{Re}(S_1)\mathrm{Im}(S_2) - \mathrm{Im}(S_1)\mathrm{Re}(S_2)) - (\mathrm{Re}(D_1)\mathrm{Im}(D_2) - \mathrm{Im}(D_1)\mathrm{Re}(D_2))}{(\mathrm{Re}(S_1)\mathrm{Im}(D_2) - \mathrm{Im}(S_1)\mathrm{Re}(D_1)) + (\mathrm{Re}(S_2)\mathrm{Im}(D_2) - \mathrm{Im}(S_2)\mathrm{Re}(D_2))}$$

$$= \frac{[S_1, S_2] - [D_1, D_2]}{[S_1, D_1] + [S_2, D_2]}$$

$$= \tan(2\theta_B)$$

$$\tag{6.66}$$

from Eq. (60). Thus, the WAL and Bahr strikes directions are *exactly* the same!

This equivalence was known to J. T. Weaver (pers. comm.), and the WAL direction is referred to as the "Bahr angle" in Weaver & Lilley (2004, bottom of left column, p. 253) and in Weaver *et al.* (2006, bottom of left column, p. 263), in papers that have been little cited as they were published in an obscure journal. Inexplicably (J. T. Weaver's word) mention of this equivalence was omitted in the Weaver *et al.* (2000) paper.

Phase-tensor strike θ_P is given by $\alpha - \beta$, where α is given by

$$\tan(2\alpha) = \frac{\Phi_{xy} + \Phi_{yx}}{\Phi_{xx} - \Phi_{yy}}$$

$$= \frac{\left(X_{yy}Y_{xy} - X_{xy}Y_{yy}\right) + \left(X_{xx}Y_{yx} - X_{yx}Y_{xx}\right)}{\left(X_{yy}Y_{xx} - X_{xy}Y_{yx}\right) - \left(X_{xx}Y_{yy} - X_{yx}Y_{xy}\right)}$$

$$= \frac{\left(X_{xx}Y_{yx} - Y_{xx}X_{yx}\right) + \left(X_{yy}Y_{xy} - Y_{yy}X_{xy}\right)}{\left(X_{yy}Y_{xx} - Y_{yy}X_{xx}\right) - \left(X_{xy}Y_{yx} - Y_{xy}X_{yx}\right)} \tag{6.67}$$

$$= \tan(2\theta_B)$$

from Eq. (65). Thus, θ_P is equal to θ_B and θ_{W2} in the case of a 2D or 3D/2D regional Earth when β is zero.

6.8 Removing distortion from magnetotelluric responses

Given the above analyses of the synthetic data presented in Section 6.6.3, how well do the various tools perform on (a) realistic synthetic data, namely far-hi described in Section 6.6.2, and (b) actual data, namely lit007 and lit008 discussed in Section 6.6.4?

6.8.1 Realistic synthetic data, far-hi

The synthetic data from site far-hi have known distortion and known regional responses, so are excellent for testing the various tools.

6.8.1.1 Conventional analysis

Conventional analysis of the synthetic data from site far-hi (Figure 6.5), using Swift skew, Swift strike and ellipticity, is shown in Figure 6.12.

- Skew remains below 0.2 at all periods.
- Swift strike appears frequency-independent, particularly at periods >1 s, at around −41° ± 2°. Taking only periods >1 s, the average is −42.1° ± 0.4°.
- Ellipticity in strike direction is below 0.25, and is particularly low at periods >1 s.

From these, it would be concluded that the MT data are from a purely 2D Earth with a regional strike direction of around −42° (or +48°), which is in fact exactly the electric field channeling azimuth (see Section 6.6.2), rather than the regional strike of +30°. Conventional analysis has failed spectacularly for these data.

6.8.1.2 Bahr analysis

Bahr analysis of the far-hi data is shown in Figure 6.13.

- Kappa (κ) is the conventional Swift skew, given by $|S_1|/|D_2| = |Z_{xx} + Z_{yy}|/|Z_{xy} - Z_{yx}|$ (Bahr, 1988, eq. (16); Simpson & Bahr, 2005, eq. (5.16)).

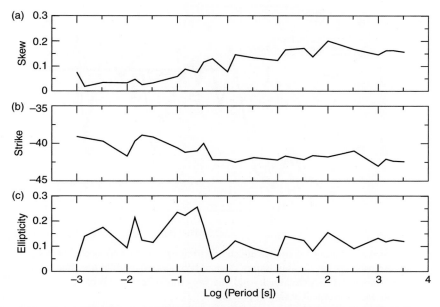

Figure 6.12. Conventional analysis of data from site far-hi: (a) Swift skew; (b) Swift strike; and (c) ellipticity in Swift strike direction.

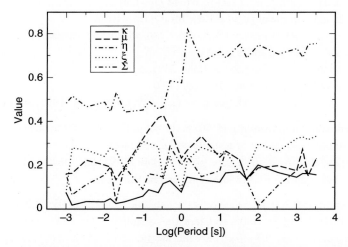

Figure 6.13. Bahr analysis of far-hi data: κ (solid line) is conventional Swift skew; μ (dashed line) is a measure of misfit to the Larsen (1975) model of distortion of responses from a 1D regional Earth; η (dashed-dotted line) is phase-sensitive skew (Section 6.7.2.4), which is a measure of three-dimensionality of the regional conductor; ξ (dotted line) is the ratio of η/κ; and Σ (dashed-double dotted line) is a measure of 2D.

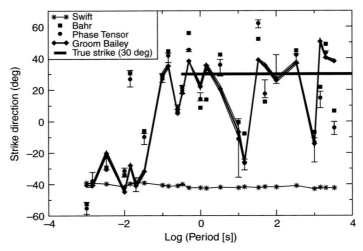

Figure 6.14. Strike directions for site far-hi data: Swift direction, stars connected by solid line; Bahr, squares; phase tensor, circles; Groom–Bailey, diamonds connected by solid line (fine solid lines either side are error estimates); true strike, thick solid line, only shown for periods of 0.3–10 000 s (MT data too distant from fault to sense it at shorter periods).

- Mu (μ) is a measure of misfit to the Larsen (1975) model of distortion of responses from a 1D regional Earth (Section 6.3.2), and is given by $(|[D_1, \ S_2]| + |[S_1, \ D_2]|)^{1/2}/|D_2|$ (Simpson & Bahr, 2005, eq. (5.21)), where $D_{1,2}$ and $S_{1,2}$ are given in Eq. (6.54) and the form [,] is taken to mean $[x, y] = \mathrm{Re}\, x\, \mathrm{Im}\, y - \mathrm{Im}\, x\, \mathrm{Re}\, y$.
- Eta (η) is the *phase-sensitive skew* (Section 6.7.2.4), which is a measure of three-dimensionality of the regional conductor, and is given by $(|[D_1, \ S_2]| + |[S_1, \ D_2]|)^{1/2}/|D_2|$ (Bahr, 1988, eq. (15) with modulus on C; Simpson & Bahr, 2005, eq. (5.32)).
- Xi (ξ) is the ratio of η/κ (Bahr, 1988, eq. (17)).
- Sigma ($\sum$) is a measure of 2D, given by $(D_1^2 + S_2^2)/D_2^2$ (Bahr, 1991, eq. (40)).

The Bahr strike direction from Eq. (6.60) is shown in Figure 6.14 (squares). There is a lot of scatter and no clear indication of a stable strike direction.

Following the classification scheme given in Bahr (1991) and the flowchart in Simpson & Bahr (2005, their figure 5.6), the following can be inferred:

- At short periods (<0.3 s), i.e. when only the galvanic response of the embedded hemisphere is sensed, $\kappa < 0.1$, but $\sum > 0.05$, so Class 1b – a simple 2D model – would be concluded.
- At intermediate periods (0.3–30 s), i.e. when the inductive effects of the 2D regional structure are strongest and are sensed, $\kappa > 0.1$, $\mu > 0.05$, $0.1 < \eta < 0.3$ and $(-\beta_1 + \beta_2)$ is not $\approx 90°$, but with unstable frequency-dependent strike, implying Class 7 – regional 3D induction.
- At the long periods (>30 s), $\kappa > 0.1$, $\mu > 0.05$ and $\eta < 0.1$, but none of the β conditions are met, hence the strike is unstable and frequency-dependent, leading to the conclusion that the data are Class 7.

These classifications are all incorrect. At short periods (<0.3 s), the MT data are distorted 1D, so are Class 2 – regional 1D model with galvanic distortion. At longer periods (>0.3 s), the data "see" the regional 2D structure and inductive effects are large, so the data are Class 5a – 3D/2D model. At periods beyond around 30 s, there are no inductive effects from the 2D fault, and the data are back to Class 2.

6.8.1.3 WAL analysis

The WAL (Weaver *et al.*, 2000) invariants, I_1 to I_7 and Q for the data from site far-hi are shown in Figure 6.15, and the WAL strike direction, θ_{W2} (Eq. (6.64)) is exactly the same as the Bahr phase-sensitive strike, θ_B, plotted in Figure 6.14 (squares, same as Bahr). The threshold values proposed by Weaver *et al.* (2000) are 0.1 for I_3 to I_7 and Q. With this threshold, then I_3 and I_4 are meaningful at all periods, and I_5 and I_6 become meaningful at periods >0.3 s, Q < 0.1 at all periods, so is not meaningful, and I_7 is poorly defined. Thus one would conclude the following:

- At periods <0.3 s, I_3 and I_4 are non-zero, and I_5, I_6 and Q are all zero, indicative of a 2D Earth. This is incorrect, as the data at short periods are 3D/1D.
- At periods >0.3 s, Q is zero, but all the other invariants are meaningful. This is indicative of a 3D Earth, which is also incorrect.

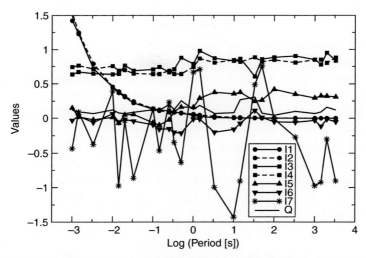

Figure 6.15. WAL invariants for site far-hi data: I_1 (dots connected by solid line) and I_2 (dots connected by dashed line) are defined as the fundamental pair by Weaver *et al.* (2000) and do not contain any intrinsic information about dimensionality; I_3 (squares connected by solid line) and I_4 (squares connected by dashed line) contain information about the strength of the 2D anisotropy, ranging from 0 in a 1D case to 1 for infinite anisotropy; I_5 (upward triangles connected by solid line), I_6 (downward triangles connected by solid line) and I_7 (stars connected by solid line) are related to galvanic distortion.

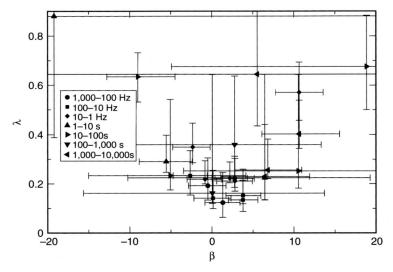

Figure 6.16. Phase-tensor dimensionality β–λ plot for site far-hi. Period ranges are: 1000–100 Hz (dots); 100–10 Hz (squares); 10–1 Hz (diamonds); 1–10 s (upward triangles); 10–100 s (rightward triangles); 100–1000 s (downward triangles); 1000–10 000 (leftward triangles). Error bars for each estimate are shown.

6.8.1.4 Phase-tensor analysis

Phase-tensor analysis of the data from site far-hi is shown in Figure 6.16 (phase-tensor dimensionality plot of β versus λ) and Figure 6.17 (phase-tensor parameters as ellipses), and the strike directions are in Figure 6.14 (black line and symbols, with errors). The errors on the strike directions have been derived using a bootstrap method with 100 realizations to define the distributions, rather than using an error propagation approach.

The phase tensors are inconsistent on a frequency-to-frequency basis, and almost uniformly have high β values above 4 at periods >0.3 s, indicative of inductive 3D responses. The dimensionality plot of β versus λ shows no strong period consistency, apart from being low at periods <0.1 s (10 Hz) (Figure 6.16). A regional strike direction of 30° would be difficult to conclude from Figure 6.14.

6.8.1.5 Groom–Bailey/McNeice–Jones analysis

The Groom–Bailey single-frequency analysis of the far-hi data are shown in Figure 6.18, and the strikes are shown in Figure 6.14 (diamonds connected by solid lines), with one-sigma intervals bounding the main response. The data all fit the distortion decomposition model, with RMS errors all <1.0. The *strike* is scattered, but is closer to the true value of 30°, especially when the inductive effects of the 2D fault are greatest at periods of 0.1–3 s

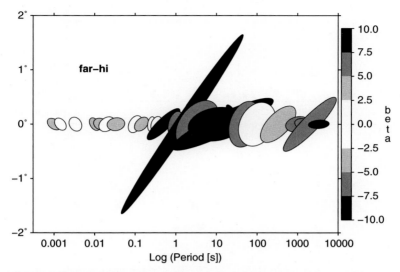

Figure 6.17. Phase-tensor ellipses of the data from site far-hi. The maximum and minimum axes of the ellipses give the amount of anisotropy. The strike of the major axis of the ellipse is given by $\alpha - \beta$, and the infill shade of gray denotes β.

and 30–300 s shown by the largest phase differences (triangles in lower panel, Figure 6.18).

As discussed in Groom *et al.* (1993), the galvanic distortion parameters are almost always more stably determined than the structural parameters (strike and regional impedances). Following the methodology described in Groom *et al.* (1993), the *shear* would be con-strained between 25° and 35°, which immediately stabilizes *twist* to the range −5° to −15° and *strike* to the range 25° to 35° (not shown).

Alternatively, applying the multifrequency extended approach of McNeice & Jones (1996, 2001) to the data searching for averages over 0.5 decade (three or four points per averaging window) yields the solutions shown in Figure 6.19. The *strike* is now far more stable, as are the *twist* and *shear*.

Undertaking a model search for the three parameters to be frequency-independent yields (not shown)

strike = 30.2° (±0.2°)
twist = −12.6° (±0.1°)
shear = 30.1° (±0.1°)2

whereas the true values are 30°, −12° and 30° respectively. This model fits the data at all periods, with an average RMS of 0.6.

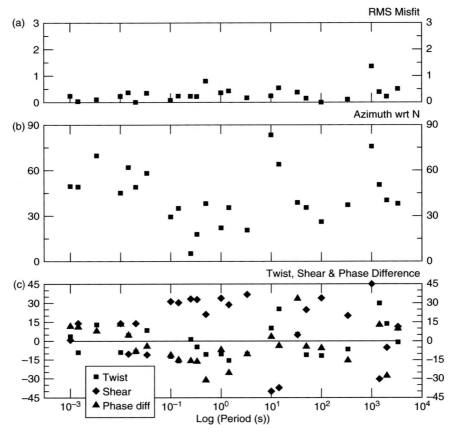

Figure 6.18. Groom–Bailey single-frequency analysis of the data from site far-hi: (a) RMS fit of the data to the GB model of distortion; (b) strike direction; (c) galvanic distortion parameters *twist* (squares) and *shear* (diamonds), plus the difference between the phases (given by $\phi_{xy} - (\phi_{yx} + 180)$) in the strike direction (triangles).

6.8.2 Actual data, lit007 and lit008

The real data from site lit008 (Figure 6.7, solid symbols) are well-behaved – the off-diagonal phases lie in their correct quadrants (ϕ_{xy} in first quadrant and ϕ_{yx} in third quadrant) and the diagonal apparent resistivities are both below the off-diagonal ones at most frequencies – whereas those of site lit007 (open symbols) are clearly strongly influenced by distortion. These two sites are located within 2 km of each other, and, at periods greater than that which yields a skin depth of 2 km, must "see" the same Earth. Given the average apparent resistivity levels at high frequencies of the two sites of 200–300 Ω m for ρ_{xy} and some 1000 Ω m for ρ_{yx}, a skin depth of 2 km is achieved for periods of 0.02–0.08 s (50–12 Hz).

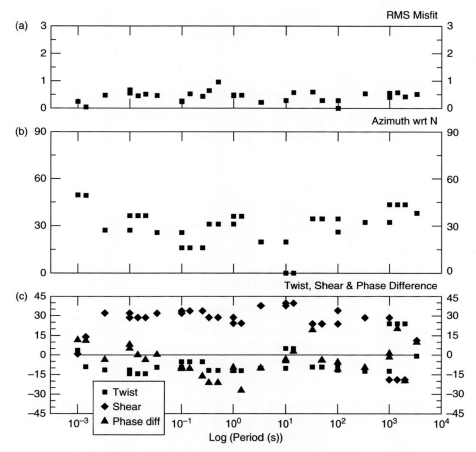

Figure 6.19. McNeice–Jones multifrequency analysis, with an averaging window of half a decade, of the data from site far-hi. Panels the same as for Figure 6.18.

6.8.2.1 Conventional analysis

Conventional analyses of the data from sites lit007 and lit008 are shown in Figure 6.20.

- For both sites, skew is >0.2 at periods >0.1 s.
- Strikes are unstable, except with a preference of around 35°–45° in the period band 1–200 s.
- Ellipticity in the strike direction is very large for lit007 in the period band 1–300 s.

The conclusions drawn would be that the data from both sites are indicative of a 3D Earth at periods >0.1 s, and that a 2D interpretation with a strike of around 40° would likely yield erroneous structure in the model.

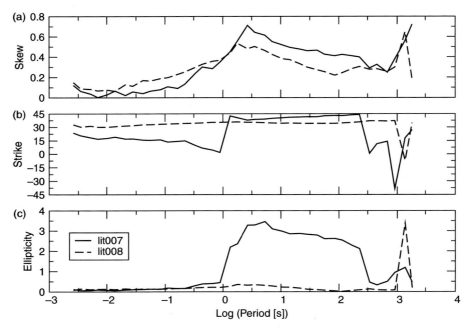

Figure 6.20. Conventional analysis of data from sites lit007 and lit008: (a) Swift skew; (b) Swift strike; and (c) ellipticity in Swift strike direction.

6.8.2.2 Bahr analyses

The Bahr analyses of the data from lit007 and lit008 are shown in Figure 6.21, and the strike directions θ_B in Figure 6.22. Following the classification scheme of Bahr (1991) and the flowchart of Simpson & Bahr (2005), almost all of the data are classified as 3D, as there is not a stable, frequency-independent strike.

6.8.2.3 WAL analyses

The WAL invariants (Weaver *et al.*, 2000) for the data from sites lit007 and lit008 are shown in Figure 6.23; the θ_{W2} (Eq. (64)) strike angles are not plotted in Figure 6.22 as they are exactly the same as θ_B. Almost all invariants I_3 to I_7 are greater than the threshold of 0.1 specified as non-zero by Weaver *et al.* (2000) and one would conclude that the data are all from a 3D Earth.

6.8.2.4 Phase-tensor analyses

Phase-tensor analyses of the lit007 and lit008 data are shown in Figures 6.24 (dimensionality plot) and 6.25, and the strike directions θ_P in Figure 6.22. There is some semblance of agreement in strike direction at periods of 10–100 s, and otherwise the phase-tensor parameters are distinctly different. The β values are large at almost all periods for both sites, indicative of inductive 3D structure. Note that, as the β values are non-zero, then θ_P is

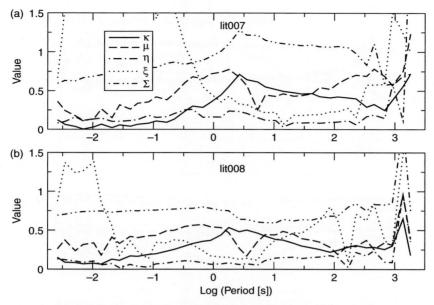

Figure 6.21. Bahr analysis of (a) lit007 and (b) lit008 data. Parameters the same as for Figure 6.13.

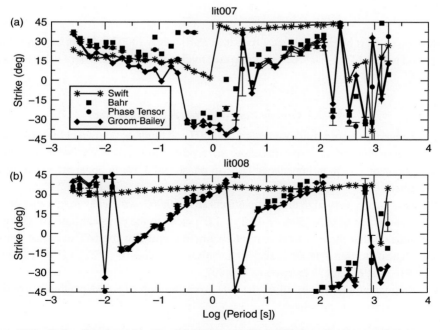

Figure 6.22. Strike directions for sites (a) lit007 and (b) lit008 data. Parameters the same as for Figure 6.14.

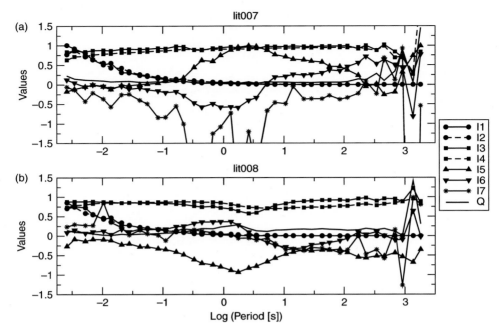

Figure 6.23. WAL invariants for sites (a) lit007 and (b) lit008 data. Parameters the same as for Figure 6.15.

close to θ_B (= θ_{W2}) but is not exactly the same. Particularly lit007 exhibits large β values in the period ranges of 0.1–10 s, compared to the smaller and more consistent values observed at lit008.

6.8.2.5 *Groom–Bailey/McNeice–Jones analyses*

Single-frequency Groom–Bailey analyses of lit007 and lit008 are shown in Figures 6.26 and 6.27, respectively, and the strike directions are repeated in Figure 6.22. For these analyses, an error floor of 1.75% in impedance (1.0° in phase and 3.5% in apparent resistivity) has been applied.

There are clearly very high distortion effects at site lit007, with an average value of *shear* in the period range of 0.2–350 s of 42.2° ± 1.0°, close to the physical limit of 45°, which would indicate that there is total current channeling and that the MT response tensor is singular. The distortion at lit008 is also high, with *shear* values of −37.1° ± 2.5° in the period range 0.01–30 s, but not as extreme as at lit007.

Multifrequency analysis of all of the data from the two sites independently yields the following distortion parameters:

lit007: *strike* = 58.3°, *twist* = −5.9°, *shear* = 42.3°, *av. rms* = 0.7
lit008: *strike* = 18.8°, *twist* = −30.2°, *shear* = 36.6°, *av. rms* = 1.0

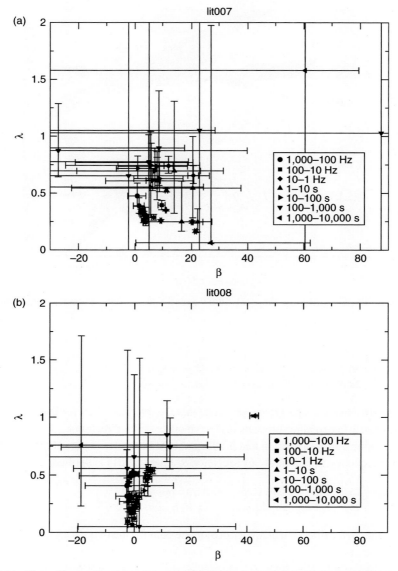

Figure 6.24. Phase-tensor dimensionality β–λ plots for sites (a) lit007 and (b) lit008. Parameters the same as for Figure 6.16.

These two are obviously not congruent – two sites some 2 km apart cannot yield regional strike directions that are 40° different. Given that they are separated by only 2 km and that a skin depth of 2 km for the average apparent resistivities of around 200–400 Ω m (Figure 6.7) is met at periods of 0.1 s and greater, the two must "see" the same regional structures at those lower frequencies. Undertaking the multifrequency analysis of all periods >0.1 s yields:

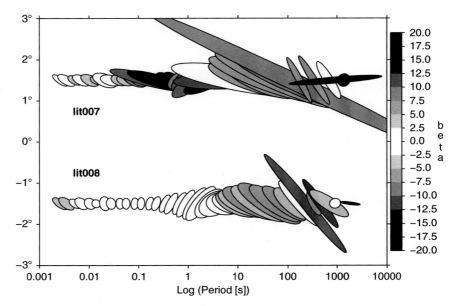

Figure 6.25. Phase tensors for sites (a) lit007 and (b) lit008 data. Parameters the same as for Figure 6.17.

lit007: *strike* = 56.7°, *twist* = 7.7°, *shear* = −42.7°, *av. rms* = 0.5
lit008: *strike* = 24.0°, *twist* = −33.7°, *shear* = 35.6°, *av. rms* = 1.0

which are also incongruent. However, given that the average RMS for lit007 is low at 0.5, there are clearly acceptable solutions that have other strike directions.

   To determine a strike direction that is acceptable to both sites at periods >0.1 s, the multi-site, multifrequency functionality of McNeice & Jones (2001) is applied, and yields

lit007: *strike* = 25.3°, *twist* = 37.0°, *shear* = 42.8°, *av. rms* = 0.6
lit008: *strike* = 25.3°, *twist* = −34.5°, *shear* = −35.3°, *av. rms* = 1.0.

Imposing this strike direction at all periods on the data for the two sites yields the following distortion parameters:

lit007: *strike* = 25.3°, *twist* = 36.6°, *shear* = 42.4°, *av. rms* = 1.0
lit008: *strike* = 25.3°, *twist* = −34.4°, *shear* = −35.1°, *av. rms* = 1.1

The MT apparent resistivity and phase curves are shown in Figure 6.28 and the RMS misfits on a frequency-by-frequency basis are contained in Figure 6.29. The distortion model fits the data from both sites acceptably except for short periods (<0.1 s) at site lit007. Thus, 3D inductive effects due to structures with lengths < 2 km are present at lit007 at short periods that are not present at lit008, but these structures become severe galvanic distorters at longer

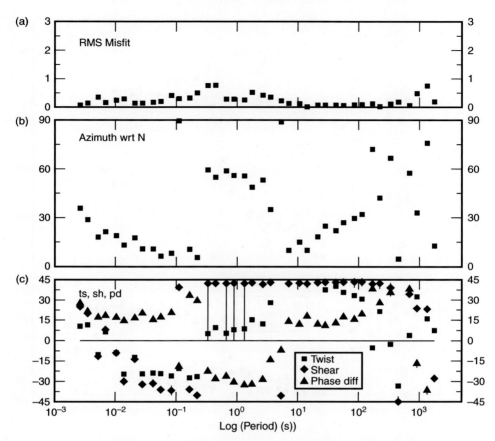

Figure 6.26. Groom–Bailey single-frequency analysis of site lit007. Parameters the same as for Figure 6.18.

periods, which makes determination of the regional responses difficult, and impossible without assistance from lit008.

6.9 Application to a one-dimensional anisotropic regional Earth

This chapter has been concerned with testing identification, appraisal and removal methods for dealing with MT response functions observed over predominantly 2D regional structures. A recent paper by Jones (2012) examines distortion of responses from an anisotropic 1D Earth.

Far less known and appreciated, even within the MT community, is a fourth form of the impedance tensor compared to the well-known 1D, 2D and 3D forms, and that is the one that the tensor assumes over a general 1D anisotropic Earth in which there are multiple layers with differing anisotropy directions. This form is given by

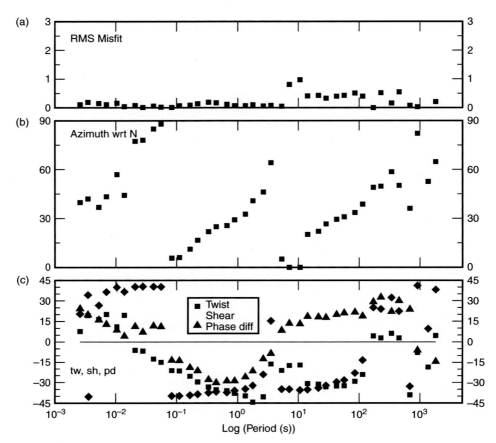

Figure 6.27. Groom–Bailey single-frequency analysis of site lit008. Parameters the same as for Figure 6.18.

$$\underline{\underline{\mathbf{Z}}}_{1Da} = \begin{bmatrix} Z_{xx} & Z_{xy} \\ Z_{yx} & -Z_{xx} \end{bmatrix} \tag{6.68}$$

of six unknowns per frequency, where the diagonal terms are equal but of opposite sign. Vozoff (1972), in his definitive review paper that helped established the MT method in the early 1970s, presents this observation (his eq. (9)) as found through computation, and Shoham & Loewenthal (1975) state that it is a consequence of the symmetry of the layer conductivity tensors. This form was shown to be valid in the general case in the theoretical development of Kováčiková & Pek (2002) given in their appendix A, and also in eqs. (30a) and (30d) of Pek & Santos (2002).

   In the specialized case where there exists only one layer within the stack of 1D layers that is anisotropic, or there are multiple anisotropic layers but they all have the same anisotropy

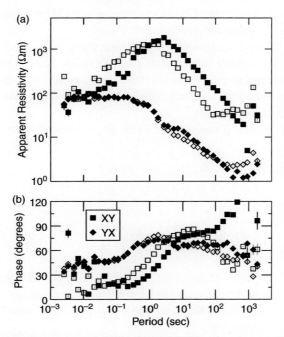

Figure 6.28. Regional 2D (a) apparent resistivity and (b) phase curves for sites lit007 (solid symbols) and lit008 (open symbols) after McNeice–Jones multi-site, multifrequency distortion decomposition: *XY*, squares; *YX*, diamonds.

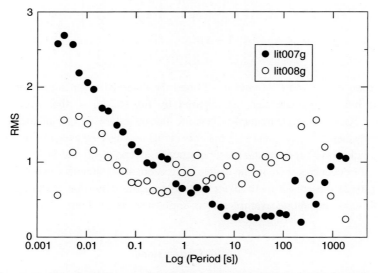

Figure 6.29. RMS misfits for the fit of the frequency-independent distortion models to the data from lit007 (solid circles) and lit008 (open circles).

directions, then the general form of Eq. (6.68) reduces to the 2D form (Eq. (6.46)) when axes are aligned parallel to one of the anisotropy directions.

In the general 2D or 3D anisotropic Earth case, there is no specific form of the MT impedance tensor; it adopts the full 3D form. For the restricted case of anisotropy in 2D with the anisotropy directions all parallel to structural strike, as used by Baba *et al.* (2006), then again when the observational coordinate system is aligned with structural/anisotropy strike, the MT tensor adopts the 2D form.

In the case of a regional 1D anisotropic Earth, the distortion can be described by

$$
\begin{aligned}
\underline{\underline{Z}}_{obs} &= \underline{\underline{C}}\,\underline{\underline{Z}}_{1Da} \\
&= g\underline{\underline{T}}\,\underline{\underline{S}}\,\underline{\underline{A}}\,\underline{\underline{Z}}_{1Da} \\
&= \underline{\underline{T}}\,\underline{\underline{S}}\,\underline{\underline{A}}\,\underline{\underline{Z}}'_{1Da} \\
&= \begin{bmatrix} 1 & -t \\ t & 1 \end{bmatrix} \begin{bmatrix} 1 & s \\ s & 1 \end{bmatrix} \begin{bmatrix} 1+a & a \\ 0 & 1-a \end{bmatrix} \underline{\underline{Z}}'_{1Da}
\end{aligned}
\tag{6.69}
$$

where $\underline{\underline{Z}}'_{1Da}$ is a scaled version of the true anisotropic 1D impedance given by $\underline{\underline{Z}}'_{1Da} = g\,\underline{\underline{Z}}_{1Da}$, as only the site gain can be absorbed into the scaled regional impedance as an indeterminable parameter. Any distortion anisotropy, a, will result in the mixing of the impedance tensor elements, viz.,

$$
\begin{aligned}
\underline{\underline{Z}}_{obs} &= \underline{\underline{C}}\,\underline{\underline{Z}}_{1Da} \\
&= \underline{\underline{A}}\,\underline{\underline{Z}}_{1Da} \\
&= \begin{bmatrix} 1+a & 0 \\ 0 & 1-a \end{bmatrix} \begin{bmatrix} Z_{xx} & Z_{xy} \\ Z_{yx} & -Z_{xx} \end{bmatrix} \\
&= \begin{bmatrix} (1+a)Z_{xx} & (1+a)Z_{xy} \\ (1-a)Z_{yx} & -(1-a)Z_{xx} \end{bmatrix}
\end{aligned}
\tag{6.70}
$$

(for distortion with s and $t = 0$ and $g = 1$) and the diagonal terms are no longer the same magnitude but of opposite sign, as required by Eq. (6.68), so that anisotropy can be recognized. Thus, at a single frequency there are nine unknowns (t, s, a and the three complex regional impedances, Z_{xx}, Z_{xy} and Z_{yx}), but only eight knowns. However, at n frequencies there are $3 + 6n$ unknowns (as t, s and a are all frequency-independent) and $8n$ knowns, thus even for a pair of frequencies there is one degree of freedom and solutions can be sought.

Jones (2012) presents a method for determining the distortion parameters in the 1D anisotropic case, and shows examples of its application and validity.

6.10 Conclusions

Conventional amplitude-based analysis methods to ascertain the dimensionality of the MT data, and, if deemed 2D, the strike direction, fail in the presence of distortion, and are unreliable even for relatively moderate noise levels that introduce in many cases unknown and unknowable bias and random errors (Chave, 2011). These methods should simply never be relied upon, even for the most precise data.

All modern distortion methods that are phase-based are able to recover the regional impedances, except for indeterminable gain factors on the apparent resistivities, when both noise and distortion are low. As either or both of these increase, some methods become more effective than others. Most approaches are virtually equivalent, as they are algebraically related. Only the Groom–Bailey type approaches differ, as they fit a model of distortion of seven parameters to the eight data. The optimum approaches are ones that treat the problem as one of statistical model-fitting, rather than algebraically manipulating the eight values in the impedance tensor. In addition, multi-site and multifrequency approaches will yield greater stability of the induction parameters, as they are often overwhelmed by the galvanic distortion parameters.

One important point, raised by Chave's (2011) examination of the distribution of Swift skew, is that the statistical properties of estimators from phase-based methods should be as exhaustively determined as that for skew.

Acknowledgements

This chapter would not have been written without the constant encouragement and badgering of my friend and colleague, Alan Chave. I tried his patience, and that of Susan Francis and Laura Clark of Cambridge University Press, for far too long through an interminable series of never-ending personal and professional delays and distractions. I truly earned and deserved the wooden spoon in this race.

I would like to thank Dick Bailey for looking at it and commenting on the importance of the uniqueness property of the Groom–Bailey decomposition approach. John Weaver provided a mild review of an earlier version of the chapter, and Alan Chave a not-so-mild review. Both helped me to focus my, at times, rambling thoughts. In particular, Alan Chave's demand for precision and rigor was imperative in a subject that has been beset and hampered by imprecision and lack of rigor.

Many, many discussions with generous colleagues over the past 20 years is embedded in the thinking presented herein. I cannot do justice to everyone, but in chronological order Ross Groom, Alan Chave, Gary McNeice, Xavier Garcia, Juanjo Ledo and Pilar Queralt all deserve special mention.

Last, but by no means least, I wish to thank all of my students and post-docs, both former and present. Their enthusiasm and enjoyment of magnetotellurics has and is keeping me enthralled in the subject.

References

Agarwal, A. K. & J. T. Weaver (2000). Magnetic distortion of the magnetotelluric tensor: a numerical study. *Earth Planets Space*, **52**, 347–353.

Agarwal, A. K., H. E. Poll, *et al.* (1993). One-dimensional and two-dimensional inversion of magnetotelluric data in continental regions. *Phys. Earth Planet. Inter.*, **81**(1–4), 155–176.

Alabi, A. O., P. A. Camfield, *et al.* (1975). North-American Central Plains conductivity anomaly. *Geophys. J. R. Astron. Soc.*, **43**(3), 815–833.

Avdeeva, A., D. B. Avdeev, *et al.* (2011). Inverting Dublin secret dataset 2 with x3Di. In *MT3DINV2: Second Three-Dimensional Magnetotelluric Inversion Workshop*, Dublin, Ireland.

Baba, K. & A. D. Chave (2005). Correction of seafloor magnetotelluric data for topographic effects during inversion. *J. Geophys. Res. – Solid Earth*, **110**(B12), 16.

Baba, K., A. D. Chave, *et al.* (2006). Mantle dynamics beneath the East Pacific Rise at 17 degrees S: insights from the Mantle Electromagnetic and Tomography (MELT) experiment. *J. Geophys. Res. – Solid Earth*, **111**(B2), 18.

Bahr, K. (1984). Elimination of local 3D distortion of the magnetotelluric tensor impedance allowing for two different phases. In *Seventh Workshop on Electromagnetic Induction in the Earth and Moon*, Ile-Ife, Nigeria.

Bahr, K. (1988). Interpretation of the magnetotelluric impedance tensor – regional induction and local telluric distortion. *J. Geophys. (Z. Geophys.)*, **62**(2), 119–127.

Bahr, K. (1991). Geological noise in magnetotelluric data: a classification of distortion types. *Phys. Earth Planet. Inter.*, **66**, 24–38.

Bailey, R. C. & R. W. Groom (1987). Decomposition of the magnetotelluric impedance tensor which is useful in the presence of channeling. In *57th Annual International Society of Exploration Geophysicists Meeting and Exposition*, Tulsa, OK.

Becken, M., O. Ritter, *et al.* (2008). Mode separation of magnetotelluric responses in three-dimensional environments. *Geophys. J. Int.*, **172**(1), 67–86.

Berdichevsky, M. N. & I. V. Dmitriev (1976a). Distortion of magnetic and electric fields by near-surface lateral inhomogeneities. *Acta Geodaet. Geophys. Montanist. Acad. Sci. Hung.*, **11**(3–4), 447–483.

Berdichevsky, M. N. & I. V. Dmitriev (1976b). Basic principles of interpretation of magnetotelluric sounding curves. In *Geoelectric and Geothermal Studies*, ed. A. Adam. KAPG Geophysical Monograph. Budapest: Akademiai Kiado, pp. 165–221.

Berdichevsky, M. N. & I. V. Dmitriev (2008). *Models and Methods of Magnetotellurics*. Berlin: Springer.

Berdichevsky, M. N., L. L. Vanyan, *et al.* (1989). Methods used in the USSR to reduce near-surface inhomogeneity effects on deep magnetotelluric sounding. *Phys. Earth Planet. Inter.* **53**(3–4), 194–206.

Bibby, H. (1986). Analysis of multi-source bipole–dipole resistivity surveys using the apparent resistivity tensor. *Geophysics*, (51), 972–983.

Bibby, H. M., T. G. Caldwell, *et al.* (2005). Determinable and non-determinable parameters of galvanic distortion in magnetotellurics. *Geophys. J. Int.*, **163**(3), 915–930.

Bostick, F. X. & H. W. Smith (1962). Investigation of large scale inhomogeneities in the Earth by the magnetotelluric method. *Proc. Inst. Radio Eng.*, **50**, 2339–2346.

Caldwell, T. G., H. M. Bibby, *et al.* (2004). The magnetotelluric phase tensor. *Geophys. J. Int.*, **158**(2), 457–469.

Caldwell, T. G., H. M. Bibby, *et al.* (2007). Comment on "The magnetotelluric phase tensor" by T. Grant Caldwell, Hugh M. Bibby and Colin Brown – Reply. *Geophys. J. Int.*, **171** (2), 567–567.

Cerv, V., J. Pek, *et al.* (2010). Bayesian approach to magnetotelluric tensor decomposition. *Ann. Geophys.*, **53**(2), 21–32.

Chakridi, R., M. Chouteau, *et al.* (1992). A simple technique for analyzing and partly removing galvanic distortion from the magnetotelluric impedance tensor – application to Abitibi and Kapuskasing data (Canada). *Geophys. J. Int.*, **108**(3), 917–929.

Chave, A. D. (2011). On the statistics of the magnetotelluric skew. *Geophys. J. Int.*, submitted.

Chave, A. D. & A. G. Jones (1997). Electric and magnetic field galvanic distortion decomposition of BC87 data. *J. Geomagn. Geoelectr.*, **49**(6), 767–789.

Chave, A. D. & J. T. Smith (1994). On electric and magnetic galvanic distortion tensor decompositions. *J. Geophys. Res. – Solid Earth*, **99**(B3), 4669–4682.

Clowes, R. M. (2010). Initiation, development, and benefits of Lithoprobe – shaping the direction of Earth science research in Canada and beyond. *Can. J. Earth Sci.*, **47**(4), 291–314.

Cochrane, N. A. & R. D. Hyndman (1974). Magnetotelluric and magnetovariational studies in Atlantic Canada. *Geophys. J. R. Astron. Soc.*, **39**(2), 385–406.

de Groot-Hedlin, C. (1995). Inversion of regional 2-D resistivity structure in the presence of galvanic scatterers. *Geophys. J. Int.*, **122**(3), 877–888.

de Groot-Hedlin, C. & S. Constable (1990). Occam inversion to generate smooth, 2-dimensional models from magnetotelluric data. *Geophysics*, **55**(12), 1613–1624.

Edwards, R. N., H. Lee, *et al.* (1978). Theory of magnetometric resistivity (MMR) methods. *Geophysics*, **43**(6), 1176–1203.

Eggers, D. E. (1982). An eigenstate formulation of the magnetotelluric impedance tensor. *Geophysics*, **47**(8), 1204–1214.

Eisel, M. & K. Bahr (1993). Electrical anisotropy in the lower crust of British Columbia – an interpretation of a magnetotelluric profile after tensor decomposition. *J. Geomagn. Geoelectr.*, **45**(9), 1115–1126.

Evans, R. L., A. G. Jones, *et al.* (2011) The electrical lithosphere beneath the Kaapvaal Craton, Southern Africa. *J. Geophys. Res. – Solid Earth*, **16**, B04105, doi:10.1029/2010JB007883.

Everett, J. E. & R. D. Hyndman (1967). Magneto-telluric investigations in south-western Australia. *Phys. Earth Planet. Inter.*, **1**(1), 49–54.

Garcia, X. & A. G. Jones (1999). Extended decomposition of MT data. In *Second Int. Symp. on Three-Dimensional Electromagnetics*, Salt Lake City, Utah.

Garcia, X. & A. G. Jones (2002). Extended decomposition of MT data. In *Three-Dimensional Electromagnetics*, ed. M. S. Zhdanov & P. E. Wannamaker. Amsterdam: Elsevier, pp. 235–250.

Gatzemeier, A. & M. Moorkamp (2005). 3D modelling of electrical anisotropy from electromagnetic array data: hypothesis testing for different upper mantle conduction mechanisms. *Phys. Earth Planet. Inter.*, **149**(3–4), 225–242.

Groom, R. W. (1988). The effects of inhomogeneities on magnetotellurics. Ph.D. thesis, Department of Physics, University of Toronto.

Groom, R. W. & K. Bahr (1992). Corrections for near surface effects: decomposition of the magnetotelluric impedance tensor and scaling corrections for regional resistivities: a tutorial. *Surv. Geophys.*, **13**(4–5), 341–379.

Groom, R. W. & R. C. Bailey (1989). Decomposition of magnetotelluric impedance tensors in the presence of local three dimensional galvanic distortion. *J. Geophys. Res.*, **94**, 1913–1925.

Groom, R. W. & R. C. Bailey (1991). Analytical investigations of the effects of near surface three dimensional galvanic scatterers on MT tensor decomposition. *Geophysics*, **56**(4), 496–518.

Groom, R. W., R. D. Kurtz, *et al.* (1993). A quantitative methodology for determining the dimensionality of conductive structure from magnetotelluric data. *Geophys. J. Int.*, **115**, 1095–1118.

Haak, V. (1972). Magnetotelluric method: the determination of transfer functions in areas with lateral variation of electrical conductivity. *Z. Geophys.*, **38**, 85–102.

Habashy, T. M., R. W. Groom, *et al.* (1993). Beyond the Born and Rytov approximations – a nonlinear approach to electromagnetic scattering. *J. Geophys. Res. – Solid Earth*, **98**(B2), 1759–1775.

Hamilton, M. P., A. G. Jones, *et al.* (2006). Electrical anisotropy of South African lithosphere compared with seismic anisotropy from shear-wave splitting analyses. *Phys. Earth Planet. Inter.*, **158**(2–4), 226–239.

Hermance, J. F. (1982). The asymptotic response of 3-dimensional basin offsets to magnetotelluric fields at long periods – the effects of current channeling. *Geophysics*, **47**(11), 1562–1573.

Honkura, Y. (1975). Perturbation of the electric current by a resistivity anomaly and its application to earthquake prediction. *J. Geomagn. Geoelectr.*, **28**, 47–57.

Jiracek, G. R. (1990). Near-surface and topographic distortions in electromagnetic induction. *Surv. Geophys.*, **11**, 163–203.

Jones, A. G. (1983). The problem of current channelling: a critical review. *Surv. Geophys.*, **6**, 79–122.

Jones, A. G. (1988a). Discussion of "A magnetotelluric investigation under the Williston Basin" by J. M. Maidens and K. V. Paulson. *Can. J. Earth Sci.*, **25**, 1132–1139.

Jones, A. G. (1988b). Static shift of magnetotelluric data and its removal in a sedimentary basin environment. *Geophysics*, **53**(7), 967–978.

Jones, A. G. (1993a). The COPROD2 dataset – tectonic setting, recorded MT data, and comparison of models. *J. Geomagn. Geoelectr.*, **45**(9), 933–955.

Jones, A. G. (1993b). Electromagnetic images of modern and ancient subduction zones. *Tectonophysics*, **219**(1–3), 29–45.

Jones, A. G. (1993c). The BC87 dataset – tectonic setting, previous EM results, and recorded MT data. *J. Geomagn. Geoelectr.*, **45**(9), 1089–1105.

Jones, A. G. (1993d). Introduction to MT-DIW1 special section. *J. Geomagn. Geoelectr.*, **45**, 931–932.

Jones, A. G. (2006). Electromagnetic interrogation of the anisotropic Earth: looking into the Earth with polarized spectacles. *Phys. Earth Planet. Inter.*, **158**(2–4), 281–291.

Jones, A. G. (2011) Three-dimensional galvanic distortion of three-dimensional regional conductivity structures: Comments on "Three-dimensional joint inversion for magnetotelluric resistivity and static shift distributions in complex media" by Y. Sasaki and M. A. Meju (2006). *J. Geophys. Res. – Solid Earth*, **116**, B12104, doi:10.1029/2011JB008665.

Jones, A. G. (2012). Distortion decomposition of the magnetotelluric impedance tensors from a one-dimensional anisotropic Earth. *Geophys. J. Int.*, in press, doi:10.1111/j.1365-246X.2012.05362.

Jones, A. G. & J. A. Craven (1990). The North American Central Plains conductivity anomaly and its correlation with gravity, magnetics, seismic, and heat flow data in the Province of Saskatchewan. *Phys. Earth Planet. Inter.*, **60**, 169–194.

Jones, A. G. & I. Dumas (1993). Electromagnetic images of a volcanic zone. *Phys. Earth Planet. Inter.*, **81**, 289–314.

Jones, A. G. & X. Garcia (2003). Okak Bay AMT data-set case study: lessons in dimensionality and scale. *Geophysics*, **68**(1), 70–91.

Jones, A. G. & G. D. Garland (1986). Preliminary interpretation of the upper crustal structure beneath Prince Edward Island. *Annales Geophysica*, **4B**, 157–164.

Jones, A. G. & R. W. Groom (1993). Strike angle determination from the magnetotelluric impedance tensor in the presence of noise and local distortion – rotate at your peril. *Geophys. J. Int.*, **113**(2), 524–534.

Jones, A. G. & R. Hutton (1979). A multi-station magnetotelluric study in southern Scotland – I. Fieldwork, data analysis and results. *Geophys. J. R. Astron. Soc.*, **56**(2), 329–349.

Jones, A. G. & G. McNeice (2002). Audio-magnetotellurics (AMT) for steeply-dipping mineral targets: importance of multi-component measurements at each site. In *Proc. 72nd SEG Meeting*, Salt Lake City, Utah, pp. 496–499.

Jones, A. G. & A. D. Schultz (1997). Introduction to MT-DIW2 Special Issue. *J. Geomagn. Geoelectr.*, **49**, 727–737.

Jones, A. G., D. I. Gough, *et al.* (1992). Electromagnetic images of regional structure in the southern Canadian cordillera. *Geophys. Res. Lett.*, **19**(24), 2373–2376.

Jones, A. G., R. W. Groom, *et al.* (1993). Decomposition and modelling of the BC87 dataset. *J. Geomagn. Geoelectr.*, **45**(9), 1127–1150.

Jupp, D. L. B. & K. Vozoff (1976). The magnetotelluric method in the exploration of sedimentary basins – discussion. *Geophysics*, **41**(2), 325–328.

Kováčiková, S. & J. Pek (2002). Generalized Riccati equations for 1-D magnetotelluric impedances over anisotropic conductors, Part I: Plane wave field model. *Earth Planets Space*, **54**(5), 473–482.

Kurtz, R. D. & G. D. Garland (1976). Magnetotelluric measurements in eastern Canada. *Geophys. J. R. Astron. Soc.*, **45**(2), 321–347.

Larsen, J. C. (1975). Low-frequency (0.1–6.0 cpd) electromagnetic study of deep mantle electrical-conductivity beneath Hawaiian islands. *Geophys. J. R. Astron. Soc.*, **43**(1), 17–46.

Larsen, J. C. (1977). Removal of local surface conductivity effects from low frequency mantle response curves. *Acta Geodaet. Geophys. Montanist. Acad. Sci. Hung.*, **12**, 183–186.

LaTorraca, G. A., T. R. Madden, *et al.* (1986). An analysis of the magnetotelluric impedance for three-dimensional conductivity structures. *Geophysics*, **51**(2), 466–466.

Ledo, J. (2005). 2-D versus 3-D magnetotelluric data interpretation. *Surv. Geophys.*, **26**(5), 511–543.

Ledo, J., P. Queralt, *et al.* (1998). Effects of galvanic distortion on magnetotelluric data over a three-dimensional regional structure. *Geophys. J. Int.*, **132**(2), 295–301.

Ledo, J., P. Queralt, *et al.* (2002). Two-dimensional interpretation of three-dimensional magnetotelluric data: an example of limitations and resolution. *Geophys. J. Int.*, **150**(1), 127–139.

Leibecker, J., A. Gatzemeier, *et al.* (2002). Evidence of electrical anisotropic structures in the lower crust and the upper mantle beneath the Rhenish Shield. *Earth Planet. Sci. Lett.*, **202**(2), 289–302.

Lezaeta, P. (2002). Confidence limit of the magnetotelluric phase sensitive skew. *Earth Planets Space*, **54**(5), 451–457.

Lezaeta, P. & V. Haak (2003). Beyond magnetotelluric decomposition: induction, current channeling, and magnetotelluric phases over 90 degrees. *J. Geophys. Res. – Solid Earth*, **108**(B6), 20.

Lilley, F. E. M. (1976). Diagrams for magnetotelluric data. *Geophysics*, **41**(4), 766–770.

Lilley, F. E. M. (1993a). Magnetotelluric analysis using Mohr circles. *Geophysics*, **58**(10), 1498–1506.

Lilley, F. E. M. (1993b). Three-dimensionality of the BC87 magnetotelluric data set studied using Mohr circles. *J. Geomagn. Geoelectr.*, **45**(9), 1107–1113.

Lilley, F. E. M. (1998a). Magnetotelluric tensor decomposition: Part I, Theory for a basic procedure. *Geophysics*, **63**(6), 1885–1897.

Lilley, F. E. M. (1998b). Magnetotelluric tensor decomposition: Part II, Examples of a basic procedure. *Geophysics*, **63**(6), 1898–1907.

Mardia, K. V. (1972). *Statistics of Directional Data*. New York: Academic Press.

Marti, A., P. Queralt, *et al.* (2005). Improving Bahr's invariant parameters using the WAL approach. *Geophys. J. Int.*, **163**(1), 38–41.

Marti, A., P. Queralt, *et al.* (2009). WALDIM: a code for the dimensionality analysis of magnetotelluric data using the rotational invariants of the magnetotelluric tensor. *Comput. Geosci.*, **35**(12), 2295–2303.

Matsuno, T., N. Seama, *et al.* (2010). Upper mantle electrical resistivity structure beneath the central Mariana subduction system. *Geochem. Geophys. Geosyst.*, **11**, Q09003, doi:10.1029/2010GC003101.

McNeice, G. & A. G. Jones (1996). Multisite, multifrequency tensor decomposition of magnetotelluric data. In *66th Society of Exploration Geophysicists Annual General Meeting*, Denver, Colorado.

McNeice, G. W. & A. G. Jones (2001). Multisite, multifrequency tensor decomposition of magnetotelluric data. *Geophysics*, **66**(1), 158–173.

Miensopust, M. P. & A. G. Jones (2011). Artefacts of isotropic inversion applied to magnetotelluric data from an anisotropic Earth. *Geophys. J. Int.*, **187**, 277–689, doi:10.1111/j.1365-246X.2011.05157.x.

Miensopust, M. P., A. G. Jones, *et al.* (2011). Lithospheric structures and Precambrian terrane boundaries in northeastern Botswana revealed through magnetotelluric profiling as part of the Southern African Magnetotelluric Experiment. *J. Geophys. Res. – Solid Earth*, **116**, B02401.

Moorkamp, M. (2007). Comment on "The magnetotelluric phase tensor" by T. Grant Caldwell, Hugh M. Bibby and Colin Brown. *Geophys. J. Int.*, **171**(2), 565–566.

Morse, P. M. & H. Feshbach (1953). *Methods of Theoretical Physics*. New York: McGraw-Hill.

Park, S. K. & R. J. Mackie (1997). Crustal structure at Nanga Parbat, northern Pakistan, from magnetotelluric soundings. *Geophys. Res. Lett.*, **24**(19), 2415–2418.

Parker, R. L. (2011) New analytic solutions for the 2-D TE mode MT problem. *Geophys. J. Int.*, **186**, 980–986, doi:10.1111/j.1365-246X.2011.05091.x.

Patro, P. K. & G. D. Egbert (2011). Application of 3-D inversion to magnetotelluric profile data from the Deccan volcanic province of western India. *Phys. Earth Planet. Inter.*, **187**, 33–46.

Pedersen, L. B. & M. Svennekjaer (1984). Extremal bias coupling in magnetotellurics. *Geophysics*, **49**(11), 1968–1978.

Pek, J. & F. A. M. Santos (2002). Magnetotelluric impedances and parametric sensitivities for 1-D anisotropic layered media. *Comput. Geosci.*, **28**(8), 939–950.

Poll, H. E., J. T. Weaver, *et al.* (1989). Calculations of voltages for magnetotelluric modelling of a region with near-surface inhomogeneities. *Phys. Earth Planet. Inter.*, **53**, 287–297.

Pracser, E. & L. Szarka (1999). A correction to Bahr's phase deviation method for tensor decomposition. *Earth Planets Space*, **51**, 1019–1022.

Queralt, P., A. G. Jones, *et al.* (2007). Electromagnetic imaging of a complex ore body: 3D forward modeling, sensitivity tests, and down-mine measurements. *Geophysics*, **72**(2), F85–F95.

Ranganayaki, R. P. (1984). An interpretive analysis of magnetotelluric data. *Geophysics*, **49** (10), 1730–1748.

Reddy, I. K. & D. Rankin (1971). Magnetotelluric measurements in central Alberta. *Geophysics*, **36**(4), 739.

Reddy, I. K. & D. Rankin (1974). Coherence functions for magnetotelluric analysis. *Geophysics*, **39**(3), 312–320.

Richards, M. L., U. Schmucker, *et al.* (1982). Entzerrung der Impedanzkurven von magnetotellurischen Messungen in der Schwabischen Alb. In *Kolloquium für Elektromagnetische Tiefenforschung*.

Roux, E., M. Moorkamp, *et al.* (2011). Joint inversion of long-period magnetotelluric data and surface-wave dispersion curves for anisotropic structure: application to data from Central Germany. *Geophys. Res. Lett.*, **38**, L05304.

Sasaki, Y. (2004). Three-dimensional inversion of static-shifted magnetotelluric data. *Earth Planets Space*, **56**(2), 239–248.

Sasaki, Y. & M. A. Meju (2006). Three-dimensional joint inversion for magnetotelluric resistivity and static shift distributions in complex media. *J. Geophys. Res. – Solid Earth*, **111**(B5), 11.

Shoham, Y. & D. Loewenthal (1975). Matrix polynomial representation of anisotropic magnetotelluric impedance tensor. *Phys. Earth Planet. Inter.*, **11**(2), 128–138.

Simpson, F. (2001). Resistance to mantle flow inferred from the electromagnetic strike of the Australian upper mantle. *Nature*, **412**(6847), 632–635.

Simpson, F. & K. Bahr (2005). *Practical Magnetotellurics*. Cambridge: Cambridge University Press.

Sims, W. E. & F. X. Bostick (1969). Methods of magnetotelluric analysis. Tech. Rep. 58, Electr. Geophys. Res. Lab., University of Texas at Austin.

Siripunvaraporn, W., G. Egbert, *et al.* (2005). Three-dimensional magnetotelluric inversion: data-space method. *Phys. Earth Planet. Inter.*, **150**(1–3), 3–14.

Smith, J. T. (1995). Understanding telluric distortion matrices. *Geophys. J. Int.*, **122**(1), 219–226.

Smith, J. T. (1997). Estimating galvanic-distortion magnetic fields in magnetotellurics. *Geophys. J. Int.*, **130**(1), 65–72.

Spitz, S. (1985). The magnetotelluric impedance tensor properties with respect to rotations. *Geophysics*, **50**(10), 1610–1617.

Sternberg, B. K., J. C. Washburne, *et al.* (1988). Correction for the static shift in magnetotellurics using transient electromagnetic soundings. *Geophysics*, **53**(11), 1459–1468.

Swift, C. M. (1967). *A magnetotelluric investigation of an electrical conductivity anomaly in the southwestern United States*. Geology and geophysics. Ph.D. thesis, Massachusetts Institute of Technology, Cambridge, MA.

Swift, C. M. (1986). A magnetotelluric investigation of an electrical conductivity anomaly in the southwestern United States. In *Magnetotelluric Methods*, ed. K. Vozoff. Tulsa: Society of Exploration Geophysicists, pp. 156–166.

Szarka, L. & M. Menvielle (1997). Analysis of rotational invariants of the magnetotelluric impedance tensor. *Geophys. J. Int.*, **129**(1), 133–142.

Utada, H. & H. Munekane (2000). On galvanic distortion of regional three-dimensional magnetotelluric impedances. *Geophys. J. Int.*, **140**(2), 385–398.

Vanyan, L. L., M. N. Berdichewski, *et al.* (1977). The study of asthenosphere of East European platform by electromagnetic sounding. *Phys. Earth Planet. Inter.*, **14**(2), P1–P2.

Vanyan, L. L., A. P. Shilovsky, *et al.* (1989). Electrical conductivity of the crust of the Siberian platform. *Phys. Earth Planet. Inter.*, **54**(1–2), 163–168.

Vozoff, K. (1972). The magnetotelluric method in the exploration of sedimentary basins. *Geophysics*, **37**(1), 98–141.

Wannamaker, P. E., G. W. Hohmann, *et al.* (1984). Magnetotelluric responses of three-dimensional bodies in layered earths. *Geophysics*, **49**(9), 1517–1533.

Ward, S. H. (1967). Electromagnetic theory for geophysical applications. In *Mining Geophysics*. Tulsa: Society of Exploration Geophysics, pp. 10–196.

Weaver, J. T. & F. E. M. Lilley (2004). Using Mohr circles to identify regional dimensionality and strike angle from distorted magnetotelluric data. *Explor. Geophys.*, **35**, 251–254.

Weaver, J. T., B. V. Le Quang, *et al.* (1985). A comparison of analytic and numerical results for a two-dimensional control model in electromagnetic induction – I. B-polarization calculations. *Geophys. J. R. Astron. Soc.*, **82**, 263–277.

Weaver, J. T., B. V. Le Quang, *et al.* (1986). A comparison of analytic and numerical results for a two-dimensional control model in electromagnetic induction – II. E-polarization calculations. *Geophys. J. R. Astron. Soc.*, **87**, 917–948.

Weaver, J. T., A. K. Agarwal, *et al.* (2000). Characterization of the magnetotelluric tensor in terms of its invariants. *Geophys. J. Int.*, **141**(2), 321–336.

Weaver, J. T., A. K. Agarwal, *et al.* (2006). The relationship between the magnetotelluric tensor invariants and the phase tensor of Caldwell, Bibby, and Brown. *Explor. Geophys.*, **37**, 261–267.

Weckmann, U., O. Ritter, *et al.* (2003). Images of the magnetotelluric apparent resistivity tensor. *Geophys. J. Int.*, **155**(2), 456–468.

Word, D. R., H. W. Smith, *et al.* (1970). *An investigation of the magnetotelluric tensor impedance method*. Tech. Rep. 82, Electr. Geophys. Res. Lab., University of Texas at Austin.

Xu, Y. S. B. T. Poe, *et al.* (1998). Electrical conductivity of olivine, wadsleyite, and ringwoodite under upper-mantle conditions. *Science*, **280**(5368), 1415–1418.

Yee, E. & K. V. Paulson (1987). The canonical decomposition and its relationship to other forms of magnetotelluric impedance tensor analysis. *J. Geophys. (Z. Geophys).*, **61**(3), 173–189.

Zhang, P., R. G. Roberts, *et al.* (1987). Magnetotelluric strike rules. *Geophysics*, **52**(3), 267–278.

Zhdanov, M. S., R. B. Smith, *et al.* (2011). Three-dimensional inversion of large-scale EarthScope magnetotelluric data based on the integral equation method: geoelectrical imaging of the Yellowstone conductive mantle plume. *Geophys. Res. Lett.*, **38**, L08307, doi:10.1029/2011GL046953.

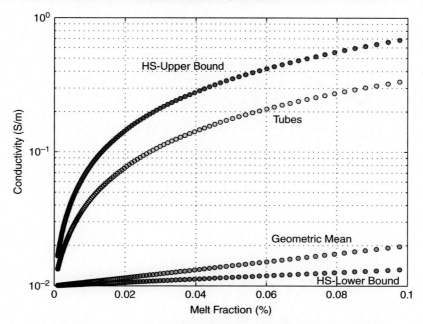

Figure 3A.5. Conductivity as a function of melt fraction for the various relationships discussed in the text (upper and lower Hashin–Shtrikmann (HS) bounds; melt in tubes; and the geometric mean of conductivities). The background conductivity is set at 0.01 S/m, essentially that of oceanic mantle on an adiabat. The melt is assigned a conductivity of 10 S/m, consistent with a hydrous basalt at high temperature (>1200°C; Pommier et al., 2010a,b). For both the HS upper bound and melt in tubes, both of which assume an interconnected melt distribution, the bulk conductivity is greatly enhanced even at low melt fractions. Typical melt fractions envisioned in the mantle beneath active spreading centers is of the order of 1–2%, but with such melt fractions an order-of-magnitude increase in bulk conductivity is possible.

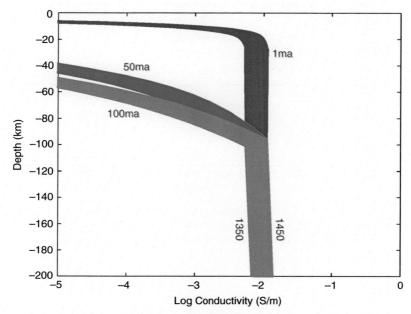

Figure 3A.8. Conductivity–depth profiles from the GDH1 thermal model calculated as in Figure 3A.7, but shown as depth profiles for three different plate ages. In the oceans below ~100 km depth, the conductivity is essentially independent of age if only thermal factors influence the structure. Variations in adiabatic temperature from 1350 to 1450°C bracket the curves as labeled. The uppermost lithosphere is predicted to be highly resistive.

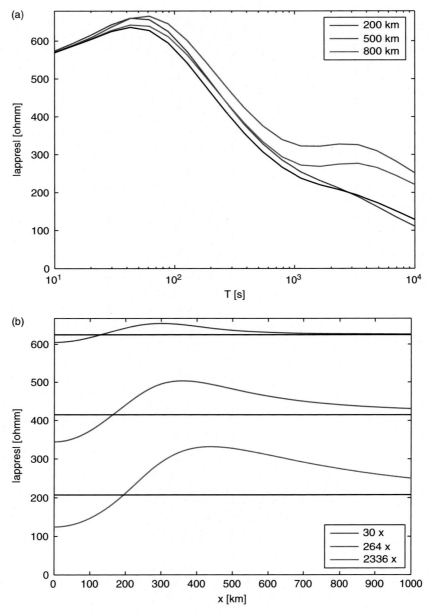

Figure 3B.12. (a) Apparent resistivity in the case of a time-harmonic Gaussian electrojet at a height of 110 km, with halfwidth 100 km, and centered at $x = 0$. (b) Apparent resistivity as a function of period at three fixed locations. The black lines show the value due to a plane-wave source.

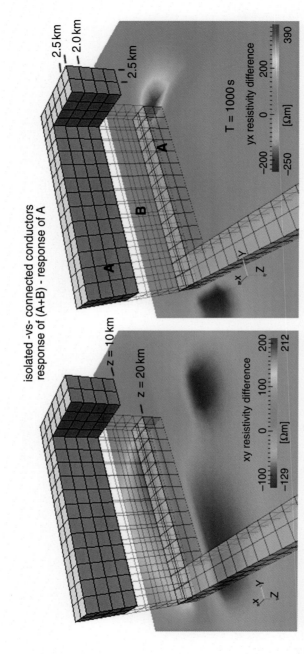

Figure 7.23. Effect on apparent resistivity of a mid-crustal electrical "short" (B) between and upper and lower crustal conductors (A). Electrical conductors A and B are all 1 Ω m dike-like structures, embedded in a 1D ground model (see text for details). Shown by the grayscale (color scale) is the difference in apparent resistivity due to the presence of block B at a frequency 0.001 Hz (T = 1000 s).

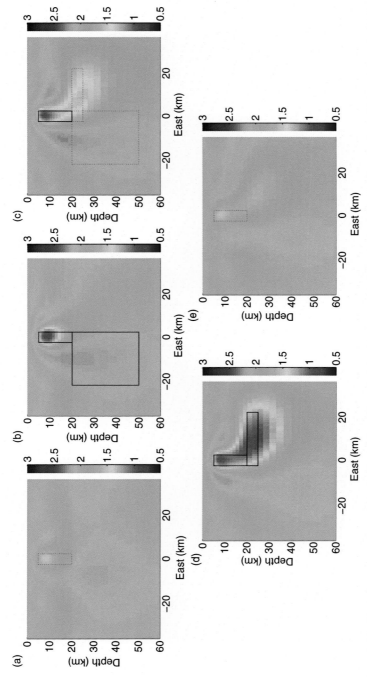

Figure 8.12. Vertical cross-sections through the 3D inversion model, with the grayscale (color scale) indicating the base 10 logarithm of resistivity: (a) $x = -20$ km; (b) $x = -10$ km; (c) $x = 0$ km; (d) $x = +10$ km; and (e) $x = +20$ km. The outlines of the bodies in the true resistivity model (Figure 8.11) are shown as solid lines, except when the cross-section is located at a discontinuity in the model, in which case the adjacent body is plotted as a dotted line.

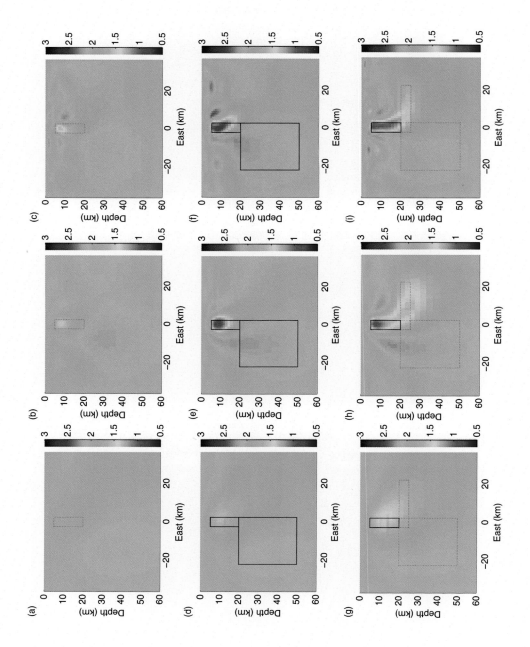

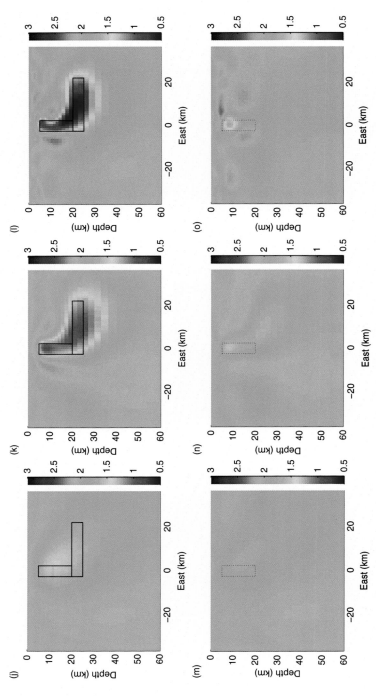

Figure 8.15. A comparison of the inversion models for the three different values of the regularization parameter (λ) shown on the trade-off curve of Figure 8.14. The first row (a)–(c) is for the vertical cross-section located at $x = -20$ km, the second row (d)–(f) is for $x = -10$ km, the third row (g)–(i) is for $x = 0$ km, the fourth row (j)–(l) is for $x = +10$ km and the final row (m)–(o) is for $x = +20$ km. The left-hand column (a)–(m) is for the model with a high value of λ that resulted in a normalized misfit of 1.55, the middle column (b)–(n) is for the optimal value of λ that resulted in a normalized misfit of 1.00, and the right-hand column (c)–(o) is for a smaller λ that resulted in a misfit of 0.94.

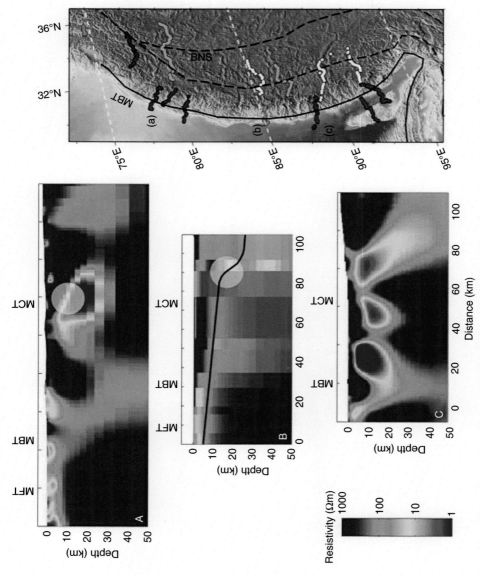

Figure 10.2. Magnetotelluric profiles crossing the central Himalaya. From Unsworth (2010). The location of each of the profiles on the left is shown in the panel on the right: (A) results of Israil *et al.* (2008), (B) results of Lemmonier *et al.* (1999) and (C) results of Patro & Harinarayana (2009). The white circle indicates a zone of seismicity.

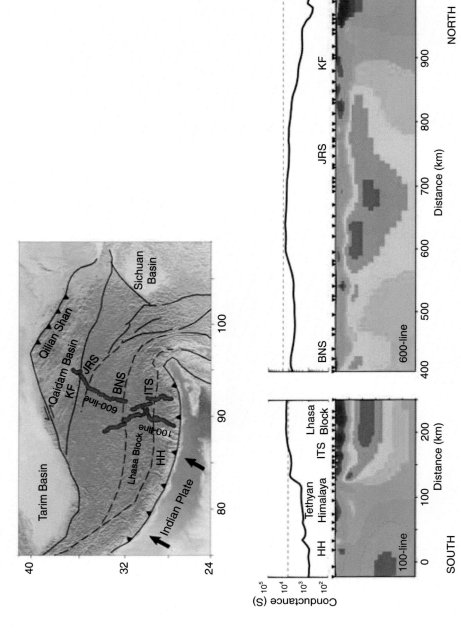

Figure 10.3. Magnetotelluric profile crossing southern Tibet. Modified from Unsworth (2010). The locations of the INDEPTH profiles are shown on the map.

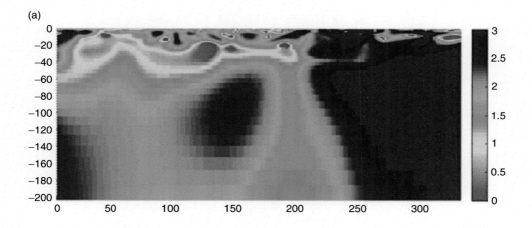

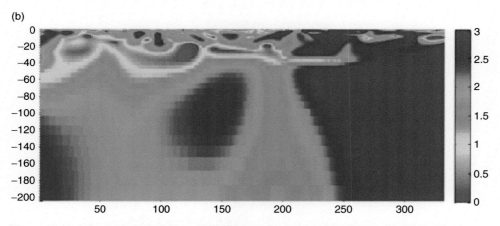

Figure 10.4. Anisotropic resistivity model of the northern part of the 600 profile (F. Le Pape, pers. comm.): (a) the *XX* (NW–SE) component of the resistivity; (b) the *YY* (NE–SW) component of the resistivity.

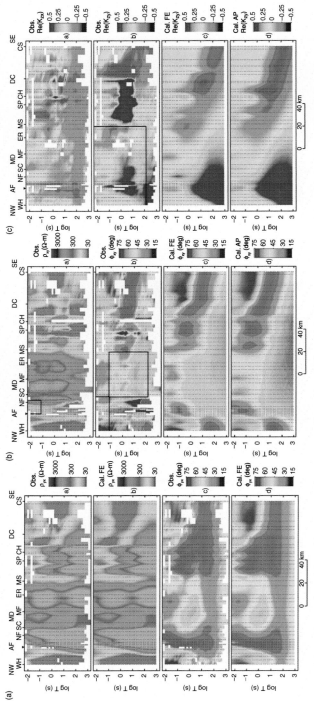

Figure 10.7. Pseudo-sections of the (a) TM, (b) TE and (c) tipper responses for the New Zealand magnetotelluric profile. Results are shown for the observed data (Obs.), the response of a 2D finite-element model (Cal. FE) and the response of a 2D *a priori* model (Cal. AP). Modified from Wannamaker *et al.* (2002). Responses from outside the boxes shown on the TE and tipper observations were excluded from the inversions. Abbreviations in addition to those defined above are: NW, northwest coast; WH, Whatanoa Delta; AF, Alpine Fault; DC, DC power line; SE, southeast coast.

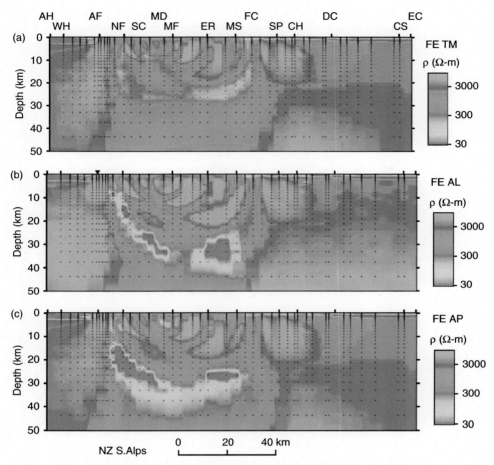

Figure 10.8. Finite-element inversion models fitted to the data. (a) Model (FE TM) fitted to the TM mode data only. (b) Model (FE AL) fitted to a dataset consisting of the TM mode data and subsets of the TE and tipper data. (c) Model (FE AP) fitted to the same subsets of the TE and tipper data using the TM model as the *a priori* model.

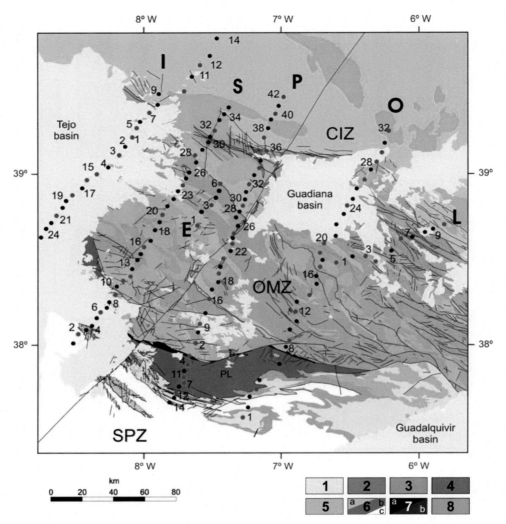

Figure 10.11. Location of the magnetotelluric sites. Modified from Muñoz *et al.* (2008). Black (red) circles show sites chosen for 3D model construction. Simplified geology is as follows: 1, Cenozoic sedimentary cover; 2, Ordovician–Silurian to Carboniferous sequences in OMZ; 3, Cambrian sequences in OMZ; 4, Upper Proterozoic in OMZ; 5, Paleozoic sequences in CIZ; 6, Paleozoic sequences in SPZ (a, Phyllitic–Quartzitic Group, Upper Devonian; b, Volcanic–Sedimentary Complex, Upper Devonian to Lower Visean; c, Flysch Group, Upper Visean to Lower Westfalian); 7, exotic terranes (a, Beja-Acebuches Ophiolite Complex; b, Pulo do Lobo Terrane); 8, intrusive bodies, mostly granitic in nature and Variscan in age.

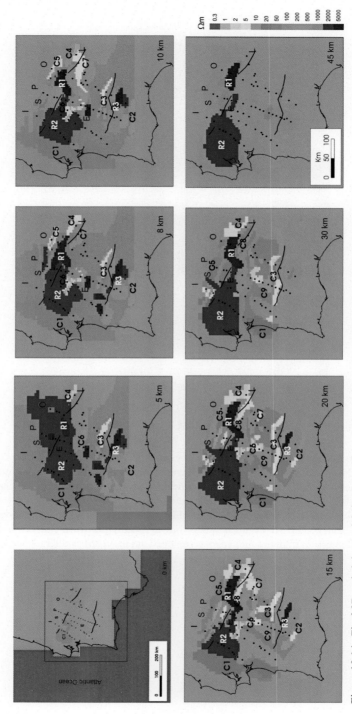

Figure 10.14. Final 3D resistivity model shown as slices at increasing depth. Modified from Muñoz *et al.* (2008).

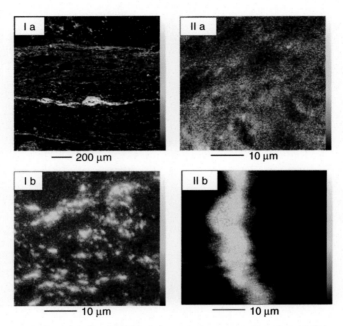

Figure 10.16. Backscattered electron microprobe images of Serie Negra black slates. Modified from Pous *et al.* (2004). The images show the concentration of carbon, with pale shade (yellow) indicating increased concentration. Additional images showing oxygen concentration (not included) allow the enhanced carbon to be attributed to graphite rather than to carbonate. (I a) Continuous bands of graphite along platy cleavage. (I b) Discrete flakes of graphite (1–5 μm diameter, 1 μm thickness). (II) Examples of graphite fault gouge: (II a) small grains (up to 1 μm) pervasively disseminated in silicate groundmass; and (II b) filling of thin veinlets (up to 10 μm in width) parallel to the fault.

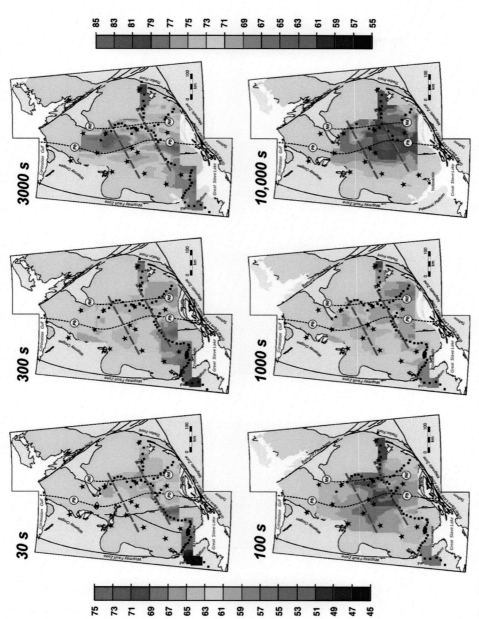

Figure 10.20. Contoured averaged phase maps for periods from 30 to 10 000 s. The data are the arithmetic (Berdichevsky) average of the two orthogonal phases, and is a rotational invariant. The phases in the shortest frequency maps (30 and 100 s) range from 45° to 75°, whereas the phases in the other four maps range from 55° to 85°. Also shown on the maps are the locations of two geochemically defined mantle domain boundaries, and crustal Pb and Nd isotope boundaries.

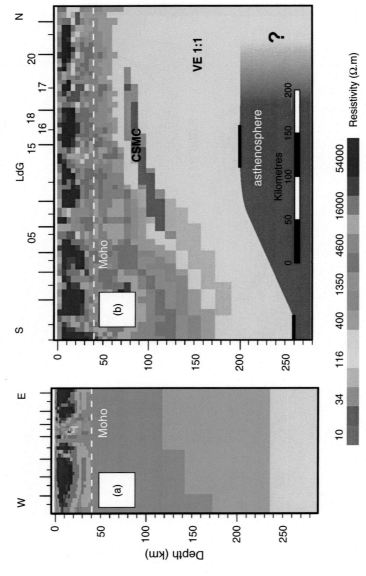

Figure 10.21. (a) Two-dimensional inversion model for sites along an east–west profile in the southwest of the Slave Craton. (b) Two-dimensional inversion model for sites along a north–south profile crossing the CSMC (along line AB shown in Figure 10.17). Modified from Jones *et al.* (2003).

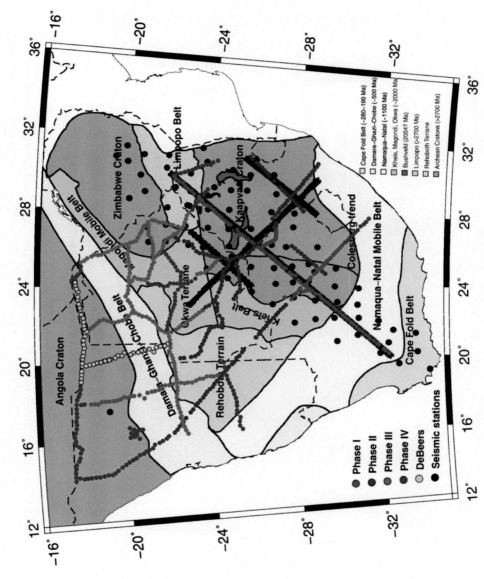

Figure 10.22. SAMTEX (dots in various shades of gray or various colors) and SASE (black dots) coverage of southern Africa. The tectonic base is from Webb (2009).

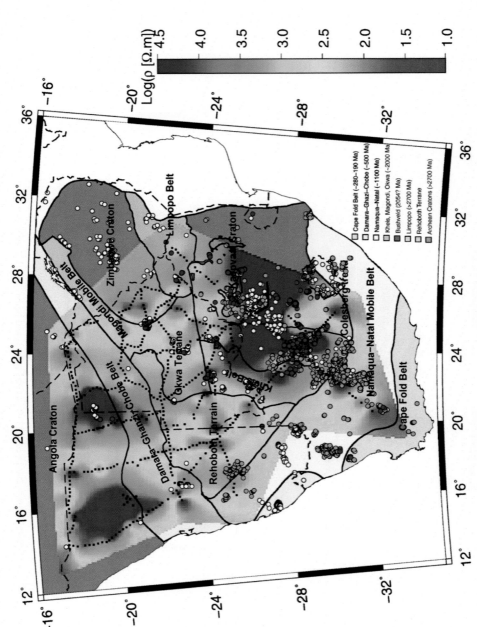

Figure 10.23. An image of the resistivity at 200 km depth based on an approximate transformation of the magnetotelluric responses from period to depth and taking the maximum resistivity. See Jones *et al.* (2009a) for details. The gray shades (various colors) are $\log_{10}$(resistivity), and the black dots show stations where data were used. Also shown on the figure are kimberlite locations: dark gray (red) means known to be diamondiferous, mid-gray (green) means known to be non-diamondiferous, and white means not defined or unknown.

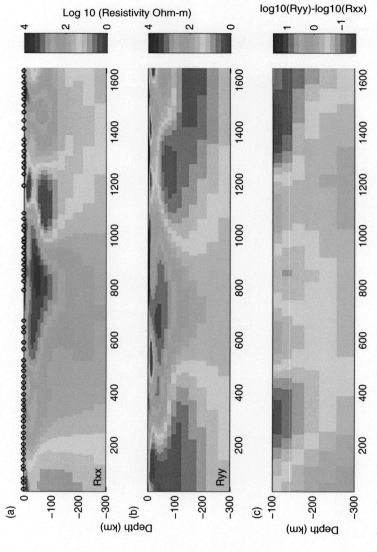

Figure 10.24. Anisotropic models for the Kaapvaal data. Conductivity in the directions (a) perpendicular to the profile (R_{xx}) and (b) parallel to the profile (R_{yy}) are shown. (c) The levels of anisotropy calculated as the difference in $\log_{10}$(resistivity) between the R_{yy} and R_{xx} models. Taken from Evans et al. (2011).

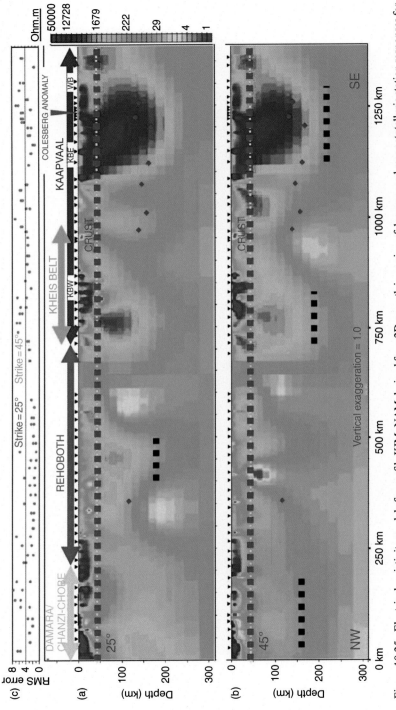

Figure 10.25. Electrical resistivity models for profile KIM–NAM derived from 2D smooth inversion of decomposed magnetotelluric station responses for (a) 25°E of N strike azimuth and (b) 45°E of N azimuth. (c) The 2D inversion RMS misfit error at each station. The surface extent of the geological terranes is shown in (a); abbreviations are used for Western Kimberley Block (KBW), Eastern Kimberley Block (KBE) and Witwatersrand Block (WB). Black dashed lines in (a) and (b) indicate the interpreted depth to the base of the lithosphere where well constrained, and dark gray (red) diamonds indicate the depth to the base of the chemically depleted lithosphere as defined by Cr/Ca-in-pyrope barometry from kimberlitic concentrates. Modified from Muller et al. (2009).

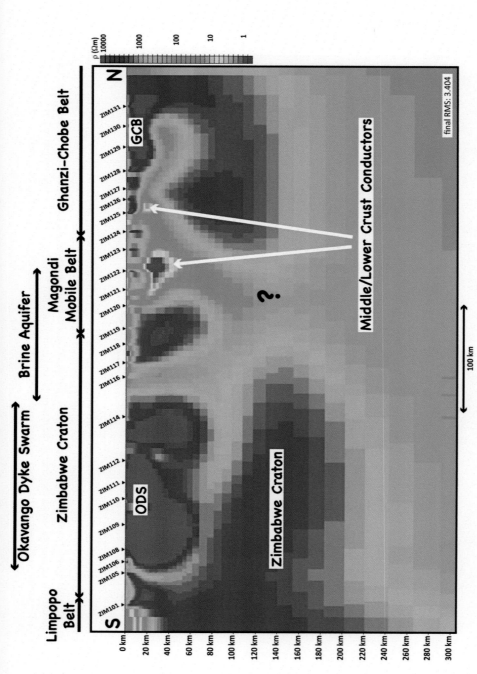

Figure 10.26. The 2D smooth inversion model (vertical exaggeration = 1.0) from the ZIM profile in relation to the known surface extent of geological terranes; taken from Miensopust *et al.* (2011). The arrows above the image of the resistivity structure show the crustal extents of the Limpopo Belt, Zimbabwe Craton, Magondi Mobile Belt and Ghanzi–Chobe Belt (GCB) with respect to magnetotelluric sites of the ZIM line; adapted from the regional-scale geological terrane boundaries based on potential field data (Webb, 2009). The extent of the Okavango Dike Swarm (ODS), known from magnetic data, is indicated, as well as an estimated extent of the brine aquifer related to the Makgadikgadi salt pan complex. The dominant resistivity features related to the main geological terranes are labeled and the question mark indicates the area of missing data coverage. Two dominant mid- to lower-crustal conductors are also apparent.

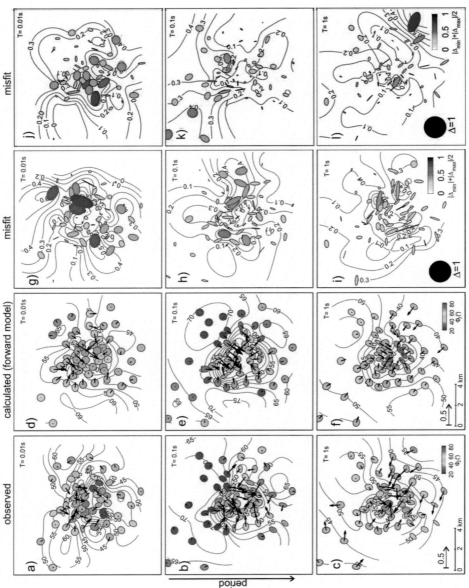

Figure 10.27. (a)–(c) Phase-tensor ellipses and induction arrows between 0.01 s and 1 s. The ellipses are normalized by Φ_{max} and the gray shades (colors) show. Induction arrows that point in the direction of increasing conductance show the strong conductivity contrast at shallow depths. (d)–(f) Calculated phase response from 3D forward modeling. (g)–(i) Tensor misfit ellipses calculated for the observed and calculated forward model phase tensors. The gray shade (color) used to fill the ellipses shows the mean of the maximum and minimum misfit. Small and light shaded (colored) ellipses indicate that the misfit is small. (j)–(l) Tensor misfit ellipses for the observed and calculated inversion model phase tensors. Modified from Heise *et al.* (2008).

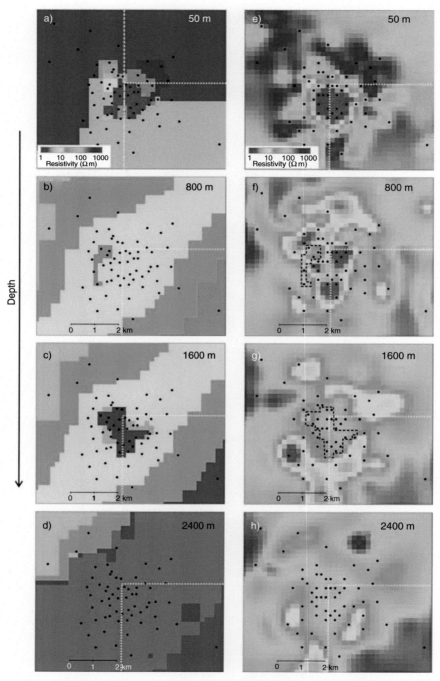

Figure 10.28. Comparison of horizontal sections of (a)–(d) the 3D forward model and (e)–(h) the 3D inverse model. Taken from Heise *et al.* (2008).

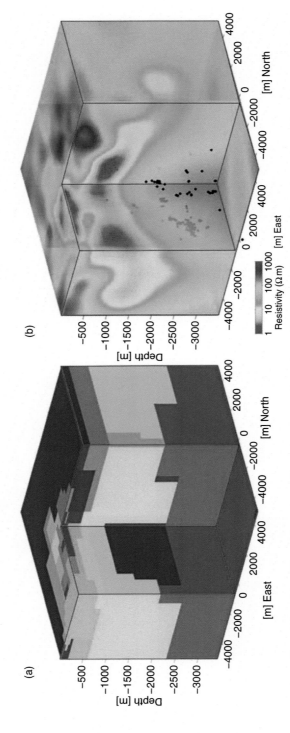

Figure 10.29. Block diagrams of (a) the 3D forward model and (b) the 3D inverse model. Earthquake hypocenters are shown in the inverse model as black dots. Taken from Heise *et al.* (2008).

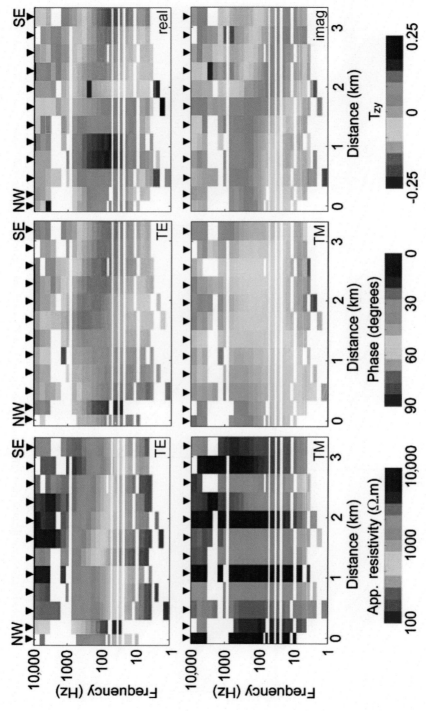

Figure 10.33. Pseudo-sections of TE, TM and induction arrow responses for profile 224. Taken from Tuncer *et al.* (2006).

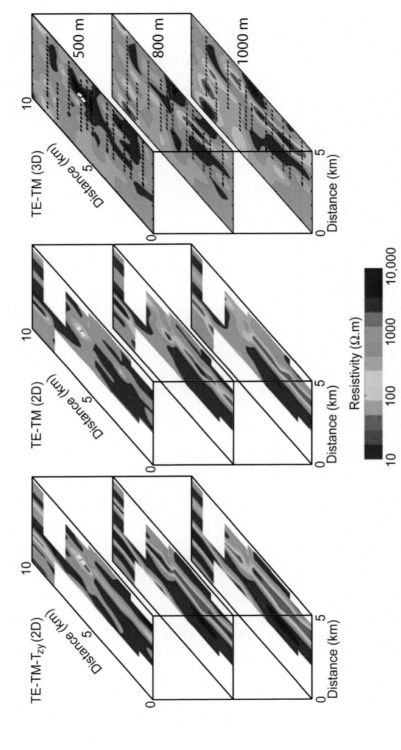

Figure 10.34. Comparison of models derived from different 2D inversions and a 3D inversion. Taken from Tuncer *et al.* (2006).

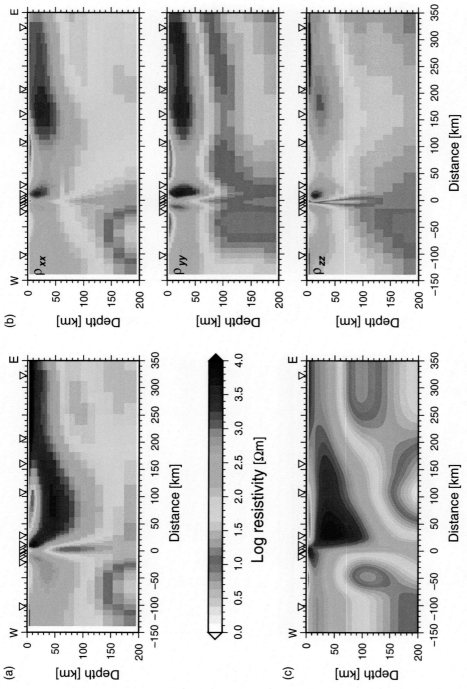

Figure 10.37. Best-fitting 2D (a) isotropic and (b) anisotropic resistivity models obtained for the combined TE and TM mode magnetotelluric responses from the MELT southern line. For the anisotropic result, the panels are along-strike (top), across-strike (middle) and vertical (bottom). The RMS misfits are 2.55 (isotropic) and 2.44 (anisotropic), and are indistinguishable at more than the 80% significance level using a two-sided F-test. The ridge axis is at 0 km and the inverted triangles show the locations of magnetotelluric sites. Taken from Baba *et al.* ((2006a).

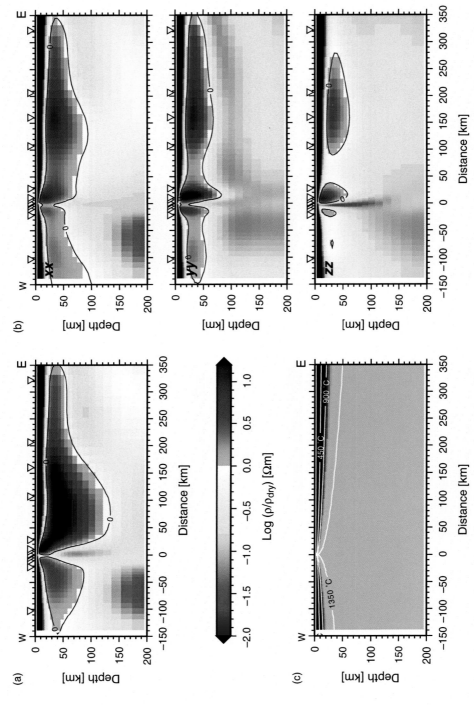

Figure 10.38. Anisotropic resistivity model after subtraction of the dry mantle reference model described in the text. Taken from Baba *et al.* 2006a.

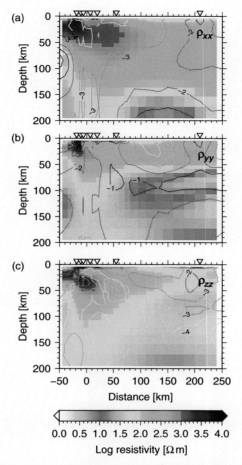

Figure 10.39. Best-fitting anisotropic resistivity model for the northern MELT line. Taken from Baba *et al*. 2006b. The ridge axis is located at 0 km, and the inverted diamonds are the locations of magnetotelluric sites. The three panels are the (a) along, (b) across and (c) vertical components of resistivity. Contours show relative sensitivity.

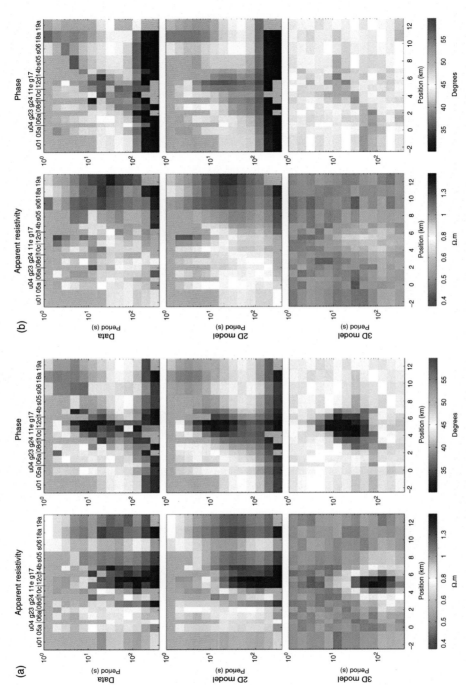

Figure 10.42. Pseudo-sections for Line A in Figure 10.40. (a) In-line electric field response. (b) Cross-line electric field response. The upper panels show observed data, the middle panels show the response from 2D inversions, and the lower panels show the response of 3D forward models. Modified from Key et al. (2006).

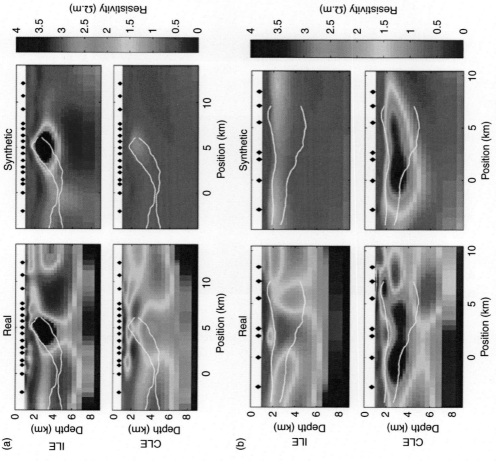

Figure 10.43. Inversion results for (a) Line A and (b) Line 1. Taken from Key *et al.* (2006). The left-hand panels correspond to inversion of the actual data, and the right-hand panels correspond to inversions of the synthetic data. The models are oriented from southwest (left) to northeast (right). The white line shows the location of the top and bottom surfaces of the salt structure as determined from 3D seismic reflection data.

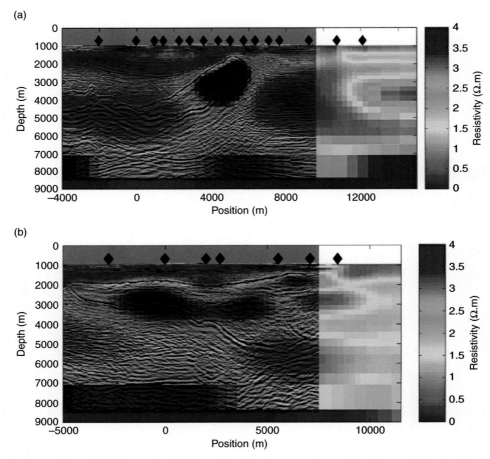

Figure 10.44. Comparison of resistivity images with seismic reflection results for the ILE model from Line A and the CLE model for Line 1. Taken from Key *et al.* (2006).

7

The two- and three-dimensional forward problems

CHESTER J. WEISS

7.1 Introduction

Without any loss of generality, this chapter concerns itself with the numerical solution
to the homogeneous "curl–curl" equations for the time-harmonic electric field $\mathbf{E}$ (see
Section 2.6.1)

$$\nabla \times \mu^{-1} \nabla \times \mathbf{E} + i\omega\hat{\sigma}\mathbf{E} = 0 \tag{7.1}$$

and magnetic field $\mathbf{H}$

$$\nabla \times \hat{\sigma}^{-1} \nabla \times \mathbf{H} + i\omega\mu\mathbf{H} = 0 \tag{7.2}$$

in a source-free medium endowed with a complex-valued electrical conductivity $\hat{\sigma}$ and real-
valued magnetic permeability μ. The convention of an $e^{i\omega t}$ time dependence ($i = \sqrt{-1}$) is
implicit in these equations, and is equivalent to Fourier-transform $\int f(t) exp(i\omega t)\, dt$ the
governing Maxwell equations. For the remainder of this chapter, it will be assumed that all
field quantities are in the frequency domain, hence are complex-valued entities, sometimes
referred to as "phasors" in the classical electromagnetics literature (Smythe, 1939; Stratton,
1941). Furthermore, the complex-valued conductivity, $\hat{\sigma} = \sigma + i\omega\varepsilon$ (with real part equal to
the ohmic conductivity σ, and imaginary part equal to the frequency–permittivity product
$\omega\varepsilon$), reflects the compatibility of these equations with the generalized convolution operator *
in the constitutive relationship for ohmic conduction currents $\mathbf{J} = \sigma * \mathbf{E}$ and magnetic
induction $\mathbf{B} = \mu * \mathbf{H}$ (Jackson, 1962). It is common to consider the material properties σ,
ε and μ to be frequency-invariant, and the effects of electric permittivity ε are sufficiently
negligible at frequencies less than a few megahertz that it can be safely ignored (see
Section 2.3).

Casual inspection of these equations suggests that analytic or quasi-analytic solutions are
accessible only in exceptional circumstances – for example, in cases where the conductivity
and permittivity are constant, or piecewise so over simple geometric boundaries that are
conformal to some convenient coordinate system. While such models may reveal interesting
aspects of the physics of EM induction and provide valuable "benchmark" solutions for
inter-code comparisons, their resemblance to actual geological systems is minimal. Hence,
solution methods for these equations are sought that can accommodate a spatially variable,

The Magnetotelluric Method: Theory and Practice, ed. Alan D. Chave and Alan G. Jones. Published by Cambridge University
Press © Cambridge University Press 2012.

three-dimensional conducting medium, and in the most general case, a medium whose conductivity is further generalized as a rank-two tensor in order to capture the electrical anisotropy of the medium (see Section 2.6.2). Observational evidence of Earth's multi-dimensional conducting subsurface is abundant, as revealed through splitting of the transverse electric and magnetic (TE and TM) modes of the magnetotelluric impedance tensor and intra-regional variations in Earth's electric strike. The numerical methods outlined in this chapter aim to capture such geological complexity and replicate its electromagnetic response through discretized numerical methods for solving the governing equations.

7.2 Numerical methods in two dimensions

With a few exceptions, such as the double quarter-space model examined by D'Erceville & Kunetz (1962) and Weaver *et al.* (1985, 1986), two-dimensional distributions of electrical conductivity in Earth, invariant along the direction of electrical strike, require the use of numerical methods to solve the Maxwell equations, and hence evaluate their magnetotelluric response. Before forging ahead and exploring the details of any particular method, the salient features of the magnetotelluric forward modeling problem in 2D will be outlined by considering the form of the governing partial differential equation, henceforth known as the differential problem (D). For an arbitrary electrical conductivity distribution $\sigma(x, z)$ throughout the Earth region Ω_e (Figure 7.1), recall that the TE mode is the mode where the electric field is non-zero only in the x direction, that of electric strike, and is spatially coupled throughout the conducting Earth purely through induction. No galvanic or charge buildup effects are present (i.e. $\nabla \cdot \mathbf{E} = 0$). Writing the electric field $\mathbf{E}^{\mathrm{TE}} = u(y, z)\hat{\mathbf{x}}$ and assuming for simplicity that the magnetic permeability μ is constant and equal to the free-space value $\mu_0 = 4\pi \times 10^{-7}$ H/m, (7.1) reduces to a simple scalar partial differential equation:

$$- \nabla \cdot k(\mathbf{r})\nabla u(\mathbf{r}) + q(\mathbf{r})u(\mathbf{r}) = 0 \quad \text{for} \quad \mathbf{r} \in \Omega = \Omega_e \cup \Omega_a \qquad (7.3)$$

where $\mathbf{r}$ is the position vector in (y, z), k is a simple constant $1/\mu_0$, $q(\mathbf{r}) = i\omega\sigma(\mathbf{r})$ and Ω is the modeling domain over which the numerical solution is obtained. Observe that, for this TE mode problem, the presence of a perfectly resistive ($\sigma = 0$) air region Ω_a poses no problems for (7.3) in terms of singular or ill-behaved quantities. This equation also possesses no source terms, either within the domain $\Omega_a \cup \Omega_e$ or on its boundary Γ, and is at present insufficient for admitting a solution. Assuming a quasi-uniform source term, a boundary condition (either strictly Dirichlet or some combination of Neumann and Dirichlet) can be imposed on Γ that is compatible with this source. This is the *total field formulation*. The "predetermined values" of the electric field at the boundary are further discussed below.

Moving on to the TM mode, the electric field is orthogonal to the strike direction, and hence the only non-zero component of the magnetic field that persists lies in the x direction. To foreshadow the underlying parallels between the formulation of the TE and TM modes, the generic function u will be recycled, trusting that its meaning (i.e. whether it is the electric or magnetic field) will become apparent from the context. Consequently, $\mathbf{H}^{\mathrm{TM}} = u(y, z)\hat{\mathbf{x}}$.

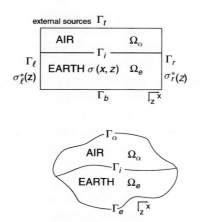

Figure 7.1. Components of the model domain for the 2D magnetotelluric problem.

However, note that the $1/\sigma$ term in (7.2) is undefined in the air region Ω_a of the model domain (Figure 7.1). There are two options to resolve this: either choose σ as something very small (numerically stable but physically insignificant), or find a way to avoid Ω_a altogether. Fortunately, the second option is easily realized, as was shown in Section 2.5.1. Consequently, (7.2) reduces to a differential equation of the same form as (7.3). The only difference is that, for the TM mode problem, the definitions of the functions k and q are correspondingly modified:

$$k(\mathbf{r}) = 1/\sigma(\mathbf{r}) \quad \text{and} \quad q = i\omega\mu_0 \tag{7.4}$$

Observe the following. Because the electric field is orthogonal to the electrical strike, the TM mode will contain galvanic terms where $\nabla \cdot \mathbf{E} \neq 0$ that correspond to jumps in the electric field. However, this is not an issue because, unlike the TE mode, the TM mode calculations are based on a single component of the magnetic field that is fully continuous throughout the computational domain. As for the air layer issue, observe that Ampere's law requires $\nabla \cdot \sigma\mathbf{E} = 0$, and therefore the normal electric field at the air–Earth interface Γ_i must be zero. This requires that $\partial_\tau u = 0$, where ∂_τ is the derivative in the y–z plane tangent to the air–Earth interface, and therefore imposing a homogeneous Dirichlet condition is sufficient to maintain the proper physics and eliminate the air region Ω_a from the calculation.

Pausing at this time to reflect upon (7.3) and the definitions of k and q for the TE and TM modes, respectively, consider what is expected from them in terms of a numerical solution and how these expectations manifest themselves as design elements of a numerical method. It is useful to bear in mind a couple of key points regarding pragmatic use of numerical methods for solving partial differential equations. First, numerical methods are *approximate* methods and hence cannot recover the exact solution no matter how hard one tries. As a corollary to this rule, the solution recovered by numerical methods is intrinsically tied to various *arbitrary* decisions made by the user, such as mesh design, polynomial order of the method, post-processing schemes and solution accuracy of whatever linear systems may arise. These issues, while important, are distinct from the numerical method itself because the method cannot be held responsible for poor choices by the end-user. Rather, the methods usually offer some guarantees of optimality (in the case of finite elements) or polynomial convergence, bearing in mind that optimality on a poorly designed numerical problem is likely of little consolation. Second, it is important to remember that an accurate solution is not required everywhere in the solution domain. Since the purpose of modeling is often to compare observables to numerically derived predictions, errors in some part of the model domain distant from the observations are irrelevant provided they do not affect the solution at those locations. The exception to this is in an experimental design study where the question of where the best places to make a measurement for resolving a given subsurface model are addressed. Lastly, the third guiding principle to bear in mind is that numerical methods require work – whether in the form of mesh design or the cost of solving a linear system – and that, generally, the less one asks of the numerical method the better. Hence, problems that can be decomposed into a part that is easy to solve analytically (like a 1D background model) and a part that is not (some complicated geo-target) are especially attractive because they only require of the numerical method that part of the solution that captures the *interaction* between the target and the background.

Suppose that the electrical conductivity $\sigma(\mathbf{r})$ can be split into two parts $\sigma_0 + \sigma'$, which are the "reference" and "secondary" conductivities, respectively, such that the solution to (7.3) with $\sigma = \sigma_0$ is cheap and easy to compute, whether through analytic solutions or whatever. Calling this solution u_0, then the total solution u is also equal to the sum of a primary and secondary field such that $u = u_0 + u'$ provided that u' satisfies

$$- \nabla{\cdot}k(r)\nabla u'(\mathbf{r}) + qu'(\mathbf{r}) = f(\mathbf{r}) \quad \text{for} \quad \mathbf{r} \in \Omega. \tag{7.5}$$

where k and q retain their usual definitions for the TE and TM modes, respectively, and the function f is now introduced as a mode-dependent source term taking the value

$$f(\mathbf{r}) = \begin{cases} -i\omega(\boldsymbol{\sigma} - \boldsymbol{\sigma}_0)u_0(\mathbf{r}), & \text{for the TE mode} \\ \nabla \cdot [(1/\boldsymbol{\sigma} - 1/\boldsymbol{\sigma}_0)\nabla u_0(\mathbf{r})], & \text{for the TM mode} \end{cases} \tag{7.6}$$

Equations (7.5) and (7.6) are known as the *secondary field formulation* because only solutions u' are sought, not the total field u. A numerical method (or, at least, a finite-difference or finite-element solution) is used only for computation of those parts of the electromagnetic fields that arise from the presence of conductivity perturbations σ' that tend to be spatially localized around the perturbations themselves rather than permeating the entire domain Ω.

7.2.1 Boundary conditions in two dimensions

At this point, the only remaining issue in the statement of the 2D magnetotelluric problem is explicitly defining the boundary conditions on the perimeter of the domain Ω. It has been argued that, for the TM mode, the solution domain is restricted to Ω_a and that setting u equal to a constant on Γ_i is sufficient for enforcement of continuity of normal electric current on the air–Earth interface. This is true for both the scattered and total field formulations in the TM mode, is independent of the choice of Ω, and hence need not be considered any further. For all remaining cases, however, it is useful to remember that the computational domain is chosen such that it has the following properties: it sufficiently encompasses the area of interest (i.e. the subsurface target); it minimizes unnecessary computation of the fields in places that are not of interest; and, if possible, it allows for simple boundary conditions that do not adversely affect the solution in parts of the domain that are not of interest. How the third point manifests itself in each of the formulations described in (7.3)–(7.5) will now be examined.

For the total field formulation, regardless of mode, the intent is to recover the complete field u throughout the domain Ω. Starting with the simplest case where the conductivity distributions on the left and right sides of the domain are a function of depth only, if the assumption that those profiles extend laterally outside of the domain is made, then the fields on the left and right sides should approximate the 1D solution corresponding to those vertical profiles. Thus, the option exists either to set a non-homogeneous Dirichlet condition on those sides of the domain, or, by realizing that a 1D solution admits no lateral variability, to set a homogeneous Neumann condition and define Ω such that its sides are vertical. These are equivalent realizations of the same assumption of a 1D Earth "stitched" to the side of the domain of interest. How, then, to deal with the top and bottom of the mesh? First, note that these are regions that are typically not of interest in the solution for geophysical interpretation purposes. Hence, the details of the solution are not of too great concern near these boundaries. For compatibility with the 1D assumption along the mesh sides, the only consistent strategy is to impose a Dirichlet condition along the top and bottom of the mesh that smoothly transitions between the field values corresponding to the 1D profiles on the mesh sides. Furthermore, whatever tapering function is used should have zero horizontal gradient at the sides of the mesh for consistency with the stitched 1D Earth assumption. A common choice is the cosine taper, but any smooth function will suffice.

In the scattered field formulation, the primary attraction is that the computational domain need only describe the fields arising from the anomalous conductivity σ' that is presumed to be spatially localized to the vicinity of σ' itself. In other words, the scattered field decays to zero, or nearly so, at some point (hopefully) not too far away from the target. Define Ω such that the scattered field is approximately zero on its perimeter, so that a homogeneous Dirichlet condition is appropriate. If this distance is so large that it requires the domain Ω to be excessively big, and therefore computationally expensive to consider, it is not unreasonable to truncate Ω to a smaller value provided that the error introduced by such a maneuver is not too large at the points in the domain that are of the greatest interest. The governing philosophy emerges: Pragmatism does not require a numerical method to give accurate answers *everywhere*, but only requires accurate answers *where it counts*. Fortunately, the diffusive electromagnetic fields in magnetotellurics decay exponentially – a property that often works favorably, limiting the influence of compromises in the boundary conditions to the vicinity of the boundary itself. Finally, in cases like those described where the left and right sides of the mesh have different vertical conductivity profiles, the boundary condition is again Dirichlet, but with a smooth lateral taper going from zero on one side (the reference side of the mesh) to the difference in u on the other side.

7.2.2 Summary of the two-dimensional magnetotelluric differential problem statement (D)

It has been shown that, regardless of mode or formulation, the essence of the 2D magnetotelluric modeling problem is seeking a solution to differential equations of the form

$$-\nabla \cdot k\nabla u(\mathbf{r}) + qu(\mathbf{r}) = f(\mathbf{r}) \quad \text{for} \quad \mathbf{r} \in \Omega, \tag{7.7}$$

with some combination of the dependent variable $u(\mathbf{r})$ and its normal derivatives $\partial_n u(\mathbf{r})$ given on Γ, where the functions k, q, u and f are defined according to Table 7.1 with the boundary conditions given in Table 7.2. Recall that, in general, the quantities in (7.7) are complex-valued, and this will add a small but not insurmountable complication to the process of obtaining a numerical solution, as will be seen later when linear solvers are discussed.

Table 7.1. *Definition of quantities in (7.7) as they apply to the 2D magnetotelluric modeling problem. The subscript zero specifies the primary field arising from a reference conductivity model σ_0, and the prime specifies the scattered field that is added to the primary one, yielding the total field.*

Mode	u	k	q	f
TE total	E_y	$1/\mu_0$	$i\omega\sigma$	0
TE scattered	E_y'	$1/\mu_0$	$i\omega\sigma$	$-i\omega(\sigma - \sigma_0)\hat{\mathbf{y}} \cdot \mathbf{E}_0(\mathbf{r})$
TM total	H_y	$1/\sigma$	$i\omega\mu_0$	0
TM scattered	H_y'	$1/\sigma$	$i\omega\mu_0$	$\nabla \cdot [(1/\sigma - 1/\sigma_0)\nabla(\hat{\mathbf{y}} \cdot \mathbf{H}_0(\mathbf{r}))]$

Table 7.2. *Definition of boundary conditions as they apply to the 2D magnetotelluric modeling problem.*

Mode	Boundary condition
TE total	Non-homogeneous Dirichlet on $\Gamma_e \cup \Gamma_a$ with lateral taper
TE scattered	Homogeneous Dirichlet on reference side of $\Gamma_e \cup \Gamma_a$ and lateral taper on alternate one
TM total	Constant on Γ_i and lateral taper on Γ_e
TM scattered	Constant total field on Γ_i and lateral taper on Γ_e

7.3 Finite differences, elements, volumes and all that

Up to this point, the focus has been on a generalized statement of the differential problem (D) for magnetotellurics, in which it has been observed that, regardless of the choice of formulation, the structure of the differential equation remains unchanged. Further, the choice of formulation and the specifics of the geo-problem itself dictate additional choices about the nature of the domain Ω, the reference model σ_0, and the boundary conditions. As a rule, these choices are driven by a compromise between solution accuracy in parts of the domain that are of interest and the computational burden (or algorithmic complexity) in obtaining said solution. However, multiple numerical methods are available, and it is worthwhile to reflect first on their taxonomy so that a better understanding of their relationships can be achieved, and, as a result, lay the groundwork to understand the solutions they generate.

In short, the three main numerical methods reviewed in this section are finite-difference, finite-element and finite-volume methods, and their distinction from one another is defined by the very first action required in seeking a numerical solution to (D) (Figure 7.2). In the finite-difference (FD) methods, the principal idea is that the differential operator is approximated by simple quotients of the differences between field values at discrete points and the distances between those points, which approaches an exact derivative for sufficiently

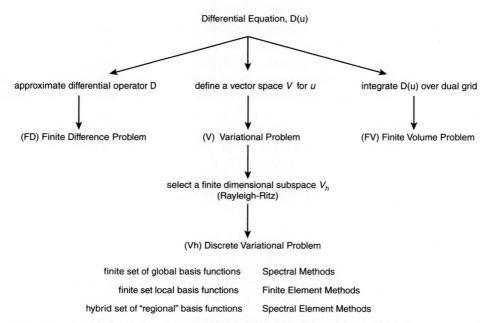

Figure 7.2. Illustration of the taxonomic relationship and procedures therein between the finite-difference (FD), finite-volume (FV) and variational (V and V_h) methods for solving the differential equation D(u).

smooth functions as the distances go to zero. Field values are evaluated at these points (nodes) only, and the continuity of the function between nodes is implicit in the formulation, but not necessarily called out or exploited when writing down the quotients. In contrast, the finite-element methods are a specific realization of a powerful and more general class of methods encompassed by the variational formulation (V), which starts off by considering the differential operator in its full expression, and instead defines a function space from which a numerical solution is drawn. This choice of function space is the distinguishing feature between finite-element (FE), spectral, spectral element and related methods, all of which derive from the variational problem itself. Finally, the finite-volume (FV) method is included for completeness as a middle ground between FE and FD methods, which starts off with a discretization much like the FE method, but then explicitly integrates (D) over the discretization, approximating the integrals through simple difference quotients as in FD.

There are two parts to a discretization. The first is a set of discrete points distributed over the domain Ω. The second is a topology – how the nodes are "connected" to one another, forming a spatially continuous tiling over Ω of simplices such as triangles or higher-order polygons in 2D (tetrahedra, etc., in 3D). Methods that are "meshless" (node clouds, but no cells or topology) have been applied elsewhere in the applied mathematics literature, but not in computational electromagnetics.

7.3.1 Finite differences (FD)

In the most common manifestation of the finite-difference method, the discretization is built on a tensor-product grid of N_y and N_z node locations in the y and z directions, respectively, thus forming a Cartesian grid of quadrilateral cells (Figure 7.3). The advantage of this grid is that it is simple to construct and economical to store in a computer because the topology is implicit and the node coordinates are completely specified by two, short, one-dimensional arrays of a much larger $N_y \times N_z$ array of all node coordinates. By defining the value of u on the grid points by $u_{i,j} = u(y_i, z_j)$, the derivatives in (7.7) can be defined as follows. At point $u(y_{i+1/2}, z_j)$, the partial derivative in the y direction is given approximately by $(u_{i+1,j} - u_{i,j})/(y_{i+1} - y_i)$. Similar quotients follow for the remaining three midpoints surrounding (y_i, z_j), thus setting up another application of this same idea to get second derivatives centered on (y_i, z_j), as required by (7.7). Specifically, this five-point stencil is written as

$$\nabla \cdot k \nabla u_{i,j} \approx \frac{2}{y_{i+1} - y_{i-1}} \left[\frac{k_{i+1/2,j} \left(u_{i+1,j} - u_{i,j} \right)}{y_{i+1} - y_i} - \frac{k_{i-1/2,j} \left(u_{i,j} - u_{i-1,j} \right)}{y_i - y_{i-1}} \right]$$

$$+ \frac{2}{z_{j+1} - z_{j-1}} \left[\frac{k_{i,j+1/2} \left(u_{i,j+1} - u_{i,j} \right)}{z_{j+1} - z_j} - \frac{k_{i,j-1/2} \left(u_{i,j} - u_{i,j-1} \right)}{z_j - z_{j-1}} \right]. \tag{7.8}$$

Notice from (7.8) that values of k are required along the same four midpoints surrounding (y_i, z_j). Defining k as piecewise constant over the cells of the discretization leads to a volume-weighted average for values along the midpoints. Furthermore, defining q over the

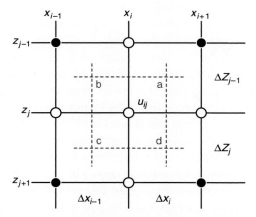

Figure 7.3. Finite-difference stencil for the 2D MT problem. Strike-aligned components of the magnetic field (TM mode) or electric field (TE mode) at the nodes (i, j) of the mesh are represented by $u_{i,j}$ node coordinates in x and z directions by x_i and z_j, respectively, and node spacing by Δ. Shown in dashed lines is a portion of the dual Voronoi grid with nodes a–d surrounding the primal grid node (i,j).

discretization leads to a similar approximation at the point (y_i, z_j), all of which leads to a discrete form of (7.7) for all indices i and j:

$$
\frac{2}{y_{i+1} - y_{i-1}} \left[\frac{k_{i+1/2,\,j}(u_{i+1,\,j} - u_{i,\,j})}{y_{i+1} - y_i} - \frac{k_{i-1/2,\,j}(u_{i,\,j} - u_{i-1,\,j})}{y_i - y_{i-1}} \right]
$$
$$
+ \frac{2}{z_{j+1} - z_{j-1}} \left[\frac{k_{i,\,j+1/2}(u_{i,\,j+1} - u_{i,\,j})}{z_{j+1} - z_j} - \frac{k_{i,\,j-1/2}(u_{i,\,j} - u_{i,\,j-1})}{z_j - z_{j-1}} \right] \qquad (7.9)
$$
$$
+ \quad q_{i,\,j} u_{i,\,j} \quad = \quad f_{i,\,j}
$$

Equation (7.9) represents a set of $N_y \times N_z$ linear equations in terms of the unknown quantities $u_{i,\,j}$ and source vector (in the linear algebra sense) $\{f_{i,\,j}\}$. Multiplying equation (7.9) by the product $\frac{1}{2}(y_{i+1} - y_{i-1})(z_{j+1} - z_{j-1})$ symmetrizes the resulting coefficient matrix and is an inexpensive calculation that will provide benefit when the linear system is solved.

Implementation of boundary conditions is the final step in setting up the finite-difference system of linear equations. Examination of the left-hand side of (7.9) reveals that nodes on the boundary Γ represent terms that are fully known, like the terms $f_{i,\,j}$ on the right-hand side. To express this fact, take the equations (i, j) for nodes *adjacent* to a boundary node, and move the corresponding boundary node term to the right-hand side by subtracting it from $f_{i,\,j}$. Then, either the boundary node terms can be culled from the vector of unknowns (since, in fact, they are known by definition) or they can be kept in the vector of unknowns by modifying the coefficient matrix with zeros in those columns, ones on the diagonal, and replacing the corresponding entries $f_{i,\,j}$ with the function value itself. The advantage of the former method is that it results in a slightly smaller linear system of equations to store and solve. The advantage of the latter is that the solution vector from the linear system contains the complete solution, boundary and all, and is not parsed across multiple data structures in the computer. Observe also that, if a Neumann condition is chosen on the vertical boundaries of Ω, they are also manifest as known constants in (7.9), and are subjected to the same treatment as described for the Dirichlet condition. Although the horizontal derivatives in (7.9) are interior to the mesh by half a cell dimension, this discrepancy is typically small enough not to propagate a significant error inward more than a few grid cells.

In summary, the finite-difference method uses repeated application of a discrete representation of the differential operator that appears in (7.7) to arrive at a complex symmetric linear system of equations. The major advantage is that the system is easy to conceptualize and set up. The price for such ease is a restriction on the mesh design. The tensor product mesh does not admit local mesh refinement in regions of interest. Rather, node densification in, for example, the y direction is propagated throughout the entire domain, leading to unnecessary mesh refinement and cells with high aspect ratios far from the region of interest.

7.3.2 Finite elements (FE) and the variational formulation (V)

In contrast to the finite-difference method, the starting point for finite elements lies in a fundamentally different direction – one in which, rather than considering an approximation to the differential operator, its full expression is maintained, while instead considering the vector space from which solutions to (D) are drawn. Such an idea lies at the heart of the variational approach from which the finite-element method obtains.

Inspection of (7.7) suggests that a solution u must reside in a vector space V that is, at a minimum, piecewise differentiable, and must further adhere to the boundary conditions on Γ. More formally, it is required that $u \in V = \{$piecewise differentiable functions on Ω; u or $\partial_n u$ given on $\Gamma\}$. An auxiliary "test" function v is introduced that also resides in a vector space $\tilde{V}$, which will be defined in a moment to maximize the usefulness of v. Multiplication of (7.7) by v and integrating over Ω yields the following relationship:

$$\int_\Omega \nabla v \cdot k \nabla u + q v u \, d\Omega - \int_\Gamma k v \partial_n u \, d\Gamma = \int_\Omega f v \, d\Omega \qquad (7.10)$$

From the first integral on the left-hand side of (7.10), it clearly must be required that v, as for u, be at least piecewise differentiable on Ω. The second integral on the left-hand side requires a bit more thought. Because v has been introduced purely for convenience, the following observation holds: setting $v = 0$ on Γ eliminates the integral and avoids the problem of working with $\partial_n u$ on Γ in cases where a Dirichlet condition is required (Table 7.2). Alternatively, for the Neumann condition, ∂_n is explicitly defined on Γ and, again, the surface integral is manageable, but now with no additional constraints on $\tilde{V}$. Regardless, the goal is elimination of unknowns in u associated with this integral. It is achieved through either benign neglect of v in the case of Neumann conditions on u, or imposition of a homogeneous Dirichlet condition on v when u itself takes on a predetermined value on the boundary. Without any loss of generality, only the Dirichlet problem will be considered in the sequel since it is an inescapable part of the magnetotelluric problem, regardless of mode or formulation (Table 7.1). Hence, the following statement of the variational problem (V) can be made: find $u \in V$ such that $a(u, v) = (f, v)$ for all $v \in \tilde{V}$, where

$$a(u, v) = \int_\Omega \nabla v \cdot k \nabla u + qvu \, d\Omega \quad \text{and} \quad (f, v) = \int_\Omega fv \, d\Omega \tag{7.11}$$

What is the usefulness of such a general statement? Consider that the exact solution u lies in V, and take some approximate solution $u_h \in V_h \subset V$ with homogeneous Dirichlet conditions. Because $v \in V_h$ may be chosen, both $a(u, v) = (f, v)$ and $a(u_h, v) = (f, v)$ for all $v \in V_h$ hold true, and upon subtraction yields $a(u - u_h, v) = 0$ for all $v \in V_h$. In other words, the error between the approximate solution u_h and the exact solution u is orthogonal to the space V_h with respect to $a(\cdot, \cdot)$. Seen another way, the approximate solution u_h is the projection with respect to $a(\cdot, \cdot)$ of the exact solution u onto the space V_h. In short, there is a guarantee that the approximate solution u_h is the closest solution possible to u in V_h when measured by a norm corresponding to the bilinear form $a(\cdot, \cdot)$ – a guarantee that is possible without explicit knowledge of u beforehand!

To move forward from the variational problem (V) to the finite-element problem (FE), the idea of considering a finite-dimensional vector space from which to draw approximate solutions u_h will be continued. Denoting the scalar basis functions of this space by $\varphi_i(\mathbf{r})$, the approximate solution $u_h(\mathbf{r}) = \mathbf{u}^\mathsf{T} \mathbf{\Phi}(\mathbf{r})$ is given by the inner product of a vector of coefficients $\mathbf{u}^\mathsf{T} = [u_1, u_2, \ldots, u_N]$, while dropping the subscript h for notational simplicity, and the vector of basis functions $\mathbf{\Phi}^\mathsf{T}(\mathbf{r}) = [\varphi_1(\mathbf{r}), \varphi_2(\mathbf{r}), \ldots, \varphi_N(\mathbf{r})]$. Letting $\mathbf{v}^\mathsf{T}$ denote the row vector of coefficients for the test function v and substituting these quantities into (7.11) yields

$$\mathbf{v}^\mathsf{T} \left[\int_\Omega k \nabla \mathbf{\Phi} \nabla \mathbf{\Phi}^\mathsf{T} + q \mathbf{\Phi} \mathbf{\Phi}^\mathsf{T} \, d\Omega \right] \mathbf{u} = \mathbf{v}^\mathsf{T} \int_\Omega \mathbf{\Phi}(\mathbf{r}) f(\mathbf{r}) d\Omega$$

which, after dropping the common prefactor $\mathbf{v}^\mathsf{T}$ on each side, gives a linear system of equations in terms of the unknown coefficients $\mathbf{u}$ independent of $\mathbf{v}$ as required by (V):

$$\left[\int_\Omega k \nabla \mathbf{\Phi} \nabla \mathbf{\Phi}^\mathsf{T} + q \mathbf{\Phi} \mathbf{\Phi}^\mathsf{T} \, d\Omega \right] u = \int_\Omega \mathbf{\Phi}(\mathbf{r}) f(\mathbf{r}) d\Omega \tag{7.12}$$

In (7.12), the quantity in brackets on the left-hand side is the coefficient matrix, whereas the right-hand side represents the source vector.

What specifically constitutes a statement of the finite-element problem at this point? The answer is the choice of $\mathbf{\Phi}$, and hence V_h. For the finite-element (FE) problem, V_h is chosen such that Ω is discretized (see above) with $j = 1, 2, \ldots, M$ elements (simplices) over which a set of $i = 1, 2, \ldots, N$ spatially localized basis functions are defined (Figure 7.4). The simplest example is the set of linear nodal basis functions

$$\varphi_i(\mathbf{r}) = \begin{cases} 1, & \mathbf{r} = \mathbf{r}_i \\ 0, & \mathbf{r} = \mathbf{r}_{i'} \quad (i' \neq i) \\ \text{linear}, & \text{otherwise} \end{cases} \tag{7.13}$$

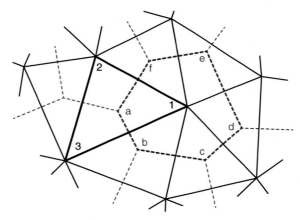

Figure 7.4. Unstructured grid used for 2D finite-element calculations (full) with corresponding Voronoi grid (dashed). Local enumeration 1–3 for nodes on an element of the primal grid is shown, as is a–f for one cell of the Voronoi grid.

which are non-zero over all elements sharing node i as a vertex, and identically zero over all remaining elements. Note that it would be legitimate to avoid a discretization altogether and choose $\varphi_i(\mathbf{r})$ as a globally extensive set of orthogonal polynomials. In that case, the variational problem (V) is known as a spectral method. Hybrid approaches – low-order expansions over elements of a coarse discretization – define the spectral element approach. The important point is the observation that these seemingly disparate methods are all rooted in the variational problem (V), and endowed with its architecture for proving useful properties such as optimality.

Getting to the specifics of finite-element methods for magnetotellurics, the scalar nodal basis just defined will be utilized as a template for specifying how to assemble the linear system in (7.12). Noting that it will be required to integrate products of $\varphi_i(\mathbf{r})$ over triangular subdomains (elements), a bit of algebraic complexity can be avoided by introducing locally enumerated, barycentric coordinates $(\lambda_1, \lambda_2, \lambda_3)$ for each element e (Figure 7.5), defined as

$$\lambda_i = \frac{S_{p,j,k}}{S_e} = \frac{\text{area of triangle}(p,j,k)}{\text{area of triangle}(1,2,3)}, \quad i = 1, 2, 3, \quad j, k \neq i$$

where the area functions $S_{p,j,k}$ are computed via the determinant

$$S_{p,j,k} = \frac{1}{2} \begin{vmatrix} y & z & 1 \\ y_j & z_j & 1 \\ y_k & z_k & 1 \end{vmatrix} \tag{7.14}$$

in the (y, z) coordinate system. Clearly, these functions λ_i take the value unity at node i, zero at nodes j, $k \neq i$, and are linear functions of position $\mathbf{r}$. Thus, the nodal basis functions are simply degree-one polynomials in the barycentric coordinates. If at some future time it is desirable to consider a higher-order polynomial basis function over the elements, then

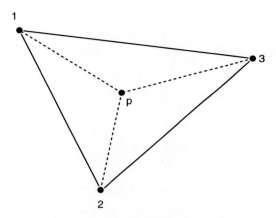

Figure 7.5. Local node enumeration, 1–3, for a given element and interior point p.

polynomials in λ are an efficient way to capture this idea. More important at present is how these coordinates help in the evaluation of the integrals in (7.12). Over a triangular element, e, with barycentric coordinates $\lambda_i(\mathbf{r})$, $i = 1, 2, 3$, it can be shown that

$$\int_e \lambda_1^l \lambda_2^m \lambda_3^n d\Omega = \frac{l!m!n!}{l + m + n + 2!} \cdot 2S_e \tag{7.15}$$

a remarkable result that obviates a load of tedious algebra!

Close inspection of the first term on the left-hand side integral of (7.12) reveals that the choice of linear basis functions in (7.13) results in simply integrating the function k over Ω. Noting that the domain Ω is the sum of elements e, then $\int_\Omega \cdots d\Omega = \sum_e \int_e \cdots d\Omega$, and attention can therefore be focused on integration over the elements themselves. Presuming that k is smooth over the element e, the integral of k over e is easily obtained by quadrature, the simplest of which is a one-point formula where $\int_e k(\mathbf{r}) \, d\Omega \approx k_e S_e$, with k_e being some average value of k over e. As a result, the 3×3 *element stiffness matrix* – an allusion to the rich history of finite-element work in the engineering mechanics literature – is defined in local node enumeration 1–3 by

$$\left[\int_e k \nabla \boldsymbol{\Phi}_e \nabla \boldsymbol{\Phi}_e^{\mathrm{T}} d\Omega \right]$$

$$= \begin{pmatrix} \int_e k \nabla \varphi_1 \cdot \nabla \varphi_1 \, d\Omega & \int_e k \nabla \varphi_1 \cdot \nabla \varphi_2 \, d\Omega & \int_e k \nabla \varphi_1 \cdot \nabla \varphi_3 \, d\Omega \\ \cdot & \int_e k \nabla \varphi_2 \cdot \nabla \varphi_2 \, d\Omega & \int_e k \nabla \varphi_2 \cdot \nabla \varphi_3 \, d\Omega \\ \cdot & \cdot & \int_e k \nabla \varphi_3 \cdot \nabla \varphi_3 \, d\Omega \end{pmatrix} \tag{7.16}$$

where $\boldsymbol{\Phi}_e \subset \boldsymbol{\Phi}$ contains only the three non-zero basis functions $\varphi_i(\mathbf{r})$ on the element e. Observing that evaluation of linear nodal basis functions $\varphi_i(y, z)$ may be written as $a_i + b_i y + c_i z$, $i = 1, 2, 3$, with values of the coefficients a_i, b_i and c_i deriving from (7.14), the element stiffness matrix in (7.16) simplifies to

$$\int_e k\nabla\boldsymbol{\Phi}_e\nabla\boldsymbol{\Phi}_e^{\mathrm{T}}d\Omega \approx \frac{k_e}{4S_e}\begin{pmatrix} b_1^2 + c_1^2 & b_1b_2 + c_1c_2 & b_1b_3 + b_1b_3 \\ \cdot & b_2^2 + c_2^2 & b_2b_3 + c_2c_3 \\ \cdot & \cdot & b_3^2 + c_3^2 \end{pmatrix} \qquad (7.17)$$

Applying (7.15) to the second term in the integrand on the left-hand side of (7.12) yields the symmetric 3×3 *element mass matrix*

$$\int_e q\boldsymbol{\Phi}_e\boldsymbol{\Phi}_e^{\mathrm{T}}d\Omega \approx \frac{q_eS_e}{12}\begin{pmatrix} 2 & 1 & 1 \\ \cdot & 2 & 1 \\ \cdot & \cdot & 2 \end{pmatrix} \qquad (7.18)$$

where, as before, q_e is some average value of q over the element e. Finally, evaluating the right-hand side of (7.12) and again applying the combinatorial formula in (7.15) yields the element source vector, which for smooth functions f is

$$\int_e \boldsymbol{\Phi}_e f\, d\Omega \approx \frac{f_eS_e}{3}\begin{pmatrix} 1 \\ 1 \\ 1 \end{pmatrix} \qquad (7.19)$$

Observe that, in the case of delta function sources at points $\mathbf{r}_s$ – seen elsewhere to arise in the use of adjoint methods for computing sensitivity matrices in the inverse problem – the right-hand side of (7.12) is trivial to evaluate, and distributes the delta function over nodes of the element where it resides:

$$\int_e \boldsymbol{\Phi}_e\delta(\mathbf{r} - \mathbf{r}_s)\, d\Omega = \begin{pmatrix} \varphi_1(\mathbf{r}_s) \\ \varphi_2(\mathbf{r}_s) \\ \varphi_3(\mathbf{r}_s) \end{pmatrix}$$

In summary, by computing element mass and stiffness matrices along with the element source vector, a complete description of the (FE) linear system of equations ensues, which, like the (FD) linear system, is complex symmetric. By decomposing integration over the whole domain Ω into a series of smaller, simple integrations over elements e, the coefficient matrix (7.17) and (7.19) can be considered element-wise, thus lending itself naturally to parallelization or storage economy when considering methods for inverting the linear system to recover $\mathbf{u}$, and hence u_h.

7.3.3 Finite volumes (FV) and the variational formulation (V)

For completeness, this discussion of numerical methods concludes with a brief overview of the finite-volume method because of its close relationship to both finite elements and finite differences, and because the idea of integrating the differential problem (D) will reappear later in the discussion of numerical methods in 3D. As with finite differences, the starting point of the finite-volume method is the construction of a discretization on Ω. There is some flexibility here: both Cartesian (Figure 7.3) and unstructured (Figure 7.4) grids may be used provided that a *dual* or Voronoi grid can be constructed. The Voronoi grid is defined as a complementary discretization to the primary grid, with nodes lying at the centroid of cells in the primary grid and connected to one another by edges that are perpendicular bisectors of the edges in the primary grid. As long as the Voronoi grid is available, integration of (7.12) over Ω can be decomposed into a sum of integrations over cells in the Voronoi grid. As before, let $u(\mathbf{r}_i)$ represent the fields for nodes $\mathbf{r}_i$ on the primary grid. Integration of (7.12) over the cell i is then relatively straightforward:

$$\int_i \nabla \cdot k \nabla u \, d\Omega = \int_{a..f} k \partial_n u \, d\Gamma \approx \sum_{\eta=a,...f} \ell_\eta k_\eta \left(u_\eta - u_i \right)$$

where ℓ_η is the length of the Voronoi edge $(\eta, \eta + 1)$ and k_η is the average value of k along this edge. Hence, this surface integral term couples nodal values u_i with their topological neighbors. The remaining integrals in (7.12) can be evaluated by low-order quadratures,

$$\int_i qu \, d\Omega \approx S_i q_i u_i \quad \text{and} \quad \int_i f \, d\Omega \approx S_i f_i$$

where S_i is the area of the *i*th Voronoi cell. As before, it can be seen that integration over the entire domain Ω yields a linear system of equations in terms of the unknown quantities u_i, and a complex symmetric coefficient matrix.

7.3.4 Numerical examples in two dimensions

While in 1D the off-diagonal elements of the magnetotelluric impedance tensor yield are equivalent, the same cannot be said of the 2D problem, owing to the presence of galvanic effects in the TM mode and their absence in the TE mode, and the breakdown of the concept of TE and TM modes in anything but an approximate sense at special locations in highly symmetric 3D problems. To further examine the nature of these effects, a simple 2D finite-element solution was constructed based on the previous discussion of linear nodal basis functions over triangular elements. Nominal node spacing in the region of interest was 20 m and the mesh was built using a constrained Delaunay triangulation (Shewchuk, 2002). The requisite derivatives for computing the elements of the impedance tensor were obtained using direct differentiation of linear basis functions, and the same mesh was used for all calculations, regardless of frequency or mode. Alternative methods for post-processing, such as local polynomial interpolants (e.g. Omeragić & Silvester, 1996) are available at a modest additional cost, including methods using surface classification schemes to accommodate kinks in the underlying functional form (e.g. Fleishman *et al.*, 2005). The latter is particularly relevant to the MT problem at hand when analyzing finite-element results on conductivity boundaries like the air–Earth interface (or sea–sediment for marine studies). However, no single post-processing scheme appears at present to have gained community-wide acceptance, barring case-specific, one-sided polynomial extrapolation from a region of relative uniformity toward the interface itself (e.g. Wannamaker & Stodt, 1987).

The models considered here are permutations of a buried rectangular block and topographic variations on a resistive 10 Ω m Earth. Perhaps the simplest of the latter is a simple step, a model that is interrogated over five decades in frequency (0.001–10.0 Hz) in both the TE (Figure 7.6) and TM (Figure 7.7) modes. Away from the step, unsurprisingly, at all frequencies in both modes, the apparent resistivity is that of the 10 Ω m half-space, and the TE and TM phases are 135° and 45°, respectively. This is the 1D result. Over this frequency band, apparent resistivities computed within a kilometer or so of the step show decisively non-1D behavior. Because the boundary condition in the TM mode is a constant magnetic field on the air–Earth interface, this behavior is due entirely to vertical derivatives of magnetic field that appear in the numerator of the apparent resistivity quotient for TM calculations. The upward deflection of apparent resistivity amplitude from the downthrown side of the step is from the "tightening" of the magnetic field isolines as they wrap upward around the step. Hence, their vertical derivative is larger (approaching infinity at the bottom corner of the step) and the apparent resistivity is likewise larger. From the upthrown side of the step, the opposite case is true. Apparent resistivity is decreased because the magnetic field isolines bend downward along the face of the step. Hence, their vertical derivative approaches zero, and so follows the apparent resistivity. For the TE mode, the vertical derivatives in the electric field manifest in the denominator of the apparent resistivity quotient and the resulting curves are a combination of effects from spatially variable electric field along the air–Earth interface along with its vertical derivative.

Given this qualitative analysis of the simple step response, the TE and TM mode responses of the ramp function (Figures 7.8 and 7.9) are easy to understand. Isolines of magnetic field wrap upward at the bottom left corner of the step, following the face of the ramp, and gently smooth out toward horizontal at the upper right corner of the ramp. Hence, an increase in the vertical derivative occurs near the downward corner of the ramp, and a decrease near the upper corner. Since the vertical derivative of the magnetic field drives the apparent resistivity calculation, the result is an increase and decrease in apparent resistivity, respectively, near these locations.

Turning attention to the problem of a buried block with no topographic effects to complicate its magnetotelluric expression, the TE mode for both conductive (Figure 7.10) and resistive (Figure 7.12) blocks demonstrates the familiar concept of magnetotelluric "depth sounding", where a frequency is used as a proxy for depth, following the characteristic decay length $\sqrt{2/\omega\mu\sigma}$. As demonstrated, high frequencies sample only the shallow part of the model, above the block inclusion, and yield apparent resistivities consistent with that of host half-space. The same result follows for low frequencies, where the presence of the block inclusion is volumetrically insignificant, leading to apparent resistivity values consistent with the 1D response for a 10 Ω m uniform Earth. Intermediate frequencies show either a low (Figure 7.10) or high (Figure 7.12) apparent resistivity at locations on the air–Earth interface over the block, and depending on the block resistivity relative to the host half-space. In the TM mode (Figures 7.11 and 7.13), however, the concept of a depth-sounding pseudo-section

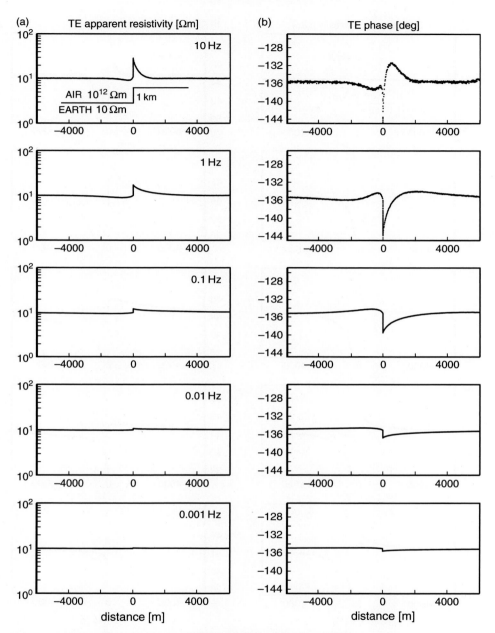

Figure 7.6. TE mode (a) apparent resistivity and (b) phase over frequencies 10, 1, 0.1, 0.01 and 0.001 Hz (top to bottom) for a 10 Ω m Earth with a 1 km high topographic step (inset, top left panel).

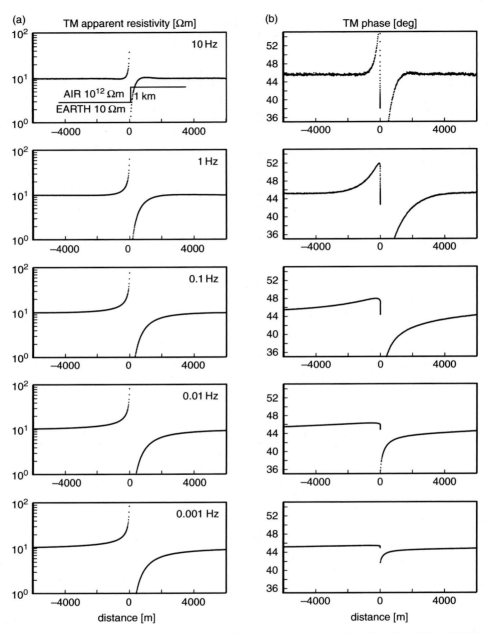

Figure 7.7. TM mode (a) apparent resistivity and (b) phase over frequencies 10, 1, 0.1, 0.01 and 0.001 Hz (top to bottom) for a 10 Ω m Earth with a 1 km high topographic step (inset, top left panel).

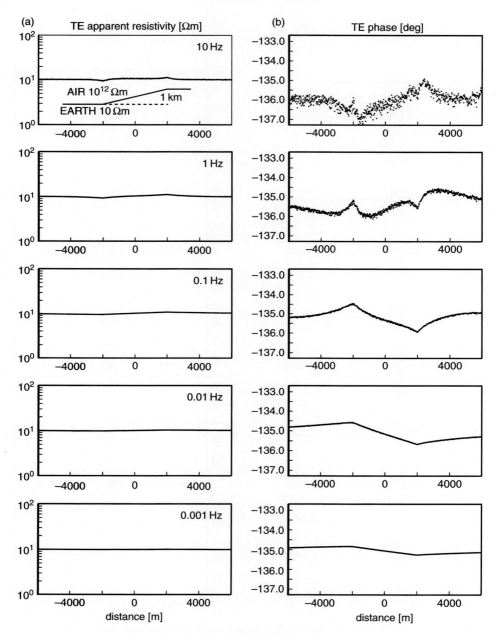

Figure 7.8. TE mode (a) apparent resistivity and (b) phase over frequencies 10, 1, 0.1, 0.01 and 0.001 Hz (top to bottom) for a 10 Ω m Earth with a 1 km high topographic ramp extending over 4 km (inset, top left panel).

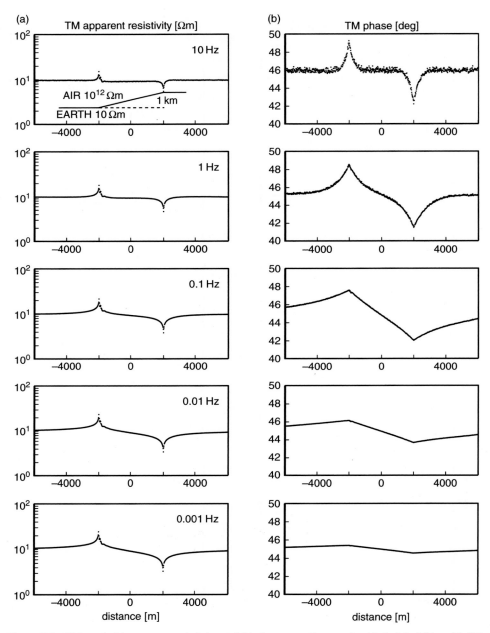

Figure 7.9. TM mode (a) apparent resistivity and (b) phase over frequencies 10, 1, 0.1, 0.01 and 0.001 Hz (top to bottom) for a 10 Ω m Earth with a 1 km high topographic ramp extending over 4 km (inset, top left panel).

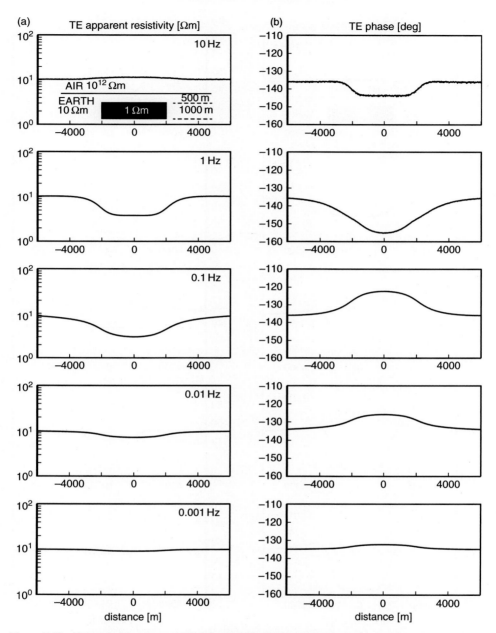

Figure 7.10. TE mode (a) apparent resistivity and (b) phase over frequencies 10, 1, 0.1, 0.01 and 0.001 Hz (top to bottom) for a 10 Ω m Earth with a conductive 1 Ω m, 4 km × 1 km block buried 500 m beneath the air–Earth interface (inset, top left panel).

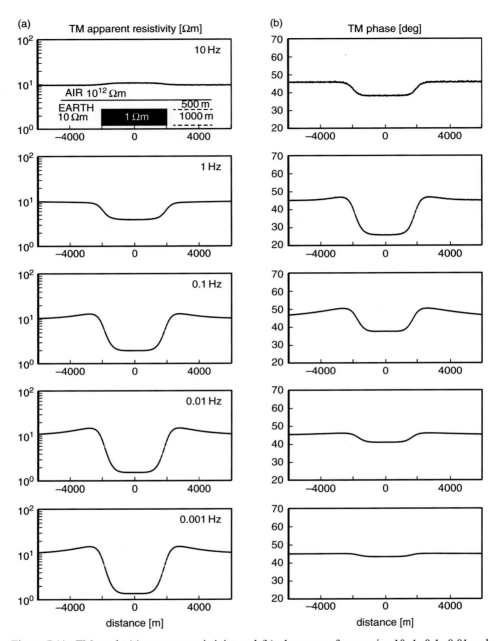

Figure 7.11. TM mode (a) apparent resistivity and (b) phase over frequencies 10, 1, 0.1, 0.01 and 0.001 Hz (top to bottom) for a 10 Ω m Earth with a conductive 1 Ω m, 4 km × 1 km block buried 500 m beneath the air–Earth interface (inset, top left panel).

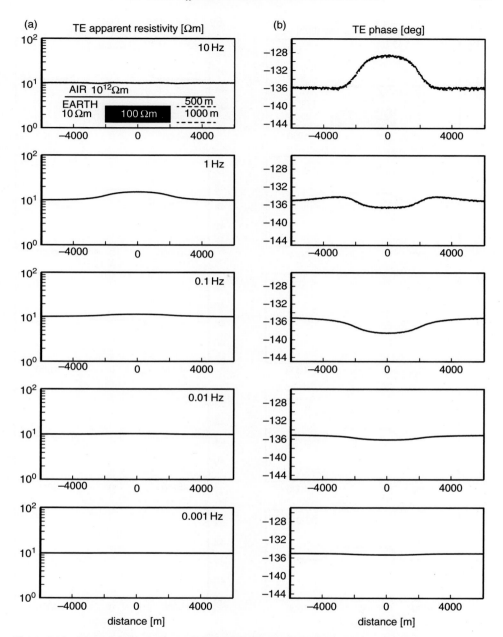

Figure 7.12. TE mode (a) apparent resistivity and (b) phase over frequencies 10, 1, 0.1, 0.01 and 0.001 Hz (top to bottom) for a 10 Ω m Earth with a resistive 100 Ω m, 4 km × 1 km block buried 500 m beneath the air–Earth interface (inset, top left panel).

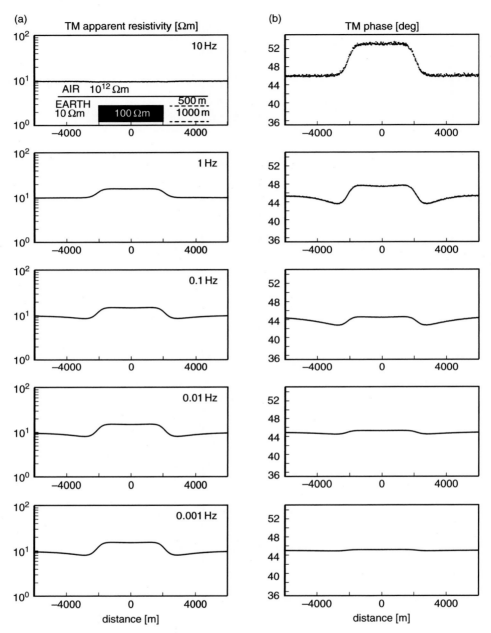

Figure 7.13. TM mode (a) apparent resistivity and (b) phase over frequencies 10, 1, 0.1, 0.01 and 0.001 Hz (top to bottom) for a 10 Ω m Earth with a resistive 100 Ω m, 4 km × 1 km block buried 500 m beneath the air–Earth interface (inset, top left panel).

breaks down. Because the TM electric currents are forced across conductivity boundaries, such as the vertical sides of the block, galvanic charges accumulate there and generate a corresponding electric field that persists for low frequencies. Hence, the "depth" estimate available from frequency sounding of the apparent resistivity amplitude in the TE mode is simply not present in the TM mode. However, at least two observations merit consideration before the TM mode is fully dismissed. First, in comparison to the TE mode, the TM mode response is generally sharper, or less smooth, giving a better estimate of the lateral position of the subsurface target. Second, the phase of the apparent resistivity does itself demonstrate some of the pseudo-section properties observed in the TE mode amplitude. The phase response of TM mode apparent resistivity shows generally 1D structure at long periods.

Although the Maxwell equations are linear in the source, they are not linear in conductivity. The induced electromagnetic fields of an Earth model with multiple subsurface targets is not simply the sum of the fields from multiple Earth models each with a single target. However, for the magnetotelluric problem, the response function *is* conveniently linear in the conductivity (see Chapter 2, Eq. ((2.70)) and following for details). This section on 2D model calculations concludes with demonstration of said linearity.

The remaining model results (Figures 7.14–7.17) are permutations of the buried block and topography models just discussed. Following theory, these 2D numerical calculations demonstrate that, indeed, the conductivity superposition principle holds for impedances – a comforting result until the realization sets in that exploiting such linearity requires solution to the Ricatti equation. The "law of conservation of evil" manifests again: no clever deed goes unpunished.

7.4 The leap to three dimensions

7.4.1 Finite-difference solutions in three dimensions

Perhaps the most popular approach (e.g. Mackie *et al.*, 1994; Newman & Alumbaugh, 1995; Siripunvaraporn *et al.*, 2002) for fully solving the magnetotelluric problem in three dimensions is finite differences, an approach facilitated by the "staggered grid" concept (Yee, 1966), in which only certain vector components of the electric or magnetic field are specified as the dependent variables in (7.1) and (7.2) at a given point instead of the complete vector field. Specifically, the 3D finite-difference mesh is composed of a tensor product grid of nodes in the three Cartesian directions, and its cells are regions of piecewise uniform electrical conductivity. The nodes of this mesh reside on conductivity boundaries, as do the edges, and these two sets of locations constitute points where the normal electric field is potentially discontinuous. The Yee grid defines the dependent variable in the FD system of equations as the single component of the electric field located at the midpoint of an edge and oriented parallel to it (Figure 7.18). Dealing with the discontinuity of the normal electric field is thus pushed off to the later, post-processing stage of evaluating the FD results, rather than polluting the formulation of the FD problem.

Defining the finite-difference stencils in 3D can be done in one of two ways: either through the differential form of the Maxwell equations, or from their integral form. Each

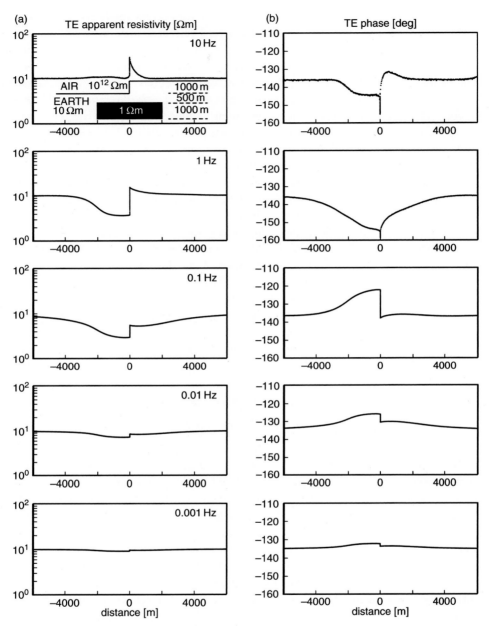

Figure 7.14. TE mode (a) apparent resistivity and (b) phase over frequencies 10, 1, 0.1, 0.01 and 0.001 Hz (top to bottom) for a 10 Ω m Earth with a conductive 1 Ω m, 4 km × 1 km block buried 500 m beneath the base of a 1 km tall topographic step (inset, top left panel).

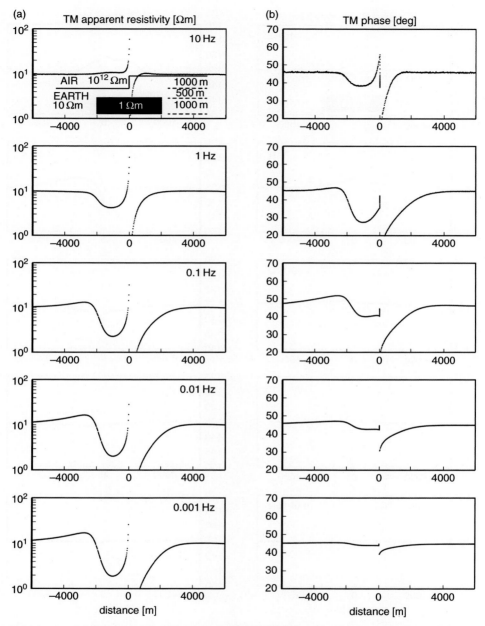

Figure 7.15. TM mode (a) apparent resistivity and (b) phase over frequencies 10, 1, 0.1, 0.01 and 0.001 Hz (top to bottom) for a 10 Ω m Earth with a conductive 1 Ω m, 4 km × 1 km block buried 500 m beneath the base of a 1 km tall topographic step (inset, top left panel).

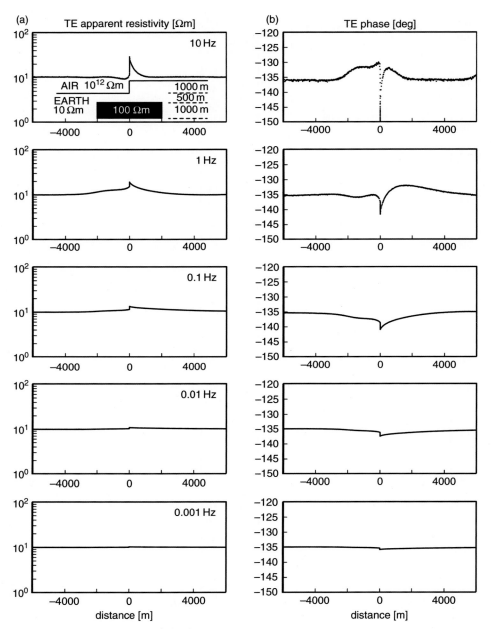

Figure 7.16. TE mode (a) apparent resistivity and (b) phase over frequencies 10, 1, 0.1, 0.01 and 0.001 Hz (top to bottom) for a 10 Ω m Earth with a resistive 100 Ω m, 4 km × 1 km block buried 500 m beneath the base of a 1 km tall topographic step (inset, top left panel).

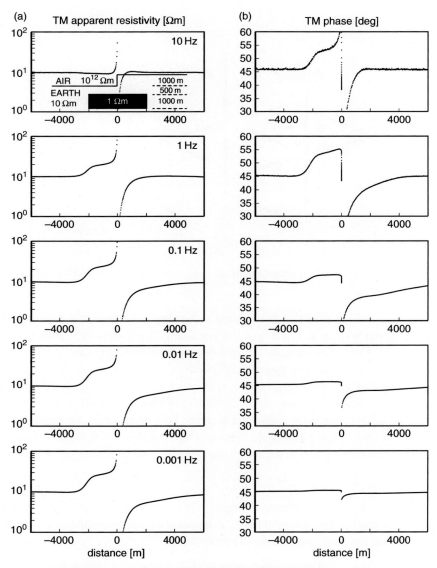

Figure 7.17. TM mode (a) apparent resistivity and (b) phase over frequencies 10, 1, 0.1, 0.01 and 0.001 Hz (top to bottom) for a 10 Ω m Earth with a resistive 100 Ω m, 4 km × 1 km block buried 500 m beneath the base of a 1 km tall topographic step (inset, top left panel).

results in an identical system of FD equations. The choice of approach is then purely a personal one that has no effect on the outcome. However, for completeness, each will be reviewed in an attempt to familiarize advocates of one approach with the advantages of the other. Both approaches start with the basic "paddlewheel" stencil that couples 12

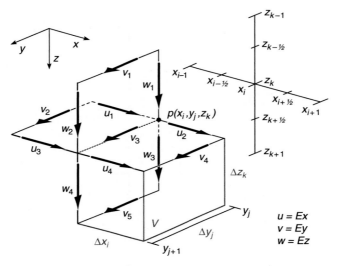

Figure 7.18. Staggered Yee grid discretization. Redrawn from Weiss & Constable (2006).

neighboring components of the electric field to the single component at the center of the stencil, thus resulting in a sparse (but not necessarily low-bandwidth) system of linear equations with 13 non-zero elements in each row of the coefficient matrix.

Adherents to the differential approach of constructing the FD stencil start by projecting the governing curl–curl vector equation (7.1) into each of the three Cartesian directions and simply writing down the requisite partial derivatives of the electric field. There is no need to consider the dual grid on whose edges the magnetic field components are similarly defined in a staggered sense. Take, for example, the y-directed stencil shown in Figure 7.18. Looking at only the curl–curl term, one is faced with approximating the following sum of derivatives:

$$\hat{\mathbf{y}} \cdot \nabla \times \nabla \times \mathbf{E} = \partial_z(\partial_y E_z - \partial_z E_y) - \partial_x(\partial_x E_y - \partial_y E_x) \tag{7.20}$$

Using the local enumeration scheme of Figure 7.18 to economize notation, it is straightforward to show that the derivatives of the E_y component are approximated by the same five-point Laplacian stencil as used for the 2D problem:

$$(\partial_{xx} + \partial_{xx})E_y = \frac{1}{\Delta \tilde{x}_i}\left(\frac{v_4 - v_3}{\Delta x_i} - \frac{v_3 - v_2}{\Delta x_{i-1}}\right) + \frac{1}{\Delta \tilde{z}_k}\left(\frac{v_5 - v_3}{\Delta z_k} - \frac{v_3 - v_1}{\Delta z_{k-1}}\right) \tag{7.21}$$

where $\Delta \tilde{x}_k = \frac{1}{2}(\Delta x_{k-1} + \Delta x_k)$ and $\Delta \tilde{z}_k = \frac{1}{2}(\Delta z_{k-1} + \Delta z_k)$. The remaining terms in equation (7.20) are dispatched in a similar way:

$$\partial_{xy}E_x = \frac{1}{\Delta \tilde{x}_i}\left(\frac{u_4 - u_2}{\Delta y_j} - \frac{u_3 - u_1}{\Delta y_j}\right) \tag{7.22}$$

and

$$\partial_{xz}E_z = \frac{1}{\Delta\tilde{z}_k}\left(\frac{w_4 - w3}{\Delta y_j} - \frac{w_2 - w_1}{\Delta y_j}\right) \tag{7.23}$$

Over a grid of $N_x \times N_y \times N_z$ nodes, there are $N_x(N_y - 1)N_z$ equations resulting from these y-directed stencils. Equations for the remaining two Cartesian directions follow in a similar fashion, resulting in a system of $(N_x - 1)N_yN_z + N_x(N_y - 1)N_z + N_xN_y(N_z - 1)$ linear equations and unknown coefficients, the values of the electric field on the mesh edges. As written, these particular finite-difference equations result in a non-symmetric coefficient matrix for the resulting linear system. The system can be symmetrized by scaling each of the x equations by the corresponding volume $\Delta\tilde{x}_i\,\Delta y_j\Delta z_k$, y edges by $\Delta x_i\Delta\tilde{y}_j\,\Delta z_k$, and $\Delta x_i\Delta y_j\Delta\tilde{z}_k$ for z.

Adherents to the integral approach start by observing that the line integral of electric field around any flap of the paddlewheel is approximately equal to the normal magnetic field at the flap's center multiplied by the area of the flap. A second line integral – this time of the magnetic field – along a rectangular path connecting the centers of the four flaps surrounding the stencil's center (Figure 7.19) relates the magnetic field back to the electric field, hence arriving at the same 13-point stencil previously derived by considering discrete differences, and with the *post hoc* symmetrization factors *mostly* in place. For example,

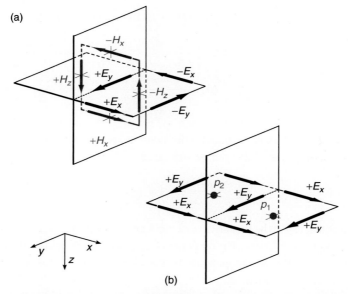

Figure 7.19. (a) Integral form of the Maxwell equations applied to the staggered grid. (b) Discrete differential form. Redrawn from Weiss & Constable (2006). Points p_1 and p_2 represent locations where first derivatives of field components are used in the discrete approximation of second derivatives.

consider again the y-directed components of (7.1). Faraday's law for each of the four flaps is approximated by

$$
i\omega\mu_0 H_z(x_{i+\frac{1}{2}}, y_{j+\frac{1}{2}}, z_k)\Delta x_i\Delta y_j = i\omega\mu_0 H_1\Delta x_i\Delta y_j
$$
$$
\approx (v_3 - v_4)\Delta y_j + (u_4 - u_2)\Delta x_i \tag{7.24a}
$$

$$
i\omega\mu_0 H_x(x_i, y_{j+\frac{1}{2}}, z_{k-\frac{1}{2}})\Delta z_{k-1}\Delta y_j = i\omega\mu_0 H_2\Delta z_{k-1}\Delta y_j
$$
$$
\approx (v_3 - v_1)\Delta y_j + (w_1 - w_2)\Delta z_{k-1} \tag{7.24b}
$$

$$
-i\omega\mu_0 H_z(x_{i-\frac{1}{2}}, y_{j+\frac{1}{2}}, z_k)\Delta x_{i-1}\Delta y_j = -i\omega\mu_0 H_3\Delta x_{i-1}\Delta y_j
$$
$$
\approx (v_3 - v_2)\Delta y_j + (u_1 - u_3)\Delta x_{i-1} \tag{7.24c}
$$

and

$$
-i\omega\mu_0 H_x(x_i, y_{j+\frac{1}{2}}, z_{k+\frac{1}{2}})\Delta z_k\Delta y_j = -i\omega\mu_0 H_4\Delta z_k\Delta y_j
$$
$$
\approx (v_3 - v_5)\Delta y_j + (w_4 - w_3)\Delta z_k \tag{7.24d}
$$

Approximating the line integral of the magnetic field around the center of the stencil according to Ampere's law results in an equivalent discrete expression of the curl–curl term as was found by substituting (7.22) and (7.23) and the five-point 2D Laplacian stencil into (7.21):

$$
i\omega\mu_0\sigma\left(x_i, y_{j+\frac{1}{2}}, z_k\right)v_3\Delta\tilde{x}_i\Delta\tilde{z}_k \approx i\omega\mu_0[(H_1 - H_3)\Delta\tilde{z}_k + (H_2 - H_4)\Delta\tilde{x}_i]
$$
$$
\approx \Delta\tilde{z}_k\left[\frac{(v_3 - v_2)}{\Delta x_{i-1}} + \frac{(v_3 - v_4)}{\Delta x_i} + \frac{(u_1 - u_2 - u_3 + u_4)}{\Delta y_i}\right] +
$$
$$
\Delta\tilde{x}_i\left[\frac{(w_1 - w_2 - w_3 + w_4)}{\Delta y_i} + \frac{(v_3 - v_5)}{\Delta z_k} + \frac{(v_3 - v_1)}{\Delta z_{k-1}}\right] \tag{7.25}
$$

Multiplication of (7.25) by Δy_j completes the symmetrization process, and the first, second, fifth and sixth quotients in (7.25) can be written as

$$
-\Delta\tilde{x}_i\Delta y_j\Delta\tilde{z}_k\left[\frac{1}{\Delta\tilde{x}_i}\left(\frac{v_4 - v_3}{\Delta x_i} - \frac{v_3 - v_2}{\Delta x_{i-1}}\right) + \frac{1}{\Delta\tilde{z}_k}\left(\frac{v_5 - v_3}{\Delta z_k} - \frac{v_3 - v_1}{\Delta z_{k-1}}\right)\right] \tag{7.26}
$$

which, as expected, is just the (negative) discrete Laplacian operator scaled by the volume of the cell surrounding the v_3 edge. The remaining terms in (7.25) constitute the cross-coupling terms between the different components of the electric field vector, which after multiplication by Δy_j may be written as

$$
\Delta\tilde{x}_i\Delta y_j\Delta\tilde{z}_k\left[\frac{(u_1 - u_2 - u_3 + u_4)}{\Delta\tilde{x}_i\Delta y_j} + \frac{(w_1 - w_2 - w_3 + w_4)}{\Delta y_j\Delta\tilde{z}_k}\right] \tag{7.27}
$$

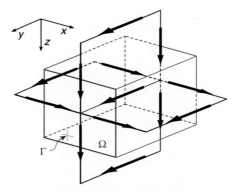

Figure 7.20. Voronoi cell for *y*-directed edge. Redrawn from Weiss & Constable (2006).

and are simply (7.22) and (7.23) scaled by the volume of the cell surrounding the v_3 edge. Hence, the two formulations are equivalent.

For completeness, one could also choose a finite-volume approach to solve (7.1) on the staggered grid. As with the 2D case, the idea is to piecewise integrate the governing partial differential equation over the cells of the Voronoi mesh (Figure 7.20), the volumes of which were the scaling factors in the discrete version of the finite volume. Transforming piecewise volume integrations over the Voronoi cells into surface integrals with outward-pointing normals **n** using the vector identity

$$\int_{\Omega} \nabla \times \nabla \times \mathbf{E} \, d\Omega = \int_{\Gamma} \boldsymbol{n} \times \nabla \times \mathbf{E} \, d\Gamma$$

it is straightforward, albeit tedious, to show that one ends up with an equivalent system of linear equations. At the end of the day, choosing the integral form or the discrete form of the finite difference system of equations or choosing the finite-volume approach leads to the *exact same coefficient matrix* for the resulting linear system of equations. Note, however, that choice of the FD integral or FV formulation leads to a more natural way of handling source terms that are not symmetric with respect to mesh geometry and hence offer a means of minimizing mesh bias through a high-order integration of the source terms.

There is one subtle difference in how the source vector is dealt with in each of the three approaches. Presuming that a secondary field formulation is being used (or, even more devious, delta function sources required for an adjoint calculation), the discrete approach offers no clear way to deal with source function structure that lies between the nodes (or edges) of the mesh. In fact, some might argue that wild source structure between the nodes (or edges) is irrelevant because it is not present where it counts: precisely at the mesh edges. Clearly this violates an implicit assumption on the regularity of the functions one is dealing with in the finite-difference formulation, but nonetheless the roadmap capturing intranode source structure is poorly defined and the algorithm developer is

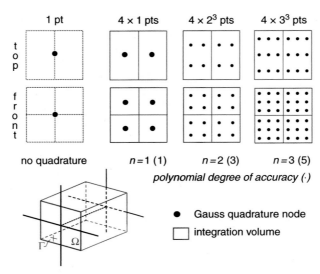

Figure 7.21. Quadrature nodes for evaluating source vector in (FV) method for solving the curl–curl equations. Redrawn from Weiss & Constable (2006).

faced with devising an *ad hoc* scheme (typically through weighted averages or expectation values) to capture such structure on the discrete locations of the mesh where the dependent variables are defined. The case is somewhat simplified in the integral formulation of finite-difference equations because the intranode structure can be captured in the single surface integral that arises from application of Ampere's law – the second step in the electric field formulation (7.1). In this regard, the finite-volume approach has an advantage because the source term is integrated volumetrically (Figure 7.21), most easily by a weighted quadrature scheme (Weiss & Constable, 2006). The end-result is that the weighted average schemes typically used in the finite-difference discretization find some theoretical foundation when looked at from the perspective of the finite-volume method.

7.4.2 Finite-element solutions in three dimensions

Much as it does in 2D, the promise of finite-element solutions for the 3D magnetotelluric problem lies in the unstructured grid upon which they are based and the opportunities that such a discretization offers to model natural topography on Earth's surface, as well as to address numerical issues such as localized mesh refinement. Hence, there is ample motivation to consider the finite-element variational (or its parent class, the variational problem). However, note that in a perfectly resistive air region, (7.1) is singular and (7.2) is undefined. For the second-order finite-difference methods described in the previous subsection, a sufficiently small but non-zero air conductivity is typically sufficient to minimize these problems, along with an *ad hoc* "divergence correction procedure" (Smith, 1996). However, the construction of well-formulated finite-element problems requires a bit more rigor. As such, two dominant approaches have emerged: work with potentials instead of fields; and define a vector space *and* formulation that is inherently divergence-free.

The simple approach recasts the governing equations in terms of vector and scalar potentials instead of the fields themselves. The consequence is that, if the potentials are well behaved and continuous, then nodal elements may yet be applied – a result that spares one from considering a more sophisticated vector space than previously discussed for 2D. On the other hand, recovering the fields requires differentiation of the potentials, and this is an operation that commonly reduces the numerical accuracy of the final result. Borrowing from the electrical engineering literature, early success in using the $(\mathbf{A} - \Phi, \Psi)$ was realized by Everett & Schultz (1996) as applied to deep Earth global geomagnetic sounding. Because the magnetic induction $\mathbf{B}$ is divergence-free, it is written in terms of the curl of a magnetic vector potential $\mathbf{A}$ in the conducting region. Hence, the electric field in Earth is given by $\mathbf{E} = i\omega(\mathbf{A} - \zeta\nabla\Phi)$, where ζ is an arbitrary constant introduced to stabilize the numerical problem and, following the finite-element literature, an $exp(-i\omega t)$ time dependence (in contrast to (7.1) and (7.2)) for reasons that will become clear momentarily. The introduction of the electric scalar potential Φ provides an additional set of unknowns whose uniqueness is set by the choice of gauge. Enforcement of the Coulomb gauge ($\nabla \cdot \mathbf{A} = 0$) in the air region results in the coupled set of differential equations (Bìrò & Preis, 1989),

$$\begin{pmatrix} -\nabla^2 - i\omega\mu_0\boldsymbol{\sigma} & i\omega\mu_0\boldsymbol{\sigma}\zeta\nabla \\ \zeta\nabla \cdot i\omega\mu_0\boldsymbol{\sigma} & -\zeta^2\nabla \cdot i\omega\mu_0\boldsymbol{\sigma}\zeta\nabla \end{pmatrix} \begin{pmatrix} \mathbf{A} \\ \Phi \end{pmatrix} = \begin{pmatrix} \mu_0\mathbf{J}_s \\ -\zeta\mu_0\nabla \cdot \mathbf{J}_s \end{pmatrix} \quad (7.28)$$

A finite-element formulation of (7.28) based on linear finite elements is possible because of the continuity of these potentials. There are 16 unknown degrees of freedom defined over a tetrahedral element: three for each of the components of $\mathbf{A}$ at each of the four nodes of the element, plus an additional four for the scalar potential Φ at each node. In the final formulation, unknowns corresponding to the three components of the vector potential are accommodated by the three Cartesian components of the differential equation resulting from substitution of both vector and scalar potentials. The fourth and remaining unknown, the scalar potential, is matched by the appropriate choice of gauge condition, thus forming a square system of equations and unknowns.

Construction of the variational problem follows the methodology already discussed for the 2D scalar case provided that a vector test function v is used to contract the first equation in (7.28) into a scalar. For the second equation in (7.28) – the scalar equation – a simple scalar test function w will suffice, as was done for 2D. Finding the bilinear form for the variational formulation starts with contraction of (7.28) with the row vector of the test functions $\mathbf{v}$ and w and subsequent integration over the model domain Ω:

$$\int_{\Omega}\left[\begin{pmatrix} \mathbf{v}^{\mathrm{T}} & w \end{pmatrix}\begin{pmatrix} -\nabla^2 - i\omega\mu_0\boldsymbol{\sigma} & i\omega\mu_0\boldsymbol{\sigma}\zeta\nabla \\ \zeta\nabla\cdot i\omega\mu_0\boldsymbol{\sigma} & -\zeta^2\nabla\cdot i\omega\mu_0\boldsymbol{\sigma}\nabla \end{pmatrix}\begin{pmatrix} \mathbf{A} \\ \Phi \end{pmatrix}\right]d\Omega$$

$$= \int_{\Omega}\left[\begin{pmatrix} \mathbf{v}^{\mathrm{T}} & w \end{pmatrix}\begin{pmatrix} \mu_0\mathbf{J}_s \\ -\zeta\mu_0\nabla\cdot\mathbf{J}_s \end{pmatrix}\right]d\Omega \qquad (7.29)$$

After integration by parts, (7.29) leads to a predictable variation on the usual statement of the variational problem, "Find $a(\mathbf{A}, \mathbf{v}|\Phi, w) = \mu_0(\mathbf{J}_s, \mathbf{v}| - \zeta\nabla\cdot\mathbf{J}_s, w)$ for all $\mathbf{v}$, w in the function space ...", where, assuming a Dirichlet boundary condition on Ω, the expanded bilinear form $a(\cdot|\cdot)$ and inner product $(\cdot|\cdot)$ are defined as follows:

$$a(\mathbf{A}, \mathbf{v}|\Phi, w) = \int_{\Omega}(\nabla\mathbf{A}\cdot\nabla\mathbf{v}) - i\omega\mu_0\boldsymbol{\sigma}(A\cdot\mathbf{v}) + \zeta^2 i\omega\mu_0\boldsymbol{\sigma}(\nabla\Phi\cdot\nabla w)$$
$$+ \zeta i\omega\mu_0\boldsymbol{\sigma}(\mathbf{v}\cdot\nabla\Phi) + \zeta i\omega\mu_0 w\nabla\cdot\boldsymbol{\sigma}\mathbf{A}\,d\Omega \qquad (7.30a)$$

and

$$(\mathbf{J}_s, v| - \zeta\nabla\cdot\mathbf{J}_s, w) = \int_{\Omega}(\mathbf{J}_s\cdot\mathbf{v}) - \zeta w\nabla\cdot\mathbf{J}_s\,d\Omega \qquad (7.30b)$$

At this point it is clear that the variational formulation for the full 3D problem of electromagnetic induction requires little more mathematical machinery than has already been utilized for 2D in Section 7.3.2. Observe that the switch to an $exp(i\omega t)$ time dependence leads to a complex symmetric global coefficient matrix for the finite-element problem. If the same Helmholtz decomposition was applied to a finite-difference formulation, such a switch is not needed to maintain matrix symmetry.

The high number of degrees of freedom per element has led to investigation of more computationally economical vector spaces for the potential formulation, namely, the introduction of the Whitney (1957) or Nédélec (1980) "edge element" (Figure 7.22). Defined as $\varphi_{ij} = \varphi_i\nabla\varphi_j - \varphi_j\nabla\varphi_i$, this class of finite elements contains only six degrees of freedom, thus reducing the total number of degrees of freedom for the $(\mathbf{A} - \Phi, \Psi)$ formulation from 16 to 10.

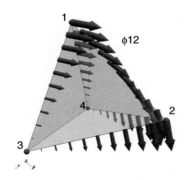

Figure 7.22. Vector edge element φ_{12} evaluated along the perimeter of the tetrahedron.

However, when applied to potential formulations, the lack of continuous normal components on the element faces requires alternative gauging strategies, including "ungauged formulations" based on building a consistent right-hand side source vector – see Bìrò (1999) for details. Again, use of a vector test function is required since one is still working with a potential formulation. In this case, the test function is drawn from the space of edge elements over the mesh rather than nodal elements. Note also that one is also required to work with the "curl–curl" equation for the vector potential. However, the null space in this term is strategically eliminated by the variational formulation in conjunction with the edge elements, rather than simply by the existence of edge elements themselves. Eliminating the null space makes finite-element discretization of the curl–curl equation into an invertible linear system.

7.4.3 Numerical examples in three dimensions

To illustrate some of the features seen in the fully 3D magnetotelluric response, a set of models was constructed and evaluated using the staggered grid finite-volume code described in Weiss & Constable (2006), modified since publication to include plane-wave sources. Drawing inspiration from the problem of imaging continental volcanic provinces, the models consist of a layered background impregnated with spatially extended conductive bodies crudely representing hypothetical "fluid-bearing" or "melt" zones. The 1D model consists of a layered structure containing a 15 km thick surface layer of 1000 Ω m material, over a 35 km thick layer of 500 Ω m, over a 100 Ω m semi-infinite half-space. In the shallow part of the model there resides a thin L-shaped (35 km on the long axis, 15 km on the short axis, in map view) upper crustal conductive zone of 1 Ω m resistivity spanning 1–10 km in

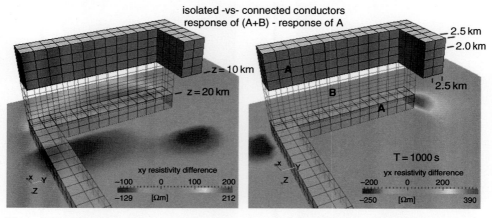

Figure 7.23. Effect on apparent resistivity of a mid-crustal electrical "short" (B) between and upper and lower crustal conductors (A). Electrical conductors A and B are all 1 Ω m dike-like structures, embedded in a 1D ground model (see text for details). Shown by the grayscale (color scale) is the difference in apparent resistivity due to the presence of block B at a frequency 0.001 Hz (*T* = 1000 s). (See plate section for color version.)

depth (upper structure "A" in Figure 7.23). This conductive target would maximally attenuate incident signals, and thus would tend to obscure structures located beneath it. In the lower crust, an additional, and electrically distinct, L-shaped body is rotated 90° in azimuth relative to the upper body. This is done to help distinguish its signature from that of the shallow conductive target. The lower body, also of 1 Ω m resistivity, extends from 20 to 50 km in depth and is 115 km × 35 km in width and length, respectively (lower structure "A" in Figure 7.23). Finally, the effect of electrical connectivity of these two distinct bodies is considered by introducing a thin 1 Ω m connecting slab between depths 20 and 50 km along the common sides of the "L" (mid-crustal structure "B" in Figure 7.23).

Models were evaluated on a regular 100 km × 100 km × 100 km grid with 2.5 km node spacing in the horizontal directions and 2.0 km spacing in the vertical. The finite-difference system of linear equations was therefore small (228 930 unknowns) and relatively quick to solve using a quasi-minimal-residual iterative solver, typically converging within a few hundred iterations (frequency of 10 Hz) to a few thousand (period of 1000 s) with a simple Jacobi preconditioner.

Figures 7.24–7.26 illustrate the apparent resistivity and phase for the off-diagonal impedance tensor elements for shallow, shallow plus deep and fully connected models,

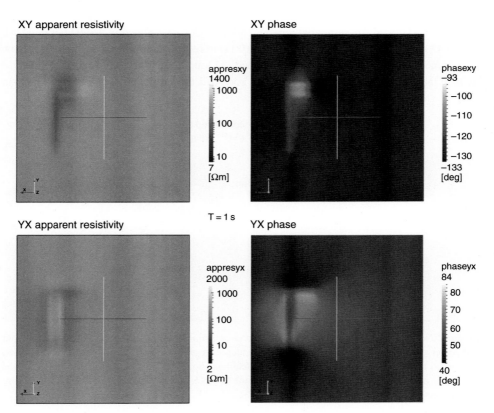

Figure 7.24. Map view of 100 km × 100 km area showing 3D MT response of shallow conductor at period *T* = 1 s.

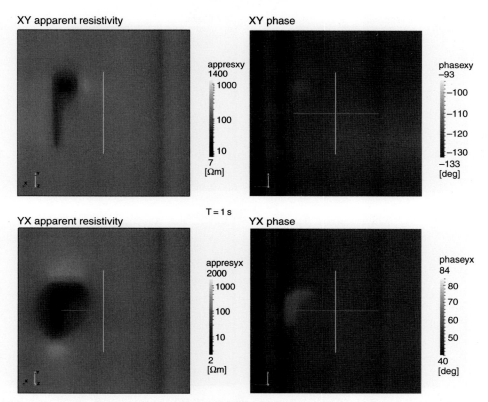

Figure 7.25. Map view of 100 km × 100 km area showing 3D MT response of shallow conductor and isolated deep conductor at period $T = 1$ s.

respectively, at a period of 1 s. Notice that the outline of the shallow conductor is clearly evident in all three model responses; however, the effect of connectivity (Figure 7.26) is clearly to sharpen up the apparent resistivity and phase of the shallow conductive target, almost to the same fidelity as that of the isolated body itself. Thus, one could conclude that the response of the isolated shallow (Figure 7.24) and fully connected (Figure 7.26) models is dominated by the upper part of the shallow conductive target, and that the resistive gap that is present in the model with distinct conductors (Figure 7.25) is significant enough to generate a response that obstructs the uppermost part of the shallow target. Hence, it is the gap in the composite conductive assemblage that generates the greatest effect in the model at this frequency rather than the presence of some conductive pathway.

Looking deeper into the pseudo-section at a longer period of 100 s, one finds that the effect of the top of the shallow conductor is minimized (Figure 7.27) and that the dominant response comes from the mid-crustal conductor (Figure 7.28) and the underlying "L"

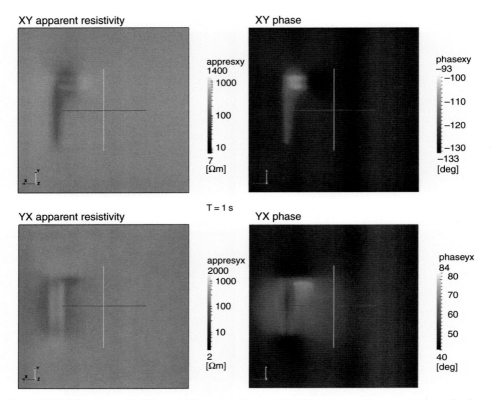

Figure 7.26. Map view of 100 km × 100 km area showing 3D MT response of trans-crustal conducting assemblage at period $T = 1$ s.

structure in the lower crust (Figure 7.29). Indeed, the presence of the mid-crustal connecting slab serves to sharpen up both the apparent resistivity and phase responses when compared to the isolated conductive bodies.

7.5 Closing remarks

It is clear that there is a well-stocked arsenal of numerical methods available to practitioners of multidimensional magnetotelluric analysis, and, fortunately, much of what is required for posing and solving the 2D and 3D problems can be learned from careful analysis of the 1D problem itself. This is comforting news considering that it is also clear from the discussion here and the abundance of examples from the literature that there is no single method that will excel in all circumstances or exploration scenarios. Hence, a bit of creativity and familiarity with the range of usable conditions that a given method may

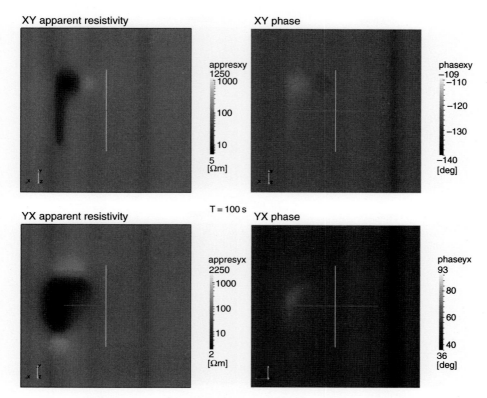

Figure 7.27. Map view of 100 km × 100 km area showing 3D MT response of shallow conductor at period $T = 100$ s.

offer is a key element for the continued advancement of computational electromagnetics in geophysics.

For the sake of brevity, a few important topics were excluded from the present discussion, such as parallelization on traditional cluster/symmetric multiprocessing platforms in addition to specialty hardware such as graphic processing units of field programmable gate arrays. Also left out was much mention of the black art of preconditioning linear systems, through either physics-based approximate solutions or more fundamental mathematical theory. Not to be ignored is the issue of algorithmic efficiency and the difficulty in weighing its relevance in the face of highly variable programming and hardware efficiency. And, finally, it is worth noting that, as the structural complexity of geophysical models increases, so it parallels the need to visualize and construct Earth models in 3D, ultimately in a time-lapse sense. A great deal of time is still spent on mesh design and model construction, and the effects of such efforts are not trivial: whether a model can be evaluated or not often depends intimately on the quality of the underlying mesh. It can be troublesome enough to drive analysts to hang up their 3D boots and stick to 2D! But if the future is 3D, then efficient methods for mesh design, visualization and

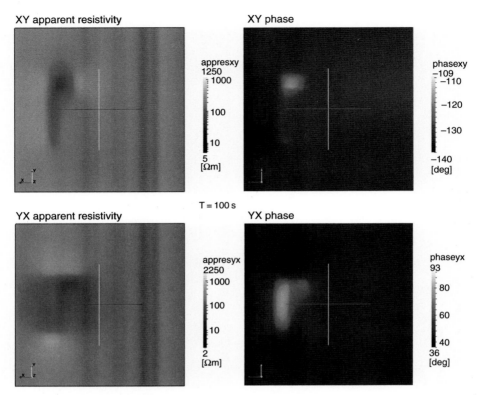

Figure 7.28. Map view of 100 km × 100 km area showing 3D MT response of shallow conductor and isolated deep conductor at period $T = 100$ s.

optimization are part of that future, too, and thus it presents an interesting opportunity for new cross-disciplinary collaborations to be realized.

Acknowledgements

I am dearly indebted to Alan Chave and Alan Jones for the opportunity to contribute to this book, and equally so for their patience in guiding me through the process. To be in the company of my esteemed co-contributors is an honor I cherish, and it would not have been possible without those who have taken the time over the years to share their numerical insights with me and to help me hone my own sense of computational acumen. There are too many to mention them all here, but special thanks are extended to Raytcho Lazarov, Mark Everett, Adam Schultz and Greg Newman. Inspiration for the 3D crustal "fluid" models came from ongoing discussions with Adam Schultz and Matt Fouch during the fall of 2010.

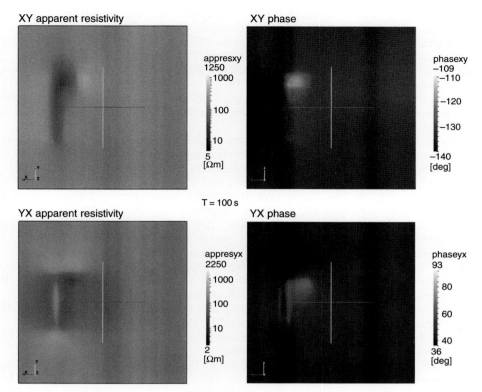

Figure 7.29. Map view of 100 km × 100 km area showing 3D MT response of trans-crustal conducting assemblage at period T = 100 s.

References

Bìrò, O. (1999). Edge element formulations of eddy current problems. *Comput. Meth. Appl. Mech. Eng.*, **169**, 391–405.

Bìrò, O. & K. Preis (1989). On the use of the magnetic vector potential in the finite element analysis of three-dimensional eddy currents. *IEEE Trans. Magn.*, **25**, 3145–3159.

D'Erceville, I. & G. Kunetz (1962). The effect of a fault on the Earth's natural electromagnetic field. *Geophysics*, **27**, 651–665.

Everett, M. E. & A. Schultz (1996). Electromagnetic induction in a heterogeneous sphere: azimuthally symmetric test calculations and the response of an undulating 660-km discontinuity. *J. Geophys. Res. – B*, **101**, 2765–2783, doi:10.1029/95JB03541.

Fleishman, S., D. Cohen-Or & C. T. Silva (2005). Moving least-squares fitting with sharp features. *ACM Trans. Graphics*, **23**, 544–552.

Jackson, J. D. (1962). *Classical Electrodynamics*. New York: John Wiley.

Mackie, R. L., J. T. Smith & T. R. Madden (1994). Three-dimensional modeling using finite difference equations: the magnetotelluric example. *Radio Sci.*, **29**, 923–935.

Nédélec, J. C. (1980). Mixed finite elements in R^3. *Numer. Math.*, **35**, 315–341.

Newman, G. A. & D. L. Alumbaugh (1995). Geophysical modeling of airborne electromagnetic responses using staggered finite differences. *Geophys. Prosp.*, **43**, 1021–1042.

Omeragić, D. & P. P. Silvester (1996). Numerical differentiation in magnetic field postprocessing. *Int. J. Numer. Model.: Electron. Netw. Devices Fields*, **9**, 99–113.

Shewchuk, J. R. (2002). Delaunay refinement algorithms for triangular mesh generation. *Comput. Geom.: Theory Appl.*, **22**, 21–74.

Siripunvaraporn, W., G. Egbert & Y. Lenbury (2002). Numerical accuracy of magnetotelluric modeling: a comparison of finite difference approaches. *Phys. Earth Planet. Inter.*, **54**, 721–725.

Smith, J. T. (1996). Conservative modeling of 3-D electromagnetic fields. Part II: Biconjugate gradient solution and an accelerator. *Geophysics*, **61**, 1319–1324.

Smythe, W. R. (1939). *Static and Dynamic Electricity.* New York: McGraw-Hill.

Stratton, J. A. (1941). *Electromagnetic Theory.* New York: McGraw-Hill.

Wannamaker, P. E. & J. A. Stodt (1987). A stable finite element solution for two-dimensional magnetotelluric modeling. *Geophys. J. Int.*, **88**, 277–296.

Weaver, J. T., B. V. Le Quang & G. Fischer (1985). A comparison of analytical and numerical results for a 2-D control model in electromagnetic induction – I. *B*-polarization calculations. *Geophys. J. R. Astron. Soc.*, **82**, 263–277.

Weaver, J. T., B. V. Le Quang & G. Fischer (1986). A comparison of analytical and numerical results for a 2-D control model in electromagnetic induction – II. *E*-polarization calculations. *Geophys. J. R. Astron. Soc.*, **87**, 917–948.

Weiss, C. J. & S. Constable (2006). Mapping thin resistors with marine EM methods: modeling and analysis in 3D. *Geophysics*, **71**, G321–332.

Whitney, H. (1957). *Geometric Integration Theory*, Princeton Mathematical Series, Vol. **21**. Princeton, NJ: Princeton University Press.

Yee, K. S. (1966). Numerical solution of initial boundary value problems involving Maxwell's equations in isotropic media. *IEEE Trans. Antennas Propag.*, **AP-14**, 302–307.

8

The inverse problem

WILLIAM L. RODI AND RANDALL L. MACKIE

8.1 Introduction

In previous chapters, various methods were considered for solving the magnetotelluric *forward* problem: predicting the MT responses that would be observed at arbitrary locations and frequencies given a hypothetical model of Earth's resistivity structure. This chapter will consider the *inverse* problem: inferring the Earth's resistivity structure on the basis of actual observations of MT responses at specific locations and frequencies. Solving the MT inverse problem involves finding one or more models of resistivity whose predicted responses match, or fit, the observed responses. Of course, the notion of "fit" must take into consideration measurement errors in the observations, computational errors in theoretical predictions, and even the appropriateness of the model itself (e.g. the use of 1D models in a 3D heterogeneous Earth). In most circumstances, a wide range of resistivity models will provide acceptable fits to a set of MT observations – that is, solutions to the inverse problem are non-unique – and it becomes necessary to impose additional constraints on the model or to find a concise characterization of the range of acceptable models.

The earliest efforts in MT inversion, as in most other types of geophysical inversion, relied on a trial-and-error process to fit observed data. At each step, forward modeling is performed on a trial resistivity model to obtain predicted responses that are compared to the observations. The discrepancy between the observed and predicted responses is used to intuit adjustments to the resistivity model for a new trial. Trial-and-error inversion is a cumbersome process, needless to say, and it often forces the restriction of resistivity models to a few free parameters (e.g. the resistivities and thicknesses of only a few planar layers) that are usually inadequate to fit field data.

The establishment of geophysical inverse theory as an active field of research in the 1960s (e.g. Backus & Gilbert, 1967), combined with the increasing availability of computers, spurred a tremendous amount of research into the development of more automated and rigorous inversion procedures for geophysical data. Because of numerical simplicity and tractability, the early efforts at inversion dealt with 1D models that allow variation in electrical conductivity as a function of depth – see Whittall & Oldenburg (1992) for a detailed review of a wide variety of 1D MT inverse methods.

The Magnetotelluric Method: Theory and Practice, ed. Alan D. Chave and Alan G. Jones. Published by Cambridge University Press © Cambridge University Press 2012.

The earliest work in nonlinear MT inversion focused on automated schemes for fitting data with simply parameterized 1D models. In particular, Wu (1968) and Patrick (1969) developed iterative schemes for finding the resistivity and thickness of a few layers that minimized a least-squares measure of the discrepancy between the observed and predicted responses. This work was followed by algorithmic modifications to improve numerical stability and convergence (e.g. Jupp & Vozoff, 1975; Shoham *et al.*, 1978; Larsen, 1981). Meanwhile, others (e.g. Parker, 1970; Oldenburg, 1980) tackled the problem of non-uniqueness in MT inverse problems by applying the concepts introduced by Backus & Gilbert (1967, 1968, 1970). In these studies, resistivity was assumed to vary continuously with depth, and non-uniqueness was characterized by the accuracy and spatial resolution of localized averages of the resistivity profile. Subsequent workers (e.g. Larsen, 1981) integrated the earlier efforts into a methodology for finding least-squares solutions and analyzing their non-uniqueness in the Backus–Gilbert sense. However, the concepts of Backus and Gilbert were not designed to provide optimal models that satisfy the data within a prescribed accuracy. Further, the Backus–Gilbert approach presumes that non-uniqueness stems only from the inability of the data to resolve details of the model function on a small spatial scale, which is an oversimplification in many geophysical inverse problems, as Jackson (1979) pointed out.

Concurrently with these early approaches, others examined exact nonlinear inverse methods that not only directly determined conductivity models but also established uniqueness and existence theorems for problems with complete and precise data and certain classes of smooth models (Bailey, 1970; Weidelt, 1972). Others established direct methods for less smooth models, including Loewenthal (1975), who formulated a direct method for models with equal penetration depth layers, and Parker (1980), who derived a complete theory for the existence of solutions for 1D conductivity models. Unlike the others, Parker also considered the more realistic situation of data comprising a finite number of possibly imprecise measurements. He set out to determine under what conditions such data are compatible with a 1D conductivity model and, most importantly, how to construct the model when it exists. Parker (1980) showed that the forward problem of electromagnetic induction has solutions for a very broad class of models, including non-smooth conductivity models. In particular, he demonstrated that, for imprecise data, conductivity models consisting of delta functions with finite conductance gave the lowest possible misfit (such as in the traditional χ^2 sense). Parker (1980) termed these models D+ models (see Chapter 4 for the mathematical details of these methods).

Electrical conductivity models consisting of delta functions are not, however, geologically plausible. Later work by Parker & Whaler (1981) showed that smoother models could be found that had χ^2 misfits close to the global minimum of the D+ solution, but were more geologically reasonable. The fact that minimum-misfit 1D models consist of sequences of delta functions with finite conductance also explains the general tendency of least-squares inversion algorithms to find oscillatory resistivity profiles to reach the same level of misfit (Chave & Booker, 1987; Smith & Booker, 1988). This suggests the usefulness of including some type of penalty on the spatial roughness of the resistivity profile, or stabilizing the solution in some other way, when fitting MT data.

While all of the early work on direct nonlinear inversions, including uniqueness and existence theorems, was of fundamental importance, the practicality and usefulness of modern inversion schemes really gained ground following the work of Constable *et al.* (1987) and Smith & Booker (1988), who introduced the concept of regularized, smoothest model inversion for 1D layered-Earth models to the magnetotelluric community. This approach attempts to fit the data within some acceptable tolerance while minimizing the amount of structure in the resistivity profile. The concept of regularizing an inverse problem to find smooth solutions is often attributed to Tikhonov & Arsenin (1977), but similar techniques were developed independently by others in the mathematical community, such as Foster (1961), Phillips (1962), Twomey (1963) and Franklin (1970).

Meanwhile, application of Bayesian inference methods to geophysics provided an alternative way to stabilize the solution to the inverse problem by specifying prior probability information on the model. While applied mainly to seismic problems (e.g. Tarantola, 2005), it has sometimes been applied to electrical resistivity problems (Yang, 1999; Malinverno, 2002) and also to the MT problem (e.g. Tarits *et al.*, 1994; Grandis *et al.*, 1999).

As computing power grew, and as the limitations of 1D approaches became better understood, regularized least-squares solutions were also being applied to 2D problems. Some prominent examples include Jiracek *et al.* (1987), Madden & Mackie (1989), Rodi (1989), Sasaki (1989), de Groot-Hedlin & Constable (1990), Smith & Booker (1991), Siripunvaraporn & Egbert (2000) and Rodi & Mackie (2001). While many of these papers investigate algorithmic shortcuts to bypass the generation and inversion of the large matrices associated with a direct application of a nonlinear least-squares approach, they all attempted to minimize data misfit in conjunction with some measure of model roughness. Practical regularized nonlinear 3D inversions were initially carried out primarily for industrial applications, but mostly using massively parallel computers, as noted by Newman & Alumbaugh (2000) and Mackie *et al.* (2001). However, regularized least-squares solutions are also being more routinely applied to smaller 3D problems. Some examples include Sasaki (2001, 2004), Siripunvaraporn *et al.* (2005), Farquharson & Craven (2009) and Siripunvaraporn & Egbert (2009). A comprehensive review of 2D and 3D inversion algorithms can be found in Avdeev (2005), and a more recent review of 3D inversion algorithms in Siripunvaraporn (2012).

8.1.1 Chapter plan

The field of geophysical inverse theory and its application to magnetotellurics comprises such a large body of work that it would be impossible to describe every major topic with reasonable depth and clarity in a single chapter of a book. Besides, there are already several excellent books on geophysical inverse theory, including ones by Menke (1989), Parker (1994), Zhdanov (2002), Tarantola (2005) and Aster *et al.* (2005). Many books on inverse problems in other fields are available as well, such as Bertero & Boccacci (1998) and Vogel (2002); and the mathematical and numerical methods used in solving geophysical inverse problems are thoroughly described in texts by Nocedal & Wright (1999), Kelley (1999) and Fletcher (1987), for example. Therefore, the goal of this chapter is to describe, within a common framework and in some detail, the methods that dominate the current practice of MT inversion.

The first part of the chapter is a brief primer on inverse theory, beginning with the mathematical formulation of the inverse problem and turning to a discussion of key concepts like nonlinearity, well-posed and ill-posed problems, and the existence, uniqueness and stability of solutions. It then develops the method of damped least-squares estimation, which is the most widely used approach to solving ill-posed, nonlinear inverse problems in geophysics, especially problems involving large numbers of observations and model parameters. This method is presented from the perspective of explaining MT observations with spatially smooth models of resistivity. The discussion pays a great deal of attention to analytical results for the linearized inverse problem, which provide a basis for iterative schemes to solve the nonlinear problem and which are often used as an approximate description of the uncertainty in nonlinear solutions.

The chapter proceeds to the discussion of numerical algorithms for computing damped least-squares solutions. Here the emphasis is on iterative, gradient-based algorithms implemented with a conjugate gradients technique. Gradient methods make direct use of the sensitivity, or Jacobian, matrix relating data and model perturbations. The Jacobian provides valuable information that greatly assists the search for a solution in large problems. Likewise, conjugate gradients facilitates the efficiency of solving large problems by avoiding costly matrix decomposition or inversion. Special attention is devoted to the description of computational shortcuts for generating or operating with the Jacobian matrix for MT data. Because of their computational advantages, similar techniques have gained prominence in a range of inverse problems involving the observation of wave fields, notably the adjoint methods developed for seismic waveform inversion (e.g. Tromp *et al.*, 2005). Also receiving wide attention in recent years are global optimization methods (e.g. Moorkamp *et al.*, 2007), which, while generally less efficient than gradient methods, address the issue of the global versus local optimality of solutions, unlike gradient methods. These are not discussed in this chapter but are treated thoroughly in Sen & Stoffa (1995).

The methods described theoretically are illustrated with a number of examples of MT inversion involving both 1D and 3D models of resistivity. The chapter closes with an overview of various extensions of damped least-squares inversion that are currently an active area of research. These extensions result from the departure from Gaussian treatments of data errors, as well as from generalization of the notion of smooth models.

8.2 Formulation of the magnetotelluric inverse problem

8.2.1 Parameterization of conductivity models

A general way to parameterize a model of Earth's resistivity is as a numerical function of position $m(\mathbf{r})$. In a 3D model, where resistivity varies in all three spatial dimensions, $\mathbf{r} = (x, y, z)$

in Cartesian coordinates. However, without changing notation, $\mathbf{r} = (y, z)$ for a 2D resistivity model and $\mathbf{r} = z$ for a 1D model. The function m may specify the resistivity at each point ($m = \rho$), its logarithm ($m = \log \rho$), the conductivity ($m = \sigma \equiv 1/\rho$), or any number of other derived quantities.

A model function is useful in analyzing an inverse problem theoretically, but computations can be performed only with a finite number of variables. In practice, then, a resistivity model is specified with a set of numerical parameters, m_j, $j = 1, \ldots, M$, which are assumed to be real-valued. In a 1D resistivity model, for example, the parameters might be the resistivities of M homogeneous planar layers, or the resistivities and thicknesses of $(M + 1)/2$ layers (allowing the deepest layer to be a half-space), or the resistivity at M fixed depths with a specified interpolation scheme defining resistivity between those depths. In a 2D or 3D model, the m_j could be the resistivities in M homogeneous blocks, or at M points, arranged on a 2D or 3D grid. As with model functions, the model parameters may be values of resistivity or some function of resistivity, such as $\log \rho$.

A number of considerations enter into the decision of how best to parameterize Earth's resistivity structure for MT inversion. One is compatibility with the physics of electromagnetic diffusion in a conductive medium. For 1D problems, Smith & Booker (1988) suggested that a model be parameterized as $m = \sigma$ since, owing to the nature of electromagnetic induction, this would give the smallest errors associated with a linearization of the forward problem. However, they also pointed out that using $m = \log \sigma$, while more nonlinear than σ, guarantees conductivity to be positive. Additionally, using logarithmic parameterization has the added benefit of more easily accommodating the wide range of conductivities in the Earth.

A second consideration is concordance with numerical algorithms for forward modeling. Exact algorithms are available to compute MT transfer functions in media comprising homogeneous, plane layers (see Chapter 2 for the layered-Earth solution). On the other hand, general 2D and 3D models require approximate modeling techniques, as discussed in Chapter 7. For example, finite-difference schemes are easily implemented with rectangular block models, while finite-element schemes can handle models parameterized with more general grids constructed from triangular or tetrahedral blocks.

Another consideration in choosing a model parameterization is the complexity of the study region where the MT data have been acquired. While resistivity undoubtedly varies in all three directions everywhere in Earth, some areas may be adequately represented by 1D or 2D models. For example, a few homogeneous plane layers might capture the major resistivity contrasts in a broad sedimentary basin with known lithology, while a 2D block model might well approximate the resistivity structure in the vicinity of a fault zone, or across a mountain front. However, the resistivity structure in a geologically complex overthrust belt, where MT is sometimes used for oil exploration, is likely to be reasonably rendered only with a fully 3D model. The appropriate dimensionality of a resistivity model may also be frequency-dependent. An example of this is a geothermal area, where 1D modeling may adequately resolve the near-surface clay cap from high-frequency MT data, but where 2D or 3D modeling is needed to fit lower-frequency data, which are most affected by deeper 3D structure.

A final, and crucial, consideration in choosing a model parameterization is the information available from the MT experiment: what the observed data can actually resolve. This pertains to the spatial sampling and aperture of the model, both vertically and laterally. As a rule of thumb, the parameterization with depth should be guided by the frequency range of the observed MT responses, while the lateral parameterization should be guided by the station geometry. Some difficult questions can arise when trying to follow such guidelines, however. For example, low-frequency MT data can be affected by the presence of an ocean even if the ocean is located far away from the measurement site. In this case, one must be careful to include the ocean in the MT models in order to correctly account for ocean-induced electric current that influences the MT response well inland. Another example occurs if MT stations lie on a line along Earth's surface (a linear traverse), where it will not be possible to resolve 3D resistivity variations, suggesting that a 2D model parameterization is appropriate (or at least the best that can be done). However, ignoring changes in resistivity along one lateral direction may be unrealistic, and doing so may bias the 2D resistivity model obtained by inversion. This dilemma is almost inescapable: there will be resistivity variations that can affect the data but that are not well determined by them. Mechanisms for coping include good experimental design and the use of *a priori*, or prior, information about Earth's resistivity to assist the data in constraining model parameters (e.g. Matsuno *et al.*, 2010).

8.2.2 *Magnetotelluric data and forward modeling functions*

As discussed in Chapter 2, magnetotelluric data are derived from measurements of the vector electric and magnetic fields at or near the Earth's surface, with raw time-domain measurements reduced by data processing to frequency-domain spectral ratios between field components. For a measurement site at the surface, the vertical electric field is zero (since the atmosphere is an insulator) and MT data for a given frequency are embodied in the second-rank tensor magnetotelluric transfer function $\mathbf{Z}$, relating the horizontal electric and magnetic field components, and the vector magnetic transfer function $\mathbf{T}$, relating the vertical component of the magnetic field to its horizontal components (see Chapter 4). A typical MT experiment determines $\mathbf{Z}$ and $\mathbf{T}$, which are complex quantities, at a finite number of observation sites and for a finite set of frequencies.

For the purpose of inversion theory, it is convenient to express the data from an MT experiment as a finite sequence of real numbers, d_i, $i = 1, ..., N$. The d_i can be defined in a variety of ways, the most obvious being as the real and imaginary parts, or amplitude and phase, of the elements of $\mathbf{Z}$ and $\mathbf{T}$ at the various sites and frequencies. However, the theoretical proprieties of $\mathbf{Z}$ and $\mathbf{T}$ under a given model parameterization might suggest alternatives. For 2D Earth models, $\mathbf{Z}$ has only two independent elements, instead of the four for 3D models, while $\mathbf{T}$ has one. Therefore, the dataset need contain at most six d_i for each measurement site and frequency. The data extracted from $\mathbf{Z}$ for 2D MT inversion commonly comprise the amplitude and phase of TE and TM apparent resistivities, but may also be based on various other rotational invariants defined in Chapter 4. For 1D models, $\mathbf{T} = 0$ and a single complex quantity is sufficient to represent $\mathbf{Z}$. The data for 1D inversion are often the amplitude and phase of the isotropic apparent resistivity, apparent conductivity, or Schmucker's $c(\omega)$ response.

For any definition of MT data, an essential ingredient in MT inversion is a set of functions that predict the theoretical values the data would have if Earth's resistivity structure were known. These functions, denoted here as F_i ($i = 1, \ldots, N$), take the model parameters m_j ($j = 1, \ldots, M$) as their arguments, such that $F_i(m_1, \ldots, m_M)$ is the theoretical prediction of d_i for a hypothetical Earth specified by the given values of the parameters. The F_i go by various names in the inversion theory literature, such as data functions, forward or forward modeling functions, and prediction functions. Moreover, the term "functional" is often used in place of "function" (for reasons mentioned below). In MT, F_i is defined implicitly by an algorithm that solves the Maxwell equations and converts the solution to the appropriate quantity defining d_i. The problem of evaluating F_i for a given model is the *forward* problem.

The mathematical properties of the forward functions will depend on the model parameterization and data representation that one chooses, and achieving desirable properties is another consideration in making these choices. For example, Smith & Booker (1988) pointed out that, among commonly used MT responses for 1D models, $-1/c$ is the most linear function of conductivity ($m = \sigma$).

8.2.3 *Statement of the inverse problem*

The generic magnetotelluric inverse problem can now be posed as follows. A sequence of real-valued magnetotelluric data, d_i, $i = 1, \ldots, N$, have been acquired in a field experiment whose objective is to determine the resistivity structure of a portion of Earth. Earth's resistivity is assumed to be adequately represented by a sequence of real-valued model parameters, m_j, $j = 1, \ldots, M$, which means that one and only one assignment of values to these parameters correctly represents the Earth. The experimental data and model parameters can be related with

$$d_i = F_i(m_1, m_2, \ldots, m_M) + e_i, \qquad i = 1, 2, \ldots, N \qquad (8.1)$$

where F_i is the forward function that theoretically predicts d_i for given values of the m_j, and e_i is the error in the prediction. The inverse problem is to solve Eq. (8.1) for the unknown parameters, m_j.

It is clear that one cannot solve Eq. (8.1) without constraining e_i, and to do so plausibly requires some understanding of where the errors come from. Some sources of error can be associated with the MT experiment itself, such as the effects of cultural and instrument noise. Other errors are incurred in modeling, such as those resulting from approximations employed in solving the Maxwell equations and from round-off error in the numerical computations. Modeling errors can also be ascribed to deficiencies in the model parameterization. For example, if one parameterizes the Earth with homogeneous blocks, the correct value of each block's resistivity can be interpreted, say, as the spatial average of the Earth's true resistivity function over the block volume. However, solving the Maxwell equations with a block-averaged resistivity function will ignore the effects of resistivity variations within blocks, and these effects become a source of modeling error.

It is reasonable to assume that experimental, or measurement, errors do not depend on the Earth's resistivity, but errors associated with modeling generally do. Allowing the e_i to depend on the m_j, however, complicates the inverse problem tremendously. While some authors have tackled aspects of model-dependent errors, such as Parker (1977), who analyzed errors due to parameterization, here it is assumed that no such dependence exists. Minimizing the model dependence of prediction errors can be added to the list of considerations in choosing a data representation, model parameterization and modeling algorithm.

8.2.3.1 *Matrix notation and vector spaces*

Equation (8.1) can be written more concisely using matrix notation, which will be introduced in two steps. First, let the m_j be the components of an M-dimensional column vector ($M \times 1$ matrix) $\mathbf{m}$:

$$\mathbf{m} = (m_1 \quad m_2 \quad \cdots \quad m_M)^{\mathrm{T}} \tag{8.2}$$

where superscript "T" denotes matrix transposition. The model parameter vector $\mathbf{m}$ can be considered to be an element of an M-dimensional, real vector space, which will be called the *model space* and denoted as $\mathcal{M}$. A real vector space is simply a set of elements that can be added to one another and multiplied by real numbers with the outcomes also being elements of the set. Using Eq. (8.2), Eq. (8.1) can be restated as

$$d_i = F_i(\mathbf{m}) + e_i, \qquad i = 1, 2, \ldots, N \tag{8.3}$$

This equation merely replaces the M scalar arguments of the forward functions with a single vector argument. Technically, re-defining the argument of a function changes it to a different function, but the reader should not be confused if this subtlety is ignored and F_i is used in both cases. A related point is that a function that transforms elements of a vector space to real numbers, as does F_i, is often designated with the special term *functional*, although it is still correct to use the generic term function.

The second step toward matrix notation is to consider the observed data and data errors to be components of N-dimensional column vectors:

$$\mathbf{d} = (d_1 \quad d_2 \quad \cdots \quad d_N)^{\mathrm{T}} \tag{8.4}$$

$$\mathbf{e} = (e_1 \quad e_2 \quad \cdots \quad e_N)^{\mathrm{T}} \tag{8.5}$$

These vectors are elements of *data space*, denoted $\mathcal{D}$, which is an N-dimensional, real vector space. The inverse problem now can be stated as solving for $\mathbf{m}$ in

$$\mathbf{d} = F(\mathbf{m}) + \mathbf{e} \tag{8.6}$$

where F, a transformation from model space to data space, is defined by

$$F(\mathbf{m}) = (F_1(\mathbf{m}) \quad F_2(\mathbf{m}) \quad \cdots \quad F_N(\mathbf{m}))^{\mathrm{T}} \tag{8.7}$$

8.2.3.2 Function spaces

The elements of a vector space can be functions as well as finite sequences of numbers. Therefore, Eqs. (8.3) and (8.6) apply equally well to inverse problems in which the Earth's resistivity is parameterized by a model function, with **m** then identified with $m(\mathbf{r})$. Not surprisingly, vector spaces containing functions as their elements are called *function spaces*. Since a function is indexed by a continuous variable ($\mathbf{r}$), instead of a finite integer (j), a function space is *infinite*-dimensional. Similarly, Eq. (8.6) also applies when the data space is a function space and **d** refers to a data function. This would be appropriate, for example, if one wanted to consider MT data at a continuum of frequencies, $d(\omega)$.

This chapter will focus on finite-dimensional, or discrete, data and models, although it will relate certain concepts to model functions in an informal way.

8.2.3.3 Inverse of what?

The reader may be wondering why solving Eq. (8.6), or (8.1) or (8.3), is called an *inverse* problem. The term refers to the transformation F. If F has an inverse, mapping data vectors to model vectors, and if $\mathbf{e} = 0$, the solution to (8.6) is given simply by

$$\mathbf{m} = F^{-1}(\mathbf{d}) \tag{8.8}$$

In this case, solving (8.6) can be equated to the task of finding F^{-1}, the inverse of F.

In the most interesting problems, however, the inverse of F does not exist, and the problem becomes that of finding an estimate of, or approximation to, **m** of the form

$$\tilde{\mathbf{m}} = G(\mathbf{d}) \tag{8.9}$$

The transformation G will presumably have some optimal properties that sanction it as a meaningful substitute for F^{-1}. However, when **e** is not zero, these properties might actually disqualify F^{-1} itself when it does exist, as will be shown below. Furthermore, many approaches to inverse problems do not treat G explicitly, and focus instead on criteria for the optimality of $\tilde{\mathbf{m}}$ and on numerical algorithms for computing $\tilde{\mathbf{m}}$.

Given the narrow utility of inverse transformations, it is tempting to avoid the word "inverse" and call Eq. (8.6) an *estimation* or *inference* problem. These terms are so general, though, they apply as well to problems in which there is no F and Eq. (8.6) is not relevant, such as many problems in medical diagnosis. The term "inverse problem" serves the good purpose of identifying problems of the form in Eq. (8.6) with solutions of the form in Eq. (8.9), regardless of whether F^{-1} exists; or whether $G = F^{-1}$ is an optimal solution when it does; or whether one chooses not to consider G in its entirety but only its one value at **d**. The knowledge that $\tilde{\mathbf{m}} = F^{-1}(\mathbf{d})$ is the ideal solution to Eq. (8.6) in the ideal situation when F^{-1} exists and $\mathbf{e} = 0$ underlies every inverse problem.

8.2.4 Linear versus nonlinear inverse problems

Much of geophysical inverse theory deals with *linear* inverse problems, in which the forward function F is a linear transformation obeying

$$F(\lambda_1 \mathbf{m}_1 + \lambda_2 \mathbf{m}_2) = \lambda_1 F(\mathbf{m}_1) + \lambda_2 F(\mathbf{m}_2) \tag{8.10}$$

for all real numbers λ_1 and λ_2 and all model vectors $\mathbf{m}_1$ and $\mathbf{m}_2$. It should be obvious from Eq. (8.7) that F is a linear transformation if and only if each forward functional F_i is linear.

For finite-dimensional model spaces, with $\mathbf{m}$ taken as a column vector, a linear forward functional can be expressed as

$$F_i(\mathbf{m}) = \mathbf{a}_i^T \mathbf{m} \tag{8.11}$$

where, like $\mathbf{m}$, $\mathbf{a}_i$ is an M-dimensional column vector. When $\mathbf{m}$ is a function, it is often possible to express F_i as an integration over the model domain (points $\mathbf{r}$ where the model function is defined):

$$F_i(\mathbf{m}) = \int dV(\mathbf{r}) \, a_i(\mathbf{r}) \, m(\mathbf{r}) \tag{8.12}$$

where $a_i(\mathbf{r})$ may be an ordinary function or a distribution and where $dV(\mathbf{r})$ denotes differential volume in the model domain (e.g. $dV(\mathbf{r}) = dx \, dy \, dz$ for a 3D model). For both finite- and infinite-dimensional model spaces one can express a linear forward transformation as

$$F(\mathbf{m}) = \mathbf{A}\mathbf{m} \tag{8.13}$$

with the right-hand side representing the particular linear operations that define the action of F. In the finite-dimensional case, $\mathbf{A}$ is simply an $N \times M$ matrix having the vectors $\mathbf{a}_i^T$ of Eq. (8.11) as its rows, and the operation of $\mathbf{A}$ on $\mathbf{m}$ is matrix multiplication.

The theory of linear inverse problems is a well-developed topic (e.g. Lines & Treitel, 1984; Menke, 1989) for which a rich toolbox of analytical and numerical methods exists. Unfortunately, magnetotelluric forward functionals – like most geophysical forward functionals – are *nonlinear*. The nonlinearity ultimately stems from the fact that the Maxwell equations involve the product of electric conductivity and the electric field. Nonlinear inverse problems are much more difficult to analyze theoretically and to solve numerically than linear ones.

Some approaches to nonlinear problems take advantage of linear methods by considering a *linearized* form of the inverse problem. This is done both in iterative schemes for computing solutions (as discussed in Section 8.5) and in the analysis of solution error (discussed in Section 8.3). Linearization is accomplished by expanding the forward function in a Taylor series around a *reference* model, $\mathbf{m}_*$:

$$F(\mathbf{m}) = F(\mathbf{m}_*) + \mathbf{A}_{\mathbf{m}_*}(\mathbf{m} - \mathbf{m}_*) + o(\|\mathbf{m} - \mathbf{m}_*\|) \tag{8.14}$$

where $\mathbf{A}_{\mathbf{m}_*}$ is a linear transformation known as the *Fréchet derivative*, or Fréchet differential, of F; $\| \ldots \|$ signifies the norm of a vector; and $o(\ldots)$ is the Landau "little-o" operator from asymptotic theory. The use of $\mathbf{m}_*$ as a subscript on $\mathbf{A}$ indicates that, in general, the Fréchet derivative depends on the reference model. For finite-dimensional data and model spaces, $\mathbf{A}_{\mathbf{m}_*}$ is a matrix of partial derivatives of the forward functionals, known as the *Jacobian* matrix, but also called the *sensitivity* matrix in geophysics:

$$[\mathbf{A}_{\mathbf{m}_*}]_{ij} = \frac{\partial F_i(\mathbf{m})}{\partial m_j}\bigg|_{\mathbf{m}=\mathbf{m}_*} \tag{8.15}$$

As long as these partial derivatives exist, Eq. (8.14) holds no matter how the norm of a model is defined. For example, a simple 2-norm (also called L_2-norm) can be assumed:

$$\|\mathbf{m}\| = \left(\sum_{j=1}^{M} (m_j)^2 \right)^{1/2} \tag{8.16}$$

When the model space is a function space, the first-order expansion in (8.14) can be written component-wise as

$$F_i(\mathbf{m}) = F_i(\mathbf{m}_*) + \int dV(\mathbf{r})\, a_i(\mathbf{r}; \mathbf{m}_*)[m(\mathbf{r}) - m_*(\mathbf{r})] + o(\|\mathbf{m} - \mathbf{m}_*\|) \tag{8.17}$$

where, in general, the Fréchet derivative $a_i(\mathbf{r}; \mathbf{m}_*)$ is a distribution. Equation (8.17) may hold for some definitions of the norm of a function but not others. It is generally more difficult to prove Fréchet differentiability for infinite-dimensional model spaces than finite-dimensional ones, although Chave (1984) and Parker (1977) proved that electromagnetic induction data are Fréchet differentiable with respect to 1D models. Some of the difficulties in proving Fréchet differentiability in 2D and 3D electromagnetic problems were pointed out by Rodi (1989).

The linearized inverse problem is created by replacing F with the first two terms of (8.14), dropping second- and higher-order terms. This first-order approximation to F will be denoted $\bar{F}$:

$$\bar{F}(\mathbf{m}) = F(\mathbf{m}_*) + \mathbf{A}_{\mathbf{m}_*}(\mathbf{m} - \mathbf{m}_*) \tag{8.18}$$

$\bar{F}$ is an *affine* transformation: a linear transformation plus a constant. In solving the inverse problem, however, an affine forward transformation is just as convenient as a linear transformation as given by (8.13). In fact, the difference between affine and linear transformations is subtle enough that the term "linear" is often applied to both.

8.2.5 *Well-posed and ill-posed inverse problems*

The notion of a well-posed problem – either a forward problem or an inverse problem – was established by Hadamard (1902). The Hadamard conditions for well-posedness are that: (1) the problem must have a solution (i.e. a solution *exists*), (2) the solution must be *unique*, and (3) the solution must be *stable* against perturbations to the problem inputs. If any of these are violated, the problem is said to be ill-posed.

To relate the Hadamard conditions to the inverse problem expressed in Eq. (8.6), define, for now, a solution to be any model $\tilde{\mathbf{m}}$ whose data predictions match the observed data:

$$F(\tilde{\mathbf{m}}) = \mathbf{d} \tag{8.19}$$

The transformation F maps a subset of vectors in model space, known as the *domain* of F, to a subset of vectors in data space, called the *range* of F. Therefore, the existence of $\tilde{\mathbf{m}}$ simply means that $\mathbf{d}$ must be in the range of F. The uniqueness of $\tilde{\mathbf{m}}$ follows if F is a *one-to-one* transformation, mapping different vectors in model space to different vectors in data space. If $\mathbf{d}$ is in the range of F and F is one-to-one, the unique solution of Eq. (8.19) is given by

$$\tilde{\mathbf{m}} = F^{-1}(\mathbf{d}) \qquad (8.20)$$

where the inverse transformation F^{-1} is defined such that its domain is the range of F. It is important to realize, however, that $\mathbf{d}$ belonging to the range of F is a very strict requirement owing to the presence of $\mathbf{e}$ in Eq. (8.6), especially when F is a nonlinear transformation.

The condition of stability pertains to the effect of $\mathbf{e}$ on $\tilde{\mathbf{m}}$. When $\mathbf{e} = 0$ (error-free data), Eqs. (8.6) and (8.20) imply that

$$\tilde{\mathbf{m}} = F^{-1}(F(\mathbf{m})) = \mathbf{m} \qquad (8.21)$$

That is, the solution $\tilde{\mathbf{m}}$ is not only unique, it is *correct*. In general, however,

$$\tilde{\mathbf{m}} - \mathbf{m} = F^{-1}(F(\mathbf{m}) + \mathbf{e}) - F^{-1}(F(\mathbf{m})) \qquad (8.22)$$

Stability is the requirement that the error in the solution, $\tilde{\mathbf{m}} - \mathbf{m}$, be bounded when $\mathbf{e}$ is bounded. More precisely, $\tilde{\mathbf{m}}$ is stable when a positive function $\varepsilon(\mu)$ exists such that $\|\tilde{\mathbf{m}} - \mathbf{m}\| < \mu$ whenever $\|\mathbf{e}\| < \varepsilon(\mu)$. This is just the definition of continuity: $\tilde{\mathbf{m}}$ is stable when F^{-1} is a continuous transformation.

Hadamard's criteria of well-posedness are difficult to satisfy in a finite-dimensional inverse problem when a solution is defined by Eq. (8.19). This becomes clear by considering the linear forward transformation in Eq. (8.13) for a given matrix $\mathbf{A}$: F has a function inverse exactly when $\mathbf{A}$ has a matrix inverse, in which case

$$F^{-1}(\mathbf{d}) = \mathbf{A}^{-1}\mathbf{d} \qquad (8.23)$$

A unique solution to the inverse problem exists if and only if $\mathbf{A}^{-1}$ exists, which can only occur in the fortuitous situation where the number of data and model parameters are equal ($N = M$). Analyzing well-posedness in the context of Eq. (8.19) is useful mainly in the theoretical analysis of inverse problems with infinite-dimensional data and model spaces, as were considered in earlier chapters. For inverse problems with either finite-dimensional data or model spaces, or both, it is more useful to base notions of well-posedness on a less restrictive definition of what constitutes the solution of an inverse problem.

8.3 Least-squares solutions

Least-squares estimation was presented in Chapter 5 in the context of estimating magnetotelluric responses. The method is also the basis for much of geophysical inverse theory. This section considers least squares in the context of finite-dimensional inverse problems in which the number of unknown parameters is less than the number of observations ($M < N$).

Such *over-determined* problems are the primary focus of classical statistical inference (e.g. Kendall & Stuart, 1977; Stuart *et al.*, 2009).

The existence of a solution to an over-determined inverse problem in the sense of Eq. (8.19) is virtually impossible when the data contain errors ($\mathbf{e} \neq 0$). The classical remedy for this is to re-define a solution to be a model whose predictions match the data as closely as possible in some pre-defined sense, rather than matching them exactly as required by (8.19). Such a solution always exists (although it may be non-unique or unstable).

A best-fitting model is most commonly defined as one minimizing an objective, or penalty, function that measures the discrepancy between $\mathbf{d}$ and $F(\mathbf{m})$. In geophysics, this function is often referred to as the *data misfit*, or just misfit. The least-squares method takes the misfit function, here denoted as Φ, to be a squared, weighted L_2-norm of the residual vector, $\mathbf{d} - F(\mathbf{m})$, written as

$$\Phi(\mathbf{m}) = (\mathbf{d} - F(\mathbf{m}))^{\mathrm{T}} \mathbf{W} (\mathbf{d} - F(\mathbf{m})) \tag{8.24}$$

where $\mathbf{W}$ is a positive definite matrix containing pre-assigned weights. Typically, $\mathbf{W}$ is chosen as a diagonal matrix, allowing $\Phi(\mathbf{m})$ to be also written as a simple sum of weighted, squared residuals:

$$\Phi(\mathbf{m}) = \sum_{i=1}^{N} w_i (d_i - F_i(\mathbf{m}))^2 \tag{8.25}$$

where $w_i = W_{ii}$.

A rationale for least-squares estimation follows from a probabilistic formulation of the inverse problem whereby the data errors are treated as random variables. Denote the joint probability density function (p.d.f.) of the errors as $f_e(e_1, e_2, \ldots, e_N)$ or, concisely, $f_e(\mathbf{e})$. Substituting Eq. (8.6) into $f_e(\mathbf{e})$, one obtains the p.d.f. of $\mathbf{d}$, which depends on $\mathbf{m}$:

$$f(\mathbf{d}; \mathbf{m}) = f_e(\mathbf{d} - F(\mathbf{m})) \tag{8.26}$$

With $\mathbf{d}$ fixed to its observed value, $f(\mathbf{d}; \mathbf{m})$ can be interpreted as a function of the unknown model vector $\mathbf{m}$, called the *likelihood* function. One of the classical statistical inference methods is *maximum likelihood* estimation, which defines an optimal estimate of $\mathbf{m}$ to be the value $\tilde{\mathbf{m}}$ that satisfies

$$f(\mathbf{d}; \tilde{\mathbf{m}}) = \max_{\mathbf{m}} f(\mathbf{d}; \mathbf{m}) \tag{8.27}$$

A heuristic justification for this criterion is that, among the family of data p.d.f.s parameterized by $\mathbf{m}$, the one for $\mathbf{m} = \tilde{\mathbf{m}}$ makes the observed data vector look most like a typical random sample from its underlying population.

One can formulate least-squares estimation as maximum likelihood estimation in the special case that the data errors are normal, or Gaussian, random variables. In this case, $f_e(\mathbf{e})$ is determined by its first and second moments. Denote the mean of the ith error as ε_i and the covariance between the ith and jth as σ_{ij}:

$$E[e_i] = \varepsilon_i \tag{8.28}$$

$$\mathrm{Cov}[e_i, e_j] = \sigma_{ij} \tag{8.29}$$

where the *expectation* operator E[...] and *covariance* operator Cov[..., ...] are defined by

$$E[e_i] = \int de_1 \int de_2 \cdots \int de_N \; e_i \; f_e(e_1, e_2, \ldots, e_N) \tag{8.30}$$

$$\mathrm{Cov}[e_i, e_j] = E[(e_i - E[e_i])(e_j - E[e_j])] \tag{8.31}$$

Equations (8.28) and (8.29) can be written in matrix notation as

$$E[\mathbf{e}] = \boldsymbol{\varepsilon} \tag{8.32}$$

$$\mathrm{Var}[\mathbf{e}] = \boldsymbol{\Sigma} \tag{8.33}$$

The *variance–covariance* matrix (or just variance matrix) $\boldsymbol{\Sigma}$ is assumed to be positive definite and therefore to have an inverse. The error p.d.f. is then given by

$$f_e(\mathbf{e}) = \frac{1}{(2\pi)^{N/2} (\det\boldsymbol{\Sigma})^{1/2}} \exp\left\{ -\tfrac{1}{2}(\mathbf{e} - \boldsymbol{\varepsilon})^{\mathrm{T}} \boldsymbol{\Sigma}^{-1}(\mathbf{e} - \boldsymbol{\varepsilon}) \right\} \tag{8.34}$$

where det denotes matrix determinant. If

$$\boldsymbol{\varepsilon} = 0 \tag{8.35}$$

and

$$\boldsymbol{\Sigma}^{-1} = \mathbf{W} \tag{8.36}$$

then

$$f(\mathbf{d}; \mathbf{m}) = \frac{(\det\mathbf{W})^{1/2}}{(2\pi)^{N/2}} \exp\left\{ -\tfrac{1}{2}\Phi(\mathbf{m}; \mathbf{d}) \right\} \tag{8.37}$$

where Φ is the least-squares objective function of Eq. (8.24) with its dependence on $\mathbf{d}$ now shown. It is clear that, under the assumption that $\mathbf{e}$ is Gaussian with first and second moments given by (8.35) and (8.36), minimizing the least-squares data misfit is equivalent to maximizing the likelihood function.

Chapter 5 discussed reasons why natural-source MT data often violate the assumption of Gaussian errors, thus detracting from the optimality of least-squares estimates. In particular, non-stationarity leads to a residual distribution that, while having a Gaussian center, has particularly long tails. Least-squares solutions can be biased in such cases, indicating the need for robust estimation techniques. Nonetheless, the assumption of Gaussian error statistics simplifies the mathematics and facilitates a detailed analysis of the inverse problem. Later in this chapter, and in Tarantola (2005) for example, departures from the Gaussian assumption are discussed.

8.3.1 Existence and uniqueness of least-squares solutions

Since the least-squares solution minimizes Φ, it satisfies

$$\Phi(\mathbf{m}) \geq \Phi(\tilde{\mathbf{m}}) \tag{8.38}$$

for every model $\mathbf{m}$ in the domain of Φ (which coincides with the domain of F). The solution exists under the weak assumption that the norm of the residual vector, $\mathbf{d} - F(\mathbf{m})$, tends to infinity as $\|\mathbf{m}\|$ tends to infinity, which can be arranged with the judicious choice of data responses and model parameters. The solution is unique under the rather strong assumption that Φ is strictly *concave*, meaning that

$$\Phi(\lambda\mathbf{m}_1 + (1-\lambda)\mathbf{m}_2) < \lambda\Phi(\mathbf{m}_1) + (1-\lambda)\Phi(\mathbf{m}_2) \tag{8.39}$$

for all models $\mathbf{m}_1$ and $\mathbf{m}_2$ and real $0 \leq \lambda \leq 1$. This inequality is depicted in Figure 8.1. If $\tilde{\mathbf{m}}$ is unique, then (8.38) can be replaced with

$$\Phi(\mathbf{m}) > \Phi(\tilde{\mathbf{m}}), \qquad \mathbf{m} \neq \tilde{\mathbf{m}} \tag{8.40}$$

Unfortunately, it is difficult to identify the properties of F that make Φ concave. Some mild assumptions about the differentiability of F, however, make it possible to establish when a model is a unique *local* least-squares solution. The model $\tilde{\mathbf{m}}$ is a unique local solution, and achieves a local minimum of Φ, when (8.40) holds for all $\mathbf{m}$ in a neighborhood of $\tilde{\mathbf{m}}$. In contrast, when (8.40) holds for all $\mathbf{m}$ in the entire model space, $\tilde{\mathbf{m}}$ can be said to be a unique *global* least-squares solution (see Figure 8.2).

Assume that F is continuously differentiable on its domain, meaning that its Jacobian at $\mathbf{m}$, denoted $\mathbf{A_m}$, exists at each $\mathbf{m}$ and is a continuous function of $\mathbf{m}$. The objective function will then also be continuously differentiable. Its partial derivatives with respect to the m_j can be collected into an M-dimensional *gradient* vector $\mathbf{g_m}$:

$$[\mathbf{g_m}]_j = \frac{\partial\Phi(\mathbf{m})}{\partial m_j} \tag{8.41}$$

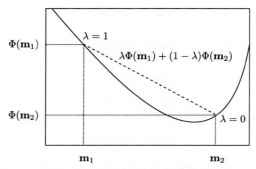

Figure 8.1. A concave function is shown as the solid line with a minimum between $\mathbf{m}_1$ and $\mathbf{m}_2$. The inequality defining a concave function means that, between any two points, the value of the function is less than a linear interpolation of the function between those two points.

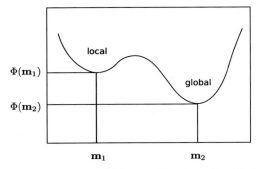

Figure 8.2. A contrived function with a local minimum $\Phi(\mathbf{m}_1)$ and a global minimum $\Phi(\mathbf{m}_2)$. Note that, at both the local and global minima, the gradient (slope) of the function is zero, and the curvature is positive.

At each $\mathbf{m}$, the vector $\mathbf{g_m}$ points in the direction of largest change of Φ. From Eq. (8.24) one can derive that

$$\mathbf{g_m} = 2\mathbf{A_m^T W}(F(\mathbf{m}) - \mathbf{d}) \tag{8.42}$$

Assume further that F is *twice* continuously differentiable, implying the existence of a matrix of second derivatives of Φ for any $\mathbf{m}$, denoted $\mathbf{H_m}$:

$$[\mathbf{H_m}]_{jk} = \frac{\partial^2 \Phi(\mathbf{m})}{\partial m_j \partial m_k} \tag{8.43}$$

The $M \times M$ matrix $\mathbf{H_m}$ is known as the *Hessian* of Φ, evaluated at $\mathbf{m}$. The Hessian of Φ describes its local curvature and is given by the formula

$$\mathbf{H_m} = 2\mathbf{A_m^T W A_m} + \mathbf{H_m^{nl}} \tag{8.44}$$

with the components of the second term given by

$$[\mathbf{H_m^{nl}}]_{jk} = 2 \sum_{i,l=1}^{N} \frac{\partial^2 F_i}{\partial m_j \partial m_k} W_{il}(F_l(\mathbf{m}) - d_l) \tag{8.45}$$

The superscript "nl" on $\mathbf{H^{nl}}$ is a reminder that this matrix is non-zero only when F is a nonlinear transformation.

Under these assumptions, a *necessary* condition for $\tilde{\mathbf{m}}$ to be a unique local minimizer of Φ is that it be a *stationary* point, or point at which $\mathbf{g_{\tilde{m}}} = 0$. The proof of this, following Kelley (1999), begins by expanding Φ in a second-order Taylor series about $\tilde{\mathbf{m}}$, yielding

$$\Phi(\mathbf{m}) = \Phi(\tilde{\mathbf{m}}) + \mathbf{g_{\tilde{m}}^T}(\mathbf{m} - \tilde{\mathbf{m}}) + \tfrac{1}{2}(\mathbf{m} - \tilde{\mathbf{m}})^T \mathbf{H_{\tilde{m}}}(\mathbf{m} - \tilde{\mathbf{m}}) + o(||\mathbf{m} - \tilde{\mathbf{m}}||^2) \tag{8.46}$$

Using this expansion and letting $\delta\mathbf{m} = \mathbf{m} - \tilde{\mathbf{m}}$, the requirement in (8.40) can be rewritten as

$$\mathbf{g_{\tilde{m}}^T}\delta\mathbf{m} + \tfrac{1}{2}\delta\mathbf{m}^T \mathbf{H_{\tilde{m}}}\delta\mathbf{m} + o(||\delta\mathbf{m}||^2) > 0, \qquad \delta\mathbf{m} \neq 0 \tag{8.47}$$

Φ will have a unique local minimum at $\tilde{\mathbf{m}}$ only if (8.47) holds for arbitrarily small $\delta\mathbf{m}$. Suppose, however, that $\mathbf{g}_{\tilde{\mathbf{m}}} \neq 0$ and $\delta\mathbf{m} = -(\|\delta\mathbf{m}\|/\|\mathbf{g}_{\tilde{\mathbf{m}}}\|)\mathbf{g}_{\tilde{\mathbf{m}}}$. The inequality in (8.47) becomes

$$-\|\mathbf{g}_{\tilde{\mathbf{m}}}\|\,\|\delta\mathbf{m}\| + \frac{1}{2}\frac{\|\delta\mathbf{m}\|^2}{\|\mathbf{g}_{\tilde{\mathbf{m}}}\|^2}\mathbf{g}_{\tilde{\mathbf{m}}}^{\mathrm{T}}\mathbf{H}_{\tilde{\mathbf{m}}}\mathbf{g}_{\tilde{\mathbf{m}}} + o(\|\delta\mathbf{m}\|^2) > 0 \tag{8.48}$$

or, multiplying by $2\|\mathbf{g}_{\tilde{\mathbf{m}}}\|^2/\|\delta\mathbf{m}\|$ and rearranging,

$$\|\delta\mathbf{m}\|\mathbf{g}_{\tilde{\mathbf{m}}}^{\mathrm{T}}\mathbf{H}_{\tilde{\mathbf{m}}}\mathbf{g}_{\tilde{\mathbf{m}}} + o(\|\delta\mathbf{m}\|) > 2\|\mathbf{g}_{\tilde{\mathbf{m}}}\|^3 \tag{8.49}$$

This clearly cannot hold for arbitrarily small $\delta\mathbf{m}$ since the right-hand side is constant while the left-hand side goes to zero as $\|\delta\mathbf{m}\| \to 0$. Therefore, by contradiction, $\mathbf{g}_{\tilde{\mathbf{m}}}$ must be zero.

Given that $\tilde{\mathbf{m}}$ is a stationary point, a *sufficient* condition for it to be a unique local solution is that the Hessian matrix $\mathbf{H}_{\tilde{\mathbf{m}}}$ be positive definite, meaning that

$$\delta\mathbf{m}^{\mathrm{T}}\mathbf{H}_{\tilde{\mathbf{m}}}\delta\mathbf{m} > 0, \qquad \delta\mathbf{m} \neq 0 \tag{8.50}$$

To prove this, drop the first term of (8.47) to obtain

$$\tfrac{1}{2}\delta\mathbf{m}^{\mathrm{T}}\mathbf{H}_{\tilde{\mathbf{m}}}\delta\mathbf{m} + o(\|\delta\mathbf{m}\|^2) > 0, \qquad \delta\mathbf{m} \neq 0 \tag{8.51}$$

Assuming $\mathbf{H}_{\tilde{\mathbf{m}}}$ is positive definite, this can be rewritten as

$$1 + \frac{o(\|\delta\mathbf{m}\|^2)}{\delta\mathbf{m}^{\mathrm{T}}\mathbf{H}_{\tilde{\mathbf{m}}}\delta\mathbf{m}} > 0, \qquad \delta\mathbf{m} \neq 0 \tag{8.52}$$

This clearly is true for arbitrarily small $\delta\mathbf{m}$ since the second term of the left-hand side approaches zero as $\|\delta\mathbf{m}\| \to 0$. This proves that (8.50) is sufficient for (8.47) to hold in a neighborhood of a stationary point $\tilde{\mathbf{m}}$.

Referring to (8.42), for $\tilde{\mathbf{m}}$ to be a stationary point it must be a solution to

$$\mathbf{A}_{\mathbf{m}}^{\mathrm{T}}\mathbf{W}(\mathbf{d} - F(\mathbf{m})) = 0 \tag{8.53}$$

This is known as the *normal equation* (or *equations*, if considered component-wise) because it requires the weighted residual vector to be normal to each column of the Jacobian.

The analysis of the previous paragraphs has shown that $\tilde{\mathbf{m}}$ is a unique, local least-squares solution if it is a point with zero gradient and positive definite Hessian. The Hessian requirement, it turns out, is sufficient but not necessary in general. These results do not, unfortunately, address the question of whether stationary points with a positive definite Hessian actually exist or, when they do, how many there are and which achieves the global minimum of Φ.

8.3.2 Stability and model uncertainty

To analyze the stability of the least-squares solution, consider it to be a function, G, of the observed data vector, as shown in Eq. (8.9): $\tilde{\mathbf{m}} = G(\mathbf{d})$. The solution $\tilde{\mathbf{m}}$ is stable in the Hadamard sense if G is a continuous function. Below, it is shown that G is in fact differentiable, which implies that it is continuous. Therefore, a finite perturbation to the data vector induces a finite perturbation to the solution model.

To relate this concept to the error in $\tilde{\mathbf{m}}$, substitute (8.6) into (8.9) to obtain

$$\tilde{\mathbf{m}} = G(F(\mathbf{m}) + \mathbf{e}) \tag{8.54}$$

If G is differentiable at $\mathbf{d} = F(\mathbf{m})$, this can be expanded as

$$\tilde{\mathbf{m}} = G(F(\mathbf{m})) + \mathbf{B}_{F(\mathbf{m})}\mathbf{e} + o(||\mathbf{e}||) \tag{8.55}$$

where the $M \times N$ matrix $\mathbf{B_d}$, for arbitrary data vector $\mathbf{d}$, denotes the Jacobian of G at $\mathbf{d}$:

$$[\mathbf{B_d}]_{ji} = \frac{\partial G_j(\mathbf{d})}{\partial d_i} \tag{8.56}$$

Now make the rather bold assumption that the solution $\tilde{\mathbf{m}}$ is correct when the data are error-free:

$$G(F(\mathbf{m})) = \mathbf{m} \tag{8.57}$$

Then (8.55) becomes

$$\tilde{\mathbf{m}} - \mathbf{m} = \mathbf{B}_{F(\mathbf{m})}\mathbf{e} + o(||\mathbf{e}||) \tag{8.58}$$

This equation ensures that the error in $\tilde{\mathbf{m}}$ can be bounded by bounding $\mathbf{e}$.

The existence of the Jacobian of G can be demonstrated by treating the stationarity condition of (8.53) as an implicit representation of G. Taking the gradient vector to be a function, Γ, of both $\mathbf{m}$ and $\mathbf{d}$, the stationarity condition can be written as

$$\Gamma(\mathbf{m}, \mathbf{d}) = 0 \tag{8.59}$$

The partial derivative of Γ with respect to its first argument ($\mathbf{m}$) is an $M \times M$ matrix that depends on $\mathbf{m}$ and $\mathbf{d}$ and will be shown as $D_1\Gamma(\mathbf{m}, \mathbf{d})$. Similarly, the partial derivative of Γ with respect to its second argument (an $M \times N$ matrix) will be denoted $D_2\Gamma(\mathbf{m}, \mathbf{d})$. Given these, the implicit function theorem (e.g. Apostol, 1957) states that

$$D_1\Gamma(G(\mathbf{d}), \mathbf{d})\mathbf{B_d} = -D_2\Gamma(G(\mathbf{d}), \mathbf{d}) \tag{8.60}$$

assuming that $D_1\Gamma$ and $D_2\Gamma$ exist. Equations (8.41) and (8.43) imply that $D_1\Gamma$ does exist and is just the Hessian of Φ:

$$D_1\Gamma(\mathbf{m}, \mathbf{d}) = \mathbf{H_m} \tag{8.61}$$

noting that $\mathbf{H_m}$ depends on $\mathbf{d}$ by virtue of $\mathbf{H_m^{nl}}$ in (8.44). Moreover, $D_2\Gamma$ can be found by differentiating the formula for the gradient in Eq. (8.42):

$$D_2\Gamma(\mathbf{m}, \mathbf{d}) = -2\mathbf{A_m^T}\mathbf{W} \tag{8.62}$$

which happens not to depend on $\mathbf{d}$ (since $\mathbf{g_m}$ is linear in $\mathbf{d}$). Equations (8.60)–(8.62) imply that, if the Hessian of Φ evaluated at the model $G(\mathbf{d})$ is positive definite, the Jacobian of G at $\mathbf{d}$ is given by

$$\mathbf{B_d} = 2\mathbf{H}_{G(\mathbf{d})}^{-1}\mathbf{A}_{G(\mathbf{d})}^{\mathrm{T}}\mathbf{W} \tag{8.63}$$

which can be evaluated with $\mathbf{d} = F(\mathbf{m})$ for the purposes of (8.58).

The notion of stability is not very useful if it is taken to be a binary property of a solution to the inverse problem. Knowing that the data errors perturb $\tilde{\mathbf{m}}$ from $\mathbf{m}$ only a finite amount does not mean that amount is small. If the elements of $\mathbf{B}_{F(\mathbf{m})}$ in Eq. (8.58) are too large, the least-squares problem can be said to be *ill-conditioned* even if well-posed in the strict sense. The expression for the Jacobian of G in (8.63) shows that how well-conditioned the least-squares problem is depends largely on how well-conditioned the Hessian of Φ is. If $\mathbf{H_m}$ is a poorly conditioned matrix, its inverse will have large elements, causing $\mathbf{B}_{F(\mathbf{m})}$ to have large elements.

In statistical inference, the concepts of stability and conditioning are encompassed within the analysis of *uncertainty*. As shown by Eq. (8.54), the estimate $\tilde{\mathbf{m}}$ is a function of both $\mathbf{m}$ (the true model) and $\mathbf{e}$. With $\mathbf{e}$ taken to be a random variable, $\tilde{\mathbf{m}}$ will be a random variable. The p.d.f. of $\tilde{\mathbf{m}}$, which depends on $\mathbf{m}$ and will be shown as $f(\tilde{\mathbf{m}}; \mathbf{m})$, describes the error, or uncertainty, in $\tilde{\mathbf{m}}$ as an estimate of $\mathbf{m}$. Most often, the first two moments of $f(\tilde{\mathbf{m}}; \mathbf{m})$ are considered in the evaluation of estimation uncertainty. The first moment, $\mathrm{E}[\tilde{\mathbf{m}}; \mathbf{m}]$, indicates the *bias* in $\tilde{\mathbf{m}}$. The estimate is said to be unbiased if

$$\mathrm{E}[\tilde{\mathbf{m}}; \mathbf{m}] = \mathbf{m} \tag{8.64}$$

When $\tilde{\mathbf{m}}$ *is* unbiased, the variance matrix of $\tilde{\mathbf{m}}$, $\mathrm{Var}[\tilde{\mathbf{m}}; \mathbf{m}]$, indicates the size of the estimation error, $\tilde{\mathbf{m}} - \mathbf{m}$. The estimation variance can be thought of as a continuous measure of stability, with lower variance indicating less sensitivity to $\mathbf{e}$ and thus greater stability of $\tilde{\mathbf{m}}$.

Unfortunately, it is difficult to implement these concepts in nonlinear inverse problems. In general, even knowing if unbiased estimates exist is problematical, not to speak of finding one and its variance. Moreover, the variance of an estimate depends on $\mathbf{m}$ in general, which greatly complicates its interpretation. Therefore, uncertainty analysis is often based on linearization of F.

8.3.3 The linearized problem

As discussed in Section 8.2.4, the linearized inverse problem replaces the forward transformation F with its first-order expansion around a given reference model, as shown in Eq. (8.18). The linearized transformation will be written here more simply as

$$\bar{F}(\mathbf{m}) = F(\mathbf{m}_*) + \mathbf{A}(\mathbf{m} - \mathbf{m}_*) \tag{8.65}$$

where it is understood that $\mathbf{A}$ is evaluated at $\mathbf{m}_*$, the reference model. For now, $\mathbf{m}_*$ can be thought of as some prior estimate of the true model. This section will show that, with some restrictions on $\mathbf{A}$, the linearized inverse problem has a unique, global least-squares solution. This solution, which can be expressed analytically in linear algebraic terms, provides a basis for iterative, numerical algorithms for computing nonlinear least-squares solutions (see Section 8.5). Additionally, in the linearized problem, the mean and variance of $\tilde{\mathbf{m}}$ can be expressed analytically and these provide an *approximate* uncertainty analysis for the non-linear problem.

The least-squares objective function for the linearized inverse problem is given by

$$\Phi(\mathbf{m}) = (\mathbf{d} - \bar{F}(\mathbf{m}))^{\mathrm{T}}\mathbf{W}(\mathbf{d} - \bar{F}(\mathbf{m})) \qquad (8.66)$$

Substituting $\bar{F}$ for F in Eqs. (8.42) and (8.44), the gradient and Hessian of Φ are found to be

$$\mathbf{g_m} = 2\mathbf{A}^{\mathrm{T}}\mathbf{W}(\bar{F}(\mathbf{m}) - \mathbf{d}) \qquad (8.67)$$

$$\mathbf{H} = 2\mathbf{A}^{\mathrm{T}}\mathbf{W}\mathbf{A} \qquad (8.68)$$

Owing to the fact that $\bar{F}$ has a constant Jacobian ($\mathbf{A}$), the objective function is quadratic in $\mathbf{m}$, while its gradient is linear in $\mathbf{m}$ and its Hessian is constant. In particular, the second term in the general expression for the Hessian, $\mathbf{H_m^{nl}}$ in Eq. (8.44), is zero since second derivatives of $\bar{F}$ are zero.

For any $\mathbf{A}$ it is easy to show that $\mathbf{H}$ is positive *semi*-definite, meaning that

$$\delta\mathbf{m}^{\mathrm{T}}\mathbf{H}\delta\mathbf{m} \geq 0 \qquad (8.69)$$

for all vectors $\delta\mathbf{m}$. This is evident from

$$\delta\mathbf{m}^{\mathrm{T}}\mathbf{H}\delta\mathbf{m} = 2\delta\mathbf{m}^{\mathrm{T}}\mathbf{A}^{\mathrm{T}}\mathbf{W}\mathbf{A}\delta\mathbf{m} = 2(\mathbf{A}\delta\mathbf{m})^{\mathrm{T}}\mathbf{W}(\mathbf{A}\delta\mathbf{m}) \qquad (8.70)$$

Since $\mathbf{W}$ is positive definite, the right-hand side cannot be negative. If $\mathbf{A}$ has a trivial null space, meaning that only $\delta\mathbf{m} = 0$ makes $\mathbf{A}\delta\mathbf{m} = 0$, the right-hand side must be positive when $\delta\mathbf{m} \neq 0$ and $\mathbf{H}$ will be positive *definite*. This occurs when the columns of $\mathbf{A}$ are linearly independent vectors, in which case $\mathbf{A}$ is said to be *full rank*.

Inserting the expression for $\bar{F}$ in (8.65) into (8.67) and setting $\mathbf{g_m}$ to zero yields the normal equation for the linearized least-squares problem:

$$\mathbf{A}^{\mathrm{T}}\mathbf{W}\mathbf{A}(\tilde{\mathbf{m}} - \mathbf{m}_*) = \mathbf{A}^{\mathrm{T}}\mathbf{W}(\mathbf{d} - F(\mathbf{m}_*)) \qquad (8.71)$$

This also results from replacing F with $\bar{F}$ in Eq. (8.53). Equation (8.71) always has a solution because the right-hand side vector, being in the range of $\mathbf{A}^{\mathrm{T}}$, is also in the range of $\mathbf{A}^{\mathrm{T}}\mathbf{W}\mathbf{A}$. Further, if $\mathbf{A}$ is full rank, the left-hand side matrix (half the Hessian) is positive definite and (8.71) has the unique solution

$$\tilde{\mathbf{m}} = \mathbf{m}_* + \mathbf{B}(\mathbf{d} - F(\mathbf{m}_*)) \qquad (8.72)$$

where

$$\mathbf{B} = (\mathbf{A}^{\mathrm{T}}\mathbf{W}\mathbf{A})^{-1}\mathbf{A}^{\mathrm{T}}\mathbf{W} \qquad (8.73)$$

This solution is unique in a global sense and achieves a global minimum of Φ. It is also worth noting that Eq. (8.73) is a special case of (8.63), with $\mathbf{B}$ here being the constant Jacobian of the affine mapping (G) defined by (8.72).

8.3.4 *Uncertainty analysis*

Estimation uncertainty is easily analyzed in the linearized inverse problem. Substituting (8.6) into (8.72), replacing F with $\bar{F}$, yields

$$\tilde{\mathbf{m}} = \mathbf{m}_* + \mathbf{B}(\bar{F}(\mathbf{m}) + \mathbf{e} - F(\mathbf{m}_*)) \tag{8.74}$$

$$= \mathbf{m}_* + \mathbf{B}(\mathbf{A}(\mathbf{m} - \mathbf{m}_*) + \mathbf{e}) \tag{8.75}$$

Since $\mathbf{BA} = \mathbf{I}$ (the $M \times M$ identity matrix),

$$\tilde{\mathbf{m}} = \mathbf{m} + \mathbf{Be} \tag{8.76}$$

which is the special case of Eq. (8.58) for linearized F. The statistical assumptions about $\mathbf{e}$ that are consistent with the definition of Φ in (8.24) are that $\mathbf{e}$ is Gaussian with moments

$$E[\mathbf{e}] = 0 \tag{8.77}$$

$$\text{Var}[\mathbf{e}] = \mathbf{W}^{-1} \tag{8.78}$$

With $\tilde{\mathbf{m}}$ as the affine function of $\mathbf{e}$ in (8.76), it follows that it, too, is Gaussian with

$$E[\tilde{\mathbf{m}}; \mathbf{m}] = \mathbf{m} \tag{8.79}$$

$$\text{Var}[\tilde{\mathbf{m}}] = \mathbf{BW}^{-1}\mathbf{B}^{\text{T}} \tag{8.80}$$

noting that the variance does not depend on $\mathbf{m}$. Inserting the expression for $\mathbf{B}$ in (8.73), the variance becomes

$$\text{Var}[\tilde{\mathbf{m}}] = \left(\mathbf{A}^{\text{T}}\mathbf{WA}\right)^{-1} = 2\mathbf{H}^{-1} \tag{8.81}$$

Equation (8.79) shows that the linear least-squares solution is an unbiased estimate of $\mathbf{m}$ and (8.81) shows that its variance is proportional to the inverse Hessian. Therefore, if the Hessian is ill-conditioned, the variance of $\tilde{\mathbf{m}}$ will be large.

A very useful tool for analyzing the variance of the solution is the method of *singular value decomposition* (SVD). Let

$$\mathbf{W}^{1/2}\mathbf{A} = \mathbf{U}\mathbf{\Lambda}\mathbf{V}^{\text{T}} = \sum_{\ell=1}^{k} \lambda_\ell \mathbf{u}_\ell \mathbf{v}_\ell^{\text{T}} \tag{8.82}$$

where k is the rank of $\mathbf{A}$. Then

$$\mathbf{A}^{\text{T}}\mathbf{WA} = \mathbf{V}\mathbf{\Lambda}^2\mathbf{V}^{\text{T}} \tag{8.83}$$

and, assuming $\mathbf{A}$ to be full rank ($k = M$),

$$\text{Var}[\tilde{\mathbf{m}}] = \mathbf{V}\mathbf{\Lambda}^{-2}\mathbf{V}^{\text{T}} = \sum_{\ell=1}^{M} \lambda_\ell^{-2} \mathbf{v}_\ell \mathbf{v}_\ell^{\text{T}} \tag{8.84}$$

The variance of $\tilde{\mathbf{m}}$ becomes larger as the singular values become smaller. The variance is also related to the *condition number* of $\mathbf{A}^T\mathbf{WA}$, which is the ratio of its largest eigenvalue to its smallest: λ_1^2/λ_M^2. The larger the condition number, the more ill-conditioned and the higher the variance.

Given the difficulty of finding the p.d.f. of $\tilde{\mathbf{m}}$, or its moments, in nonlinear inverse problems, the linearized uncertainty results are often used as an approximation in such problems. In particular, a commonly used approach to nonlinear problems is to compute $\tilde{\mathbf{m}}$ – applying a numerical algorithm to minimize Φ (see Section 8.5) – and then infer the uncertainty in $\tilde{\mathbf{m}}$ based on linearized analysis, using $\tilde{\mathbf{m}}$ as the reference model:

$$\text{Var}[\tilde{\mathbf{m}}] \approx \left(\mathbf{A}_{\tilde{\mathbf{m}}}^T\mathbf{WA}_{\tilde{\mathbf{m}}}\right)^{-1} \tag{8.85}$$

It is just assumed that $\tilde{\mathbf{m}}$ is unbiased.

8.4 Damped least-squares and smooth models

The previous section showed that a least-squares solution to the inverse problem is well-posed if the Hessian matrix of the objective function is positive definite at the solution model. In practice, this sufficient condition is a prerequisite for the application of least squares unless one can rely on specific properties of F to overrule it. Based on linear analysis, the condition translates to the Jacobian matrix being full rank: it must not contain linearly dependent columns. This cannot be true if the number of unknown parameters M exceeds the number of observed data N, and is not guaranteed even when $M \leq N$. Moreover, the Hessian being positive definite accomplishes only the most permissive notion of well-posedness. The variance of the least-squares estimate – a measure of its difference from the true model – will depend on how well-conditioned the Hessian matrix is. If columns of the Jacobian are nearly parallel, the variance will be large.

It is certainly possible to coerce an inverse problem to be well-conditioned by restricting the number of parameters, e.g. parameterizing Earth's structure as a 1D model with relatively few layers. However, doing so raises the issue of whether the parameterization, unless based on geological information, taints the inversion results. The difficulty lies with the implicit assumption that one element of the model space is the true model, capable of explaining observations from all possible experiments, no matter how complex or numerous. If the model space is too small, no element of the space will have this status.

Such concerns gave impetus in geophysics to the approach of parameterizing Earth models with one or more spatial functions, elevating the model space to a function space of infinite dimension. In practice, this approach is executed with finely discretized models, such as 1D Earth models having a large number of thin layers, or 2D and 3D models defined on dense sampling grids; in any case, this makes $M \gg N$. The inversion theory needed for this approach necessarily goes beyond classical statistical inference theory.

Several mathematical methods have been developed for defining well-posed solutions to inverse problems involving model and/or data spaces of large or infinite dimension. These include (citing a flagship reference for each) the Backus–Gilbert method (Backus & Gilbert, 1970), Tikhonov regularization (Tikhonov & Arsenin, 1977), Bayesian inference (Tarantola, 2005), stochastic inversion (Franklin, 1970), ridge regression (Hoerl & Kennard, 1970), generalized spline smoothing (Wahba, 1990) and kriging (Deutsch & Journel, 1998). Some of these methods specifically deal with infinite-dimensional model and data spaces (Tikhonov regularization and stochastic inversion), while others allow only the model space to be infinite-dimensional (Backus–Gilbert method, spline smoothing and kriging). Bayesian inference requires both spaces to be finite-dimensional but still allows $M > N$.

This section will focus on methods that are formulated as an extension of least-squares estimation, i.e. as *damped* least squares. In the list above, these include Tikhonov regularization and spline smoothing. Bayesian inference, applied in a particular way, also fits this description. A damped least-squares estimate of **m** can be defined as a constrained least-squares estimate, with the constraint specified as

$$\Omega(\mathbf{m}) \leq \mu \tag{8.86}$$

where $\mu > 0$ and Ω is a positive-valued function known as a *stabilizing functional*, in the terminology of Tikhonov regularization. The stabilizing functional is typically designed to penalize an undesirable property of **m**, such as oscillations that do not contribute significantly to fitting **d**. The model minimizing the least-squares objective function Φ of Eq. (8.24), subject to (8.86), is called a *regularized* solution to the inverse problem. The method of Lagrange multipliers can be used to convert this constrained optimization problem to the unconstrained problem of minimizing the augmented objective function given by

$$\Psi(\mathbf{m}) = \Phi(\mathbf{m}) + \lambda\,\Omega(\mathbf{m}) \tag{8.87}$$

for some $\lambda > 0$ that is determined by μ. The variable λ is known as the *regularization parameter*.

The theory of Tikhonov regularization is particularly concerned with inverse problems in which the data space and model space are infinite-dimensional function spaces, and is developed mainly for linear problems, i.e.

$$\mathbf{d} = \mathbf{Am} + \mathbf{e} \tag{8.88}$$

In many such problems, **A** has an inverse, yielding $\tilde{\mathbf{m}} = \mathbf{A}^{-1}\mathbf{d}$ as a unique solution, but $\mathbf{A}^{-1}$ is not a continuous transformation from data to model space. Therefore, the solution is unstable. The theory developed by Tikhonov & Arsenin (1977) and others, such as Morozov (1993), shows that, when Ω is defined properly, a regularized solution is stable and approaches the true model as **e** and λ tend to zero. The method of spline smoothing is similar to Tikhonov regularization but is concerned primarily with problems having a finite number of data (N), which are the focus of this chapter. In such problems, **A** in (8.88) does not have an inverse and the stabilizing functional takes on the task of achieving solution uniqueness in addition to stability. Furthermore, spline methods are developed more within a statistical inference framework, especially with regard to optimal choice of the regularization parameter.

The remainder of this section discusses the practical application of damped least squares to finite-dimensional inverse problems, avoiding the mathematical intricacy involved with function spaces. This said, the physical meaning of the stabilizing functional and its effect on solutions to the inverse problem are rooted in concepts relating to model functions, which are therefore considered at a more informal level, starting with the next subsection.

8.4.1 Stabilizing functionals

In applications to geophysical inverse problems, the stabilizing functional is typically chosen to be a measure of the spatial *roughness* of the unknown model. The constraint in Eq. (8.86) then has the effect of isolating an optimally smooth model as the unique solution of the inverse problem. Spatial roughness, in turn, is associated with derivatives of the model function ($m(\mathbf{r})$ in this chapter). In particular, the stabilizing function is often given in the form

$$\Omega(\mathbf{m}) = \int_V dV \ |Lm|^2 \tag{8.89}$$

where V is the domain of the model function, L is a differential operator, and $|\ldots|$ denotes the absolute value of a number or magnitude of a vector. For example, for a 1D model function $m(z)$ defined over $0 \leq z \leq a$, setting $L = d/dz$ gives

$$\Omega(\mathbf{m}) = \int_0^a dz \ \left(\frac{dm}{dz}\right)^2 \tag{8.90}$$

which is just the L_2-norm of dm/dz. Setting L to the second-derivative operator sets Ω to the L_2-norm of d^2m/dz^2:

$$\Omega(\mathbf{m}) = \int_0^a dz \ \left(\frac{d^2m}{dz^2}\right)^2 \tag{8.91}$$

The extension of Eqs. (8.90) and (8.91) to 2D or 3D models are, respectively,

$$\Omega(\mathbf{m}) = \int_V dV |\nabla m|^2 = \int_V dV \ \nabla m \cdot \nabla m \tag{8.92}$$

and

$$\Omega(\mathbf{m}) = \int_V dV \ (\nabla^2 m)^2 \tag{8.93}$$

Alternatively, the stabilizing function can be written as

$$\Omega(\mathbf{m}) = \int_V dV \ mKm \tag{8.94}$$

for an appropriate differential operator K. For example, integrating the right-hand side of (8.90) by parts yields

$$\int_0^a dV \left(\frac{dm}{dz}\right)^2 = m(a)\frac{dm}{dz}(a) - m(0)\frac{dm}{dz}(0) - \int_0^a dz\, m\frac{d^2m}{dz^2} \qquad (8.95)$$

If the space of model functions is restricted to either Dirichlet boundary conditions ($m = 0$) or Neumann conditions ($dm/dz = 0$), the first two terms of the right-hand side vanish, so that

$$\Omega(\mathbf{m}) = -\int_0^a dz\, m\frac{d^2m}{dz^2} \qquad (8.96)$$

i.e. $K = -d^2/dz^2$. For (8.91), integrating by parts twice reveals that $K = d^4/dz^4$ assuming, say, $dm/dz = d^3m/dz^3 = 0$ at the end-points. Jumping to Eq. (8.93), the right-hand side integral can be evaluated using Green's second identity, which states that for twice differentiable functions $f(\mathbf{r})$ and $g(\mathbf{r})$,

$$\int_V dV\left[f\nabla^2 g - g\nabla^2 f\right] = \int_S dS\left[f\frac{\partial g}{\partial n} - g\frac{\partial f}{\partial n}\right] \qquad (8.97)$$

where S is the boundary of the model domain and $\partial/\partial n$ denotes the directional derivative along the outward normal to the boundary. If $f = \nabla^2 m$ and $g = m$, then

$$\int_V dV\left(\nabla^2 m\right)^2 = \int_V dV\, m\nabla^2\nabla^2 m + \int_S dS\left[\nabla^2 m\frac{\partial m}{\partial n} - m\frac{\partial \nabla^2 m}{\partial n}\right] \qquad (8.98)$$

$\nabla^2\nabla^2$ (the Laplacian of the Laplacian) is the biharmonic operator, often shown as ∇^4. The boundary conditions $\partial m/\partial n = \partial\nabla^2 m/\partial n = 0$ remove the second integral, and the stabilizing function of Eq. (8.93) can be written as Eq. (8.94) with $K = \nabla^4$.

8.4.1.1 Finite-dimensional models

When $\mathbf{m}$ is a finite-dimensional model, as required for numerical application, the concept of model roughness can be applied in an approximate way. If the M components of $\mathbf{m}$ can be related to a continuous function $m(\mathbf{r})$ – e.g. as samples of the function on a grid, or as coefficients for a set of basis functions – it will generally be possible to construct a symmetric, positive semi-definite matrix $\mathbf{K}$ such that

$$\Omega(\mathbf{m}) = \mathbf{m}^T\mathbf{K}\mathbf{m} \approx \int_V dV(\mathbf{r})\, m(\mathbf{r})Km(\mathbf{r}) \qquad (8.99)$$

For sampled models, $\mathbf{K}$ will represent a finite-difference approximation to the differential operator K. Alternatively, one can devise matrices $\mathbf{L}$ and $\mathbf{V}$ such that

$$\Omega(\mathbf{m}) = \mathbf{m}^T\mathbf{L}^T\mathbf{V}\mathbf{L}\mathbf{m} \approx \int_V dV(\mathbf{r})\, |Lm(\mathbf{r})|^2 \qquad (8.100)$$

where $\mathbf{L}$ is a finite-difference approximation to L and $\mathbf{V}$ contains coefficients for numerical integration. This implicitly results in

$$\mathbf{K} = \mathbf{L}^T \mathbf{V} \mathbf{L} \tag{8.101}$$

For example, in the case of a 1D Earth model, let the components of the finite-dimensional model vector $\mathbf{m}$ be samples of a model function at given depths z_j: $m_j = m(z_j)$. To approximate the stabilizing function of Eq. (8.90), the matrix $\mathbf{L}$ can be defined such that $\mathbf{L}\mathbf{m}$ approximates samples of dm/dz. Using simple first differences to approximate derivatives, $\mathbf{L}$ becomes the $(M-1) \times M$ matrix

$$\mathbf{L} = \begin{pmatrix} -\frac{1}{h_1} & \frac{1}{h_1} & 0 & \cdots & \\ 0 & -\frac{1}{h_2} & \frac{1}{h_2} & 0 & \cdots \\ & & \ddots & \ddots & \\ & \cdots & 0 & -\frac{1}{h_{M-1}} & \frac{1}{h_{M-1}} \end{pmatrix} \tag{8.102}$$

where $h_j = z_{j+1} - z_j$. The integration in (8.90) can be approximated with a rectangle-rule sum, which is accomplished by setting $\mathbf{V}$ to the diagonal matrix

$$\mathbf{V} = \begin{pmatrix} h_1 & 0 & \cdots & \\ 0 & h_2 & 0 & \cdots \\ & & \ddots & \\ & \cdots & 0 & h_{M-1} \end{pmatrix} \tag{8.103}$$

Equation (8.101) then yields the $M \times M$ tridiagonal matrix

$$\mathbf{K} = \begin{pmatrix} \frac{1}{h_1} & -\frac{1}{h_1} & 0 & \cdots & & \\ -\frac{1}{h_1} & \frac{1}{h_1}+\frac{1}{h_2} & -\frac{1}{h_2} & 0 & \cdots & \\ 0 & -\frac{1}{h_2} & \frac{1}{h_2}+\frac{1}{h_3} & -\frac{1}{h_3} & 0 & \cdots \\ & & \ddots & \ddots & \ddots & \\ & \cdots & 0 & -\frac{1}{h_{M-2}} & \frac{1}{h_{M-2}}+\frac{1}{h_{M-1}} & -\frac{1}{h_{M-1}} \\ & & \cdots & 0 & -\frac{1}{h_{M-1}} & \frac{1}{h_{M-1}} \end{pmatrix} \tag{8.104}$$

and the numerical approximation to Eq. (8.90) becomes

$$\int dz \left(\frac{dm}{dz}\right)^2 \approx \sum_{j=1}^{M-1} (z_{j+1}-z_j)\left(\frac{m_{j+1}-m_j}{z_{j+1}-z_j}\right)^2 = \sum_{j=1}^{M-1} \frac{(m_{j+1}-m_j)^2}{z_{j+1}-z_j} \tag{8.105}$$

For the remainder of this section, the stabilizing functional will be taken as

$$\Omega(\mathbf{m}) = (\mathbf{m}-\mathbf{m}_0)^T \mathbf{K}(\mathbf{m}-\mathbf{m}_0) \tag{8.106}$$

This is the form given in (8.99) with the introduction of a default model, $\mathbf{m}_0$. In some applications, it may be meaningful to constrain the roughness of the solution model relative to a default, although it is more common to set $\mathbf{m}_0 = 0$. The main reason to include $\mathbf{m}_0$ is that it facilitates the comparison of smoothing to Bayesian inference, presented later in Section 8.4.6.

8.4.2 The nonlinear problem

Most of the earlier analysis of ordinary (undamped) least squares is relevant to damped least squares with some modification to account for the additional term in the objective function. Substituting Eqs. (8.24) and (8.106) into Eq. (8.87), the damped least-squares objective functional becomes

$$\Psi(\mathbf{m}) = (\mathbf{d} - F(\mathbf{m}))^T \mathbf{W}(\mathbf{d} - F(\mathbf{m})) + \lambda(\mathbf{m} - \mathbf{m}_0)^T \mathbf{K}(\mathbf{m} - \mathbf{m}_0) \tag{8.107}$$

The gradient and Hessian of Ψ are given by

$$\mathbf{g_m} = 2\mathbf{A_m^T W}(F(\mathbf{m}) - \mathbf{d}) + 2\lambda\mathbf{K}(\mathbf{m} - \mathbf{m}_0) \tag{8.108}$$

and

$$\mathbf{H_m} = 2\mathbf{A_m^T W A_m} + \mathbf{H_m^{nl}} + 2\lambda\mathbf{K} \tag{8.109}$$

These expressions are just those of Eqs. (8.42) and (8.44), the gradient and Hessian of Φ, with an additional term in each arising from the stabilizing functional. Analogous reasoning to that used in Section 8.3.1 shows that a model $\tilde{\mathbf{m}}$ is a unique, local damped least-squares solution if $\mathbf{g_{\tilde{m}}} = 0$ (stationarity) and $\mathbf{H_{\tilde{m}}}$ is positive definite. Of crucial importance is the fact that the contribution of $\lambda\mathbf{K}$ to the Hessian makes it more likely that the Hessian is positive definite since $\mathbf{K}$ is positive semi-definite, and more so as λ increases.

The analysis of stability and uncertainty in damped least squares proceeds somewhat differently from ordinary least squares (Section 8.3.2). While, as before, little can be said for the general nonlinear problem, Eq. (8.55), relating $\tilde{\mathbf{m}}$ to $\mathbf{m}$ and $\mathbf{e}$, still applies, with $\mathbf{B_d}$ still given by Eq. (8.63). The damped least-squares solution is thus stable in the sense of Hadamard. However, a side effect of augmenting the objective function is that it is no longer reasonable to assume that $G(F(\mathbf{m})) = \mathbf{m}$, with the implication that $\tilde{\mathbf{m}}$ is not correct when $\mathbf{e} = 0$ and, therefore, is a biased estimate of $\mathbf{m}$. On the other hand, one might suspect that the variance of $\tilde{\mathbf{m}}$, while very difficult to characterize in the nonlinear case, is reduced by inclusion of the stabilizing functional. These statements will be abundantly clear in the linearized inverse problem.

8.4.3 The linearized problem

For the linearized problem, Ψ is the quadratic function that results from replacing F with the first-order approximation $\bar{F}$, defined in Eq. (8.65). The gradient and Hessian of Ψ become

$$\begin{aligned}\mathbf{g_m} &= 2\mathbf{A^T W}(\bar{F}(\mathbf{m}) - \mathbf{d}) + 2\lambda\mathbf{K}(\mathbf{m} - \mathbf{m}_0) \\ &= 2\mathbf{A^T W}(F(\mathbf{m}_*) + \mathbf{A}(\mathbf{m} - \mathbf{m}_*) - \mathbf{d}) + 2\lambda\mathbf{K}(\mathbf{m} - \mathbf{m}_0)\end{aligned} \tag{8.110}$$

and

$$\mathbf{H} = 2\mathbf{A^T W A} + 2\lambda\mathbf{K} \tag{8.111}$$

while the normal equation for $\tilde{\mathbf{m}}$ is

$$\mathbf{A}^T\mathbf{W}\mathbf{A}(\tilde{\mathbf{m}} - \mathbf{m}_*) + \lambda\mathbf{K}(\tilde{\mathbf{m}} - \mathbf{m}_0) = \mathbf{A}^T\mathbf{W}(\mathbf{d} - F(\mathbf{m}_*)) \tag{8.112}$$

Two useful expressions for the solution to (8.112) can be found as follows. First, adding $\lambda\mathbf{K}(\mathbf{m}_0 - \mathbf{m}_*)$ to both sides, the normal equation becomes

$$\mathbf{A}^T\mathbf{W}\mathbf{A}(\tilde{\mathbf{m}} - \mathbf{m}_*) + \lambda\mathbf{K}(\tilde{\mathbf{m}} - \mathbf{m}_*) = \mathbf{A}^T\mathbf{W}(\mathbf{d} - F(\mathbf{m}_*)) + \lambda\mathbf{K}(\mathbf{m}_0 - \mathbf{m}_*) \tag{8.113}$$

whose solution is

$$\tilde{\mathbf{m}} = \mathbf{m}_* + \left(\mathbf{A}^T\mathbf{W}\mathbf{A} + \lambda\mathbf{K}\right)^{-1}\left[\mathbf{A}^T\mathbf{W}(\mathbf{d} - F(\mathbf{m}_*)) + \lambda\mathbf{K}(\mathbf{m}_0 - \mathbf{m}_*)\right] \tag{8.114}$$

The inverse matrix in this expression exists exactly when the Hessian of Ψ is positive definite. The other useful expression results if $\mathbf{A}^T\mathbf{W}\mathbf{A}(\mathbf{m}_* - \mathbf{m}_0)$ is added to both sides of Eq. (8.112), yielding

$$\begin{aligned} \mathbf{A}^T\mathbf{W}\mathbf{A}(\tilde{\mathbf{m}} - \mathbf{m}_0) + \lambda\mathbf{K}(\tilde{\mathbf{m}} - \mathbf{m}_0) &= \mathbf{A}^T\mathbf{W}(\mathbf{d} - F(\mathbf{m}_*) + \mathbf{A}(\mathbf{m}_* - \mathbf{m}_0)) \\ &= \mathbf{A}^T\mathbf{W}(\mathbf{d} - \bar{F}(\mathbf{m}_0)) \end{aligned} \tag{8.115}$$

This leads naturally to the expression

$$\tilde{\mathbf{m}} = \mathbf{m}_0 + \mathbf{B}(\mathbf{d} - \bar{F}(\mathbf{m}_0)) \tag{8.116}$$

where

$$\mathbf{B} = \left(\mathbf{A}^T\mathbf{W}\mathbf{A} + \lambda\mathbf{K}\right)^{-1}\mathbf{A}^T\mathbf{W} \tag{8.117}$$

Equation (8.116) portrays the damped least-squares solution model as a perturbation to a default model $\mathbf{m}_0$ as opposed to the reference model $\mathbf{m}_*$ as in Eq. (8.114). Both expressions are merely a consequence of the general relation

$$\mathbf{g}_\mathbf{m} = \mathbf{g}_{\mathbf{m}_1} + \mathbf{H}(\mathbf{m} - \mathbf{m}_1) \tag{8.118}$$

which holds for any $\mathbf{m}_1$ since the gradient is a linear function of $\mathbf{m}$. The normal equation becomes

$$\mathbf{g}_{\mathbf{m}_1} + \mathbf{H}(\tilde{\mathbf{m}} - \mathbf{m}_1) = 0 \tag{8.119}$$

leading to

$$\tilde{\mathbf{m}} = \mathbf{m}_1 - \mathbf{H}^{-1}\mathbf{g}_{\mathbf{m}_1} \tag{8.120}$$

which is Eq. (8.114) when $\mathbf{m}_1 = \mathbf{m}_*$ and (8.116) when $\mathbf{m}_1 = \mathbf{m}_0$.

For the case when $\mathbf{K}$ is positive definite, an alternative expression for $\mathbf{B}$ is available as (e.g. Foster, 1961; Tarantola & Valette, 1982)

$$\mathbf{B} = \mathbf{K}^{-1}\mathbf{A}^T\left(\mathbf{A}\mathbf{K}^{-1}\mathbf{A}^T + \lambda\mathbf{W}^{-1}\right)^{-1} \tag{8.121}$$

This expression for $\mathbf{B}$, containing the inverse of a matrix on the data space, can be computed more efficiently than (8.117) when $N < M$. Equation (8.121) can be derived directly as follows:

$$\begin{aligned}
\mathbf{B}\left(\mathbf{A}\mathbf{K}^{-1}\mathbf{A}^{\mathrm{T}} + \lambda\mathbf{W}^{-1}\right) &= \left(\mathbf{A}^{\mathrm{T}}\mathbf{W}\mathbf{A} + \lambda\mathbf{K}\right)^{-1}\mathbf{A}^{\mathrm{T}}\mathbf{W}\left(\mathbf{A}\mathbf{K}^{-1}\mathbf{A}^{\mathrm{T}} + \lambda\mathbf{W}^{-1}\right) \\
&= \left(\mathbf{A}^{\mathrm{T}}\mathbf{W}\mathbf{A} + \lambda\mathbf{K}\right)^{-1}\left(\mathbf{A}^{\mathrm{T}}\mathbf{W}\mathbf{A}\mathbf{K}^{-1}\mathbf{A}^{\mathrm{T}} + \lambda\mathbf{A}^{\mathrm{T}}\right) \\
&= \left(\mathbf{A}^{\mathrm{T}}\mathbf{W}\mathbf{A} + \lambda\mathbf{K}\right)^{-1}\left(\mathbf{A}^{\mathrm{T}}\mathbf{W}\mathbf{A} + \lambda\mathbf{K}\right)\mathbf{K}^{-1}\mathbf{A}^{\mathrm{T}} \\
&= \mathbf{K}^{-1}\mathbf{A}^{\mathrm{T}}
\end{aligned} \tag{8.122}$$

which implies (8.121).

8.4.4 Uncertainty analysis

Substituting (8.6) into (8.116)), replacing F with $\bar{F}$, yields

$$\begin{aligned}
\tilde{\mathbf{m}} &= \mathbf{m}_0 + \mathbf{B}(\bar{F}(\mathbf{m}) + \mathbf{e} - \bar{F}(\mathbf{m}_0)) \\
&= \mathbf{m}_0 + \mathbf{B}(\mathbf{A}(\mathbf{m} - \mathbf{m}_0) + \mathbf{e}) \\
&= \mathbf{R}\mathbf{m} + (\mathbf{I} - \mathbf{R})\mathbf{m}_0 + \mathbf{B}\mathbf{e}
\end{aligned} \tag{8.123}$$

where

$$\mathbf{R} = \mathbf{B}\mathbf{A} \tag{8.124}$$

$\mathbf{R}$ is known as the *resolution matrix* (or resolution *operator* for infinite-dimensional models). In ordinary least squares, $\mathbf{R}$ is just the identity matrix, but with damping it is not:

$$\mathbf{R} = \left(\mathbf{A}^{\mathrm{T}}\mathbf{W}\mathbf{A} + \lambda\mathbf{K}\right)^{-1}\mathbf{A}^{\mathrm{T}}\mathbf{W}\mathbf{A} \tag{8.125}$$

As a consequence, Eq. (8.123) states that the estimate $\tilde{\mathbf{m}}$ is a *filtered* version of the true model $\mathbf{m}$ plus an error term due to $\mathbf{e}$. In statistical terms, using (8.77) and (8.78) for the moments of $\mathbf{e}$, the mean and variance of $\tilde{\mathbf{m}}$ become

$$E[\tilde{\mathbf{m}}\,;\mathbf{m}] = \mathbf{R}\mathbf{m} + (\mathbf{I} - \mathbf{R})\mathbf{m}_0 \tag{8.126}$$

$$\begin{aligned}
\mathrm{Var}[\tilde{\mathbf{m}}] &= \mathbf{B}\mathbf{W}^{-1}\mathbf{B}^{\mathrm{T}} \\
&= \left(\mathbf{A}^{\mathrm{T}}\mathbf{W}\mathbf{A} + \lambda\mathbf{K}\right)^{-1}\mathbf{A}^{\mathrm{T}}\mathbf{W}\mathbf{A}\left(\mathbf{A}^{\mathrm{T}}\mathbf{W}\mathbf{A} + \lambda\mathbf{K}\right)^{-1}
\end{aligned} \tag{8.127}$$

Equation (8.126) shows that $\tilde{\mathbf{m}}$ is a *biased* estimate of $\mathbf{m}$ when the resolution matrix is not the identity.

Several authors have discussed the role of the regularization parameter in regulating a trade-off between the bias and variance of $\tilde{\mathbf{m}}$. As λ is decreased, $\mathbf{R}$ becomes closer to the identify matrix, reducing the bias in the estimate, while $\mathrm{Var}[\tilde{\mathbf{m}}]$ increases. The reader is pointed to Jackson (1972), in the geophysical literature, and Marquardt (1970), in the statistics literature, for more complete discussions of the bias–variance trade-off in the damped least-squares setting. However, it is important to note that the topic was introduced to geophysics by Backus & Gilbert (1968, 1970) in a somewhat different context from least squares. In the Backus–Gilbert formulation, which dealt with 1D model functions, the resolution operator was seen as a spatial low-pass filter, with estimation bias being associated with the filter width. This interpretation is compatible with choosing the stabilizer Ω in damped least squares to be a measure of spatial roughness, thereby enforcing spatial smoothness on $\tilde{\mathbf{m}}$. Moreover, the association of resolution with low-pass filtering is most valid if the default model $\mathbf{m}_0$ is set to zero, in which case the last term of (8.126) vanishes, giving simply

$$E[\tilde{\mathbf{m}}; \mathbf{m}] = \mathbf{Rm} \qquad (8.128)$$

However, even with these choices, the action of the resolution matrix will not necessarily be that of a low-pass filter, as pointed out by Jackson (1979). MT data, in particular, can be sensitive to high-wavenumber components of Earth's resistivity function, especially for 2D and 3D models.

As in ordinary least squares, linearized uncertainty analysis is often applied to the non-linear inverse problem as an approximation. Specifically, after the nonlinear least-squares solution model is found, the resolution and variance matrices are computed based on linearization around that solution. That is, Eqs. (8.125) and (8.127) are evaluated with $\mathbf{A}$ set to $\mathbf{A}_{\tilde{\mathbf{m}}}$.

8.4.5 *Choosing the regularization parameter*

Damped least-squares estimation requires specification of the regularization parameter, λ, in the objective function of Eq. (8.87). This parameter controls the relative importance of minimizing the data misfit, $\Phi(\mathbf{m})$, and the stabilizing functional, $\Omega(\mathbf{m})$, which in the case of smooth model inversion is the amount of spatial roughness in the model. Larger values of λ will give more weight to minimizing the model roughness, resulting in smoother models yielding poorer fits to the data. From a statistical point of view, λ controls the bias in the solution model, which is measured by the resolution matrix, versus its sensitivity to the data errors, which is measured by the model variance matrix. As λ is increased, the variance is reduced but resolution is degraded. Theoretically, the models for all λ are valid solutions to the inverse problem and they, together with a description of the misfit–roughness and variance–resolution trade-offs, constitute a complete solution of the inverse problem. In practice, however, there is usually a need to select a single model as the solution. The question arises then as to which value of λ yields the best solution.

In MT applications, and geophysical applications in general, the selection of λ has been based primarily on the trade-off between data misfit and model roughness. The goal is to find the "corner" of the trade-off curve, $\Phi(\tilde{\mathbf{m}})$ versus $\Omega(\tilde{\mathbf{m}})$, that is generated as λ is varied. The corner is the point where decreasing λ achieves only a small decrease in Φ at the expense of a large increase in Ω, while increasing λ is equally unsatisfying in the opposite sense. The *L-curve* method (Hansen, 1992) defines the corner of the trade-off curve mathematically as the point of maximum curvature. The larger goal, however, is to find a model that is geologically reasonable and whose data residuals are consistent with an understanding of the errors in the data. Therefore, the selection of λ is often based on visual inspection of $\tilde{\mathbf{m}}$ and its predicted data, not just on the values of Φ and Ω.

Another approach commonly used in MT is to target a specified value of the data misfit, namely, to select λ such that

$$\Phi(\tilde{\mathbf{m}}) = \phi \tag{8.129}$$

The target misfit value, ϕ, is chosen in accordance with statistical assumptions or bounds on $\mathbf{e}$. This is known as the *discrepancy* method in the Tikhonov regularization literature (e.g. Morozov, 1993), in which a major concern is *convergence*: whether, and how fast, the error in $\tilde{\mathbf{m}}$ (compared to the true model) goes to zero as the variance of $\mathbf{e}$, or a bound on $||\mathbf{e}||$, is reduced to zero. The literature on spline smoothing, which deals specifically with finite-dimensional data and infinite-dimensional models, further addresses convergence as $N \to \infty$.

For a fixed dataset and fixed assumptions about $\mathbf{e}$, the primary concern in the discrepancy method is how to set the target misfit. Statistically, it is reasonable to set ϕ to the expected value of $\Phi(\tilde{\mathbf{m}})$, in which case (8.129) becomes

$$\Phi(\tilde{\mathbf{m}}) = E[\Phi(\tilde{\mathbf{m}})] \tag{8.130}$$

This criterion, in effect, prevents the solution from over-fitting the data (λ too small) or under-fitting them. Unfortunately, the right-hand side of (8.130) depends not only on λ but also on the true model $\mathbf{m}$. This is clear from the linearized inverse problem, in which it follows from Eq. (8.123) that

$$\mathbf{d} - \bar{F}(\tilde{\mathbf{m}}) = (\mathbf{I} - \mathbf{S})(\mathbf{A}(\mathbf{m} - \mathbf{m}_0) + \mathbf{e}) \tag{8.131}$$

where the $N \times N$ matrix $\mathbf{S}$, known as the *influence* matrix (e.g. Vogel, 2002) or *information density* matrix (Jackson, 1972; Wiggins, 1972), is given by

$$\mathbf{S} = \mathbf{AB} \tag{8.132}$$

Forming the data misfit from the residual vector in (8.131), and taking the expected value of the result, will leave a dependence on $\mathbf{m}$ (the true model), rendering (8.130) unsolvable. For this reason, (8.130) is often replaced with

$$\Phi(\tilde{\mathbf{m}}) = E[\Phi(\mathbf{m})] \tag{8.133}$$

Since $\Phi(\mathbf{m}) = \mathbf{e}^{\mathrm{T}}\mathbf{We}$ and $E[\mathbf{e}^{\mathrm{T}}\mathbf{We}] = N$ (based on the statistical assumptions of (8.77) and (8.78)), Eq. (8.133) becomes simply

$$\Phi(\tilde{\mathbf{m}}) = N \tag{8.134}$$

Recalling Eq. (8.25), this states that the weighted root mean square (RMS) residual produced by $\tilde{\mathbf{m}}$, given by $(\Phi(\tilde{\mathbf{m}})/N)^{1/2}$, should be unity.

The discrepancy method, along with most other methods for choosing λ, can impose a considerable computational burden on an inversion. To solve Eq. (8.134) by trial and error, for example, $\tilde{\mathbf{m}}$ will have to be calculated for many values of λ. A much more efficient scheme is described by Constable *et al.* (1987), de Groot-Hedlin & Constable (1990) and Parker (1994). Their algorithm minimizes the damped least-squares objective function (Ψ) and solves (8.134) simultaneously by adjusting λ on each step of an iterative minimization algorithm.

Another approach to selecting the regularization parameter, which to date has received little attention in MT inversion, is to consider the prediction capability of the inversion model: how well $\tilde{\mathbf{m}}$ can predict functionals of the true model. Methods that follow this approach include (ordinary) *cross-validation*, *generalized cross-validation* and the *unbiased predictive risk estimator* method. These methods are described in books by Wahba (1990), Bertero & Boccacci (1998) and Vogel (2002). The following brief description of cross-validation conveys the essential idea of the prediction-error approach.

Cross-validation considers the prediction errors, p_i, in the N data functionals used in deriving $\tilde{\mathbf{m}}$:

$$p_i = F_i(\tilde{\mathbf{m}}) - F_i(\mathbf{m}) \tag{8.135}$$

The objective is to find the value of λ that minimizes the p_i in some sense. Since the true model, $\mathbf{m}$, is not known, the prediction errors are approximated as

$$p_i \approx F_i(\tilde{\mathbf{m}}_{(i)}) - d_i \tag{8.136}$$

This uses the observation d_i as a proxy for $F_i(\mathbf{m})$ and, to avoid circularity, replaces $\tilde{\mathbf{m}}$ with the *leave-one-out* solution of the inverse problem, $\tilde{\mathbf{m}}_{(i)}$, obtained by removing the ith observation from $\mathbf{d}$. Cross-validation chooses λ as the value minimizing the function

$$CV = \sum_{i=1}^{N} w_i(d_i - F_i(\tilde{\mathbf{m}}_{(i)}))^2 \tag{8.137}$$

which is just the weighted sum of squared prediction errors as they are given in (8.136). Nominally, the calculation of CV for a single value of λ requires the minimization of N damped least-squares objective functions, with each datum left out in turn. This computational burden can be avoided, however, if linearization of the data functionals is used, in which case CV can be expressed as (Wahba, 1990)

$$CV = \sum_{i=1}^{N} \frac{w_i}{(1 - S_{ii})^2} (d_i - F_i(\tilde{\mathbf{m}}))^2 \tag{8.138}$$

where S_{ii} is the ith diagonal element of the influence matrix. This expression for CV involves only the inversion model ($\tilde{\mathbf{m}}$) obtained with the full data vector. On the other hand, it requires the computation of S_{ii}, which can be problematical in large inverse problems.

There are many variations on cross-validation, such as leaving out multiple data per trial instead of one. Another variation is *generalized* cross-validation (Wahba, 1990), which like (8.138) avoids repeated inversion of data subsets. An advantage of cross-validation methods (ordinary and generalized) is that, unlike the discrepancy method, they do not require the data variances ($\sigma_i^2 \equiv 1/w_i$) to be known in an absolute sense: if the w_i are scaled by a constant factor, the value of λ that minimizes CV will scale by the same factor, resulting in the same $\tilde{\mathbf{m}}$.

8.4.6 Comparison to Bayesian inference

Bayesian inference applies to finite-dimensional model and data spaces (N, $M < \infty$). In this approach, the unknown model parameters, like the data errors, are treated as random variables. To stabilize an ill-conditioned inverse problem, information about $\mathbf{m}$ is added in the form of a *prior* probability density function on $\mathbf{m}$, denoted $f(\mathbf{m})$. The Bayesian solution to the inverse problem is embodied in the *posterior* p.d.f. of $\mathbf{m}$, which is the *conditional* p.d. f. of $\mathbf{m}$ given $\mathbf{d}$:

$$f(\mathbf{m}|\mathbf{d}) = \frac{f(\mathbf{d}, \mathbf{m})}{f(\mathbf{d})} \tag{8.139}$$

where $f(\mathbf{d})$ is the p.d.f. of $\mathbf{d}$ (*marginal* with respect to $\mathbf{m}$) and $f(\mathbf{d}, \mathbf{m})$ is the *joint* p.d.f. of $\mathbf{d}$ and $\mathbf{m}$. Typically these distributions are not given directly but, using the laws of probability, are expressed as

$$f(\mathbf{d}, \mathbf{m}) = f(\mathbf{d}|\mathbf{m}) f(\mathbf{m}) \tag{8.140}$$

$$f(\mathbf{d}) = \int_{\mathcal{M}} dV(\mathbf{m}) f(\mathbf{d}, \mathbf{m}) \tag{8.141}$$

where $\mathcal{M}$ is the model space and $dV(\mathbf{m})$ is shorthand for $dm_1\, dm_2 \cdots dm_M$. The conditional p.d.f. of $\mathbf{d}$ given $\mathbf{m}$, $f(\mathbf{d}\,|\,\mathbf{m})$, is the same function denoted earlier as $f(\mathbf{d}; \mathbf{m})$ when $\mathbf{m}$ was treated as a deterministic variable. Recalling Eq. (8.26), it is given by

$$f(\mathbf{d}|\mathbf{m}) = f_e(\mathbf{d} - F(\mathbf{m})) \tag{8.142}$$

Using Eqs. (8.140) and (8.141) in (8.139) yields

$$f(\mathbf{m}|\mathbf{d}) = \frac{f(\mathbf{d}|\mathbf{m}) f(\mathbf{m})}{\int_{\mathcal{M}} dV(\mathbf{m}) f(\mathbf{d}|\mathbf{m}) f(\mathbf{m})} \tag{8.143}$$

This is Bayes' theorem for vector random variables.

Bayesian inference is a general method that does not require that the prior p.d.f.s of $\mathbf{e}$ and $\mathbf{m}$ or the forward transformation F be restricted to any particular form. However, it is not necessarily simple to compute $f(\mathbf{m}\,|\,\mathbf{d})$ or even its moments. The computational challenge is to evaluate the integral comprising the denominator of (8.143). When the number of model parameters is small, it may be possible to perform this integral numerically. Otherwise, other strategies for characterizing the posterior p.d.f. of $\mathbf{m}$ must be employed. For example, random sampling techniques, such as Markov chain Monte Carlo (MCMC; Sambridge & Mosegaard, 2002), generate an ensemble of models that constitute a random sampling from $f(\mathbf{m}\,|\,\mathbf{d})$. Another tractable approach is to calculate the mode of the posterior p.d.f., i.e. the model that maximizes $f(\mathbf{m}\,|\,\mathbf{d})$. The posterior mode is known as the *maximum a posteriori* (MAP) estimate and satisfies

$$f(\tilde{\mathbf{m}}|\mathbf{d}) = \max_{\mathbf{m}} f(\mathbf{m}|\mathbf{d}) \tag{8.144}$$

The damped least-squares estimate minimizing the objective function Ψ of Eq. (8.107) is also a Bayesian MAP estimate if the prior distributions of e and m are chosen appropriately. As assumed earlier in this chapter, let $f_e(e)$ be a Gaussian p.d.f. with first and second moments given by Eqs. (8.77) and (8.78). Then, recalling (8.142)

$$f(\mathbf{d}|\mathbf{m}) = \frac{(\det \mathbf{W})^{1/2}}{(2\pi)^{N/2}} \exp\left\{-\tfrac{1}{2}(\mathbf{d} - F(\mathbf{m}))^{\mathsf{T}} \mathbf{W}(\mathbf{d} - F(\mathbf{m}))\right\} \qquad (8.145)$$

Let $f(\mathbf{m})$ also be Gaussian and assign its first and second moments as

$$\mathrm{E}[\mathbf{m}] = \mathbf{m}_0 \qquad (8.146)$$

$$\mathrm{Var}[\mathbf{m}] = \lambda^{-1}\mathbf{K}^{-1} \qquad (8.147)$$

where $\mathbf{K}$ must now be positive definite. Then

$$f(\mathbf{m}) = \frac{\lambda^{M/2}(\det \mathbf{K})^{1/2}}{(2\pi)^{M/2}} \exp\left\{-\tfrac{1}{2}\lambda(\mathbf{m} - \mathbf{m}_0)^{\mathsf{T}} \mathbf{K}(\mathbf{m} - \mathbf{m}_0)\right\} \qquad (8.148)$$

Substituting these Gaussian p.d.f.s into (8.143), taking note of (8.107),

$$f(\mathbf{m}|\mathbf{d}) = \frac{\exp\left\{-\tfrac{1}{2}\Psi(\mathbf{m};\mathbf{d})\right\}}{\int_{\mathcal{M}} dV(\mathbf{m}) \exp\left\{-\tfrac{1}{2}\Psi(\mathbf{m};\mathbf{d})\right\}} \qquad (8.149)$$

where the dependence of the objective function on $\mathbf{d}$ is now shown. Equation (8.149) shows that the damped least-squares estimate of $\mathbf{m}$, which minimizes Ψ, maximizes the posterior p.d.f. and is thus also the MAP estimate.

Equation (8.149) also implies that, for the linearized inverse problem, the posterior p.d.f. of $\mathbf{m}$ is a Gaussian distribution. This follows from the fact that, when F is replaced by $\bar{F}$ in (8.107), Ψ becomes a quadratic function, which can be written as

$$\Psi(\mathbf{m};\mathbf{d}) = \Psi(\tilde{\mathbf{m}};\mathbf{d}) + (\mathbf{m} - \tilde{\mathbf{m}})^{\mathsf{T}}(\mathbf{A}^{\mathsf{T}}\mathbf{W}\mathbf{A} + \lambda\mathbf{K})(\mathbf{m} - \tilde{\mathbf{m}}) \qquad (8.150)$$

where $\tilde{\mathbf{m}}$ is the damped least-squares estimate. This can be proved knowing only that $\tilde{\mathbf{m}}$ minimizes Ψ and that the Hessian of Ψ is Eq. (8.111). Given this formula for Ψ, it follows from Eq. (8.149) that

$$\mathrm{E}[\mathbf{m}|\mathbf{d}] = \tilde{\mathbf{m}} \qquad (8.151)$$

$$\mathrm{Var}[\mathbf{m}|\mathbf{d}] = (\mathbf{A}^{\mathsf{T}}\mathbf{W}\mathbf{A} + \lambda\mathbf{K})^{-1} \qquad (8.152)$$

That is, the posterior mean of $\mathbf{m}$ is the damped least-squares model. It can be shown that the posterior variance is *larger* than the estimation variance given in Eq. (8.127). In fact,

$$\mathrm{Var}[\mathbf{m}|\mathbf{d}] = \mathrm{Var}[\tilde{\mathbf{m}};\mathbf{m}] + \lambda^{-1}(\mathbf{I} - \mathbf{R})\mathbf{K}^{-1}(\mathbf{I} - \mathbf{R})^{\mathsf{T}} \qquad (8.153)$$

In the Bayesian approach, the posterior variance of $\mathbf{m}$ accounts for the two sources of error considered in the deterministic approach: the variance resulting from data errors and the bias introduced by damping.

8.5 Minimization algorithms

Algorithms for minimizing the damped, nonlinear least-squares objective function given by Eq. (8.107) will be described, and in particular minimization methods specifically designed for the situation where F is a nonlinear function of $\mathbf{m}$ and thus Ψ is non-quadratic in $\mathbf{m}$.

Gradient or *descent* methods start with an initial guess of the solution, taken here to be $\mathbf{m}_0$, and generate a sequence of updated models $\mathbf{m}_\ell$, $\ell = 1, 2, \ldots$, with the goal of achieving $\Psi(\mathbf{m}_{\ell+1}) < \Psi(\mathbf{m}_\ell)$. Model $\mathbf{m}_{\ell+1}$ is based on the local behavior of Ψ in the neighborhood of model $\mathbf{m}_\ell$, as characterized by the gradient and, usually, Hessian of Ψ evaluated at $\mathbf{m}_\ell$. A convergence criterion (more properly, a *stopping* criterion, since it is often difficult to prove that a model sequence actually converges) selects the last iterate $\mathbf{m}_L$ as the solution of the problem.

Gradient methods perform *local*, as opposed to *global*, optimization. This means that they do not sample the model space away from the trajectory of the model sequence they generate in an attempt to find models, perhaps far from $\mathbf{m}_0$, that yield values of Ψ even smaller than $\Psi(\mathbf{m}_L)$. On the other hand, global optimization methods *do* attempt to sample the entire model space with some sort of adaptive search scheme. Such methods include grid search, genetic algorithms and simulated annealing (see Sambridge & Mosegaard (2002) for an overview). The present focus will be on gradient methods because of their greater efficiency compared to global search methods. By using the information from the gradient and Hessian of the objective function, they test many fewer models than global search methods typically do. This is very important in large inverse problems like 2D and 3D MT. The efficiency of gradient methods comes at the expense of a far greater likelihood that the globally optimal solution will be missed. This can be mitigated to some extent if one repeats the gradient algorithm with several different initial models, which can still be more efficient than a global search (Alumbaugh, 2000).

8.5.1 Newton's method

Newton's method generates the model sequence, $\mathbf{m}_\ell$, based on local quadratic approximations to Ψ. Let $\mathbf{g}_\ell$ and $\mathbf{H}_\ell$, respectively, denote the gradient and Hessian of Ψ evaluated at the ℓth model:

$$\mathbf{g}_\ell \equiv \mathbf{g}_{\mathbf{m}_\ell} \tag{8.154}$$

$$\mathbf{H}_\ell \equiv \mathbf{H}_{\mathbf{m}_\ell} \tag{8.155}$$

It should be noted that this is the exact Hessian from (8.109), including the second-derivative terms. Let $\bar{\Psi}_\ell$ be the second-order Taylor series expansion of Ψ around $\mathbf{m}_\ell$:

$$\bar{\Psi}_\ell(\mathbf{m}) \equiv \Psi(\mathbf{m}_\ell) + \mathbf{g}_\ell^{\mathrm{T}}(\mathbf{m} - \mathbf{m}_\ell) + \tfrac{1}{2}(\mathbf{m} - \mathbf{m}_\ell)^{\mathrm{T}}\mathbf{H}_\ell(\mathbf{m} - \mathbf{m}_\ell) \tag{8.156}$$

Newton's method finds $\mathbf{m}_{\ell+1}$ as a stationary point of $\bar{\Psi}_\ell$. Setting the gradient of $\bar{\Psi}_\ell$ to zero implies that

$$\mathbf{g}_\ell + \mathbf{H}_\ell(\mathbf{m}_{\ell+1} - \mathbf{m}_\ell) = 0 \tag{8.157}$$

or, assuming the Hessian is non-singular, that

$$\mathbf{m}_{\ell+1} = \mathbf{m}_\ell - \mathbf{H}_\ell^{-1}\mathbf{g}_\ell \tag{8.158}$$

If F were linear, the model sequence would converge at $\ell = 1$ with $\mathbf{m}_1$ being the solution $\tilde{\mathbf{m}}$ given by Eqs. (8.116) and (8.117). With nonlinear F, Eq. (8.158) must be iterated until $\mathbf{g}_\ell$ is satisfactorily small.

However, the Newton iteration does not necessarily converge if the Hessian matrix is not well-conditioned at some or all of the model iterates. The presence of the second term in Eq. (8.109), which involves second derivatives of F, poses a particular threat to convergence since, unlike the first and third terms, it may not even be positive semi-definite. In any case, convergence of Newton's method guarantees only that the final model $\mathbf{m}_L$ is a stationary point of Ψ, not necessarily a minimum, and, even if so, not necessarily a global minimum. However, if $\mathbf{g}_L$ is satisfactorily close to zero and $\mathbf{H}_L$ is positive definite, the solution does achieve a local minimum of Ψ. Haber (2000) showed that, for certain 1D MT problems, Newton's method reached a solution in fewer iterations than the more common variant, the Gauss–Newton method described next. Whether this result holds more generally is unclear.

8.5.2 Gauss–Newton method

The Gauss–Newton method is similar to Newton's method but uses a different quadratic approximation to the objective function Ψ that results from linearization of F around $\mathbf{m}_\ell$. In it, F is replaced with the first-order approximation, denoted by $\bar{F}_\ell$ where $\bar{F}_\ell$ is the third term in (8.159), obtained by setting $\mathbf{m}_* = \mathbf{m}_\ell$, in (8.18)

$$\bar{\Psi}_\ell(\mathbf{m}) = (\mathbf{d} - \bar{F}_\ell(\mathbf{m}))^\mathrm{T}\mathbf{W}(\mathbf{d} - \bar{F}_\ell(\mathbf{m})) + \lambda(\mathbf{m} - \mathbf{m}_0)^\mathrm{T}\mathbf{K}(\mathbf{m} - \mathbf{m}_0) \tag{8.159}$$

where $\bar{F}_\ell$ is short-hand for $\bar{F}_{\mathbf{m}_\ell}$. Since $\bar{F}_\ell(\mathbf{m}) = F(\mathbf{m}_\ell)$ at $\mathbf{m} = \mathbf{m}_\ell$, the gradient of $\bar{\Psi}_\ell$ at $\mathbf{m} = \mathbf{m}_\ell$ is the same as the gradient of Ψ at $\mathbf{m}_\ell$, denoted $\mathbf{g}_\ell$ in Section 8.5.1. The Hessian of $\bar{\Psi}_\ell$ at $\mathbf{m}_\ell$ is given by

$$\bar{\mathbf{H}}_\ell = 2\mathbf{A}_\ell^\mathrm{T}\mathbf{W}\mathbf{A}_\ell + 2\lambda\mathbf{K} \tag{8.160}$$

(where $A_\ell \equiv \mathbf{A}_{\mathbf{m}_\ell}$), which differs from the Hessian of Ψ by excluding the term $\mathbf{H}_\mathbf{m}^{\mathrm{nl}}$ in (8.109). Since $\bar{\Psi}_\ell$ is a quadratic function of $\mathbf{m}$, Eq. (8.159) must be equivalent to

$$\bar{\Psi}_\ell(\mathbf{m}) = \Psi(\mathbf{m}_\ell) + \mathbf{g}_\ell^\mathrm{T}(\mathbf{m} - \mathbf{m}_\ell) + \tfrac{1}{2}(\mathbf{m} - \mathbf{m}_\ell)^\mathrm{T}\bar{\mathbf{H}}_\ell(\mathbf{m} - \mathbf{m}_\ell) \tag{8.161}$$

Stated concisely, the Gauss–Newton method generates $\mathbf{m}_{\ell+1}$ as the minimum point of $\bar{\Psi}_\ell$. Applying the results of Section 8.4.3 yields

$$\mathbf{m}_{\ell+1} = \mathbf{m}_\ell - \bar{\mathbf{H}}_\ell^{-1}\mathbf{g}_\ell \tag{8.162}$$

which can also be derived by setting the gradient of $\bar{\Psi}_\ell$ in Eq. (8.161) to zero. The Gauss–Newton method differs from Newton's method simply by replacing $\mathbf{H}_\ell$, the exact Hessian of Ψ, with $\bar{\mathbf{H}}_\ell$ or, equivalently, by ignoring the second derivatives of F. Aside from the obvious computational savings, excluding these second derivatives ensures that $\bar{\mathbf{H}}_\ell$ is a positive definite matrix, tending to lead to a more stable model sequence. However, the Gauss–Newton method still cannot guarantee convergence to a global minimum of Ψ.

The results of Section 8.4.3 allow (8.162) to be written as (see Eq. (8.114))

$$\mathbf{m}_{\ell+1} = \mathbf{m}_\ell + \left(\mathbf{A}_\ell^{\mathrm{T}}\mathbf{W}\mathbf{A}_\ell + \lambda\mathbf{K}\right)^{-1}\left[\mathbf{A}_\ell^{\mathrm{T}}\mathbf{W}(\mathbf{d} - F(\mathbf{m}_\ell)) + \lambda\mathbf{K}(\mathbf{m}_0 - \mathbf{m}_\ell)\right] \qquad (8.163)$$

The explicit evaluation of this expression entails the inversion of the $M \times M$ Hessian matrix. When $\mathbf{K}$ is positive definite, Eqs. (8.116) and (8.121) provide the alternative formula

$$\mathbf{m}_{\ell+1} = \mathbf{m}_0 + \mathbf{K}^{-1}\mathbf{A}_\ell^{\mathrm{T}}\left(\mathbf{A}_\ell\mathbf{K}^{-1}\mathbf{A}_\ell^{\mathrm{T}} + \lambda\mathbf{W}^{-1}\right)^{-1}(\mathbf{d} - \bar{F}_\ell(\mathbf{m}_0)) \qquad (8.164)$$

which involves the inverse of an $N \times N$ matrix. Siripunvaraporn & Egbert (2000) and Siripunvaraporn *et al.* (2005) exploited this formula in their *data space* inversion method as a more efficient approach to 2D and 3D inversion problems involving many more model parameters than data ($M \gg N$).

8.5.3 Levenberg–Marquardt method

Levenberg (1944) and Marquardt (1963) proposed a modification of the Gauss–Newton method that further stabilizes the model sequence. The Levenberg–Marquardt model update minimizes an augmented objective function given by

$$\bar{\Psi}_\ell^\varepsilon(\mathbf{m}) = \bar{\Psi}_\ell(\mathbf{m}) + \varepsilon_\ell(\mathbf{m} - \mathbf{m}_\ell)^{\mathrm{T}}\mathbf{D}_\ell(\mathbf{m} - \mathbf{m}_\ell) \qquad (8.165)$$

where $\bar{\Psi}_\ell$ is the Gauss–Newton approximation to the objective function (Eq. (8.159)), $\mathbf{D}_\ell$ is a diagonal, positive definite damping matrix, and $\varepsilon_\ell > 0$. Reasoning by analogy with Eq. (8.162), the updating scheme becomes

$$\mathbf{m}_{\ell+1} = \mathbf{m}_\ell - (\bar{\mathbf{H}}_\ell + 2\varepsilon_\ell\mathbf{D}_\ell)^{-1}\mathbf{g}_\ell \qquad (8.166)$$

Damping the Hessian has the effect of shortening and redirecting the model increment $\mathbf{m}_{\ell+1} - \mathbf{m}_\ell$ with the aim of keeping it within a "domain of validity" for the quadratic approximation to Ψ. The damping reduces the sensitivity of $\mathbf{m}_{\ell+1}$ to the smallest eigenvalues of $\bar{\mathbf{H}}_\ell$ that are probably not predictive of the behavior of Ψ away from $\mathbf{m}_\ell$. The damping also suppresses computational error, which can result when $\bar{\mathbf{H}}$ is poorly conditioned.

The best choices for ε_ℓ and $\mathbf{D}_\ell$ are not obvious, but both Levenberg and Marquardt propose setting $\mathbf{D}_\ell$ to the diagonal of $\bar{\mathbf{H}}_\ell$ as one possibility. Marquardt (1963) proposes a decision tree for selecting ε_ℓ that forces Ψ to decrease monotonically: $\Psi(\mathbf{m}_{\ell+1}) < \Psi(\mathbf{m}_\ell)$. The reader is directed to the original papers and to Pujol (2007) for a more detailed discussion of Levenberg's and Marquardt's algorithms.

The damping used in the Levenberg–Marquardt iteration is not to be confused with the stabilizing function that contributes to Ψ given by the second term of Eq. (8.107) involving λ and **K**. Whereas minimizing the stabilizing function is an integral part of the objective in solving the nonlinear inverse problem, ε_ℓ and $\mathbf{D}_\ell$ are merely numerical devices for finding the solution. To emphasize this, note that, the larger λ is, the more quadratic Ψ becomes, and the less Levenberg–Marquardt damping is needed to successfully minimize Ψ.

8.5.4 Model updates by conjugate gradients

The formulas for the Gauss–Newton and Levenberg–Marquardt model updates (Eqs. (8.162) and (8.166), respectively) can be evaluated by inverting the Hessian (or damped Hessian) matrix and applying the inverse to the gradient vector. Algorithms for inverting positive definite matrices are readily available for this purpose. However, Eqs. (8.162) and (8.166) do not need to be read so literally; instead, they can be interpreted to say that the updated model is a solution to the linear system

$$(\bar{\mathbf{H}}_\ell + \varepsilon_\ell \mathbf{D}_\ell)(\mathbf{m} - \mathbf{m}_\ell) = -\mathbf{g}_\ell \qquad (8.167)$$

(with $\varepsilon_\ell = 0$ for Gauss–Newton). An algorithm that solves this system will calculate $\mathbf{m} = \mathbf{m}_\ell$. For example, Rodi & Mackie (2001) implemented the Gauss–Newton method using components from the LINPACK software library (Dongarra *et al.*, 1979) for factoring $\bar{\mathbf{H}}_\ell + \varepsilon_\ell \mathbf{D}_\ell$ with Gaussian elimination and for solving triangular linear systems. This approach and explicit matrix inversion are both direct solution methods, and have similar computational requirements. Each requires the generation and storage of the Hessian matrix, which contains $M(M+1)/2$ independent elements, and on the order of M^3 computations either to invert or to factor the Hessian. The computer memory and CPU usage with direct methods can be excessive in 2D and 3D MT inversion.

Alternatively, one can calculate $\mathbf{m}_{\ell+1}$ with iterative methods for solving linear systems or, equivalently, for minimizing quadratic functions (i.e. $\bar{\Psi}_\ell^\varepsilon$). This section will describe one such method: *conjugate gradients*, introduced by Hestenes & Stiefel (1952). The conjugate gradients (CG) method can be applied to the MT inverse problem with less memory usage and fewer computations than direct solution methods.

The CG algorithm will be described as it applies to Eq. (8.167) with $\varepsilon_\ell = 0$ (the Gauss–Newton system). For each fixed ℓ, CG generates a sequence of models, $\mathbf{m}_{\ell,k}$, $k = 0, 1, 2, \ldots,$ K, where it is convenient to set the initial model as $\mathbf{m}_{\ell,0} = \mathbf{m}_\ell$ and where the final model is taken as the solution, such that $\mathbf{m}_{\ell+1} = \mathbf{m}_{\ell,K}$. The sequence is generated as

$$\mathbf{m}_{\ell,k+1} = \mathbf{m}_{\ell,k} + \alpha_k \mathbf{p}_k, \qquad k = 0, 1, \ldots, K-1 \qquad (8.168)$$

where the vector $\mathbf{p}_k$ is a *search direction* in model space and the scalar α_k is a *step size*. (The sequences $\mathbf{p}_k$ and α_k are different for different ℓ, but this will be omitted from the notation.) Given $\mathbf{m}_{\ell,k}$ and $\mathbf{p}_k$, α_k is chosen to minimize $\bar{\Psi}_\ell(\mathbf{m}_{\ell,k} + \alpha \mathbf{p}_k)$ with respect to α. Substituting into Eq. (8.161) and simplifying yields

$$\bar{\Psi}_\ell(\mathbf{m}_{\ell,k} + \alpha\mathbf{p}_k) = \bar{\Psi}_\ell(\mathbf{m}_{\ell,k}) + \alpha\mathbf{g}_{\ell,k}^\mathrm{T}\mathbf{p}_k + \tfrac{1}{2}\alpha^2\mathbf{p}_k^\mathrm{T}\bar{\mathbf{H}}_\ell\,\mathbf{p}_k \tag{8.169}$$

where $\mathbf{g}_{\ell,k}$ denotes the gradient of $\bar{\Psi}_\ell$ at $\mathbf{m}_{\ell,k}$ and is given by

$$\mathbf{g}_{\ell,k} = \mathbf{g}_\ell + \bar{\mathbf{H}}_\ell(\mathbf{m}_{\ell,k} - \mathbf{m}_\ell) \tag{8.170}$$

Differentiating with respect to α and equating to zero, this quadratic function of α is minimized by

$$\alpha_k = -\frac{\mathbf{g}_{\ell,k}^\mathrm{T}\mathbf{p}_k}{\mathbf{p}_k^\mathrm{T}\bar{\mathbf{H}}_\ell\,\mathbf{p}_k} \tag{8.171}$$

As the model sequence is generated, the gradient vectors $\mathbf{g}_{\ell,k}$ can be computed with Eq. (8.170), but it is more efficient to iterate them as

$$\mathbf{g}_{\ell,k+1} = \mathbf{g}_{\ell,k} + \alpha_k\,\bar{\mathbf{H}}_\ell\,\mathbf{p}_k, \qquad k = 0, 1, \dots, K - 2 \tag{8.172}$$

The sequence is initialized with $\mathbf{g}_{\ell,0} = \mathbf{g}_\ell$, the gradient at $\mathbf{m}_\ell$.

CG generates the search directions as follows. The initial search direction is taken to be

$$\mathbf{p}_0 = \mathbf{C}\mathbf{g}_\ell \tag{8.173}$$

where $\mathbf{C}$ is a positive definite matrix known as a *preconditioner*. Subsequent directions are iterated as

$$\mathbf{p}_{k+1} = \mathbf{C}\mathbf{g}_{\ell,k+1} + \beta_k\mathbf{p}_k, \qquad k = 0, 1, \dots, K - 2 \tag{8.174}$$

where

$$\beta_k = \frac{\mathbf{g}_{\ell,k+1}^\mathrm{T}\mathbf{C}\mathbf{g}_{\ell,k+1}}{\mathbf{g}_{\ell,k}^\mathrm{T}\mathbf{C}\mathbf{g}_{\ell,k}} \tag{8.175}$$

This formula guarantees a number of interesting properties of the conjugate gradients sequence, which are proved by Hestenes & Stiefel (1952). Paramount among them are the relations,

$$\mathbf{p}_k^\mathrm{T}\bar{\mathbf{H}}_\ell\,\mathbf{p}_{k'} = 0 \tag{8.176}$$

$$\mathbf{g}_{\ell,k}^\mathrm{T}\mathbf{C}\mathbf{g}_{\ell,k'} = 0, \qquad k \neq k' \tag{8.177}$$

Equation (8.176) states that the search directions are mutually *conjugate* with respect to $\bar{\mathbf{H}}_\ell$, while (8.177) states that the gradient vectors are mutually conjugate with respect to the preconditioner. These properties guarantee that the current search direction is orthogonal to previous gradients:

$$\mathbf{g}_{\ell,k'}^\mathrm{T}\mathbf{p}_k = 0, \qquad k' < k \tag{8.178}$$

8.5.4.1 Stopping criterion

A reasonable stopping criterion for CG is that $\mathbf{g}_{\ell,k}$ be small, for example as measured by the size of $\mathbf{g}_{\ell,k}^{\mathrm{T}}\mathbf{C}\mathbf{g}_{\ell,k}$. With infinite-precision calculations, $\mathbf{g}_{\ell,k}$ will be zero after at most M steps (i.e. $K \leq M$). For less-than-perfect preconditioning, it will generally take exactly M steps to achieve $\mathbf{g}_{\ell,k} = 0$. Because of this, CG is sometimes considered to be a direct solution method requiring order M^3 calculations just like matrix inversion. However, CG often reaches a good solution in fewer than M steps, in which case it is properly thought of as an iterative method. The number of steps required for a desired solution accuracy depends mainly on how good the initial guess is, the effectiveness of the preconditioner and the condition number of $\bar{\mathbf{H}}_{\ell}$.

8.5.4.2 Computational requirements

The main computational effort in the CG algorithm (assuming an efficient preconditioner is used) is involved in the calculation of $\bar{\mathbf{H}}_{\ell}\,\mathbf{p}_k$. If $\bar{\mathbf{H}}_{\ell}$ is calculated and stored prior to the CG iteration, then the number of computations needed to perform this matrix–vector product on each CG iteration, if done efficiently, will be proportional to the number of non-zero elements in $\bar{\mathbf{H}}_{\ell}$. For this reason, conjugate gradients is particularly well-suited for problems in which $\bar{\mathbf{H}}_{\ell}$ is a sparse matrix. In addition to CPU time, memory usage is reduced when $\bar{\mathbf{H}}_{\ell}$ is sparse.

Examining Eq. (8.160), the sparseness of $\bar{\mathbf{H}}_{\ell}$ depends on the sparseness of matrices $\mathbf{A}_{\ell}$ and $\mathbf{K}$. The regularization matrix $\mathbf{K}$ is often a sparse matrix by design, as when it represents a finite-difference approximation to a differential operator (see Section 8.4.1). However, while the Jacobian may be sparse for some geophysical inverse problems (e.g. travel-time tomography), $\mathbf{A}_{\ell}$ is not sparse for the MT problem.

Operation on the search direction p_k with the first term of $\bar{\mathbf{H}}_{\ell}$ in (8.160) factors into the multiplication $\mathbf{A}_{\ell}\mathbf{p}_k \equiv \mathbf{f}$ followed by the multiplication $\mathbf{A}_{\ell}^{\mathrm{T}}\mathbf{W}\mathbf{f}$. The obvious way to implement these calculations is to calculate and store $\mathbf{A}_{\ell}$ prior to the CG iteration, and then operate with $\mathbf{A}_{\ell}$ and $\mathbf{A}_{\ell}^{\mathrm{T}}$ as needed on each CG iteration. Unfortunately, in 2D and 3D MT inversion, generating $\mathbf{A}_{\ell}$ can dominate the computations needed for the entire CG iteration (see Section 8.6.2). Moreover, $\mathbf{A}_{\ell}$ is typically a large matrix, and its storage will dominate the memory requirements of the CG algorithm. However, it can be shown that applications of $\mathbf{A}_{\ell}$ and $\mathbf{A}_{\ell}^{\mathrm{T}}$ can be performed implicitly, avoiding the computation and storage of $\mathbf{A}_{\ell}$. This is discussed in Section 8.6.2.

8.5.4.3 Truncated conjugate gradients

Even with efficient $\mathbf{A}_{\ell}$ and $\mathbf{A}_{\ell}^{\mathrm{T}}$ computations, the number of CG steps required to ensure convergence of the CG iteration is an important consideration, since, if the number of steps is too large, the computation time will be prohibitive. For example, Mackie & Madden (1993) did not use a stopping criterion that seeks convergence of the CG iteration, but rather they placed a strict limit on the number of steps K. The limit was chosen to ensure that the CPU time needed by the CG iteration would be less than that needed to generate $\mathbf{A}_{\ell}$. In so doing, their algorithm did not solve Eq. (8.167) completely, making it different from and faster than the Gauss–Newton method.

In fact, this is an example of *truncated* conjugate gradients, a general method described by Kelley (1999). CG truncation can be thought of as an alternative to Levenberg–Marquardt damping as a device for constraining the Gauss–Newton model update. In this view, the computational saving ensuing from small K does not necessarily slow, and in fact may speed up, the convergence of the model sequence $\mathbf{m}_\ell$. This is because the incomplete model update resulting from truncating the CG iteration is less likely to be deleteriously affected by the nonlinearity of F. However, if K is made too small, this benefit will be outweighed by the negative impact of incomplete solution of the CG iteration.

8.5.4.4 Preconditioning

Generally speaking, the CG iteration will converge faster when the preconditioner $\mathbf{C}$ is a good approximation to the inverse of the Hessian matrix. In fact, when $\mathbf{C} = \bar{\mathbf{H}}_\ell^{-1}$, the solution is reached in one step. On the other hand, if the construction of $\mathbf{C}$ or operations with $\mathbf{C}$ involve significant computation, the total computation time for the CG iteration can increase, even if fewer steps are done. Therefore, considerable care is needed to derive an effective preconditioner for a particular type of problem.

Rodi & Mackie (2001) used a preconditioner of the form

$$\mathbf{C} = (\mathbf{M} + \lambda\mathbf{K})^{-1} \tag{8.179}$$

with $\mathbf{M}$ chosen to be a scaled version of the identity matrix such that the scaling approximated $\mathbf{A}_\ell^T\mathbf{W}\mathbf{A}_\ell$ in some sense. With $\mathbf{M}$ and $\mathbf{K}$ both sparse, it becomes very efficient to apply $\mathbf{C}$ indirectly to a vector $\mathbf{g}$ by solving

$$(\mathbf{M} + \lambda\mathbf{K})\mathbf{h} = \mathbf{g} \tag{8.180}$$

to obtain $\mathbf{h} = \mathbf{C}\mathbf{g}$. Newman & Boggs (2004) and Mackie *et al.* (2007) found that using an approximation to the diagonal of $\mathbf{A}_\ell^T\mathbf{W}\mathbf{A}_\ell$ works well, although, in 1D and 2D algorithms, one could use the true diagonal of $\mathbf{A}_\ell^T\mathbf{W}\mathbf{A}_\ell$. Farquharson & Oldenburg (1996) and Han *et al.* (2008) discuss other approximations to $\mathbf{A}_\ell$ for use in 3D MT inversion algorithms.

8.5.5 Nonlinear conjugate gradients

The method of *nonlinear* conjugate gradients (NLCG) is similar to linear CG just described, but directly minimizes a non-quadratic objective function, abandoning the paradigm of repeated linearized inversion. The NLCG method is well described in the optimization literature (e.g. Kelley, 1999). A particular version of the method due to Rodi & Mackie (2001) will be described here, as applied to the minimization of Ψ in Eq. (8.107). Newman & Alumbaugh (2000) also used the method of NLCG to solve the 3D MT inverse problem, but with differences in the line search and preconditioning approaches.

The model sequence for NLCG will be denoted as $\mathbf{m}_\ell$, $\ell = 0, 1, 2, \ldots$, as for the Gauss–Newton sequence. The sequence is generated in the same fashion as for linear CG, Eq. (8.168), but the step size α_ℓ is chosen to minimize the non-quadratic objective function:

$$\Psi(\mathbf{m}_\ell + \alpha_\ell \mathbf{p}_\ell) = \min_\alpha \Psi(\mathbf{m}_\ell + \alpha \mathbf{p}_\ell) \tag{8.181}$$

Solving this univariate minimization problem is referred to as a *line search* for α.

The search directions are iterated analogously to linear CG:

$$\begin{aligned} \mathbf{p}_0 &= \mathbf{C}_0 \mathbf{g}_0 \\ \mathbf{p}_{\ell+1} &= \mathbf{C}_{\ell+1} \mathbf{g}_{\ell+1} + \beta_\ell \mathbf{p}_\ell, \qquad \ell = 0, 1, \ldots \end{aligned} \tag{8.182}$$

where, using the Polak–Ribiere version of NLCG (Polak, 1971),

$$\beta_\ell = \frac{\mathbf{g}_{\ell+1}^{\mathrm{T}} \mathbf{C}_{\ell+1}(\mathbf{g}_{\ell+1} - \mathbf{g}_\ell)}{\mathbf{g}_\ell^{\mathrm{T}} \mathbf{C}_\ell \mathbf{g}_\ell} \tag{8.183}$$

Note that the preconditioner $\mathbf{C}_\ell$ is allowed to vary during the iteration. Similar to linear CG, the preconditioner (Section 8.5.4) plays an important role in the convergence of the NLCG algorithm. Unlike linear CG, the search directions are not necessarily conjugate with respect to some fixed matrix, as in (8.176), but they do satisfy the weaker condition

$$\mathbf{p}_\ell^{\mathrm{T}}(\mathbf{g}_\ell - \mathbf{g}_{\ell-1}) = 0, \qquad \ell > 0 \tag{8.184}$$

8.5.5.1 Line search

The line search problem (8.181) is not quadratic and requires some type of iterative scheme to solve it. Since only a single unknown is involved, it is tempting to attack the problem as one of global optimization: finding a global minimum of Ψ with respect to α. Doing so would gain one advantage over the Gauss–Newton method, which makes no attempt to distinguish local from global minima. However, global optimization potentially leads to many forward problem calculations per NLCG step. Concerned with the computational intensity of 2D and 3D MT forward modeling, Rodi & Mackie (2001) presented a line search algorithm that does not attempt global minimization, but rather puts a premium on computational efficiency. Their line search algorithm is a univariate version of the Gauss–Newton method with certain modifications. To describe it, denote the univariate function to be minimized as ψ:

$$\psi(\alpha) \equiv \Psi(\mathbf{m}_\ell + \alpha \mathbf{p}_\ell) \tag{8.185}$$

The line search generates a minimizing sequence $\alpha_{\ell,k}$, $k = 0, 1, \ldots, K$, with $\alpha_{\ell,0} = 0$ and with $\alpha_{\ell,K}$ taken as the solution α_ℓ in (8.181). For each step of the line search, define the quantities

$$\mathbf{m}_{\ell,k} = \mathbf{m}_\ell + \alpha_{\ell,k} \mathbf{p}_\ell \tag{8.186}$$

$$\mathbf{A}_{\ell,k} = \mathbf{A}_{\mathbf{m}_{\ell,k}} \tag{8.187}$$

A Gauss–Newton approximation to ψ at the kth step is given by

$$\bar{\psi}_k(\alpha) = \bar{\Psi}_{\ell,k}(\mathbf{m}_{\ell,k} + \alpha\mathbf{p}_\ell) \tag{8.188}$$

where $\bar{\Psi}_{\ell,k}$ is the Gauss–Newton approximation to Ψ referenced to $\mathbf{m}_{\ell,k}$. It is clear that $\bar{\psi}_{\ell,k}$ is a quadratic function of α since $\bar{\Psi}_{\ell k}$ is quadratic. The first and second derivatives of $\bar{\psi}_k$ at the kth iterate are given by

$$\bar{\psi}'_k(\alpha_{\ell k}) = \mathbf{g}_{\ell,k}^T\mathbf{p}_\ell \tag{8.189}$$

$$\bar{\psi}''_k(\alpha_{\ell k}) = \mathbf{p}_\ell^T\,\bar{\mathbf{H}}_{\ell,k}\,\mathbf{p}_\ell \tag{8.190}$$

where, as indicated by their subscripts, the gradient and approximate Hessian are evaluated at $\mathbf{m}_{\ell,k}$. Therefore, the Gauss–Newton update for α is given by

$$\alpha_{\ell,k+1} = \alpha_{\ell,k} - \frac{\mathbf{g}_{\ell,k}^T\mathbf{p}_\ell}{\mathbf{p}_\ell^T\,\bar{\mathbf{H}}_{\ell,k}\,\mathbf{p}_\ell} \tag{8.191}$$

This can be augmented with tests to avoid a local maximum of $\psi(\alpha)$. Additionally, if consecutive trials of α bracket a local minimum of ψ, Gauss–Newton updating is abandoned in favor of a scheme based on cubic interpolation of ψ.

After the line search converges or is otherwise stopped, the best α tested during the search, corresponding to the smallest ψ is taken as α_ℓ for the purpose of updating the model as $\mathbf{m}_{\ell+1} = \mathbf{m}_\ell + \alpha_\ell\mathbf{p}_\ell$. If the line search does not converge satisfactorily, the NLCG iteration is restarted from $\mathbf{m}_\ell$ with the search direction set to $\mathbf{p}_\ell = \mathbf{C}_\ell\mathbf{g}_\ell$, *in lieu* of Eq. (8.182), thus breaking conjugacy with previous directions.

8.5.5.2 Computational requirements

The dominant computations in the NLCG algorithm are very similar to those of linear CG, and can be reduced to operations with $\mathbf{A}_{\ell,k}$ and its transpose on appropriate vectors (see Section 8.6.2). The Jacobian operations are used somewhat differently, however. The quantity $\mathbf{p}_\ell^T\,\bar{\mathbf{H}}_{\ell,k}\,\mathbf{p}_\ell$ in (8.191) is obtained from $\mathbf{A}_{\ell,k}\mathbf{p}_\ell$, while $\mathbf{g}_{\ell,k}$ is obtained from $\mathbf{A}_{\ell,k}^T\mathbf{W}(\mathbf{d} - F(\mathbf{m}_{\ell,k}))$. This difference accommodates NLCG's need for the gradient of the non-quadratic Ψ for each ℓ and k. However, an extra demand of NLCG is the need to compute the nonlinear forward function F for each ℓ and k, not just for each ℓ as in the previous section. The memory requirements of linear and nonlinear CG are also similar, scaling linearly with N and M.

Under normal circumstances, the model updates $(\mathbf{m}_{\ell+1} - \mathbf{m}_\ell)$ in the later stages of the NLCG iteration will be sufficiently small to make the quadratic approximations to Ψ and ψ very accurate. When this happens, the Gauss–Newton line search will converge in one step $(K = 1)$ and the NLCG model recursion in can be rewritten as

$$\mathbf{m}_{\ell+1} = \mathbf{m}_\ell - \frac{\mathbf{g}_\ell^T\mathbf{p}_\ell}{\mathbf{p}_\ell^T\,\bar{\mathbf{H}}_\ell\,\mathbf{p}_\ell}\mathbf{p}_\ell \tag{8.192}$$

$$\mathbf{p}_{\ell+1} = \mathbf{C}_{\ell+1}\mathbf{g}_{\ell+1} + \frac{\mathbf{g}_{\ell+1}^T\mathbf{C}_{\ell+1}(\mathbf{g}_{\ell+1} - \mathbf{g}_\ell)}{\mathbf{g}_\ell^T\mathbf{C}_\ell\mathbf{g}_\ell}\mathbf{p}_\ell \tag{8.193}$$

This is the same recursion as linear CG, applied to the main model sequence. Therefore, if the linear regime is detected, the NLCG iteration can be completed as one final (untruncated) Gauss–Newton iteration. Additionally, the evaluation of F for each ℓ can be avoided, as in linear CG.

8.5.6 Subspace methods

The subspace method (Oldenburg & Ellis, 1993; Oldenburg *et al.*, 1993; Oldenburg, 1994; Oldenburg & Li, 1994) represents the model space by a subspace that is much smaller in size, yet is still representative of the entire model space. Let $\mathbf{v}_i$ ($i = 1, \ldots, q$) be basis vectors for a q-dimensional subspace of $\mathbf{m}$. A linearized model perturbation given by $(\mathbf{m} - \mathbf{m}_\ell) = \sum \alpha_i \mathbf{v}_i = \mathbf{V}\boldsymbol{\alpha}$ is sought. Substituting this into Eq. (8.161) and setting the gradient with respect to $\boldsymbol{\alpha}$ to zero results in a Gauss–Newton algorithm in which the gradient and the Hessian are modified from their previous forms and given by

$$\mathbf{g}_q = \mathbf{V}^{\mathrm{T}}\mathbf{g}_{\mathrm{m}_\ell} \tag{8.194}$$

and

$$\mathbf{H}_q = \mathbf{V}^{\mathrm{T}}\mathbf{H}_\ell \mathbf{V} \tag{8.195}$$

The gradient is now of length q and the Hessian has dimensions of $q \times q$ where $q \ll M$ and M is the original number of model parameters. The advantages of this approach are clear: the inverse of the Hessian is much easier to obtain and less computationally demanding. The disadvantage is that the reduced subspace may miss parts of the model space that are important in finding the solution. Therefore, it is critical that the subspace vectors be chosen carefully, and the interested reader is referred to Oldenburg & Ellis (1993) and Oldenburg *et al.* (1993) for discussion.

8.6 Derivatives of the forward functions

In previous sections, the partial derivatives of the forward modeling functions – the elements of the Jacobian matrix $\mathbf{A}$ – were seen to play a central role in the solution of an inverse problem because they contain information about how changes in the model parameters affect the predicted data. This section describes techniques for computing these partial derivatives for MT data.

As was discussed in Section 8.2.2, a wide variety of MT responses can be used as data for inversion, each having its own forward function. These include the components of the complex MT transfer function, components of the complex vertical magnetic transfer function, complex apparent resistivity or conductivity, the determinant of the MT transfer function, and the *c*-response. However an MT datum is defined, it is ultimately some function of the electric and magnetic induction vectors measured at a point in the model domain, i.e. $\mathbf{E}(\mathbf{r}_0)$ and $\mathbf{B}(\mathbf{r}_0)$ for some $\mathbf{r}_0$. This section will focus on the derivatives of the components of $\mathbf{E}(\mathbf{r}_0)$ and $\mathbf{B}(\mathbf{r}_0)$, knowing that the chain rule can be used to obtain the derivatives of an MT forward function.

There are many approaches for obtaining the derivatives of the electric and magnetic fields. These will not all be described here; the reader is referred to McGillivray & Oldenburg (1990) for a detailed description of common approaches. Before proceeding to numerical techniques, however, it is instructive to consider the theoretical sensitivity of electric and magnetic fields to the conductivity of a medium, as implied by the Maxwell equations.

8.6.1 Theoretical sensitivity distribution

The derivation presented here closely follows the Green's function approach of Rodi (1989) and the "auxiliary field" solution of McGillivray *et al.* (1994). It is also closely related to the more general adjoint approach of Lanczos (1961), which is described for the MT problem by Madden (1990).

Although the Maxwell equations were discussed in detail in Chapter 2, they are repeated here in a slightly different form, which includes impressed magnetic current sources. While non-physical, this alternative way of characterizing sources provides symmetry to the equations. Assuming an $e^{i\omega t}$ time dependence of the fields, the Maxwell equations can be written as

$$\nabla \times \mathbf{E}(\mathbf{r}) = -i\omega \mathbf{B}(\mathbf{r}) + \mathbf{M}_s(\mathbf{r}) \tag{8.196}$$

$$\nabla \times \mathbf{B}(\mathbf{r}) = \mu_0 [\mathbf{r}\mathbf{E}(\mathbf{r}) + \mathbf{J}_s(\mathbf{r})] \tag{8.197}$$

where $\mathbf{M}_s$ and $\mathbf{J}_s$ are impressed magnetic and current densities, and $\mathbf{E}$ and $\mathbf{B}$ are the electric field and magnetic induction vectors. These equations are assumed to hold in a volume V bounded by a surface S. A very general form of boundary conditions on S is given by McGillivray *et al.* (1994) as

$$\alpha(\mathbf{n} \times \mathbf{U}) + \beta(\mathbf{n} \times \mathbf{n} \times \nabla \times \mathbf{U}) = \mathbf{S} \tag{8.198}$$

where $\mathbf{U}$ can be either $\mathbf{E}$ or $\mathbf{B}$; $\mathbf{n}$ is the unit vector outwardly normal to the boundary; α and β are constants; and $\mathbf{S}$ is the magnetic or electric current on the boundary. This formulation allows one to specify boundary conditions on either the tangential electric or magnetic field, or their ratio. The MT electric and magnetic fields solve Eqs. (8.196) and (8.197) in the volume V together with the boundary conditions (8.198) on S. The goal is to find the first-order sensitivity of the components of $\mathbf{E}(\mathbf{r}_0)$ and $\mathbf{B}(\mathbf{r}_0)$ to the conductivity function σ, where $\mathbf{r}_0$ is an observation point. The various field components can be summarized as a *complex-valued*, scalar datum d expressed as

$$d = \mathbf{a} \cdot \mathbf{E}(\mathbf{r}_0) + \mathbf{b} \cdot \mathbf{B}(\mathbf{r}_0) \tag{8.199}$$

where **a** and **b** are given "selection" vectors. For example, setting **a** = **i** (the unit vector in the x direction) and **b** = 0 gives $d = E_x(\mathbf{r}_0)$. Accordingly, setting **a** and **b** to other unit vectors yields each electric and magnetic field component.

Suppose that the conductivity is perturbed by $\delta\sigma$, inducing perturbations $\delta\mathbf{E}$ and $\delta\mathbf{B}$ to **E** and **B**, respectively, and

$$\delta d = \mathbf{a} \cdot \delta\mathbf{E}(\mathbf{r}_0) + \mathbf{b} \cdot \delta\mathbf{B}(\mathbf{r}_0) \tag{8.200}$$

to the datum d. Substituting $\sigma + \delta\sigma$, $\mathbf{E} + \delta\mathbf{E}$ and $\mathbf{B} + \delta\mathbf{B}$ into the Maxwell equations, dropping second-order terms such as $\delta\sigma\delta\mathbf{E}$, and subtracting the Maxwell equations for **E** and **B** yields

$$\nabla \times \delta\mathbf{E} = -i\omega\delta\mathbf{B} \tag{8.201}$$

$$\nabla \times \delta\mathbf{B} = \mu_0[\sigma\delta\mathbf{E} + \delta\sigma\mathbf{E}] \tag{8.202}$$

Further, since **S** in (8.198) is not perturbed, the field perturbations satisfy the homogeneous boundary conditions

$$\alpha(\mathbf{n} \times \delta\mathbf{U}) + \beta(\mathbf{n} \times \mathbf{n} \times \nabla \times \delta\mathbf{U}) = 0 \tag{8.203}$$

Thus, to first order in $\delta\sigma$, the perturbations $\delta\mathbf{E}$ and $\delta\mathbf{B}$ solve the Maxwell equations with a current source of $\delta\sigma\mathbf{E}$, zero magnetic source and homogeneous boundary conditions.

Now assume the existence of Green's functions $\mathbf{E}_G(\mathbf{r})$ and $\mathbf{B}_G(\mathbf{r})$ satisfying

$$\nabla \times \mathbf{E}_G(\mathbf{r}) = -i\omega\mathbf{B}_G(\mathbf{r}) + \mu_0\mathbf{b}\delta(\mathbf{r} - \mathbf{r}_0) \tag{8.204}$$

$$\nabla \times \mathbf{B}_G(\mathbf{r}) = \mu_0[\sigma\mathbf{E}_G(\mathbf{r}) + \mathbf{a}\delta(\mathbf{r} - \mathbf{r}_0)] \tag{8.205}$$

and the homogeneous boundary conditions

$$\alpha(\mathbf{n} \times \mathbf{U}_G) + \beta(\mathbf{n} \times \mathbf{n} \times \nabla \times \delta\mathbf{U}_G) = 0 \tag{8.206}$$

The Green's functions are the electromagnetic responses to a superposition of point current and magnetic sources at the point $\mathbf{r}_0$. This is termed the "auxiliary problem" by McGillivray *et al.* (1994), although they use a different notation.

Using the vector identity $\nabla \cdot (\mathbf{E} \times \mathbf{B}) = \mathbf{B} \cdot \nabla \times \mathbf{E} - \mathbf{E} \cdot \nabla \times \mathbf{B}$ and Eqs. (8.201), (8.202), (8.204) and (8.205), one can derive that

$$\frac{1}{\mu_0}\nabla \cdot (\mathbf{E}_G \times \delta\mathbf{B} - \delta\mathbf{E} \times \mathbf{B}_G)$$
$$= \delta\mathbf{E} \cdot \mathbf{a}\delta(\mathbf{r} - \mathbf{r}_0) + \delta\mathbf{B} \cdot \mathbf{b}\delta(\mathbf{r} - \mathbf{r}_0) - \delta\sigma \, \mathbf{E}_G \cdot \mathbf{E} \tag{8.207}$$

Integrating over the domain V and using the divergence theorem gives

$$\frac{1}{\mu_0}\int_S (\mathbf{E}_G \times \delta\mathbf{B} - \delta\mathbf{E} \times \mathbf{B}_G) \cdot \mathbf{n} \, dS$$
$$= \delta\mathbf{E}(\mathbf{r}_0) \cdot \mathbf{a} + \delta\mathbf{B}(\mathbf{r}_0) \cdot \mathbf{b} - \int_V \delta\sigma \, \mathbf{E}_G \cdot \mathbf{E} \, dV \tag{8.208}$$

McGillivray *et al.* (1994) proved that, under the boundary conditions (8.203) and (8.206), the left-hand side of this equation is zero. The first two terms on the right-hand side are recognized from Eq. (8.200) to be δd. Therefore, to first order in $\delta\sigma$,

$$\delta d = \int_V \delta\sigma \mathbf{E}_G \cdot \mathbf{E} \, dV \tag{8.209}$$

This can also be written as

$$\delta d = \int_V a(\mathbf{r})\delta\sigma(\mathbf{r}) \, dV(\mathbf{r}) \tag{8.210}$$

where

$$a(\mathbf{r}) = \mathbf{E}(\mathbf{r}) \cdot \mathbf{E}_G(\mathbf{r}) \tag{8.211}$$

Recalling Eq. (8.17), $a(\mathbf{r})$ is the sensitivity distribution for the datum d. At each point $\mathbf{r}$, the sensitivity value is given by the inner product between $\mathbf{E}$, the electric field portion of the MT response, and $\mathbf{E}_G$, the electric field due to point current and/or magnetic sources at the observation site $\mathbf{r}_0$. It is important to note that $a(\mathbf{r})$ will be a proper function, rather than a distribution, as long as $\mathbf{E}(\mathbf{r})$ and $\mathbf{E}_G(\mathbf{r})$ are functions, which will be the case if the medium conductivity, $\sigma(\mathbf{r})$, is well-behaved. Rodi (1989) points out, however, that, in 2D and 3D media, $\mathbf{E}_G$ is singular (blows up) at $\mathbf{r}_0$; therefore, the sensitivity function is singular at $\mathbf{r}_0$.

The sensitivity functions derived above provide a basis for the numerical computation of partial derivatives for inversion. A partial derivative with respect to a discrete model parameter is given by integrating $a(\mathbf{r})$ over some volume weighted by an interpolation function. The needed Green's functions can be solved for analytically in 1D models or numerically in 2D and 3D models using the same techniques applied to computing the MT fields. However, it is also possible to derive expressions for partial derivatives within the framework of numerical solutions of the Maxwell equations, as the next section shows.

8.6.2 Numerical techniques

One can compute exact partial derivatives of the data predicted by numerical solution of the Maxwell equations, as described by Rodi (1976), Jupp & Vozoff (1977), McGillivray & Oldenburg (1990), Haber *et al.* (2000) and Rodi & Mackie (2001). The advantage of this approach is its conceptual simplicity compared to the more complicated mathematics that are involved in numerically computing Green's functions.

The Gauss–Newton and related methods require the computation of each element of the Jacobian, or sensitivity, matrix $\mathbf{A}$. By contrast, the nonlinear conjugate gradients algorithm and other similar gradient descent methods employ $\mathbf{A}$ only in the computation of the quantities $\mathbf{Ap}$ and $\mathbf{A}^{\mathrm{T}}\mathbf{q}$ for specific vectors $\mathbf{p}$ and $\mathbf{q}$. The computation of all these quantities will be demonstrated in the next subsections. As in Section 8.6.1, the numerical techniques will be developed for the case of complex-valued data and, accordingly, partial derivatives. The relationship of the complex-valued Jacobian and related quantities to the real quantities used in inversion theory is spelled out in detail in the appendix of Rodi & Mackie (2001).

In Chapter 7, numerical techniques for solving the Maxwell equations were described that generally required solution of a complex linear system of equations, given by

$$\mathbf{D}(\mathbf{m})\mathbf{v}(\mathbf{m}) = \mathbf{s}(\mathbf{m}) \tag{8.212}$$

where $\mathbf{D}$ is the matrix that results from a finite-difference or finite-element numerical discretization of the Maxwell equations; $\mathbf{v}$ is the vector of parameterized electric and/or magnetic fields; and $\mathbf{s}$ is the right-hand side vector representing known boundary values and sources terms. All of these quantities, in general, depend on the vector of model parameters, $\mathbf{m}$. For the purposes of this discussion, it is convenient to let (8.212) include all relevant source polarizations and frequencies simultaneously. If the system is partitioned by polarization and frequency, $\mathbf{D}$ will be a block-diagonal matrix. Each complex MT datum for the inverse problem will then be some nonlinear function of $\mathbf{v}$, depending on which MT response defines the datum. For simplicity here, each datum will be taken simply as an electric or magnetic field component at an observation site, as in the previous section. The data can then be written as

$$d_i = \mathbf{b}_i(\mathbf{m})^{\mathrm{T}}\mathbf{v}(\mathbf{m}) \tag{8.213}$$

where the components of the complex vector $\mathbf{b}_i$ are coefficients chosen to perform, as needed, spatial interpolation to the observation site, and numerical differentiation or integration to convert values of $\mathbf{E}$ to $\mathbf{B}$ or vice versa. The vectors $\mathbf{b}_i$ may depend on the conductivity model itself, i.e. $\mathbf{m}$.

8.6.2.1 Computation of the complex Jacobian

Taking partial derivatives of Eq. (8.213), abbreviating $\partial/\partial m_j$ as ∂_j, the elements of the complex Jacobian are seen to be

$$A_{ij} = (\partial_j \mathbf{b}_i)^{\mathrm{T}}\mathbf{v} + \mathbf{b}_i^{\mathrm{T}}\partial_j\mathbf{v} \tag{8.214}$$

Differentiating Eq. (8.212) gives

$$(\partial_j \mathbf{D})\mathbf{v} + \mathbf{D}\partial_j\mathbf{v} = \partial_j\mathbf{s} \tag{8.215}$$

Solving this equation for $\partial_j\mathbf{v}$ and substituting into the previous equation yields

$$A_{ij} = (\partial_j \mathbf{b}_i)^{\mathrm{T}}\mathbf{v} + \mathbf{b}_i^{\mathrm{T}}\mathbf{D}^{-1}[\partial_j\mathbf{s} - (\partial_j\mathbf{D})\mathbf{v}] \tag{8.216}$$

Rodi (1976) described two algorithms for calculating the components of $\mathbf{A}$ based on Eq. (8.216). Both assume that Eq. (8.212) has previously been solved for $\mathbf{v}$. Both also assume that $\partial_j\mathbf{D}$, $\partial_j\mathbf{s}$ and $\partial_j\mathbf{b}_j$ are given, since they can typically be evaluated with analytical formulas, as can $\mathbf{D}$, $\mathbf{s}$ and $\mathbf{b}_j$. The first algorithm fills out $\mathbf{A}$ one column at a time by solving the linear systems (for $j = 1, \ldots, M$)

$$\mathbf{D}\mathbf{y}_j = \partial_j\mathbf{s} - (\partial_j\mathbf{D})\mathbf{v} \tag{8.217}$$

yielding solution vectors $\mathbf{y}_j = \mathbf{D}^{-1}\left[\partial_j\mathbf{s} - (\partial_j\mathbf{D})\mathbf{v}\right]$. Given $\mathbf{y}_j$, the jth column of $\mathbf{A}$ is then calculated as

$$A_{ij} = (\partial_j\mathbf{b}_i)^\mathrm{T}\mathbf{v} + \mathbf{b}_i^\mathrm{T}\mathbf{y}_j \tag{8.218}$$

for $i = 1, \ldots, N$. The second algorithm generates $\mathbf{A}$ row by row by solving the systems

$$\mathbf{D}\mathbf{u}_i = \mathbf{b}_i \tag{8.219}$$

implying that $\mathbf{u}_i = \mathbf{D}^{-1}\mathbf{b}_i$. Assuming $\mathbf{D}$ is symmetric, and therefore $\mathbf{b}_i^\mathrm{T}\mathbf{D}^{-1} = (\mathbf{D}^{-1}\mathbf{b}_i)^\mathrm{T}$, it follows that

$$A_{ij} = (\partial_j\mathbf{b}_i)^\mathrm{T}\mathbf{v} + \mathbf{u}_i^\mathrm{T}\left[\partial_j\mathbf{s} - (\partial_j\mathbf{D})\mathbf{v}\right] \tag{8.220}$$

The matrices $\partial_j\mathbf{D}$ are sparse since $\mathbf{D}$ itself is sparse and its elements typically depend on only a few of the m_j. The vectors $\mathbf{b}_i$, $\partial_j\mathbf{b}_i$ and $\partial_j\mathbf{s}$ are likewise sparse, or zero. Therefore, in either algorithm, construction of the right-hand side vectors for the forward problems and evaluation of either Eq. (8.218) or Eq. (8.220) take relatively little computation. The major computational effort in each algorithm is in solution of the appropriate set of forward problems, Eq. (8.217) or Eq. (8.219). Which algorithm is more efficient will depend on the relative sizes of N and M.

8.6.2.2 *Operations with the complex Jacobian*

In the NLCG method and other gradient descent methods, it is not necessary to form and store the Jacobian explicitly; rather, one needs only the result of the Jacobian or its transpose operating on a particular vector. It will now be shown that each of these computations is equivalent to solving one forward problem with an appropriate set of sources distributed either in the model volume (for operations with the Jacobian) or among the observation sites (for operations with its transpose).

Starting with $\mathbf{Ap}$, Eq. (8.216) implies that

$$\sum_j A_{ij}p_j = \sum_j p_j(\partial_j\mathbf{b}_i)^\mathrm{T}\mathbf{v} + \sum_j p_j\mathbf{b}_i^\mathrm{T}\mathbf{D}^{-1}\left[\partial_j\mathbf{s} - (\partial_j\mathbf{D})\mathbf{v}\right] \tag{8.221}$$

Therefore, if $\mathbf{t}$ satisfies

$$\mathbf{D}\mathbf{t} = \sum_j p_j\left[\partial_j\mathbf{s} - (\partial_j\mathbf{D})\mathbf{v}\right] \tag{8.222}$$

Eq. (8.221) becomes

$$\sum_j A_{ij}p_j = \sum_j p_j(\partial_j\mathbf{b}_i)^\mathrm{T}\mathbf{v} + \mathbf{b}_i^\mathrm{T}\mathbf{t} \tag{8.223}$$

Thus, the algorithm for $\mathbf{Ap}$ is to solve the single forward problem, (8.222), for $\mathbf{t}$ and then evaluate (8.223) for each i.

The method for $\mathbf{A}^\mathrm{T}\mathbf{q}$ can be derived analogously. Again, from (8.216), one has

$$\sum_i A_{ij} q_i = \sum_i q_i (\partial_j \mathbf{b}_i)^{\mathrm{T}} \mathbf{v} + \sum_i q_i \mathbf{b}_i^{\mathrm{T}} \mathbf{D}^{-1} [\partial_j \mathbf{s} - (\partial_j \mathbf{D}) \mathbf{v}] \qquad (8.224)$$

Letting $\mathbf{z}$ satisfy

$$\mathbf{D} \mathbf{z} = \sum_i q_i \mathbf{b}_i \qquad (8.225)$$

and assuming $\mathbf{D}^{\mathrm{T}} = \mathbf{D}$, it follows that

$$\sum_i A_{ij} q_i = \sum_i q_i (\partial_j \mathbf{b}_i)^{\mathrm{T}} \mathbf{v} + \mathbf{z}^{\mathrm{T}} [\partial_j \mathbf{s} - (\partial_j \mathbf{D}) \mathbf{v}] \qquad (8.226)$$

This time, solving the forward problem, (8.225), for $\mathbf{z}$ and then evaluating (8.226) for each j yields $\mathbf{A}^{\mathrm{T}} \mathbf{q}$.

The sums on the right-hand side of both Eqs. (8.223) and (8.226) require relatively little effort to calculate. The major computation in each of these algorithms is the solution of one forward problem: for $\mathbf{t}$ in (8.222) or $\mathbf{z}$ in Eq. (8.225).

8.7 Examples

8.7.1 One-dimensional models

This section illustrates the concepts that have been described earlier, beginning with 1D inversion for simplicity. A 1D conductivity model is represented as a stack of homogeneous plane layers, letting the logarithm of each layer conductivity ($\log \sigma$) comprise the elements of the model vector $\mathbf{m}$. Parameterization with $\log \sigma$ makes MT forward functionals more nonlinear than with σ, but ensures positivity of the conductivity. Each component of the data vector is either the log amplitude or phase of the complex apparent conductivity, σ_A, for a given frequency. The complex apparent conductivity is defined as

$$\sigma_A = -i \omega \mu Y^2(0) \qquad (8.227)$$

where $Y(0)$ is the complex admittance at $z = 0$ (Earth's surface); $Y(0)$ can be computed analytically for a layered model using the recursive algorithm of Madden & Nelson (1964). The recursion formula is similar to the one for the poloidal mode (PM) response function derived from first principles in Chapter 2, but at the zero-wavenumber limit. The recursion begins at the bottom half-space of the model and propagates the admittance upward through the layers to the surface. Partial derivatives of $\log \sigma_A$ with respect to the layer conductivities are computed analytically with top-down recursion formulas derived by Rodi (1989).

A 1D inversion of synthetic data is performed using a Gauss–Newton algorithm that minimizes the damped least-squares objective function given in Eq. (8.107). The stabilizing function is taken as a measure of spatial roughness in the form of a gradient norm, similar to that given in Eq. (8.90) but incorporating a depth transformation $f(z)$:

$$\Omega(\mathbf{m}) = \int \left(\frac{dm(z)}{df(z)} \right)^2 df(z) = \int \left(\frac{df(z)}{dz} \right)^{-1} \left(\frac{dm(z)}{dz} \right)^2 dz \qquad (8.228)$$

Smith & Booker (1988) found that $f(z) = \log(z + z_0)$ provided a uniform fit to MT data at all frequencies compared to other choices, including $f(z) = z$. The logarithmic transformation yields

$$\Omega(\mathbf{m}) = \int (z + z_0) \left(\frac{dm(z)}{dz}\right)^2 dz \qquad (8.229)$$

While the choice of z_0 is somewhat arbitrary, following Smith & Booker (1988), it is set to half the inductive scale length at the highest frequency. The inductive scale length (Weidelt, 1972) is the mean depth of the inductive current system, and at the highest frequency would represent the minimum depth that would reasonably be expected to contain resolvable conductivity variations.

The synthetic model under consideration is the 1D model used by Whittall & Oldenburg (1992) for their review of 1D MT inverse methods. This geologically plausible model comprises four layers over a half-space. The true model is shown as the thin solid line in Figure 8.3. Synthetic complex apparent conductivity data were generated for the model at 14 logarithmically spaced frequencies from 1000 to 0.0003 Hz. This frequency range is typical for a wide-band MT sounding, and logarithmically spaced frequencies allow efficient sampling of the data as a function of frequency. In order to quantify misfit and simulate

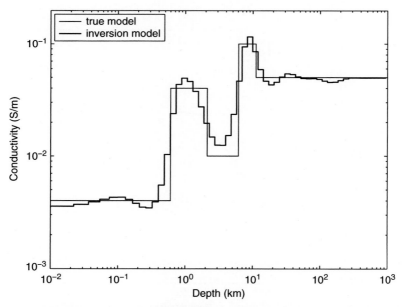

Figure 8.3. The 1D model from which synthetic complex apparent conductivity data were generated is shown as the thin solid line. Synthetic data were computed at 14 frequencies spaced logarithmically from 1000 to 0.0003 Hz with 2% random Gaussian noise added prior to inversion. The result of a damped least-squares inversion that resulted in a weighted RMS of the data residuals of 1.0 is shown as the thick solid line. The inversion model has 54 layers increasing logarithmically with depth.

real data, 2% random Gaussian noise was added to the real and imaginary parts of log σ_A prior to inversion. The starting model comprised 54 layers with thicknesses increasing logarithmically with depth. The first layer was 10 m in thickness, and each layer increased by a constant factor of 1.2. The use of logarithmically increasing layer thicknesses compensates in a reasonable way for the range of structural scales necessary to fit data spanning several decades of frequency without the use of inordinately large models (e.g. Oldenburg, 1980).

The regularization parameter λ was determined using the discrepancy method described in Section 8.4.5. That is, the objective function was minimized for several values of λ with the goal of selecting that value resulting in a weighted RMS of the data residuals of unity. The result of the inversion using this value of λ and starting from a uniform half-space of 20 Ω m is shown in Figure 8.3, where the resistivity of each layer in the inverse model is plotted as the heavy solid line. Because roughness is penalized, only a smoothed version of the true model is recovered. However, all of the main features and structures of the model are apparent. The amplitude of the apparent conductivity predicted by this inversion model is shown in Figure 8.4 plotted against the observed data for comparison. Although not plotted, equally good data fits are observed for the phase of the apparent conductivity.

In Section 8.3.1, it was shown that a model was a unique local least-squares solution if the gradient at that point was zero and the Hessian was positive definite. The inversion result

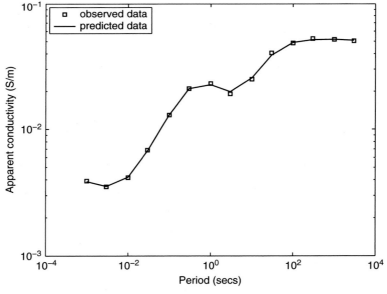

Figure 8.4. The observed and predicted response for the 1D model of Figure 8.3. The amplitude of the complex apparent conductivity is plotted for the observed and predicted responses. The observed responses are plotted as open squares, while the predicted response is plotted as a solid line. Because of their small size, error bars representing the added 2% random Gaussian noise are not plotted.

shown in Figure 8.3 is the model that results when the Gauss–Newton algorithm reduces the objective function to the point where it is unable to make any further progress. At this point, the gradient is zero and a stationary point has been reached where the objective function will not change any further. Inversions beginning from different starting models, that is, uniform half-spaces of resistivity other than 20 Ω m as in this example, resulted in the same final inversion model.

It is instructive to look at the history of the objective function during the Gauss–Newton iteration. Figure 8.5 is a plot of the total objective function Ψ, as defined by Eq. (8.107), versus iteration number. The separate terms of the objective function are also plotted: the term related to the least-squares measure of data misfit, and the term related to the model roughness. The total objective function Ψ is reduced very quickly, and after six iterations it remains unchanged. Note that the model roughness is not plotted for the very first iteration because the algorithm was initiated with a uniform model (every layer with the same conductivity), and the gradient model norm applied to a uniform model is zero. The model roughness increases to its maximum value at the third iteration as structure is put into the model, but then both model roughness and data misfit are reduced and become constant after six iterations.

The effect of the regularization parameter λ is shown in Figure 8.6 where the square root of model roughness Ω defined by Eq. (8.229) has been plotted against the normalized data

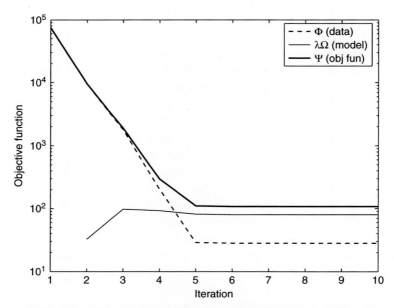

Figure 8.5. The objective function versus iteration is shown for the inversion of Figure 8.3. The total objective function Ψ as defined by Eq. (8.107) is plotted as the thick solid line, and the data misfit Φ defined by Eq. (8.25) is the dashed line. The model roughness Ω, as defined by Eq. (8.229), multiplied by the regularization parameter λ is plotted as the thin solid line.

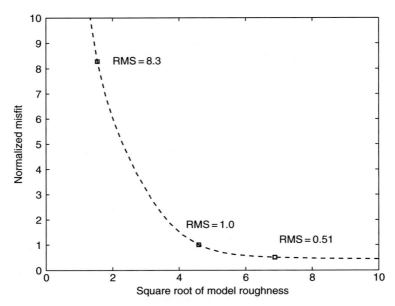

Figure 8.6. The trade-off curve showing the inverse relationship between the normalized data misfit $(\Phi/N)^{1/2}$ and the square root of model roughness defined by Eq. (8.229). Three points are noted, whose models are shown in Figure 8.7.

misfit $(\Phi/N)^{1/2}$. The curve was generated by running the inversion to convergence (i.e. where it reached a stationary point) for several values of λ to determine the final value of Φ and Ω for each λ. The important concept to take from this plot is the inverse relationship between the data misfit and model roughness as λ is changed. As λ is increased, a larger penalty is placed on model roughness, and the inversion model becomes smoother but yields a poorer data fit as a consequence. This is illustrated by the point labeled RMS = 8.3 obtained with a very high value for λ. As λ is decreased, model roughness can increase, and data misfit decreases. The second point labeled RMS = 1.0 corresponds to the model that fits the data to within its correctly prescribed errors, with the minimum amount of structure. Reducing λ to lower values allows for more model structure and smaller data misfit, such as the point labeled RMS = 0.51, at the expense of over-fitting noise in the data. In the limit of $\lambda \to 0$, the model would be expected to approach the D+ model described in Section 4.2.6. These three models are plotted in Figure 8.7 for comparison. High values of λ significantly over-smooth the inversion result, while lower values of λ result in lower data misfits but at the expense of extraneous and unphysical structure.

It is straightforward to apply linearized uncertainty analysis, as described in Section 8.4.4, to this example inverse problem. Specifically, the inversion model from Figure 8.3 is taken as the reference model and, using a linearization around that model, the resolution and variance matrices are computed from Eqs. (8.125) and (8.127). The resolution matrix, shown in Figure 8.8, can be thought of as a filter through which the inversion model is

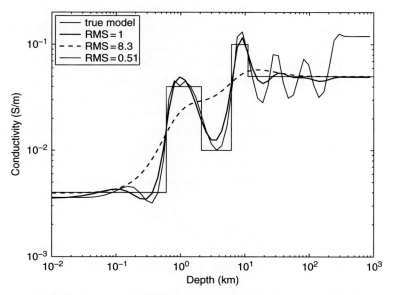

Figure 8.7. Plot showing the inversion models for three different values of the regularization parameter λ, which resulted in the misfit value shown. These models were the points described in the previous trade-off plot, Figure 8.6. The true model is shown as the thin solid line. The model resulting from a high value of λ is shown as the dashed line. The model with the optimal amount of smoothing is shown as the thick solid line, and the model with too little smoothing is shown by the medium solid line.

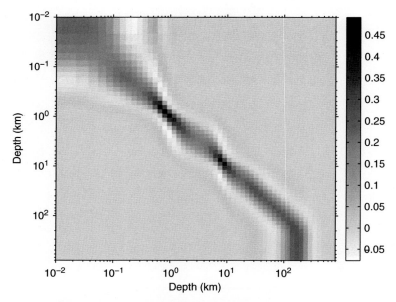

Figure 8.8. The resolution matrix, Eq. (8.125), for the inversion model of Figure 8.3 plotted as a function of the depth in kilometers.

obtained from the true model. Well-resolved model parameters correspond to rows of the resolution matrix that are sharply peaked on the diagonal, with small off-diagonal elements. The figure shows that, at depths less than 100 m, the resolution matrix is broad, indicating that the inversion model is poorly resolved. This is consistent with the fact that these depths are smaller than the inductive scale length at the highest frequency. The figure also shows that the inversion model is poorly resolved at depths greater than 100 km, and in fact is an extrapolation of the model at shallower depths. Between 100 m and 100 km, the model is reasonably well resolved, although the scale of local averaging is on the order of the depth itself. The best resolved depths are where the conductivity values of the inversion model are highest. While it is true that 1D MT data are more sensitive to conductive than to resistive layers, this may in part be an artifact of the nonlinearity brought on by a logarithmic parameterization of conductivity.

The variance matrix of the inversion model in Figure 8.3 is shown in two parts. Figure 8.9 shows the standard deviation of the model parameters as a function of depth, $\sqrt{\mathrm{Var}[m(z)]}$, obtained from the diagonal elements of the variance matrix. The off-diagonal elements of the variance matrix are displayed in Figure 8.10, normalized as correlation coefficients. That is, Figure 8.10 is an image, in z–z' space, of

$$\frac{\mathrm{Cov}[m(z), m(z')]}{\sqrt{\mathrm{Var}[m(z)]\mathrm{Var}[m(z')]}} \tag{8.230}$$

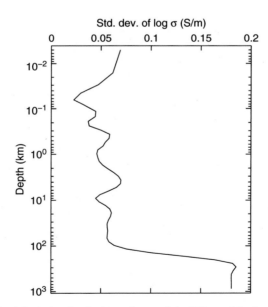

Figure 8.9. The standard deviation for the inversion model of Figure 8.3 plotted as a function of the depth in kilometers.

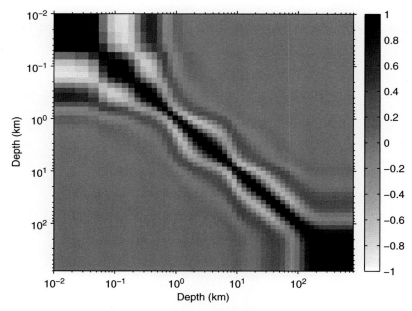

Figure 8.10. The correlation matrix, Eq. (8.230), for the inversion model of Figure 8.3 plotted as a function of the depth in kilometers.

whose values range from -1 to $+1$. Figure 8.9 shows that, at depths less than 100 km, the standard deviation of log σ is less that 0.1, meaning that σ has a relative error less than 10%. The correlation matrix shown in Figure 8.10 displays similar characteristics as the resolution matrix. At depths of poor resolution ($z < 100$ m and $z > 100$ km), the model estimates are highly correlated between depths. In the depth range of good resolution, the correlation drops with depth separation over a scale comparable to the local resolution scale.

8.7.2 Three-dimensional models

As will be shown in Chapter 10, 1D and 2D inversion algorithms have been quite successful for interpreting real MT datasets, but nonetheless come with inherent problems that are not present in 3D inversions. For example, when interpreting MT data along a profile with a 2D algorithm, it is assumed *a priori* that the geology under the profile is actually close to 2D. Furthermore, it must be decided how to rotate the data (e.g. in the direction of the profile, or to maximize some component of the MT transfer function, or to fit a distortion model as in Chapter 6). These issues do not arise in 3D inversion, and, while it is other potential problems for 3D inversions (such as the large computer resources needed), 3D inversion is clearly more in harmony with the 3D nature of geological structures.

Therefore, the next example illustrates these inversion concepts using a dataset comprising synthetic model responses for the 3D Dublin Test Model 1 (DTM1) (Miensopust *et al.*,

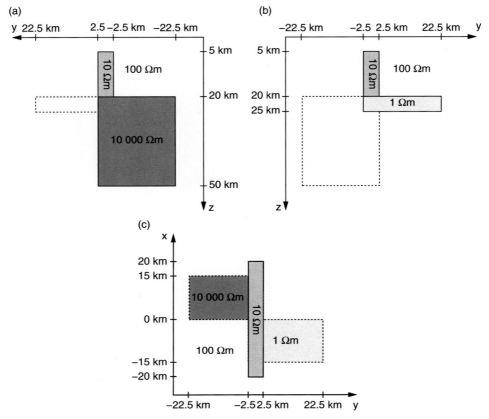

Figure 8.11. The Dublin Test Model 1 (DTM1) from Miensopust *et al.* (2008). The model comprises three blocks in a 100 Ω m homogeneous half-space: (a) south-facing section view; (b) north-facing section view; and (c) plan view. The resistivities of the blocks are: $\rho_1 = 10\ \Omega$ m, $\rho_2 = 1\ \Omega$ m and $\rho_3 = 10\ 000\ \Omega$ m, as indicated. Figure courtesy of Marion Miensopust.

2008). The model is shown in Figure 8.11 and was designed to compare 3D MT forward modeling and inversion codes on a challenging model encompassing a large range of resistivities (1–10 000 Ω m) with both inductive and galvanic effects. Therefore, it serves as a useful reference model to test inversion algorithms. For this example, synthetic MT and vertical magnetic transfer functions were computed at 11 frequencies logarithmically spaced from 10 to 0.001 Hz using the finite-difference algorithm described in Mackie *et al.* (1994, 2007). Prior to inversion, 2% random Gaussian noise was added to the synthetic transfer functions. Data at 59 stations, spaced 5 km apart and located on three east–west profiles, at $x = +15$ km, $x = 0$ km and $x = -15$ km, and one north–south profile, at $y = 0$ km, were used in the inversion. To eliminate bias, different model grids were used for computing the forward synthetic responses and the inversion model.

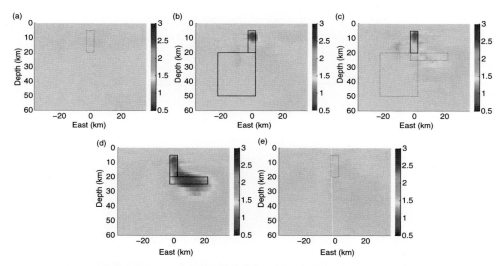

Figure 8.12. Vertical cross-sections through the 3D inversion model, with the grayscale (color scale) indicating the base 10 logarithm of resistivity: (a) $x = -20$ km; (b) $x = -10$ km; (c) $x = 0$ km; (d) $x = +10$ km; and (e) $x = +20$ km. The outlines of the bodies in the true resistivity model (Figure 8.11) are shown as solid lines, except when the cross-section is located at a discontinuity in the model, in which case the adjacent body is plotted as a dotted line. (See plate section for color version.)

Inversions of the synthetic data for 3D models were carried out using a NLCG algorithm that has been used extensively in industrial applications (Mackie *et al.*, 2001). This algorithm minimizes a damped least-squares objective function with a stabilizing function that approximates the L_2-norm of the Laplacian, Eq. (8.93), weighted as a function of the logarithm of depth. As discussed in Section 8.5.4, the preconditioner used in this NLCG algorithm is an approximation to the diagonal part of $\mathbf{A}_\ell^T \mathbf{W} \mathbf{A}_\ell$, which is computationally feasible for large 3D problems.

Similar to the 1D example, the regularization parameter was determined by running the algorithm to convergence for several values, and choosing that value resulting in a weighted RMS of the data residuals of 1.0. The result of this inversion is shown in Figure 8.12 for vertical cross-sections at $x = -20$, -10, 0, $+10$ and $+20$ km, respectively. Also shown superimposed for comparison are the outlines of the true resistivity model. For those cross-sections located at boundaries between adjacent bodies (e.g. the cross-section at $x = -20$ km is located where the 10 Ω m body ends), the adjacent body is outlined by a dotted line to make clear that the cross-section is located at a discontinuity in the true model. Since this inversion penalizes roughness, only a smoothed version of the true model is recovered. Although the data were sparsely sampled over the 3D model, the inversion recovered the main features and structures of the true model. The conductive bodies were better resolved than the resistive body, but this is because in general MT data are less sensitive to resistive bodies than conductive bodies.

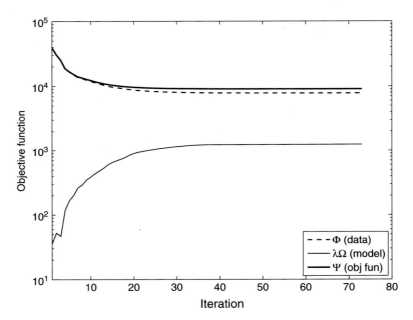

Figure 8.13. The objective function versus iteration is shown for the inversion result of Figure 8.12. The total objective function Ψ as defined by Eq. (8.107) is plotted as the thick solid line; the data misfit Φ defined by Eq. (8.25) is the dashed line; and the model roughness times the regularization parameter ($\lambda\Omega$) is the thin solid line.

Figure 8.13 shows the progression of the 3D inversion as a function of iteration. The behavior of the 3D inversion is similar to that of 1D in that the data misfit and total objective function are reduced in the early iterations while the model roughness increases as structure is added. For this 3D example, more inversion iterations were required to reach convergence compared to the 1D example. This is mostly due to the use of only a rough approximation to the Hessian (**M**) as a preconditioner in the NLCG algorithm, although the higher-dimensional model space may also be playing a role. However, experience with 2D inversion (Rodi & Mackie, 2001) indicates that the use of an approximation to the Hessian, while computationally efficient, increases the number of iterations needed to reach convergence.

The effect of the regularization parameter λ in this 3D inversion is shown in Figure 8.14, where the square root of model roughness is plotted against the normalized data misfit $(\Phi/N)^{1/2}$. The curve was generated by running the inversion to convergence (i.e. where it reached a stationary point) for several values of λ to determine the final values of Φ and Ω for each λ. Similar to the 1D example, an inverse relationship between model roughness and data misfit is observed: higher values of λ yield smoother models but at the expense of data fit. Likewise, lower values of λ fit the data better, but, if λ is too low, data can be over-fitted, resulting in extraneous and unphysical structure in the inversion model. The three models corresponding to the points labeled in Figure 8.14 are compared in Figure 8.15. The first

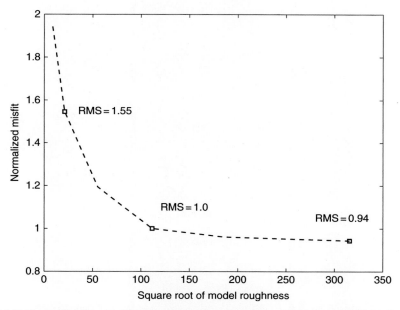

Figure 8.14. The trade-off curve showing the inverse relationship between the normalized data misfit $(\Phi/N)^{1/2}$ and the square root of model roughness. Three points are noted, whose models are shown in Figure 8.15.

column represents the inversion model obtained with a high value of λ and is therefore overly smooth. The second column is the inverse model that had a normalized data misfit of 1.0, while the third (right-hand) column is the model obtained with a low value of λ and therefore having extraneous structure and unphysical resistivity variations.

8.8 Beyond least squares

Thus far, only approaches to solving underdetermined, nonlinear inverse problems based on the method of damped least squares have been discussed. In these methods, a solution is taken to be the model vector minimizing the objective function in Eq. (8.87) where the function Φ is the measure of data misfit given by Eq. (8.24) and Ω is the stabilizing function (stabilizer) given in Eq. (8.106). The functional form of Φ arises from the assumption that the data errors have a Gaussian probability distribution. From a Bayesian point of view, Ω corresponds to the assumption that the prior probability distribution of the model parameter vector $\mathbf{m}$ is Gaussian. In a deterministic framework, the same stabilizing function is interpreted as a measure of the spatial roughness of $\mathbf{m}$, considered to be a discrete parameterization of a model function $m(\mathbf{r})$.

The function Φ is a quadratic function of the data residual vector, $\mathbf{d} - F(\mathbf{m})$, and in fact is the square of a weighted L_2-norm of the residual vector, as is apparent from Eq. (8.25).

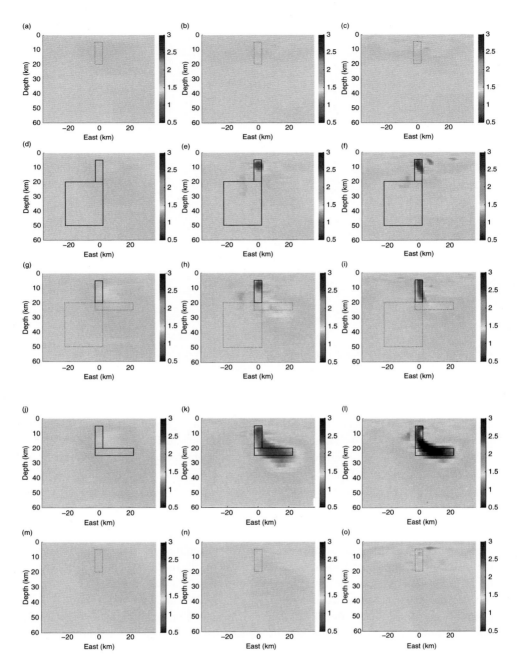

Figure 8.15. A comparison of the inversion models for the three different values of the regularization parameter (λ) shown on the trade-off curve of Figure 8.14. The first row (a)–(c) is for the vertical cross-section located at $x = -20$ km, the second row (d)–(f) is for $x = -10$ km, the third row (g)–(i) is for $x = 0$ km, the fourth row (j)–(l) is for $x = +10$ km and the final row (m)–(o) is for $x = +20$ km. The left-hand column (a)–(m) is for the model with a high value of λ that resulted in a normalized misfit of 1.55, the middle column (b)–(n) is for the optimal value of λ that resulted in a normalized misfit of 1.00, and the right-hand column (c)–(o) is for a smaller λ that resulted in a misfit of 0.94. (See plate section for color version.)

Likewise, the stabilizing functional is the square of a weighted L_2-norm of **m**, and is also quadratic. In this section, the use of non-quadratic functions for Φ and Ω will be considered, thus departing from pure least squares.

8.8.1 Non-Gaussian data errors

The assumption of Gaussian statistics is often taken to be a reasonable assumption, even if tenuous, owing to the central limit theorem (e.g. Feller, 1971), which states that the mean of a sufficiently large number of random variables will be normally distributed. Gaussian statistics applies reasonably well to estimates of the MT transfer function if a bounded influence estimator is used to estimate it, as described in Chapter 5. However, the assumption of Gaussian statistics does not apply to quantities derived from the MT transfer function, such as apparent resistivity or phase. Consequently, there is no inherent reason, other than departing from the quadratic functional form of $\Phi(\mathbf{m})$, that one is not at liberty to consider other distributions describing the data errors (e.g. Constable, 1988).

For example, in seismic event location (e.g. Billings *et al.*, 1994; Rodi, 2006), the *generalized Gaussian* distribution has been used and is given by (for zero mean errors)

$$f_e(\mathbf{e}) = K_p \prod_{i=1}^{N} \exp\left\{ -\frac{1}{p} \left| \frac{e_i}{\sigma_i} \right|^p \right\} \tag{8.231}$$

where p is the order of the distribution Σ_i is a standard error and K_p is a constant. For $p = 2$, the p.d.f. is Gaussian, given by Eq. (8.34). The generalized Gaussian leads to a misfit function that is the L_p-norm of the data residuals, raised to the power p:

$$\Phi(\mathbf{m}) = \sum_{i=1}^{N} \left| \frac{d_i - F_i(\mathbf{m})}{\sigma_i} \right|^p \tag{8.232}$$

An important case of the generalized Gaussian distribution is the Laplace distribution, obtained with $p = 1$, which leads to the L_1-norm of the data residual vector for the misfit functional (e.g. Chave *et al.*, 1987). The L_1-norm is important because it is less sensitive than the L_2-norm to large data residuals, or *outliers*, and thus yields solutions less affected by large errors in the data (Claerbout & Muir, 1973). However, other norms exist that also provide robust measures of data misfit besides the L_1-norm (see Chapter 5).

8.8.2 Non-quadratic stabilizers

Section 8.4.1 considered stabilizing functions that were squared L_2-norms of $Lm(\mathbf{r})$, where L was some differential operator (e.g. $L = \nabla^2$). Minimizing the spatial derivatives of the model function (gradients, curvature or higher derivatives) is very effective in stabilizing an ill-conditioned inverse problem. The use of the L_2-norm places a particularly high penalty on large model derivatives relative to small ones, leading to very smooth solutions.

The L_2-norm also makes the stabilizing function quadratic, enhancing the effectiveness of gradient-based minimization algorithms.

However, smooth models of Earth structure are often unrealistic. Portions of Earth do have rapid spatial variations, including discontinuities, in physical properties, especially in the shallow crust. In fact, gradients and contrasts in properties are often the target of a geophysical investigation, as in resource exploration and engineering studies that seek features like oil reservoirs, faults and the basement under a sedimentary sequence. Strong smoothness constraints then obscure important information. Stabilizers that are not L_2-norms of spatial derivatives may be more appropriate for such applications.

One extension of the stabilizer of Section 8.4 is the replacement of the L_2-norm with the L_p-norm (raised to the power p):

$$\Omega(\mathbf{m}) = \int dV |Lm(\mathbf{r})|^p \qquad (8.233)$$

As p is decreased from 2, the stabilizer becomes relatively less sensitive to the larger values of Lm, allowing larger gradients in the solution for $m(\mathbf{r})$ to be accommodated. The L_1-norm, in particular, admits the construction of piecewise constant models (Farquharson & Oldenburg, 1998). For $p = 1$ and $L = \nabla$, the stabilizing function is called the *total variation* of $m(\mathbf{r})$ and is given by

$$\Omega(\mathbf{m}) = \int dV |\nabla m(\mathbf{r})| \qquad (8.234)$$

This stabilizing function measures the difference between the peak and trough values of m, being insensitive to the shape of m between the extrema. It has been used extensively for image restoration, noise removal, and de-blurring (e.g. Rudin *et al.*, 1992; Acar & Vogel, 1994).

A problem with stabilizers based on the L_1-norm is that their gradient is not defined when Lm is zero. This is remedied with a modification of L_p stabilizers proposed by Ekblom (1973):

$$\Omega(\mathbf{m}) = \int dV \left(|Lm(\mathbf{r})|^2 + \varepsilon^2 \right)^{p/2} \qquad (8.235)$$

The small constant ε removes problems at $Lm = 0$, and for $p = 1$ and $L = \nabla$ it yields a modification of the total variation that behaves quadratically for small Lm.

Other stabilizing functionals are a significant departure from L_p-norms, such as the *minimum support* stabilizers first considered by Last & Kubik (1983) and later by Portniaguine & Zhdanov (1999a). These take the form

$$\Omega(\mathbf{m}) = \int dV \frac{|Lm(\mathbf{r})|^2}{|Lm(\mathbf{r})|^2 + \varepsilon^2} \qquad (8.236)$$

This stabilizer behaves quadratically when $|Lm| \ll \varepsilon$ but assigns nearly equal penalty to large values of $|Lm|$. In the limit $\varepsilon \to 0$, $\Omega(\mathbf{m})$ becomes the volume over which

$|Lm| \neq 0$, i.e. the volume of the *support* of $|Lm|$. With $L = \nabla$, for example, the minimum support stabilizer will seek the smallest volume over which a solution model has a large gradient without penalizing the actual size of the gradient.

8.8.3 Minimization algorithms

When the data misfit functional and the model stabilizer depart from L_2-norms, they introduce non-quadratic dependence of Ψ on **m** over and above that caused by the non-linearity of the forward function, F. To the extent that this happens, the gradient-based minimization methods considered in Section 8.5 become less effective. For example, L_1-norms are non-smooth and non-differentiable functions, and special methods are available for minimizing them (see e.g. Fletcher, 1987). The usual approach to L_1 minimization converts the problem to the optimization of a linear objective function subject to linear equality and inequality constraints that is solved by linear programming (Menke, 1989; Tarantola, 2005).

Another method that can be used to solve L_1-norm problems is the method of iteratively re-weighted least squares (IRLS) (see e.g. Yarlagadda *et al.*, 1985; Scales *et al.*, 1988; Wolke & Schwetlick, 1988; Wolke, 1992), which solves a sequence of weighted least-squares problems. This sequence of problems converges to a solution that is the L_1-norm solution. Additionally, the IRLS algorithm can be used to solve minimization problems that involve other norms, including the non-quadratic measures of model roughness and data misfit considered above. The M-estimators described in Chapter 5 are another example of IRLS.

In the IRLS method, the non-quadratic stabilizing function is written in the form of Eq. (8.100) but with **V** dependent on the model $m(\mathbf{r})$. The quadratic approximation comes from taking **V** to be a constant based on the model values from the previous iteration, and carrying out, for example, a Gauss–Newton iteration to solve for the model update. On the other hand, minimization of these non-quadratic functions can be carried out directly using the NLCG algorithm discussed in Section 8.5.5 without having to resort to the quadratic approximations of the IRLS method.

8.8.4 Examples

The effect of non-quadratic stabilizers is illustrated with the 1D synthetic example analyzed in Section 8.7.1. Inversions were carried out with a NLCG algorithm to minimize the damped least-squares objective function, with the stabilizing function given by either the Ekblom L_p stabilizing function, Eq. (8.235), or the minimum support stabilizing function, Eq. (8.236). In both cases, the differential operator L was defined to be $L = d/dz$.

The result of an inversion using the Ekblom L_p-norm with $p = 1$ is shown in Figure 8.16 along with the true model. For $p = 1$ and large values of Lm, this stabilizing function behaves like the total variation norm, which is relatively less sensitive to large gradients in the model

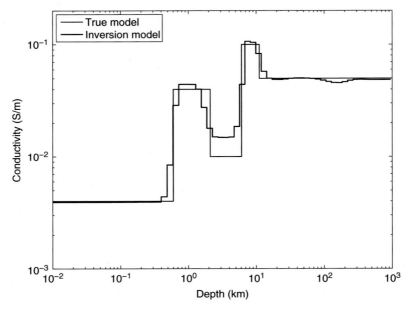

Figure 8.16. The 1D model resulting from an inversion using the Ekblom L_p-norm, Eq. (8.235), for p = 1 is plotted as the thick solid line. The true model (thin solid line) and synthetic data are the same as those analyzed in Section 8.7.1.

than when $p = 2$. At small values of Lm, the stabilizing function behaves quadratically. The result of an inversion using the minimum support norm is shown in Figure 8.17 along with the true model. In both figures, it is clear that the non-quadratic stabilizers admit solutions with larger gradients than are possible using standard quadratic L_2-norm-based stabilizing functions.

8.8.5 Sharp boundary inversion

As a further departure from the framework of unconstrained damped least-squares inverse methods, there exist alternative inverse methods for deriving models with sharp boundaries from magnetotelluric data that do not involve the non-quadratic stabilizing functions just discussed. These methods proceed, for example, by parameterizing 2D models with laterally variable layers such that the layer interfaces are described by a set of nodes with linear interpolation between nodes and conductivities within each layer also described by a set of linear interpolations between nodes (Smith *et al.*, 1999). The inversion can be stabilized by minimizing the lateral roughness of the layer interfaces and the conductivity within each interface. Similarly, de Groot-Hedlin & Constable (2004) present a sharp boundary inversion algorithm that solves for a fixed number of layers of uniform resistivity but variable thickness. The algorithm is stabilized by minimizing variations in

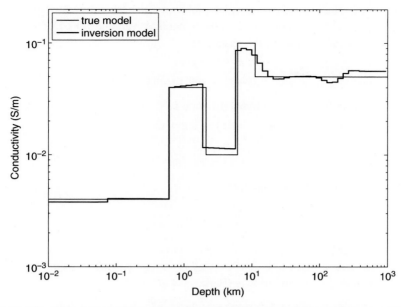

Figure 8.17. The 1D model resulting from an inversion using the minimum support stabilizer, Eq. (8.236), is plotted as the thick solid line. The true model (thin solid line) and synthetic data are the same as those analyzed in Section 8.7.1.

boundary depths, and optionally the resistivity contrast between layers. Another related class of algorithms are parametric inversion algorithms (Portniaguine & Zhdanov, 1999b) that solve for a few simple parameters relating to simple geometric bodies. For example, localized ore bodies could be imaged by inverting for a simple cube, or spheroid, with the parameters of the inversion set to be the depth and radius, or other parameters as appropriate to describing the geometric shape.

8.8.6 Model bounds

Unconstrained optimization problems that allow the model parameters to take on any value have been emphasized in this chapter. However, it is often desirable to put bounds on the model parameters that may come from an understanding of the physical properties, or knowledge about what values those properties should attain in any particular problem. Because the electrical resistivity has to be greater than zero, it is useful to enforce positivity on the model parameters. Furthermore, a reasonable range of resistivity values may be known (e.g. from well logs). Reducing the allowable model space will reduce the range of models that can fit the data, and should help in producing more physically realistic models.

Let $\rho(\mathbf{r})$ parameterize a model of Earth's resistivity on which bounds will be set, and let $m(\mathbf{r})$ be an unbounded function related to $\rho(\mathbf{r})$ by a transformation $\rho(\mathbf{r}) = M(m(\mathbf{r}))$ and its inverse

$m(\mathbf{r}) = M^{-1}(\rho(\mathbf{r}))$. The original minimization problem, which is a constrained minimization on $\rho(\mathbf{r})$, can be mapped to a minimization on the unbounded variable $m(\mathbf{r})$. Derivatives must be modified and this can be done by the chain rule

$$\frac{\partial}{\partial m(\mathbf{r})} = \frac{\partial \rho(\mathbf{r})}{\partial m(\mathbf{r})} \frac{\partial}{\partial \rho(\mathbf{r})} \tag{8.237}$$

Positivity can be enforced by the transformation pair

$$m(\mathbf{r}) = \log \rho(\mathbf{r}) \tag{8.238}$$

$$\rho(\mathbf{r}) = \exp(m(\mathbf{r})) \tag{8.239}$$

Model derivatives must then be modified by

$$\frac{\partial}{\partial m(\mathbf{r})} = \rho \frac{\partial}{\partial \rho(\mathbf{r})} \tag{8.240}$$

If upper and lower bounds on the resistivity are to be enforced, i.e. $\rho_l \leq \rho \leq \rho_u$, then other transformations can be used to turn a bounded inverse problem into an unbounded inverse problem. One useful transformation is (Barbosa *et al.*, 1999)

$$m(\mathbf{r}) = -\ln\frac{\rho_u - \rho}{\rho - \rho_l} \tag{8.241}$$

which has an inverse transformation of

$$\rho(\mathbf{r}) = \frac{\rho_u - \rho_l}{1 + \exp(-m(\mathbf{r}))} + \rho_l \tag{8.242}$$

As $m(\mathbf{r}) \to \infty$, $\rho(\mathbf{r}) \to \rho_u$, and as $m(\mathbf{r}) \to -\infty$, $\rho(\mathbf{r}) \to \rho_l$, thus keeping $\rho(\mathbf{r})$ bounded. As before, the derivative terms must be modified for this transformation, which becomes

$$\frac{\partial}{\partial m(\mathbf{r})} = \frac{\partial}{\partial \rho(\mathbf{r})} \frac{(\rho_u - \rho)(\rho - \rho_l)}{(\rho_u - \rho_l)} \tag{8.243}$$

Other useful transformations can be found in Habashy & Abubakar (2004) and Zhdanov & Tolstaya (2004).

There are bounded constrained minimization methods in the optimization literature that have not been discussed in this section. The reader is referred to Nocedal & Wright (1999) for a good discussion of these methods.

References

Acar, R. & C. R. Vogel (1994). Analysis of bounded variation penalty methods for ill-posed problem. *Inverse Problems*, **10**, 1217–1229.

Alumbaugh, D. L. (2000). Linearized and nonlinear parameter variance estimation for two-dimensional electromagnetic induction inversion. *Inverse Problems*, **16**, 1323–1341.

Apostol, T. M. (1957). *Mathematical Analysis, A Modern Approach to Advanced Calculus*. Reading, MA: Addison-Wesley.

Aster, R. C., B. Borchers & C. H. Thurber (2005). *Parameter Estimation and Inverse Problems*. London: Elsevier Academic Press.

Avdeev, D. B. (2005). Three-dimensional electromagnetic modelling and inversion from theory to application. *Surv. Geophys.*, **26**, 767–799.

Backus, G. E. & F. Gilbert (1967). Numerical application of a formalism for geophysical inverse problems. *Geophys. J. R. Astron. Soc.*, **13**, 247–276.

Backus, G. E. & F. Gilbert (1968). The resolving power of gross Earth data. *Geophys. J. R. Astron. Soc.*, **16**, 169–205.

Backus, G. E. & F. Gilbert (1970). Uniqueness in the inversion of inaccurate gross Earth data. *Phil. Trans. R. Soc.*, **A266**, 123–192.

Bailey, R. C. (1970). Inversion of the geomagnetic induction problem. *Proc. R. Soc. Lond.*, **A315**, 185–194.

Barbosa, V. C. F., J. B. C. Silva & W. E. Medeiros (1999). Gravity inversion of a discontinuous relief stabilized by weighted smoothness constraints on depth. *Geophysics*, **64**, 1429–1437.

Bertero, M. & P. Boccacci (1998). *Introduction to Inverse Problems in Imaging*. Bristol: Institute of Physics.

Billings, S. D., M. S. Sambridge & B. L. N. Kennett (1994). Errors in hypocenter location: picking, model and magnitude dependence. *Bull. Seismol. Soc. Am.*, **84**, 1978–1990.

Chave, A. D. (1984). The Fréchet derivatives of electromagnetic induction. *J. Geophys. Res.*, **89**, 3373–3380.

Chave, A. D. & J. R. Booker (1987). Electromagnetic induction studies. *Rev. Geophys.*, **25**, 989–1003.

Chave, A. D., D. J. Thomson & M. E. Ander (1987). On the robust estimation of power spectra, coherences, and transfer functions. *J. Geophys. Res.*, **92**, 633–648.

Claerbout, J. F. & F. Muir (1973). Robust modeling with erratic data. *Geophysics*, **38**, 826–844.

Constable, C. G. (1988). Parameter estimation in non-Gaussian noise. *Geophys. J.*, **94**, 131–142.

Constable, S. C., R. L. Parker & C. G. Constable (1987). Occam's inversion: a practical algorithm for generating smooth models from electromagnetic sounding data. *Geophysics*, **52**, 289–300.

de Groot-Hedlin, C. & S. Constable (1990). Occam's inversion to generate smooth, two-dimensional models from magnetotelluric data. *Geophysics*, **55**, 1613–1624.

de Groot-Hedlin, C. & S. Constable (2004). Inversion of magnetotelluric data for 2D structure with sharp resistivity contrasts. *Geophysics*, **69**, 78–86.

Deutsch, C. V. & A. G. Journel (1998). *GSLIB: Geostatistical Software Library and User's Guide*, 2nd edn. New York: Oxford University Press.

Dongarra, J. J., J. R. Bunch, C. B. Moler & G. W. Stewart (1979). *LINPACK: Users' Guide*. Philadelphia: Society for Industrial and Applied Mathematics.

Ekblom, H. (1973). Calculation of linear, best Lp-approximations. *BIT*, **13**, 292–300.

Farquharson, C. G. & J. A. Craven (2009). Three-dimensional inversion of magnetotelluric data for mineral exploration: an example from the McArthur River uranium deposit, Saskatchewan, Canada. *J. Appl. Geophys.*, **68**, 450–458.

Farquharson, C. G. & D. W. Oldenburg (1996). Approximate sensitivities for the electromagnetic inverse problem. *Geophys. J. Int.*, **126**, 235–252.

Farquharson, C. G. & D. W. Oldenburg (1998). Non-linear inversion using general measures of data misfit and model structure. *Geophys. J. Int.*, **134**, 213–227.

Feller, W. (1971). *An Introduction to Probability Theory and its Applications*, Vol. 2, 3rd edn. New York: John Wiley.

Fletcher, R. (1987). *Practical Methods of Optimization*, 2nd edn. New York: John Wiley.

Foster, M. (1961). An application of the Wiener–Kolmogorov smoothing theory to matrix inversion. *J. Soc. Indust. Appl. Math.*, **9**, 387–392.

Franklin, J. N. (1970). Well-posed stochastic extensions of ill-posed linear problems. *J. Math. Anal. Appl.*, **31**, 682–716.

Grandis, H., M. Menvielle & M. Roussignol (1999). Bayesian inversion with Markov chains – I. The magnetotelluric one-dimensional case. *Geophys. J. Int.*, **138**, 757–768.

Habashy, T. M. & A. Abubakar (2004). A general framework for constraint minimization for the inversion of electromagnetic measurements. *Prog. Electromagn. Res.*, **46**, 265–312.

Haber, E., U. M. Ascher & D. Oldenburg (2000). On optimization techniques for solving nonlinear inverse problems. *Inverse Problems*, **16**, 1263–1280.

Hadamard, J. (1902). Sur les problèmes aux dérivées partielles et leur signification physique. *Princeton University Bulletin*, 49–52.

Han, N., M. J. Nam, H. E. Kim, T. J. Lee, Y. Song & J. H. Suh (2008). Efficient three-dimensional inversion of magnetotelluric data using approximate sensitivities. *Geophys. J. Int.*, **175**, 477–485.

Hansen, P. C. (1992). Analysis of discrete ill-posed problems by means of the L-curve. *SIAM Rev.*, **34**(4), 561–580.

Hestenes, M. R. & E. Stiefel (1952). Methods of conjugate gradients for solving linear systems. *J. Res. Natl Bur. Stand.*, **49**, 409–436.

Hoerl, A. E. & R. W. Kennard (1970). Ridge regression: biased estimation for nonorthogonal problems. *Technometrics*, **12**, 55–67.

Jackson, D. D. (1972). Interpretation of inaccurate, insufficient, and inconsistent data. *Geophys. J. R. Astron. Soc.*, **28**, 97–109.

Jackson, D. D. (1979). The use of *a priori* data to resolve non-uniqueness in linear inversion. *Geophys. J. R. Astron. Soc.*, **57**, 137–157.

Jiracek, G. R., W. L. Rodi & L. L. Vanyan (1987). Implications of magnetotelluric modeling for the deep crustal environment in the Rio Grande rift. *Phys. Earth Planet. Inter.*, **45**, 179–192.

Jupp, D. L. B. & K. Vozoff (1975). Stable iterative methods for the inversion of geophysical data. *Geophys. J. R. Astron. Soc.*, **42**, 957–976.

Jupp, D. L. B. & K. Vozoff (1977). Two-dimensional magnetotelluric inversion. *Geophys. J. R. Astron. Soc.*, **50**, 333–352.

Kelley, C. T. (1999). *Iterative Methods for Optimization*. Philadelphia: Society for Industrial and Applied Mathematics.

Kendall, M. G. & A. Stuart (1977). *The Advanced Theory of Statistics, vol. 2, Inference and Relationship*, 4th edn. New York: Macmillan.

Lanczos, C. (1961). *Linear Differential Operators*. Princeton, NJ: Van Nostrand.

Larsen, J. C. (1981). A new technique for layered earth magnetotelluric inversion. *Geophysics*, **46**, 1247–1257.

Last, B. J. & K. Kubik (1983). Compact gravity inversion. *Geophysics*, **48**, 713–721.

Levenberg, K. (1944). A method for the solution of certain non-linear problems in least squares. *Q. Appl. Math.*, **2**, 164–168.

Lines, L. R. & S. Treitel (1984). Tutorial: a review of least-squares inversion and its application to geophysical problems. *Geophys. Prosp.*, **32**, 159–186.

Loewenthal, D. (1975). Theoretical uniqueness of the magnetotelluric inverse problem for equal penetration discretizable models. *Geophys. J. R. Astron. Soc.*, **43**, 897–903.

Mackie, R. L. & T. R. Madden (1993). Three-dimensional magnetotelluric inversion using conjugate gradients. *Geophys. J. Int.*, **115**, 215–229.

Mackie, R. L., T. J. Smith & T. R. Madden (1994). Three dimensional electromagnetic modeling and inversion using conjugate gradients. *Geophysics*, **60**, 923–935.

Mackie, R. L., W. Rodi & M. D. Watts (2001). 3-D magnetotelluric inversion for resource exploration. In *Expanded Abstracts of the SEG International Exposition and Annual Meeting*, San Antonio, Texas.

Mackie, R. L., M. D. Watts & W. Rodi (2007). Joint 3D inversion of marine CSEM and MT data. In *Expanded Abstracts of the SEG International Exposition and Annual Meeting*, San Antonio, Texas.

Madden, T. R. (1990). Inversion of low-frequency electromagnetic data. In *Oceanographic and Geophysical Tomography*, ed. Y. Desaubies, A. Tarantola & J. Zinn-Justin. Amsterdam: Elsevier Science.

Madden, T. R. & R. L. Mackie (1989). Three-dimensional magnetotelluric modeling and inversion. *Proc. IEEE*, **77**, 318–333.

Madden, T. R. & P. Nelson (1964). A defense of Cagniard's magnetotelluric method. *Geophys. Lab*. ONR NR-371-401, Final Report, MIT, Cambridge, MA.

Malinverno, A. (2002). Parsimonious Bayesian Markov chain Monte Carlo inversion in a nonlinear geophysical problem. *Geophys. J. Int.*, **151**, 675–688.

Marquardt, D. W. (1963). An algorithm for least-squares estimation of nonlinear parameters. *J. Soc. Indust. Appl. Math.*, **11**, 431–441.

Marquardt, D. W. (1970). Generalized inverses, ridge regression, biased linear estimation, and nonlinear estimation. *Technometrics*, **12**, 591–612.

Matsuno, T., N. Seama, R. L. Evans, *et al*. (2010). Upper mantle electrical resistivity structure beneath the central Mariana subduction system. *Geochem. Geophys. Geosyst.*, **11**, Q09003, doi:10.1029/2010GC003101.

McGillivray, P. R. & D. W. Oldenburg (1990). Methods for calculating Fréchet derivatives and sensitivities for the nonlinear inverse problem: a comparative study. *Geophys. Prosp.*, **38**, 499–524.

McGillivray, P. R., D. W. Oldenburg, R. G. Ellis & T. M. Habashy (1994). Calculation of sensitivities for the frequency domain electromagnetic problem. *Geophys. J. Int.*, **116**, 1–4.

Menke, W. (1989). *Geophysical Data Analysis: Discrete Inverse Theory*. San Diego: Academic Press.

Miensopust, M. P., P. Queralt & A. G. Jones (2008). MT 3D inversion – a recapitulation of a successful workshop. In *Proc. 19th Electromagnetic Induction in the Earth Workshop*, Beijing, China.

Moorkamp, M., A. G. Jones & D. W. Eaton (2007). Joint inversion of teleseismic receiver functions and magnetotelluric data using a genetic algorithm: Are seismic velocities and electrical conductivities compatible? *Geophys. Res. Lett.*, **34**, L16311, doi:10.1029/2007GL030519.

Morozov, V. A. (1993). *Regularization Methods for Ill-Posed Problems*, Engl. edn, ed. M. Stessin. Boca Raton, FL: CRC Press.

Newman, G. A. & D. L. Alumbaugh (2000). Three-dimensional magnetotelluric inversion using non-linear conjugate gradients. *Geophys. J. Int.*, **140**, 410–424.

Newman, G. A. & P. T. Boggs (2004). Solution accelerators for large-scale three-dimensional electromagnetic inverse problems. *Inverse Problems*, **20**, S151–S170.

Nocedal, J. & S. J. Wright (1999). *Numerical Optimization*, Springer Series in Operations Research. New York: Springer.

Oldenburg, D. W. (1980). One-dimensional inversion of natural source magnetotelluric observations. *Geophysics*, **44**, 1218–1244.

Oldenburg, D. W. (1994). Practical strategies for the solution of large-scale electromagnetic inverse problems. *Radio Sci.*, **29**, 1081–1099.

Oldenburg, D. W. & R. G. Ellis (1993). Efficient inversion of magnetotelluric data in two dimensions. *Phys. Earth Planet. Inter.*, **81**, 177–200.

Oldenburg, D. W. & Y. Li (1994). Subspace linear inverse method. *Inverse Problems*, **10**, 915–935.

Oldenburg, D. W., P. R. McGillivray & R. G. Ellis (1993). Generalized subspace methods for large scale inverse problems. *Geophys. J. Int.*, **114**, 12–20.

Parker, R. L. (1970). The inverse problem of electrical conductivity in the mantle. *Geophys. J. R. Astron. Soc.*, **22**, 121–138.

Parker, R. L. (1977). Linear inference and underparameterized models. *Rev. Geophys. Space Phys.*, **15**, 446–456.

Parker, R. L. (1980). The inverse problem of electromagnetic induction: existence and construction of solutions based on incomplete data. *J. Geophys. Res.*, **85**, 4421–4428.

Parker, R. L. (1994). *Geophysical Inverse Theory*. Princeton, NJ: Princeton University Press.

Parker, R. L. & K. A. Whaler (1981). Numerical methods for establishing solutions to the inverse problem of electromagnetic induction. *J. Geophys. Res.*, **86**, 9574–9584.

Patrick, F. W. (1969). *Magnetotelluric modeling techniques*. Ph.D. thesis, University of Texas, Austin.

Phillips, D. L. (1962). A technique for the numerical solution of certain integral equations of the first kind. *J. Assoc. Comput. Mach.*, **9**, 84–97.

Polak, E. (1971). *Computational Methods in Optimization: A Unified Approach*. New York: Academic Press.

Portniaguine, O. & M. S. Zhdanov (1999a). Focusing geophysical inversion images. *Geophysics*, **64**, 874–887.

Portniaguine, O. & M. S. Zhdanov (1999b). Parameter estimation for 3-D geoelectromagnetic inverse problems. In *Three-Dimensional Electromagnetics*, ed. M. Oristaglio & B. Spies. Tulsa: Society of Exploration Geophysics, pp. 222–232.

Pujol, J. (2007). The solution of nonlinear inverse problems and the Levenberg–Marquardt method. *Geophysics*, **72**, W1–W16.

Rodi, W. L. (1976). A technique for improving the accuracy of finite element solutions for magnetotelluric data. *Geophys. J. R. Astron. Soc.*, **44**, 483–506.

Rodi, W. L. (1989). *Regularization and Backus–Gilbert estimation in nonlinear inverse problems: application to magnetotellurics and surface waves*. Ph.D. thesis, Pennsylvania State University.

Rodi, W. (2006). Grid-search event location with non-Gaussian error models. *Phys. Earth Planet. Inter.*, **158**, 55–66.

Rodi, W. & R. L. Mackie (2001). Nonlinear conjugate gradients algorithm for 2-D magnetotelluric inversion. *Geophysics*, **66**, 174–187.

Rudin, L. I., S. Osher & E. Fatemi (1992). Nonlinear total variation based noise removal algorithms. *Physica D*, **60**, 259–268.

Sambridge, M. & K. Mosegaard (2002). Monte Carlo methods in geophysical inverse problems. *Rev. Geophys.*, **40**(3), 1009, doi:10.1029/2000RG000089.

Sasaki, Y. (1989). Two-dimensional joint inversion of magnetotelluric and dipole–dipole resistivity data. *Geophysics*, **54**, 254–262.

Sasaki, Y. (2001). Full 3-D inversion of electromagnetic data on PC. *J. Appl. Geophys.*, **46**, 45–54.

Sasaki, Y. (2004). Three-dimensional inversion of static-shifted magnetotelluric data. *Earth Planets Space*, **56**, 239–248.

Scales, J. A., A. Gersztenkorn & S. Treitel (1988). Fast l_p solution of large, sparse, linear systems: application to seismic travel time tomography. *J. Comput. Phys.*, **75**, 314–333.

Sen, M. & P. L. Stoffa (1995). *Global Optimization Methods in Geophysical Inversion*. New York: Elsevier.

Shoham, Y., A. Ginzburg & F. Abramovici (1978). Crustal structure in central Israel from the inversion of magnetotelluric data. *J. Geophys. Res.*, **83**, 4431–4440.

Siripunvaraporn, W. (2012). Three-dimensional magnetotelluric inversion: an introductory guide for developers and users. *Surv. Geophys.*, **33**, 5–27, doi:10.1007/s10712–011–9122–6.

Siripunvaraporn, W. & G. Egbert (2000). An efficient data-subspace inversion method for 2-D magnetotelluric data. *Geophysics*, **65**, 791–803.

Siripunvaraporn, W. & G. Egbert (2009). WSINV3DMT: vertical magnetic field transfer function inversion and parallel implementation. *Phys. Earth Planet. Inter.*, **173**, 317–329.

Siripunvaraporn, W., G. Egbert, Y. Lenbury & M. Uyeshima (2005). Three-dimensional magnetotelluric inversion: data-space method. *Phys. Earth Planet. Inter.*, **150**, 3–14.

Smith, J. T. & Booker J. R. (1988). Magnetotelluric inversion for minimum structure. *Geophysics*, **53**, 1565–1576.

Smith, J. T. & Booker, J. R. (1991). Rapid inversion of two- and three-dimensional magnetotelluric data. *J. Geophys. Res.*, **96**, 3905–3922.

Smith, J. T., M. Hoversten, E. Gasperikova & F. Morrison (1999). Sharp boundary inversion of 2-D magnetotelluric data. *Geophys. Prosp.*, **47**, 469–486.

Stuart, A., K. Ord & S. Arnold (2009). *Kendall's Advanced Theory of Statistics, Vol. 2A, Classical Inference and the Linear Model*, 6th edn. New York: John Wiley.

Tarantola, A. (2005). *Inverse Problem Theory*. Philadelphia: Society for Industrial and Applied Mathematics.

Tarantola, A. & B. Valette (1982). Generalized nonlinear inverse problems solved using the least squares criterion. *Rev. Geophys. Space Phys.*, **20**, 219–232.

Tarits, P., V. Jouanne, M. Menvielle & M. Roussignol (1994). Bayesian statistics of non-linear inverse problems: example of the magnetotelluric 1-D inverse problem. *Geophys. J. Int.*, **119**, 353–368.

Tikhonov, A. N. & V. Y. Arsenin (1977). *Solutions of Ill-Posed Problems*. Washington, DC: Winston.

Tromp, J., C. Tape & Q. Liu (2005). Seismic tomography, adjoint methods, time reversal and banana–doughnut kernels. *Geophys. J. Int.*, **160**, 195–216.

Twomey, S. (1963). On the numerical solution of Fredholm integral equations of the first kind by the inversion of the linear system produced by quadrature. *J. Assoc. Comput. Mach.*, **10**, 97–101.

Vogel, C. R. (2002). *Computational Methods for Inverse Problems*. Philadelphia: Society for Industrial and Applied Mathematics.

Wahba, G. (1990). *Spline Models for Observational Data*. Philadelphia: Society for Industrial and Applied Mathematics.

Weidelt, P. (1972). The inverse problem of geomagnetic induction. *Z. Geophys.*, **38**, 257–289.

Whittall, K. P. & D. W. Oldenburg (1992). *Inversion of Magnetotelluric Data for a One-Dimensional Conductivity.* Geophysical Monograph Series, No. 5. Tulsa: Society of Exploration Geophysicists.

Wiggins, R. A. (1972). The general linear inverse problem: implication of surface waves and free oscillations for earth structure. *Rev. Geophys. Space Phys.*, **10**, 251–285.

Wolke, R. (1992). Iteratively reweighted least squares: a comparison of several single step algorithms for linear models. *BIT*, **32**, 506–524.

Wolke, R. & H. Schwetlick (1988). Iteratively reweighted least squares: algorithms, convergence analysis, and numerical comparisons. *SIAM J. Sci. Stat. Comput.*, **9**, 907–921.

Wu, F. T. (1968). The inverse problem of magnetotelluric sounding. *Geophysics*, **33**, 972–979.

Yang, X. (1999). *Stochastic inversion of 3-D ERT data.* Ph.D. thesis, University of Arizona.

Yarlagadda, R., B. Bednar & T. L. Watt (1985). Fast algorithms for L_p deconvolution. *IEEE Trans. Acoust. Speech Signal Process.*, **ASSP-33**, 174–182.

Zhdanov, M. S. (2002). *Geophysical Inverse Theory and Regularization Problems.* Amsterdam: Elsevier Science.

Zhdanov, M. & E. Tolstaya (2004). Minimum support nonlinear parameterization in the solution of a 3D magnetotelluric inverse problem. *Inverse Problems*, **20**, 937–952.

9

Instrumentation and field procedures

IAN J. FERGUSON

9.1 Overview of magnetotelluric recording

9.1.1 Requirements of magnetotelluric instrumentation

In most MT surveys, it is essential to acquire high-quality data over a broad period range. The MT response represents a spatial average of the electrical properties of the underlying Earth, and the contribution to the response of a particular target may be relatively small, perhaps creating only a few degrees of phase change or 10% or so change in apparent resistivity. Thus, it is often necessary to define the MT response or impedance with an uncertainty of less than a few percent. This objective requires time series of very high fidelity, necessitating modern high-quality instrumentation, thoughtful site selection and careful installation, and appropriate choices of site layout and data filtering. Some MT surveys focus only on determining the location or coarse spatial delineation of a conductive feature, allowing some trade-off between data quality and the number of acquired sites. However, even for these surveys, acquisition of high-quality data can usually provide additional information, such as improved resolution of the depth or shape of the target.

The components of an MT recording system can be divided into three parts: electrometers for sensing electric fields (or, more correctly, potential differences); magnetometers for sensing the magnetic field (or, more correctly, the magnetic induction); and recording/timing units for controlling the timing, digitization, filtering and recording of the data. Figure 9.1 shows typical spectra of the natural MT magnetic and electric field components in land and seafloor environments (see also Chapter 2). The data have an extremely broad period range, with large variations in signal level at different periods. MT instrumentation must have sufficient sensitivity, dynamic range and filtering capabilities to be able to measure such signals.

The magnetic field signal levels in land MT surveys will often be at a comparable level to those shown in Figure 9.1, with similar levels for the two orthogonal horizontal components, and the signal level for the vertical component typically a factor of 2–50 smaller. However, on occasion during quiet geomagnetic periods, signal levels may be a factor of 100 lower, placing more stringent constraints on the equipment needed to record the signals. There is considerable variation in electric field signal levels, and as shown by Figure 9.1 the signal levels at a more conductive site may be several orders of magnitude lower than over a

The Magnetotelluric Method: Theory and Practice, ed. Alan D. Chave and Alan G. Jones. Published by Cambridge University Press © Cambridge University Press 2012.

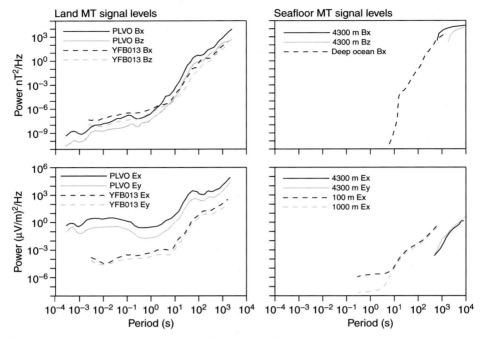

Figure 9.1. Power spectra of magnetic and electric field signals. Land data are from a relatively resistive region in central Canada (PLVO) and a relatively conductive site in southwestern Australia (YFB013). Seafloor data are for a site in 4300 m deep water in the Pacific Ocean off California. Based on Chave & Filloux (1984). Also shown are the typical horizontal magnetic field spectrum for the deep ocean at short periods (Filloux, 1973) and typical horizontal electric field spectra in 100 m and 1000 m deep ocean water (Hoversten *et al.*, 2000).

resistive region. On the seafloor, both magnetic and electric signal levels are much smaller than on land, owing to exponential decay. The magnetic field is reduced even further due to the presence of the highly resistive subsurface (the magnetic field on the surface of an infinite insulator from a uniform source is zero). In the deep ocean, MT studies are restricted to periods larger than 10–100 s; and in the more shallow ocean, to periods longer than 1 s.

9.1.2 Categories of magnetotelluric recording systems

It is possible to classify the instrumentation used in natural-source land MT surveys into one of three groups based on the period range of data acquired and the specific equipment used in the survey:

- Long-period magnetotelluric (LMT) instruments typically record data over a period range of 1 s to >10 000 s, and ultra-long-period MT surveys extend this range to 100 000 s and beyond.

- Broadband MT (BBMT) instruments typically record data over a period range from around 0.001 s to more than 1000 s.
- Audio-frequency MT (AMT) instruments typically record data over the period range of 0.0001 s to 0.1 s.

Three additional classifications of MT equipment can be defined on the basis of the source of the recorded signals and the environment of deployment:

- Controlled-source AMT (CSAMT) instrumentation uses an artificial electromagnetic source to create the signal, and usually operates in a period range of 10^{-4} s to 8 s.
- Radio-frequency MT (RMT) instrumentation uses radio-frequency transmitters, and measures MT signals for a period range of 10^{-6} to 10^{-4} s.
- Seafloor MT (SFMT) instrumentation, with which MT data are collected within the BBMT and LMT period ranges, uses equipment deployed on the seafloor.

In this chapter the main focus will be on land AMT, BBMT and LMT instrumentation, but some information will be included on other equipment. A number of review papers and textbooks discuss land MT instrumentation (e.g. Kaufman & Keller, 1981; Vozoff, 1991; Simpson & Bahr, 2005). Other reviews have focused on either magnetometers or electrodes, such as the extensive investigation of electrode performance in Perrier *et al.* (1997). Some of the most significant changes to MT instrumentation that have occurred since the earlier reviews include the ubiquitous use of Global Positioning System (GPS) timing, introduction of low-power, low-noise, ring–core fluxgate magnetometers, and the availability of inexpensive solid-state data storage. Reviews and descriptions of marine MT instrumentation include Filloux (1987), Chave *et al.* (1991) and Constable *et al.* (1998). Zonge & Hughes (1991) provide detailed information on CSAMT instrumentation, and Tezkan (1999) and Pedersen *et al.* (2005) describe RMT instrumentation.

Different equipment is used for recording in different period ranges. LMT recordings require magnetometers sensitive to low-frequency signals, such as fluxgate sensors with low power consumption, and non-polarizing metal–metal ion electrodes with limited long-period drift. It is often desirable for LMT systems to be relatively inexpensive to allow the acquisition of multiple systems that can be deployed simultaneously for long durations (days to months). BBMT systems use equipment sensitive to higher-frequency signals (such as induction coil magnetometers), use higher sampling frequencies, typically have higher power consumption and often permit more complex on-board signal filtering than LMT systems. AMT and BBMT systems are more expensive than LMT systems and are deployed for a shorter duration – a few hours (AMT) to a few days (BBMT). In the AMT period range, it is common for induction coil sensors to be used, but because of the increased sensitivity of these coils at high frequencies, it is often most convenient to use smaller and more portable coils than in BBMT surveys. Because of decreased electrode noise levels at higher frequencies, it is possible to use cheaper and more portable polarizing metal rod electrodes in AMT surveys. At AMT frequencies, the signals typically exceed noise levels in only short time windows, so some AMT systems record only selected segments of the time series.

In CSAMT the signal is produced by a long grounded dipole, with a length of some hundreds of meters for small-scale features (groundwater, geothermal, mining) to some kilometers for the crustal scale, or by a horizontal loop source (Zonge & Hughes, 1991). In order for the resultant electromagnetic fields to be modeled as spatially uniform MT fields, the recording sites must be more than four or five skin depths in the near-surface substrate away from the source. Sensors used in CSAMT surveys are generally similar to those used in natural-source AMT surveys. However, the recording system is usually synchronized with the transmitter, permitting the transmission, acquisition and stacking of data using a pre-programmed time sequence of frequencies. The shape of the source signal is typically sinusoidal, but can be otherwise. The RMT method evolved from the geophysical very-low-frequency (VLF) method. It uses the same basic equipment as AMT (induction coil magnetic sensors and electrometers). However, the equipment is generally smaller in form factor than AMT equipment. Smaller coil sensors can be used for the higher-frequency signals that are measured, and electrode lines may be shorter than in AMT surveys (e.g. Tezkan 1999).

The absorption of shorter-period signals by the conductive ocean water limits the lower end of the period range for SFMT surveys to around 1 s in ocean water of 1 km depth and around 100 s in the deep ocean. SFMT surveys therefore use similar magnetic sensors as land BBMT and LMT surveys. Recordings in shallow water use induction coil sensors for magnetic field measurements (Constable *et al.*, 1998) whereas those in the deep ocean tend to use similar sensors as LMT land surveys such as fluxgates (Chave *et al.*, 1991). Electrometers for marine surveys most commonly use low-noise Ag–AgCl electrodes and a short-arm salt-bridge apparatus rather than a long cable.

9.2 Magnetotelluric instrumentation: electrometers

The role of the electrometer, or telluric sensor, in MT systems is measurement of the electric field in two horizontal orthogonal directions. In land systems, this measurement is made by determining the potential difference between pairs of grounded electrodes spaced a known distance apart. For most installations, the electrode separation is between 25 m (AMT) and 100 m (BBMT), and the pairs of electrodes are typically laid in either the magnetic or geographic north–south and east–west directions. The electrode lines are usually arranged as either an "X", with the recording unit near the center, or an "L", with one electrode shared between the orthogonal dipoles, with the recording unit near the common electrode. Most MT systems employ an additional ground electrode installed near the recording unit that serves as the ground or earth electrode for the electronics in the recording unit, and may be used to form a reference voltage for the signal in an "L" configuration. Most land MT surveys employ porous-pot Cu–CuSO$_4$, Pb–PbCl$_2$ or Ag–AgCl electrodes, but some high-frequency MT surveys employ metal rod electrodes. For high-frequency problems, such as RMT or AMT at high frequencies, capacitive electrodes may hold some advantages, but they are not routinely used in MT. In marine MT systems, the electric field may be measured between the ends of a salt-bridge system consisting of open tubes connected to Ag–AgCl electrodes located closer to the instrument.

9.2.1 Physical principles

In general, there exists a potential difference (or voltage) between any two points in an electric field. In MT, the electric field is determined by a measurement of the potential difference between such points. In electromagnetic studies, the electric field may be expressed in terms of the vector and scalar potentials as

$$\mathbf{E} = -\nabla\phi - \partial_t\mathbf{A} \tag{9.1}$$

where

$$\mathbf{B} = \nabla \times \mathbf{A} \tag{9.2}$$

The second term in (9.1), related to the time derivative of the magnetic field, means that the electric field is non-conservative, and the potential difference between two points therefore depends on the path between the points. Defining the measurement path using a vector $\mathbf{s}$, the potential difference $\phi_{12} = \phi(P_1) - \phi(P_2)$ will be related to the electric field by

$$\phi_{12} = \int_{P_1}^{P_2} \mathbf{E} \cdot d\mathbf{s} \tag{9.3}$$

For a straight-line path between the points defined by vector $\mathbf{r}$ (extending from P_1 toward P_2) and a uniform electric field, the potential is given by

$$\phi_{12} = \mathbf{E} \cdot \mathbf{r} \tag{9.4}$$

In this situation the potential difference ϕ between two points separated horizontally by distance l is related to the electric field E in the same direction by

$$E = \frac{\phi}{l} \tag{9.5}$$

In SI units, ϕ is given in volts (V), l in meters (m) and E in volts per meter (V/m).

In order for a measurement of the potential difference between two electrodes oriented in a particular direction to provide a measure of the electric field in that direction, it is necessary for the wire joining the electrodes to be straight (e.g. Swift, 1967; Gómez-Treviño, 1987). Errors arising from non-straight wires will be greater at higher frequencies, and can involve a few percent in apparent resistivity or a few degrees in phase when the average deviation from a straight wire becomes a small fraction of the electromagnetic skin depth (Gómez-Treviño, 1987). In the extreme case of a large circular loop in the wire, the time variations in the vertical magnetic field will induce a spurious potential proportional to the area enclosed by the loop. Therefore, as a field procedure, there should not be spare electrode wire lying in loops. Any spare wire should be laid backward and forward across itself so as to reduce to zero the area of the loop.

Measurement of the potential difference between a pair of grounded electrodes may be thought to be a straightforward operation, but it involves measuring two grounded "earths" using a moderately complicated electrochemical circuit, and it is important to be aware of the salient physical and chemical processes to ensure accurate measurements. Similar measurements are made in other fields such as induced polarization (IP; Merriam, 2007) and biomedical monitoring (e.g. Tallgren *et al.*, 2005). A grounded electrode forms an electrochemical half-cell with a corresponding half-cell potential and impedance. In general, the impedance is a complex quantity, and includes both a real component, the resistance, and an imaginary component, the reactance, that has a contribution from the electrode capacitance. Accurate measurement of the time variations in the telluric potential requires the electrode half-cell potential to be stable over time in varying ambient physical and chemical conditions, and that the electrode impedance be small.

An electrode forms a contact between a metallic conductor, the electrode wire, and an ionic conductor, the saline porewater in the soil, and therefore undergoes electrochemical reactions at each boundary. The electrical response depends on the electrical and chemical properties of the electrode itself, the chemistry of the aqueous fluids in the soil and, less importantly, the mineral constituents of the soil. Two types of electrode can be defined on the basis of the current flow across the electrode–soil interface. A non-polarizing electrode is one in which there is free exchange of charge across the interface, or a Faradaic process. A polarizing electrode is one in which there is no transfer of charge across the electrode–electrolyte interface, and hence only displacement current occurs. In practice, most electrodes lie between these extremes. The porous-pot Cu–$CuSO_4$, Pb–$PbCl_2$ or Ag–$AgCl$ electrodes used in MT surveys are non-polarizing, whereas the metal rod electrodes used in some high-frequency MT surveys are polarizing.

If a metal is placed in an electrolyte solution, an ion–electron exchange will occur in which metal ions enter the solution and electrolyte ions combine with the metal of the electrode. The result is a charge distribution at the metal–electrolyte interface called an electric double layer, which creates an associated half-cell potential. A non-zero potential will occur between two electrodes of different composition in a uniform electrolyte, or two identical electrodes in an electrolyte of varying composition. Even when two electrodes of the same metal are placed in the electrolyte, impurities in the metals cause an irregular time-varying potential. These effects may be lessened by encasing the metal in a layer of a corresponding salt, such as silver in silver chloride. One objective of fieldwork is the reduction to a minimum of the number of interfaces between metals and electrolytes, and between electrolytes of different salts.

Figure 9.2 shows an equivalent circuit for a potential measurement by an electrode. The electrode response can be approximated by an electrode, or Faradaic, impedance Z_{fa} and a half-cell potential $E_{1/2}$. The Faradaic impedance can be described by an equivalent circuit including a Faradaic or charge transfer resistance R_{fa}, associated with the transfer of ions at

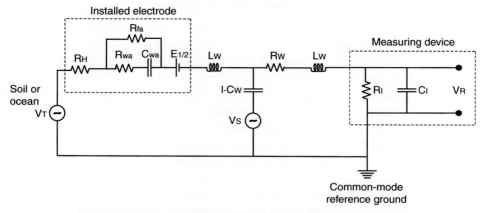

Figure 9.2. Circuit representation of a telluric measurement. Modified from Zonge & Hughes (1985). V_T is the telluric potential in the soil or ocean water, V_S is the potential sensed by the electrode and V_R is the measured potential. The electrode has a half-cell potential $E_{1/2}$, Warburg resistance R_{wa}, Warburg capacitance C_{wa}, Faradaic resistance R_{fa} and electrolyte resistance R_H. The resistance, inductance and distributed capacitance of the wire are R_W, L_W and C_W, and the equivalent resistance and capacitance of the measuring device are R_I and C_I.

the metal interface, in parallel with a Warburg impedance Z_{wa}, associated with the supply of ions to the interface through the process of diffusion. The Warburg impedance can be divided into a Warburg resistance R_{wa} and a Warburg capacitance C_{wa}. As the double layer involves a charge distribution separated by a very small distance, the capacitance may be relatively high. The combined effect of these contributions leads to a frequency dependence of the real part of the impedance, the electrode resistance. At low frequencies the impedance tends to the Faradaic resistance, whereas at high frequencies it tends to the parallel addition of Faradaic and Warburg resistances. There is an additional resistance R_H associated with the transfer of ions through the electrolyte. The specific values of the resistances and capacitances depend on the electrolyte chemistry, electrode chemistry and surface area, and the temperature and current density crossing the electrode–electrolyte boundary (e.g. Merriam, 2007). The requirement for low electrode resistance is aided by using an electrode with an appropriately large surface area.

The electrical properties of other parts of the measuring circuit are also important to telluric measurements. The impedance of the voltmeter used to measure the potential difference should greatly exceed the resistance between the electrodes in order to ensure minimal current flow through the measuring circuit and minimal disruption of the electro-chemical conditions. The wire connecting the electrodes should have sufficiently low resistance and inductance not to affect the measured response. Finally, appropriate corrections to the response must be made for any preamplifiers, filters or other electronic components included in the measuring circuit.

9.2.2 Magnetotelluric electrodes

Electrodes are the most critical component in the measurement of the telluric response, and indeed are often the weakest component of any MT system. If reasonable care is taken with all other aspects of the electric measurements, then it is electrode noise that limits the quality of the electric field measurements (Petiau & Dupis, 1980). Two days of MT recording with good electrodes will produce better data than two weeks of recording with poor electrodes. Most porous-pot electrodes used in MT surveys are based on Pb–PbCl$_2$, Ag–AgCl, Cu–CuSO$_4$ or Cd–CdCl$_2$ electrochemistry, with the metal ion either absorbed in plaster, kaolinite, bentonite or a synthetic substance, or present in the form of a liquid electrolyte (e.g. Petiau & Dupis, 1980; Perrier *et al.*, 1997). Contact with the soil is made either through the absorbing material or through a contact interface such as a ceramic or wooden plug. Metal electrodes used in MT surveys include steel, iron, brass and aluminum rods, and lead and aluminum plates.

The quality of MT electrodes can be defined in terms of the ability to accurately record time variations of telluric potential over a broad range of frequencies and for long durations. The first factor contributing to accuracy is the stability of the electrode half-cell potential under both constant and varying temperature and moisture conditions. The second factor is the requirement for low electrode impedance, including both a low electrode contact resistance and a low electrode capacitance. Electrodes must also have appropriate mechanical attributes to allow their transport and deployment in a range of field environments.

9.2.2.1 Half-cell potential

Time variations in electrode half-cell potential occurring under constant ambient conditions include long-term drift and spontaneous changes. These variations are attributed to electro-chemical reactions within the electrode or at its interface. Causes of long-term noise include deposition of metal ion salt on the metal interface, physical degradation of metallic surfaces (e.g. Tallgren *et al.*, 2005) and spontaneous changes, that can be as large as 20 mV, due to reaction charge release (Clerc *et al.*, 1998; Lu & Macnae, 1998). The time–frequency characteristics of the noise are different for distinct types of electrodes, and the electro-chemical processes are not fully understood (Chave *et al.*, 1991; Lu & Macnae, 1998; Tallgren *et al.*, 2005). For at least some electrodes, the level of noise increases with rising self-potential of the electrodes (Petiau, 2000).

The self-noise of MT electrodes increases with period and, at long periods (>100 s), is much larger for metal electrodes than for metal–metal ion electrodes (Figure 9.3). Tests of metal MT and IP electrodes show that stainless steel provides the lowest noise levels, followed by iron, copper, lead and brass with intermediate noise levels, and aluminum and graphite with the highest noise levels (Petiau & Dupis, 1980; Lu & Macnae, 1998; LaBrecque & Daily, 2007). At 1000 s period, metal electrode noise (defined in terms of power in V^2/Hz) is higher than metal–metal ion electrode noise by a factor of about 3 for stainless steel, 30 for iron and 300 for aluminum (Lu & Macnae, 1998). The relative performance of metal electrodes improves at shorter periods. At 1 s period, poor metal

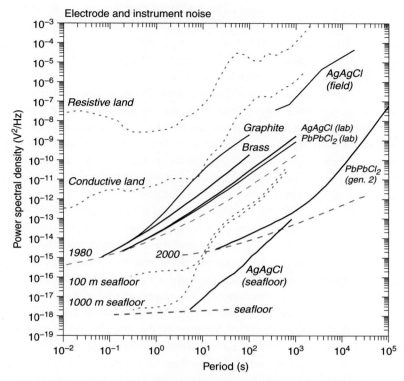

Figure 9.3. Comparison of the noise levels of electrodes and telluric recording instrumentation with signal levels. Short-dashed gray lines are the telluric potential for a resistive and conductive land site (PLVO and YFB013 from Figure 9.1) for 100 m long dipoles, and for 100 m and 1000 m deep seafloor sites with 10 m long dipoles (Hoversten *et al.*, 2000). Long-dashed gray lines show the level of instrument noise for 1980 technology (Petiau & Dupis, 1980), 2000 technology (Petiau, 2000) and seafloor MT systems (Hoversten *et al.*, 2000). Solid black lines show electrode noise levels. Laboratory results for graphite, brass, Ag–AgCl and Pb–PbCl$_2$ electrodes are for first-generation electrodes reported in Petiau & Dupis (1980). Laboratory results for a second-generation Pb–PbCl$_2$ electrode are from Petiau (2000). Noise levels are also shown for Ag–AgCl electrodes at a land field site (Junge, 1990) and at a seafloor site (Hoversten *et al.*, 2000).

electrodes are less than a factor of 10 noisier than metal–metal ion electrodes (Petiau & Dupis, 1980), and at 0.2 s there is little difference between the noise of stainless steel, iron and metal–metal ion electrodes (Lu & Macnae, 1998).

Pb–PbCl$_2$, Ag–AgCl and Cu–CuSO$_4$ systems presently provide the lowest noise levels for land MT metal–metal ion electrodes, with the preparation of the lowest noise levels requiring careful control of electrolyte chemistry (e.g. Filloux, 1973; Junge, 1990; Petiau, 2000; Korepanov *et al.*, 2008a). Silver has the lowest potential difference between the metal and its salt, with lead next and then copper. Therefore, Ag–AgCl electrodes will have the lowest noise levels, but their cost makes them prohibitive for land use. Marine electrodes are usually Ag–AgCl electrodes because of the additional requirement for the metal ion to be reasonably insoluble in the ocean water (Filloux, 1987; Webb *et al.*, 1985) and because of the larger cost of the rest of the MT system. Good-quality metal–metal ion electrodes have noise of 10^{-14} V^2/Hz at 1 s period, 10^{-9} V^2/Hz at 100 s period and 10^{-5} V^2/Hz at 10^5 s period, corresponding to mean peak-to-peak fluctuations (assuming bandwidth equal to frequency) of 0.3 µV, 3 µV and 100 µV, respectively (Petiau & Dupis, 1980). Exceptional-quality electrodes may have noise power levels that are a factor of 50 or more lower than these values (Petiau, 2000), and poor electrodes may have noise levels a factor of 10 or more higher (Lu & Macnae, 1998, with correction). The variation in electrode potential of carefully prepared electrodes of the same design has an approximately Gaussian distribution (Filloux, 1973). Selection of pairs of electrodes with similar self-potential can minimize the magnitude of the differential potential during recordings.

The frequency variation in the electrode noise and telluric spectra means that electrode noise will have the greatest impact on the signal-to-noise ratio at periods of 1–10 s (Figure 9.3). At very short periods, noise will be dominated by amplifier and other instrumental sources rather than by the electrodes. In practice, changes in ambient physical and chemical conditions mean that electrode noise is higher in land field deployments than in the laboratory, as seen by comparison of the responses of Ag–AgCl electrodes in the laboratory and field in Figure 9.3. In contrast, electrode noise level in the seafloor environment is lower than in the laboratory because of the almost isothermal and isosaline conditions (Constable *et al.*, 1998). The "cleanest" land-based electric field measurements made to date were recorded in Carty Lake, northern Ontario, Canada, in 1991–93 (Schultz *et al.*, 1993) and provided excellent responses to periods of some days (>200 000 s). The electrodes were placed at the base of a lake, in over 15 m of water, where they were chemically and thermally stable.

The temperature dependence of the electrodes is most important in long-duration surveys that expose electrodes to the largest temperature variation. It arises because of the temperature sensitivity of the chemical reactions at the electrode interface. Perrier *et al.* (1997) found that temperature sensitivity for first-generation non-polarizing electrodes was 30 to 200 µV/°C for Pb–$PbCl_2$ electrodes and 160 to 300 µV/°C for Ag–AgCl electrodes. An extreme case of temperature dependence occurs when the ground freezes. At the onset of freezing, large changes in potential are observed as the salinity of the remaining unfrozen pore fluid increases. At even colder temperatures, the potential and contact resistance of the electrodes increase considerably, as both the ground and possibly the electrode freeze. Electrodes used in MT deployments on ice have included copper (Beblo & Liebig, 1990) and titanium sheets (Wannamaker *et al.*, 2004).

Changes in the chemistry and moisture levels of the soil will produce long-term changes in the electrode potential. Following the installation of electrodes, there is usually a period of electrode drift with a time scale of minutes to tens of minutes to hours as the chemistry of the electrode–soil system reaches equilibrium (e.g. Petiau & Dupis, 1980). During subsequent recordings, drift may be caused by natural changes in the ambient moisture levels, such as after a heavy rainfall, and salinity of the soil or ocean water. The sensitivity to salinity change is 500 μV/% of salinity change for Ag–AgCl electrodes (Filloux, 1973).

Undesirable electrode drift also occurs because of aging and deterioration of electrodes. For metal–metal ion electrodes, these processes include changes in the concentration of metal ions and corrosion of the metal wire within the electrode or exposed wire in the vicinity of the electrode. Metal electrodes may also undergo corrosion, causing large magnitude drift. For metal–metal ion electrodes, the drift is observed to be lowest for Pb–$PbCl_2$ and Ag–AgCl electrodes and higher for Cu–$CuSO_4$ electrodes, but the construction of the electrode can be more of a determining factor than the chosen chemical system (Perrier *et al.*, 1997). For good-quality electrodes, drift should be less than about 10 mV/month (Perrier *et al.*, 1997), but for poor electrodes it may be a factor of 5 or more higher. For metal electrodes, the magnitude of the drift appears to depend more on the electrode surface area than on the particular metal involved.

9.2.2.2 Electrode impedance

Accurate measurement of telluric potentials requires electrode contact impedance to be small. Large values of electrode contact resistance can couple with the distributed capacitance of the dipole wires to alter the shorter-period telluric response. Zonge & Hughes (1985) showed that as a rule of thumb this effect can become significant if the product LfR_c of dipole wire length L (km), frequency f (kHz) and electrode contact resistance R_c (kΩ) exceeds 2. It is possible to overcome this problem using a high-impedance amplifier located near the electrode and by shielding the lead wires from the amplifier to the receiver (Zonge & Hughes, 1985; Wannamaker *et al.*, 2004). A high contact resistance will result in the earth electrode recording system acting as an *RC* low-pass circuit. The high-frequency electric field amplitude will be strongly attenuated, causing a decrease in apparent resistivity curves that increases with increasing frequency and a corresponding distortion of the phase response. An extreme example of this effect can be seen in the MT responses from site baf-001 (Figure 9.4) recorded on Baffin Island in Canada by Evans *et al.* (2005). The site was located on rocks from which the Barnes Icecap Glacier had retreated only a few years before such that there was not any lichen on the rocks. The measured contact resistance was in excess of 1 MΩ. The estimated apparent resistivity curves begin to attenuate at periods less than 0.1 s, and by 0.03 s are more than two orders of magnitude below their correct level of $>10^4$ Ω m.

The contact resistance of a metal electrode depends on the shape and size of the electrode and on the resistivity of the material in which it is buried. For rod-shaped electrodes, the Faradaic contribution to the contact resistance can be estimated using the result for the contact resistance of a cylindrical rod of length L and diameter a in a half-space of resistivity ρ,

$$R = \frac{\rho}{2\pi L}\left[\ln\left(\frac{2L}{a}\right) - 1\right] \tag{9.6}$$

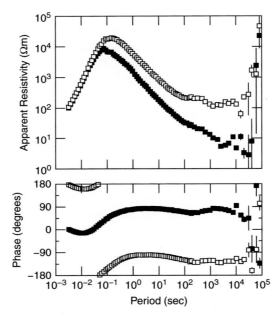

Figure 9.4. MT data recorded at site baf001 on Baffin Island, northern Canada, by Evans *et al.* (2005). Solid symbols are the *xy* component and open symbols are the *yx* component. Note the strong attenuation of the apparent resistivity curves with decreasing period (increasing frequency) from about 0.1 s and lower due to the high contact resistance causing the earth electrode recording system to act as an *RC* low-pass filter.

(Sunde, 1949). For example, a 1 cm diameter rod buried to 50 cm depth in a half-space of 1000 Ω m will have a contact resistance of 1150 Ω. For the dimensions of metal electrodes used in most land MT surveys, the contact resistance will be between 0.75 and 2 times the numerical value of the resistivity. In resistive soils such as dry sand, with resistivity exceeding 5000 Ω m, the electrode contact resistance may be sufficiently high that $L f R_c$ exceeds 2 for shorter-period (<0.001 s) signals and, in the absence of appropriate measures, the MT response will be distorted. For non-polarizing metal–metal ion electrodes, Constable *et al.* (1998) report a value of 6 Ω for Ag–AgCl marine electrodes, and values reported for land surveys vary from around 20 to 500 Ω. Zonge & Hughes (1985) report a value of 300 Ω for Cu–CuSO$_4$ electrodes. Part of the observed variation in contact resistance is due to differences in electrode chemistry, size and geometry, and part is due to the variation of the host material.

In most land MT surveys, metal–metal ion electrodes are installed in a small hole back-filled with saturated clay or soil. As the desired measurement is the electric potential in the surrounding undisturbed material, the circuit representing the measurement requires the addition in series of a resistance corresponding to the saturated material (Figure 9.2). There will be contributions to the contact resistance from both the electrode contact resistance, the internal resistance of the saturated material, and the contact resistance of the zone formed by the saturated soil. It is again important for the total contact resistance to be as small as possible, and it may be minimized by the addition of salt to the saturating liquid in the hole or by increasing its dimensions. A common practice is to use NaCl salt to form a mud solution in the electrode hole to reduce contact resistance, but the NaCl forms an interface with the electrode salt ($CuSO_4$, $PbCl_2$ or AgCl), and can become a noise source. For a hole with an average linear dimension of about 50 cm, the numerical value of the contact resistance with the surrounding undisturbed soil will be of the same order of magnitude as the resistivity of the surrounding soil (for example, as calculated using equation (9.6)). For resistive soils, this value may be 1000–10 000 Ω, considerably larger than the actual electrode resistance. For the larger values in this range, there is a risk of distortion of the MT response.

In many MT surveys, contact resistance is assessed by measuring the resistance between pairs of electrodes in their recording configuration. The resistance measurement is sometimes called an electrode contact resistance, but in fact corresponds to the response of a pair of electrodes, the material in the deployment holes and a smaller contribution from the remainder of the material between the electrodes. Such a measurement should be carefully distinguished from the actual contact resistance of the electrodes, which is best measured using electrodes deployed side by side in the recording environment.

Significant values of the imaginary component of the electrode impedance, arising from the capacitance of the electrode, can affect shorter-period MT responses. A significant capacitive component has been observed in the contact impedance of metal IP electrodes (Merriam, 2007; LaBrecque & Daily, 2007) and metal biomedical electrodes (Tallgren *et al.*, 2005). There have been few studies appropriate for the period ranges, current densities and physical environments of MT surveys. Wannamaker *et al.* (2004) suggested that some features of the observed response of an Antarctic MT system may be explained by complex contact resistance between titanium electrodes and the surrounding dielectric fern. However, in this environment, the real component of the contact impedance is extremely high (around 0.2 MΩ), although not as high as on the Baffin Island example (Figure 9.4). It appears probable that, in most situations relevant to MT surveys, distortion due to the combined effect of large contact resistance and capacitance of the dipole wires will greatly exceed distortion due to electrode capacitance.

9.2.3 Other components of magnetotelluric electrometers

9.2.3.1 Dipole wires and salt bridges

The wire connecting the electrodes to the measuring device should be insulated (using material with a much greater resistance than the electrode contact resistance) to prevent both shorting of the measurement and corrosion. Some researchers have used coaxial wires with an external electrical conducting shield grounded at the central point of the electrode array to reduce noise. Such efforts are usually not required (Kaufman & Keller, 1981), although they can result in cleaner signals that are less affected by motion-induced signals (i.e. shaking in the wind). The wire should have a sufficiently low resistance R_W and self-inductance L_W not to affect the measured electric fields. In the case of extreme electrode contact resistance, such as on ice surfaces, the capacitance of the wire has a measurable effect (Zonge & Hughes, 1985; Wannamaker *et al.*, 2004) and the configuration of the recording must be modified from the standard method to avoid distortion of the measured response.

Marine MT recordings have used both wires connecting electrodes and salt-bridge systems to measure the telluric signals. In a salt bridge, the field is measured between the open ends of non-conducting tubes extending laterally from the instrument (Filloux, 1973). The finite conductivity of the ocean water in the tubes, relative to copper wire, limits the length of the tubes to several tens of meters. With a salt-bridge system, the electrodes may be located on-board the main recording unit.

9.2.3.2 Amplifiers and filters

There is no absolute ground reference in telluric recordings, so the amplifiers used in recording systems are differential types designed to multiply the voltage difference between two channels by an appropriate gain factor. A typical circuit diagram for measuring telluric signals can be found in the classic paper of Trigg (1972) that formed the basis of many early MT systems. There are typically three electrodes for each recording direction, two main electrodes and the circuit ground electrode. Some MT systems amplify the differential signal between the main electrode relative to the ground before combining the two signals, while other systems amplify the differential signal between the two main electrodes. In the former case, it is important to use a ground electrode of comparable quality to the four main electrodes in order to minimize the risk of drift or other electrode noise from the ground electrode producing saturation. It is essential for the amplifier to have a very high common mode rejection (ratio of amplification of differential signal to amplification of common signal) in order to allow proper recording of large, rapidly changing signals affecting both inputs, e.g. due to lightning strikes in the survey region (Vozoff, 1991).

Amplifiers used in telluric measurements must have a number of other characteristics. They should have appropriate impedance characteristics matched to the electrode impedance, the response should be linear over the range of input voltages and the required frequency band, noise levels should be low relative to electrode noise, temperature dependence should be low, and power consumption should be limited. Electronic devices produce a number of types of noise, including Johnson noise generated by thermal fluctuations of charge carriers in resistive materials, and noise with a $1/f$ power spectrum produced by semiconductors. Spectra of geomagnetic and telluric signals can fall off more steeply than $1/f$ (e.g. Figure 9.1), so $1/f$ noise can be a problem in MT recordings. It can be minimized by using a chopping amplifier in which the output is obtained from a modified input signal that is reversed in sign at an appropriately short period and integrated. The chopping amplifier will also reduce output drift and remove the DC offset between the input channels, allowing measurement of the long-period telluric response, but there is a trade-off between the noise reduction and the frequency bandwidth of the output. Some seafloor telluric systems have incorporated a mechanical chopper in which a series of insulated valves reverse the connection between the ends of the salt-bridge arms and the electrodes (Filloux, 1973, 1987; Chave *et al.*, 1991).

Chopping amplifiers and filters used in telluric circuitry will affect the frequency response of the recording, and require appropriate system calibration to be performed for subsequent correction of the output. In the absence of a chopper, the telluric recordings may use a low-cut filter to remove longer-period signals. This mode of recording is often referred to as AC coupling, and removes the DC offset between the electrodes. It has the important advantage of removing long-period electrode drift from the recorded signal, making it less likely that the maximum input range of the system (often around several hundred millivolts) will be exceeded. Since the advent of 24-bit analog-to-digital (A/D) conversion, this method is less common than it once was, as the full dynamic range can usually be accommodated within 24 bits. Most telluric recordings will incorporate a low-pass anti-aliasing filter to remove high-frequency signals prior to digitization of the signal. LMT recordings will often be switchable between DC coupling and AC coupling utilizing filters with a corner period of 10 000 to 30 000 s. The corner period of the low-pass filter will be about 1 s (e.g. Tomczyk, 2007). The induction coil magnetic sensors used in BBMT and AMT recordings provide AC coupling, so the corresponding telluric recordings are usually AC-coupled using a corner frequency matched to the magnetic response. The corner period for the low-pass filter is matched to the sampling frequency. In very resistive sounding locations, it is important to be aware of possible response distortion arising because of the interaction of filter components and the electrical characteristics of the rest of the measurement circuit. See Figure 9.4 and the discussion above.

9.2.3.3 Lightning protection

For land telluric recordings it is necessary to protect the system against spike overvoltages of up to 1000 V that may be induced by lightning strikes at or near the recording site. This may be done using diodes and varistors (Tomczyk, 2007). In telluric systems that amplify the differential signal between the main electrodes rather than between the main electrodes and ground separately, it is possible to use a ground electrode having a higher noise level but a better capacity to inject current into the ground, such as a large metal spike. Such an electrode will provide superior lightning protection to a porous-pot electrode. Burial of the electrode wires may also reduce the risk of direct lightning strikes on the recording system.

9.2.4 Field deployment of electrometers

9.2.4.1 Geometry

For convenience, the electrode dipoles in land surveys are usually aligned in the magnetic or geographic north–south and east–west directions, with the two dipoles crossing near their midpoint, and the ground electrode and recording unit are located nearby. However, L-shaped configurations are used when space is limited on two sides of a recording site, and it is desirable to have the recording unit near the corner of the space. Compasses and surveying methods are used to ensure that the electric dipoles are aligned to within 1° or better of the required azimuth. The lengths of electrode lines are measured with an accuracy of 5–10 cm. It is essential for the electrode wires to remain stationary during data acquisition, so the wires should be weighted or buried to minimize movement. Wind-induced movement of the wires will induce a spurious signal that has a dominant period of 0.1–10 s as the wires move through the Earth's static magnetic field.

Seafloor recordings require considerable attention to aspects of instrument housing and methods for deployment and recovery of instruments (Filloux, 1973). The deployment of reasonably straight electrode wires in marine surveys requires a significant logistical effort. In contrast, the polypropylene tubes used in salt-bridge systems allow orthogonal horizontal electric field measurements to be made more easily. A similar system of tubes can also be used with electrodes located at the end of each arm and the electrode wires running through the tubes (Constable *et al.* 1998).

9.2.4.2 *Installation of electrodes*

To obtain the most accurate telluric recording, it is necessary for deployment of the electrodes to appropriately match the environmental conditions. In land surveys, electrodes are optimally deployed in moist, salted, kaolinitic mud or bentonite. The saline pore fluid helps to reduce the overall contact resistance, and the clays help to retain the moisture. In soils with an intermediate moisture level, long-term drift can be reduced by deploying electrodes within non-metallic buckets – sometimes called the Russian bucket configuration (Lu & Macnae, 1998). The bucket helps to reduce moisture loss and variation along with changes in the porewater chemistry in the vicinity of the electrode. In very dry soils, it may be better to install the electrodes directly in the soil and to allow the hydroscopic properties of the salted clay to attract moisture from the surrounding soil. In wet soils, no bucket is required, as adequate moisture is present, and it is sometimes best not to use salt, as advection of water around the electrode after rainfall can change the salt concentration, leading to changes in electrode potential (Perrier *et al.*, 1997). For long-duration telluric recordings, electrodes should be deployed at least 1 m deep to minimize temperature-induced potential variations.

Some environments call for more specialized electrode deployments. The diurnal temperature variation can be particularly problematic in desert regions, where it may be up to 30°C, exacerbated by electrode burial in very dry, loose sand. For typical soils, the diurnal variation is attenuated by approximately one order of magnitude per meter of soil, so a 20°C diurnal variation will cause 2°C variation at the depth of most electrodes. However, in desert regions, the thermal conductivity of sand is such that it requires 2 m of burial depth for an order-of-magnitude attenuation, so electrodes buried one meter in sand will experience 10°C thermal variation during a typical 30°C diurnal variation.

In areas in which the ground freezes in winter, long-duration recordings require that electrodes be installed at depths exceeding the limit of freezing. Researchers have also had some success using electrodes deployed in a Russian bucket with a commercial antifreeze additive added to the bentonite mixture. In areas of permafrost or very shallow bedrock, it is sometimes possible to ground one or more of the electrodes in a small pond or lake. An effective method for achieving this is to wrap the electrode with a clay mixture, place it inside a flexible permeable material, such as a nylon stocking, and install it on the lake floor (e.g. McNeice & Jones, 1998).

Environmental regulations are requiring increased consideration of the environmental impact of MT surveys. In addition, lead is defined as a toxic substance, and special permission is required to freight Pb–PbCl electrodes by air in many countries. In such situations, it is preferable to use less toxic Ag–AgCl or Cu–CuSO$_4$ electrodes rather than Pb- or Cd-based electrodes.

9.3 Magnetotelluric instrumentation: magnetometers

The role of the magnetometer in MT measurements is measurement of the time variations of the magnetic field **H** or, more correctly, the magnetic induction **B**, in either the two horizontal directions or all three orthogonal directions. In modern MT systems, the measurements are most commonly made with induction coil sensors at RMT, AMT and BBMT frequencies and with fluxgate magnetometers at LMT frequencies. A potentially exciting new development is low-power, millimeter-scale atomic magnetometers with femtotesla sensitivity (Kitching *et al.*, 2009). These instruments provide a measurement of the vector magnetic induction **B**.

9.3.1 Induction coil sensors

The principle of operation of induction coil magnetometers, which are also known as search coil magnetometers, involves the induction of an electromotive force (EMF) in a coil by time variation of the component of the magnetic field parallel to the coil axis (e.g. Kaufman & Keller, 1981; Tumanski, 2007). The EMF induced in an idealized coil of negligible resistance, capacitance and inductance by a time-varying magnetic induction **B** is given by

$$\text{EMF} = nA\partial_t(\mathbf{a} \cdot \mathbf{B}) \tag{9.7}$$

where n is the number of turns of wire in the coil, A is the area of the coil, and **a** is the unit vector parallel to its axis. It is evident that the EMF rises with increasing frequency, and hence the sensitivity of induction coils decreases at long periods. The response depends on the direction of the magnetic field variations, so a vector measurement of the field requires three orthogonal sensors.

The response of a real coil will depend on the lumped resistance R_c, capacitance C_c and inductance L_c of the coil (Figure 9.5). If amplification of signals occurs near the coil, the resistance, capacitance and inductance of lead wires may be assumed to be negligible. The combination of the capacitance and the resistance of the induction coil creates a low-pass filter in which the response is reduced above a resonant frequency ω_{max}. If the input resistance of the coil is much less than that of the measuring device R_i, the corner frequency will be given by

$$\omega_{max} = \sqrt{\frac{1}{L_c C_c}} \qquad (9.8)$$

(Kaufman & Keller, 1981). In order for the rollover frequency to be high, it is desirable for the capacitance – which depends on the geometry of the coil, the diameter of the wire, the characteristics of the insulation and the sequence of winding of the turns – to be as low as possible. To minimize noise levels, it is also desirable for the resistance – which depends on the gauge of the wire and the metal used – to be as low as possible. In general, both the resistance and capacitance increase approximately linearly with the length of wire. The sensitivity increases with the size of the coil, but is limited by practical consideration of the weight and length of the coil.

The sensitivity of induction coil sensors can be increased by the addition of magnetically permeable material within its core. In this case, the response of a coil at frequencies less than the resonant frequency becomes

$$\text{EMF} = -k_c n A \partial_t (\mathbf{a} \cdot \mathbf{B}) \qquad (9.9)$$

where k_c is the relative magnetic permeability of the core. Materials such as permalloy and supermalloy have extremely high magnetic permeability and the capacity to greatly amplify the response. However, because of the geometrically controlled demagnetization process, the effective magnetic permeability of the core will be less than the true material permeability. Demagnetization is minimized when the magnetic material has a long rod-like form and the coil is wrapped in the form of a solenoid around the rod. The primary advantage of including the permeable material in a sensor is thus to reduce the cross-section of a coil without decreasing the effective area $k_c A$. This change permits the use of a smaller amount of wire, and therefore provides lower resistance and capacitance (Kaufman & Keller, 1981).

The characteristics of the coil can be adjusted by varying the winding and the electric and magnetic properties of the coil. Increasing the number of turns while keeping the length of wire constant will not change the output voltage, but will shift the response toward lower frequencies. Increasing the effective area $k_c A$ will increase the output, but will also shift the response toward lower frequencies. Overall, these effects mean that coils designed to record only RMT or AMT frequencies will be smaller and lighter than those designed to collect BBMT frequencies.

The form of the induction coil response also depends on the characteristics of the measuring electronics. For example, if the ratio of the coil resistance to the load resistance R_c/R_i is significant, the response will have a flat peak with corner frequencies

$$\omega_{low} = \frac{R_c + R_i}{L_c} \quad \text{and} \quad \omega_{high} = \frac{1}{R_i C_c} \qquad (9.10)$$

(Tumanski, 2007). Preamplifiers in MT systems must be well matched to the coils to provide a relatively flat response in the target period band, and in modern MT systems are constructed as

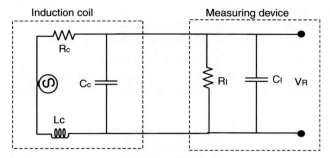

Figure 9.5. The electrical circuit equivalent to an induction coil sensor.

part of the coils (e.g. Vozoff, 1991). The amplifiers may also include a chopper or integrating transducer that will further alter the form of the response (Tumanski, 2007).

The maximum sensitivity of common MT induction coil sensors is from about 0.1 V/nT for smaller RMT and AMT coils rising to 1–10 V/nT for larger BBMT coils. The amplitude response is relatively flat over the target period band (Figure 9.6). The phase difference of the output and the input signals passes through zero in the target period band, with a negative phase response occurring at shorter periods and a positive phase response occurring at longer periods. Inclusion of integrating circuitry reduces the fall-off in the long-period response and reduces the maximum phase difference at these periods from 180° to 90°. The spectrum of natural magnetic field signals increases with increasing period, so measurement of the time derivative of the field, as provided by induction coils, automatically minimizes the influence of large longer-period signals. There has been considerable effort made in recent years to produce coils with broad frequency ranges to minimize the number of different coils needed for broadband recordings. However, a relatively narrow bandwidth can sometimes be considered advantageous for minimizing the effects of noise outside the frequency range of interest. An additional type of induction coil sensor in common use is the air loop. This is a large-diameter multi-turn loop deployed flat on the ground and used to measure the vertical field in locations where a coil cannot be installed. The response of such a loop has similar characteristics to other induction coils, but generally exhibits a lower overall sensitivity (Figure 9.6).

The noise in induction coil magnetometers arises from the coil itself and from the associated electronic circuitry, including the amplifier. The major source of noise from the coil itself at higher frequencies is the thermal resistance and Johnson noise of the wiring in the coil. This noise has a white spectrum and increases with the resistance of the wire. At high frequencies, the current noise in the amplifier becomes the most significant source of noise; and at very low frequencies, the noise spectrum is dominated by $1/f$ noise from semiconductors in the electronic circuitry. Figure 9.6 shows typical noise spectra of induction coil sensors. A good induction coil sensor has a noise level of less than 10^{-8} nT^2/Hz at a period of 1 s, corresponding to mean peak-to-peak fluctuations of less than a picotesla. The characteristics of an induction coil may change with time due to aging of the magnetic material in coils with high-permeability cores, changes of the permeability as a function of Earth's field strength, changes in the core material due to changes in temperature and moisture content, and changes in geometry (e.g. bending, distortion). Coils should be calibrated regularly to account for the long-term changes. They should be calibrated at temperatures comparable to that where they will be used, and for long-duration recordings, the temperature dependence of the response should be as small as possible.

9.3.2 Fluxgate sensors

Fluxgate sensors consist of an excitation coil carrying an alternating current wound around a magnetically permeable body, and a detector coil to measure the response. The ambient geomagnetic field alters the point at which magnetic saturation occurs in the output signal time sequence. For measurement directions parallel and antiparallel to the Earth's field, there will be an asymmetry in the measured response that can be related to the strength of the field. In parallel-rod fluxgate magnetometers, the sensor consists of parallel rods wound as solenoids in alternating directions, and the instrument measures the magnetic field component parallel to the rods. In ring-core fluxgate magnetometers, the sensor consists of circular metallic ribbons wound as an excitation coil and located within an external detector coil. The component of the magnetic field in the plane of the ring and parallel to the axis of the detector coil is measured (Burger, 1972). It is possible to increase the gain of fluxgate magnetometers by configuring them as null instruments. A compensating solenoid is used to provide negative feedback to offset the baseline level of Earth's field in the direction of measurement (Andersen *et al.*, 1988; Narod & Bennest, 1990). If the fluxgate output is nulled completely, the current in the compensating solenoid is a measure of the magnetic field.

The sensitivity of fluxgate sensors is flat with zero phase at periods significantly longer than that of the measurement cycle, and the amplitude response of sensors used in geophysics is commonly in the range 0.5–10 mV/nT (Figure 9.6). This value is restricted by dynamic range constraints of the amplifier and recording electronics rather than by the sensor itself. A comparison of the sensitivity of fluxgate and induction coil sensors shows that induction coil sensors will generally provide greater sensitivity at periods of less than about 10^3 s, and fluxgate sensors will have superior sensitivity at longer periods (Figure 9.6).

Sources of instrumental noise in a fluxgate magnetometer can be divided into sensor noise, thermal drift and long-term drift (Narod & Bennest, 1990). Sensor noise arises in large part because ferromagnetic materials undergo magnetization changes in a series of small, discrete Barkhausen steps caused by magnetostriction. This noise can be minimized by an appropriate choice of core material. Narod *et al.* (1985) and other researchers have investigated the optimal materials for reducing this noise in ring-core sensors, and have found that metallic glasses consisting of alloys of Fe, Co, Si and B have suitable properties. Cores of such material show a $1/f$ spectral characteristic, and most ring cores have noise levels between 2×10^{-5} and 8×10^{-5} nT^2/Hz at 1 Hz (Figure 9.6). The typical noise level of fluxgate sensors is centered around this range. At 1 s period, the noise from fluxgate sensors is several orders of magnitude larger than that of induction coils, but this difference decreases significantly at longer periods. In practice, LMT fluxgate instruments are usually deployed for longer durations than BBMT induction coil instruments, and the resulting improvement in the signal-to-noise ratio means that, depending on signal levels, LMT instruments can provide reasonable MT responses to periods as short as a few tens of seconds at mid-latitudes, and a few seconds close to the equatorial and auroral electrojets.

Thermal sensitivity arises mainly from electronic circuitry, including the voltage reference (Narod & Bennest, 1990). In observatory environments where power consumption is not a critical factor, the thermal sensitivity can be reduced to 3 ppm/deg, which corresponds to 0.2 nT/deg for a 60 000 nT field. Field-based instruments designed to use lower-power components may have corresponding values of 10 ppm/deg and 0.5 nT/deg. In comparison, the sensor itself typically has a sensitivity of better than 0.1 nT/deg. Long-term drift arises from aging of sensor windings and other electronic components. Engineering data suggest that this drift may be around 50 ppm/yr or 3 nT/yr in a 60 000 nT field, but instrument tests have suggested that values of 1 nT/yr may be possible.

9.3.3 Additional types of magnetometer sensors

Several additional types of magnetometers are used in MT surveys or have been used in the past (e.g. Kaufman & Keller, 1981). A large volume of land LMT data have been collected over the past four decades using torsion-fiber magnetometers that are based on observing or nulling the deflection of suspended magnets. A large proportion of the extant deep seafloor marine MT data were also acquired using very carefully constructed torsion-fiber magnetometers (Filloux, 1987).

Superconducting quantum interference device (SQUID) magnetometers were used in the late 1970s and early 1980s (Clarke, 1977, and references therein). SQUID magnetometers use the Josephson junction effect in superconductors to measure the magnetic field. Their noise level is comparable to or better than that of many induction coils in the 10^{-2} to 10^2 s period range (e.g. Vozoff, 1991). Low-temperature SQUID magnetometers require a supply of liquid helium, so the instruments are not particularly portable. Their use in MT surveys was abandoned by the mid-1980s because of the difficulty of obtaining liquid helium in remote locations. Newer high-temperature SQUID magnetometers require only liquid nitrogen rather than liquid helium, providing greater portability than low-temperature SQUIDs, but they have higher noise levels (e.g. Bick *et al.*, 1999; Nabighian & Macnae, 2005; Nabighian *et al.*, 2005). High-temperature radio-frequency SQUIDs were developed in the late 1990s that are showing promise (e.g. Fagaly, 2006); however, they are an order of magnitude greater in cost than three high-quality coils, so will likely only be used in special cases where very low noise is essential.

9.3.4 Other components of magnetometers

The noise associated with 50 or 60 Hz power line signals and their harmonics occurs within the AMT and BBMT frequency ranges. This noise is usually dominated by the primary frequency and odd harmonics that arise from the departure of the waveform from a sinusoidal form (e.g. due to symmetric clipping of peaks). Even harmonics are usually, but not always, of smaller magnitude than odd harmonics, and will occur in the case of unbalanced signals when the waveform is asymmetric about its zero level. Power line noise is usually removed using comb filters acting on the primary frequency and its harmonics. In some MT systems, only odd harmonics are included in the filter, which limits the application of the system to regions with balanced power grids. Low-pass filters are used to remove high-frequency noise, such as VLF radio-frequency signals, from the recorded magnetic data. The roll-off of the filters may be weaker for BBMT and LMT recordings. For AMT recordings made within about 100 km of a VLF transmitter, it is important for the filters to have a steep roll-off in order to record signals at periods of, or smaller than, 10^{-4} s.

9.3.5 Field deployment of magnetometers

In most land MT surveys, the magnetic sensors are buried at a depth of between 30 cm and 1 m to minimize the effects of temperature variation, microseismic shake and wind noise. The components of two sensors are usually aligned in the magnetic or geographic north and east directions, with the third sensor aligned vertically in three-component recordings. Compasses, levels and surveying methods should be used to ensure that the sensors are aligned to within better than $1°$ of the required azimuth. In LMT and BBMT surveys, it is also critical for the sensors to be installed in firm soil to minimize long-term drift resulting from changes in sensor orientation, such as slow changes from soil settling, or rapid ones from "traffic" (pedestrian and vehicular) in the area. An instrument rotation of $0.1°$ can produce spurious signals of up to 100 nT in a 60 000 nT field. In seafloor surveys, the vertical orientation of the instrument is controlled using weights attached to the instrument. If the total static magnetic field is measured by the instrument, the horizontal orientation of the instrument can be determined relative to magnetic north using the recorded data. However, if only magnetic field variations are measured, the orientation must be established using a compass system.

In all types of magnetometers, care must be taken to minimize magnetic field noise at the sensors created by the recording electronics and by current in the wiring connecting the sensors and recording units. This noise is of greatest concern in MT systems in which the sensors are located close to the recording unit (e.g. in some seafloor MT systems; Hoversten *et al.*, 2000). In older systems, without solid-state memory, noise also arose from writing to hard disks and magnetic tapes.

Most MT instruments include temperature sensors to record temperature variations in the recording unit and/or the sensors. These data permit evaluation of possible temperature effects in the data, but, because of the varying temperature dependence of the different components of the system, they are not normally used to make corrections to the data.

9.4 Magnetotelluric data recording

9.4.1 Digitization and dynamic range

As with other equipment, the number of bits used in the digitization and recording of the data affects the dynamic range of the instruments. Modern broadband MT systems have 24-bit digital resolution, providing a dynamic range (defined in terms of signal power) of over 130 dB. Older digital systems are 16-bit or less, with a maximum dynamic range of approximately 90 dB. The analog paper recording systems of the 1970s and earlier had an equivalent resolution of 12 bits or less.

For LMT magnetic recordings, the increased dynamic range of 24-bit systems provides the very important advantage of being able to record the total static magnetic field without the need to remove a baseline value for each component. For a maximum total field value of 100 000 nT, a 24-bit system allows recordings using a least count of 6 pT. This value corresponds to a noise level of 5×10^{-6} nT^2/Hz at a period of 1 s, which is just lower than the inherent noise of good fluxgate sensors (Figure 9.6), and therefore provides optimum data collection. In contrast, a 16-bit system using the same least count might record maximum magnetic field fluctuations of about 400 nT. In general, 16-bit systems use a somewhat larger least count and a maximum recording range of around 1000 nT.

In BBMT and AMT systems, AC coupling of the induction coil magnetometers means that the large magnetic static field is no longer a constraint in the dynamic range consideration. Recording of the time derivative of the magnetic field also flattens the spectrum, so that the variation of the spectrum, in $(nT/s)^2/Hz$, is less than several decades. In this situation, the larger dynamic range of 24-bit systems allows the accurate recording of narrowband noise signals superimposed on the background, along with accurate recordings during both magnetically disturbed and quiet times.

Similar considerations apply to the dynamic range of electric field recordings. From Figure 9.3, the dynamic range of land LMT recordings should be defined so that the accuracy of recordings at 1 s period is 10^{-15} V^2/Hz or better, corresponding to fluctuations of 0.1 μV (assuming bandwidth equal to frequency). For a least count of 0.1 μV, a 24-bit system provides a maximum value of 1700 mV, which is large enough to accommodate most observed variations in electrode self-potential. In contrast, a 16-bit system with this least count would allow accurate recording of variations up to 6 mV, so, to allow for reasonable variations in electrode self-potential, it is necessary to use a larger least count. In order to allow for variations in the level of the electric field from site to site (e.g. Figure 9.1), many MT systems include a gain setting for the electric field recordings that adjusts the level of signal amplification. For seafloor recordings, it is necessary to use a smaller least count and record the fields with greater accuracy since the electric field and electrode noise are both lower than in land recordings.

9.4.2 Data acquisition control and storage

In modern MT instruments, data acquisition is controlled by an on-board microcomputer. A range of microcomputer chips are employed. Older-generation microcomputer chips offer more flexible program control on input and output and lower power consumption, whereas newer chips provide greater ease of data transfer with field computers. For each acquisition run, the program to be executed and/or a table of acquisition parameters is downloaded to the MT equipment from a field computer or from a memory card or other storage device.

One of the most important advances in MT instrumentation in the past two decades has been the introduction of GPS-synchronized data acquisition. All commercial land MT systems now use a clock system synchronized to GPS. This system provides a time accuracy of several hundred nanoseconds (±340 nanoseconds to be precise), and allows effective remote reference recording to be done at AMT frequencies without any need for a wired connection between instruments. Seafloor MT surveys employ GPS signals to synchronize timing of the instruments at the start and end of deployments. In older seafloor equipment, a crystal oscillator is used to control the timing during each deployment. The recent development of a low-power (100 mW), chip-mounted, rubidium clock, with three orders of magnitude better stability than a quartz oscillator, has greatly increased the accuracy of timing. The timing drift should be less than a few milliseconds over seafloor BBMT deployments to permit determination of accurate MT phase responses at 0.1 s period, and it should be less than about 100 ms in seafloor LMT recordings in order to permit accurate phase responses at 10 s period. In both cases the maximum drift should be on the order of 1 ms/day.

MT data are now recorded with solid-state memory such as SD memory cards, and are retrieved by either removal of the memory cards or by uploading the data to a field computer. The progressive decrease in the physical size and cost of storage media has allowed a major increase in the collected data volume. Many BBMT and AMT surveys now involve several hundred megabytes or more of MT time series at each site. LMT surveys with 1–10 Hz sampling and recording durations of up to one month involve data collection of close to one megabyte per site. Data are generally recorded in binary format to minimize their volume. Commercial systems often use a proprietary data format requiring specialized instrument software from the company for decoding.

9.4.3 Sampling rates, frequency windows and recording strategies

Most LMT instruments record continuous time series with a fixed sample interval typically between 0.1 and 1 s. There is little to be gained from using shorter sampling periods. Typical signal levels in the MT dead band around 1 s period are lower than the sensitivity of most

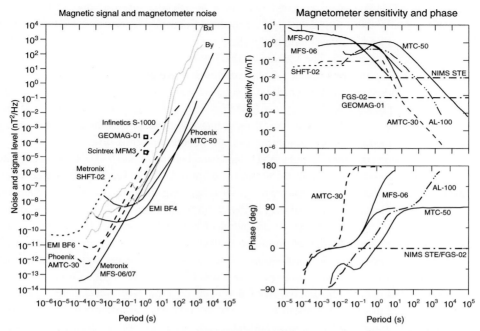

Figure 9.6. Relative noise level, sensitivity and phase response of some induction coil and fluxgate magnetometer sensors. Results are shown for BBMT induction coil sensors (solid lines), AMT induction coil sensors (long-dashed lines), RMT induction coil sensors (short-dashed line), air-loop induction coil sensors (double dot-dashed line) and fluxgate sensors (single dot-dashed line). Curves that split the branch with lower sensitivity and higher phase at long periods correspond to recordings without a chopper, and the other branch corresponds to recordings with a chopper. Magnetic field signals levels recorded at PLVO in Canada (Figure 9.1) are shown in gray for comparison with magnetometer noise levels. Responses are shown: for Phoenix Geophysics AMTC-30, MTC-50 and airloop AL-100 sensors (Phoenix Ltd website); for Metronix (Cooper Tools) MFS-06 and MFS-07 BBMT sensors, SHFT-02 RMT sensor and FGS-02 with GEOMAG-01 fluxgate sensor (Metronix website); for EMI BF4 and BF6 sensors (EMI Inc., 2002); for a noise level corresponding to the geometric mean of 100 Infinetics S-1000 fluxgate sensors (Narod & Bennest, 1990); for the sensitivity of a NIMS STE fluxgate sensor (B. Narod, 2000, pers. comm.); and for a Scintrex MFM3 design fluxgate (Edwards *et al.*, 1985).

fluxgate sensors (Figure 9.6). The use of recording durations of several weeks or more and robust data processing will improve the MT responses from fluxgate systems (Booker & Narod, 2002), but the lower limit of useful responses is still typically observed to be between 2 and 10 s period. Induction coil systems must be used if accurate responses are required at periods of less than about 2 s, and more typically 20 s in mid-latitudes during normal solar activity.

AMT and BBMT systems typically use more complicated recording arrangements involving two or more frequency ranges. Recording in each frequency range uses different base-frequency sampling rates and recording windows, and sometimes different amplifier and A/D boards. Acquisition configurations vary between instrument manufacturers. As an example, the Phoenix MTU-A system collects AMT data in a low-frequency band with 150 Hz sampling and continuous time series, an intermediate-frequency band with 2.4 kHz sampling in a series of 1 s long windows, and a high-frequency band with 24 kHz data recorded in a series of 0.1 s long windows. The Metronix AD-07 system allows simultaneous acquisition of LMT and BBMT data. For seafloor BBMT recordings, there is no need to record at periods shorter than 0.1 s because of the attenuation of signals at these periods by the conductive ocean. Hoversten *et al.* (2000) describe a seafloor BBMT system with a sampling frequency of 25 Hz. With this sampling frequency and typical recording durations of less than 48 hours using modern storage media, it is straightforward to collect continuous time series.

The background signal levels in the AMT dead band between 1 and 5 kHz are typically one or two orders of magnitude less than the sensitivity of most AMT field sensors (Garcia & Jones, 2002). Various strategies have been proposed for optimizing the temporal range of collected AMT data, including acquisition of data in nighttime hours when AMT signals are stronger (Garcia & Jones, 2002) and threshold triggering of buffered recordings (Korepanov *et al.*, 2008b). However, threshold triggering must be applied with care, as there is the risk of focusing acquisition on large-amplitude noise signals or on signals from nearby sources that are not quasi-uniform. Garcia & Jones (2005) propose a hybrid method, with acquisition of telluric-only stations during the night, and full AMT during the daytime. New wavelet-based processing methods can, in some circumstances, pick out the very weak signal among the dominant noise (Garcia & Jones, 2008), but these specialist approaches are not universally applicable.

In CSAMT and RMT recordings, the MT response is measured at a series of discrete periods. The MT instrument is tuned to the required frequency by hand or through automated electronic control, and then the amplitude of the magnetic and electric field components and their phase differences are recorded. Some earlier RMT instruments provide only the amplitude response and no phase information (e.g. Tezkan, 1999). The components of the MT response can be computed directly by the microprocessor on the CSAMT and RMT instruments, removing the need for a permanent recording of the time series (Pfaffhuber, 2001).

9.4.4 Power requirements and batteries

Most commercial land BBMT and AMT systems have typical power consumptions of between 5 and 15 W, and draw around 1 A from a 12 V supply. They are typically operated for one to two days using one or two high-quality wet-cell car batteries with a capacity of 45 A h or greater. In contrast, LMT systems use a maximum of 1–2 W, drawing 0.1 A, and must be able to record without service visits for one week or longer (e.g. Narod *et al.*, 2001; Korepanov *et al.*, 2008b). These systems usually require one wet-cell battery per week of recording. Typically, batteries at LMT installations are larger (up to 120 A h), and are the marine deep-cycle type to give maximum performance. The use of solar panels to power LMT systems is becoming more common, particularly in long-duration and remote recording locations in which weekly service visits would be difficult or expensive. Solar power has its own set of attendant problems, including noise generation and security issues.

For seafloor BBMT recordings, the lower sampling rates compared to land BBMT recordings allow lower-power instruments (<1 W) to be constructed (Hoversten *et al.*, 2000). With battery volume and weight being of greater concern in seafloor surveys, these systems may be powered by banks of rechargeable batteries, providing recording duration of up to a week, and sets of lithium batteries, providing recording durations of up to a month. Seafloor LMT instruments with low power consumption can use lithium battery packs to achieve required recording durations of multiple months to several years.

Steady improvement in battery technology over the past two decades has enabled the use of smaller and lighter power sources to achieve the required recording duration. It has also become possible to use sealed batteries that are certified for transport by airfreight, which is an important consideration for surveys in remote locations using lithium batteries. Further improvements in battery, solar panel and battery charger technology will likely lead to steady gains in MT recording duration.

9.4.5 *Telemetry and distributed acquisition systems*

Radio telemetry has been used quite commonly in AMT and BBMT studies to allow a central recording unit to acquire data from distributed sensors (e.g. Clarke *et al.*, 1983; Romo *et al.*, 1997). Radio telemetry was particularly useful, prior to the availability of GPS synchronization, for recording synchronous data from a main MT site and a remote reference site. Telemetry also enables separation of magnetic and telluric sensors. For example, in recordings in the Slave Province in northern Canada, telluric sensors were deployed through ice-covered lakes onto the lake floor, whereas magnetic sensors had to be located on the lake shore in order to reduce vibration noise (McNeice & Jones, 1998). Satellite and Internet telemetry enable LMT and observatory-style MT recordings to be made for very long durations without the need for site visits to download data. Such recordings also permit real-time monitoring of the data and the status of the recording equipment (e.g. Eaton *et al.*, 2005; Schultz *et al.*, 2008). This telemetry also permits the provision of very distant remote reference data for shorter-period surveys (e.g. Yamashita & Fox, 2008).

Telemetry also enables the collection of more than the normal four or five magnetic or electric field components per site. In most land locations, spatial variation of the horizontal components of the magnetic field is much more gradual than the spatial variations in the electric field (Jiracek, 1990), and it is most efficient to record using more broadly spaced magnetic sensors and more densely spaced two-channel electric field sensors. A similar approach can be applied to vertical magnetic field recordings (e.g. Fox *et al.*, 2008). Telemetry allows such measurements to be made using a smaller number of MT instruments.

In some recent geophysical acquisition systems, such as MIMDAS and the Quantec Geoscience Titan 24 system, MT measurements are combined with DC resistivity and IP measurements (e.g. Sheard, 1998; Garner & Thiel, 2000; Nabighian & Macnae, 2005; Goldie, 2007). Data are collected simultaneously at multiple sites using a distributed acquisition system in which 24-bit data acquisition units for each sensor are networked to a central recording unit. The systems enable collection of data for several IP and DC array types, and provide increased depth of resolution for these methods.

9.4.6 *Magnetotelluric instrument calibration and instrument noise evaluation*

MT equipment must be accurately calibrated. For BBMT and AMT surveys, it is necessary to conduct a frequency-dependent calibration. For LMT surveys, the system response often has a negligible period dependence in the period range of interest, so it is possible to calibrate at a single period such as the DC response. For LMT systems with an analog high-pass filter, the calibration is often based on the theoretical response of the filter.

MT instrument calibration first requires that the recording unit respond correctly to input signals. This involves checking that a known reference voltage input, representing a sensor output, is recorded at the correct level. The second part of the calibration involves defining the response for each sensor including any associated preamplifiers and filters. Magnetic induction coil sensors are usually calibrated using a secondary coil that creates an accurately known magnetic field. The calibration is done for a series of discrete frequencies and the observations are fitted with either an appropriate response function or an equivalent series of filter poles. For the most reliable results, it is of critical importance to ensure that the data are fitted appropriately well by the response function and that the range of periods used in the calibration fully covers the period range of the MT response.

For BBMT and AMT systems, and LMT systems with significant period dependence of the response, the calibrations are usually applied to the frequency-domain response during data processing. It is rare for fully calibrated time series to be computed. However, for LMT data for which the instrument response is flat, the recorded time series will correspond to true electric and magnetic field variations once the frequency-independent calibration has been applied.

It is important for MT systems to be tested for instrument noise levels prior to their use in an MT survey. These tests can be done at several levels of sophistication. First, the system can be used to make recordings in an area in which electromagnetic noise sources are at a relatively low level. Spectra derived from the recorded time series can be examined for indications of noise, such as spectra that are asymptotic to a white or $1/f$ form at high frequencies, providing an indication of random noise or electronic noise that is limiting the response. Note that it is very important to apply appropriate windowing during these tests to minimize spectral leakage, and it is important to consider the absolute limit of resolution defined by the instrument least count and sampling frequency. The spectra can be examined for other sources of narrowband noise such as at the basic frequency of operation of the associated microprocessor and its harmonics.

Second, the instrument response can be compared with that of a second instrument. In side-by-side coil tests, the difference between the spectra of time series recorded by closely spaced parallel coils is evaluated and provides an indication of the frequency-dependent noise level of the coils. A similar approach can be used for electrodes, although it is more common for the noise to be assessed using a DC approach in which the voltage is measured between electrodes placed side by side on cloth or paper moistened with a saline fluid.

Finally, the overall response of an MT system can be assessed by comparing the full response measured by two instruments at the same location. Although this approach does not provide information about which components are causing the noise, it has the advantage of testing the overall system at one time. Ideally, the tests should be done on a side-by-side basis in order to ensure comparable signal levels for the two instruments.

9.4.7 Common magnetotelluric site layout errors

Problems associated with the layout of an MT site can lead to errors in the computed response. The errors may be quite subtle, such as errors of a few percent associated with slightly incorrect sensor orientation, or very obvious, such as a 180° phase change when electrodes are connected in reverse.

Magnetic sensors and electrode lines should be installed to an angular accuracy of better than 1°. Care must be taken to use geomagnetic or geographic coordinates systematically, and in the second case to set the declination correctly on all compasses used in the field. For large-scale surveys in which there is significant variation in magnetic declination across the survey, it is safest to use magnetic coordinates throughout. Installation of a site using a consistently incorrect angular reference will result in a corresponding error in the geoelectric strike azimuth determined for the site. Installation of a site with a differential angular error between the magnetic and electric fields will result in an apparent galvanic distortion, with the Swift skew angle estimate equal to the differential angular error. Errors in the measured length of electrode lines will map to the proportional errors in the response magnitude. Because the apparent resistivity is proportional to the square of the MT response, its percentage errors will be approximately twice as large. The phase response will not be affected.

Even the most experienced MT practitioners occasionally install a site with an incorrect magnetic sensor orientation or electrode configuration. A common error involves installing coils or fluxgate sensors 180° from the correct orientation, and can be recognized by a 180° phase shift in the response. This problem can easily be corrected during data processing. It is also possible to recognize when electric field lines are connected with reversed polarity. Although it is possible to identify, it is impossible to correct data for a situation in which one north or south electrode line is switched with one east or west line. In this situation the electric field is measured between two parallel or antiparallel dipoles, and one component of the electric field is therefore not measured. The situation is recognized in data from galvanic distortion in which there is a Groom–Bailey shear angle of +45° or −45°.

9.5 Magnetotelluric field procedure: site selection

There are a number of factors involved in the selection of a good MT site, including the following:

- its distance from the planned location in a profile or array;
- appropriate physical conditions for site installation;
- distance of the site from sources of electromagnetic noise;
- distance of the site from natural structures other than the survey target (geological noise);
- distance of the site from man-made structures;
- availability of permission from the land-owner to install a site; and
- security of the site from interference by humans and animals.

Even for experienced MT practitioners, finding an MT site that completely satisfies all of these criteria is nearly impossible, so in most situations the location chosen for a site represents a compromise between the listed factors. It is therefore important to have a good understanding of how these aspects affect the measured MT responses.

9.5.1 Physical requirements of magnetotelluric sites

A good land MT survey site should be relatively flat to enable accurate positioning of sensors and minimize potentials caused by the electrokinetic effect of percolating water. If possible, the site should be fairly free of large trees that can cause ground vibration during strong winds, restrict the layout of the electric lines, and possibly cause bioelectric potentials. However, competing requirements, such as the need to install the MT equipment in a shaded environment in very hot areas, or the need to increase site security, may make deployments near large trees more desirable.

The soil at the site should be comparatively thick. It should be possible to bury magnetic and electric sensors to at least 30 cm depth, and preferably to 1 m depth for BBMT acquisition, to minimize temperature and moisture variations and the direct effect of wind-induced motion on the sensors. This requirement often makes it difficult to find suitable sites for co-located MT and teleseismic measurements, as seismic sites are ideally located on, or very close to, bedrock. The soil should be uniform over the survey area to reduce the effect of natural electrode potential variations and local distortion of the electric field. It should also provide appropriately low electrode contact resistance: loam and clay soils will usually provide low resistance, but it is sometimes difficult to obtain comparable values in clean sands. Finally, the soil should ideally have an intermediate moisture content that is sufficiently low to allow the stable installation of sensors, but sufficiently high to ensure that there is no excessive drying of electrodes.

9.5.2 Electromagnetic noise

MT sites should be located as far as possible from sources of electromagnetic noise. Modern MT equipment with effective filters and large dynamic range, and improved data processing methods including remote referencing, bounded influence methods and effective windowing in spectral analysis, have increased the tolerance to noise. However, high levels of noise may still degrade responses because of amplifier saturation, increased stochastic uncertainty on spectral estimates due to a decreased signal-to-noise ratio, and spectral leakage from narrowband noise sources. Sources of noise affecting MT surveys were reviewed in detail by Szarka (1988) and Junge (1996), and other studies have examined the effect of particular sources of electromagnetic noise on MT measurements.

In modern studies, the increased volume of collected MT data has meant, by necessity, that detailed examination of MT time series and responses for electromagnetic noise has become less common. Many surveys now rely only on robust remote reference response function estimation to reduce the effect of noise and bias, and utilize the observation of large data misfit in inversion models to identify site responses strongly affected by them. However, it remains important to examine data from at least representative sites for noise (e.g. Szarka, 1988). Visualization methods such as those described by Weckmann *et al.* (2005) can assist in this process, and multivariate data processing methods such as those described in Chapter 5 or by Egbert (2002) can help to identify pervasive quasi-uniform noise. In the absence of detailed noise studies, it becomes more important to be aware of possible sources of electromagnetic noise and their effects on the data.

In land MT surveys, the main sources of electromagnetic noise are artificial sources, such as electrified railways, power systems and electric fences. The levels of magnetic field noise at a given distance from a particular source will tend to be larger in terranes with resistive surface materials (Szarka, 1988), such as granitic terranes with thin glaciated soils, and lower in terranes with conductive surface materials, such as basaltic terranes with tropical regolith. However, electric field noise may persist to a greater distance from a source in conductive areas. The effects of the noise will be greatest when signal levels are lower, such as during quiet geomagnetic periods, and in areas of conductive lithosphere, where the electric fields are smaller. Different instruments will exhibit different tolerances to noise. For example, an instrument with a large dynamic range will be able to accommodate higher noise levels without saturation. Vozoff (1991) recommends locating sites at least 1 km from active sources, such as railways and power lines, but, as will be discussed below, much larger separations are needed for some sources of noise, and the remote reference site must be chosen with great care.

9.5.2.1 Power lines

The most pervasive source of cultural electromagnetic noise in MT surveys is the 50 or 60 Hz noise from AC power transmission and distribution systems.

Most AC power transmission systems are three-phase systems in which separate conductors carry signals separated by one-third of a cycle. In lower-voltage systems, there may also be a neutral line. In a perfectly balanced system, the phase currents sum to zero, hence canceling each other out, and there will be no active electromagnetic noise produced in the Earth (e.g. Ádám *et al.*, 1986; Szarka, 1988). However, most transmission systems are unbalanced, resulting in a net current called a zero sequence current that is common to all three conductors. The corresponding return current will flow on the neutral line, or, in the absence of a neutral line, as a ground return following the path of the overhead wires, and will create noise for MT surveys. Connections from the power grid to houses or other facilities may be single-phase and cause more significant electromagnetic noise. Single-wire

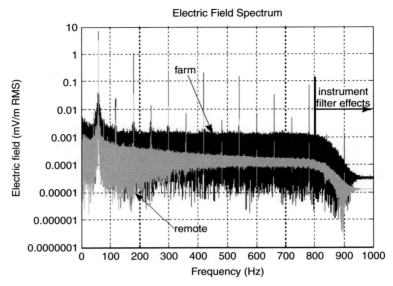

Figure 9.7. Electric field noise due to 60 Hz AC power lines on a farm and at a remote site for comparison (Pellerin *et al.*, 2004). The data at the two sites were recorded simultaneously with a 2000 Hz sampling rate.

earth return (SWER) systems, in which the return AC current flows through the earth, may be used for supplying single-phase power from a grid to rural or remote areas.

Distortion of the power line signal from a sinusoidal form, and imbalance of three-phase signals, results in additional noise at harmonics of the base frequency (Figure 9.7). Such effects can be caused by the nonlinear response of conversion elements such as transformers, imbalance in the phase currents, asymmetric loading of the network and geomagnetic effects on the power system. It is common for strong signals to be observed at 10 or more harmonics (e.g. Szarka, 1988; Pellerin *et al.*, 2004), and significant effects may extend to frequencies above 1 kHz, where they become even more important due to the lower signal level in the AMT dead band (Hoover *et al.*, 1978). The odd harmonics (e.g. 180 Hz, 300 Hz, etc., in a 60 Hz system) are often larger than the even harmonics. Power lines may also carry subharmonics of the base frequency (e.g. Junge, 1996; Trad & Travassos, 2000).

Variations in the magnitude of electromagnetic noise due to a time-varying load will cause modulation and broadening of spectral peaks, and in some networks the frequency itself is affected by the changes in load (Szarka, 1988). Pellerin *et al.* (2003) describe an example from a farm, where increasing load causes not only an increase in the magnitude of the harmonic peaks in the electric and magnetic fields, but also, because of spectral leakage, an increase in the background spectrum between the peaks.

High-voltage DC (HVDC) transmission lines transfer large amounts of power over large distances more efficiently than AC systems. In bipole HVDC systems, each of two conductors is at high potential relative to the ground, but with opposite polarity. In monopole

HVDC systems, the return current flows in the earth between ground electrodes at the converter stations. Bipole systems, without a third metallic conductor, may also operate in this mode if one line fails or is removed from the circuit for maintenance. The earth return current in monopole HVDC systems, and currents arising from imbalance of the currents in bipole systems, will produce electromagnetic noise in MT surveys. Strong electric fields may arise within tens of kilometers of the grounding electrodes. Noise may also arise from transient changes in the current levels carried by the HVDC systems. DC power lines may also carry AC signals that can affect MT surveys. Hoover *et al.* (1978) describe the noise arising from 60 Hz AC signals carried by a 400 kV DC transmission line in Nevada, USA, and Ferguson *et al.* (1999) describe noise arising from a broadband centred at 4 Hz, signal carried by a 500 kV bipole system in Manitoba, Canada.

Power line noise is usually observed most strongly in MT data in the electric field parallel to the line and in the vertical magnetic field. Figure 9.8 shows the fields beneath a three-phase AC line. The response is more complex than that of a single-phase system because the magnetic field would be tangential to the current. For the three-phase system, a strong magnetic field occurs within 50 m of the power line due to the asymmetric position of the wires (Szarka, 1988). At greater distances, the magnetic field is approximately vertical, and both the electric and magnetic fields drop off more slowly than the response close to the lines. Hoover *et al.* (1978) report a decrease in the electric field response of AC harmonics of 36 dB (a factor of about 4000) between sites 300 m to 7 km from the power line, and Chaize & Lavergne (1970) show that there is a decrease in the magnetic field magnitude due to transients in a 225 kV, 50 Hz AC line of 12 dB over about the same distance range. Trad & Travassos (2000) report the presence of AC power line noise, as well as its harmonics and subharmonics, more than 15 km from the source. This noise is interpreted to have

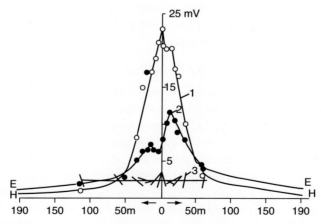

Figure 9.8. Electromagnetic noise crossing a three-phase power line (Takács, 1979, reported in Szarka, 1988). Profile 1 is the total magnetic field, profile 2 is the horizontal electric field, and profile 3 shows the orientation of the total magnetic field.

propagated through resistive Precambrian basement. Wannamaker *et al.* (2002) show the noise from a DC power line extending to distances of at least 40 km from the line.

In the presence of low-amplitude AC power line signals, the MT response often exhibits erratic values within an octave of the 50 or 60 Hz base frequency and possibly its first few harmonics. At sites containing large-magnitude noise that is not removed by remote referencing, the responses can show near-field effects. The near-field effects represent the response of an electric dipole, and include phase values close to $0°$ or $180°$, and apparent resistivity responses having a $45°$ slope on a log apparent resistivity versus log frequency plot (e.g. Qian & Pedersen, 1991; Wannamaker *et al.*, 2002; Pellerin *et al.*, 2003). For noise with a broad bandwidth (e.g. due to transients on a power line), this effect can be particularly strong in the MT or AMT dead bands, where the MT signal strength is lower (Iliceto & Santarato, 1999). The effect of power line noise on the MT tipper response can also be large, with the real component of the tipper exceeding unity. The effect of the reversal in the real component of the tipper caused by power line noise has the opposite sense to that of a conductor in the same location (Wannamaker *et al.*, 2002).

AC power line noise can be significantly reduced in MT measurements using notch filters. However, the MT site should still be located at a sufficient distance from the power line that the noise does not cause saturation of electronic components. Its effect on the computed responses is minimized if there is a remote reference site that is relatively uncontaminated. However, because of the ubiquitous nature of the noise, it is common to experience difficulty in finding a suitable remote site. In such situations, various filtering methods, including digital notch filtering, time-domain delay line periodic filtering and wavelet filtering, can produce significant improvement in the responses (e.g. Hoover *et al.*, 1978; Szarka, 1988; Trad & Travassos, 2000; Pellerin *et al.*, 2003). DC power line noise can also be reduced by processing the data using a good remote reference, but once again the remote site must be chosen carefully. In northern Manitoba, Canada, it was found that the remote site needed to be more than several tens of kilometers from DC power lines to effectively reduce the noise. Because of the broadband nature of the noise caused by DC power lines, appropriate filtering methods must be used.

9.5.2.2 Railways

A significant proportion of the world's rail transport is driven by electric traction, and electrified railway systems can cause major problems for MT surveys. Urban and local electric railway systems tend to use DC power systems, whereas long-distance systems use either DC or AC power systems. In DC systems, currents flow in a circuit that involves overhead power lines or a third rail, the rail vehicles, the main two rails and the earth. If the rails were perfectly insulated, the currents in the rails would balance those in the power lines, and the field would cancel at distances of more than a few kilometers. However, significant leakage occurs between the rails and the earth, and the resulting noise in MT measurements is caused by time variations in the current system due to train movement between grounding points on the tracks (Szarka, 1988; Iliceto & Santarato, 1999; Pádua *et al.*, 2002; Pirjola *et al.*, 2007; Lowes, 2009). AC power systems typically use single-phase currents and

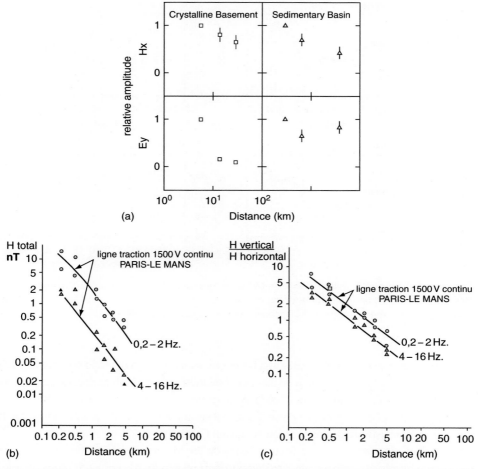

Figure 9.9. Distance extent of DC railway noise. (a) Relative decay of magnetic and electric field noise from a 1500 V DC railway in southeastern Brazil (Pádua *et al.*, 2002). (b) Magnetic field decay from 1500 V DC railway in France for two different frequency ranges, and (c) orientation of the magnetic field noise (slightly modified from Chaize & Lavergne, 1970).

frequencies of 50 or 60 Hz, or integer fractions of these frequencies (e.g. $16\frac{2}{3}$ Hz and 25 Hz). The resulting noise will include a strong component at the operational frequency and its harmonics, as well as other components due to the time variation of the currents.

The noise from electric railways can extend hundreds of kilometers from the source (e.g. Egbert, 1997) depending on the upper crustal resistivity. Figure 9.9 shows observations of the levels of DC railway noise from Brazil and France. Magnetic field noise is measured to distances of more than 120 km from rail lines in resistive crystalline terrains (Pádua *et al.*, 2002), and is present in both the vertical and horizontal components. Electric field noise

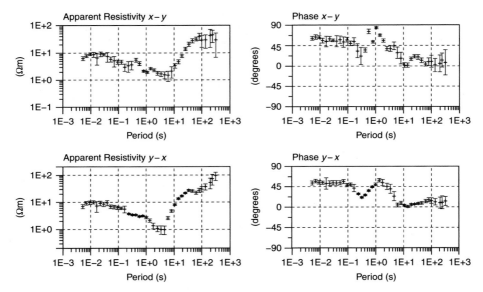

Figure 9.10. Effect of DC railway noise on MT responses (modified from Iliceto & Santarato, 1999). Both off-diagonal components show strong near-field noise effects in the MT dead band with phases approaching zero and the slope of the apparent resistivity approaching 45°.

decays with distance from the source at a faster rate than the magnetic field noise in resistive terrains, but persists for very large distances in conductive rocks such as sedimentary basins adjacent to the source (Pádua *et al.*, 2002). A more extreme example is seen in central South Africa, where DC trains produce noise that is coherent over hundreds of kilometers due to the very high resistivity of the Archean Kaapvaal Craton.

The effect of broadband noise from railways on the MT responses is a typical near-field response with phase approaching zero and apparent resistivity having a 45° slope on a log apparent resistivity versus log frequency plot (e.g. Egbert, 1997; Iliceto & Santarato, 1999; Oettinger *et al.*, 2001). As shown in Figure 9.10, the effect is typically most evident in the MT dead band. A variety of processing schemes have been applied to MT data to reduce the effects of railway noise, including robust and multi-site spectral analyses (Egbert, 1997, 2002; Oettinger *et al.*, 2001).

MT surveys in the vicinity of electrified railways must be planned very carefully. In order to obtain unbiased responses, particularly in the dead bands, it will be necessary to employ either a specialized multi-site acquisition and processing scheme (Egbert, 1997, 2002; Oettinger *et al.*, 2001) or a remote reference at sufficient distance not to be contaminated by the railway signals. The remote reference may need to be more than 100 km from the main site, and in the case of South Africa more than 500 km. Non-electrified railways may also carry electrical signals for controlling level crossings within a particular range of track or within a particular distance from a train. It is recommended that MT sites be located at least 1 km from such railways, and ideally the remote reference should be located at least 10 km away.

9.5.2.3 Pipelines

A cathodic protection system is usually used to make the surface of a pipeline the cathode of an electrochemical cell to control corrosion. In passive cathodic protection systems, the metal infrastructure to be protected is simply connected electrically to a more easily corroded metal. In contrast, larger-scale infrastructure such as pipelines use an active system of electrical protection, creating a more significant noise source for MT surveys (Szarka, 1988; Junge, 1996). The active method uses a cathodic protection rectifier attached to the AC power grid to apply a pulsed potential to the pipeline. The output of the rectifier is typically rated at 10–50 A and 50 V, but the actual current strength is controlled by the desired potential between the pipeline and the soil, and will therefore vary during geomagnetic disturbances or in the presence of other current sources (Szarka, 1988). The parasitic electromagnetic fields created by the cathodic protection system may include both periodic and broadband components (Figure 9.11).

Noise from cathodic protection systems has been observed in a number of electromagnetic surveys in land and marine environments (Szarka, 1988). On land, the signals appear to have a typical magnitude of around 10 nT near the pipeline, and decreases as the linear response of a line current with distance. In general, the electromagnetic noise reaches relatively low levels at a distance of 500 m from the pipeline, and it should be possible to

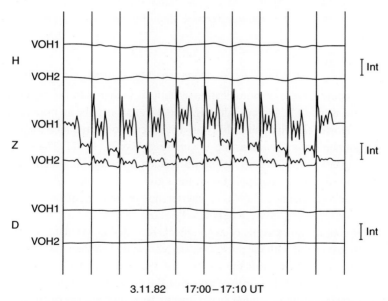

Figure 9.11. Magnetic field variations measured at sites 90 m (VOH1) and 490 m (VOH2) from a pipeline with a cathodic protection system (Junge, 1996). The figure shows 10 minutes of data.

obtain good MT data in these locations, especially if a good remote reference is available (Junge, 1996; Türkoğlu *et al.*, 2009). However, some studies do report larger noise effects (e.g. Ádám *et al.*, 1986).

Another source of noise associated with pipelines is high-voltage test signals applied to them to test the integrity of their coating. Oettinger *et al.* (2001) indicate that such signals can be observed up to tens of kilometers from the pipeline. The signals observed by Oettinger *et al.* (2001) consisted of 3 s pulses applied every 30 s. They cause broadband noise with typical near-field effects on MT responses.

9.5.2.4 Electric fences

Electric fences have caused problems for MT surveys in many different locations (e.g. Ádám *et al.*, 1986; Ingham *et al.*, 2001; Gowan *et al.*, 2009). A modern electric fence applies a high-voltage pulse, possibly as short as 10 μs, to a fence wire every one or two seconds. One terminal from the pulse energizer is connected to the wire and the second to a ground rod. An animal completing the circuit between the fence and the ground receives a shock. Noise will arise from the basic circuit, and its magnitude may vary with time if vegetation or moisture short the circuit.

The spectrum of electric fence noise may consist of both narrow spectral lines and broader-band components with a near-field response. The noise may be at least partially coherent between sites near electric fences on different farms, and it has been observed that a remote reference site on a farm 10 km from the main site may fail to remove the electric fence noise from the MT response. Ideally, MT sites should be located at least 1 km away from electric fences, and the remote reference should be at least 3–5 km from the nearest electric fence.

9.5.2.5 Radio communication systems

The base frequency of radio communication systems is higher than the frequency range of interest in conventional MT surveys, although these signals may be an important source for RMT surveys. However, a number of MT surveys have encountered increased noise at sites near radio communication broadcasting towers (e.g. Vozoff, 1991). The effects of radio transmission signals on MT recordings could include noise on the magnetic and electric fields due to low-frequency components in the radio signals, such as those produced by modulation effects (e.g. Junge, 1996), or interference of radio-frequency signals with the electronics of the recordings system (e.g. Clarke *et al.*, 1983).

Junge (1996) reports an example of radar signals recorded in MT data. The MT data will contain signals associated with the pulse frequency (usually less than 100 Hz) as well as the sweep frequency associated with the rotation of the antenna, but not the actual radio-frequency signals (Figure 9.12). It is of note that the noise can be represented in terms of signals from a magnetic dipole source, in contrast to the earlier sources of noise, which could be represented as electric dipole sources.

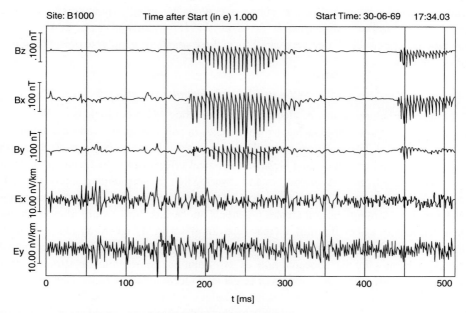

Figure 9.12. Radar sweeps recorded in MT data (Junge, 1996).

Care must be taken when applying low-pass filters to remove low-frequency communication signals from MT recordings. A number of VLF communication signals have frequencies around 20 kHz, and the corresponding filter drop-off must be sharp in order to avoid any attenuation of high-frequency AMT signals.

It is recommended that, where possible, MT sites be located more than 5 km from large radio communication sources, especially those in military establishments, and particularly in resistive terrains. A high-quality remote reference site should also be used in data processing.

9.5.2.6 Vehicles and other sources of electromagnetic noise

Relatively slow movement of ferric objects near MT recording sites can create magnetic field noise. The most common source is motor vehicles traveling on roads near MT sites, but marine measurements have also recorded electric field perturbations due to passing ships in shallow water (Junge, 1996). MT sites should be located at least 250–500 m from major roads and highways, and 100 m from smaller roads, in order to avoid significant effects. In general, high-quality remote reference data can completely remove noise due to roads.

A variety of other electromagnetic equipment can affect MT surveys. In rural areas, pumps and generators can affect MT recordings made within a distance of several kilometers. Soundings made in urban areas and in the vicinity of power generation stations, military establishments and active mining operations are usually quite noisy (e.g. Ádám *et al.*, 1986; Szarka, 1988; Qian & Pedersen, 1991; Weckmann *et al.*, 2005). Remote reference sites must be chosen very carefully in these situations in order to obtain reasonable MT responses.

9.5.2.7 Self-potential and electrokinetic effects

An important natural source of electrical noise affecting MT recordings is spurious variations in electric potential associated with changes in soil chemistry, water movement and biological factors. These sources cause electric potential differences between electrodes that are superimposed on the desired signals. Biological effects can be reduced by deploying electrodes away from trees and bushes and ensuring that the surrounding soil is reasonably free of organic material. The other sources of potential tend to be strongest after rainfall. Water moving through the soil creates potentials on the order of tens of millivolts through the electrokinetic effect. It is difficult to reduce this effect on MT recordings, but selection of sites on flat ground with the electrodes in the same type of soil will help to minimize the signals. Changes in the concentration of salt solutions surrounding electrodes creates potential changes of the same size as or larger than electrokinetic signals. Deeper burial of electrodes and use of the Russian bucket technique will help to minimize their influence. Other sources such as corroding bodies adjacent to the electrodes will also cause potentials that form noise on electric field recordings.

9.5.2.8 Motionally induced electromagnetic noise in marine magnetotelluric surveys

In marine MT surveys, a primary source of noise is motionally induced electromagnetic fields caused by the movement of ocean water through the Earth's static magnetic field. In both the electric and magnetic fields, the noise may have a magnitude comparable to or greater than the level of MT signals. See Chapter 2 for a detailed description of motionally induced electromagnetic fields.

9.5.3 Geological noise

Geological noise is defined as the effect on the MT response of spatial changes in electrical resistivity due to geological sources other than the survey target. Sources of geological noise with importance for land MT surveys include soil and regolith layers, swamps, topography and coastlines; and sources of geological noise with importance for marine MT surveys include the sediment layer, bathymetry and coastlines. The response from these features can affect the MT response, and may obscure the response of the target.

9.5.3.1 Soil and regolith

In most regions of the world, there is a surface layer of soil and weathered rock that affects electromagnetic measurements. This layer may vary from a thin soil in recently glaciated terranes, with an integrated conductance of only 0.01–0.1 S, to the thick regolith formed in tropical climates, with an integrated conductance of tens of siemens (Palacky, 1988). In most MT surveys, this layer is not the target of the survey and is a source of geological noise. It has a number of effects.

1. A conductive near-surface layer has the positive effect of decreasing electrode contact resistance. In thick conductive soils, the contact resistance is typically less than 500 Ω, whereas in resistive soils such as clean sands it may exceed 10 000 Ω. It also has the positive effect of more rapidly attenuating electromagnetic noise with distance from the source.

2. A near-surface layer containing inhomogeneities in resistivity can have the negative effect of causing galvanic distortion of the measured electric field (e.g. Jiracek, 1990; Torres-Verdín & Bostick, 1992; Groom *et al.*, 1993). This distortion can also affect the tipper response (Zhang *et al.*, 1993). In cases of very strong distortion, the horizontal magnetic field may also be affected at periods of less than a few thousand seconds (Chave & Smith, 1994). Chapter 6 discusses galvanic distortion in more detail.

3. A very conductive or thick near-surface layer will make it more difficult to resolve underlying targets because of electromagnetic induction in the layer. One effect is strong attenuation of the electromagnetic fields within it. A near-surface conductive layer can also have a significant inductive effect on the MT response to periods as long as 0.1–1 s. Depending on the uniformity of the layer, these effects may be 1D, 2D or 3D. For most resistivity structures, the response will mask the resolution of underlying zones that have higher resistivity (e.g. Zhang & Pedersen, 1991). As a general rule of thumb in MT surveys, it is not possible to properly resolve layers with a conductance less than that of an overlying layer.

9.5.3.2 Other near-surface conductive structures

A number of near-surface features may create similar effects to those noted above. For MT surveys targeting the crust, features such as swamps, saline aquifers, salt lakes and valleys containing deep sediment deposits can create conductive surface layers. Smaller features can cause galvanic distortion of the MT data, and larger features can reduce the resolution of the underlying resistivity structure. In larger-scale surveys, surface sedimentary basins may reduce the resolution of resistive parts of the underlying crust. Finally, in surveys targeting the mantle, conductive layers within the crust may limit the resolution of mantle features. For example, Jones (2006) shows that good crustal conductors can completely attenuate the TE mode of broadband MT data within the crust.

9.5.3.3 Topography

Steep topography surrounding an MT recording site can perturb the impedance (e.g. Ádám *et al.*, 1986; Wannamaker *et al.*, 1986; Jiracek, 1990). The geometry of the topography creates a lateral resistivity contrast formed between the air and the earth that allows charge accumulation to occur and cause a galvanic perturbation of the MT component with electric fields perpendicular to the topography (the TM component). The effect leads to decreased apparent resistivity values on hill tops and increased apparent resistivity values on valley floors. For a case in which there is a 100 m elevation change, a wavelength of 2400 m and a uniform resistivity of 100 Ω m in the subsurface, the hill top apparent resistivity will be decreased to approximately 80 Ω m and the valley floor apparent resistivity will be increased to approximately 120 Ω m (Jiracek, 1990). There may also be indirect effects associated with topography, such as the presence of a thick conductive layer in the valley floor. This layer may have a stronger effect on the TE mode response than on the TM response.

9.5.3.4 Effect of coastlines on land magnetotelluric surveys

The most prominent effect of a nearby coastline on land MT data is observed in the tipper or induction arrow response. In land surveys conducted near oceans, the localized electric current flow within the ocean creates a vertical magnetic field that can be observed in the magnetic field transfer function at distances of up to several tens of kilometers from the coast at short periods (<0.01 s) to several hundred kilometers from the coast at longer periods (>100 s). This effect is known as the coast effect (e.g. Parkinson, 1962).

The presence of the oceans will also affect both the TE and TM MT response. The effects of the coast on the TE mode, involving electric fields parallel to the coastline, will include changes to the electric field and the perpendicular horizontal magnetic field (as well as the vertical magnetic field). The lateral extent of these effects will depend to first order on the electromagnetic skin depth in the underlying material. In general, at distances of more than several tens of kilometers from the coast, the TE response will be affected to a smaller degree than the induction arrow response. For the TM mode, involving electric fields perpendicular to the coast, the effect will be mainly in the electric field, and will be caused by charge accumulation on the conductivity contrast at the boundary. The lateral extent of these effects will depend on the horizontal adjustment distance. This distance scale was originally defined by Berdichevsky & Dmitriev (1976) and Ranganayaki & Madden (1980). For a sequence of conductive rocks overlying resistive rocks, it is equal to the square root of the ratio of the vertically integrated conductivity of the conductive rocks to the vertically integrated resistivity of the resistive rocks.

In the vicinity of good local conductors located within a skin depth or adjustment distance of the coast, the effect of the resistivity contrast at the coast can become more complicated. For example, Jones & Garcia (2003) show how electromagnetic coupling can occur between the ocean and a conductor located 50 km from the coast.

9.5.3.5 Marine sediments

In marine surveys, the layer of unconsolidated seafloor sediments forms a conductor that can limit the resolution of the underlying crust and mantle. The conductivity of ocean sediments varies from around 2.5 S/m for unlithified, seawater-saturated sediments to 0.01 S/m for indurated materials (e.g. Fainberg, 1980; Chave & Cox, 1982). In many parts of the ocean, the layer of sediments is less than a few hundred meters thick, and the conductance of the sediment layer is of the order of 10 S. Globally, the conductance of the sediments is typically less than 10% that of the oceans (Kuvshinov *et al.*, 2002). However, there are some areas in which the integrated conductance of the sediments is much higher. Lilley *et al.* (1993) determined that the conductance of sediments in the Tasman Sea, off the southeast coast of Australia, exceeds 500 S, and Kuvshinov *et al.* (2002) note that, in the Gulf of Mexico, Arctic Ocean, Black Sea and Caspian Sea, the conductance of the sediments is comparable to that of the ocean layer. In these situations, the resolution of the MT method is typically limited to the detection of deeper conductive layers in the mantle or strong localized mantle conductors.

9.5.3.6 *Effect of coastlines and bathymetry on marine magnetotelluric surveys*

The coastline can be very important in marine MT surveys. For the TM response, the adjustment distance over which the continent affects the MT response has been observed in some marine surveys to be of the order of 1000 km (e.g. Ferguson *et al.*, 1990; Heinson & Lilley, 1993) at passive continental margins. This distance is reduced by the presence of conductive paths to the asthenosphere, such as in regions of active subduction (Wannamaker *et al.*, 1989; Tarits *et al.*, 1993), and in three-dimensional situations (Heinson & Constable, 1992; Constable & Heinson, 1993). Beyond several tens of kilometers from the coastline, the effect of the coast has a galvanic form with a frequency-independent decrease in the apparent resistivity response and no change to the phase response. The coastline also creates an anomaly in the seafloor TE response due to enhancement of the sheet current in the ocean adjacent to coastlines (e.g. Fischer *et al.*, 1978; Constable *et al.*, 2009). This enhancement extends several tens of kilometers from the coastline at long periods, and causes strong attenuation of the seafloor horizontal magnetic field perpendicular to the coastline, and in some cases a reversal of the sign of this component. These effects lead to strong lateral variations in both the impedance and induction vector responses (e.g. Ferguson, 1988; Constable *et al.*, 2009). Ocean bathymetry can also have a severe effect on marine MT measurements (Baba & Chave, 2005; Baba *et al.*, 2006; Constable *et al.*, 2009), and may involve a frequency-dependent distortion of the regional response. The distortion may involve coupling with mantle structures, requiring the use of sophisticated methods for its removal (Baba & Chave, 2005).

9.5.4 *Artificial resistivity structures*

In addition to natural resistivity variations, there are a variety of artificial conductive structures that may impact MT data even when unpowered, including pipelines, railways and metallic fences located near MT sites (Szarka, 1988). The significance of these features on the response will depend on the geometry and integrated conductance of the bodies relative to the distance from them and the background conductivity variations in the Earth. Where possible, MT sites should be located more than a kilometer from large-scale, well-grounded features such as railways and pipelines, more than 500 m from well-grounded metallic fences, and more than 100 m from smaller objects such as poorly grounded fences and metal tanks (e.g. Vozoff, 1991). In conductive areas, it is possible to be closer than these recommended distances, and in very resistive areas it is desirable to be further away. In all cases, it is important to note in field records the proximity of such features and to examine the MT response carefully for their possible effects. For example, a strongly distorted response with a local strike parallel to a nearby pipeline would provide a warning that the feature may have affected the measured response.

9.6 Fieldwork

There are a number of detailed instructions on conducting MT fieldwork available in instrument field manuals from companies and in university, government and company field manuals. Marine MT surveys require much more specialized planning and procedures. In the following description, the focus is on land BBMT and LMT surveys.

9.6.1 Survey planning and arrangement

9.6.1.1 Long-lead-time planning

Any MT survey must be planned carefully to ensure that the results optimally meet the objectives of the project. Fundamental survey parameters that must be defined well in advance of the survey are the required site distribution and spacing, frequency range, recording duration and remote reference site selection. Forward MT modeling of expected resistivity structures can help to ensure that the survey objectives are met. The planning must also consider the budget, time scale, human resources and instrument resources available, as well as major logistical constraints such as site access and weather conditions.

There are two main styles of MT data acquisition: contracted acquisition by an MT company, and "in-house" acquisition. In contracted data acquisition, it is advisable to formulate a detailed contract for the work. The contract should document aspects including:

- safety requirements;
- technical specifications of the equipment;
- MT site requirements, such as minimum electrode line length, accuracy of coil and electrode line azimuths, or depth of electrodes;
- the duties to be performed by the contractor (e.g. whether instrument calibration, site scouting, site permitting and site clearing are included);
- the amount of MT data processing to be done and the processing algorithms to be applied; and
- the expected deliverables from the survey.

It is critical to define the requirements on the MT responses, such as the minimum recording duration at each site, target maximum data errors over a specified frequency range, or a combination of these two aspects, and who is responsible for approving the completion of acquisition at a site. It is also critical to define how aspects such as noise and loss of productivity due to weather conditions or low signal levels are to be handled. As with other contracted geophysical surveys, it is generally effective for a person from the client organization to spend some time on the actual survey ensuring that the specified conditions of the survey are met. There needs to be daily communication between the client and the contracted company to discuss the completion of sites and methods for dealing with unforeseen situations. In-house MT surveys may be conducted by mineral exploration companies, government departments and universities. Similar considerations must be made to those listed for the contracted surveys. In all cases, it is very strongly recommended that the MT survey involve one or more people with previous MT experience. Acquisition of good-quality MT data requires attention to many details. Inexperienced operators and field personnel with rented MT equipment will generally not obtain high-quality MT data.

There are several important aspects of planning that should be considered in either contracted or in-house surveys. The first is the nature of the field crew or crews. An MT field team usually requires a minimum of two people and a field vehicle. A team of two provides for efficient site servicing and recovery, and reasonably efficient site installation in easy terrain. A team of three will install sites faster than a team of two, but any additional people may not lead to increased efficiency. It is important for an MT survey to have a clearly identified field chief with responsibility for making decisions and for interacting with the client in contracted surveys. A second important aspect of survey planning is defining the degree of data processing to be done in the field. Most AMT and BBMT surveys will involve the field processing of data files to at least the level of the Society of Exploration Geophysicists Electronic Data Interchange (edi) format (White, 1988). In-field examination of the MT data allows re-collection of data from faulty installations or noisy sites as well as updating of the survey in response to the available data. However, this processing requires an operator with appropriate expertise, and will typically require a minimum of one to two hours of work per MT site. A third important aspect of planning is defining the requirements and use of the remote reference site. With larger field crews, it may be possible to use one very good remote site, located within easy access of the accommodation base, for the whole surveys. With smaller crews and larger surveys, it may be necessary to remote reference adjacent pairs of sites.

9.6.1.2 Short-lead-time planning

Closer to the start of the survey, it is important to define additional aspects of site and recording configurations. For the site configuration, it is necessary to define the electrode line length, whether an "X" configuration or "L" configuration will be used, whether electrode lines must be fully buried, and so on. The MT recording parameters to be defined include the recording times and durations for each frequency band, the gain settings, the use of AC or DC coupling (high-pass filtering), and the configuration of any high-frequency (low-pass) filters. Some of these aspects may need to be adjusted during the survey in response to the measured data.

There are also some additional parameters that should be set for the whole survey. These include: (a) whether azimuths are to be recorded relative to magnetic north or geographic north, (b) whether GPS locations will be recorded in latitude–longitude or Universal Transverse Mercator (UTM) format, and which datum and UTM zone will be used for UTM coordinates, and (c) whether recording times will be defined in terms of the local time zone or Coordinated Universal Time (UTC).

Prior to the MT survey, it is important to design a working schedule indicating which sites are to be scouted, installed, serviced and recovered on each day of the survey. This schedule is particularly important if the survey involves multiple field teams and vehicles and/or long distances between sites. Inevitably, the schedule must be adjusted during the survey for unforeseen circumstances, including breakdown of vehicles, failure to properly install an MT site or unexpected noise at a selected site.

9.6.2 Site selection and permitting

The factors involved in MT site selection were discussed above. MT site selection and permitting can be integrated with site installation, but this approach can have drawbacks. It sometimes takes an unexpectedly long time to permit an MT site, for example, if land-owners cannot be found or if documentation must be prepared. This can lead to delays in installation, and may cause problems if it is critical to install a second site on the same day for simultaneous recording.

Modern technology has enhanced the site selection process. High-resolution digital aerial photography and satellite imagery now allow site selection to be started ahead of the field-work. Georeferenced satellite images, aerial photographs and topographic maps can be loaded onto a field computer and used to guide site selection in the field. This is particularly useful in rough or forested terrains where it is difficult to find flat and cleared areas. A real-time GPS system can also be used to track the vehicle position on these images or a topographic map.

The job of obtaining permission to use a site for MT recordings may take a significant amount of time. For example, permitting sites in national or state parks and nature reserves may require the preparation of appropriate documentation, including a statement describing the environmental impact of the recordings. Permitting sites on private land can also be time-consuming because of the necessity of finding the owner of the land, finding a suitable time to talk with them, and taking the time to explain the nature of MT studies. Many land-owners are very accommodating, but some will not be. It is useful to provide to each land-owner a short illustrated document explaining the objectives of the survey, the basics of MT and the environmental impact of the work, as well as listing contact details to enable the owner to contact the survey lead if necessary.

9.6.3 Required equipment and supplies

9.6.3.1 Magnetotelluric recording units and magnetic sensors

Prior to the MT survey, it is necessary to have in place an appropriate number of functioning MT recording units and magnetic sensors. For each MT recording unit, there will usually be a GPS antenna and GPS connecting cable, one or two battery connection cables, and one to three magnetic sensor cables. For many MT systems, each recording unit will also have a cable to connect to a field computer and a dedicated memory device. In many surveys, the equipment taken to the field includes a spare set of magnetic sensors, a spare set of connecting cables, a spare GPS antenna, additional storage media and instrument manuals. The equipment should be checked for correct operation before leaving for distant field locations. It is particularly important to check the communications between the field computer and the recording unit and for correct operation of the GPS.

9.6.3.2 Electrodes, electrode wire and batteries

As mentioned in Section 9.2.2, it is critical to use high-quality electrodes in order to get good MT data. The electrodes should be stored in moist conditions between surveys, and transported to, and during, the survey in the same manner. Prior to the survey, the potential between pairs of electrodes should be checked by placing the electrodes next to each other on damp paper towels and measuring the differential potential. Using a number of electrodes, it is possible to identify any with large offsets, and these should be discarded from the survey. Spare electrodes should be taken to the field and used to replace any electrodes that develop high potential and/or high resistance.

Electrode wires are usually pre-cut to lengths 5 m to 10 m longer than the ground-to-electrode distance used in the survey. They may be hand-coiled or coiled on lightweight cable reels. During the MT survey, each cable may be kept connected to an electrode in order to avoid an unnecessary amount of disconnection and re-splicing of cables. It is generally preferable for the cables and electrodes to be disconnected between surveys to allow more efficient storage of both and to minimize the risk of hidden corrosion developing on the electrode–cable join. Prior to an MT survey, the wires should be checked for breaks in the insulation and corrosion on the exposed ends. The insulation should be repaired using electrical tape and/or heat-shrink tubing, and any corroded parts should be removed. Any cable with a significant number of breaks should be discarded.

Most MT surveys use rechargeable 12 V wet-cell batteries. Solar panel and marine batteries can provide better performance than automobile batteries. The batteries should be stored appropriately between surveys, and it is advisable to run them through a discharge–recharge cycle several times prior to the survey to assess their performance. It is important to have one or more fully reliable battery chargers for use during the actual MT survey.

9.6.3.3 Field computer

The field computer should be a robust laptop or netbook that is able to run on battery power for a reasonable duration and has good screen visibility in bright sunlight. It must be capable of running the software needed to interface with the MT recording unit, and it must have either appropriate ports to connect to the recording unit or alternative wireless facilities. It is often advantageous for the field computer to be capable of MT data processing to enable on-site processing. Flash drives or external hard drives can be used to supplement the internal storage of the computer. Every MT survey should have at least one backup field computer. It is also valuable to have a small inverter to enable the computer to be run from a 12 V battery at an MT site if necessary.

The lifespan of computers is much shorter than the lifespan of MT systems. As a result, many current MT systems require older computer operating systems to run the MT software, and still need either serial or parallel ports to communicate with the recording unit. It is possible to run virtual operating systems on newer computers to accommodate the software requirements, but it is becoming more difficult to replace computers with the required ports. In order to avoid these issues, it is advisable to maintain several dedicated field computers matched to a particular generation of MT equipment.

9.6.3.4 Navigation and communication equipment

Each MT survey requires appropriate geological and topographic maps and GPS systems for locating sites and recording site locations. A digital camera is useful for photographing each recording site. Cellular phones or satellite phones are necessary for making logistic arrangements from the field and for communicating between different vehicles and field teams. Walkie-talkies are useful for communicating between vehicles traveling together and for installation of MT sites with large electrode spacings.

9.6.3.5 Tools and meters

There are a number of tools required for MT site installation, servicing and retrieval, as follows.

- *Surveying tools.* Electric dipoles are usually installed using a tripod-mounted compass and measuring tapes. A good hand-held compass and a small level are used to install magnetic sensors.
- *Electric meters and tools.* The electrode resistance is commonly measured in MT surveys using an AC ohmmeter or an analog ohmmeter, whereas electrode DC voltages and AC noise levels are usually measured with a digital voltmeter. The voltmeter can also be used to measure battery voltage. Each member of an MT team should carry a wire stripper and wire cutter. Additional electrical tools and supplies will be needed if repairs to the MT equipment are expected during the survey.
- *Digging tools.* The basic digging equipment for an MT survey consists of shovels and spades, and in more rocky terrain a pry bar (crowbar). Hand augers are ideal for installing vertical coils; motorized augers often provide minimal advantage over hand augers. Hoes or trenching tools are very useful for installing air loops and burying electrode lines. Finally, a small wooden mallet is very useful for tamping soil around vertical coils and LMT magnetic sensors.

9.6.3.6 Additional equipment

On most MT surveys, it is useful to have one plastic or canvas tarpaulin for each deployed site as well as a number of spares. The tarpaulins are used to wrap the recording units during deployment to help control the temperature of the instrument as well as reduce its visibility.

A supply of large water containers is needed for carrying fresh water and salt water for deployments.

9.6.3.7 Consumable supplies

Consumable items for installing electrodes include up to 10 liters of water per electrode, salt, and either bentonite or cat litter. In some environments it is also useful to carry bags of soil or sand purchased from garden supply stores. Electrical insulation tape is required for repairing and joining wires. Plastic or garbage bags may be used for wrapping the magnetic sensors during their deployment, and disposable rubber gloves can be used to help seal connections between magnetic sensors and sensor cables. It is useful to enclose a one-page laminated document in each MT recording unit during deployment, noting contact details for the survey and the purpose of the measurements. Flagging or colored tape is often used in MT surveys for marking sites and for marking discharged batteries or faulty equipment.

9.6.4 Instrument calibration

Methods for MT equipment calibration were discussed in Section 9.4.6. All equipment should be calibrated at the start of the survey and, in longer surveys, again at the end. It is also necessary to calibrate any new instruments or magnetic sensors added to the equipment during the survey. If a piece of equipment is possibly damaged by an event such as being dropped or water infiltration, it should be recalibrated. The value of the MT data collected in the main survey hinges on the quality of the calibration, so this phase of the survey must be done with great care. It is useful for the MT equipment to be calibrated at either the survey location or in a well-studied location where previous MT measurements or calibrations have been made. The calibration will often be done at the site of the first MT station as long as that site has low noise levels and is in a good physical environment.

There are a number of factors that need to be considered during the calibration. For some systems, it is important for a particular sensor or set of sensors to be calibrated with a particular recording unit; and in other cases, it is also desirable for the same connecting cable(s) to be used with the same device throughout the calibration and survey. The calibration will produce response files that may need to be available during the data recording, and will need to be available during the data processing. It is critical that these files be handled very carefully, including ensuring that they are correctly associated with particular instruments and sensors, that they are correctly associated with a particular calibration event and that backup copies are made. It is useful to collate a full list of all instrument and sensor serial numbers during the calibration.

9.6.5 Site installation

MT site installation involves deciding on the layout, installing the MT equipment, making measurements of resistances and voltages, programming the MT instrument for data acquisition and recording all of the details.

Site layout commences with a decision by the layout chief on the optimal location for the middle of an "X" electrode configuration or the corner of an "L" electrode configuration. This decision is usually made based on the availability of space to install sufficiently long electrode lines, and the availability of a good environment near the center of the array for sensor burial. Depending on the site, other factors to be considered include the visibility from nearby roads, the presence of shade and proximity to large trees. Once the central point is established, the electrode lines can be sited in using the compass and measuring tapes. The layout of the magnetic sensors must be planned carefully so that all sensors are at the recommended distance from the recording unit, from each other, and from the electrode lines. These requirements are easier to accommodate in an "X" electrode configuration. At this point, the layout of the site and its GPS location should be recorded in the logbook or layout sheet. The serial numbers of the recording unit and sensor to be deployed should also be recorded.

The installation of equipment proceeds with the layout chief setting up the recording unit while the other team members install the electrodes and magnetic sensors.

The electrodes are then buried and, if needed, the electrode lead is spliced into the electrode wire. The join should be completely sealed and be able to withstand mechanical forces on the wire such as those caused by a foraging animal. Wires are often knotted to the handle of the electrode to ensure that there will be no direct force on the connection between the electrode lead and the electrode. The electrode wires are then run in relatively straight lines from the electrodes to the recording unit. They should be as close to the ground as possible, and either weighted down at short intervals or buried. Any slack wire should be folded back along the main part of the wire in a tight "S" form so that it does not lead to spurious measurements (Section 9.2.1). It is a common convention in MT surveys for a series of knots to be tied in the recording unit end of the wire immediately after they are strung in from the electrode. Zero knots denotes the ground electrode, one knot the north electrode, two knots the east electrode, three knots the south electrode, and four knots the west electrode. This process minimizes the chance of incorrect connection of the electrodes.

The induction coil or fluxgate sensor units may be wrapped in plastic to seal them during burial. A rubber glove can be used to complete the seal onto the connecting lead. The sensors are buried at the recommended depth using a compass and a level for accurate positioning. For fluxgate units and vertical coils, the soil around the device should be tamped down during the refilling of the hole to minimize movement of device. Care must be taken with the installation of coil and fluxgate sensor units to ensure that they are not oriented 180° from the proper azimuth. In induction coil studies, the cables from the north, east and vertical coils may be marked by one, two or three rings of tape in order to help ensure correct connections.

At the recording unit, the layout chief checks and records the battery voltage, and then sets up the recording unit, connecting the cables from the coils, the battery, the GPS and the computer. The electrode lines are usually connected after the voltage and resistance measurements. The recording instrument is powered up, communication with the field computer is established and the GPS lock is checked.

Voltage measurements should be made before the resistance measurements, as they cause less perturbation to the electrochemical circuit formed by the electrodes, wires and voltmeter. They should be made with a digital voltmeter in order to obtain accurate readings in the 1–100 mV range typically encountered in MT surveys. A high-input impedance voltmeter minimizes perturbation of the electrochemical circuit. The DC voltage is measured between each main electrode and the ground electrode, and then between the pairs of east and west, and north and south, electrodes, and the values are recorded. Any electrode exhibiting anomalously large values (greater than 25 mV for a good set of electrodes) may need re-installation or replacement, but large potentials on an electrode may also be caused by a variation in soil type or differential groundwater movement near that electrode. It may be necessary to accommodate high DC voltages using a decreased gain setting or AC coupling on the MT instrument. Next, the AC voltage is measured between the pairs of east and west, and north and south, electrodes, and values are recorded. Large values (greater than 200 mV) over a 100 m dipole indicate the presence of strong AC noise sources.

Contact resistance is determined by measuring the resistance between pairs of electrodes. An AC ohmmeter provides a measurement that is not perturbed by the self-potential of the electrodes and avoids charging of the electrodes. In the absence of an AC ohmmeter, an analog ohmmeter provides a more robust measurement of the resistance than a digital one because the larger current it uses is less perturbed by the self-potential of the electrodes. If a digital DC ohmmeter is used, each measurement should be made twice, with the meter positive and negative leads reversed between them, and the results should be averaged. The resistance readings should be taken as quickly as possible to reduce the impact of the current applied by the meter. The contact resistance is measured between each main electrode and the ground electrode, and then between the pairs of east and west, and north and south, electrodes, and the values are recorded. If the contact resistance has a magnitude comparable to the input impedance of the recording unit, it can only be measured accurately with the wires disconnected from the MT instrument.

A significant effort should be made to reduce the resistance of electrodes with high contact resistance values (greater than 10–20 kΩ) because of the potential for distortion of the measured MT response and increased noise levels (Section 9.2.2). Methods that can be used include burial of the electrode at greater depth, addition of conducting soil from elsewhere to the electrode hole, replacement of a poor electrode and/or use of two parallel electrodes.

The data recording acquisition parameters are usually programmed by the layout chief. For many installations, the setup includes running a short recording, perhaps of 5 minutes length, and examining the results. Any signal saturation in the data may indicate the need to reduce gain settings or to turn on a low-pass filter. The operator then programs the recording start and end times and other acquisition parameters before leaving the system in a ready-to-record state. In some MT systems a pre-prepared set of parameters can be downloaded to the acquisition unit.

It is of paramount importance to keep accurate notes of all details of the installation, including the site layout and the recording parameters. These notes may help to recover data from an incorrect installation or to understand problems in a calculated MT response. Most MT surveys record a standard set of recorded information, and it can be useful to

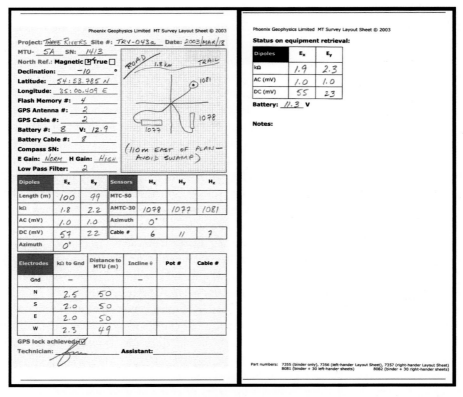

Figure 9.13. MT layout sheet used by Phoenix Geophysics Ltd (modified from Phoenix Geophysics, 2004).

have pre-prepared forms on which to write the information. Figure 9.13 shows an example of the layout form used by a commercial MT company.

9.6.6 Site servicing

It is common practice to visit and service a BBMT site approximately midway through the recording session and an LMT site every one or two weeks. In older surveys, this practice was driven by the need to replace discharged batteries and the limited storage capacity of the computer media. With solar panels and enhanced computer storage, these factors are no longer as critical. However, site visits can still be very valuable. They allow a fresh examination of the layout of the MT site, a check that the acquisition was programmed correctly during the installation, a check for human or animal interference at the site, a check on the performance of the electrodes and the collection of an initial batch of data. The analysis of these data may suggest modification of the site or the survey due to the presence of unexpected noise in a region, and it is useful for it to be available as early as possible.

The key component of servicing an MT site is the uploading of all data. In some MT systems, it is critical to turn off the instrument in a defined order to correctly recover data. The correct turn-off sequence will ensure that header information is written correctly and files are closed. In older systems, the data were uploaded by cable, and in modern systems they may be transferred on a flash drive. The MT data represent the only return for the combined effort of the MT site location, permitting and installation, and it is critical that they are handled with care! It is recommended that the data be backed up immediately in the field so that at all times there are two copies of the data on independent storage media.

During upload of the data, the other MT crew members should check the electrode lines, if they are not fully buried, for animal bites and cuts. Any problems should be repaired with electrical tape, and if necessary part or all of the wire should be replaced. The voltages and resistances should be remeasured and recorded, and any electrodes showing large increases in potential or resistivity should be replaced. A fully charged battery should be substituted for the original, and the voltage of both should be recorded. It is a useful convention to mark the discharged battery and any problem electrodes removed from the site with a piece of colored or flagging tape.

The final component of a service visit is reprogramming of the recording parameters in the MT instrument for the next recording interval.

9.6.7 Site retrieval

MT site retrieval includes the same components as the first part of a service visit, but ends with the collection of all of the MT equipment and restoration of the site to its original condition.

Data upload is again the key component of retrieval, especially if no data were collected during service visits. It is important to do a quick examination of the data before any equipment is removed from the site. This may involve examining time series and perhaps processing a short segment of the data. At the very least, the size of the data files must be checked to ensure that the data were recorded correctly. If there are problems with the recording, it may be necessary, depending on the availability of results from an earlier service visit, to extend the length of recording at the site. The voltages and resistances should be measured and recorded again during the site retrieval. This information may allow the identification of a problem electrode before it is installed at a new site, and it may be useful in understanding a poor MT response.

After it is established that the MT data from the site are adequate, the MT equipment can be unburied and collected. Electrodes and magnetic sensors should be unburied carefully to avoid damaging them. Electrode lines should be cabled up carefully, especially if they are hand-coiled, to avoid twisting and kinking the wires. Finally, the site should be returned to as close to its original state as possible. Holes should be refilled and any flagging tape used to identify the site or sensor locations should be removed.

References

Ádám, A., L. Szarka, J. Verö, A. Wallner & R. Gutdeutsch (1986). Magnetotellurics (MT) in mountains – noise, topographic and crustal inhomogeneity effects. *Phys. Earth Planet. Inter.*, **42**, 165–177.

Andersen, F., B. B. Boerner, K. Harding, A. G. Jones, R. D. Kurtz, J. Parmelee & D. Trigg (1988). LIMS: long period intelligent magnetotelluric system. In *Proc. 9th Workshop on EM Induction in the Earth*, Sochi, USSR.

Baba, K. & A. D. Chave (2005). Correction of seafloor magnetotelluric data for topographic effects during inversion. *J. Geophys. Res.*, **110**, B12105, doi:10.1029/2004JB003463.

Baba, K., A. D. Chave, R. L. Evans, G. Hirth & R. L. Mackie (2006). Mantle dynamics beneath the East Pacific Rise at 17S: insights from the Mantle Electromagnetic and Tomography (MELT) experiment. *J. Geophys. Res.*, **111**, B02101, doi:10.1029/2004JB003598, 2006.

Beblo, M. & V. Liebig (1990). Magnetotelluric measurements in Antarctica. *Phys. Earth Planet. Inter.*, **60**, 89–99.

Berdichevsky, M. N. & V. I. Dmitriev (1976). Basic principles of interpretation of magnetotelluric sounding curves. In *Geoelectric and Geothermal Studies*, ed. A. Ádám. KAPG Geophysical Monograph. Budapest: Akademiai Kiado, pp. 165–221.

Bick, M., G. Panaitov, N. Wolters, Y. Zhang, H. Bousack, A. I. Braginski, U. Kalberkamp, H. Burkhardt & U. Matzander (1999). A HTS rf SQUID vector magnetometer for geophysical exploration. *IEEE Trans. Appl. Supercond.*, **9**, 3780–3785.

Booker, J. R. & B. Narod (2002). Enhancing deadband magnetotelluric responses from long period array data. *EOS, Trans. AGU*, Abstracts of Fall Meeting, 2002.

Burger, J. R. (1972). The theoretical output of a ring core fluxgate sensor. *IEEE Trans. Magn.*, MAG-8, 791–796.

Chaize, L. & M. Lavergne (1970). Signal et bruit en magnétotellurique. *Geophys. Prosp.*, **18**, 64–87.

Chave, A. D. & C. S. Cox (1982). Controlled electromagnetic sources for measuring electrical conductivity beneath the oceans. 1. Forward problem and model study. *J. Geophys. Res.*, **87**, 5327–5338.

Chave, A. D. & J. H. Filloux (1984). Electromagnetic induction fields in the deep ocean off California: oceanic and ionospheric sources. *Geophys. J. R. Astron. Soc.*, **77**, 143–171.

Chave, A. D. & J. T. Smith (1994). On electric and magnetic galvanic distortion tensor decompositions. *J. Geophys. Res.*, **99**, 4669–4682.

Chave, A. D., S. C. Constable & R. N. Edwards (1991). Electrical exploration methods for the seafloor. In *Electromagnetic Methods in Applied Geophysics*, Vol. **2**, *Applications*, ed. M. N. Nabighian. Tulsa: Society of Exploration Geophysicists, pp. 931–972.

Clarke, J. (1977). Superconducting quantum interference devices for low frequency measurements. In *Superconductor Applications: Solids and Machines*, Vol. **3**, ed. B. B. Schwartz & S. Foner. New York: Plenum, pp. 67–124.

Clarke, J., T. D. Gamble, W. M. Goubau, R. H. Koch & R. F. Miracky (1983). Remote-reference magnetotellurics: equipment and procedures. *Geophys. Prosp.*, **31**, 149–170.

Clerc, G., G. Petiau & F. Perrier (1998). *The Garchy 1995–1996 electrode experiment, Technical Report, CNRS, Garchy, France.*

Constable, S. C. & G. S. Heinson (1993). In defence of a resistive oceanic upper mantle: reply to a comment by Tarits, Chave and Schultz. *Geophys. J. Int.*, **114**, 717–723.

Constable, S. C., A. S. Orange, G. M. Hoversten & H. F. Morrison (1998). Marine magneto-tellurics for petroleum exploration. Part 1: a sea-floor equipment system. *Geophysics*, **63**, 816–825.

Constable, S. C., K. Key & L. Lewis (2009). Mapping offshore sedimentary structure using electromagnetic methods and terrain effects in marine magnetotelluric data. *Geophys. J. Int.*, **176**, 431–442.

Eaton, D. W., J. Adams, I. Asudeh, G. M. Atkinson, M. G. Bostock, J. F. Cassidy, I. J. Ferguson, S. Samson, D. B. Snyder, K. F. Tiampo & M. J. Unsworth (2005). Geophysical arrays to investigate lithosphere and earthquake hazards in Canada. *EOS, Trans. AGU*, **86**, 169–173.

Edwards, R. N., L. K. Law, P. A. Wolfgram, D. C. Nobes, D. F. Trigg & J. M. DeLaurier (1985). First results of the MOSES experiment: sea sediment thickness and conductivity determination, Bute Inlet, British Columbia, by magnetometric offshore electrical sounding. *Geophysics*, **50**, 153–160.

Egbert, G. D. (1997). Robust multiple station magnetotelluric data processing. *Geophys. J. Int.*, **130**, 475–496.

Egbert, G. D. (2002). Processing and interpretation of electromagnetic induction array data. *Surv. Geophys.*, **23**, 207–249.

EMI Inc. (2002). *MT24/LF system operation and maintenance manual, version 1.0*, ElectroMagnetic Instruments, Inc., Richmond, CA.

Evans, S., A. G. Jones, J. Spratt & J. Katsube (2005). Central Baffin electromagnetic experiment (CBEX) maps the NACP in the Canadian arctic. *Phys. Earth Planet. Inter.*, **150**, 107–122.

Fagaly, R. L. (2006). Superconducting quantum interference device instruments and applications. *Rev. Sci. Instrum.*, **77**, 101101, doi:10.1063/1.2354545.

Fainberg, E. B. (1980). Electromagnetic induction in the world ocean. *Geophys. Surv.*, **4**, 157–171.

Ferguson, I. J. (1988). *The Tasman Project of magnetotelluric exploration*. Ph.D. thesis, Australian National University.

Ferguson, I. J., F. E. M. Lilley & J. H. Filloux (1990). Geomagnetic induction in the Tasman Sea and electrical conductivity structure beneath the Tasman seafloor. *Geophys. J. Int.*, **102**, 299–312.

Ferguson, I. J., A. G. Jones, Yu Sheng, X. Wu & I. Shiozaki (1999). Geoelectric response and crustal electrical-conductivity structure of the Flin Flon Belt, Trans-Hudson Orogen, Canada. *Can. J. Earth Sci.*, **36**, 1917–1938.

Filloux, J. H. (1973). Techniques and instrumentation for study of natural electromagnetic induction at sea. *Phys. Earth Planet. Inter.*, **7**, 323–338.

Filloux, J. H. (1987). Instrumentation and experimental methods for oceanic studies. In *Geomagnetism*, Vol. **1**, ed. J. A. Jacobs. New York: Academic Press, pp. 143–247.

Fischer, G., P. A. Schnegg & K. D. Usadel (1978). Electromagnetic response of an ocean-coast model to E-polarization induction. *Geophys. J. R. Astron. Soc.*, **53**, 599–616.

Fox, L., O. Ingerov, A. Golyashov, I. Ingerov & A. Kolin (2008). High-sensitivity EM prospecting technique based on measurement of three magnetic components of the natural EM field. In *Proc. 19th Workshop on EM Induction in the Earth*, Beijing, China.

Garcia, X. & A. G. Jones (2002). Atmospheric sources for audio-magnetotelluric (AMT) sounding. *Geophysics*, **67**, 448–458.

Garcia, X. & A. G. Jones (2005). A new methodology for the acquisition and processing of audio-magnetotelluric (AMT) data in the AMT dead-band. *Geophysics*, **70**, 119–126.

Garcia, X. & A. G. Jones (2008). Robust processing of magnetotelluric data in the AMT dead band using the continuous wavelet transform. *Geophysics*, **73**, F223–F234.

Garner, S. J. & D. V. Thiel (2000). Broadband (ULF-VLF) surface impedance measurements using MIMDAS. *Expl. Geophys.*, **31**, 173–178.

Goldie, M. (2007). A comparison between conventional and distributed acquisition induced polarization surveys for gold exploration in Nevada. *The Leading Edge*, **26**, 180–183, doi:10.1190/1.2542448.

Gómez-Treviño, E. (1987). Should the electric line be straight in magnetotelluric surveys? *Geophys. Prosp.*, **35**, 920–923.

Gowan, E. J., I. J. Ferguson, A. G. Jones & J. A. Craven (2009). Geoelectric structure of the northeastern Williston basin and underlying Precambrian lithosphere. *Can. J. Earth Sci.*, **46**, 441–464.

Groom, R. W., R. D. Kurtz, A. G. Jones & D. E. Boerner (1993). A quantitative methodology to extract regional magnetotelluric impedances and determine the dimension of the conductivity structure. *Geophys. J. Int.*, **115**, 1095–1118.

Heinson, G. S. & S. C. Constable (1992). The electrical conductivity of the oceanic upper mantle. *Geophys. J. Int.*, **110**, 159–179.

Heinson, G. S. & F. E. M. Lilley (1993). An application of thin-sheet electromagnetic modelling to the Tasman Sea. *Phys. Earth Planet. Inter.*, **81**, 231–251.

Hoover, D. B., C. L. Long & R. M. Senterfit (1978). Some results from audiomagnetotelluric investigations in geothermal areas. *Geophysics*, **43**, 1501–1414.

Hoversten, G. M., S. C. Constable & H. F. Morrison (2000). Marine magnetotellurics for base-of-salt mapping: Gulf of Mexico field test at the Gemini structure. *Geophysics*, **65**, 1476–1488.

Iliceto, V. & G. Santarato (1999). On the interference of man-made EM fields in the magnetotelluric "dead band". *Geophys. Prosp.*, **47**, 707–719.

Ingham, M., K. Whaler & D. McKnight (2001). Magnetotelluric sounding of the Hikurangi Margin, New Zealand. *Geophys. J. Int.*, **144**, 343–355.

Jiracek, G. R. (1990). Near-surface and topographic distortions in electromagnetic induction. *Surv. Geophys.*, **11**, 163–203.

Jones, A. G. (2006). Electromagnetic interrogation of the anisotropic Earth: looking into the Earth with polarized spectacles. *Phys. Earth Planet. Inter.*, **158**, 281–291.

Jones, A. G. & X. Garcia (2003). Okak Bay AMT dataset case study: lessons in dimensionality and scale. *Geophysics*, **68**, 70–91.

Junge, A. (1990). A new telluric KCl probe using Filloux's AgAgCl electrode. *Pure Appl. Geophys.*, **134**, 589–598.

Junge, A. (1996). Characterization and correction for cultural noise. *Surv. Geophys.*, **17**, 361–391.

Kaufman, A. A. & G. Keller (1981). *The Magnetotelluric Sounding Method*. Amsterdam: Elsevier.

Kitching J., S. Knappe, W. C. Griffith, J. Preusser, V. Gerginov, P. D. D. Schwindt, V. Shah & R. Jimenez-Martinez (2009). Uncooled, millimeter-scale atomic magnetometers with femtotesla sensitivity. *IEEE Sensors*, **1–3**, 1844–1847.

Korepanov, V., V. Glemba, V. Loshevsky & O. Svenson (2008a). Super-low noise non-polarized electrodes. In *Proc. 19th Workshop on EM Induction in the Earth*, Beijing, China.

Korepanov, V., Y. Klymovych & V. Regubenko (2008b). Recent advances in magnetotelluric instrumentation and field results. In *Proc. 19th Workshop on EM Induction in the Earth*, Beijing, China.

Kuvshinov, A. V., N. Olsen, D. B. Avdeev & O. V. Pankratov (2002). Electromagnetic induction in the oceans and the anomalous behaviour of coastal *C*-responses for periods up to 20 days. *Geophys. Res. Lett.*, **29**, 1595, doi:10.1029/2001GL014409.

LaBrecque, D. & W. Daily (2007). Assessment of measurement errors for galvanic-resistivity electrodes of different composition. *Geophysics*, **73**, F55–F64.

Lilley, F. E. M., J. H. Filloux, P. J. Mulhearn & I. J. Ferguson (1993). Magnetic signals from an ocean eddy. *J. Geomagn. Geoelectr.*, **45**, 403–422.

Lowes, F. J. (2009). DC railways and the magnetic fields they produce-the geomagnetic context. *Earth Planets Space*, **61**(8), i–xv.

Lu, K. & J. Macnae (1998). The international campaign on intercomparison between electrodes for geoelectrical measurements. *Explor. Geophys.*, **29**, 484–488.

McNeice, G. W. & A. G. Jones (1998). Magnetotellurics in the frozen north: measurements on lake ice. In *Proc. 14th Workshop on EM Induction*, Sinaia, Romania.

Merriam, J. B. (2007). Induced polarization and surface electrochemistry. *Geophysics*, **4**, F157–F166.

Nabighian, M. N. & J. C. Macnae (2005). Electrical and EM methods, 1980–2005. *The Leading Edge*, **24**, Suppl., S42–S45, doi:10.1190/1.2112391.

Nabighian, M. N., V. J. S. Grauch, R. O. Hansen, T. R. LaFehr, Y. Li, J. W. Peirce, J. D. Phillips & M. E. Ruder (2005). The historical development of the magnetic method in exploration. *Geophysics*, **70**, 33ND–61ND, doi:10.1190/1.2133784.

Narod, B. B. & J. R. Bennest (1990). Ring–core fluxgate magnetometers for use as observatory variometers. *Phys. Earth Planet. Inter.*, **59**, 23–28.

Narod, B. B., J. R. Bennest, J. O. Strom-Olsen, F. Nezil & R. A. Dunlap (1985). An evaluation of the noise performance of Fe, Co, Si, and B amorphous alloys in ring–core fluxgate magnetometers. *Can. J. Phys.*, **63**, 1468–1472.

Narod, B. B., J. R. Bennest & J. R. Booker (2001). The low power long period magneto-telluric system, *EOS Trans. AGU*, **82**, Fall Meet. Suppl., Abstract GP21A-0247.

Oettinger, G., V. Haak & J. C. Larsen (2001). Noise reduction in magnetotelluric time-series with a new signal–noise separation method and its application to a field experiment in the Saxonian Granulite Massif. *Geophys. J. Int.*, **146**, 659–669.

Pádua, M. B., A. L. Padilha & Í. Vitorello (2002). Disturbances on magnetotelluric data due to DC electrified railway: a case study from southeastern Brazil. *Earth Planets Space*, **54**, 591–596.

Palacky, G. J. (1988). Resistivity characteristics of geologic targets. In *Electromagnetic Methods in Applied Geophysics*, Vol. **1**, *Theory*, ed. M. N. Nabighian. Tulsa: Society of Exploration Geophysicists, pp. 53–129.

Parkinson, W. D. (1962). The influence of continents and oceans on geomagnetic variations. *Geophys. J. R. Astron. Soc.*, **6**, 441–449.

Pedersen, L. B., M. Bastani & L. Dynesius (2005). Groundwater exploration using combined controlled-source and radiomagnetotelluric techniques. *Geophysics*, **70**, G8–G15, doi:10.1190/1.1852774.

Pellerin, L., D. Alumbaugh & N. Cueva (2003). Characterization of cultural noise in the AMT band. In *ASEG 16th Geophysics Conference and Exhibition*, Extended Abstracts, Brisbane, Australia.

Pellerin, L., D. Alumbaugh, D. J. Reinemann & P. D. Thompson (2004). Power line induced current in the Earth determined by magnetotelluric techniques. *Appl. Eng. Agric.*, **20**, 703–706.

Perrier, F. E., G. Petiau, G. Clerc, V. Bogorodsky, E. Erkul, L. Jouniaux, D. Lesmes, J. Macnae, J. M. Meunier, D. Morgan, D. Nascimento, G. Oettinger, G. Schwarz,

H. Toh, M. J. Valiant, K. Vozoff & O. Yazici-Çakin (1997). A one-year systematic study of electrodes for long period measurements of the electric field in geophysical environments. *J. Geomagn. Geoelectr.*, **49**, 1677–1696.

Petiau, G. (2000). Second generation of lead–lead chloride electrodes for geophysical applications. *Pure Appl. Geophys.*, **157**, 357–382.

Petiau, G. & A. Dupis (1980). Noise, temperature coefficient, and long time stability of electrodes for telluric observations. *Geophys. Prosp.*, **28**, 792–804.

Pfaffhuber, A. (2001). *Development and test of a controlled source MT method in the frequency range 1 to 50 kHz*. Diploma thesis, Technical University Berlin.

Phoenix Geophysics (2004). *V5 System 2000 MTU/MTU – a user guide, version 1.4.* Phoenix Geophysics Ltd, Toronto, Canada.

Pirjola, R., L. Newitt, D. Boteler, L. Trichtchenko, P. Fernberg, L. McKee, D. Danskin & G. Jansen van Beek (2007). Modelling the disturbance caused by a dc-electrified railway to geomagnetic measurements. *Earth Planets Space*, **59**, 943–949.

Qian, W. & L. B. Pedersen (1991). Industrial interference magnetotellurics: an example from the Tangshan area, China. *Geophysics*, **56**, 265–273.

Ranganayaki, R. P. & T. R. Madden (1980). Generalized thin sheet analysis in magnetotellurics: an extension of Price's analysis. *Geophys. J. R. Astron. Soc.*, **60**, 445–457.

Romo, J. M., C. Flores, R. Vega, R. Vazquez, M. Perez Flores, E. Gómez-Treviño, S. J. Esparza, J. E. Quijanot & V. H. Garcia (1997). A closely-spaced magnetotelluric study of the Ahuachapan–Chipilapa Geothermal Field, El Salvador. *Geothermics*, **26**, 627–656.

Schultz, A., R. D. Kurtz, A. D. Chave & A. G. Jones (1993). Conductivity discontinuities in the upper mantle beneath a stable craton. *Geophys. Res. Lett.*, **20**, 2941–2944.

Schultz, A., P. Bedrosian, R. Evans, G. Egbert, A. Kelbert, K. Mickus, D. Livelybrooks, S. Park, P. Patro, T. Peery, P. Wannamaker, M. Unsworth, C. Weiss & B. Woodward (2008). An emerging view of the crust and mantle of tectonic North America from EMScope: a mid-term progress review of Earthscope's magnetotelluric program. *EOS, Trans. AGU*, **89**, Fall Meeting Suppl., Abstract GP51C-07.

Sheard, N. (1998). MIMDAS: a new direction in geophysics. In *ASEG Geophysics Conference and Exhibition*.

Simpson, F. & K. Bahr (2005). *Practical Magnetotellurics*. Cambridge: Cambridge University Press.

Sunde, E. D. (1949). *Earth Conduction Effects in Transmission Systems*. New York: Van Nostrand.

Swift, C. M., Jr. (1967), *A magnetotelluric investigation of an electrical conductivity anomaly in the southwestern United States*. Ph.D. thesis, Massachusetts Institute of Technology.

Szarka, L. (1988). Geophysical aspects of man-made electromagnetic noise in the Earth – a review. *Surv. Geophys.*, **9**, 287–318.

Takács, E. (1979). Investigations into geophysical application of electromagnetic field of 50 Hz electric power transmission lines. *Magyar Geofizik*, **20**, 121–127 (in Hungarian).

Tallgren, P., S. Vanhatalo, K. Kaila & J. Voipo (2005). Evaluation of commercially available electrodes and gels for recording of show EEG potentials. *Clin. Neurophysiol.*, **116**, 799–806.

Tarits, P., A. D. Chave & A. Schultz (1993). Comment on "The electrical conductivity of the oceanic upper mantle" by G. Heinson and S. Constable. *Geophys. J. Int.*, **114**, 711–716.

Tezkan, B. (1999). A review of environmental applications of quasistationary electromagnetic techniques. *Surv. Geophys.*, **20**, 279–308.

Tomczyk, S. (2007). A new model of telluric amplifier for magnetotelluric soundings. *Publ. Inst. Geophys. Pol. Acad. Sci.*, **C-99**, 398.

Torres-Verdín, C. & F. X. Bostick, Jr. (1992). Principles of spatial surface electric field filtering in magnetotellurics: electromagnetic array profiling (EMAP). *Geophysics*, **57**, 603–622.

Trad, D. O. & J. M. Travassos (2000). Wavelet filtering of magnetotelluric data. *Geophysics*, **65**, 482–491.

Trigg, D. F. (1972). An amplifier and filter system for telluric signals. *Publ. Earth Phys. Branch*, 44, 1–5, Department of Energy, Mines and Resources, Canada.

Tumanski, S. (2007). Induction coil sensors – a review. *Meas. Sci. Technol.*, **18**, R31–R46, doi:10.1088/0957-0233/18/3/R01.

Türkoğlu, E., M. Unsworth & D. Pana (2009). Deep electrical structure of northern Alberta (Canada): implications for diamond exploration. *Can. J. Earth Sci.*, **46**, 139–154.

Vozoff, K. (1991). The magnetotelluric method. In *Electromagnetic Methods in Applied Geophysics*, Vol. **2**, *Applications*, ed. M. N. Nabighian. Tulsa: Society of Exploration Geophysicists, pp. 641–711.

Wannamaker, P. E., J. A. Stodt & L. Rijo (1986). Two-dimensional topographic responses in magnetotellurics modeled using finite elements. *Geophysics*, **51**, 2131–2144.

Wannamaker, P. E., J. R. Booker, A. G. Jones, A. D. Chave, J. H. Filloux, H. S. Waff & L. K. Law (1989). Resistivity cross-section through the Juan de Fuca Ridge subduction system and its tectonic implications. *J. Geophys. Res.*, **94**, 14 127–14 144.

Wannamaker, P. E., G. R. Jiracek, J. A. Stodt, T. G. Caldwell, V. M. Gonzalez, J. D. McKnight & A. D. Porter (2002). Fluid generation and pathways beneath an active compressional orogen, the New Zealand Southern Alps, inferred from magneto-telluric data. *J. Geophys. Res.*, **107**, 2117, doi:10.1029/2001JB000186.

Wannamaker, P. E., J. A. Stodt, L. Pellerin, S. L. Olsen & D. B. Hall (2004). Structure and thermal regime beneath the South Pole region, East Antarctica, from magnetotelluric measurements. *Geophys. J. Int.*, **157**, 36–54.

Webb, S. C., S. C. Constable, C. S. Cox & T. K. Deaton (1985). A seafloor electric field instrument. *J. Geomagn. Geoelectr.*, **37**, 1115–1129.

Weckmann, U., A. Magunia & O. Ritter (2005). Effective noise separation for magneto-telluric single site data processing using a frequency domain selection scheme. *Geophys. J. Int.*, **161**, 635–652.

White, D. E. (1988). *MT/EMAP Data Interchange Standard*. Tulsa: Society of Exploration Geophysicists.

Yamashita, M. & L. Fox (2008). Introduction of automated stationary MT systems. In *Proc. 19th Workshop on EM Induction in the Earth*, Beijing, China.

Zhang, P. & L. B. Pedersen (1991). Can magnetotelluric data resolve lower-crustal con-ductors? *Phys. Earth Planet. Inter.*, **65**, 248–254.

Zhang, P., L. B. Pedersen, M. Mareschal & M. Chouteau (1993). Channeling contribution to tipper vectors: a magnetic equivalent to electrical distortion. *Geophys. J. Int.*, **113**, 693–700.

Zonge, K. L. & L. J. Hughes (1985). The effect of electrode contact resistance on electric field measurements. In *55th Annual International Society of Exploration Geophysicists Meeting*, Tulsa, OK, Expanded Abstracts, Contribution MIN.1.5, pp. 231–234.

Zonge, K. L. & L. J. Hughes (1991). Controlled source audiofrequency magnetotellurics. In *Electromagnetic Methods in Applied Geophysics*, Vol. **2**, *Applications*, ed. M. N. Nabighian. Tulsa: Society of Exploration Geophysicists, pp. 713–798.

10

Case histories and geological applications

IAN J. FERGUSON, ALAN G. JONES AND ALAN D. CHAVE

10.1 Introduction

This chapter describes selected modern applications of magnetotelluric surveys. The examples included illustrate many of the analysis and modeling methods described in Chapters 2–8, as well as the geological information that can be derived from magnetotelluric studies. The exemplars are divided into crustal studies, lithospheric mantle investigations, resource exploration applications and marine surveys. There are many excellent descriptions of magnetotellurics in the published literature, and those described in this chapter represent just a sample of the information that is available.

10.2 Magnetotelluric studies of continental crust

This section examines the application of magnetotellurics to define structures and processes in the continental crust. The examples presented are all large-scale projects, involving multiple university and government groups, and, in two of the examples, multiple phases of data acquisition. The studies demonstrate the importance of very careful dimensionality analysis in crustal studies, application of multidimensional modeling and the incorporation of anisotropic conductors. They also demonstrate the interpretation of electrical resistivity in terms of geology.

There have been an extremely large number of magnetotelluric studies of crustal targets over recent decades; see reviews by Heinson (1990), Jones (1992), Brown (1994), Korja (2007) and Unsworth (2010). These investigations have provided sufficient information to define the vertically averaged resistivity of continental crust, including its variation with age (Haak & Hutton, 1986; Jones, 1992). They have also provided constraints on resistivity structures of particular tectonic and geological features, including: convergent, divergent and strike-slip plate margins; faults and shear zones; partially molten rocks; intrusive and extrusive rocks; sedimentary basins; and different types of metamorphic terranes. Chapter 3 provides an extensive overview of typical values for, and factors affecting, the resistivity of the geological components of continental and oceanic crust.

Magnetotellurics has been applied in many older geological settings to define the tectonic structure in three dimensions, and to make inferences regarding past tectonic processes. The

The Magnetotelluric Method: Theory and Practice, ed. Alan D. Chave and Alan G. Jones. Published by Cambridge University Press © Cambridge University Press 2012.

interpretation of the resistivity structure in these settings must be done in light of changes in the thermal and fluid state of the crust since the termination of the active processes that produced them. However, the signatures of past tectonic events are often recorded by compositional variation in rocks, as well as by changes in grain size and grain shape. For example, the presence and geometry of conductive graphitic and sulfidic metasedimentary rocks can help to resolve the three-dimensional structure of Proterozoic suture zones (Korja & Hjelt, 1993; Boerner *et al.*, 1999; Gowan *et al.*, 2009), and define the concentration and alignment of these conductive minerals during deformational processes (Jones *et al.*, 1997, 2005a).

There have been numerous magnetotelluric studies of Precambrian crust over different regions of the world. Large-scale magnetotelluric surveys conducted in Archean and Proterozoic regions of Canada in the Lithoprobe program (e.g. Boerner *et al.*, 2000a,b; Jones *et al.*, 2005a, b; Ferguson *et al.*, 2005), in Scandinavia (e.g. Korja & Hjelt, 1993; Korja *et al.*, 2002; Korja, 2007), in India (e.g. Patro & Sarma, 2009), in Australia (e.g. Lilley *et al.*, 2003; Selway *et al.*, 2009) and elsewhere record crustal geological structures that resemble those produced in younger tectonic terranes.

The resistivity structure of older Phanerozoic crust may image compositional and grain changes recording past tectonization. In younger Phanerozoic settings, where tectonic processes ceased in the last 100 Ma, surveys may record thermal and fluid signatures that are directly associated with tectonic events. For example, studies in the southern Canadian Cordillera image enhanced conductivity in the lower crust in a region that underwent extension between 59 and 46 Ma (Ledo & Jones, 2001). The enhanced conductivity is interpreted as mainly due to interconnected aqueous fluids that have been trapped for the last 46 Ma.

Some of the most exciting new applications of magnetotellurics in crustal studies come from areas of active plate tectonics. By mapping the distribution of partial melt and saline fluids through their resistivity signature, magnetotellurics provides constraints both on the geometry of interacting lithospheric plates and on some of the fundamental geological processes that affect their behavior. Fluids and melt are responsible for significant rheological weakening of crustal materials, so magnetotellurics can in principle directly image zones of active deformation. In addition, the distribution of fluids reflects temperature and composition variations, and, near the surface, the presence of increased porosity associated with ductile deformation.

10.2.1 Imaging of the India–Asia collision

Magnetotelluric results from the India–Asia plate interaction zone are providing important information on tectonic processes associated with continent–continent collision. Unsworth (2010) provides a comprehensive review. The initial collision between the subcontinents of India and Asia occurred between 45 and 70 Ma along the Indus–Tsangpo suture (Figure 10.1). The Tibetan Plateau grew as convergence proceeded, producing a steady-state situation with a modeled maximum elevation of 5 km. Continuing collision resulted in

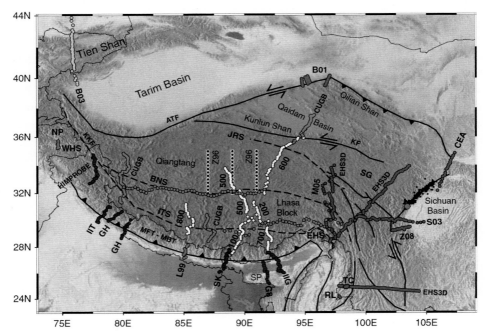

Figure 10.1. Magnetotelluric surveys and tectonic features of the India–Asia collisional zone from Unsworth (2010). The labeled lines of gray circles show the locations and names of MT profiles. Defined features are: SP, Shillong Plateau; NP, Nana Parbat; ATF, Altyn Tagh Fault; KF, Kunlun Fault; KKF, Karakorum Fault; JRS, Jinsha River Suture; BNS, Bangong–Nuijiang Suture; ITS, Indus–Tsangpo Suture; EHS, eastern Himalayan syntaxis; MFT, Main Frontal Thrust; MBT, Main Boundary Thrust; WHS, western Himalayan syntaxis. The locations of the May 12, 2008, Wenchuan earthquake and its aftershocks are shown by the single larger gray circle, just west of the Sichuan Basin, and the black dots, respectively.

lateral growth of the Tibetan Plateau to the northeast. Tectonically, this was manifest as thrust faulting at low elevations and collapse at higher elevations as the plateau underwent east–west extension and normal faulting. Extension produced north–south trending rift zones in southern Tibet. GPS data indicate that active convergence is now accommodated on the Main Boundary Thrust (MBT) and Main Frontal Thrust (MFT; see Figure 10.1), but geological studies indicate that earlier thrusting occurred farther north on the Main Central Thrust (MCT; see Figure 10.2 for location). Seismic studies in the central Himalaya have imaged a strong reflector that dips north at 10–15°, and can be extrapolated to the surface at the MBT and MFT. Tectonic processes at the eastern and western end of the Himalaya (the eastern and western syntaxes) are distinct from those in the central Himalaya. In these regions, rapid uplift is coupled to erosion and deformation. The large-scale tectonics is also influenced by the presence of strong crustal blocks. There is minimal deformation within the Tarim Basin to the north of the Plateau, but a high rate

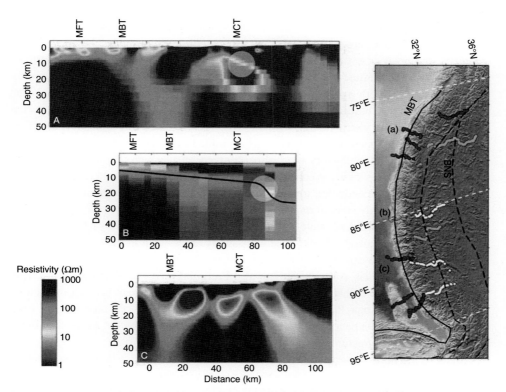

Figure 10.2. Magnetotelluric profiles crossing the central Himalaya from Unsworth (2010). The location of each of the profiles on the left is shown in the panel on the right: (A) results of Israil *et al.* (2008), (B) results of Lemmonier *et al.* (1999) and (C) results of Patro & Harinarayana (2009). The white circle indicates a zone of seismicity. (See plate section for color version.)

of deformation occurs in the Tien Shan farther to the northwest. The Sichuan Basin has also been interpreted as a strong crustal block that causes localization of surrounding deformation.

Magnetotelluric studies have imaged a number of features associated with these tectonic structures and processes. Surveys in the central Himalaya and the adjacent region to the south have imaged the unconsolidated sediments of the Indian Foreland as a low-resistivity feature (Figure 10.2). Northward-dipping conductors suggest the presence of underthrust sedimentary rocks. The magnetotelluric profiles all show a conductor close to the location where the strong crustal seismic reflector steepens significantly near the surface expression of the MCT. The zone is closely associated with significant seismicity at 10–20 km depth, and has been interpreted as the effect of crustal fluids.

Farther to the north, magnetotelluric studies in southern Tibet include the INDEPTH project profiles (Figure 10.1; Chen *et al.*, 1996; Li *et al.*, 2003; Unsworth *et al.*, 2004, 2005;

Spratt *et al.*, 2005; Solon *et al.*, 2005). Data were acquired using Phoenix (V5 and MTU-5) and EMI systems for broadband magnetotellurics (BBMT) and LIMS and NIMS for long-period studies (see Chapter 9). INDEPTH, along with earlier studies, has defined the geometry and areal extent of a significant mid-crustal conductor (Figure 10.3). North of the Indus–Tsangpo Suture (ITS, Figure 10.1 and 10.3), the top of the conductor is approximately coincident with a negative-polarity seismic reflection, suggesting that it is due to crustal fluids, and, based on the expected crustal temperature, it is probable that partial melt is present. Early interpretations of the INDEPTH 100 and 200 profiles (Figure 10.1) suggested that partial melt is present because the profile lies within an east–west extensional rift zone, but similar conductors were also observed on other profiles outside the rifts. The reduced crustal resistivity is consistent with a factor of 10 reduction in crustal strength. The continuity of the conductor, implied in Figure 10.3 and in Unsworth *et al.* (2005), is a consequence of overly smooth modeling by Spratt *et al.* (2005).

Similar resistivity structures have been imaged in the northwest Himalaya, but the amount of melt required to explain the mid-crustal resistivity is less than in the central Himalaya. The difference may be explained by reduced convergence rates that produce slower crustal thickening and less melt. A conductor has also been observed in the western Himalayan syntaxis in the Nanga Parbat region (Figure 10.1), but is non-existent beneath the rapidly uplifting peak of Nanga Parbat (Park & Mackie, 1997, 2000). Its absence has been interpreted to indicate a paucity of fluids, and has been explained by the removal of fluids through a network of fractures created by rapid deformation.

Farther to the north, magnetotelluric studies conducted on the elevated Tibetan Plateau (e.g. in the Qiangtang terrane in Figure 10.1) reveal a northern continuation of the mid-crustal conductor. Results from the INDEPTH 600 line suggest that at 94°E the conductor terminates close to the surface location of the Kunlun Fault (Figures 10.1 and 10.3). Recent reanalysis of data from the northern part of the 600 line crossing the Kunlun Fault using newer algorithms has yielded more robust regional response function estimates that have been modeled using a newer 2D anisotropic inversion code (Baba *et al.*, 2006a). The new anisotropic model in Figure 10.4 (F. Le Pape, pers. comm.) is distinctly different from the smoother model of Unsworth *et al.* (2004) in Figure 10.3. In particular, and of rheological importance, the conductor is restricted to the mid-crust, and does not penetrate into the deep crust and uppermost mantle as implied in Figure 10.3. Also significant is the implication that the conductors are connected in a fault-perpendicular NE–SW direction (*YY*) but not in a fault-parallel NW–SE direction (*XX*). This suggests that partial melt extends NE of the Kunlun Fault in penetrative fingers.

One of the important roles of magnetotelluric studies in the Himalaya is assessment of models of crustal flow in eastern Tibet (Unsworth, 2010). Teleseismic studies have revealed a spatially extensive low-velocity layer at mid-crustal depths that can be attributed to a weak fluid-rich layer. The mid-crustal high-conductivity layer observed using magnetotellurics can be interpreted as a weakened layer. The high conductivities observed in northern Tibet on the INDEPTH profiles and in the Qiangtang terrane may be due to channel flow (Zhang *et al.*, 1996; Unsworth *et al.*, 2004). An extensive magnetotelluric

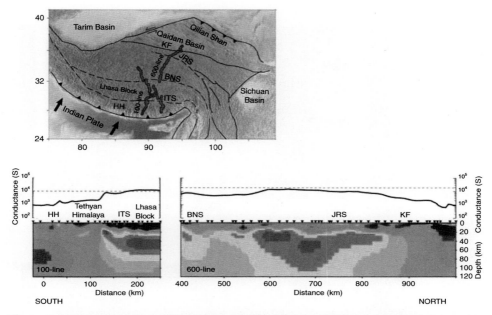

Figure 10.3. Magnetotelluric profile crossing southern Tibet. Modified from Unsworth (2010). The locations of the INDEPTH profiles are shown on the map. (See plate section for color version.)

survey in eastern Tibet shows that zones of low resistivity can be traced over lateral distances of more than 100 km, supporting this interpretation (Bai *et al.*, 2010). Geodynamic models suggest that the channel flow should be diverted around the Sichuan Basin, and a magnetotelluric study on the west side of the Sichuan Basin (Sun *et al.*, 2003) shows that the mid-crustal conductor terminates in this region, in agreement with the model.

10.2.2 Imaging of fluids in an oblique compressional orogen in the Southern Alps, New Zealand

Wannamaker *et al.* (2002) describe the study of an oblique continent–continent collision in the Southern Alps of New Zealand. The Alpine Fault marks the surface expression of the plate boundary between the Pacific and Australian plates, and extends over most of the length of South Island, New Zealand (Figure 10.5). Tectonic activity occurs in a narrow zone, with much of the strain occurring within 50 km of the Alpine Fault, and is focused mainly to its southeast. Most of the compressional deformation and growth of the Southern Alps has occurred since 6–7 Ma. Approximately 70 km of shortening has occurred in the study area, and uplift (up to 10 mm/yr) and erosion rates are presently among the highest in the world.

(a)

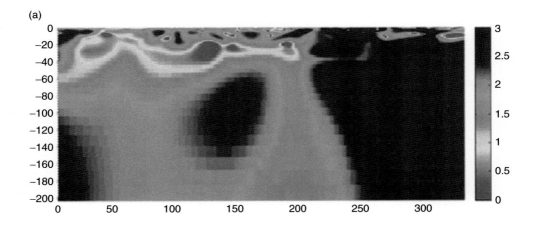

(b)

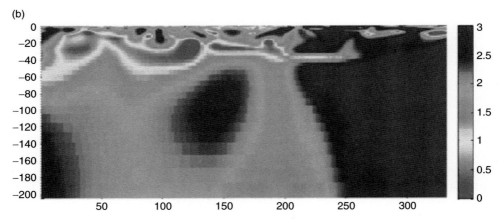

Figure 10.4. Anisotropic resistivity model of the northern part of the 600 profile (F. Le Pape, pers. comm.): (a) the *XX* (NW–SE) component of the resistivity; (b) the *YY* (NE–SW) component of the resistivity. (See plate section for color version.)

Magnetotelluric study of the orogen complements an active seismic survey, with the electromagnetic profile set parallel to two seismic profiles (Figure 10.5). The magnetotelluric survey comprised 41 sites on the main line, and additional site pairs (WF and MC) were used to assess the continuity of results along-strike. Recordings were made using University of Utah dual-site systems, and employed EMI induction coils and Pb–PbCl$_2$ electrodes separated by 200 m. The recording period range was typically 0.005 to 1000 s. Data processing used the remote reference method, with the remote data coming either from other sites or from mountain sites deployed by helicopter. Significant noise sources included a NE–SW DC transmission line (Figure 10.5), and, in the coastal areas, electric fences that degraded the data over the 0.3–3 s period range.

Analysis of geoelectric strike directions determined using the Groom–Bailey method (see Chapter 6) and based on the middle of the period range indicated an optimal value of N40°E (Figure 10.6). Figure 10.7 shows pseudo-sections of the 2D TE and TM mode responses for a coordinate system aligned parallel and perpendicular with the estimated strike. Low values of the TM mode apparent resistivity (ρ_{yx}) in the northwest and southeast parts of the island are attributed to 1–3 km of conductive Cenozoic sediments. Relatively high TM mode apparent resistivity in the Stew Point–Coal Hill area (SP, CH in Figure 10.5 and 10.7) is attributed to an intrusive formation, and relatively high apparent resistivity to the southeast of the Alpine Fault is due to high metamorphic-grade schist. The high value of the TM mode phase (ϕ_{yx}) at periods of 0.1–20 s farther to the southeast of the Alpine Fault, and a

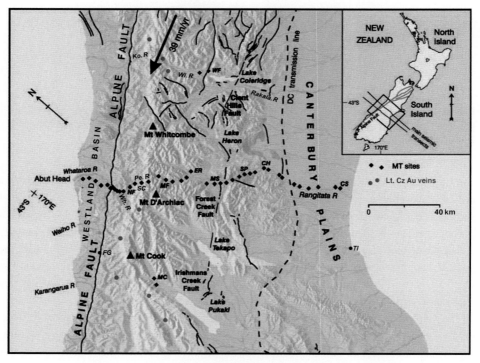

Figure 10.5. Map of the central-western part of South Island, New Zealand, showing magnetotelluric sites (diamonds) and geographic features from Wannamaker *et al.* (2002). The inset shows the location of two seismic profiles. Initials refer to Nolan's Flat (NF), Scone Hut (SC), Mistake Flat (MF), Erewhon Station (ER), Mesopotamia Station (MS), Stew Point (SP), Coal Hill (CH), Coldstream Station (CS), Wilberforce magnetotelluric sites (WF) and Mount Cook (MC) magnetotelluric sites. The Upper Whataroa, Perth, Kokatahi and Wilberforce rivers are denoted by Wh. R, Pe. R, Ko. R and Wi. R, respectively.

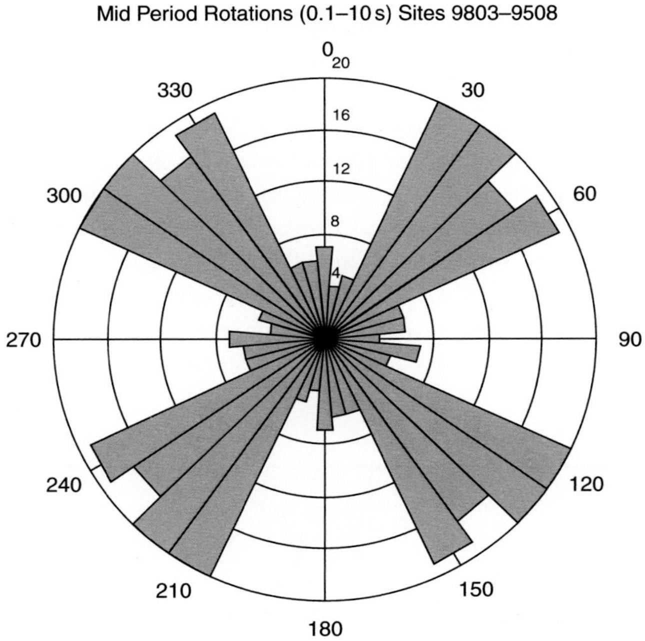

Figure 10.6. Rose diagram of Groom–Bailey strike directions for the part of the profile from SC to ER (Figure 10.5) that displays enhanced lower crustal conductivity. From Wannamaker *et al.* (2002). Results account for the 90° ambiguity in strike determination.

corresponding minimum in the TM mode apparent resistivity, indicate the presence of a crustal-scale conductive zone under the Southern Alps. The TE mode response shows similar behavior to the TM mode at short periods. At longer periods, the TE responses differ significantly from the TM ones, and in some areas cannot be fitted by spatially consistent tensor decompositions. Because of conductive surface rocks and the ocean, the amplitude of the long-period TE mode electric field is smaller than the TM mode values. The small TE magnitude response is susceptible to minor distortion or inductive processes that combine it with a component of the larger-magnitude TM response.

The vertical magnetic field transfer functions show that the response for short to intermediate periods in the component parallel to the assumed strike direction (K_{zx}) is quite weak, supporting the choice of strike direction. However, at long periods this component shows a spatially-consistent negative response, indicating the presence of a conductor to the northeast. For the component of the transfer function perpendicular to the assumed strike (K_{zy}) at periods >100 s, there is a strong response corresponding to the coast effect in the northwest part of the profile and a weaker coast effect at the southeast end of the profile. There is also a strong response centered on periods of 1–10 s associated with the crustal conductor beneath the Southern Alps. Finally, there is a reversal in the response centered at a

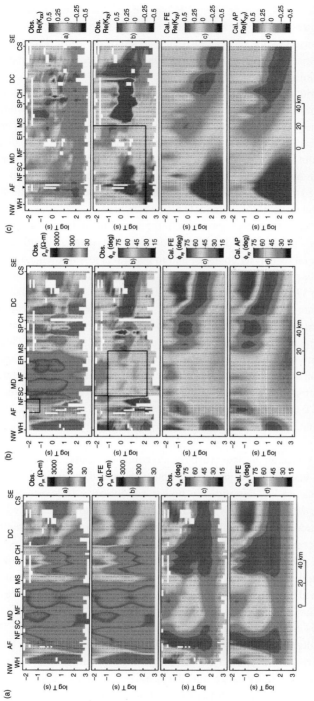

Figure 10.7. Pseudo-sections of the (a) TM, (b) TE and (c) tipper responses for the New Zealand magnetotelluric profile. Results are shown for the observed data (Obs.), the response of a 2D finite-element model (Cal. FE) and the response of a 2D *a priori* model (Cal. AP). Modified from Wannamaker *et al.* (2002). Responses from outside the boxes shown on the TE and tipper observations were excluded from the inversions. Abbreviations in addition to those defined above are: NW, northwest coast; WH, Whatanoa Delta; AF, Alpine Fault; DC, DC power line; CS, southeast coast. (See plate section for color version.)

period of 1 s beneath the DC power line. This has the opposite sense to that for a crustal conductor (i.e. the corresponding Parkinson-convention induction arrows point away from the feature), and is caused by the power line. Comparison of the vertical magnetic field and magnetotelluric responses shows that the TE mode is also affected by this signal, but the TM mode appears to be uninfluenced.

The magnetotelluric data were modeled using 2D inversion algorithms with careful selection of data to minimize the effect of 3D elements. The inversion algorithms utilized were a Gauss–Newton finite-element method that damps variations from an *a priori* structure, and a nonlinear conjugate gradient algorithm that dampens the spatial curvature of the resistivity model against that of an *a priori* model. The initial model consisted of a 2000 Ω m half-space modified by the inclusion of structures representing the ocean and marine sediments and 1–2 km of the land structure based on 1D inversion of the short-period (<0.1 s) data. Only the finite-element models are shown here.

Figure 10.8 shows 2D models that optimally fit the data. The first model is an inversion of the TM mode data labeled FE TM. The TM mode response is more sensitive than the TE mode response to lateral resistivity changes, and in general is less influenced by 3D effects. The normalized root mean square (RMS) misfit of this model is 3.7. The second model is a joint inversion of the TM, TE and tipper data defined in Figure 10.7 and is labeled FE AL. The TE mode data are included to help identify anisotropy and narrow conductors that would not be defined by the TM mode data, and the shorter-period TE mode data are included to improve resolution of the near-surface part of the Alpine Schist. The tipper data have a different sensitivity to finite-strike-length features than the TE mode magnetotelluric data, so their inclusion and consideration of the resulting misfit allows identification of such components. The normalized RMS misfit for the FE AL model is 4.3, with the individual misfits for the TM, TE and tipper data being 4.1, 6.1 and 5.4. The final model shown is an inversion of the TE and tipper data defined above using the TM mode FE AP model as the *a priori* one.

All of the models reveal an arcuate crustal-scale conductor extending downward from a surface location southeast of the Alpine Fault, and then extending horizontally at depth to the southeast. The feature has significantly higher conductance in inversion models that include the TE mode data, and is more continuous in the inversion model in which the TM mode data were included only through the *a priori* model. The higher conductivity of the lower crustal conductor seen in the models based on TE mode data suggests that the conductivity may be greater in the along-strike direction than in the cross-strike direction sensed by the TM mode currents. In addition, the large TE and tipper misfit of the FE AL model can be explained by a 3D structure in which the along-strike conductance of the deep conductor is higher than that along the magnetotelluric profile.

Figure 10.9 shows a geological interpretation of the resistivity results. The highest conductivity in the deep conductor lies within the crustal root zone, and the enhanced conductivity is inferred to be caused by release of highly saline fluid during prograde metamorphism of material (graywacke and the Mesozoic oceanic crust on which it sits) advected to deeper regions of the crust. For fluids of 4 to >20 wt% salt, the decreased lower

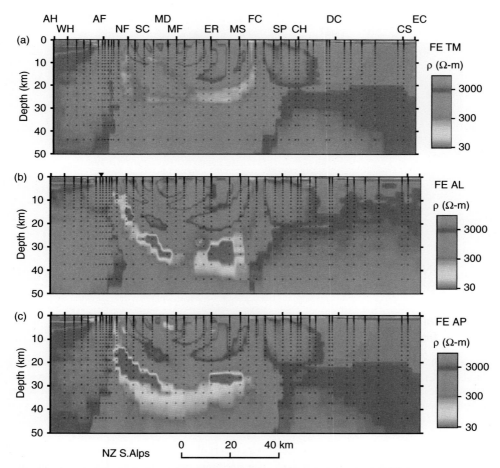

Figure 10.8. Finite-element inversion models fitted to the data. (a) Model (FE TM) fitted to the TM mode data only. (b) Model (FE AL) fitted to a dataset consisting of the TM mode data and subsets of the TE and tipper data. (c) Model (FE AP) fitted to the same subsets of the TE and tipper data using the TM model as the *a priori* model. (See plate section for color version.)

crustal resistivity observed in the TM mode model implies nominal porosity of 0.02–0.2% assuming grain-edge or grain-face fluid interconnection, and the lower resistivity in the TE mode models implies porosity of up to 1% for efficiently connected pores. It is not possible to estimate the true porosity because syntectonic fluid flow is expected to concentrate in faults and shear zones, and a regionally interconnected shear structure is postulated to dominate the bulk conductivity.

In the interpreted quartzo-feldspathic graywacke composition of the lower crust, fluids are expected to play an important role in controlling deformation. Fluids will increase diffusion creep deformation, creating additional weakening and enhancing bulk transport.

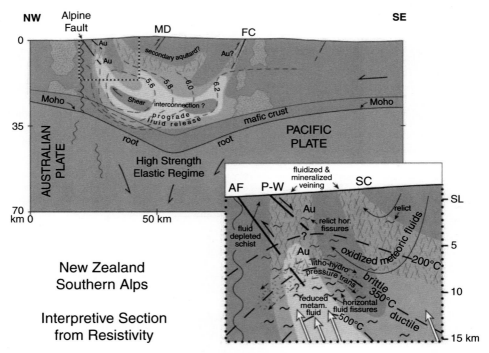

Figure 10.9. Geological interpretation of resistivity results and additional constraints from Wannamaker *et al.* (2002). Shading represents resistivity, with lighter shades indicating more conductive regions. Dashed lines represent P-wave velocity. Criss-cross patterns represent fault fracture mesh, and coarse stippling represents Late Cretaceous plutonic intrusive rocks.

Thus, the high-conductivity zone in the lower crust is expected to be a zone of weakness and a possible host of regional detachment. The northwestern of the two highest conducting zones corresponds to the fluid-rich dilatant region modeled in coupled deformation–fluid flow studies, and the southwestern zone may represent backthrusting.

The rise of the conductor to the northwest is spatially correlated with the position of the Alpine Fault, but the magnetotelluric data are unable to resolve the true form of the associated detachment. Shallowing of the conductor is interpreted to represent exhumation of high-grade schist and uplift of the brittle–ductile transition. This transition is believed to represent a permeability contrast and the upper depth bound of high lower crustal conductivity. Uplift would help to preserve free fluid at a higher than normal level in the crust by impeding retrograde reactions. The absence of a conductor right at the Alpine Fault is probably due to release of fluids through a fault-fracture mesh.

The authors also considered whether the conductors could be explained by the presence of graphite. Graphite occurs in particular lithologies in the metamorphosed graywackes making up much of the crust. However, the measured geometries of the conductors are inconsistent with the location of the graphitic lithologies, and it would be necessary for the graphite to have been mobilized and re-deposited to explain them. In this alternative interpretation, fluids once again play a key role in controlling the location of the conductors, but the observed conductivity results from a combination of fluids and a graphite solid phase.

10.2.3 Three-dimensional imaging of the Ossa Morena Zone of the Variscan fold–thrust belt

Excellent studies involving extensive data collection, detailed 2D inversion, 3D modeling and comprehensive geological interpretation have been conducted in the Betic-Rif orogen (Martí *et al.*, 2009; Rosell *et al.*, 2011) and in the southwest Iberian Variscides (Pous *et al.*, 2004; Muñoz *et al.*, 2008), both located in Spain. The latter studies are reviewed here to illustrate the application of magnetotellurics in an older tectonic setting.

The Variscan fold–thrust belt has extensive exposure in southwest Iberia. The southern part of the zone includes three continental blocks, the Ossa Morena Zone (OMZ), the Central Iberian Zone (CIZ) and the South Portuguese Zone (SPZ), separated by suture zones containing rocks of oceanic chemical affinity (Figure 10.10). The Guadiana Basin is a Neogene shallow terrestrial basin containing clastic sediments that partially overlies the boundary between the OMZ and CIZ. Magnetotelluric studies have been used to define the 3D geometry of these units, and in particular the OMZ. Pous *et al.* (2004) describe 2D modeling of profiles crossing the OMZ, and Muñoz *et al.* (2008) describe 3D modeling of an expanded dataset centered on the OMZ.

The OMZ includes a variably metamorphosed Proterozoic basement that locally includes high-pressure metamorphic rocks (Figure 10.11). Volcanic sedimentary sequences were added to the basement during two major stages of accretion. The first of these, during the Upper Proterozoic Cadomian Orogeny, imposed important controls on the crustal geometry near the northern margin of the OMZ. The second stage occurred during the Variscan Orogen in the Upper Devonian to Carboniferous eras, and led to juxtaposition of the OMZ with other terranes that control the crustal geometry near the its southern margin.

Figure 10.11 shows the site locations of the magnetotelluric dataset in the OMZ, along with the subset used for the 3D modeling study. The complete database consists of measurements from 148 sites covering the period range 0.004 s to 4000 s, along with vertical magnetic field transfer functions at 88 sites for the same period range. In earlier studies, the data from profiles S, P, O and L were analyzed using 2D techniques. The resistivity structures along all of the profiles exhibit a preferred strike direction between N100°E and N125°E. The data from new profile I was found to be inconsistent with a 2D model, suggesting the presence of off-line conductors to the east and west of the profile. In addition, induction vectors indicate that a conductor lies to the west that is not explained by the presence of the Atlantic Ocean, requiring a conductor beneath the Tejo Basin (Figure 10.11).

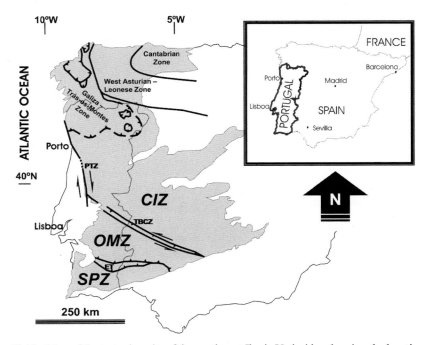

Figure 10.10. Map of the tectonic units of the southwest Iberia Variscides showing the location of the Ossa Morena Zone (OMZ). Taken from Muñoz *et al.* (2008). Gray-shaded area represents the pre-Mesozoic basement in the Iberian Peninsula; CIZ, Central Iberian Zone; SPZ, South Portuguese Zone; PTZ, Porto–Tomar shear zone; TBCS, Tomar–Badajoz–Córdoba shear belt; ET, exotic terranes. Thin black lines represent the traces of thrust zones and strike-slip fault zones.

The data from the OMZ were therefore re-modeled using a 3D trial-and-error forward approach.

An initial phase of 3D modeling was carried out to evaluate the influence of the surrounding ocean on the magnetotelluric and tipper responses. The model consisted of a 1D Earth with a superimposed 3D conductor representing the ocean. The mantle resistivity in the 1D Earth was chosen to be high to maximize the effect of the ocean on the data. Figures 10.12 and 10.13 show the resulting responses. At 20 s, the ocean has an effect only on the induction vectors in the far south of the study area. At 100 s, the ocean effect is more significant, with the induction vector response at the majority of the magnetotelluric sites showing the influence of the Atlantic Ocean to the south and west. As noted above, the model did not fully explain the observed induction vectors, thus indicating the presence of additional structures with the continental crust. The main effect of the ocean on the magnetotelluric responses is a decrease in apparent resistivity and an increase in phase in the longer-period (>10 s) response in the component with electric fields parallel to the coastline. However, there are also smaller changes in the other response elements.

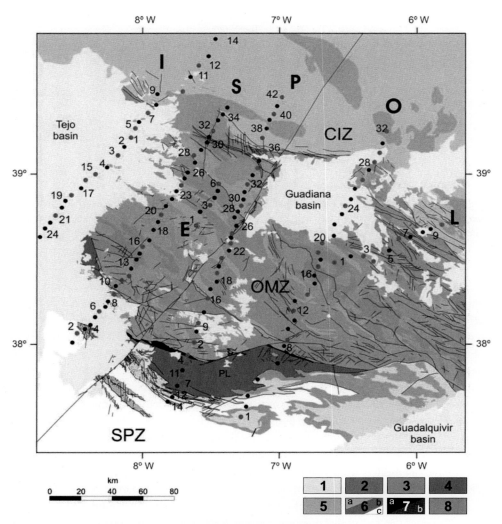

Figure 10.11. Location of the magnetotelluric sites. Modified from Muñoz *et al.* (2008). Dark gray (red) circles show sites chosen for 3D model construction. Simplified geology is as follows: 1, Cenozoic sedimentary cover; 2, Ordovician–Silurian to Carboniferous sequences in OMZ; 3, Cambrian sequences in OMZ; 4, Upper Proterozoic in OMZ; 5, Paleozoic sequences in CIZ; 6, Paleozoic sequences in SPZ (a, Phyllitic–Quartzitic Group, Upper Devonian; b, Volcanic–Sedimentary Complex, Upper Devonian to Lower Visean; c, Flysch Group, Upper Visean to Lower Westfalian); 7, exotic terranes (a, Beja-Acebuches Ophiolite Complex; b, Pulo do Lobo Terrane); 8, intrusive bodies, mostly granitic in nature and Variscan in age. (See plate section for color version.)

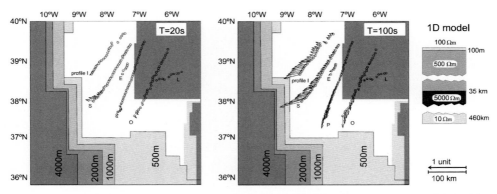

Figure 10.12. Influence of the ocean on induction vectors. Modified from Muñoz *et al.* (2008). The real induction vector response (Parkinson convention) is shown for a 1D model with superimposed oceans of 0.3 Ω m and the depth shown in the figure.

The 3D modeling used a set of 61 carefully selected magnetotelluric sites (Figure 10.11) chosen to be representative of the regional response and to minimize galvanic distortion effects. The starting model was built using 2D modeling results and induction vector data as a guide. The mesh size was determined by successively refining it until stable results were obtained, and consisted of 80 (N–S) × 77 (E–W) × 76 (vertical) cells, corresponding to an average cell size of 4 km × 4 km in the horizontal plane and increasing exponentially with depth from 25 m at the surface to 5 km at 50 km depth. Several layers of uniform conductivity were included at the base of the model. The modeling minimized the misfit to the apparent resistivity and phase of the off-diagonal components of the impedance tensor at all sites, and included the diagonal components for profile I, as well as the real and imaginary components of the geomagnetic transfer functions.

A total of 84 models were examined in the process of determining a reasonable data fit. The misfit was calculated assuming error floors of 10% for apparent resistivity, 5° for phase and 0.05 for the geomagnetic transfer functions, yielding a total RMS misfit of 4.58. The misfit on individual components was 5.42 on the apparent resistivity of the off-diagonal components, 2.79 for the phases of the off-diagonal components, 14.96 for the apparent resistivity of the diagonal components, and 2.86 for the magnetic transfer functions. The misfit was distributed fairly evenly across the study area. Figure 10.14 shows the final resistivity model. The forward modeling included thorough testing of the continuity of the various conductors and resistors, ensuring that their final form is robust.

The distribution of the conductive and resistive zones has provided significant three-dimensional information on the regional geology and tectonics. Some of the conductors, such as C3 and C8, correspond to the OMZ boundaries. The present discussion will focus on the observed enlargement and coalescence of the conductors that occurs at 15–20 km depth

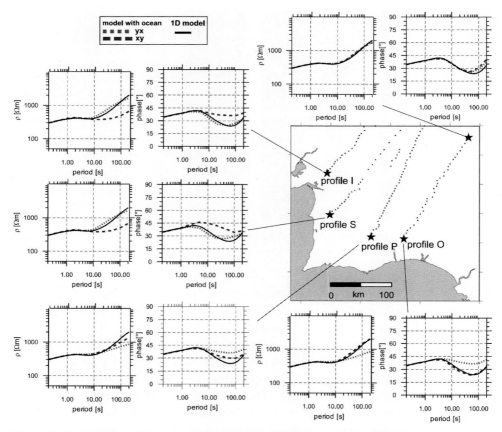

Figure 10.13. Influence of the ocean on the magnetotelluric response. Modified from Muñoz *et al.* (2008). The off-diagonal components of the responses are shown for the model in Figure 10.12. *X* refers to the north and *Y* to the east polarizations.

and extend to around 30 km. This pattern can be seen more clearly in Figure 10.15. At 20 km, the depth at which the conductive structure extends more than 200 km east–west and 185 km north–south, there are merging conductors within the feature associated with internal domains of the OMZ (e.g. C1, C7 and C9) and those associated with its boundaries (C3 and C8). At slightly shallower depths (15–20 km), the feature is semi-continuous, but broken by resistive structures such as R1 and deep, sub-vertical tectonic structures such as C6. The conductivity decreases with depth within the zone.

The extensive conductive zone correlates spatially with a 2 s thick, reflective band in seismics called the Iberian Reflective Body (IRB). Taken together, the resistivity and seismic data support the presence of an important décollement that has developed immediately above, or coinciding with, the top of the OMZ granulite basement. This

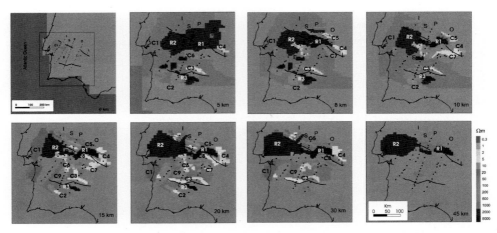

Figure 10.14. Final 3D resistivity model shown as slices at increasing depth. Modified from Muñoz *et al.* (2008). (See plate section for color version.)

interpretation is also supported by the study of Pous *et al.* (2004), who conducted laboratory measurements on graphite in the Precambrian Serie Negra rocks from within the OMZ. This work used backscattered electron imaging to define the distribution of carbon within representative samples of the Serie Negra black slates (Figure 10.16). The results demonstrate directly that penetrative deformation can enhance the connectivity of graphite present in the rocks by concentrating it in veinlets and producing finely disseminated graphite in fault gouge. The conductivity may be further enhanced by shearing, which would improve the connectivity of the graphite phase due to its high plasticity.

The mid-crustal décollement has important implications for the tectonics of the OMZ region. It is interpreted to lie at the top of the granulitic basement and the overlying pile of metasedimentary rocks. The resistivity results suggest that tectonic units to the north and south of the OMZ (CIZ and SPZ) have an identical sub-horizontal structure at the same depth, although this is not as well resolved in the other units. In general, sub-horizontal shear zones, acting as décollements, are interpreted to develop as a consequence of compression in rheologically stratified media. During orogenesis, the pattern of strain is partitioned vertically by decoupling the shallow, more deformed, upper crust from the lower crust, and the lower crust may be decoupled from the mantle in a similar way. This pattern of strain proportioning is expected to be a common feature of orogens involving oblique continental collision.

The specific tectonics of the décollement in the OMZ may vary from the north to the south. In the north, where the northern OMZ boundary represents a reactivated Cadomian suture, the strain partitioning would have been controlled during the Variscan orogen by zones of pre-existing weakness. This could have caused decoupling at the base of the metasedimentary pile

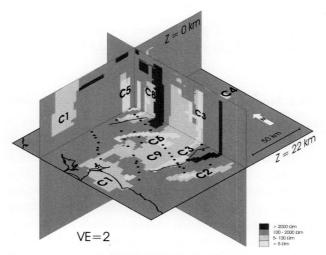

Figure 10.15. Synthesized view of the 3D resistivity model from Muñoz *et al*. (2008). The number of divisions in the resistivity scale has been reduced so that conductors with similar resistivity (i.e. C1, C3, C6 and C9) are merged.

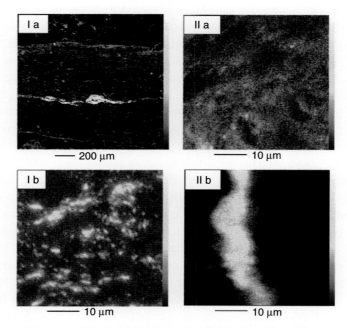

Figure 10.16. Backscattered electron microprobe images of Serie Negra black slates. Modified from Pous *et al*. (2004). The images show the concentration of carbon, with pale shade (yellow) indicating increased concentration. Additional images showing oxygen concentration (not included) allow the enhanced carbon to be attributed to graphite rather than to carbonate. (I a) Continuous bands of graphite along platy cleavage. (I b) Discrete flakes of graphite (1–5 μm diameter, 1 μm thickness). (II) Examples of graphite fault gouge: (II a) small grains (up to 1 μm) pervasively disseminated in silicate groundmass; and (II b) filling of thin veinlets (up to 10 μm in width) parallel to the fault. (See plate section for color version.)

rather than pure shear thickening of the upper crust. Farther south, the mid-crustal structures are due to mechanical heterogeneity generated by the OMZ–SPZ continental collision.

10.3 Magnetotelluric studies of the continental mantle

This section examines applications of magnetotellurics to the study of structures and processes in the continental mantle. Chapter 3 reviews the resistivity structure of continental mantle. It was shown that temperature is a major factor in controlling mantle resistivity, whereas bulk composition is expected to have a smaller influence (Jones *et al.*, 2009a). Other aspects of composition, such as the presence of fluids or melt (Glover *et al.*, 2000), carbon (Duba & Shankland, 1982; Mareschal *et al.*, 1995), sulfides (Ducea & Park, 2000), hydrous phases (Boerner *et al.*, 1999) and hydrogen diffusion (Karato, 1990; Hirth *et al.*, 2000), may have a significant influence on mantle resistivity, but the full influence of these parameters is still under debate. It has also been suggested that grain size may influence continental mantle resistivity (Spratt *et al.*, 2009).

Magnetotellurics has been used in continental mantle studies to define both vertical and lateral variations in resistivity. To resolve resistivity in the asthenosphere and at greater depths, magnetotelluric data are sometimes integrated with geomagnetic depth sounding methods (e.g. Schultz *et al.*, 1993; Bahr *et al.*, 1993). Targets have included very large-scale features, such as the lithosphere–asthenosphere boundary beneath stable cratons (e.g. Muller *et al.*, 2009), and smaller-scale features such as resistivity structure in suture zones (e.g. Wu *et al.*, 2005; Spratt *et al.*, 2009) and subducting plates (e.g. Kurtz *et al.*, 1986; Wannamaker *et al.*, 1989; Ichiki *et al.*, 2000; Brasse *et al.*, 2002; Ledo *et al.*, 2004; Jödicke *et al.*, 2006; Soyer & Unsworth, 2006). Magnetotelluric studies of the continental mantle have usually been carried out for scientific purposes, but there are also important applications in strategies for diamond exploration (Jones & Craven, 2004; Jones *et al.*, 2009b; Türkoğlu *et al.*, 2009). Another important target for mantle magnetotelluric surveys has been resistivity anisotropy (Ji *et al.*, 1996; Bahr & Duba, 2000; Bahr & Simpson, 2002; Frederiksen *et al.*, 2006; Hamilton *et al.*, 2006; Roux *et al.*, 2011). Resistivity anisotropy in the lithospheric mantle can reflect deformational processes such as shear, and may record the direction of past motion of the lithospheric plate. Within the asthenosphere, resistivity anisotropy is controlled mainly by present-day plate motion direction.

The case studies considered below illustrate large-scale, long-period magnetotelluric studies of Precambrian continental mantle. To resolve resistivity at depths of several hundred kilometers, it is necessary to use very large arrays and recording times that are sufficiently long to yield accurate responses to periods of 10^4 s or longer. At the same time, it is necessary to use reasonably dense arrays and determine the shorter-period magnetotelluric response to define the resistivity structure of the overlying crust and the effect that it has on the long-period response. The resulting magnetotelluric analyses involve large datasets and broad period ranges, and require careful selection of remote reference sites and thorough dimensionality and strike analyses.

10.3.1 Slave Craton

The Archean Slave Craton is located in northwestern North America. It is one of a number of Archean cratons welded together by Proterozoic orogenic belts that form the central nucleus of North America (Hoffman, 1988). The craton includes the oldest known rocks on Earth, with a sample of Acasta gniess from the Anton complex in the southwestern part of the craton having been dated at 4.03 Ga. The Anton complex is part of a contiguous Mesoarchean basement complex in the western half of the craton. Deep-probing magnetotelluric soundings were conducted in the Slave Craton as a component of the Canadian Lithoprobe Slave and Northern Cordillera Lithospheric Evolution (SNORCLE) transect, and later as collaborative studies between investigators at the Geological Survey of Canada and Woods Hole Oceanographic Institution. Data were collected using three fundamentally different approaches: conventional land recordings, winter ice road recordings and lake-bottom recordings over a full year (Figure 10.17).

The first magnetotelluric soundings done in the Slave Craton in the SNORCLE project were conventional land measurements collected along an all-weather road spanning the most accessible part of the region. Broadband soundings were completed at sites spaced approximately 10 km apart, and long-period soundings were collected at every second broadband site. The broadband soundings were done using Phoenix V5 instrumentation and recordings of two-day duration. Electric lines were all approximately 100 m long. The long-period data were acquired using LIMS instruments and the same electrode holes as for the broadband surveys to minimize static shift. The soundings were of either one-, two- or three-week duration, with the longer recordings spaced evenly along the profile. Winter magnetotelluric recordings were made along the seasonal ice road connecting the northern parts of the Slave Craton with more developed regions to the south. Recordings were done in temperatures as low as $-30°C$ using frozen lakes near the ice road, and employed Phoenix V5 and GMS-06 instruments (Jones *et al.*, 2003). Electrode lines crossed the lake surface and terminated with electrodes lowered to the lake floor through holes drilled in the ice. Magnetometers were deployed adjacent to the lakes because movement of the frozen lake ice introduced noise into the recordings at 10–100 s periods. Sites in the remote northern part of the Slave Craton were installed by Twin Otter float plane, and consisted of lake-bottom recordings done using low-power continental shelf magnetotelluric instrumentation. Electrode chemistry was adjusted to suit fresh-water installations. These sites were very long-period deployments with durations of one year (Jones *et al.*, 2003).

Data collected in these surveys were analyzed using several types of robust algorithms employing multiple remote references to reduce the influence of non-uniform source fields at the subauroral to auroral latitudes of the survey. Tensor decomposition methods including multi-site, multifrequency Groom–Bailey decomposition were also employed.

The magnetotelluric surveys have revealed a number of important features of the lithospheric mantle of the Slave Craton. The results from the Anton complex in the southwest corner of the craton reveal the presence of extremely resistive crust (>40 000 Ω m), enabling the first definitive electrical imaging of the Mohorovičić (moho) discontinuity (Jones & Ferguson,

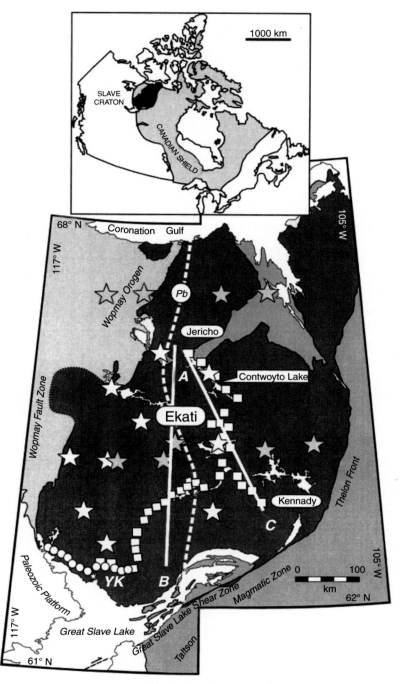

Figure 10.17. Map of the study area. Modified from Jones *et al.* (2001). White circles represent all-weather road sites, white squares represent winter road sites and stars represent lake-bottom sites. Also shown is the north-trending crustal Pb isotope line. YK = Yellowknife.

2001). To establish the depth of the electrical moho, smoothed 1D inversions were carried out that include a discontinuity at a series of specified depths. The best-fitting model was derived for a discontinuity at 35.5 km depth, providing excellent agreement with seismic reflection, seismic refraction and teleseismic studies on the SNORCLE profile, which indicate a moho depth of 36 km. The magnetotelluric results also provide a rare accurate measurement of the resistivity of the uppermost continental mantle of 4000 Ω m (Figure 10.18).

Jones *et al.* (2003) combined the magnetotelluric responses from 138 sites in the Slave Craton to yield a spatially averaged craton response to which they fitted a five-layer 1D model extending from the surface to 300 km depth (Figure 10.19). The model has a resistivity of ~1000 Ω m in the upper 6 km, underlain by very resistive crust (>5000 Ω m). The craton average for the resistivity of the crust and underlying mantle (1100 Ω m) are lower than observed in the Anton complex. The model includes a decrease in mantle resistivity to 300 Ω m at 135 km and a significant decrease to 75 Ω m at 260 km. The 260 km change is interpreted to be the spatially averaged depth to the base of the Slave lithosphere; this value is weighted toward the southern part of the craton where most of the sites lie. Figure 10.19 compares the Slave result with an accurately determined 1D resistivity model from the middle of the Archean Superior Craton in central Canada (Schultz *et al.*, 1993). Remarkably, the responses merge at a period of about 300 s, and indicate a similar thickness for the lithosphere beneath the two cratons. Hirth *et al.* (2000) interpreted the resistivity of the deeper components of the Superior lithosphere to be the consequence of a low (relative to oceanic lithosphere) water content, and suggested that it may be stabilized against convective instabilities by a concomitant high viscosity. The similarity of the Superior and Slave resistivity structures suggests that a similar mechanism may pertain beneath the latter. The uppermost 10 km of the crust in the central Superior Craton is more resistive than the spatially averaged result for the Slave Craton, but the rest of the Superior crust is significantly more conductive than for the Slave Craton.

Maps of the phase response across the Slave Craton provide an excellent overview of resistivity variation (Figure 10.20). Relatively low phase in the extreme southwest of the craton at the shortest periods indicates the limited lateral extent of the extremely resistive crustal rocks of the Anton complex. There is a significant increase in the phase response in the central Slave in the 30 s and 100 s maps, providing a suggestion of a conductor in the mantle. This feature has been named the Central Slave Mantle Conductor (CSMC), and lies mainly within a specific geochemically defined mantle domain boundary. The high phases farther to the northwest at 300 s and 1000 s indicate that the conductor dips in this direction. The CSMC also correlates spatially with the locus of diamond occurrence in the Slave.

Two-dimensional inversion was done for a number of profiles in the Slave Craton. Figure 10.21 shows results for a 150 km long east–west profile in the southwest and a north–south profile crossing the CSMC. The inversions assume a geoelectric strike direction perpendicular to each profile. This assumption will not introduce spurious features into the models, as the resistivity structure changes very slowly along each profile, and the response at each site is close to a 1D form (Jones *et al.*, 2003). The inversions used the nonlinear conjugate gradient

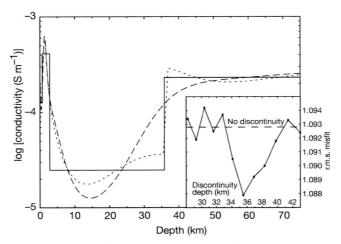

Figure 10.18. One-dimensional conductivity–depth models that fit the azimuthally averaged impedances from SNORCLE site 106 in the Anton complex. Taken from Jones & Ferguson (2001). Solid line: best-fitting layered-Earth model. Dashed line: best-fitting continuous model. Dotted line: best-fitting continuous model with a permitted step change at 35.5 km. Inset: RMS misfit for best-fitting continuous models varying the depth to the permitted step change.

code of Rodi & Mackie (2001), fitting the TE and TM responses simultaneously to within 5° in phase and 25% in apparent resistivity to allow for static shift.

The results for the north–south profile provide an excellent image of the CSMC. The top of the mantle conductor is at a depth of 80–100 km. It has an internal resistivity of <30 Ω m, and sensitivity studies show that, for a resistivity of 15 Ω m, the body must be at least 15 km thick (Jones *et al.*, 2001). Kimberlite studies have defined a unique two-layer mantle lithosphere in the central Slave, with an ultra-depleted harzburgitic layer separated from a less depleted layer at 140–150 km depth. There is excellent spatial correlation of the ultra-depleted layer with the lateral extent and depth of the CSMC. Jones *et al.* (2001) interpret the CSMC as due to the presence of carbon in the mantle. At depths below 145 km, which delineates the graphite–diamond stability boundary, carbon would exist as graphite and contribute to the observed CSMC anomaly. However, at greater depths, carbon would be present as resistive diamond. The presence of the CSMC provides constraints on tectonic interpretations of the formation of the Slave Craton. The structure was possibly emplaced during lithospheric subduction and the trapping of overlying oceanic mantle at 2630–2620 Ma.

10.3.2 *Kaapvaal Craton*

The Kaapvaal Craton and neighboring Archean cratons in Angola and Zimbabwe, and their stitching belts, the Namaqua–Natal Mobile Belt, Limpopo Mobile Belt, Rehoboth Terrane, Damara–Ghanzi Chobe Belt, Kheis Belt and Okwa Terrane, have recently been studied

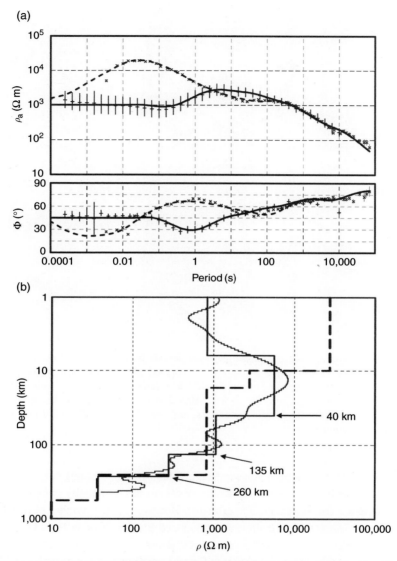

Figure 10.19. (a) Spatially averaged Slave magnetotelluric responses (+) with 95% error bounds and the layered model response from Jones *et al.* (2003). The apparent resistivity response has been corrected for static shift as discussed in Jones *et al.* (2003). Also shown is the response (×) from the central Superior Craton obtained by Schultz *et al.* (1993). (b) The 1D layered resistivity model (thick solid line) and smooth resistivity model (thin solid line) that best fit the responses and the 1D layered resistivity model for the Superior Craton (dashed line).

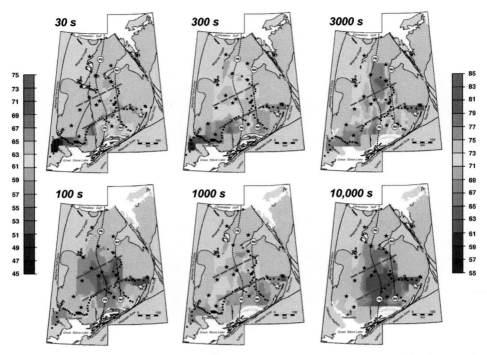

Figure 10.20. Contoured averaged phase maps for periods from 30 to 10 000 s. The data are the arithmetic (Berdichevsky) average of the two orthogonal phases, and is a rotational invariant. The phases in the shortest frequency maps (30 and 100 s) range from 45° to 75°, whereas the phases in the other four maps range from 55° to 85°. Also shown on the maps are the locations of two geochemically defined mantle domain boundaries, and crustal Pb and Nd isotope boundaries. (See plate section for color version.)

during the SAMTEX (Southern African Magnetotelluric Experiment) project. The project was initially conceived to consist of two acquisition phases spatially coincident with the SASE (Southern African Seismic Experiment). However, as the number of partners grew from the initial four (Dublin Institute for Advanced Studies, Woods Hole Oceanographic Institution, Canadian Geological Survey and De Beers) to eleven during the course of the project, it was expanded from two phases to four to cover much of South Africa and Namibia and all of Botswana (Figure 10.22). The SAMTEX dataset comprises over 750 sites in an area of approximately two million square kilometers, and thus represents the largest regional-scale magnetotelluric study conducted to date.

   Results from SAMTEX have been presented as both depth images and formal models. The depth images were produced using the Niblett–Bostick transformation (see Chapter 4) to determine the rotationally variant maximum and minimum resistivities at a given depth (Jones *et al.*, 2009a). An example of depth imaging is shown in Figure 10.23, where the

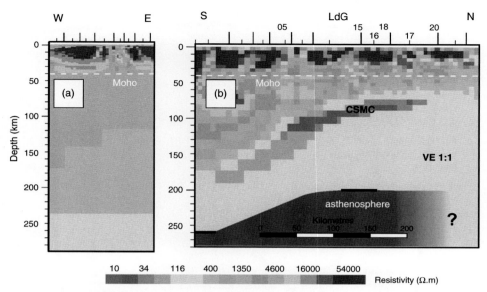

Figure 10.21. (a) Two-dimensional inversion model for sites along an east–west profile in the southwest of the Slave Craton. (b) Two-dimensional inversion model for sites along a north–south profile crossing the CSMC (along line AB shown in Figure 10.17). Modified from Jones *et al.* (2003). (See plate section for color version.)

maximum resistivity at each station is smoothed and contoured – see Jones *et al.* (2009a) for details. Also shown on the figure are the locations of publicly known kimberlites, coded in dark gray (red) for diamondiferous kimberlites, mid-gray (green) for non-diamondiferous kimberlites, and white for those whose diamond potential is not reported in the open literature. Diamondiferous kimberlites are evidence for lithosphere that is at least 20 km thicker than the graphite–diamond stability boundary of around 150 km beneath cratons. Non-diamondiferous kimberlites are suggestive of lithosphere that is thinner than that. There is a remarkable spatial coincidence between the locations of diamondiferous kimberlites and the edges of cratons, suggesting either that the kimberlite magmas are unable to penetrate through the thick roots at their center or that the initiation of kimberlitic eruptive magmas occurs preferentially at depths shallower than the thickest roots.

Two-dimensional models were obtained using Rodi & Mackie's (2001) code, as implemented in the Schlumberger WinGLink system, plus a newer 2D anisotropy code presented in Baba *et al.* (2006a). Models from three profiles are presented here: one for the main SW–NE KAP03 profile (mid-gray/blue Phase I dots on Figure 10.22), one for the SE–NW KIM–NAM profile (mid-gray/blue and dark gray/purple Phases I and II dots on Figure 10.22), and one for the N–S ZIM profile (mid-gray/green Phase III dots in NE Botswana). The anisotropic model of the KAP03 profile (Figure 10.24) reveals complex structures within the lithospheric mantle. Large variations in maximum resistivity at depths to 200–250 km relate directly to age and

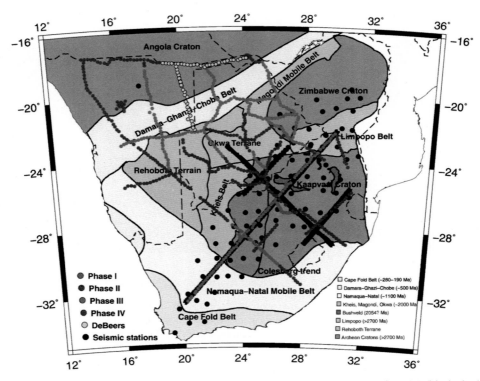

Figure 10.22. SAMTEX (dots in various shades of gray or various colors) and SASE (black dots) coverage of southern Africa. The tectonic base is from Webb (2009). (See plate section for color version.)

tectonic provenance of surface structures. Within the central portions of the Kaapvaal Craton, there are regions of resistive lithosphere about 230 km thick, in agreement with estimates from xenolith thermobarometry (Rudnick & Nyblade, 1999) and SASE seismic surface-wave tomography (Li & Burke, 2006), and in reasonable agreement with the 260 km thickness for the Slave and Superior lithospheres discussed above (Figure 10.19), but thinner than the value inferred from seismic body-wave tomography in SASE (Fouch *et al.*, 2004). The magnetotelluric data are unable to discriminate between a completely dry or a slightly "damp" (water at a few hundred parts per million) structure within the transitional region at the base of the lithosphere. However, the structure of the uppermost ~150 km of lithosphere is consistent with enhanced, but still low, conductivities reported for hydrous olivine and orthopyroxene at levels of water consistent with that in Kaapvaal xenoliths. The electrical lithosphere around the Kimberley and Premier diamond mines is thinner than the maximum craton thickness found between Kimberley and Johannesburg/Pretoria. The mantle beneath the Bushveld Complex is highly conducting at depths of around 60 km. Possible explanations for these high conductivities include graphite or sulfide and/or iron metals associated with the Bushveld magmatic event. The authors suggest that one of these conductive phases (most likely melt-related

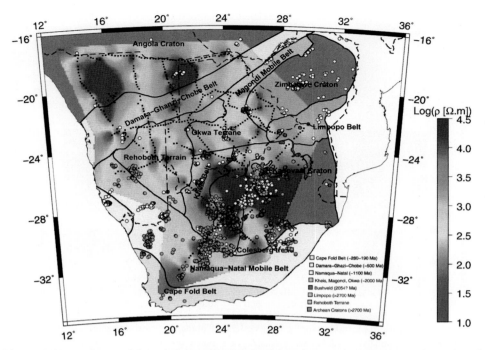

Figure 10.23. An image of the resistivity at 200 km depth based on an approximate transformation of the magnetotelluric responses from period to depth and taking the maximum resistivity. See Jones *et al.* (2009a) for details. The gray shades (various colors) are $\log_{10}$(resistivity), and the black dots show stations where data were used. Also shown on the figure are kimberlite locations: dark gray (red) means known to be diamondiferous, mid-gray (green) means known to be non-diamondiferous, and white means not defined or unknown. (See plate section for color version.)

sulfides) could electrically connect iron-rich garnets in a garnet-rich eclogitic composition associated with a relict subduction slab.

Two models obtained from the data along the perpendicular KIM–NAM profile are shown in Figure 10.25, one for the data defined in a coordinate system with strike N25°E, and the other for the data in the system N45°E (Muller *et al.*, 2009). Inversion parameters were as follows: simultaneous inversion for phase and apparent resistivity for both TE and TM modes, and for static shift, with smoothing factor $\tau = 3.0$ and error floors for phase and apparent resistivity set to 5% and 10%, respectively. Although available at some stations, tipper data were not used. The value of the parameter controlling the smoothness of the model was chosen based on an L-curve of model RMS error versus τ. The 2D inversion RMS misfit errors are the root mean square differences between the observed and calculated magnetotelluric responses at each site normalized by an error measure incorporating both "observational" and "2D decomposition" errors. Because "decomposition" errors are

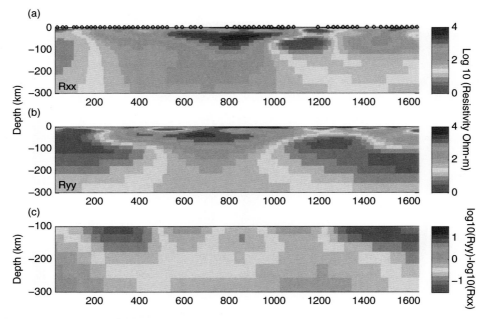

Figure 10.24. Anisotropic models for the Kaapvaal data. Conductivity in the directions (a) perpendicular to the profile (R_{xx}) and (b) parallel to the profile (R_{yy}) are shown. (c) The levels of anisotropy calculated as the difference in $\log_{10}$(resistivity) between the R_{yy} and R_{xx} models. Taken from Evans *et al.* (2011). (See plate section for color version.)

incorporated into the magnetotelluric data error, the final 2D models only match the magnetotelluric data observations to the extent that they are two-dimensional.

The 1400 km long KIM–NAM profile across the western part of the Archean Kaapvaal Craton, the Proterozoic Rehoboth Terrane and the Late Proterozoic/Early Phanerozoic Ghanzi–Chobe/Damara Belt reveals significant lateral heterogeneity in the electrical resistivity structure of the southern African lithosphere. The lithospheric structures of the Rehoboth Terrane and Ghanzi–Chobe/Damara Belt have not been imaged previously by geophysical methods. Temperature has the primary control on the resistivity of mantle minerals, and the magnetotelluric-derived lithospheric thicknesses therefore provide a very reasonable proxy for the "thermal" thickness of the lithosphere (i.e. the thickness defined by the intersection of a conductive geotherm with the mantle adiabat), allowing approximate present-day geotherms to be calculated. This indicates the following present-day average lithospheric thicknesses, to a precision of about ± 20 km, for each of the terranes traversed (inferred geotherms in brackets): Eastern Kimberley Block of the Kaapvaal Craton, 220 km (41 mW/m^2); Western Kimberley Block, 190 km (44 mW/m^2); Rehoboth Terrane, 180 km (45 mW/m^2); and Ghanzi–Chobe/Damara Belt, 160 km (48 mW/m^2).

A clear relationship between the electrical resistivity structure of the lithosphere and the tectonic stabilization age of the terrane is evident. Good agreement between the inferred present-day lithospheric geotherms and surface heat flow measurements indicate that the latter are strongly controlled by variation in lithospheric thickness. A significant difference in lithospheric thickness is observed between the Eastern and Western Kimberley Blocks, and is consistent with seismic tomography images of the Kaapvaal Craton. The present-day lithospheric thickness, and reduced intrusion into the diamond stability field, accounts for the absence of diamondiferous kimberlites in the Gibeon and Gordonia kimberlite fields in the Rehoboth Terrane. However, previously published mantle xenolith *P–T* arrays from the Gibeon, Gordonia and Kimberley fields suggest that the Rehoboth Terrane has equilibrated to a cooler conductive paleogeotherm (40–42 mW/m^2), very similar to that of the Eastern Kimberley Block of the Kaapvaal Craton, at some time prior to the Mesozoic eruption of the kimberlites. The timing and nature of both the thermal equilibration of the Rehoboth Terrane, and the subsequent lithospheric heating/thinning event required to account for its present-day lithospheric structure, are not well constrained. A model combining the penetration of heat-transporting magmas into the lithosphere with associated chemical refertilization at an early stage of Mesozoic activation appears to be the most plausible way to account for both the present-day lithospheric structure of the Rehoboth Terrane and an earlier, cooler paleogeotherm. However, problems remain for the isostatic response of the model. Based on a compilation of xenocryst Cr/Ca-in-pyrope barometry observations, the extent of depleted mantle in the Rehoboth Terrane is significantly reduced with respect to the Eastern Kimberley Block: 117 km versus 138–167 km. It appears likely that the depletion depth in both terranes, at least in the vicinity of kimberlite eruption, is explained by refertilization of the lower lithospheric mantle.

Finally, within the framework of SAMTEX, a focused study was undertaken to gain improved knowledge of the lithospheric geometries and structures of the westerly extension of the Zimbabwe Craton into Botswana, with the overarching aim of increasing understanding of southern African tectonics (Miensopust *et al.*, 2011). The area of interest is located in northeastern Botswana where Kalahari Desert sands cover most of the geological terranes, and very little is known about lithospheric structure and thickness. Some of the regional-scale terrane boundary locations, defined based on potential field data, are insufficiently accurate for local studies. Investigation of the NNW–SSE oriented, 600 km long ZIM profile (Figure 10.26) crossing the Zimbabwe Craton, Magondi Mobile Belt and Ghanzi–Chobe Belt showed that the Zimbabwe Craton is characterized by thick (~220 km) resistive lithosphere, consistent with geochemical and geothermal estimates from kimberlite samples of the nearby Orapa and Letlhakane pipes (~175 km west of the profile). The lithospheric mantle of the Ghanzi–Chobe belt is resistive, but its lithosphere is only about 180 km thick. At crustal depths, a northward-dipping boundary between the Ghanzi–Chobe and Magondi belts was identified, and two mid- to lower-crustal conductors were discovered in the Magondi belt. The crustal terrane boundary between the Magondi and Ghanzi–Chobe belts was found to be located farther to the north, and the southwestern

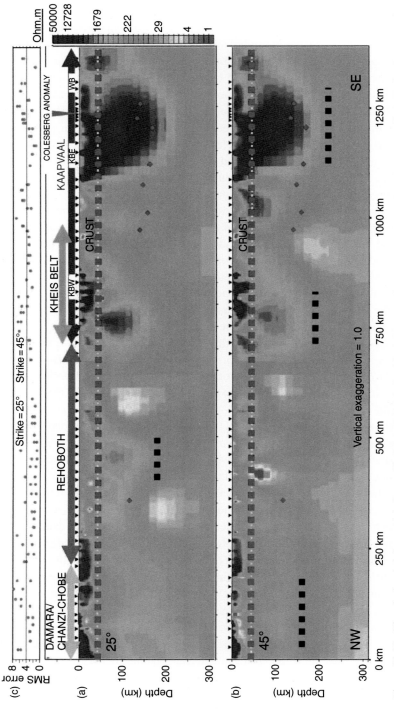

Figure 10.25. Electrical resistivity models for profile KIM–NAM derived from 2D smooth inversion of decomposed magnetotelluric station responses for (a) 25°E of N strike azimuth and (b) 45°E of N azimuth. (c) The 2D inversion RMS misfit error at each station. The surface extent of the geological terranes is shown in (a); abbreviations are used for Western Kimberley Block (KBW), Eastern Kimberley Block (KBE) and Witwatersrand Block (WB). Black dashed lines in (a) and (b) indicate the interpreted depth to the base of the lithosphere where well constrained, and dark gray (red) diamonds indicate the depth to the base of the chemically depleted lithosphere as defined by Cr/Ca-in-pyrope barometry from kimberlitic concentrates. Modified from Muller et al. (2009). (See plate section for color version.)

boundary of the Zimbabwe Craton might be further to the west than previously inferred from regional potential field data.

10.4 Applied magnetotelluric studies

There are numerous applications of magnetotellurics to targets of economic and societal interest, including mineral exploration, petroleum exploration, geothermal exploration, groundwater exploration and salinity evaluation, environmental and engineering assessment, and earthquake and volcano monitoring.

Focusing on surveys of economic targets, magnetotellurics – and particularly audiomagnetotellurics – has been utilized in mineral exploration; see reviews by Meju (2002) and Sheard *et al.* (2005). Magnetotellurics may be used to delineate the host geological structure (Boerner *et al.*, 1993; Routh & Oldenburg, 1999; Dennis *et al.*, 2011), to reveal small- and large-scale alteration zones and hydrothermal zones associated with the fluid processes producing a deposit (Heinson *et al.*, 2006), or to directly image an ore body (Livelybrooks *et al.*, 1996; Phillips *et al.*, 2001; Jones & Garcia, 2003; Queralt *et al.*, 2007). In diamond exploration, deep-penetrating magnetotelluric surveys have been applied both to define areas with a lithospheric geometry suitable for diamond formation (Jones & Craven, 2004; Jones *et al.* 2009b; Türkoğlu *et al.*, 2009) and to image kimberlites (e.g. Pettit, 2009). Magnetotellurics has been applied to petroleum exploration in situations where access is difficult for conventional seismic reflection methods, and for cases in which near-surface geological features, including extrusive rocks, carbonate reefs and salt structures, reduce the resolution of seismic reflection data (Orange, 1989; Christopherson, 1991; Withers *et al.*, 1994; Sheard *et al.*, 2005; Unsworth, 2005). Magnetotellurics can be used both to indirectly detect resistive hydrocarbons when they displace conductive saline pore fluids, and to image structures that are potential traps. Pellerin *et al.* (1996) examine the resolution of resistivity targets in a conceptual geothermal system by geophysical methods, including magnetotellurics. Targets suitable for magnetotellurics include a conductive clay cap and possibly a more resistive underlying reservoir.

The magnetotelluric method has also been valuable in environmental, groundwater and engineering applications. For example, modern audiomagnetotelluric surveys, including radio-frequency surveys, have provided 3D delineation of waste pits (Newman *et al.*, 2000; Bastani & Pedersen, 2001), and definition of subsurface geological structures and hydrogeology at potential waste sites (Unsworth *et al.*, 2000). For many near-surface environmental targets, alternative electromagnetic methods are required to achieve the required spatial resolution – see, for example, the reviews by Nobes (1996), Tezkan (1999) and Pellerin (2002). However, magnetotelluric surveys can offer good spatial resolution when high site density and high-frequency recording is applied. Most recently, audiomagnetotelluric methods have been evaluated as a potential method to monitor carbon dioxide sequestration in subsurface aquifers. There are also wide-ranging applications of magnetotelluric surveys to groundwater studies, including aquifer delineation (Pedersen *et al.*, 2005; Falgàs *et al.*, 2011), assessment of aquifer quality (Sainato *et al.*, 2000) and delineation of

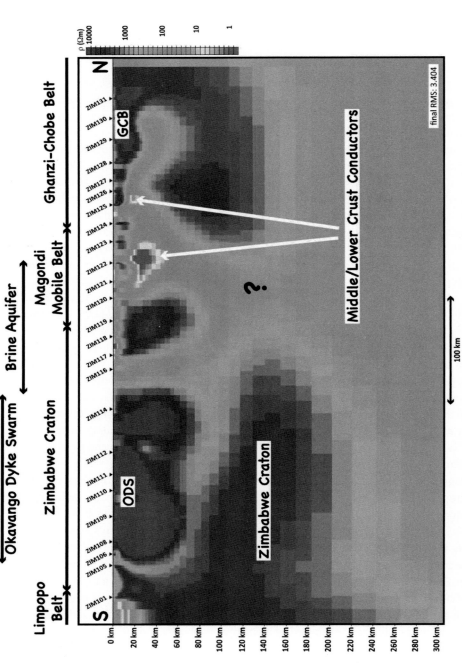

Figure 10.26. The 2D smooth inversion model (vertical exaggeration = 1.0) from the ZIM profile in relation to the known surface extent of geological terranes; taken from Miensopust *et al.* (2011). The arrows above the image of the resistivity structure show the crustal extents of the Limpopo Belt, Zimbabwe Craton, Magondi Mobile Belt and Ghanzi–Chobe Belt (GCB) with respect to magnetotelluric sites of the ZIM line; adapted from the regional-scale geological terrane boundaries based on potential field data (Webb, 2009). The extent of the Okavango Dike Swarm (ODS), known from magnetic data, is indicated, as well as an estimated extent of the brine aquifer related to the Makgadikgadi salt pan complex. The dominant resistivity features related to the main geological terranes are labeled and the question mark indicates the area of missing data coverage. Two dominant mid- to lower-crustal conductors are also apparent. (See plate section for color version.)

salt water intrusion (Falgàs *et al.*, 2005). Linde *et al.* (2006) and Slater (2007) review the relationship between electrical properties and hydrological parameters. Other applications of magnetotellurics to engineering surveys include the determination of suitable locations to ground electrodes in DC power systems (Manglik *et al.*, 2009) and modeling of geomagnetically induced currents and potentials in power systems (Pulkkinen *et al.*, 2007).

Finally, magnetotellurics has important applications for investigating and monitoring earthquake zones and volcanoes. In imaging applications, the method provides information on the distribution of aqueous fluids and melt. Seismic hypocenters associated with earthquake aftershock sequences are usually associated with resistive regions (Honkura *et al.*, 2000; Ogawa *et al.*, 2001). In monitoring applications, magnetotelluric surveys can be used to measure time variations of subsurface resistivity structure (e.g. Park, 1991), and to isolate normal magnetotelluric signals from the magnetic and electric field fluctuations arising from earthquakes or volcanoes. Johnston (1997) and Uyeshima (2010) review these aspects.

The two case studies below are included to illustrate modern applications of magnetotellurics to economic targets. They demonstrate that high magnetotelluric site density and careful dimensionality and modeling analysis are essential for near-surface studies.

10.4.1 Geothermal investigation

The magnetotelluric method has been successfully applied to evaluate geothermal resources around the world. As an exemplar, Heise *et al.* (2008) describe a geothermal study in New Zealand where cutting edge analyses were required, including 3D modeling of phase-tensor responses and 3D inverse modeling of tensor impedance data. The outcome provides an illustrative example of the phase tensor in magnetotelluric analysis, and shows the value of modeling this parameter in strongly 3D environments.

The geological setting is the Taupo Volcanic Zone of New Zealand in an active continental rift setting, of which the uppermost 1–3 km comprises a mixture of rhyolitic lavas, ignimbrites and volcaniclastic sediments. The principal target was the Rotokawa geothermal field located adjacent to the southeast margin of the Taupo Volcanic Zone in a region where basement rocks have been down-faulted along major SW–NE striking faults. The Taupo Volcanic Zone contains most of New Zealand's high-temperature (>200°C) geothermal systems. The Rotokawa geothermal field produces 33 MW of electric power from four production wells. Hydrological problems due to extraction of hot water from shallow parts of the field can be reduced by including high-temperature geothermal fluids from the deeper (>1500 m) parts of the hydrothermal system, motivating the study of Heise *et al.* (2008).

DC resistivity surveys have provided extensive information on the shallow parts of the hydrothermal system, and reveal low-resistivity areas within the geothermal field at Rotokawa. Deeper-penetrating bipole–dipole DC surveys show that the deeper parts of the system are more resistive. The decreased near-surface resistivity is attributed to a combination of hydrothermal alteration of young volcanics, producing conductive smectite clay, and the presence of high-temperature, saline fluid. At a greater depth, the resistivity increases due to a decrease in pore space and a transition of the alteration product to less conductive illitic and chloritic clay. Outside the geothermal system, the resistivity structure consists of a layer of resistive young volcanics overlying more conductive older volcanics and a resistive graywacke basement.

The magnetotelluric survey consisted of 64 broadband soundings with a period range of 3×10^{-4} s ($\sim$3000 Hz) to 2000 s. The sites form an irregular grid over an area of approximately 10 km $\times$ 10 km. Site spacing ranged between 2 and 4 km at the outer part of the survey area, and decreased to 200–500 m within the central part of the field, as defined by the shallow conductive zone detected by DC resistivity surveys. Figure 10.27 shows the phase tensors and induction vectors determined from the data. The gray shade (color) of the ellipses provides a measure of the invariant phase averaged over polarization direction. The higher phases observed at short periods suggest the presence of high-conductivity material within the shallow part of the hydrothermal system beneath a more resistive surface cap. Lower phases at longer periods indicate a decrease in resistivity at depth. Inward-pointing, real induction vectors are also consistent with the presence of a conductive region within the geothermal field. The orientations of the major and minor axes of the ellipses are a measure of the direction of maximum and minimum phase, respectively. Spatial variation of the phase tensors and induction vectors provide evidence for localized resistivity changes within the geothermal field.

The data were modeled using two approaches. In the first, the observed phase tensors were fitted using trial-and-error 3D forward modeling. The phase tensors provide information on spatial variation in resistivity and the resistivity gradient, but not the absolute level of resistivity, so it was necessary to include additional information to constrain the result. This came from transient electromagnetic soundings done at each site, and from DC resistivity surveys. These results, as well as those from long-period transient (LOTEM) soundings and regional magnetotelluric studies, were used to create a preliminary 3D model of the region. The phase-tensor response was then computed by forward modeling using a 3D finite-difference code. The fit of the model phase tensors to the observed ones was assessed using maps of the responses at different periods. The resistivity model was then adjusted on the basis of the observed misfit until satisfactory agreement was obtained.

Figure 10.27 compares the observed and model tensors at three periods, and also shows the misfit tensor that parameterizes the difference between the responses. At short periods, the phase tensor is fitted quite well, and the misfit tensors have a random alignment. However, at 1 s period, the phase tensors in the southeast show a consistent NW–SE alignment perpendicular to the southeast margin of the volcanic zone. This suggests that the conductance of the material overlying the basement has been underestimated in the model, permitting inappropriately large signal penetration.

Three-dimensional inversion of the impedance tensor data was based on the WSINV3D magnetotelluric code of Siripunavaraporn *et al.* (2005). Prior to inversion, distortion and static shift effects were removed from the data, and the result was interpolated to two estimates per decade and assigned 5% errors. After six iterations, the misfit reached a normalized RMS value of 5.77. Figure 10.27 shows the misfit ellipses for three periods and Figure 10.28 shows the model. The model results underestimate the phase at the highest frequency, and provided

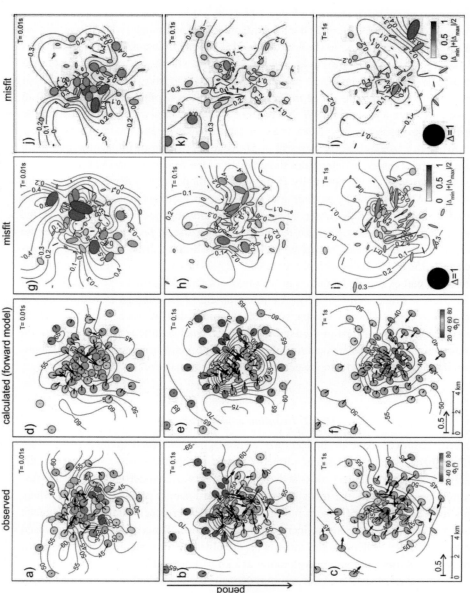

Figure 10.27. (a)–(c) Phase-tensor ellipses and induction arrows between 0.01 s and 1 s. The ellipses are normalized by Φ_{max} and the gray shades (colors) show . Induction arrows that point in the direction of increasing conductance show the strong conductivity contrast at shallow depths. (d)–(f) Calculated phase response from 3D forward modeling. (g)–(i) Tensor misfit ellipses calculated for the observed and calculated forward model phase tensors. The gray shade (color) used to fill the ellipses shows the mean of the maximum and minimum misfit. Small and light shaded (colored) ellipses indicate that the misfit is small. (j)–(l) Tensor misfit ellipses for the observed and calculated inversion model phase tensors. Modified from Heise et al. (2008). (See plate section for color version.)

a much poorer fit than the 3D forward model. This problem occurred because it was not possible to represent a 15 m thick, resistive surface layer adequately. At longer periods, the fit to the phase-tensor data is superior to that in the forward model, and the consistent misfit in the southeast of the study area is no longer observed.

Figures 10.28 and 10.29 compare the forward and inversion models, which show similar spatial variations in resistivity. At shallow depths (15–200 m), both models have resistivities of 1–5 Ω m within the geothermal field, in good agreement with the DC resistivity data. The region outside the geothermal field has resistive young volcanic rocks (300 Ω m) that become more conductive (30–10 Ω m) below about 600 m depth. The 800 and 1600 m depth sections show a prominent SW–NE band of 8 Ω m material that was included in the forward model to represent conductive volcanic rocks. The inversion model shows that the correct resistivity is significantly lower (1–8 Ω m) at the margin of the geothermal field. Because the phase response determines the gradient of the resistivity, and not the absolute value, overestimation of the resistivity in the forward model at this depth is reflected by inappropriately high resistivity extending to greater depth. Corresponding underestimation of the conductance causes the observed misfit in the forward model phase tensors in the southeast. One of the most significant features is the increase in resistivity at depth in the geothermal zone. This increase begins at ~1200 m, where drill log data show that the temperature exceeds 250°C. Smectite, a highly conductive clay, is unstable at these temperatures. The most resistive part of the model occurs in the middle of the geothermal zone, and correlates with the highest temperatures. Northwest-dipping basement is observed in both models at around 1600 m depth. Through comparison with seismicity data, it was possible to correlate the resistive region with a zone of fluid injection. The magnetotelluric results in combination with seismics appear to have identified the core of the geothermal system where high-temperature fluid is being fed into the geothermal system from deeper levels.

10.4.2 Uranium exploration

Tuncer *et al.* (2006) describe an application of audiomagnetotellurics to uranium exploration in the Proterozoic Athabasca Basin of northern Canada. Uranium deposits are found near the base of Proterozoic sedimentary rocks at their unconformity with underlying Archean rocks. As exploration progressed, the focus transitioned toward deeper parts of the basin, where deposits may be found at 500 m or more, and hence deeper-penetrating electromagnetic methods, such as magnetotellurics, increased in importance. Tuncer *et al.* (2006) focused on the McArthur River deposit and mine, with the objective of assessing the potential of audiomagnetotellurics in uranium exploration at a known deposit.

The formation and location of the uranium deposits are controlled by the unconformity and by cross-cutting faults. The faults are often graphitic, forming a good target for electromagnetic exploration. Alteration zones around the deposits contain quartz dissolution, and illitic clay formation represents a secondary resistivity target. Figure 10.30 shows the resistivity model for deposits in this area. The ore body lies beneath 525 m of resistive,

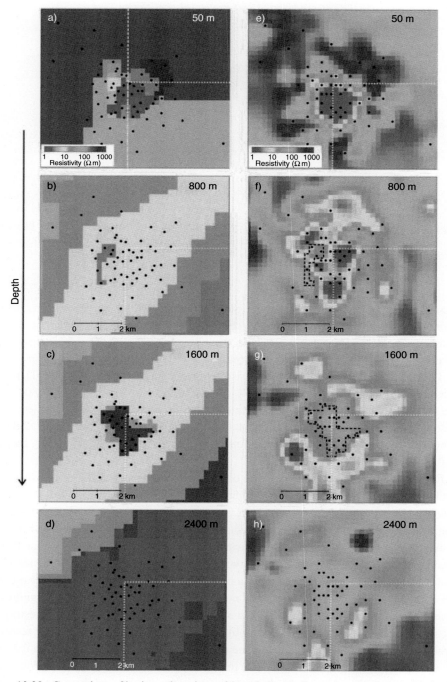

Figure 10.28. Comparison of horizontal sections of (a)–(d) the 3D forward model and (e)–(h) the 3D inverse model. Taken from Heise *et al.* (2008). (See plate section for color version.)

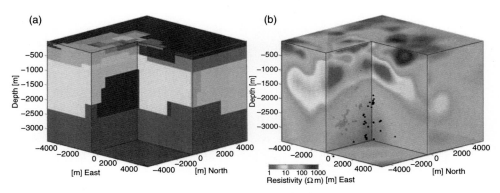

Figure 10.29. Block diagrams of (a) the 3D forward model and (b) the 3D inverse model. Earthquake hypocenters are shown in the inverse model as black dots. Taken from Heise *et al.* (2008). (See plate section for color version.)

sandstone-dominated sedimentary rocks of the Athabasca Group and 25 m of moderately resistive overburden. The underlying Archean gneisses of the Wollaston Group are very resistive. The ore body has lower resistivity than the host rocks, and lies at the intersection of a very conductive fault and the unconformity.

The audiomagnetotelluric survey consisted of soundings at 132 stations on 11 profiles crossing the P2 North fault at the deposit (Figure 10.31). The line spacing is ~800 m and the station spacing is ~300 m. Data were collected using Metronix audiomagnetotelluric systems with EMI BF-6 and BF-10 induction coils. The time series were processed using robust algorithms, yielding usable audiomagnetotelluric data over the frequency range 10 200–3 Hz.

Figure 10.32 shows maps of the strike direction determined using the extended Groom–Bailey tensor decomposition method of McNeice & Jones (2001). For the frequency range 1000–1 Hz, a consistent strike direction parallel to the fault is defined. The result was compared to an analysis of synthetic data derived from a simple 3D model of the fault in which there is an offset in the conductors. The synthetic data display strikes that are predominantly parallel to the fault, but with increased complexity near the ends of the fault segments. Figure 10.32 also shows a map of the misfit of the Groom–Bailey model to the data over the 1000–1 Hz frequency range. There are several regions of significantly increased misfit near the ends of the fault segments for the synthetic data, suggesting that increases in the actual data are caused by analogous 3D components of the resistivity structure. The final responses shown in the right-hand panels of Figure 10.32 are induction arrows for a period of 100 s. Although the real (reversed) induction arrows point dominantly toward the fault, there is considerable complexity in the dataset, and there are some zones 1 km or more in size in which the arrows are sub-parallel to the fault. In contrast, the arrows for the synthetic dataset remain perpendicular to the fault even near the ends of the two

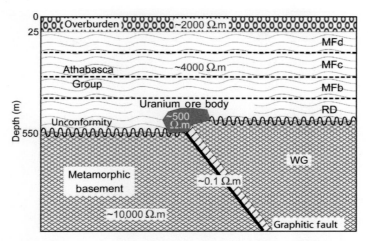

Figure 10.30. Generic resistivity model of a uranium deposit. Modified from Tuncer *et al.* (2006). The simplified model does not include an alteration zone around the ore body. WG = Wollaston Group.

segments. This suggests that the true resistivity structure contains additional features not present in the synthetic model, or that features from outside the survey area are having an effect on the observed induction arrow response.

Pseudo-sections of the TE and TM mode apparent resistivity and phase, and of the induction arrow responses, are shown in Figure 10.33. A conductor associated with the fault is indicated by enhanced apparent resistivity in the center of the profile at frequencies of 300–3 Hz, and by an increased phase response at this location for periods of 1000–100 Hz. There is also a clear reversal of the real and imaginary induction arrows. In contrast to the obvious response in the TE mode, for the TM mode there is only a weak indication for a conductor through slightly enhanced phase values.

Two-dimensional inversions of the data from each profile were performed using the nonlinear conjugate gradient algorithm of Rodi & Mackie (2001). Error floors of 20%, 5% and 0.025 were used for the apparent resistivity, phase and induction arrow, respectively. The large error floor on resistivity is one way of addressing static shift in the data. Figure 10.34 shows the results for horizontal slices at 500 m, 800 m and 1000 m depth. These are similar for inversions with and without the induction arrows included, although there are some differences in the vicinity of the McArthur River mine, which may be due to degradation of the data by cultural noise. The result shows a clear linear conductor associated with the fault, and suggests the presence of a second conductive fault to the northwest of the main conductor. The top of the conductor is in the basement at a depth of 500 m, and extends to at least 2000 m depth.

Because of the potential presence of 3D features in the resistivity model, the audiomagnetotelluric data were re-inverted using the 3D algorithm of Siripunavaraporn *et al.* (2005).

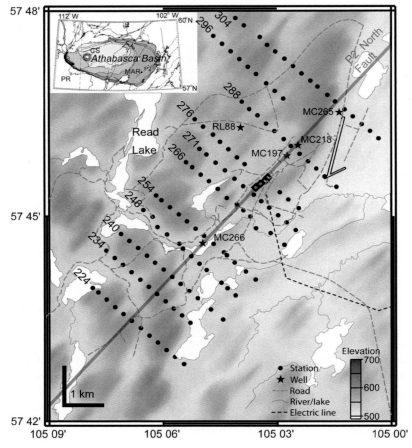

Figure 10.31. Map of audiomagnetotelluric survey sites. The location of the McArthur River deposit is denoted in the inset by MAR. Taken from Tuncer *et al.* (2006).

The inversion utilized only the off-diagonal components of the impedance tensor over 16 frequencies from 151 sites and an error floor of 5%. The RMS misfit of the final model was 1.38. The model is shown in Figure 10.34. The main conductor in the model is similar to that in the 2D models, but the 3D model suggests greater complexity to the northwest.

Figure 10.35 contains a geological interpretation of the 2D model for the southernmost profile. The conductor defines the geometry of the northeast-dipping graphitic basement fault. In addition, both the 2D and 3D models reveal a resistive zone above the fault intersection with the unconformity that can be interpreted as silicification associated with mineralization. Tuncer *et al.* (2006) completed additional sensitivity tests on the 2D and 3D inversion models, and validated the model using borehole resistivity logs. They concluded that, although the audiomagnetotelluric data are sensitive to a resistive zone above the locus

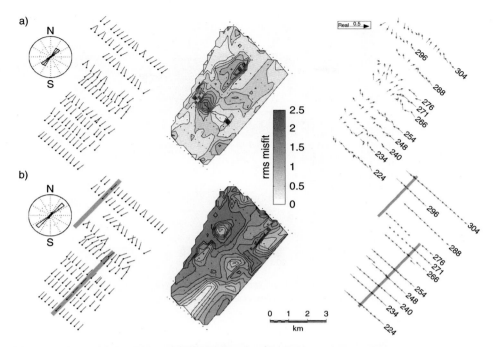

Figure 10.32. Dimensionality and strike indicators for (a) actual audiomagnetotelluric data and (b) synthetic audiomagnetotelluric data corresponding to a simple 3D resistivity model in which there are two fault segments. These are denoted by the gray rectangles on the figure. The left-hand panels show the strike direction constrained over the frequency range 1–1000 s. The middle panels show the misfit of the Groom–Bailey model over this frequency range. The right-hand panels show the real induction arrow response for a frequency of 100 Hz plotted using the Parkinson convention, so as to point toward increased conductivity. Modified from Tuncer *et al.* (2006).

of mineralization, such a region may also be due to regularization effects in inversion, and so careful analysis is essential.

10.5 Marine magnetotelluric studies

Magnetotellurics has been used in the marine environment to examine a range of tectonic targets. Pioneering seafloor magnetotelluric studies by J. H. Filloux (e.g. Filloux, 1980) revealed spatial variability in the electrical resistivity of the oceanic lithosphere that correlated with plate age. Oldenburg (1981) demonstrated that the magnetotelluric results could be explained by an increasing thickness of the oceanic lithosphere with increasing age. Seafloor magnetotelluric soundings have been applied to subduction systems (Wannamaker *et al.*, 1989; Worzewski *et al.*, 2010; Matsuno *et al.*, 2010), intra-plate hot spots (Nolasco *et al.*, 1998) and passive continental margins (Kellett *et al.*, 1991). The marine magneto-telluric method has been used to image the structure beneath mid-ocean ridges (Filloux,

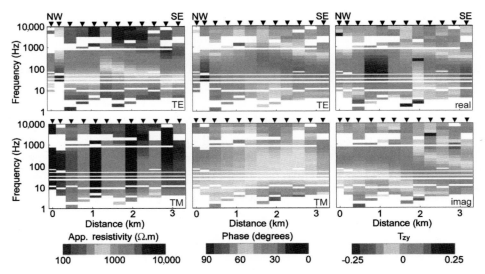

Figure 10.33. Pseudo-sections of TE, TM and induction arrow responses for profile 224. Taken from Tuncer *et al.* (2006). (See plate section for color version.)

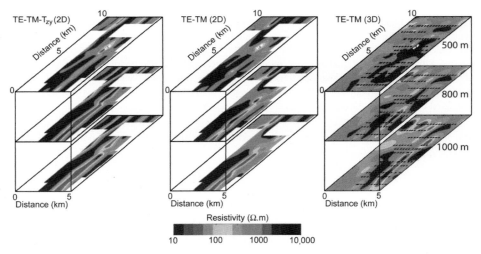

Figure 10.34. Comparison of models derived from different 2D inversions and a 3D inversion. Taken from Tuncer *et al.* (2006). (See plate section for color version.)

1981, 1982; Evans *et al.*, 1999; Heinson *et al.*, 2000; Key & Constable, 2002; Baba *et al.*, 2006a,b) and back-arc basins (Matsuno *et al.*, 2010). The marine magnetotelluric method has also been applied at shallow to intermediate depths for petroleum exploration (Hoversten *et al.*, 2000; Key *et al.*, 2006).

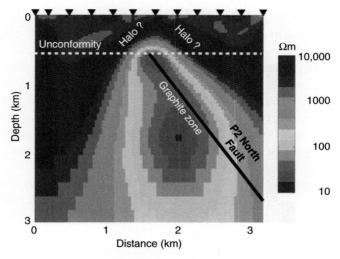

Figure 10.35. Geological interpretation of the 2D inversion model for profile 224. Modified from Tuncer *et al.* (2006). Darker shades (colors) in the middle of the cross-section indicate increasing conductivity, whereas darker shades (colors) at the edges and top indicate increasing resistivity.

10.5.1 Imaging of the East Pacific Rise

The results of the Mantle Electromagnetic and Tomography (MELT) experiment that imaged the southern East Pacific Rise (EPR) between the Pacific and Nazca plates will be described (Evans *et al.*, 1999; Baba *et al.*, 2006a, b), with emphasis on the last two papers.

The MELT experiment combined electromagnetic and seismic observations to determine the geometry of the region of partial melting, the pattern of upwelling and the melt concentration beneath a fast-spreading mid-ocean ridge. The study was principally carried out at 17°S on the EPR (Figure 10.36), where the spreading rate is ~150 mm/yr, with a secondary magnetotelluric line at 15°45′S. The former crosses the ridge crest at a melt-rich segment, while the latter locale is melt-starved. The spreading rate is asymmetric, with faster absolute plate motion on the Pacific plate leading to slow westward migration of the ridge axis. The seismic study suggested that melt generation and transport occur over a broad region of the mantle, and provided no evidence for a narrow column of melt below the ridge axis.

The magnetotelluric survey involved 47 seafloor instruments from four countries deployed at 32 stations on two east–west lines crossing the EPR. It was conducted between June 1996 and June 1997. The magnetotelluric sites extended 200 km on either side of the axis, with site spacing increasing from 3 km at the ridge axis to 75 km at the end of the profiles. The magnetotelluric responses were strongly affected by the rugged seafloor bathymetry, although they appear to be 2D when aligned with the strike of the ridge axis, with the TE mode electric field along-strike and the TM mode across-strike. Bathymetric effects were corrected using the method of Baba & Chave (2005) which models the 3D bathymetry and removes its influence, reducing the response to that for an ocean of uniform depth. The result was then cleaned of outliers using the rho+ method (see Section 4.2.7) and inverted for the resistivity structure. Because of coupling between the topography and mantle structure, topographic stripping must be recomputed for each model until convergence is achieved, although coupling proved to be weak in practice.

There were strong suggestions of anisotropy from preliminary isotropic inversions of the southern line data, and hence Baba *et al.* (2006a) inverted for a diagonal conductivity tensor using a 2D algorithm based on Rodi & Mackie (2001). Both the model smoothness and degree of anisotropy are regularized. A suite of models was generated as a function of the smoothness and degree of anisotropy, as well as with different starting models. The resulting misfits are nearly identical for the preferred isotropic and anisotropic models in Figure 10.37, both of which utilize the combined TE and TM mode data.

The isotropic model in Figure 10.37a has many features in common with the isotropic model of Evans *et al.* (1999), especially a structure that is asymmetric across the ridge axis, but the thickness of the high-resistivity zone at the ridge crest and its eastward thinning are geologically implausible. The anisotropic model in Figure 10.37b shows a structure that is also asymmetric across the ridge axis, with higher conductivity to the west, especially in the upper 60 km. In addition, it has several key features that either were not seen in earlier isotropic solutions, or could not be clearly identified owing to the assumption of isotropy. In particular, the model contains: (1) a highly conductive region in the across-ridge (y)

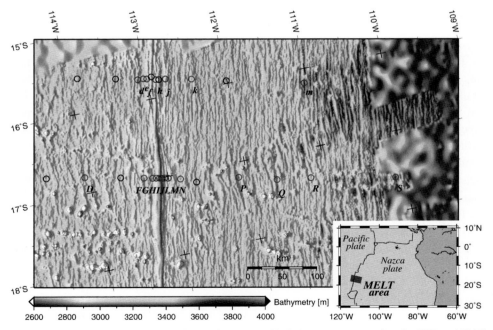

Figure 10.36. Map showing the two lines of magnetotelluric instruments crossing the EPR at 15°45′ and 17°S. Taken from Baba *et al.* (2006a). Note the rugged topography.

direction below 60 km to the east of the rise; (2) a resistive–conductive boundary that is flat-lying and independent of distance from the ridge axis; and (3) a narrow, vertically and along-strike aligned conductive sheet in the vertical conductivity located beneath the rise axis that is also apparent in the across-ridge conductivity. The first and second features are consistent with other geological and geophysical observations, and are more readily interpretable than the corresponding features in the isotropic models. The third feature, a sheet-like conductor beneath the rise axis in $\rho_{zz,}$ is probably the result of the high sensitivity of magnetotelluric data to even weak vertical electrical connection between the highly conductive ocean and deep mantle. A vertical conductor would cause electric currents to turn from the base of the crust deeper into the mantle (or vice versa), and hence can strongly influence the TM mode.

Baba *et al.* (2006a) carried out a series of tests for the significance of major features, including a formal analysis of the sensitivity matrix that showed that the data are most sensitive to the conductive region east of the ridge axis and below 60 km where the resistivity is anisotropic, being higher in the direction of plate spreading (ρ_{yy}). The data are also sensitive to the along-strike resistivity in the same region, suggesting that the detected anisotropy is real. The upper 60 km of the mantle is more resistive than at greater depths. Evans *et al.* (2005) describe additional sensitivity tests that suggest that this region is also more isotropic than below.

The contribution of temperature to conductivity can be assessed through comparison of the derived structure to that for dry isotropic peridotite based on laboratory measurements (Xu *et al.*, 2000) with a thermal structure based on a model of a cooling half-space with a basal potential temperature of 1350°C and a mantle adiabat of 0.3°C/km. The differences between the log resistivity of the observed structure and the dry model are shown for the preferred anisotropic model in Figure 10.38. The effect of anisotropy on the electrical conductivity of dry olivine is weak (approximately a factor of 2 between the *a* and *c* axes, with the latter being higher), and hence the use of an isotropic reference model with the anisotropic inversion result does not introduce major interpretation errors.

The most persistent, large-scale feature of the anisotropic inversion is the consistently higher conductivity in the direction of plate spreading below 60 km depth that occurs from very near the ridge axis to the edge of the model, together with the overlying resistive, electrically isotropic upper lithosphere (Figure 10.37). These features are more apparent in Figure 10.38, and are suggestive of a 60 km thick region depleted of volatiles by plate accretion overlying a more fertile zone. However, the near-constant 60 km depth of the conductive feature and its existence very near the ridge crest are inconsistent with a purely thermal origin (Evans *et al.*, 2005).

A region of enhanced conductivity to the east of the ridge from 70 to 120 km depth is a required feature of the magnetotelluric data. For the anisotropic model (Figure 10.37), higher conductivity is seen in the spreading (*y*) direction, and is about a factor of 4 above the along-axis (*x*) value. However, even the lower (along-axis) conductivity is 2–3 times above the dry peridotite value, while the across-axis value is over a decade higher. Unreasonable (>1500°C) potential temperatures are required to explain any of these observations with thermal structure alone.

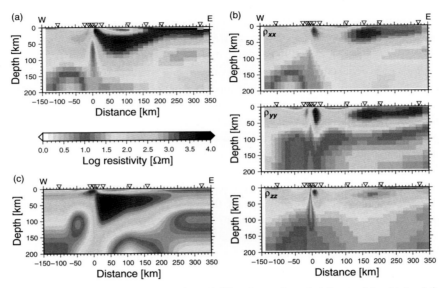

Figure 10.37. Best-fitting 2D (a) isotropic and (b) anisotropic resistivity models obtained for the combined TE and TM mode magnetotelluric responses from the MELT southern line. For the anisotropic result, the panels are along-strike (top), across-strike (middle) and vertical (bottom). The RMS misfits are 2.55 (isotropic) and 2.44 (anisotropic), and are indistinguishable at more than the 80% significance level using a two-sided F-test. The ridge axis is at 0 km and the inverted triangles show the locations of magnetotelluric sites. Taken from Baba *et al.* (2006a). (See plate section for color version.)

Evans *et al.* (2005) present laterally averaged conductivity profiles derived from the anisotropic model in Figure 10.37, and compare them to wet-mantle conductivity estimates. Their calculations assumed that all of the hydrogen is contributing to conductivity enhancement, and thus represent the maximum influence of hydrogen on conductivity. They showed that anisotropic conductivity in the across-axis direction is similar to that predicted from the Nernst–Einstein relation for hydrogen diffusion parallel to the *a* axis or [100] direction in olivine containing 1000 H/10^6 Si (Karato, 1990). The conductivity in the along-axis direction is significantly smaller even in light of uncertainty in the hydrogen diffusion data. Evans *et al.* (2005) concluded that anisotropy of conductivity can be explained if there is strong lattice-preferred orientation of olivine due to flow in the mantle away from the spreading ridge. The observed factor of 4 anisotropy is compatible with about 50% alignment of olivine in the [100] direction. The off-axis anisotropic structure continues almost undisturbed beneath the ridge axis where melting is active (see Figure 10.38). As a result, a wet-mantle interpretation, probably including a small amount of melt near the spreading center, is preferred over alternatives, including the presence of melt off-axis. These conclusions are also compatible with the basin-scale bounds in conductivity given by Lizarralde *et al.* (1995), suggesting that a wet region below ~60 km is a major feature of the oceanic mantle.

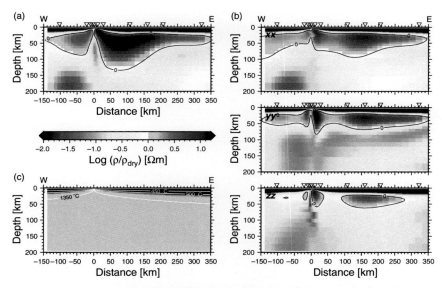

Figure 10.38. Anisotropic resistivity model after subtraction of the dry mantle reference model described in the text. Taken from Baba *et al.* (2006a). (See plate section for color version.)

In the vicinity of the rise axis, above a depth of ~70 km where the conductivity is higher than predicted for a dry peridotite mantle, the conductivity enhancement is probably a consequence of partial melting. Inference that conductivity in the vertical direction beneath the ridge is different from that in the two horizontal directions immediately suggests differences in the degree of melt interconnection, with melt better connected vertically. However, Baba *et al.* (2006a) showed that the conductivity in the vertical component model is not well defined, precluding a tight constraint. Using a parallel bound for aligned melt sheets, Baba *et al.* (2006a) showed that the anisotropic model predicts that melt will concentrate in the narrow region beneath the ridge. The lower bound on the average melt fraction in the central melt zone is ~4% for the z direction and no more than 1% for the x and y directions.

Baba *et al.* (2006b) used a similar methodology to model the MELT northern line of Figure 10.36. Their best-fitting anisotropic model is shown in Figure 10.39. The result provides no evidence for a conductive region immediately beneath the ridge, in contrast to the model previously obtained beneath the EPR at 17°S. This observation can be explained by differences in current melt production along the ridge, consistent with other geophysical observations. It is interesting to note that Matsuno *et al.* (2010) also found no evidence for melt beneath the axis of the back-arc spreading center in the Marianas region.

10.5.2 Marine petroleum exploration

The magnetotelluric method has also been applied in marine petroleum investigations (e.g. Hoversten *et al.*, 2000; Key *et al.*, 2006). Magnetotellurics has important applications

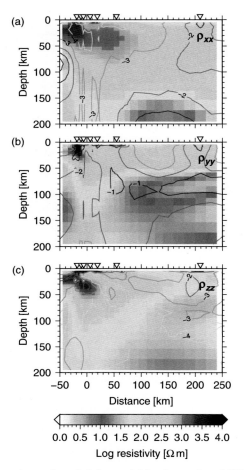

Figure 10.39. Best-fitting anisotropic resistivity model for the northern MELT line. Taken from Baba *et al.* (2006b). The ridge axis is located at 0 km, and the inverted diamonds are the locations of magnetotelluric sites. The three panels are the (a) along, (b) across and (c) vertical components of resistivity. Contours show relative sensitivity. (See plate section for color version.)

when salt deposits reduce the effective resolution of seismic reflection data, providing good resolution in relatively shallow ocean settings (<1500 m) for which high-frequency electromagnetic source signals can penetrate to the seafloor. Synthetic studies have shown that magnetotelluric data in the period range 1 to 1000 s can provide information on salt geometry by, for example, discriminating between shallow and deeply rooted salt structures.

Key *et al.* (2006) describe an application of magnetotellurics to image a complex salt body at the Gemini Prospect in the Gulf of Mexico. The geometry of the salt body is known reasonably well from seismic data. The study extends earlier work described in Hoversten

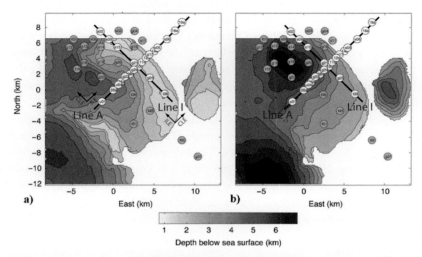

Figure 10.40. Geometry of the salt body and location of magnetotelluric stations. Modified from Key *et al.* (2006). (a) Contours showing the top-of-salt surface. (b) Contours showing the base-of-salt section. CLE and ILE refer to electric fields oriented across and parallel to the lines, respectively. Note that ILE for Line A and CLE for Line 1 correspond to electric fields perpendicular to the ridge-like structure in the top-of-salt structure.

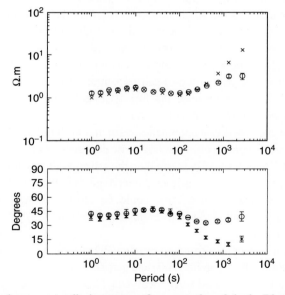

Figure 10.41. Exemplar magnetotelluric response from over the salt body. Black symbols represent the off-diagonal impedance component with the electric field oriented to the southeast; and gray symbols are for the electric field oriented to the northeast. See text for discussion. Taken from Key *et al.* (2006).

et al. (1998, 2000). The tectonic setting contains salt structures that contribute to the deformation of sediments, and may lead to the formation of sub-salt petroleum reservoirs. The site lies 200 km southeast of New Orleans in 1 km deep water. The salt forms a complex structure lying mostly between 1 and 5 km beneath the seafloor. The top-of-salt surface has a ridge-like structure with a NW–SE trend, and the base of the salt has a basin-like shape with a maximum depth exceeding 6 km. The Gemini sub-salt petroleum prospect consists of a gas sand at about 4 km depth on the southeastern edge of the salt structure.

The magnetotelluric survey consists of 42 sites spread over an area of about 10 km × 15 km (Figure 10.40), including a dense profile (Line A) perpendicular to the NW–SE trending ridge on the top of the salt surface. The site spacing is 1–3 km outside this profile, and includes a Line 1 lying parallel to the ridge that crosses a strongly 3D part of the salt structure. Data were collected over several surveys, each of which involved simultaneous recording of magnetotelluric data at multiple sites for durations of 1–2 days. Robust multi-station transfer function estimation yielded impedance tensor estimates for the period range of 1 to 3000 s.

Figure 10.41 shows an exemplar magnetotelluric response from a site located over a deep, thick portion of the structure. The gentle rise in resistivity at a period of approximately 10 s is the signature of the salt, with resistivity >10 Ω m, which constitutes a resistive target within the seawater-saturated sediments whose resistivity is around 0.5 Ω m. The split in off-diagonal impedance components at longer periods is attributed either to 2D or 3D structures at greater depth than the base of the salt, or to the effect of the topography of the nearby continental shelf. Figure 10.42 shows a pseudo-section of the data for electric currents flowing parallel to the ridge structure. The data do not exhibit significant static shift effects, but do show some inter-site variability. Higher apparent resistivity (up to 1.4 Ω m) and low phase (down to 30°) is observed at sites above the shallowest parts of the salt. Dimensionality analysis shows that the structure ranges from 1D (<20 s) to 2D (20–250 s), with the strike dominated by the top-of-salt structure. At periods longer than 250 s, the structure becomes 3D in response to the full 3D distribution of the salt as well as the surrounding basement and regional bathymetry.

Modeling of the data used two approaches restricted to 2D inversion, but, to assess the influence of 3D structure on the outcome, synthetic data generated by 3D forward modeling were also inverted. The 3D model was based on the seismic salt structure with random noise added to the response to match the estimated uncertainties in the actual data. The 3D model revealed the parts of the responses that could be most accurately inverted using 2D approaches and, for example, led to the exclusion of data with periods longer than 250 s from the final inversions.

Figure 10.43 shows the results from independent 2D inversion of the synthetic data for the modes involving current flow parallel (ILE) and perpendicular (CLE) to the profile. Its major features are similar to those based on the real data. However, the real data model contains additional smaller-scale structures inferred to be due to features not included in the 3D forward model. Inversion of the synthetic data shows that optimal results are obtained for the ILE data for Line A and CLE data for Line 1, which represent the TM mode with respect to

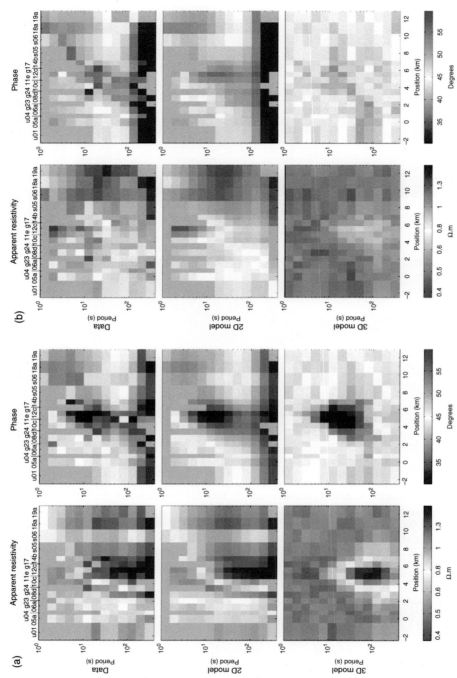

Figure 10.42. Pseudo-sections for Line A in Figure 10.40. (a) In-line electric field response. (b) Cross-line electric field response. The upper panels show observed data, the middle panels show the response from 2D inversions, and the lower panels show the response of 3D forward models. Modified from Key *et al.* (2006). (See plate section for color version.)

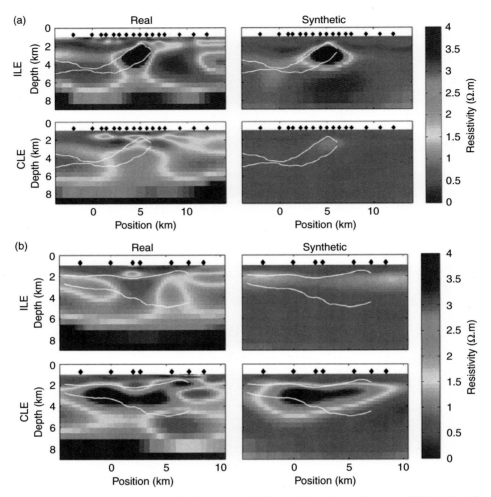

Figure 10.43. Inversion results for (a) Line A and (b) Line 1. Taken from Key *et al.* (2006). The left-hand panels correspond to inversion of the actual data, and the right-hand panels correspond to inversions of the synthetic data. The models are oriented from southwest (left) to northeast (right). The white line shows the location of the top and bottom surfaces of the salt structure as determined from 3D seismic reflection data. (See plate section for color version.)

the linear ridge on the top of the salt structure. This mode resolves the geometry of the thicker and shallower parts of the resistive salt structure. It does not resolve deeper and thinner parts of the salt, and is able to provide only a minimum estimate of the salt resistivity. The TE mode provides only a weak indication of the presence of salt.

Inversion of the real and synthetic data is able to resolve the top-of-salt surface in thicker parts of the salt body quite accurately. As shown by Figure 10.44, where the magnetotelluric

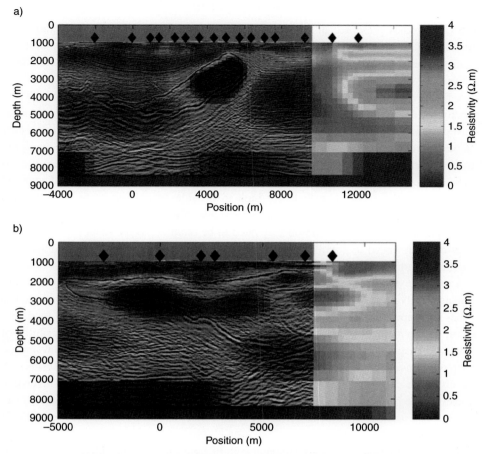

Figure 10.44. Comparison of resistivity images with seismic reflection results for the ILE model from Line A and the CLE model for Line 1. Taken from Key *et al.* (2006). (See plate section for color version.)

results are compared with seismic reflection sections, the surface is well imaged at locations shallower than 4–5 km depth. The synthetic and real data inversions also resolve the base-of-salt surface to within a depth range and lateral position of several kilometers. The results have good sensitivity to the more conductive zone beneath the salt, and are able to confirm that it is not rooted into the basement along either of the two profiles. This is important, as there are several parts of the seismic reflection image in which interpretation of the base-of-salt reflector is ambiguous. Inversion of the actual data revealed a number of other features correlating spatially with seismic reflections, such as the overhanging salt structure near the end of Line 1 and the resistive zone extending out of the end of the seismic reflection image on Line A. These provide additional constraints on the interpretation of the geophysical data.

10.6 Conclusions

The case studies presented in this chapter demonstrate the important contributions that magnetotelluric studies have made in geological and applied studies. A number of common themes are evident from these studies, and can be used to help design future magnetotelluric surveys.

A first overarching theme is that excellent magnetotelluric data quality and high site density are required for optimal resolution of most targets. The magnetotelluric responses must provide accurate phase and apparent resistivity responses for all four response tensor elements over a sufficiently broad period range to resolve both the target and the overlying resistivity structures. Comparison of the case studies with those from several decades ago shows that one of the most significant differences lies in the density and layout of acquired data. Site density must be sufficiently high to provide spatial resolution of the target structures while still allowing for some sites to be discarded from the analysis. In particular, sites with strong galvanic distortion may need to be omitted. At the same time, magneto-telluric surveys need to provide a sufficiently wide aperture to minimize the influence of conductivity structures adjacent to the primary target.

A second overarching theme of the case studies is a requirement for detailed distortion, dimensionality and strike analyses to ensure success. Application of modern tools such as tensor decomposition methods, phase-tensor examination and the application of modeling and inversion methods of appropriate dimensionality has greatly improved the reliability of resistivity images derived from magnetotelluric surveys. In some of the more recent studies, anisotropic inversion has also been shown to yield improved resistivity images. Although 3D inversion codes are now available, most studies continue to rely on 2D inversion. However, the case histories show that in many cases careful 3D forward modeling can be used in association with 2D inversion (e.g. Tuncer *et al.*, 2006) to produce reliable resistivity images.

A final overarching theme of the magnetotelluric case studies is the enhancement to their value from integrating electrical models with other geophysical and geological information. Magnetotelluric surveys can provide information on the salinity of crustal fluids, the temperature of the lithospheric mantle, and the percentage of melt near mid-ocean ridges, rather than just multidimensional images of the subsurface resistivity.

References

Baba, K. & A. D. Chave (2005). Correction of seafloor magnetotelluric data for topographic effects during inversion. *J. Geophys. Res.*, **110**, B12105, doi:10.1029/2004JB003463.

Baba, K., A. D. Chave, R. L. Evans, G. Hirth & R. L. Mackie (2006a). Mantle dynamics beneath the East Pacific Rise at 17S: insights from the Mantle Electromagnetic and Tomography (MELT) experiment. *J. Geophys. Res.*, **111**, B02101, doi:10.1029/2004JB003598.

Baba, K., P. Tarits, A. D. Chave, R. L. Evans, G. Hirth & R. L. Mackie (2006b). Electrical structure beneath the northern MELT line on the East Pacific Rise at 15°45′S. *Geophys. Res. Lett.*, **33**, L22301, doi:10.1029/2006GL027538.

Bahr, K. & A. Duba (2000). Is the asthenosphere electrically anisotropic? *Earth Planet. Sci. Lett.*, **178**, 87–95.

Bahr, K. & F. Simpson (2002). Electrical anisotropy below slow- and fast-moving plates: paleoflow in the upper mantle. *Science*, **295**, 1270–1272, doi:10.1126/science.1066161.

Bahr, K., N. Olsen & T. J. Shankland (1993). On the combination of the magnetotelluric and the geomagnetic depth sounding method for resolving an electrical conductivity increase at 400 km depth. *Geophys. Res. Lett.*, **20**, 2937–2940.

Bai, D., M. J. Unsworth, M. A. Meju, X. Ma, J. Teng, X. Kong, Y. Sun, J. Sun, L. Wang, C. Jiang, C. Zhao, P. Xiao & M. Liu (2010). Crustal deformation of the eastern Tibetan plateau revealed by magnetotelluric imaging. *Nature, Geosci.*, **3**, 358–362, doi:10.1038/NGEO830.

Bastani, M. & L. B. Pedersen (2001). Estimation of magnetotelluric transfer functions from radio transmitters. *Geophysics*, **66**, 1038–1051, doi:10.1190/1.1487051.

Boerner, D. E., J. A. Wright, J. G. Thurlow & L. E. Reed (1993). Tensor CSAMT studies at the Buchans Mine in central Newfoundland. *Geophysics*, **58**, 12–19.

Boerner, D., R. D. Kurtz, J. A. Craven, G. M. Ross, F. W. Jones & W. J. Davis (1999). Electrical conductivity in the Precambrian lithosphere of western Canada. *Science*, **283**, 668–670.

Boerner, D., R. D. Kurtz & J. A. Craven (2000a). A summary of electromagnetic studies on the Abitibi–Grenville transect. *Can. J. Earth Sci.*, **37**, 427–437.

Boerner, D., R. D. Kurtz, J. A. Craven, G. M. Ross & F. W. Jones (2000b). A synthesis of electomagnetic studies in the LITHOPROBE Alberta Basement Transect: constraints on Paleoproterozoic indentation tectonics. *Can. J. Earth Sci.*, **37**, 1509–1534.

Brasse, H., P. Lezaeta, V. Rath, K, Schwalenberg, W. Soyer & V. Haak (2002).The Bolivian Altiplano conductivity anomaly. *J. Geophys. Res.*, **107**, 2096, doi:10.1029/2001JB000391.

Brown, C. (1994). Tectonic interpretation of regional conductivity anomalies. *Surv. Geophys.*, **15**, 123–157.

Chen, L., J. R. Booker, A. G. Jones, N. Wu, M. J. Unsworth, W. Wei & H. Tan (1996). Electrically conductive crust in Southern Tibet from INDEPTH magnetotelluric surveying. *Science*, **274**, 1694–1696.

Christopherson, K. R. (1991). Applications of magnetotellurics to petroleum exploration in Papua New Guinea: a model for frontier areas. *Geophysics*, **56**, 21–27.

Dennis, Z. R., D. H. Moore & J. P. Cull (2011). Magnetotelluric survey for undercover structural mapping, Central Victoria. *Aust. J. Earth Sci.*, **58**, 33–47.

Duba, A. L. & T. J. Shankland (1982). Free carbon and electrical conductivity in the earth's mantle. *Geophys. Res. Lett.*, **9**, 1271–1274.

Ducea, M. N. & S. K. Park (2000). Enhanced mantle conductivity from sulfide minerals, southern Sierra Nevada, California. *Geophys. Res. Lett.*, **27**, 2405–2408.

Evans, R. L., P. Tarits, A. D. Chave, A. White, G. Heinson, J. H. Filloux, H. Toh, N. Seama, H. Utada, J. R. Booker & M. J. Unsworth (1999). Asymmetric mantle electrical structure beneath the East Pacific Rise at 17°S. *Science*, **286**, 752–756.

Evans, R. L., G. Hirth, K. Baba, D. Forsyth, A. Chave & R. Mackie (2005). Geophysical controls from the MELT area for compositional controls on oceanic plates. *Nature*, **437**, 249–252.

Evans, R. L., A. G. Jones, X. Garcia, M. Muller, M. Hamilton, S. Evans, C. J. S. Fourie, J. Spratt, S. Webb, H. Jelsma & D. Hutchins (2011).Electrical lithosphere beneath the Kaapvaal Craton, southern Africa. *J. Geophys. Res.*, **116**, B04105, doi:10.1029/2010JB007883.

Falgàs, E., J. Ledo, T. Teixidó, A. Gabàs, F. Ribera, C. Arango, P. Queralt, J. L. Plata, F. Rubio, J. A. Peña, A. Martí & A. Marcuello (2005). Geophysical characterization of a Mediterranean costal aquifer: Baixa Tordera fluvial deltaic aquifer unit. *Groundwater Saline Intrusion*, **15**, 395–404.

Falgàs, E., J. Ledo, B. Benjumea, P. Queralt, A. Marcuello, T. Teixidó & A. Martí (2011). Integrating hydrogeological and geophysical methods for the characterization of a deltaic aquifer system. *Surv. Geophys.*, **32**, 857–873, doi:10.1007/s10712–011–9126–2.

Ferguson, I. J., J. A. Craven, R. D. Kurtz, D. E. Boerner, R. C. Bailey, X. Wu, M. R. Orellana, J. Spratt, G. Wennberg & M. Norton (2005). Geoelectric responses of Archean lithosphere in the western Superior Province, central Canada. *Phys. Earth Planet. Inter.*, **150**, 123–143.

Filloux, J. H. (1980). Magnetotelluric soundings over the northeast Pacific may reveal spatial dependence of depth and conductance of the asthenosphere. *Earth Planet. Sci. Lett.*, **46**, 244–252.

Filloux, J. H. (1981). Magnetotelluric exploration of the North Pacific: progress report and preliminary soundings near a spreading ridge. *Phys. Earth Planet. Inter.*, **25**, 187–195.

Filloux, J. H. (1982). Magnetotelluric experiment over the ROSE area. *J. Geophys. Res.*, **87**, 8364–8378.

Fouch, M. J., D. E. James, J. C. Van Decar, S. van der Lee & Kaapvaal Seismic Group (2004). Mantle seismic structure beneath the Kaapvaal and Zimbabwe Cratons. *S. Afr. J. Geol.*, **107**, 33–44.

Frederiksen, A. W, I. J. Ferguson, D. Eaton, S.-K. Miong & E. Gowan (2006). Mantle fabric at multiple scales across an Archean–Proterozoic boundary, Grenville Front, Canada. *Phys. Earth Planet. Inter.*, **158**, 240–263.

Glover, P. W. J., J. Pous, P. Queralt, J.-A. Munoz, M. Liesa & M. J. Hole (2000). Integrated two-dimensional lithospheric conductivity modeling in the Pyrenees using field-scale and laboratory measurements. *Earth Planet. Sci. Lett.*, **178**, 59–72.

Gowan, E. J., I. J. Ferguson, A. G. Jones & J. A. Craven (2009). Geoelectric structure of the northeastern Williston basin and underlying Precambrian lithosphere. *Can. J. Earth Sci.*, **46**, 441–464.

Haak, V. & V. R. S. Hutton (1986). Electrical resistivity in continental lower crust. In *The Nature of the Lower Continental Crust*, ed. J. B. Dawson, D. A. Carswell, J. Hall & K. H. Wedepohl. Special Publication 24. London: Geological Society, pp. 35–49.

Hamilton, M. P., A. G. Jones, R. L. Evans, S. Evans, C. J. S. Fourie, X. Garcia, A. Mountford, J. E. Spratt & SAMTEX Team (2006). Electrical anisotropy of South African lithosphere compared with seismic anisotropy from shear-wave splitting analyses. *Phys. Earth Planet. Inter.*, **158**, 226–239.

Heinson, G. S. (1990). Electromagnetic studies of the lithosphere and asthenosphere. *Surv. Geophys.*, **20**, 229–255.

Heinson, G. S., S. C. Constable & A. White (2000). Episodic melt transport at mid-ocean ridges inferred from magnetotelluric sounding. *Geophys. Res. Lett.*, **27**, 2317–2320.

Heinson, G. S., N. G. Direen & R. M. Gill (2006). Magnetotelluric evidence for a deep-crustal mineralizing system beneath the Olympic Dam iron oxide copper–gold deposit, southern Australia. *Geology*, **34**, 573–576, doi:10.1130/G22222.1.

Heise, W., T. G. Caldwell, H. M. Bibby & C. Bannister (2008). Three-dimensional modeling of magnetotelluric data from the Rotokawa geothermal field, Taupo Volcanic Zone, New Zealand. *Geophys. J. Int.*, **172**, 740–750.

Hirth, J. G., R. L. Evans & A. D. Chave (2000). Comparison of continental and oceanic mantle electrical conductivity: Is the Archaean lithosphere dry? *Geochem. Geophys. Geosyst.*, **1**(12), 1030, doi:10.1029/2000GC000048.

Hoffman, P. (1988). United plates of America, the birth of a craton: early Proterozoic assembly and growth of Proto–Laurentia. *Annu. Rev. Earth Planet. Sci.*, **16**, 543–603.

Honkura, Y., A. M. Işikara, N. Oshiman, A. Ito, B. Üçer, S. Bariş, M. K. Tunçer, M. Matsushima, R. Pektaş, C. Çelik, S. B. Tank, F. Takahashi, M. Nakanishi, R. Yoshimura, Y. Ikeda & T. Komut (2000). Preliminary results of multidisciplinary observations before, during and after the Kocaeli (Izmit) earthquake in the western part of the North Anatolian Fault Zone. *Earth Planets Space*, **52**, 293–298.

Hoversten, M. G., H. F. Morrison & S. C. Constable (1998). Marine magnetotellurics for petroleum exploration, part II: Numerical analysis of subsalt resolution. *Geophysics*, **63**, 826–840.

Hoversten, M. G., S. C. Constable & H. F. Morrison (2000). Marine magnetotellurics for base-of-salt mapping: Gulf of Mexico field test at the Gemini structure. *Geophysics*, **65**, 1476–1488.

Ichiki, M., N. Sumitomo & T. Kagiyama (2000). Resistivity structure of high-angle sub-duction zone in the southern Kyushu district, southwestern Japan. *Earth Planets Space*, **52**, 539–548.

Israil, M., D. K. Tyagi, P. K. Gupta & S. Niwas (2008). Magnetotelluric investigations for imaging electrical structure of Garwhal Himalayan corridor, Uttarakhand, India. *J. Earth Syst. Sci.*, **117**, 189–200.

Ji, S., S. Rondenay, M. Mareschal & G. Senechal (1996). Obliquity between seismic and electrical anisotropies as a potential indicator of movement sense for ductile shear zones in the upper mantle. *Geology*, **24**, 1033–1036.

Jödicke, H., A. Jording, L. Ferrari, J. Arzate, K. Mezger & L. Rüpke (2006).Fluid release from the subducted Cocos plate and partial melting of the crust deduced from magneto-telluric studies in southern Mexico: implications for the generation of volcanism and subduction dynamics. *J. Geophys. Res.*, **111**, B08102, doi:10.1029/2005JB003739.

Johnston, M. J. S. (1997). Review of electric and magnetic fields accompanying seismic and volcanic activity. *Surv. Geophys.*, **18**, 441–475.

Jones, A. G. (1992). Electrical conductivity of the continental lower crust. In *Continental Lower Crust*, ed. D. M. Fountain, R. J. Arculus & R. W. Kay. Amsterdam: Elsevier, pp. 81–143.

Jones, A. G. & J. A. Craven (2004). Area selection for diamond exploration using deep probing electromagnetic surveying. *Lithos*, **77**, 765–782.

Jones, A. G. & I. J. Ferguson (2001). The electric moho. *Nature*, **409**, 331–333.

Jones, A. G. & X. Garcia (2003). The Okak Bay MT dataset case study: a lesson in dimensionality and scale. *Geophysics*, **68**, 70–91.

Jones, A. G., J. Katsube & P. Schwann (1997). The longest conductivity anomaly in the world explained: sulphides in fold hinges causing very high electrical anisotropy. *J. Geomagn. Geoelectr.*, **49**, 1619–1629.

Jones, A. G., I. J. Ferguson, A. D. Chave, R. L. Evans & G. W. McNeice (2001). Electric lithosphere of the Slave Craton. *Geology*, **29**, 423–426.

Jones, A. G., P. Lezaeta, I. J. Ferguson, A. D. Chave, R. L. Evans, X. Garcia & J. Spratt (2003). The electrical structure of the Slave Craton. *Lithos*, **71**, 505–527.

Jones, A. G., J. Ledo & I. J. Ferguson (2005a). Electromagnetic images of the Trans-Hudson orogen: the North American Central Plains (NACP) anomaly revealed. *Can J. Earth Sci.*, **42**, 457–478.

Jones, A. G., J. Ledo, I. J. Ferguson, N. Grant, G. McNeice, J. Spratt, C. Farquharson, B. Roberts, G. Wennberg, L. Wolynec & X. Wu (2005b). The electrical resistivity structure of Archean to Tertiary lithosphere along 3200 km of SNORCLE profiles, northwestern Canada. *Can. J. Earth Sci.*, **42**, 1257–1275.

Jones, A. G., R. L. Evans & D. W. Eaton (2009a). Velocity–conductivity relationships for mantle mineral assemblages in Archean cratonic lithosphere based on extremal bounds. *Lithos*, **109**, 131–143.

Jones, A. G., R. L. Evans, M. R. Muller, M. P. Hamilton, M. P. Miensopust, X. Garcia, P. Cole, T. Ngwisanyi, D. Hutchins, C. J. S Fourie, H. Jelsma, T. Aravanis, W. Pettit, S. Webb, J. Wasborg & SAMTEX Team (2009b). Area selection for diamonds using magnetotellurics: examples from southern Africa. *Lithos*, **112S**, 83–92.

Karato, S. (1990). The role of hydrogen in the electrical conductivity of the upper mantle. *Nature*, **347**, 272–273.

Kellett, R. L., F. E. M. Lilley & A. White (1991). A two-dimensional interpretation of the geomagnetic coast effect of southeast Australia, observed on land and seafloor. *Tectonophysics*, **192**, 367–382.

Key, K. W. & S. C. Constable (2002). Broadband marine MT exploration of the East Pacific Rise at 9°50′N. *Geophys. Res. Lett.*, **29**, 2054–2057, doi:10.1029/2002GL016035.

Key, K. W., S. C. Constable & C. J. Weiss (2006). Mapping 3D salt using the 2D marine magnetotelluric method: case study from Gemini Prospect, Gulf of Mexico. *Geophysics*, **71**, B17–B27.

Korja, T. (2007). How is the European lithosphere imaged by magnetotellurics? *Surv. Geophys.*, **28**, 239–272.

Korja T. & S.-E. Hjelt (1993) Electromagnetic studies in the Fennoscandian Shield – electrical conductivity of Precambrian crust. *Phys. Earth Planet. Inter.*, **81**, 107–138.

Korja, T., M. Engels, A. A. Zhamaletdinov, A. A. Kovtun, N. A. Palshin, M. Y. Smirnov, A. D. Tokarev, V. E. Asming, L. L. Vanyan, I. L. Vardaniants and **BEAR Working Group** (2002). Crustal conductivity in Fennoscandia – a compilation of a database on crustal conductance in the Fennoscandian Shield. *Earth Planets Space*, **54**, 535–558.

Kurtz, R. D., J. M. De Laurier & J. C. Gupta (1986). A magnetotelluric sounding across Vancouver Island detects the subducting Juan de Fuca plate. *Nature*, **321**, 596–599.

Ledo, J. & A. G. Jones (2001). Regional electrical resistivity structure of the southern Canadian Cordillera and its physical interpretation. *J. Geophys. Res.*, **106**, 30 755–30 769.

Ledo, J., A. G. Jones, I. J. Ferguson & L. Wolynec (2004).Lithospheric structure of the Yukon, northern Canadian Cordillera obtained from magnetotelluric data. *J. Geophys. Res.*, **109**, B04410, doi:10.1029/2003JB002516.

Lemmonier, C., G. Marquis, F. Perrier, J. P. Avouac, G. Chitrakar, B. Kafle, S. Sapkota, U. Gautam, D. Tiwari & M. Bano (1999). Electrical structure of the Himalaya of Central Nepal: high conductivity around the mid-crustal ramp along the MHT. *Geophys. Res. Lett.*, **26**, 3261–3264.

Li, A. & K. Burke (2006).Upper mantle structure of southern Africa from Rayleigh wave tomography. *J. Geophys. Res.*, **111**, B10303, doi:10.1029/2006JB004321.

Li, S., M. J. Unsworth, J. R. Booker, W. Wei, H. Tan & A. G. Jones (2003). Partial melt or aqueous fluid in the midcrust of Southern Tibet? Constraints from INDEPTH magneto-telluric data. *Geophys. J. Int.*, **153**, 289–304.

Lilley, F. E. M., L. J. Wang, F. H. Chamalaun & I. J. Ferguson (2003). Carpentaria electrical conductivity anomaly, Queensland, as a major structure in the Australian Plate. In *Evolution and Dynamics of the Australian Plate*, ed. R. R. Hillis & R. D. Muller. Geol. Soc. Aust. Spec. Pub. 22, and Geol. Soc. Am. Spec. Pap. **372**, pp. 141–156.

Linde, N., J. Chen, M. B. Kowalsky & S. Hubbard (2006). Hydrogeophysical parameter estimation approaches for field scale characterization. In *Applied Hydrogeophysics*, ed. H. Vereecken *et al.* Dordrecht: Springer, pp. 9–44.

Livelybrooks, D., M. Mareschal, E. Blais & J. T. Smith (1996). Magnetotelluric delineation of the Trillabelle massive sulfide body in Sudbury, Ontario. *Geophysics*, **61**, 970–986.

Lizarralde, D., A. D. Chave, J. G. Hirth & A. Schultz (1995). Long period magnetotelluric study using Hawaii-to-California submarine cable data: implications for mantle conductivity. *J. Geophys. Res.*, **100**, 17 837–17 854.

Manglik, A., S. K. Verma, R. P. Sasmal & D. Muralidharan (2009). Application of magnetotelluric technique in selection of Earth electrode sites for HVDC transmission systems: an example from NE India. *Earth Sci. India*, **2**, 249–257.

Mareschal, M., R. L. Kellett, R. D. Kurtz, J. N. Ludden & R. C. Bailey (1995). Archean cratonic roots, mantle shear zones and deep electrical anisotropy. *Nature*, **375**, 134–137.

Martí, A., P. Queralt, E. Roca, J. Ledo & J. Galindo-Zaldívar (2009).Geodynamic implications for the formation of the Betic-Rif orogen from magnetotelluric studies. *J. Geophys. Res.*, **114**, B01103, doi:10.1029/2007JB005564.

Matsuno, T., N. Seama, R. L. Evans, A. D. Chave, K. Baba, A. White, T. Goto, G. Heinson, G. Boren, A. Yoneda & H. Utada (2010).Upper mantle electrical resistivity structure beneath the central Mariana subduction system. *Geochem. Geophys. Geosyst.*, **11**, Q09003, doi:10.1029/2010GC003101.

McNeice, G. M. & A. G. Jones (2001). Multisite, multifrequency tensor decomposition of magnetotelluric data. *Geophysics*, **66**, 158–173.

Meju, M. A. (2002). Geoelectromagnetic exploration for natural resources: models, case studies and challenges. *Surv. Geophys.*, **23**, 133–205.

Miensopust, M. P., A. G. Jones, M. R. Muller, X. Garcia & R. L. Evans (2011).Lithospheric structures and Precambrian terrane boundaries in northeastern Botswana revealed through magnetotelluric profiling as part of the Southern African Magnetotelluric Experiment. *J. Geophys. Res.*, **116**, B02401, doi:10.1029/2010JB007740.

Muller, M. R., A. G. Jones, R. L. Evans, H. S. Grütter, C. Hatton, X. Garcia, M. P. Hamilton, M. P. Miensopust, P. Cole, T. Ngwisany, D. Hutchins, C. J. Fourie, H. A. Jelsma, S. F. Evans, T. Aravanis, W. Pettit, S. J. Webb, J. Wasborg & SAMTEX Team (2009). Lithospheric structure, evolution and diamond prospectivity of the Rehoboth Terrane and western Kaapvaal Craton, southern Africa: constraints from broadband magnetotellurics. *Lithos*, **112S**, 93–105.

Muñoz, G., A. Mateus, J. Pous, W. Heise, F. Monteiro Santos & E. Almeida (2008). Unraveling middle-crust conductive layers in Paleozoic Orogens through 3D modeling of magnetotelluric data: the Ossa-Morena Zone case study (SW Iberian Variscides). *J. Geophys. Res.*, **113**, B06106, doi:10.1029/2007JB004987.

Newman, G. A., S. Recher, B. Tezkan & F. M. Neubauer (2000). 3D inversion of a scalar radio magnetotelluric field data set. *Geophysics*, **68**, 791–802.

Nobes, D. C. (1996). Troubled waters: environmental applications of electrical and electromagnetic methods. *Surv. Geophys.*, **17**, 393–454.

Nolasco, R., P. Tarits, J. H. Filloux & A. D. Chave (1998). Magnetotelluric imaging of the Society Islands hotspot. *J. Geophys. Res.*, **103**, 30 287–30 309, doi:10.1029/98JB02129.

Ogawa, Y., M. Mishina, T. Goto, H. Satoh, N. Oshiman, T. Kasaya, Y. Takahashi, T. Nishitani, S. Sakanaka, M. Uyeshima, Y. Takahashi, Y. Honkura & M. Matsushima (2001). Magnetotelluric imaging of fluids in intraplate earthquake zones, NE Japan back arc. *Geophys. Res. Lett.*, **28**, 3741–3744.

Oldenburg, D. W. (1981). Conductivity structure of oceanic upper mantle beneath the Pacific plate. *Geophys. J. R. Astron. Soc.*, **65**, 359–394.

Orange, A. S. (1989). Magnetotelluric exploration for hydrocarbons. *Proc. IEEE*, **77**, 287–317.

Park, S. K. (1991). Monitoring changes of resistivity prior to earthquakes in Parkfield, California with telluric arrays. *J. Geophys. Res.*, **96**, 14 211–14 237.

Park, S. K. & R. J. Mackie (1997). Crustal structure at Nanga Parbat, northern Pakistan, from magnetotelluric soundings. *Geophys. Res. Lett.*, **24**, 2415–2418.

Park, S. K. & R. L. Mackie (2000). Resistive (dry?) lower crust in an active orogen, Nanga Parbat, northern Pakistan. *Tectonophysics*, **316**, 359–380.

Patro, P. K. & T. Harinarayana (2009). Deep geoelectric structure of the Sikkim Himalayas (NE India) using magnetotelluric studies. *Phys. Earth Planet. Inter.*, **173**, 171–176.

Patro, P. K. & S. V. S. Sarma (2009). Lithospheric electrical imaging of the Deccan trap covered region of western India. *J. Geophys. Res.*, **114**, B01102, doi:10.1029/2007JB005572.

Pedersen, L. B., M. Bastani & L. Dynesius (2005). Groundwater exploration using combined controlled-source and radiomagnetotelluric techniques. *Geophysics*, **70**, G8–G15, doi:10.1190/1.1852774.

Pellerin, L. (2002). Applications of electrical and electromagnetic methods for environmental and geotechnical investigations. *Surv. Geophys.*, **23**, 101–132.

Pellerin, L., J. M. Johnston & G. W. Hohmann (1996). A numerical evaluation of electromagnetic methods in geothermal exploration. *Geophysics*, **61**, 121–130.

Pettit, W. (2009). Geophysical signatures of some recently discovered large (>40 ha) kimberlite pipes on the Alto Cuilo concession in northeastern Angola. *Lithos*, **112S**, 106–115.

Phillips, N., D. Oldenburg, J. Chen, Y. Li & P. Routh (2001). Cost effectiveness of geophysical inversions in mineral exploration: applications at San Nicolas. *The Leading Edge*, **20**, 1351–1360.

Pous, J., G. Muñoz, W. Heise, J. C. Melgarejo & C. Quesada (2004). Electromagnetic imaging of Variscan crustal structures in SW Iberia: the role of interconnected graphite. *Earth Planet. Sci. Lett.*, **217**, 435–450.

Pulkkinen, A. R. Pirjola & A. Viljanen (2007). Determination of ground conductivity and system parameters for optimal modeling of geomagnetically induced current flow in technological systems. *Earth Planets Space*, **59**, 999–1006.

Queralt, P., A. G. Jones & J. Ledo (2007). Electromagnetic imaging of a complex ore body: 3D forward modeling, sensitivity tests, and down-mine measurements. *Geophysics*, **72**, F85–F95, doi:10.1190/1.2437105.

Rodi, W. & R. Mackie (2001). Nonlinear conjugate gradients algorithm for 2D magnetotelluric inversion. *Geophysics*, **66**, 174–187.

Rosell, O., A. Martí, A. Marcuello, J. Ledo, P. Queralt, E. Roca & J. Campanyà (2011). Deep electrical resistivity structure of the northern Gibraltar Arc (western Mediterranean): evidence of lithospheric slab break-off. *Terra Nova*, **23**, 179–186, doi:10.1111/j.1365-3121.2011.00996.x.

Routh, P. & D. Oldenburg (1999). Inversion of controlled source audio-frequency magnetotelluric data for a horizontally layered earth. *Geophysics*, **64**, 1689–1697.

Roux, E., M. Moorkamp, A. G. Jones, M. Bischoff, B. Endrun, S. Lebedev & T. Meier (2011). Joint inversion of long-period magnetotelluric data and surface-wave dispersion curves for anisotropic structure: application to data from Central Germany. *Geophys. Res. Lett.*, **38**, L05304, doi:10.1029/2010GL046358.

Rudnick, R. L. & A. Nyblade (1999). The thickness and heat production of Archean lithosphere: constraints from xenoliths, thermobarometry and surface heat flow. In *Mantle Petrology: Field Observations and High Pressure Experimentation, A Tribute to F. R. Boyd*. Geochem. Soc. Spec. Publ. **6**, pp. 3–12.

Sainato, C., M. C. Pomposiello, A. Landini, G. Galindo & H. Malleville (2000). Hydrogeological sections of the Pergamino basin (Buenos Aires province, Argentina): audiomagnetotelluric and geochemical results. *Rev. Bras. Geofis.*, **18**, doi:10.1590/S0102–261X2000000200007.

Schultz, A., R. D. Kurtz, A. D. Chave & A. G. Jones (1993). Conductivity discontinuities in the upper mantle beneath a stable craton. *Geophys. Res. Lett.*, **20**, 2941–2944.

Selway, K., M. Hand, G. S. Heinson & J. L. Payne (2009). Magnetotelluric constraints on subduction polarity: reversing reconstruction models for Proterozoic Australia. *Geology*, **37**, 799–802, doi:10.1130/G30175A.1.

Sheard, S. N., T. J. Ritchie, K. R. Christopherson & E. Brand (2005). Mining, environmental, petroleum, and engineering industry applications of electromagnetic techniques in geophysics. *Surv. Geophys.*, **26**, 653–669.

Siripunavaraporn, W., G. Egbert, Y. Lenbury & M. Uyeshima (2005). Three-dimensional magnetotelluric inversion: data-space method. *Phys. Earth Planet. Inter.*, **150**, 3–14.

Slater, L. (2007). Near surface electrical characterization of hydraulic conductivity: from petrophysical properties to aquifer geometries – a review. *Surv. Geophys.*, **28**, 169–197, doi:10.1007/s10712-007-9022-y.

Solon, K., A. G. Jones, K. D. Nelson, M. J. Unsworth, W. Wei, H. Tan, S. Jin, M. Deng, J. R. Booker, S. Li & P. A. Bedrosian (2005).Structure of the crust in the vicinity of the Banggong–Nujiang Suture, central Tibet from INDEPTH magnetotelluric data. *J. Geophys. Res.*, **110**, B10102, doi:10.1029/2003JB002405.

Soyer, W. & M. Unsworth (2006). Deep electrical structure of the northern Cascadia (British Columbia, Canada) subduction zone: implications for the distribution of fluids. *Geology*, **34**, 53–56, doi:10.1130/G21951.1.

Spratt, J. E., A. G. Jones, K. D. Nelson, M. J. Unsworth & INDEPTH MT Team (2005). Crustal structure of the India–Asia collision zone, southern Tibet, from INDEPTH MT investigations. *Phys. Earth Planet. Inter.*, **150**, 227–237.

Spratt, J. E., A. G. Jones, V. A. Jackson, L. Collins & A. Avdeeva (2009). Lithospheric geometry of the Wopmay orogen from a Slave Craton to Bear Province magnetotelluric transect. *J. Geophys. Res.*, **114**, B01101, doi:10.1029/2007JB005326.

Sun, J., G. Jin, D. Bai & L. Wang (2003). Sounding of electrical structure of the crust and upper mantle along the eastern border of Qinghai–Tibet Plateau and its tectonic significance. *Sci. China*, **46**, 243–253.

Tezkan, B. (1999). A review of environmental applications of quasi-stationary electromagnetic techniques. *Surv. Geophys.*, **20**, 279–308.

Tuncer, V., M. J. Unsworth, W. Siripunavaraporn & J. A. Craven (2006). Exploration for unconformity-type uranium deposits with audiomagnetotelluric data: a case study from the McArthur River mine, Saskatchean, Canada. *Geophysics*, **71**, B201–B209.

Türkoğlu, E., M. Unsworth & D. Pana (2009). Deep electrical structure of northern Alberta (Canada): implications for diamond exploration. *Can. J. Earth Sci.*, **46**, 139–154.

Unsworth, M. J. (2005). New developments in conventional hydrocarbon exploration with electromagnetic methods. *CSEG Recorder*, **30**, 35–38.

Unsworth, M. J. (2010). Magnetotelluric studies of active continent–continent collisions. *Surv. Geophys.*, **30**, 137–161, doi:10.1007/s10712-009-9086-y.

Unsworth, M. J., X. Lu & M. D. Watts (2000). CSAMT exploration at Sellafield: characterization of a potential radioactive waste disposal site. *Geophysics*, **65**, 1070–1079.

Unsworth, M. J., W. Wei, A. G. Jones, S. Li, P. A. Bedrosian, J. R. Booker, S. Jin & M. Deng (2004).Crustal and upper mantle structure of Northern Tibet imaged with magnetotelluric data. *J. Geophys. Res.*, **109**, B02403, doi:10.1029/2002JB002305.

Unsworth, M. J., A. G. Jones, W. Wei, G. Marquis, S. Gokarn & J. Spratt (2005). Crustal rheology of the Himalaya and Southern Tibet inferred from magnetotelluric data. *Nature*, **438**, 78–81, doi:10.1038/nature04154.

Uyeshima, M. (2010). EM monitoring of crustal processes including the use of the network-MT observations. *Surv. Geophys.*, **28**, 199–237, doi:10.1007/s10712-007-9023-x.

Wannamaker, P. E., J. R. Booker, A. G. Jones, A. D. Chave, J. H. Filloux, H. S. Waff & L. K. Law (1989). Resistivity cross section through the Juan de Fuca subduction system and its tectonic implications. *J. Geophys. Res.*, **94**, 14 127–14 144.

Wannamaker, P. E., G. R. Jiracek, J. A. Stodt, T. G. Caldwell, V. M. Gonzalez, J. D. McKnight & A. D. Porter (2002).Fluid generation and pathways beneath an active compressional orogen, the New Zealand Southern Alps, inferred from magnetotelluric data. *J. Geophys. Res.*, **107**, 2117, doi:10.1029/2001JB000186.

Webb, S. J. (2009). Southern African tectonics from potential field interpretation. Ph.D. thesis, University of the Witswatersrand, Johannesburg.

Withers, R., D. Eggers, T. Fox & T. Crebs (1999). A case study of integrated hydrocarbon exploration through basalt. *Geophysics*, **59**, 1666–1679.

Worzewski, T., M. Jegen, H. Kopp, H. Brasse & W. Taylor Castillo (2010). Magnetotelluric image of the fluid cycle in the Costa Rican subduction zone. *Nature, Geosci.*, **4**, 108–111, doi:10.1038/ngeo1041.

Wu, X., I. J. Ferguson & A. G. Jones (2005). Geoelectric structure of the Proterozoic Wopmay Orogen and adjacent terranes, Northwest Territories, Canada. *Can. J. Earth Sci.*, **42**, 955–981.

Xu, Y., T. J. Shankland & B. T. Poe (2000). Laboratory-based electrical conductivity in the earth's mantle. *J. Geophys. Res.*, **105**, 27 865–27 875.

Zhang, S., S. Wei & J. Wang (1996). Magnetotelluric sounding in the Qiangtang basin of Xizang (Tibet). *J. China Univ. Geosci.*, **21**, 198–202.

Index

Printed in the United States
By Bookmasters